The Reticuloendothelial System

A COMPREHENSIVE TREATISE

Volume 9

Hypersensitivity

The Reticuloendothelial System

A COMPREHENSIVE TREATISE

General Editors:
Herman Friedman, *University of South Florida, Tampa, Florida*
Mario R. Escobar, *Medical College of Virginia, Richmond, Virginia*
and
Sherwood M. Reichard, *Medical College of Georgia, Augusta, Georgia*

MORPHOLOGY
Edited by Ian Carr and W. T. Daems

BIOCHEMISTRY AND METABOLISM
Edited by Anthony J. Sbarra and Robert R. Strauss

PHYLOGENY AND ONTOGENY
Edited by Nicholas Cohen and M. Michael Sigel

IMMUNOPATHOLOGY
Edited by Noel R. Rose and Benjamin V. Siegel

CANCER
Edited by Ronald B. Herberman and Herman Friedman

IMMUNOLOGY
Edited by Joseph A. Bellanti and Herbert B. Herscowitz

PHYSIOLOGY (In two parts)
Edited by Sherwood M. Reichard and James P. Filkins

PHARMACOLOGY
Edited by John Hadden and Andor Szentivanyi

HYPERSENSITIVITY
Edited by S. Michael Phillips and Mario R. Escobar

INFECTION
Edited by John P. Utz and Mario R. Escobar

The Reticuloendothelial System

A COMPREHENSIVE TREATISE

Volume 9
Hypersensitivity

Edited by

S. MICHAEL PHILLIPS
The University of Pennsylvania
Philadelphia, Pennsylvania

and

MARIO R. ESCOBAR
Medical College of Virginia
Richmond, Virginia

PLENUM PRESS • NEW YORK AND LONDON

Library of Congress Cataloging in Publication Data

(Revised for vol. 9)

The Reticuloendothelial system.

Includes bibliographies and indexes.
Contents: v. 1. Morphology—v. 2. Biochemistry and metabolism—[etc.]—v. 9. Hypersensitivity.
1. Reticuloendothelial system—Collected works. 2. Macrophages—Collected works. I. Friedman, Herman, 1931- . II. Reichard, Sherwood M. [DNLM: 1. Reticuloendothelial system. WH650 R437]
QP115.R47 599'.029 79-25933
ISBN 978-1-4684-5160-3 ISBN 978-1-4684-5158-0 (eBook)
DOI 10.1007/978-1-4684-5158-0

Softcover reprint of the hardcover 1st edition 1986

A Division of Plenum Publishing Corporation
233 Spring Street, New York, N.Y. 10013

Contributors

PHILIP W. ASKENASE • Department of Medicine, Yale University School of Medicine, New Haven, Connecticut

MICHAEL K. BACH • Department of Hypersensitivity Diseases Research, The Upjohn Company, Kalamazoo, Michigan

ANDRÉ CAPRON • Centre d'Immunologie et de Biologie Parasitaire, INSERM U-167, CNRS 624, Institut Pasteur, Lille Cedex, France

MICHAEL D. CLAYMAN • Renal Electrolyte Section, Department of Medicine, University of Pennsylvania School of Medicine, Philadelphia, Pennsylvania

MARION C. COHEN • Department of Pathology, University of Connecticut Health Center, Farmington, Connecticut

STANLEY COHEN • Department of Pathology, University of Connecticut Health Center, Farmington, Connecticut

JEAN-PAUL DESSAINT • Centre d'Immunologie et de Biologie Parasitaire, INSERM U-167, CNRS 624, Institut Pasteur, Lille Cedex, France

RICHARD L. EDWARDS • Department of Medicine, University of Connecticut School of Medicine, Farmington, Connecticut, and the Veterans Administration Medical Center, Newington, Connecticut

VEETA A. EWAN • Department of Medicine, University of Connecticut School of Medicine, Farmington, Connecticut, and the Veterans Administration Medical Center, Newington, Connecticut

STEPHEN J. GALLI • Departments of Pathology, Beth Israel Hospital and Harvard Medical School, and the Charles A. Dana Research Institute, Beth Israel Hospital, Boston, Massachusetts

CARL S. GOLDSTEIN • Renal Electrolyte Section, Department of Medicine, University of Pennsylvania School of Medicine, Philadelphia, Pennsylvania

JANICE K. GUTOWSKI • Department of Pathology, University of Connecticut Health Center, Farmington, Connecticut

JUDITH HEAD • Departments of Cell Biology, Pathology, and Internal Medicine, University of Texas Health Science Center, Dallas, Texas

MICHEL D. KAZATCHKINE • INSERM U-28 and Unité d'Immunopathologie, Hôpital Broussais, Paris, France

JAMES W. KAZURA • Division of Geographic Medicine, Department of Medicine, Case Western Reserve University and University Hospitals, Cleveland, Ohio

ANTHONY KULCZYCKI, JR. • Department of Internal Medicine, Division of Allergy and Immunology, Washington University School of Medicine, St. Louis, Missouri

GRAHAM F. MITCHELL • Immunoparasitology Unit, The Walter and Eliza Hall Institute of Medical Research, Melbourne, Victoria, Australia

ERIC G. NEILSON • Renal Electrolyte Section, Department of Medicine, University of Pennsylvania School of Medicine, Philadelphia, Pennsylvania

URS E. NYDEGGER • Central Hematology Laboratory Inselspital, Bern, Switzerland

FREDERICK R. RICKLES • Department of Medicine, University of Connecticut School of Medicine, Farmington, Connecticut, and the Veterans Administration Medical Center, Newington, Connecticut

MICHAEL R. SIMON • Medical and Research Services, Veterans Administration Medical Center, Allen Park, Michigan, and Department of Medicine, Wayne State University School of Medicine, Detroit, Michigan

JOAN STEIN-STREILEIN • Departments of Cell Biology, Pathology, and Internal Medicine, University of Texas Health Science Center, Dallas, Texas

J. WAYNE STREILEIN • Departments of Cell Biology, Pathology, and Internal Medicine, University of Texas Health Science Center, Dallas, Texas

LEON WEISS • Laboratory of Experimental Hematology and Cell Biology, University of Pennsylvania School of Veterinary Medicine, Philadelphia, Pennsylvania

PAUL R. WOOD • Department of Microbiology, University of Melbourne, Parkville, Victoria, Australia

BURTON ZWEIMAN • Allergy and Immunology Section, University of Pennsylvania School of Medicine, Philadelphia, Pennsylvania

Foreword

This comprehensive treatise on the reticuloendothelial system is a project jointly shared by individual members of the Reticuloendothelial (RE) Society and biomedical scientists in general who are interested in the intricate system of cells and molecular moieties derived from those cells which constitute the RES. It may now be more fashionable in some quarters to consider these cells as part of what is called the mononuclear phagocytic system or the lymphoreticular system. Nevertheless, because of historical developments and current interest in the subject by investigators from many diverse areas, it seems advantageous to present in one comprehensive treatise current information and knowledge concerning basic aspects of the RES, such as morphology, biochemistry, phylogeny and ontogeny, physiology, and pharmacology as well as clinical areas including immunopathology, cancer, infectious diseases, allergy, and hypersensitivity. It is anticipated that, by presenting information concerning these apparently heterogeneous topics under the unifying umbrella of the RES, attention will be focused on the similarities as well as interactions among the cell types constituting the RES from the viewpoint of various disciplines. The treatise editors and their editorial board, consisting predominantly of the editors of individual volumes, are extremely grateful for the enthusiastic cooperation and enormous task undertaken by members of the biomedical community in general and especially by members of the American as well as European and Japanese Reticuloendothelial Societies. The assistance, cooperation, and great support from the editorial staff of Plenum Press are also valued greatly. It is hoped that this unique treatise, the first to offer a fully comprehensive treatment of our knowledge concerning the RES, will provide a unified framework for evaluating what is known and what still has to be investigated in this actively growing field. The various volumes of this treatise provide extensive in-depth and integrated information on classical as well as experimental aspects of the RES. It is expected that these volumes will serve as a major reference for day-to-day examination of various subjects dealing with the RES from many different viewpoints.

Herman Friedman
Mario R. Escobar
Sherwood M. Reichard

Preface

All of the immunologic phenomena observed by the end of the 19th century supported the view that they were related to defense mechanisms of the host. This view was, however, being disputed by experimental tests which brought about a short period of confusion. Karl Landsteiner's findings and, especially, the discovery of anaphylaxis by Charles Richet and Paul Portier in 1902 heralded in the very beginning of this century the birth of the domain which we now call Hypersensitivity. This triggered a series of clinical investigations which, not infrequently, generated a number of misconceptions. However, the crescendo of progress that has ensued in the general field of immunology during the last 25 years has brought forth new understanding and pragmatic approaches to the patient with allergic disease. Indeed, one working in the field of immunology senses a major revolution of immunobiologic thinking, much of which has relevance to the area of Hypersensitivity.

In the present volume of this comprehensive treatise on the Reticuloendothelial System (RES) we strive to bring together relevant contributions from experts in the field. The further understanding of those mechanisms that have been elucidated by *in vitro* experiments and the use of appropriate animal models, has in numerous instances paved the way towards a better interpretation of results from human studies and has reached the point of clinical application.

Dr. Weiss, in the first chapter of this volume, reviews the numerous activities of the hematopoietic system, which include diverse migratory pathways, cell sorting, cell recognition, cell differentiation, and cell functions. These activities require a complex regulation at every level of organization and function. Dr. Zweiman elaborates on the leukocyte–vascular endothelium interactions which are important to the understanding of the circulation of leukocytes and their role in localized immune and inflammatory responses. Drs. Streilein, Stein-Streilein, and Head show how the so-called regional spheres of immunologic influence, including the gastrointestinal, skin, lung, and reproductive systems, may prove to be crucial to our understanding of specific types of immune reactions and the pathogenesis of certain human inflammatory disorders. These first three chapters permitted tracing the origin and fate of various constituents of the RES in terms of structural distribution.

Dr. Kulczycki describes the role of the IgE receptor in the release and/or production of mediators from mast cells, and basophils, and Fc receptors of high and lower affinity, and/or less specificity for IgE. Drs. Dessaint and Capron lucidly present their own *in vitro* experiments on anaphylactic antibody-dependent eosinophil or macrophage cytotoxicity as they relate to receptor-mediated mechanisms of activation. These observations were made not only in human and experimental schistosomiasis, but also in several models of experimental filariasis. Confirmation was obtained by the direct evidence of the *in vivo* relevance of anaphylactic antibodies complexed to parasite antigens. They show how the interaction of phagocytic cells (e.g., macrophages and eosinophils) with anaphylactic antibodies in human defense against helminth parasites, reveals that anaphylactic antibodies in the form of immune complexes, besides their well-known function in immediate hypersensitivity, can also have an alternate function in protective immunity. Dr. Simon recounts the multiplicity of immunologic abnormalities associated with specific human allergic diseases and their relation to mediators of immediate hypersensitivity. Dr. Kazatchkine and Nydegger expand the concept of immune complexes and other effector components through an examination of mechanisms of complement activation. These authors discuss effector events both in the circulation and at the site of the tissue lesion. Dr. Kazura addresses another aspect of effector function, effector cells and antibody classes participating in antibody-dependent cell-mediated cytotoxicity reactions (ADCC) and potential clinical applications. Drs. Cohen, Gutowski, and Cohen discuss the multiple biological properties of the inflammatory lymphokines both *in vitro* and *in vivo*. Drs. Edwards, Ewan, and Rickles highlight the importance of coagulation in inflammation with particular reference to monocyte/macrophage procoagulants (PCA's) and their regulation. This regulation is dependent upon a complex network of interacting pathways. Dr. Bach reviews the areas of active research involving the leukotrienes and their role in human disease. Drs. Galli and Askenase focus on the current understanding of the mechanisms regulating the cutaneous basophil hypersensitivity (CBH) and other delayed onset reactions containing large numbers of basophils. In this connection, they present a valid argument for the need to reassess the nomenclature of CBH.

The topics which we have introduced above have dealt mainly with the cellular components and the specific release of mediators of hypersensitivity reactions, their distribution, role and mechanisms of action, as well as their interactions. The last two chapters address some of the biochemical consequences of these interactions. Drs. Clayman, Goldstein, and Neilson review the role which immunogenetics, the RES, and the cellular influence on the development of antigen-recognition and immunoregulation have in renal injury. They consider events associated with responses to antigens of both renal and nonrenal origin. Finally, Drs. Mitchell and Wood deal with the extracellular macrophage-related events in the immunology of parasitism. They emphasize the role of the macrophage in, or from, parasitized hosts as a regulator of antiparasite immune responses and as an effector cell of host resistance involving mediator release. They present their studies pertaining to the immunoregulatory role of parasitized macrophages in murine cutaneous leishmaniasis and examine the role of macrophage products in the control of infection with plasmodia and related intraerythrocytic protozoa.

In creating this volume we realized that the subject matter was somewhat arbitrary and could not be all-inclusive. We are most grateful to all the authors for their enthusiastic response to our request to review current knowledge on the general or specific aspects of the RES and Hypersensitivity.

We owe a very special debt of gratitude to Dr. Peter Abramoff who, more than anyone else, was instrumental in the initial formulation of the concepts and selection of the contributors for this volume. The expert and generous assistance given to us by Mr. Kirk Jensen from Plenum Publishing Corporation and Ms. Charlotte Phillips from the Medical College of Virginia are sincerely appreciated.

S. Michael Phillips
Mario R. Escobar

Contents

3. Regional Specialization in Antigen Presentation

J. WAYNE STREILEIN, JOAN STEIN-STREILEIN, and JUDITH HEAD

4. Structure and Expression of IgE Receptors

Anthony Kulczycki, Jr.

5. Interaction of Phagocytic Cells with Immune Complexes of Anaphylactic Antibodies

Jean-Paul Dessaint and André Capron

6. Leukocyte Function in Human Allergic Disease

Michael R. Simon

7. Complement-Mediated Injury

Michel D. Kazatchkine and Urs E. Nydegger

10. Macrophage Procoagulants, Fibrin Deposition, and the Inflammatory Response

RICHARD L. EDWARDS, VEETA A. EWAN, and FREDERICK R. RICKLES

11. Leukotrienes

MICHAEL K. BACH

1

Cellular Regulation in Hematopoiesis

LEON WEISS

1. INTRODUCTION

The hematopoietic system, which includes the reticuloendothelial system and the immune system, maintains blood cells in circulatory and recirculatory phases and supports large-scale cell migration to specific, diverse, and distant tissues where rapid cell turnover, cell differentiation, and cell functions occur (Metcalf and Moore, 1971). Regulation in this system is intricate, extensive, and efficient, effected through the actions of hormones and other systemic factors, local humoral factors, and cell surface molecules on blood cells, stromal cells, and the vasculature, which respond in tight sequences.

2. THE NATURE OF THE HEMATOPOIETIC SYSTEM AS EXEMPLIFIED BY THE LIFE CYCLE OF MACROPHAGES

The nature of the hematopoietic system may be illustrated by the life cycle of macrophages. [This topic has, of course, been covered extensively elsewhere in these volumes (Carr and Daems, 1980) but certain points will be presented that are useful for this discussion.] Macrophages originate in the adult from pluripotent stem cells (CFU-S and, likely, a less differentiated cell type) in bone marrow. Under appropriate circumstances, CFU-S differentiate into stem cells of more restricted potency, CFU-GM, capable of producing (Cohen and Sigel, 1981; Reichard and Filkins, 1981; Goldberg and Rabinovitch, 1983; Lucas, 1983) both monocytes and neutrophils. Under appropriate circumstances, which include a lower concentration of colony-stimulating factors (CSF), CFU-GM differentiation into promonocytes is favored (Lord, 1983; Metcalf and Moore, 1971). Unlike the erythrocytic and neutrophilic lineages, which hold large numbers of reticulo-

LEON WEISS • Laboratory of Experimental Hematology and Cell Biology, University of Pennsylvania School of Veterinary Medicine, Philadelphia, Pennsylvania 19104.

cytes and metamyelocytes in ready reserve in the bone marrow, promonocytes differentiate quickly and enter the blood of the venous sinuses as monocytes or as small inactive macrophages. Whether monocytes enter the systemic circulation directly or, like reticulocytes, eosinophils (in the rat) (Spry, 1971), and T cells freshly released from the thymus (Sprent, 1977), are first conditioned in the spleen for a few hours to a few days is not known. But monocytes do leave the circulation and enter the serous cavities, the connective tissue spaces—indeed, virtually any of the perivascular tissues—and remain as monocytes or differentiate into macrophages. As extensively documented in these volumes, macrophages have diverse functions. Macrophages in different locations may look different and behave distinctively. Perhaps there are subgroups of macrophages. For example, within the red pulp of the spleen, macrophages may lie within cords or pulp spaces (Simon, 1980; Weiss, 1983d). They may lie perivascularly, extending processes through the wall of venous sinuses or pulp veins into the lumen, thereby surveilling the blood cells that travel the pulp spaces, from terminal arterial vessels into veins. Macrophages may form tight clusters held within the interstices of a distinctive reticulum and, in some species, associate with antigen-presenting cells and surround terminal arterial vessels as *periarterial macrophage sheaths* (ellipsoids) (Blue and Weiss, 1981a; Tablin and Weiss, 1983). Within the liver there are "conventional" macrophages in the perilobular connective tissues and beneath the capsule as well as a population of phagocytic cells intercalated in the wall of vascular sinusoids as Kupffer cells (Wisse, 1980). There are subgroups of macrophages—Kupffer cells, fetal pulmonary macrophages (Sorokin, 1977), and, perhaps, the macrophages of the periarterial macrophage sheaths of the spleen—renewed in large part by local proliferation and therefore not as dependent as other groups of macrophages upon renewal by cells emigrated from bone marrow. There is a steady large-scale movement of macrophages from the serous cavities, where many of them collect (Daems, 1980) into lymph nodes and into blood via lymphatic vessels.

Macrophages may die as macrophages but they may live out a cycle in which they differentiate into epithelioid cells and into multinucleate giant cells of which the Langerhans foreign giant body cells and osteoclasts are subsets. The functions of epithelioid cells and multinucleate giant cells, inferred from their high concentrations of mitochondria and richly folded plasma membrane, depend upon their capacity for active transport rather than, as with their precursors, upon phagocytosis or protein synthesis (Sutton and Weiss, 1966).

Macrophages attend the differentiation and proliferation of a number of blood cell types. Perhaps the best known arrangement is the erythroblastic islet. Here a central macrophage lies surrounded by tiers of erythroblasts, placed just outside the wall of a vascular sinus in bone marrow (Berman, 1967; Bessis, 1958; Weiss, 1983a). Those erythroblasts closest to the perikaryon of the macrophage are the least mature, those that form the peripheral ring are the most mature. Further, those erythroid cells that lie against the wall of the vascular sinus are the most mature (orthochromatic erythroblasts and reticulocytes) and postmitotic, while those faced toward the hematopoietic compartment tend to be less mature and mitotic. The islet macrophage sends out slender sheetlike, multiply branched processes that enclose each of the erythroblasts, reticulocytes, and separated nuclear poles in the islet. Islet macrophages phagocytize both separated nuclear poles (which appear as erythroblasts become erythrocytes) and whole imperfect erythroblasts that represent ineffective erythropoiesis. In bone marrow the appropriate combination of macrophage, erythroid stem cells and the wall of the vascular sinus (the latter including both adventitial reticular cell and endothelial cell) and, perhaps, lymphocytes would appear to contribute a cellular microenvironment needed for erythropoiesis (Brookoff and Weiss, 1983a,b). From the distribution of erythroid colonies in marrow reconstituting after irradiation, moreover, it would appear that erythropoietic microenvironments are preferentially located periarterially in the bone marrow (Lambertsen and Weiss, 1983). The importance of macrophages in creating an erythroid microenvironment is underlined by the failure of chronically hypertransfused mice to become erythropoietic when bled (Brookoff and Weiss, 1983b). One consequence of hypertransfusion in these mice is a marked depletion of bone marrow macrophages, a depletion that may be sufficient to prevent erythropoiesis. Macrophages are known to secrete factors regulating hematopoiesis—colony-stimulating factors (CSF) that enhance hematopoiesis and prostaglandins that inhibit hematopoiesis. It is not known whether the macrophages of erythroblastic islets secrete such factors, but the morphological arrangements suggest that they do. They may have other functions, e.g., helping reticulocytes move across the wall of the vascular sinus, en route from the hematopoietic perivascular tissue into the circulation (Weiss, 1976). While the association of macrophages and erythroblasts is best known, early recognized by Bessis (1958), macrophages may similarly attend other lines of hematopoiesis, as in the heightened eosinophilopoiesis induced by *Ascaris* infection (Sakai *et al.*, 1981). Macrophages have at least two other locations in marrow. One is against the adventitial surface of vascular sinuses, with a large phagocytic process extending through the wall into the lumen. A second is within the lumen of the vascular sinus (Weiss, 1983e).

It would appear likely that bone marrow macrophages, as other macrophages, develop from bone marrow monocytes. Whether they reenter the marrow as monocytes or macrophages from the circulating blood or simply differentiate from marrow monocytes without a circulatory phase is not known.

Experimental studies establishing the migratory patterns of hematopoietic cells, the structure of stromal cells, the composition of cell clusters associated with hematopoiesis and blood cell movement and function, assays that define various stem cells, and the identification of soluble factors with substantive hematopoietic effects have set the stage for an understanding of the cellular basis for the regulation of hematopoiesis.

3. MIGRATORY CHARACTER OF THE HEMATOPOIETIC SYSTEM

The hematopoietic system is a migratory system of cells, beyond the obvious circulation of blood cells (Metcalf and Moore, 1971). Hematopoietic stem cells of different levels of differentiation and other cells that had been considered tissue-bound, have migratory phases as part of their life cycle. Understanding of this dynamic state was advanced by the experimental studies of Gowans (1961) and his colleagues demonstrating the migration of lymphocytes from blood to lymph through lymph nodes, the presence of a recirculating pool of lymphocytes (Gowans and Knight, 1964; Marchesi and Gowans, 1964), and the immunological competence of lymphocytes. There has followed a good deal of work establishing the migratory properties of differentiated cells of the immune system (lymphocytes, macrophages, and antigen-presenting cells) and hematopoietic stem cells. In the embryo, hematopoietic cells originate from migratory cells of the middle cell mass that find their way to the yolk sac (Metcalf and Moore, 1971). Hematopoietic cells migrate into the specialized hematopoietic tissues (liver, bone marrow, thymus, spleen, lymph nodes) as they become established, find suitable loci, and thereby become set up for differentiation, proliferation, and storage (Weiss, 1983b). In this system, where hematopoietic cells migrate in selective rigorously maintained pathways throughout the body, regulation is complex, involving not only the control of stem cell differentiation as occurs in confined loci as, say, of the keratinocytes in the epithelial columns of epidermis (Potten, 1983), but also the control of cell homing, cell sequestration, cell sorting, and cell storage, as occurs in cells migrating great distances to selective tissues.

4. HEMATOPOIETIC MICROENVIRONMENTS

There are within the various hematopoietic tissues a variety of distinctive zones each of which appears capable of supporting preferentially a given type of hematopoiesis. This capacity is expressive of distinctive hematopoietic inductive microenvironments (HIM) defined by Curry and Trentin (1967; Curry *et al.*, 1967; Trentin, 1970; La Pushin and Trentin, 1977) and rooted in analysis of the hematopoietic colonies that appear in mouse bone marrow and spleen after an infusion of hematopoietic stem cells into mice hematopoietically depleted by lethal irradiation. Bone marrow colonies are preponderantly neutrophilic, spleen colonies erythroblastic, indicating tissue differences in HIM. Transplantation experiments indicate that these HIM are sessile, therefore likely having stromal cells as an essential element (see below). Within bone marrow, pluripotent stem cells collect preferentially in subosteal regions, erythropoiesis occurs preferentially away from subosteum, in periarterial zones, and megakaryocyte precursors and megakaryopoiesis throughout the marrow at the walls of vascular sinuses (Lambertsen and Weiss, 1983; Young and Weiss, 1983). T cells are located in the periarterial lymphatic sheath of spleen, and the diffuse cortex of lymph nodes and B cells in lymphatic nodules (de Sousa, 1981; Weiss, 1983b). Erythropoiesis in the spleen is preferentially located in the red pulp beneath the capsule and along trabeculae. This preferential distribution of hematopoiesis indicates that there are loci, niches, or HIM which are discrete sites capable of inducing and supporting a given type of hematopoiesis.

The morphological properties of these loci have not been established, although there is circumstantial or indirect evidence of their structure. As indicated above, the HIM is sessile, unlike migratory hematopoietic cells, and therefore likely dependent upon stromal cells. Trentin (1970) transplanted discrete bits of marrow into the spleen and then carried out the spleen colony assay for CFU-S. Colonies in the splenic tissue were preponderantly erythroid. Colonies in the marrow tissue were preponderantly myeloid. In those colonies that lay over the spleen–marrow boundary, that part of the colony in bone marrow was predominantly myeloblastic, that in spleen erythroblastic. The bits of marrow evidently carried their HIM with them. Further evidence for the stromal nature of HIM comes from analysis of the congenital hypoplastic anemias of the Sl/Sl^d and W/W^v strains of mice (Bernstein *et al.*, 1968; Bernstein, 1980; McCuskey and Meineke, 1973). The W/W^v anemia is due to a deficiency in hematopoietic stem cells and may be ameliorated by infusion of stem cells. But the Sl/Sl^d animal has normal numbers of potent stem cells and can, in fact, donate stem cells to "cure" the W/W^v. The Sl/Sl^d defect can be effectively treated not by a suspension of hematopoietic cells, but only by a transplant that includes stroma. The Sl/Sl^d defect therefore is in the microenvironment that induces hematopoiesis.

A number of stromal cells, viz. thymic epithelium, bone-lining cells, reticular cells, antigen-presenting cells, macrophages, and a number of blood cells and their precursors, viz. stem cells, lymphocytes, neutrophils, may have a hand in hematopoietic regulation and would, therefore, contribute to the HIM. I will continue this discussion by turning to hematopoietic stem cells and then to cell types important in regulating hematopoiesis and to the regulatory factors.

5. HEMATOPOIETIC STEM CELLS

The hematopoietic microenvironment appears to control the differentiation of pluripotent stem cells and perhaps their immediate progeny. For example, the differentiation of a pluripotent stem cell (CFU-S and/or its precursor) into the earliest known red cell precursors (Axelrod *et al.*, 1974; McLeod *et al.*, 1974; Hora and Ogawa, 1977; Adamson and Torok-Storb, 1978), *burst forming unit–erythroid* (BFU-E), likely depends upon the CFU-S entering an erythroid niche or microenvironment and being controlled by local factors produced there. BFU-E, which possess few if any erythropoietin receptors, cannot respond to erythropoietin. They depend upon being in an erythroid niche for differentiation. Indeed, BFU-E, as part of their biological role, can circulate and seek out suitable erythroid niches and proliferate and differentiate into erythropoietin-responsive cells there. It is the more differentiated stem cells and hematopoietic cells that are largely regulated by hormones or humoral factors. In the red cell system, it is stem cells differentiated from BFU-E and their progeny [the *colony forming unit–erythroid* (CFU-E) and basophilic and polychromatophilic erythroblasts] that are erythropoietin-responsive. These cells have differentiated to the degree that their surface is well supplied with erythropoietin receptors. Responding to the stimulation of erythropoietin and, likely, of other less well-defined factors, they proliferate and differentiate.

Morphology has not adequately served the hematopoietic system. It has not been possible to identify early hematopoietic stem cells by their structure. When this becomes possible, a microscopist will be able to spot pluripotent stem cells and characterize their microenvironment quite neatly by describing the cellular and extracellular structures that immediately surround them, readily amplifying the morphology by chemical information (using cytochemical methods) and kinetics (using autoradiographic data). Although early stem cells cannot yet be recognized, it has become possible, by means of experimental animals and tissue cultures, to assay the numbers of different types of these early stem cells in a tissue. By counting hematopoietic colonies in the spleen of a lethally irradiated mouse given a tissue (in the form of a cell suspension) containing an unknown number of stem cells, one can assay the number of pluripotent stem cells in that tissue because, as has been shown by karyotype analysis of uniquely damaged stem cell chromosomes (Becker *et al.*, 1963; Wu *et al.*, 1967), each spleen colony is a clone, derived from a pluripotent stem cell. This assay does not, of course, give directly a measure of the full number of pluripotent stem cells since many of the stem cells go to tissues other than the spleen (i.e., bone marrow) and not all stem cells need/can form colonies (see below). This *spleen colony assay*, introduced in the late 1950s (Till and McCulloch, 1961), was the first of the assays for pluripotent stem cells—the colony forming unit–spleen (CFU-S)—and is still in use. Tissue culture assays have been developed which quantitate the numbers of stem cells of restricted potency, derived from the CFU-S, committed to one or more lines of hematopoiesis. Such restricted cells may be defined as stem cells because, remaining capable of further proliferation and differentiation, they provide more differentiated cells and renew themselves. Each of these tissue culture methods provides certain distinctive factors and conditions which favor the proliferation and differentiation of one or two blood cell lines. In assays of erythroid stem cells, for example, erythropoietin is an essential ingredient of the medium because the assays depend upon the development of colonies of erythropoietin-responsive cells. By the use of these cultures the existence of two erythroid precursors, BFU-E and CFU-E, can be inferred. The CFU-E is the prodigious "workhorse" of the erythropoietic system because, being the first of the erythroid stem cells fully responsive to erythropoietin, its numbers may be induced to increase under erythroid stress which generates erythropoietin. This has been shown in clinical and experimental anemias, as the hemolytic anemia due to the greatly increased hemolysis of spectrin-deficient erythrocytes. Regarding leukocyte stem cell assay, tissue culture methods have permitted recognition and quantitation of a stem cell capable of differentiating into either neutrophils or monocytes [the *colony forming unit–granulocytic and monocytic* (CFU-GM)], of a stem cell giving rise to eosinophils (CFU-Eo), of one giving rise to megakaryocytes (CFU-Meg), and, perhaps, of one from which basophils spring (CFU-Bas) (Potten, 1983; Metcalf, 1981; Till, 1982). There are also tissue culture systems that permit the assay of pluripotent stem cells, complementing the spleen colony assay. These cultures show, moreover, associations of the vasculature, stromal cells, and hematopoietic cells similar to those present in intact marrow (Allen, 1981; Dexter, 1982).

By means of the above assays, various types of stem cells have been shown to exist and have been quantitated. Each of the assays reveals the antecedent presence of a stem cell by the colony (clone) it produces, but does not reveal the stem cell itself. The assays give no information at all on the structure of the stem cell. They do not provide a population of pure stem cells which may be analyzed or studied. Many factors are needed to allow the growth of colonies from a stem cell and only if these factors are supplied will colonies grow and can the presence of stem cells be inferred. As indicated above, many stem cells must go uncounted in any assay because it is not likely that the conditions will be supplied for each of them to produce a colony. A direct visualization of a specific stem cell continues to elude us. van Bekkum *et al.* (1971; Dicke *et al.*, 1973), however, have glimpsed pluripotent stem cells in experiments in which the CFU-S in bone marrow were greatly enriched (as determined, of course, by spleen colony assay) by differential centrifugation (empirically, stem cells possess a characteristic midrange density) and pretreatment of the donor with antimitotic drugs (pluripotent stem cells tend to be in G_0 and thereby protected from the actions of these drugs, while their progeny are in the mitotic cycle and thereby vulnerable). Correlated with the increased numbers of CFU-S were increased numbers of a morphological cell type, the *candidate stem cell.* Not surprisingly, this cell type lacks the pigment, granules, and nuclear changes of differentiated erythrocytes and granulocytes, and looks like a lymphocyte. These (and more differentiated stem cells) have been classed among the null lymphocytes (Hood *et al.*, 1978), i.e., among those lymphocytes that are neither T cells nor B cells. But many scientists in the hematological community, still aware of the costly polemical monophyletic versus polyphyletic controversies, have been reluctant to classify pluripotent stem cells as lymphocytes and therefore their designation, as any distinctive individual characteristics they may have, has not yet been achieved.

6. CELL TYPES ASSOCIATED WITH HEMATOPOIETIC REGULATION

While it is not yet possible to accurately and specifically identify early hematopoietic stem cells in histological sections, it is possible, nonetheless, to address the question of the structure of hematopoietic microenvironments. These microenvironments undoubtedly differ in structure and in potency, depending upon their location. Microenvironments may be created by the close association of several cell types of which stromal cells may be the most prominent.

6.1. THYMIC EPITHELIUM

One of the most distinctive and, perhaps, decipherable of microenvironments is in the cortex of the thymus. Here, large sessile epithelial cells of different conformation are found (Bearman *et al.*, 1978; Kendall, 1981; Singh, 1981; Weiss, 1983b). A major conformation is as branched cells (epithelial–reticular cells) which form a cortical meshwork in whose interstices developing T cells lie.

A second major conformation is as flattened cells that form a simple pavement or boundary about the perimeter of the thymic lobule. Blood vessels penetrating the lobule carry in with them an investment of these boundary epithelial cells, thereby providing the outermost layer of the blood–thymus barrier. Epithelial cells also make up thymic corpuscles and the rather solid epithelial tissue of the medulla. Further, epithelial cells may form cysts and residues of the embryonic branchial epithelium. Which cortical epithelial cell types contribute most to the developing T cell's microenvironment is not known, but the epithelial reticular cells would appear well arranged to induce T-cell differentiation (Jordan and Robinson, 1981). They form a system of communicating locules with their broad sheetlike processes. Processes from contiguous cells meet and are held together by desmosomes. The epithelial reticular processes envelop developing T cells and macrophages. When epithelial reticular cells are isolated from a disrupted thymus, they may form closed epithelial sacs filled with lymphocytes. Epithelial reticular cells possess the cytoplasmic organelles characteristic of secretory cells. In tissue cultures of thymic epithelium in which the epithelial reticular cells are likely a major component, *thymosin* and other factors that stimulate T-cell proliferation and differentiation are produced. Epithelial reticular cells thus have the characteristics of "nurse cells" and may well secrete those low-molecular-weight polypeptides that regulate T lymphopoiesis in the thymus (Bach and Papiernik, 1981; Dardenne and Bach, 1981).

Epithelial reticular cells may have a regulatory role in addition to inducing differentiation and proliferation in T-cell development. Among the T cells generated in the thymus are those which may attack "self" and thereby cause autoimmune damage. If permitted to proliferate and form clones, the so-called "forbidden clones," these cells might well cause great damage. The thymus likely surveils developing T cells and causes those that would injure "self" to be destroyed or rendered unable to proliferate. Developing T cells and the epithelial reticular cells that enclose and support them in the thymic cortex have broad areas of plasma membrane in close apposition. It is possible that by this membranous interaction the thymic epithelial cells impose a test on the developing T cells. Only those congenial to self are permitted to proliferate; those hostile to self are suppressed. It is becoming evident that the idea of "self" may be expressed by proteins on the cell surface (called antigens because they are revealed by immunoglobulin reagents) coded in the major histocompatibility complex (MHC). Rouse and his colleagues (Rouse *et al.*, 1979; van Ewijk *et al.*, 1980; Rouse and Weissman, 1981) have demonstrated by electron microscopic cytochemistry the presence of certain MHC antigens on the surface of mouse thymic epithelial cells and corresponding antigens on the surface of apposed lymphocytes. The epithelial reticular cells and associated T cells in the cortex were positive for Ia antigen while the epithelial cells of the medulla displayed K antigen on their surface. This pattern of staining needs to be studied more fully to assess the possibility that the MHC antigens on the surface of epithelial cells and of lymphocytes mediate the interactions between these cell types which control the development of T cells hostile to self. Jenkinson (1981) has studied MHC antigens in the developing thymus of normal and nude mice.

While our discussion has concerned the epithelial cells, macrophages, (Jenkinson, 1981), antigen-presenting cells, developing T lymphocytes, and other cells are also present in the thymus and may well contribute to the hematopoietic microenvironment there (Janossy *et al.*, 1980, 1981). Discussion of cell types other than epithelial cells follows.

6.2. BONE-LINING CELLS

As concerns the control of hematopoiesis in bone marrow, the finding that ectopic marrow, as under the renal capsule or in the rectus sheath, typically develops only after bone develops, and develops within bone, suggests that bone induces the marrow type of hematopoiesis. Granulocytopoiesis is the type of hematopoiesis most favored in mammalian bone marrows, since, as discussed above, the marrow is preponderantly granulocytic. Granulocytopoiesis in cultures of bone marrow, moreover, occurs in association with the distinctive fibroblastic adipocyte–reticular cell adventitial to the vascular sinuses. Erythropoiesis, in contrast, occurs in association with macrophages. These associations are discussed more fully below.

Those cells lining the inside surface of bone, the bone-lining cells or endosteum (osteoblasts, osteoclasts, fibroblasts, pluripotent hematopoietic stem cells, macrophages, osteocytes, osteoprogenitor cells, reticular cells, and cells as yet unclassified), must be assessed with regard to hematopoietic inductive capacity (Deldar and Weiss, 1983, 1985; McManus and Weiss, 1984). Fibroblastic cells (including osteoblasts, reticular cells, and fibroblasts) are consistently present in this lining layer. There is an important body of experiments in developmental biology initiated by Grobstein's (1954, 1975) demonstrations that the development of the salivary gland depends upon an interaction of its mesenchymal (fibroblastic) capsule and its epithelium: the fibroblastic cells induce the differentiation of other cells (Slavkin and Greulich, 1975). Salivary gland fibroblasts, moreover, have a more powerful effect than remote fibroblasts. The likelihood that bone, or cells associated with bone, induces hematopoiesis is supported by the findings that CFU-S are concentrated subosteally, CSF are produced there in high concentration, and undifferentiated hematopoietic colonies are preferentially found subosteally in endocloned irradiated mouse bone marrow (Chan and Metcalf, 1972; Lord *et al.*, 1975; Gong, 1978; Deldar and Weiss, 1983, 1985). The inducing capacity of fibroblasts on hematopoiesis has been most explicitly addressed by Friedenstein *et al.* (1974).

6.3. RETICULAR CELLS

Within the marrow, and within lymph nodes and the pulp of the spleen, the disposition and structure of reticular cells suggest that the reticular meshwork or reticulum which they form provides the locules or spaces within which lie the microenvironments that induce hematopoiesis (Pictet *et al.*, 1969; Chen and

Weiss, 1972; Friedenstein *et al.*, 1974; Weiss, 1974a,b, 1976, 1980; La Pushin and Trentin, 1977; Sakai *et al.*, 1981; Blue and Weiss, 1981a,b,c; Tablin and Weiss, 1983). Reticular cells are a fibroblastic stromal cell type that form a meshwork that spans the bone marrow, supporting blood vessels, nerves, and hematopoietic cells. Reticular cells lie upon reticular fibers, silver-impregnable fibers made of collagen type IV, which they presumably produce. The reticular meshwork therefore has a cellular component consisting of reticular cells (cellular reticulum) and a fibrous component consisting of reticular fibers (fibrillar reticulum). Reticular cells are branched or dendritic stromal cells. Other dendritic stromal cells, discussed below, are macrophages and interdigitating cells of the Langerhans system. Reticular cells along with bone-lining cells in marrow and the other dendritic stromal cells, would appear to be important components of the hematopoietic microenvironment. In the heyday of the monophyletic–polyphyletic dichotomy, it was widely held that reticular cells forming the supporting reticular meshwork of hematopoietic tissues would, in hematopoietic stress, round up and come off the reticulum to become hemocytoblasts (pluripotent hematopoietic cells). But convincing experimental support for this source of stem cells was not forthcoming. As evident from kinetic studies using long-term administration of [^{3}H]thymidine, reticular cells turn over slowly, even in hematopoietic stress. Karyotype studies, moreover, have failed to establish a clonal relationship between hematopoietic stem cells and fibroblasts. There is, further, a good deal of evidence that pluripotent stem cells are, even in embryos and in hematopoietic stress, small circulating cells which can be classified, as discussed above, among the null lymphocytes. But while the hematopoietic stroma does not itself provide the hematopoietic stem cells and the blood cells, there is strong and increasing evidence that it plays a part in the regulation of hematopoiesis. The root of this evidence is in Grobstein's (1954, 1975) demonstration, discussed above, that the development of the salivary gland is controlled by its mesenchyme. A large body of experimental embryological evidence drawn from diverse systems (reviewed by Weiss, 1976) including the hematopoietic system (Friedenstein *et al.*, 1974) has underscored the value of the stroma in hematopoietic regulation. Evidence for this role of stroma has been buttressed by morphological studies, biochemical studies, and tissue cultures. Some of this work is reviewed below.

There is a close physical association between reticular cells and the hematopoietic cells held in their locules (and between the hematopoietic cells themselves). Junctional complexes, many held by an intercellular matrical adhesive, are abundant. While not readily visible in conventional electron micrographs, these complexes may be shown by fixing the tissue in hypertonic solutions, which draw the cells apart. The cells remain together only at the complexes, which therefore stand out boldly. These complexes likely come and go. By holding cells in place for an appropriate period of time, they may provide the positions and cellular associations that underlie those cellular interactions needed for cellular differentiation and function (Sakai and Weiss, unpublished observations). The MHC cell surface antigens may be important in all hematopoietic tissues, as discussed above for the thymus, in mediating the cell recognition and the phenomena dependent upon cell recognition, fundamental in this system in which migratory cells move into a meshwork of sessile stromal cells and are sorted out and function there.

Reticular cells may well be involved not only in sorting out and sequestering migratory cells, but in further aspects of the regulation of hematopoiesis. How much, or how distinctively, the reticular cells contribute vis-à-vis bone-lining cells is not clear. In addition, other stromal cells (as antigen-presenting cells and macrophages) and the blood cells themselves (as lymphocytes, neutrophils, and, of course, monocytes) may contribute to the regulation of hematopoiesis by participating in the creation of HIM. Close associations, including junctional complexes, occur among all of these cell types (Sakai and Weiss, unpublished observations).

A subset of reticular cells contributing to the reticulum forms the outermost coat of blood vessels—they are the *adventitial reticular cells*—and branches into the reticulum that surrounds and supports the blood vessels (Weiss, 1976). In bone marrow the venous sinuses are those large-bored venules (postcapillary in location) into which newly produced blood cells are delivered. Adventitial reticular cells, covering the outside wall of the vascular sinuses as they do, control access to these vessels and, therefore, the delivery of blood cells to the circulation. These cells have additional functions. They may accumulate lipid and become the adipocytes of the marrow. Adipocytes have a long understood function in the control of marrow volume, appearing and taking up space as hematopoiesis wanes. They aromatize testosterone to estrogen with hematopoietic consequences (testosterone drives erythropoiesis and estrogen stabilizes bone). Reticular cells and the adipocytes into which they may modulate, induce granulocytopoiesis. Reticular cells, under hematopoietic stress, may become "activated" in bone marrow, spleen, and, in all likelihood, other hematopoietic organs. These activated cells are dark, branched cells whose branches are closely associated with hematopoietic cells. They may fuse to form complex multinucleate syncytial reticula pervading the pulp of the spleen and the hematopoietic cords of the marrow. The nuclei of these activated cells regularly display several well-developed nucleoli. A salient characteristic is an extraordinarily rich system of endoplasmic reticulum, continuous with the perinuclear cisternae. This cell type, or likely its subtypes, is found in a number of cases of experimentally intensified hematopoiesis (La Pushin and Trentin, 1977), in the heightened erythropoiesis in spectrin-deficient animals (Brookoff *et al.*, 1982) and in malaria (Weiss, 1983e; Weiss, Geduldig, and Weidanz, 1986), in the abortive erythropoietic response in hypertransfused then bled mice (Brookoff *et al.*, 1982), and in the heightened eosinophilopoiesis in the course of ascariasis (Sakai *et al.*, 1981; Brookoff *et al.*, 1982). This activated reticular cell or its subtypes may well be a concomitant of heightened or intense hematopoiesis. Its role would appear to include the selective trapping or sequestration of migratory cells. These cells may be hematopoietic stem cells which, after being trapped, will go on to differentiation and proliferation. The unique capacity of the spleen to control malaria appears to depend upon these activated reticular cells trapping parasitized erythrocytes (PE), setting them up for destruction by macrophages.

Indeed, the resolution of malaria, "crisis," appears to depend upon activation of reticular cells in the red pulp of the spleen followed by large-scale trapping of PE, sweeping them out of the circulation (Weiss, Geduldig, and Weidanz, 1986).

Further support for a hematopoietic inductive role for reticular cells comes from the histiotypic long-term cultures of bone marrow studied by Dexter (1982) and Allen (1981). Adipose cells in these cultures appeared to possess an inductive role in granulocytopoiesis (neutrophilopoiesis), indicated by the presence of colonies of granulocytes on broad cytoplasmic processes of the adipocytes.

6.4. MACROPHAGES

Reticular cells are prototypic stromal cells, being sessile fibroblastic cells that support the parenchyma (hematopoietic cells) and the vasculature. But the distinction between stromal and hematopoietic cells has, by usage, become blurred. Macrophages within hematopoietic tissues are regularly classified as stromal cells, but they are not sessile, and derived as they are from monocytes, they are linked to blood cells (Friedman, 1981). Macrophages regulate hematopoiesis. Macrophages, as discussed at the outset of this chapter, produce factors that stimulate hematopoiesis (CSF) and inhibit it (the prostaglandins) as shown in tissue culture assays. Macrophages form clusters with erythroblasts in erythropoietic tissues. Results from experimental hypertransfusion and bleeding suggest that macrophages are essential for erythropoiesis (Gordon *et al.*, 1980; Brookoff *et al.*, 1982). The roles of macrophages in immune reactions constitute regulation of lymphocytopoiesis, both B and T. (See Section 2 on life cycle of macrophages.)

6.5. ANTIGEN-PRESENTING CELLS

Interdigitating cells which are part of the Langerhans antigen-presenting cell system are similar to macrophages in that they are produced in bone marrow and travel in the blood as a blood cell, one that may be classified as a subtype of monocyte or as its own type, a *veiled cell* possessing extensive, branched veil-like processes. Veiled cells and their descendents exhibit a complex life cycle, and are different from the monocyte-macrophage both in cytology (presence of distinctive Birbeck granules in antigen-presenting cells) and in the absence of significant phagocytic capacity (Balfour *et al.*, 1981; Humphrey, 1981; Hoefsmit *et al.*, 1980).

Antigen-presenting cells regulate immune responses by their capacity to bring antigen and lymphocytes together. Indeed, Sakai and Weiss (unpublished observations) have found a noteworthy cell association in spleen (and, presumably, in lymph nodes), viz., that of antigen-presenting cell (identifiable in this tissue phase as an interdigitating cell) and T lymphocytes, wherein the T cells surround the interdigitating cell forming a rosette, all of the cells associated by junctional complexes. This assemblage is present in the marginal zone and the white pulp of the spleen. The arrangement suggests that the antigen-presenting cell, antigen, and lymphocytes may assemble in the marginal zone and travel as an ensemble to a site (in white pulp) where, stimulated by antigen, the lymphocytes may differentiate further, expressing an immunological function. But it is possible that antigen-presenting cells may facilitate lymphocyte movement by clustering lymphocytes and moving with them as an ensemble *without* presenting antigen. There may be a disassociation of antigen-presenting and migratory functions. Interdigitating cells are present in the thymic cortex. Here, protected by a thymus–blood barrier and surrounded by immature T cells, it is not likely that they would be concerned with antigen presentation. Perhaps they function in the segregation and movement of developing T cells through the cortex of the thymus, steps which may be necessary for T-cell maturation (Brelinska and Pilgrim, 1983; Milicevic *et al.*, 1983). They may also convoy lymphocytes through peripheral lymphatic tissue. The follicular dendritic cell may be the B cell counterpart to the T cell's interdigitating cell.

6.6. T LYMPHOCYTES AND OTHER BLOOD CELLS

As evident from the veiled cell–Langerhans cell–interdigitating cell life cycle, blood cells have a tissue phase as well as a circulating phase, and in their tissue phase, they may have important roles in regulating hematopoiesis. As a further example of the regulation of hematopoiesis by blood cells, T cells are needed for the heightened eosinophilopoiesis associated with ascariasis or trichinosis (Beeson and Bass, 1977), and, in fact, appear as a component of eosinophilopoietic clusters in bone marrow (Sakai *et al.*, 1981). Additionally, the role of subsets of T lymphocytes in helping (T_H cells) or in suppressing (T_S cells) the differentiation of B lymphocytes is well known (Hood *et al.*, 1978). While T cells are not required to mount the heightened erythropoietic response to bleeding [nude mice readily display such compensatory erythropoiesis (Zimmerman and Weiss, unpublished results)], Nathans and his colleagues (1978) and Mangan and Desforges (1980) have shown that a fetal variety of BFU-E is enhanced by coculturing with T cells. The Diamond–Blackfan aplastic anemia, moreover, appears due to T cells suppressing erythropoiesis (Hoffman *et al.*, 1976, 1977). The influence of T cells extends beyond the hematopoietic system. In the host reaction to *Nippostrongylus brasiliensis* intestinal infection in rodents, Love *et al.* (1974) showed that the differentiation of host intestinal epithelial stem cells produces more mucous cells and fewer absorptive cells. T cells from infected rats can induce a similar mucosal shift in normal rats. It may well be that T lymphocytes have a general regulating role in cell differentiation and proliferation, perhaps dependent upon an antigenic intermediary.

Blood cells are involved in their own regulation. Neutrophils produce lactoferrin which by binding iron makes it unavailable to bacteria, whose growth is thus curtailed (Baggiolini *et al.*, 1970; Lisiewicz, 1980). But lactoferrin also inhibits the proliferation of neutrophils. A nice feedback loop exists here: Neutrophils are produced in response to a bacterial infection. After bacterial growth is inhibited by lactoferrin, lactoferrin is no longer used in bacterial control and is available for neutrophil control. Leukocytes are being found to produce a fair number of factors, including interleukins, which control proliferation, motility, and activation of leukocytes of the same and of different types (Parker and O'Brien, 1983). The level of circulating red cells controls the output of erythropoietin, and therefore the level of red cells in the circulation and in the bone marrow. Thus, regulation in the hematopoietic system is likely a function of every cell in this system, stromal and hematopoietic, modified and unmodified, tissue-bound and circulating.

7. VASCULATURE IN HEMATOPOIETIC REGULATION

The vasculature, often linked with the stroma, is another component of hematopoietic tissues important in regulation. As discussed above, the adventitial layer of blood vessels branches out to become part of the reticular meshwork and the adventitial reticular cells of vascular sinuses may undergo transformation into adipocytes. Thereby a kinship of vasculature with reticulum is established. Mature megakaryocytes lie against the wall of vascular sinuses in bone marrow, delivering their platelets into the lumen through mural apertures. The likelihood that the vascular sinuses provide a niche for the differentiation of megakaryocyte precursors has been reinforced by cytochemical study of mouse megakaryocytes using the rodent megakaryocyte marker, acetylcholinesterase. Small, lymphocyte-like acetylcholinesterase-positive megakaryocyte precursors lie at the wall of the vascular sinus (Young and Weiss, 1983). Recognition between lymphocytes and the high endothelium of postcapillary venules in the diffuse cortex of lymph nodes makes possible the essential lymph node loop by the lymphocytes of the recirculating lymphocyte pool (RLP) (Butcher and Weissman, 1980; de Sousa, 1981; Woodruff *et al.*, 1977; Woodruff and Kuttner, 1980). T cells, and B cells, circulating in blood through the vasculature of lymph nodes, reach the postcapillary venule and adhere to its endothelium. The lymphocytes then pass through the vascular wall in a distinctive way and enter the parenchyma of the node where they are sorted into T- and B-cell zones. If they do not engage in an immune response, the lymphocytes of the RLP move about for a number of hours within their respective zones and leave the node passing through the medulla and into efferent lymph. Lymphocytes may be carried by lymph into more central nodes and, from the most central nodes, into the blood via a thoracic duct. The propensity of T cells for the high endothelium of postcapillary venules in lymph nodes is so marked that after flooding a frozen section of lymph node with a suspension of lymphocytes, large numbers of the lymphocytes adhere to the sectioned postcapillary venular endothelium (Woodruff and Kuttner, 1980). A monoclonal antibody reagent has revealed a lymphocyte surface receptor specific for this adherence to the venular endothelial surface (see Section 8).

8. REGULATORY ARRANGEMENTS IN HEMATOPOIESIS: REGULATORY FACTORS

Having characterized cellular associations linked with regulation of the hematopoietic system, I will now deal with the nature of the regulating factors. There are many systemic factors which have a secondary but important effect upon the hematopoietic system. Androgens, for example, support erythropoiesis and have been used therapeutically in certain refractory congenital aplastic anemias such as the Diamond–Blackfan syndrome. Estrogen supports bone stroma which may affect hematopoiesis. Thus, the role of marrow adipocytes in aromatizing testosterone to estrogen, presented above, has consequences in the regulation of hematopoiesis. Glucocorticosteroids cause the destruction of thymic lymphocytes. There are systemic factors whose primary actions are upon the hematopoietic system. Erythropoietin induces erythropoietin-responsive cells, i.e., those erythroid cells that have developed erythropoietin receptors and remain in the mitotic pool (see Section 5), to proliferate and differentiate. It is this pool of erythroid cells that increases in erythroid stress, as in the hemolytic anemias due to spectrin-deficient erythrocytes and in malaria, and in the erythropoiesis induced by bleeding (Brookoff *et al.*, 1982). Erythropoietin is the best characterized of the hematopoietic hormones. Its counterparts, leukopoietin and thrombopoietin, are less well defined. Their existence is inferred, as was erythropoietin's, by experimental assay. For example, the serum of platelet-depleted individuals may contain a factor (or factors) that can induce heightened megakaryocytopoiesis even in individuals hypertransfused with platelets. These systemic factors will undoubtedly be better defined by further development of tissue culture assay systems for the various hematopoietic stem cells. Erythropoietin and, presumably, the other poietins affect only those relatively mature stem cells that are both committed to a blood cell lineage and differentiated to the level where appropriate cell surface receptors have appeared. Erythropoietin has little or no direct effect on BFU-E and, of course, no direct effect upon CFU-S. The factors inducing CFU-S to make a commitment to a given cell line are likely elaborated and confined to those microenvironments dedicated to inducing the CFU-S into that cell line. These factors may be cell-surface membrane proteins and short-range transient humoral factors, in contrast to stable humoral systemic hormones, like erythropoietin.

The high importance of locally produced, and locally confined, short-lived factors having local control of tissue function has become evident in many systems. Such local control is a condition, for example, of chemical neurotransmission and of many of the elements in the bradykinin and complement cascades. The decarboxylation of histidine with the formation of histamine in the basophil–mast cell system is, in a complementary way, not unlike the deamination of histidine by keratinocytes to form UV-blocking uronic acid, each reaction controlled locally. The enteroendocrine system, some of it neurally connected, is important in the local control of motility and of absorption in the gut. The idea of *paracrine* secretion, signifying local influence in contrast to the systemic influence of *endocrine* secretion, and the delineation of the APUD system, acronym for *a*mine *p*recursor *u*ptake and *d*ecarboxylation, have focused attention on such local control (Weiss, 1983c).

The implications of local control include significant consistent cellular associations and interactions, and the idea of microenvironments. In the hematopoietic system the importance of such factors has been highlighted by the delineation of lymphokines and the interleukins, low-molecular-weight compounds that have many functions. Among other actions, they control the movement and activation of macrophages, effect specific and nonspecific cellular toxicity, exercise chemotropism for eosinophils, basophils, and other cell types, provide T-cell replacing factors (TRF) and B-cell growth factors, and mediate the interactions of lymphocyte subtypes in both antibody formation and cellular immunity (Parker and O'Brien, 1983). The production of such factors is not restricted to lymphocytes and the terms *monokines,* referring to those produced by monocytes-macrophages, and *cytokines,* recognizing the capacity of cells in general to produce such factors, have been advanced. The several CSF elaborated by monocytes and macrophages which stimulate neutrophil and monocyte production (CSF-GM), eosinophil production (CSF-Eo), and megakaryocyte production (CSF-Meg) may be taken as examples of monokines (Metcalf, 1981). Presumably, factors elaborated by bone-lining cells and reticular cells which affect a hematopoietic microenvironment would include compounds classifiable as cytokines. Knowledge of factors regulating the activities of the blood cells is, at this writing, greatly expanding with efforts made to rationally name and classify compounds with regard to cell source, target cell, actions, and chemical structure. The groupings, many overlapping, include lymphokines, monokines, cytokines, arachadonic acid derivatives (monohydroxyeicosatetraenoic acids—HETE's, leukotrienes, and prostaglandins), interleukins, interferons, autocoids, complement and its associated factors, lysosomes, kinins, antimicrobial substances (superoxides, cationic proteins, neutral proteases, lactoferrins, and peptides). These factors are discussed more fully in other contributions to this and to the remaining volumes in this series. Such local factors appear, in concert with systemic factors, to play a part in the control of erythropoiesis, and other cell differentiations closely and distantly related to hematopoiesis.

Local and systemic humoral factors also work in conjunction with such quintessentially local compounds—as cell surface molecules, i.e., antigens, receptors, or markers. Examples are virtually without limit. Reference has already been made to the interaction of erythropoietin and cell surface receptors for erythropoietin. The capacity for the utilization of iron depends upon transferrin and upon transferrin receptors in the plasma membrane of erythroblasts and other iron-requiring cells. Phagocytosis is receptor-mediated. The pathological absence of cell surface receptors, as those for testosterone, causes, even with all the other elements of the system in place, grave, pervasive deficiency disease. Receptors may be engaged by local or systemic humoral factors, as indicated above, or with the receptors of vicinal cells. In the latter case, the initiation of cellular interactions which may be powerful, specific, and quite confined, need not require humoral factors: the interactions may depend entirely upon compounds in the plasma membrane. There may be no cellular interaction more intimate, more restricted, or more tightly regulated than that effected primarily or entirely through molecules at the cell surface bounded in the plasma membrane. An example of a hematopoietic interaction dependent on cell surface molecules is the activation of a T cell by an antigen-presenting cell. Two different receptors on the T cell must be satisfied, the antigen or "nonself" receptor and the "self" receptor. The antigen receptor is engaged by the antigen held on the plasma membrane of the antigen-presenting cell. The "self" receptor may be satisfied by MHC antigens on the surface of the antigen-presenting cell engaging MHC antigens on the surface of the T cells. The adherence of T cells to the high endothelium of postcapillary venules [or high endothelial venule (HEV)] in lymph nodes preparatory to their crossing the vessel and entering the parenchyma of the node, appears mediated by cell surface molecules. This has been demonstrated in rats by monoclonal antibody prepared against T-cell surfaces. (The T cells were harvested from a peripheral lymph node in a lymphoma 38C-13 in which the lymphoma cells bind HEV.) By screening for antibodies which distinguish peripheral node HEV-binding lymphoid cells from nonbinding Peyer's patch lymphocytes, Butcher and Weissman (1980) and Gallatin *et al.* (1983) isolated a monoclonal antibody, MEL-14, which binds determinants on or at the lymphocyte surface. MEL-14 blocks lymphoma cell adherence to the HEV, blocks normal lymphocyte cell binding, and blocks lymphocyte homing *in vivo.* Here, then, is an instance of endothelium–lymphocyte recognition upon which an essential loop of the recirculating lymphocyte pool depends and, therefore, upon which the immunological and other lymphocyte-related functions of lymph nodes depend.

9. CONCLUSION

The diverse activities of the hematopoietic system, which include extensive diverse migratory pathways, cell sorting, cell recognition, cell differentiation, and cell functions, require diverse complex regulation at every level of organization and function. Systemic hormones and factors, regional and local factors, and cell surface molecules interact, in concert with interactions of hematopoietic and stromal cells, and vasculature, and nerves within the hematopoietic tissues and of cells and tissues outside the hematopoietic system.

REFERENCES

Adamson, J. W., and Torok-Storb, L. N., 1978, Analysis of erythropoiesis by erythroid colony formation in culture, *Blood Cells* **4**:89.

Allen, T. D., 1981, Haemopoietic microenvironment in vitro: Ultrastructural aspects, in: *Symposium on Microenvironments and Cell Differentiation* (R. Porter and J. Whelan, eds.), pp. 38–67, Pitman Press, London.

Axelrod, A. A., McLeod, D. L., Shreeves, M. M., and Heath, D. S., 1974, Properties of cells that produce erythrocytic colonies in vitro, *in: Hemopoiesis in Culture* (W. A. Robinson, ed.), U.S. Government Printing Office, Wahington, D.C.

Bach, J. F., and Papiernik, M., 1981, Cellular and molecular signals, in T cell differentiation, in: Symposium on Microenvironments and Cell Differentiation (R. Porter and J. Whelan, eds.), pp. 215–235, Pitman Press, London.

Baggiolini, M., de Duve, C., Masson, P. L., and Heremans, J. F., 1970, Association of lactoferrin with specific granules in rabbit heterophil leukocytes, *J. Exp. Med.* **131:**559–570.

Balfour, B. M., Drexhage, H. A., Kamperdyk, E. W. A., and Hoefsmit, E. C. M., 1981, Antigen-presenting cells, including Langerhans cells, veiled cells and interdigitating cells, in: *Symposium on Microenvironments and Cell Differentiation* (R. Porter and J. Whelan, eds.), pp. 22–37, Pitman Press, London.

Bearman, R. M., Levine, G. D., and Bensch, K. G., 1978, The ultrastructure of the normal human thymus: A study of 36 cases, *Anat. Rec.* **190:**755–781.

Becker, A. J., McCulloch, E. A., and Till, J. E., 1963, Cytological demonstration of the clonal nature of spleen colonies derived from mouse marrow cells, *Nature (London)* **197:**452.

Beeson, P. B., and Bass, D. A., 1977, *The Eosinophil*, Saunders, Philadelphia.

Berman, J., 1967, The ultrastructure of erythroblastic islands and reticular cells in mouse bone marrow, *J. Ultrastruct. Res.* **17:**391.

Bernstein, S. E., 1980, Inherited hemolytic disease in mice: A review and update, *Lab. Anim. Sci.* **30:**197.

Bernstein, S. E., Russell, E. S., and Keighley, G., 1968, Two hereditary mouse anemias (Sl/Sl^d and W/W^v) deficient in response to erythropoietin, *Ann. N.Y. Acad. Sci.* **149:**475.

Bessis, M., 1958, L'ilot erythroblastique, unite fonctionelle de la moelle osseuse, *Rev. Hematol.* **13:**8.

Blue, J., and Weiss, L., 1981a, Periarterial macrophage sheaths (ellipsoids) in cat spleen: An electron microscope study, *Am. J. Anat.* **161:**115–134.

Blue, J., and Weiss, L., 1981b, Vascular pathways in non-sinusal red pulp: An electron microscope study of the cat spleen, *Am. J. Anat.* **161:**135–168.

Blue, J., and Weiss, L., 1981c, Species variation in the structure and function of the marginal zone: An electron microscope study of the cat spleen, *Am. J. Anat.* **161:**169–187.

Brelinska, R., and Pilgrim, C., 1983, Macrophages and interdigitating cells; their relationship to migrating lymphocytes in the white pulp of rat spleen, *Cell Tissue Res.* **233:**671–688.

Brookoff, D., and Weiss, L., 1983a, Adipocyte development and the loss of erythropoietic capacity in the bone marrow of mice after sustained hypertransfusion, *Blood* **60:**1337.

Brookoff, D., and Weiss, L., 1983b, Ultrastructure of the bone marrow in three murine strains with non-malignant lymphoproliferative syndromes: NZB×W, BXSB, and MRL/lpr, *Anat. Rec.* **207:**451–459.

Brookoff, D., Maggio-Price, L., Bernstein, S., and Weiss, L., 1982, Erythropoiesis in ha/ha and sph/sph mice, mutants which produce spectrin-deficient erythrocytes, *Blood* **59:**646–651.

Butcher, E. C., and Weissman, I. L., 1980, Cellular, genetic and evolutionary aspects of lymphocyte interactions with high-endothelial venules, *Ciba Found. Symp.* **71:**265–286.

Carr, I., and Daems, W. T. (eds.), 1980, *The Reticuloendothelial System: A Comprehensive Treatise*, Volume 1, Plenum Press, New York.

Chan, S. H., and Metcalf, D., 1972, Local production of colony-stimulating factor within the bone marrow: Role of nonhematopoietic cells, *Blood* **40:**646–653.

Chen, L. T., and Weiss, L., 1972, Electron microscopy of the red pulp of human spleen, *Am. J. Anat.* **134:**425–458.

Cohen, N., and Sigel, M. M. (eds.), 1981, *The Reticuloendothelial System: A Comprehensive Treatise*, Volume 3, Plenum Press, New York.

Curry, J. L., and Trentin, J. J., 1967, Hematopoietic spleen colony studies. I. Growth and differentiation, *Dev. Biol.* **15:**395–413.

Curry, J. L., Trentin, J. J., and Wolf, N. S., 1967, Hematopoietic colony studies. II. Erythropoiesis, *J. Exp. Med.* **125:**703–720.

Daems, W. T., 1980, Periarterial macrophages, in: *The Reticuloendothelial System: A Comprehensive Treatise*, Volume 1 (I. Carr and W. T. Daems, eds.), pp. 57–126, Plenum Press, New York.

Dardenne, M., and Bach, J. F., 1981, Thymic hormones, in: *The Thymus Gland* (M. D. Kendall, ed.), pp. 113–132, Academic Press, New York.

Deldar, A., and Weiss, L., 1983, Scanning and transmission electron microscopic studies on canine

bone marrow: The bone lining cells and delivery of blood cells to the circulation, in: *ASCVP 34th Annual Meeting,* San Antonio, pp. 16–17.
Deldar, A., and Weiss, L., 1985, Bone lining cells and hematopoiesis: An electron microscopic study of canine bone marrow, *Anat. Rec.* **213:**187–201.
de Sousa, M., 1981, *Lymphocyte Circulation: Experimental and Clinical Aspects,* Wiley, New York.
Dexter, T. M., 1981, Self-renewing hematopoietic progenitor cells and the factors controlling proliferation and differentiation, in: *Symposium on Microenvironments and Cell Differentiation* (R. Porter and J. Whelan, eds.), pp. 22–37, Pitman Press, London.
Dexter, T. M., 1982, Stromal cell associated haemopoiesis, *J. Cell. Physiol.* **87:**94.
Dicke, K. A., van Noord, M. J., and van Bekkum, D. W., 1973, Attempts at morphological identification of the hematopoietic stem cell in rodents and primate, in: *Haemopoietic Stem Cells.* pp. 53–62. Excerpta Medica, Amsterdam.
Friedenstein, A. J., Chailakhjan, R. K., Latsinik, N. V., Pansyuk, A. F., and Keiliss-Borok, V., 1974, Stromal cells responsible for transferring the microenvironment of hematopoietic tissues, *Transplantation* **17:**331–340.
Friedman, P. S., 1981, The immunobiology of Langerhans cells, *Immunol. Today* **2:**124–128.
Gallatin, W. M., Weissman, I., and Butcher, E. C., 1983, Lymphocyte recirculation: A molecular basis for the recognition of vascular endothelium, in: *Intercellular Communication in Leucocyte Function* (J. W. Parker and R. L. O'Brien, eds.), pp. 645–648, Wiley, New York.
Goldberg, B., and Rabinovitch, M., 1983, The connective tissue, in: *Histology,* 5th ed. (L. Weiss, ed.), pp. 160–170, Elsevier, Amsterdam.
Gong, J. K., 1978, Endosteal marrow: A rich source of hematopoietic stem cells, *Science* **199:** 1443.
Gordon, I. K., Miller, W. J., Branda, R. F., Zanjani, E. D., and Jacob, H. S., 1980, Regulation of erythroid colony formation by bone marrow macrophages, *Blood* **55:**1047.
Gowans, J. L., 1961, Biological activity of the lymphocyte, *Ciba Found. Study Group* **10:**58–59.
Gowans, J. L., and Knight, E. J., 1964, The route of recirculation of lymphocytes in the rat, *Proc. R. Soc. London Ser. B* **159:**257.
Grobstein, C., 1954, Tissue interaction in the morphogenesis of mouse embryonic rudiments in vitro, in: *Aspects of Synthesis and Order in Growth,* (E. G. Butler, ed.), pp. 233–256, Princeton University Press, Princeton, N.J.
Grobstein, C., 1975, Developmental role of intercellular matrix: Retrospection and prospection, in: *Extracellular Matrix Influences on Gene Expression* (H. Slavkin and R. C. Greulich, eds.), pp. 9–16, Academic Press, New York.
Hoefsmit, E. C. M., Kamperdyk, E. W. A., Hendricks, H. R., Beelen, R. H. J., and Balfour, B. M., 1980, Lymph node macrophages, in: *The Reticuloendothelial System: A Comprehensive Treatise,* Volume 1 (I. Carr and W. T. Daems, eds.), pp. 417–468, Plenum Press, New York.
Hoffman, R., Zanjani, E. D., Zalusky, J. V. R., Lutton, J. D., and Wasserman, L. R., 1976, Diamond–Blackfan syndrome: Lymphocyte-mediated suppression of erythropoiesis, *Science* **193:**899.
Hoffman, R., Zanjani, E. D., Lutton, J. D., Zalusky, J. V. R., and Wasserman, L. R., 1977, Suppression of erythroid-colony formation by lymphocytes from patients with aplastic anemia, *N. Engl. J. Med.* **296:**10.
Hood, L., Weissman, I. L., and Wood, W. B., 1978, *Immunology,* (J. Gillen, ed.), Benjamin–Cummings, Menlo Park, Calif.
Hora, H., and Ogawa, M., 1977, Erythropoietic precursors in mice with erythropoietic stimulation and suppression, *Exp. Hematol. (Copenhagen)* **5:**141.
Humphrey, J. H., 1981, Differentiation of function among antigen-presenting cells, in: *Symposium on Microenvironments and Cell Differentiation* (R. Porter and J. Whelan, eds.), pp. 302–321, Pitman Press, London.
Janossy, G., Thomas, J. A., and Bollum, F. J., 1980, The human thymic microenvironment: An immunohistological study, *J. Immunol.* **125:**202.
Janossy, G., Thomas, J. A., Goldstein, G., and Bollum, F. J., 1981, The thymic microenvironment, in: *Symposium on Microenvironments and Cell Differentiation* (R. Porter and J. Whelan, eds.), pp. 193–214, Pitman Press, London.
Jenkinson, E. J., 1981, Expression and function of major histocompatibility complex antigens in the

developing thymus: Studies on normal and nude mice, in: *Symposium on Microenvironments and Cell Differentiation* (R. Porter and J. Whelan, eds.), pp. 178–192, Pitman Press, London.

Jordan, R. K., and Robinson, J. H., 1981, T lymphocyte differentiation, in: *The Thymus Gland* (M. D. Kendall, ed.), pp. 151–168, Academic Press, New York.

Kendall, M. D., 1981, The cells of the thymus, in: *The Thymus Gland* (M. D. Kendall, ed.), pp. 63–84, Academic Press, New York.

Lambertsen, R. H., and Weiss, L., 1983, Studies on the organization and regeneration of bone marrow: Origin, growth, and differentiation of endocloned hematopoietic colonies, *Am. J. Anat.* **166**:369.

La Pushin, R. W., and Trentin, J. J., 1977, Identification of distinctive stromal elements in erythroid and neutrophil granuloid spleen colonies: Light and electron microscopic study, *Exp. Hematol. (Copenhagen)* **5**:505.

Lisiewicz, J., 1980, *Human Neutrophils,* Charles Press, Bowie, Md.

Lord, B. I., 1983, Haemopoietic stem cells, in: *Stem Cells: Their Identification and Characterization* (C. S. Potten, ed.), pp. 118–154, Churchill Livingstone, Edinburgh.

Lord, B. I., Testa, N. G., and Hendry, J. H., 1975, The relative spatial distributions of CFU_s and CFU_c in the normal mouse femur, *Blood* **46**:65.

Love, R. J., Kelly, D., and Dineen, J. K., 1974, Nippostrongylus brasiliensis: Effects of immunity of the pre-intestinal and intestinal larval stages of the parasite, *Int. J. Parasitol.* **4**:183–191.

Lucas, D. O., 1983, Macrophage ontogeny and function, in: *Intercellular Communication in Lymphocyte Function* (J. W. Parker and R. L. O'Brien, eds.), pp. 259–262, Wiley, New York.

McCuskey, R. S., and Meineke, H. A., 1973, Studies of the hemopoietic microenvironment. III. Differences in the splenic microvascular system and stroma between Sl/Sl^d and W/W^v anemic mice, *Am. J. Anat.* **137**:187.

McLeod, D. L., Shreeves, M. M., and Axelrod, A. A., 1974, Improved plasma culture system for production of erythrocytic colonies in vitro: Quantitative assay method for CFU-E, *Blood* **44**:517.

McManus, P., and Weiss, L., 1984, Busulfan-induced chronic bone marrow failure: Changes in cortical bone marrow, stromal cells and adherent cell colonies, *Blood* **64**:1036–1041.

Mangan, K. F., and Desforges, F., 1980, The role of T lymphocytes and monocytes in the regulation of human erythropoietic peripheral blood burst forming units, *Exp. Hematol. (Copenhagen)* **8(6)**:717–727.

Marchesi, V. T., and Gowans, J. L., 1964, The migration of lymphocytes through the endothelium of venules in lymph nodes: An electron microscopic study, *Proc. R. Soc. London Ser. B* **159**:283.

Metcalf, D., 1981, The clonal culture in vitro of human leukemic cells, in: *The Leukemic Cell* (D. Catovsky, ed.), pp. 220–251, Churchill Livingstone, Edinburgh.

Metcalf, D., and Moore, M. A. S., 1971, *Haemopoietic Cells,* Elsevier, Amsterdam.

Milicevic, N. M., Milicevic, Z., Piletic, O., Mujovie, S., and Ninkov, V., 1983, Reactivity of thymic metallophillic cells during the regeneration after the application of cyclophosphamide, *J. Reticuloendothel. Soc.* **34**:501–507.

Nathans, D. G., Chess, L., Hillman, D. G., Clarke, B., Breard, J., Merler, E., and Housman, D. E., 1978, Human erythroid burst-forming unit: T cell requirement for proliferation in vitro, *J. Exp. Med.* **147**:324.

Parker, J. W., and O'Brien, R. L. (eds.), 1983, *Intercellular Communication in Leucocyte Function,* Wiley, New York.

Pictet, R., Orci, L., Forssmann, W. G., and Firardier, L., 1969, An electron microscope study of the perfusion-fixed spleen, *Z. Zellforsch. Mikrosk. Anat.* **96**:372–399.

Potten, C. S., 1983, Stem cells in epidermis from the back of the mouse, in: *Stem Cells: Their Identification and Characterization* (C. S. Potten, ed.), pp. 200–232, Churchill Livingstone, Edinburgh.

Reichard, S. M., and Filkins, J. P. (eds.), 1981, *The Reticuloendothelial System: A Comprehensive Treatise,* Volume 4, Plenum Press, New York.

Rouse, R. V., and Weissman, I. L., 1981, Microanatomy of the thymus: Its relationship to T cell differentiation, in: *Symposium on Microenvironments and Cell Differentiation* (R. Porter and J. Whelan, eds.), pp. 161–177, Pitman Press, London.

Rouse, R. V., van Ewijk, W., Jones, P. P., and Weissman, I. L., 1979, Expression of MHC antigens by mouse thymic dendritic cells, *J. Immunol.* **122**:2508.
Sakai, N., Johnstone, C., and Weiss, L., 1981, Bone marrow cells associated with heightened eosinophilopoiesis: An electron microscope study of murine bone marrow stimulated by Ascaris suum, *Am. J. Anat.* **161**:11.
Simon, G. T., 1980, Splenic macrophages, in: *The Reticuloendothelial System: A Comprehensive Treatise*, Volume 1 (I. Carr and W. T. Daems, eds.), pp. 469–498, Plenum Press, New York.
Singh, T., 1981, The ultrastructure of epithelial reticular cells, in: *The Thymus Gland* (M. D. Kendall, ed.), pp. 133–150, Academic Press, New York.
Slavkin, H. C., and Greulich, R. C. (eds.), 1975, *Extracellular Matrix Influences on Gene Expression*, Academic Press, New York.
Sorokin, S., 1977, Phagocytes in the lungs: Decadence, general behavior and phylogeny, in: *Respiratory Defense Mechanisms* (J. D. Brain, D. F. Proctor, and L. Reid, eds.), p. 711, Dekker, New York.
Sprent, J., 1977, Recirculating lymphocytes, in: *The Lymphocyte: Structure and Function* (J. J. Marchalonis, ed.), pp. 93–101, Dekker, New York.
Spry, C. J. F., 1971, Mechanisms of eosinophilia. VI. Eosinophil mobilization, *Cell Tissue Kinet.* **4**:365.
Sutton, J. L., and Weiss, L., 1966, Transformation of monocytes in tissue culture into macrophages, epithelioid cells, and multinucleated giant cells: An electron microscope study, *J. Cell Biol.* **28**:303.
Tablin, F., and Weiss, L., 1983, The equine spleen: An electron microscope analysis, *Am. J. Anat.* **166**:393.
Till, J. E., 1982, Stem cells in differentiation and neoplasia, *J. Cell. Physiol.* **1**:3–11.
Till, J. E., and McCulloch, E. A., 1961, A direct measurement of the radiation sensitivity of normal mouse bone marrow cells, *Radiat. Res.* **14**:213.
Trentin, J. J., 1970, Influence of hematopoietic organ (hematopoietic inductive microenvironments) on stem cell differentiation, in: *Regulation of Hematopoiesis* (A. S. Gordon, ed.), pp. 161–186, Appleton–Century–Crofts, New York.
van Bekkum, D. W., van Noord, M. J., Maat, B., and Dicke, K. A., 1971, Attempts at identification of hematopoietic stem cell in mouse, *Blood* **38**:547.
van Ewijk, W., Rouse, R. V., and Weissman, I. L., 1980, Distribution of H-2 microenvironments in the mouse thymus: Immuno-electron microscopic identification of I-A and H-2k bearing cells, *J. Histochem. Cytochem.* **28**:1089.
Weiss, L., 1974a, A scanning electron microscope study of the spleen, *Blood* **43**:665.
Weiss, L., 1974b, The white pulp of the spleen: The relationships of arterial vessels, reticulum, and free cells in the periarterial lymphatic sheath, *Bull. Johns Hopkins Hosp.* **115**:99.
Weiss, L., 1976, The hematopoietic microenvironment of the bone marrow: An ultrastructural study of the stroma in normal and eosinophilic rats, *Anat. Rec.* **186**:161.
Weiss, L., 1980, The hematopoietic microenvironment of bone marrow: An ultrastructural study of the interactions of blood cells, stroma and blood vessels, in: *Blood Cells and Vessel Walls: Functional Interactions* (R. Porter and J. Whelan, eds.), pp. 3–19, Excerpta Medica, Amsterdam.
Weiss, L., 1983a, The red pulp of the spleen: Structural basis of blood flow, in: *Clinics in Haematology* (E. Lewis and W. Wolvey, eds.), pp. 375–393, Saunders, Philadelphia.
Weiss, L., 1983b, The life cycle of blood cells: Bone marrow, in: *Histology*, 5th ed. (L. Weiss, ed.), pp. 474–509, Elsevier, Amsterdam.
Weiss, L., 1983c, The thymus; lymphatic vessels and lymph nodes; the spleen, in: *Histology*, 5th ed. (L. Weiss, ed.), pp. 510–568, Elsevier, Amsterdam.
Weiss, L. (ed.), 1983d, *Histology*, 5th ed., Elsevier, Amsterdam.
Weiss, L., 1983e, Hematopoietic tissue in malaria: Facilitation of erythrocytic recycling by bone marrow in *Plasmodium berghei*-infected mice, *Am. J. Parasitol.* **69**:307.
Weiss, L., Geduldig, U., and Weidanz, L., 1986, The spleen in malaria: the mechanism of crisis, *Am. J. Anat.* (in press).
Wisse, E., 1980, On the fine structure and function of rat liver Kupffer cells, in: *The Reticuloendothelial System: A Comprehensive Treatise*, Volume 1 (I. Carr and W. T. Daems, eds.), pp. 363–380, Plenum Press, New York.

Woodruff, J. J., and Kuttner, B. J., 1980, Adherence of lymphocyte to the high endothelium of lymph nodes in vitro, in: *Blood Cells and Vessel Walls: Functional Interactions* (R. Gordon, ed.), pp. 243–263, Elsevier/North Holland.

Woodruff, J. J., Katz, I. M., Lucas, L. E., and Stamper, H. B., 1977, An in vitro model of lymphocyte homing. II. Membrane and cytoplasmic events involved in lymphocyte adherence to specialized high endothelial venules of lymph nodes, *J. Immunol.* **119:**1603.

Wu, A. M., Till, J. E., Siminovitch, L., and McCulloch, E. A., 1967, A cytological study of the capacity for differentiation of normal haemopoietic colony-forming cells, *J. Cell. Physiol.* **69:**177.

Young, K. M., and Weiss, L., 1983, Ultrastructural characterization of small acetylcholinesterase-positive cells in murine bone marrow, *Blood* **62**(Suppl. 1):5a.

2

Endothelial Reactions

BURTON ZWEIMAN

1. INTRODUCTION

Hypersensitivity reactions are frequently characterized by prominent exudation of the fluid and/or cellular components of the blood into the reaction site(s). The endothelium of normal vessels plays a very important role in the passage of both fluid and cells from the vascular compartment in normal and diseased states (reviewed by Thorgeirsson and Robertson, 1978). In essence, the endothelium in some vessels acts as a barrier that selectively regulates transfer of substances of various molecular sizes between the circulating blood and the surrounding tissues.

In this chapter I will review briefly some current knowledge about the structure of vascular endothelium and the various ways fluid components of the blood pass through these structures. I will also review information about the interaction of blood leukocytes with the vascular endothelium. It has long been suggested that this interaction is an important determinant of the degree of passage of leukocytes from the blood into various tissues. Such interactions are of obvious importance in the relationship of inflammatory cells to the reticuloendothelial system. Certain aspects of the roles of leukocyte–endothelium interactions will have been covered in other discussions in this series. In this chapter I will concentrate on what is known about the role of vascular endothelial cells in certain inflammatory responses as a component discussion of this volume dealing with the RES and hypersensitivity.

2. STRUCTURE OF VASCULAR ENDOTHELIUM

The vascular endothelium consists of a single layer of flattened, fairly uniform elongated cells, about 30 × 10 μm in dimension, with the long dimension in the direction of blood flow (Thorgeirsson and Robertson, 1978). There are

BURTON ZWEIMAN • Allergy and Immunology Section, University of Pennsylvania School of Medicine, Philadelphia, Pennsylvania 19104.

some differences in endothelial cell structures in different types of vessels and in different areas of the body. I will limit this brief description to the capillaries and small venules, since that is most relevant to inflammatory responses. Endothelial cells tend to be thickest at the area of the nucleus, thinning to diameters of less than 0.1 μm at the cell periphery. They are coated with a carbohydrate-rich glycocalyx, which may play a role in leukocyte adherence (discussed later). In addition to the usual array of intracytoplasmic components, endothelial cells contain a rather unique structure called the Weibel–Palade body (Weibel and Palade, 1964). Plasmalemmal vesicles (diameter 600–700 Å) lie within the cytoplasm or open onto the cell surface, more in capillaries than in other types of vessels (Simionescu *et al.*, 1976). The interendothelial junctions also vary in different regions of the body, a continuous endothelium being found in arterioles, veins, and muscular arteries and a fenestrated endothelium in certain visceral capillaries (Simionescu *et al.*, 1976). The junctions of capillaries and venules tend to be less organized than those of arteries, generally without communicating junctions and more devoid of junctional particles. These characteristics may be important in determining the predominant role of capillaries and venules in the passage of components from the blood to tissues (see Sections 3 and 4).

3. THE NORMAL ENDOTHELIUM AS A SELECTIVE BARRIER FOR FLUIDS

Plasmalemmal vesicles and intercellular junctions are pathways for the initial passage of water-soluble molecules from the vascular lumen to tissues (Thorgeirsson and Robertson, 1978). There appear to be two different pore sizes found in capillary fenestrae and plasmalemmal vesicles. In the postcapillary venule there is less evidence for transport through the cytoplasm. However, the loosely organized interendothelial junction is quite permeable to certain aqueous solutions. This may explain the importance of this venule in the pathophysiology of inflammation.

Limited information to date has shown differences in permeability to fluids in capillaries from different organ systems. For example, the capillaries of the central nervous system are particularly resistant to the passage of macromolecules, resulting in the effective "blood–brain barrier."

4. NORMAL LEUKOCYTE–ENDOTHELIUM INTERACTIONS

It will be helpful to first briefly review some of the knowledge about normal leukocyte–endothelium interactions. Leukocytes passing through capillaries and small venules at relatively slow rates frequently touch and perhaps adhere to the endothelium transiently (Atherton, 1972). Normally, there is little persistent adherence of leukocytes to vascular endothelium, as compared to other surfaces, possibly due to the prominent elaboration by the endothelium of prostacyclin (PGI_2), a prostaglandin with antiaggregating activity (Moncada *et al.*, 1977). Although endothelial cells synthesize both factor VIII and von Willebrand factor, procoagulant activity has not been demonstrated in the secreted material (Jaffe *et al.*, 1973). This degree of adherence does not normally result in prominent emigration of the lymphocytes from the vessels, except in certain sites, particularly the postcapillary venules of lymphoid tissue (Gowans and Knight, 1964). Such vessels are characterized by pseudocolumnar "high endothelial cells" (Gesner and Ginsberg, 1964). These cells are recognizable as cuboidal, containing pale open nuclei, copious pyroninophilic cytoplasm with numerous organelles and a thick basement membrane (Freemont, 1982). Considerable interest exists as to why lymphocyte emigration is so prominent in postcapillary venules, frequently called "high endothelial venules" (HEV) (Butcher and Weissman, 1980): (1) it is in such sites that the first decrease in wall shear stress occurs, allowing more opportunity for adhesion; (2) it has been thought for some time by at least some investigators that the presence of specific acceptors or other structural characteristics of the HEV lead to more frequent and prolonged adhesion of the circulating lymphocytes (Anderson and Anderson, 1976).

However, other studies suggest that the picture is more complex. (1) In species with HEV in lymph nodes, the high endothelial cells may actually occur as a response to the emigrating lymphocytes; the high endothelial cells are absent in neonatally thymectomized animals and appear following reconstitution of such animals with small lymphocytes (Goldschneider and MacGregor, 1968). It appears that a trypsin-sensitive component of the lymphocyte is important in this interaction (Ford *et al.*, 1976). (2) It was thought that carbohydrate components of the membrane were also important because of altered *in vivo* homing patterns of neuraminidase-treated lymphocytes. However, (a) subsequent studies (Ford *et al.*, 1976) have shown that these desialylized lymphocytes are preferentially trapped in the liver. (b) There can be ample emigration of lymphocytes into lymph nodes of species in which this tissue is not characterized by HEV (reviewed by Ford, 1975; Gowans and Steer, 1980). (c) Treatment of endothelial cells with a variety of agents, including some fixatives, has a variable effect on *in vitro* lymphocyte adhesion, sometimes being quite modest (DeBono, 1976; Stamper and Woodruff, 1976). This has suggested to some investigators a more passive role of the vascular endothelial cell in such adherence.

Once leukocytes adhere to the endothelium, they spread and then rapidly penetrate the vascular wall, mainly between endothelial cells (Schoefl, 1972). It was initially thought that the vascular endothelium played a rather passive role in this process, with the propensity to passage through venules related to their relatively loose intercellular junctions (Simionescu *et al.*, 1976). However, more recent studies (Schoefl, 1972; reviewed by Simon, 1980) suggest that endothelial cells play a more active role. They retract in response to certain inflammatory agents, such as histamine-type mediators. This contractibility of the endothelial cells is due to the presence of large quantities of contractile proteins within the cell. Such contraction leads to the formation of large intercellular gaps, permitting the passage of macromolecules and leukocytes. However, the elaborate basement membrane seen in some postcapillary venules may act as a selective barrier to plasma proteins while lymphocytes can insinuate their way through this membrane (Ford, 1980). Although it is still unclear whether increased adhesion or increased migration through the vascular wall by the lymphocyte plays the major role, the end result is increased passage into lymphoid tissue. Subsequent localization within the tissue and emigration through lymphatics play key roles in immune responses and lymphocyte recirculation.

A fascinating and provocative aspect of lymphocyte–endothelium interaction relates to the known differences in the array of lymphocytes in different lymphoid organs. For example, Peyer's patches in the intestine contain a large percentage of B cells and a minority of T cells whereas this is not the case in peripheral nodes (Gowans and Steer, 1980; Butcher *et al.*, 1982). Following intravenous infusion, lymphocytes obtained from Peyer's patches migrate preferentially into the Peyer's patch and lamina propria areas of the small intestine, possibly playing a role in local immunity to enteric immunogens. Butcher *et al.* (1982) have presented findings suggesting that different patterns of lymphocyte subpopulations in different lymphoid tissues may be related to preferential binding of a lymphocyte subpopulation to the HEV of these different tissues. Activated lymphocytes (lymphoblasts) may also bind in altered patterns.

In contrast, only a minute percentage of circulating lymphocytes emigrate through venules in nonlymphoid tissue (Ford, 1975). This situation is altered appreciably in certain inflammatory responses, as will be seen later.

Granulocytes within the vascular compartment appear to be within two functional components: circulating and marginated pools. These two pools are normally of approximately equal size, with dynamic exchange of cells between them leading to rapid equilibration. Epinephrine and prominent exercise will "demarginate" granulocytes possibly because of transient increases in blood flow through small vessels. Infection may lead to increased margination but does not necessarily lead to increased emigration. It is of note that the patterns of movement between circulating and marginated compartments may differ for two types of granulocytes, neutrophils and eosinophils, during situations such as certain infections and corticosteroid administration.

It is unclear whether there are specific receptors in endothelial cells for granulocytes although it is known that general systemic activation of complement results in a nonselective adhesion of neutrophils to endothelium. Heat-aggregated IgG induces increased adherence of neutrophils to cultured endothelial cells (Hashimoto and Hurd, 1981). Herpes virus-infected but not uninfected human venous endothelial cells express receptors for the Fc portion of human IgG and for C3 bound in immune complexes (Cines *et al.*, 1982). Such receptors have also been noted on injured pulmonary artery endothelial cells (Ryan *et al.*, 1981). Clq receptors have also been demonstrated on cultured human venous endothelial cells (Andrews *et al.*, 1981).

However, it is not clear how these receptors detected *in vitro* relate to the *in vivo* interaction of intravascular leukocytes with vascular endothelium. It is also unclear whether glycoprotein inhibitors of adherence, such as GP110, modulate *in vivo* adherence of granulocytes to vascular endothelium. Proteolytic enzymes released locally may well help in the invasion of small vessel walls by neutrophils (Cochrane and Aikin, 1966), although they do not appear to be the only factors involved in penetration.

5. *IN VITRO* STUDIES OF LEUKOCYTE–ENDOTHELIUM INTERACTIONS

In vitro models have been developed to investigate normal and altered interaction between blood leukocytes and endothelial cells. Endothelial cells obtained from vessels such as human umbilical veins, bovine or porcine arteries can be cultured under appropriate conditions (as reviewed by Gimbrone, 1976). They assume a monolayer structure with cells containing factor VIII and Weibel–Palade inclusion bodies similar to those found in vascular endothelial cells *in vivo*.

The interaction of blood leukocytes with such monolayers has mainly been investigated using stationary cultures. In these studies, semiquantitative assessments have shown that up to 20% of lymphocytes adhere over the first several hours, decreasing over the succeeding 10 hr, presumably due to detachment (DoBono, 1976). Using radiolabeled lymphocytes, it has been found that the percentage of normal lymphocytes that adhere generally increases progressively over the first 4 hr to about 10%, apparently not limited by the number of suspended lymphocytes in concentrations found normally in the circulation (Zweiman *et al.*, 1982). There is increased adherence of those lymphocytes from donors sensitized to histocompatibility antigens which can be found on endothelial cells (DeBono, 1979; see also below).

In more recent studies, it has been found (Zweiman *et al.*, 1983) that the Ia-positive cells in blood mononuclear cell populations are more likely to adhere to endothelial monolayers. In contrast, the percentages of T3 (mature T), T4 (putative helper T), and T8 (putative suppressor/cytotoxic T) cells were reduced in the endothelium-adherent cells. The relative increase in adherent Ia-positive cells could only be partially explained by the presence of monocytes in such populations.

Concerns have been raised that the normal shearing forces found even in the relatively slow flow through small venules may affect leukocyte adhesion in ways not assessed in stationary cultures. To examine this, Beesley *et al.* (1978) carried out studies of *in vitro* adherence of bovine leukocytes to monolayers of bovine vascular endothelial cells under conditions where the leukocytes were in continuous motion over the monolayers. They found that about 2–10% of the leukocytes adhered generally within the first 10 min. Leukocytes adhered preferentially to endothelial monolayers when compared to smooth muscle cells, adventitial fibroblasts, or serum-coated glass. Adherence by granulocytes and mononuclear cells was similar, as was the adherence to endothelial monolayers of aortic or venous origin (DeBono, 1976). DeBono found that the adherence of pig lymphocytes to endothelial monolayers was reduced about 50% when incubation was carried out under flow rather than static condition. Nevertheless, about 10% of the lymphocytes adhered.

However, in a recently described model (McIntyre *et al.*, 1982), leukocytes did not appear to adhere to cultured vascular endothelial cells which were in a monolayer lining glass capillary tubes through which blood was perfused. Somewhat surprisingly, there was avid adherence of leukocytes to monolayers of smooth muscle cells lining such capillary tubes.

Using another *in vitro* model, Stamper and Woodruff (1976, 1977) investigated the adherence at 7°C of various rodent lymphoid cells to fixed sections of lymph nodes. Cells of thoracic duct, nodes, and spleen (but not bone marrow or thymus) attached in significant numbers, preferentially to the postcapillary venules, the same area thought to be the site of major emigration of lymphocytes from the blood into the lymphoid tissue. They did not find such lymphoid cell adherence to similarly prepared sections of nonlymphoid tissue. This has raised the question again as to whether there are specific acceptors (like the glycoprotein surface material referred to earlier) or signals transmitted from the HEV of the lymph node postcapillary venules to the lymphocytes. Although the matter is unsettled, Woodruff's group feels that the endothelial cell plays a relatively passive role in this interaction. However, they have subsequently reported that the adherence of recirculating lymphoid cells to these venules is increased by prior treatment of the latter with factors released from incubated lymphocytes or from lymph. Another factor in the lymph appears to inhibit lymphocyte adherence, possibly by binding to receptors or acceptors on the endothelial cell.

The relevance of findings using the *in vitro* models to *in vivo* events is uncertain. However, the logistic difficulties in studying such interactions *in vivo* leave us with the hope that current and emerging *in vitro* approaches will allow at least reliable estimates of the normal and abnormal patterns of *in vivo* adherence of leukocytes.

6. A POSSIBLE ROLE FOR ENDOTHELIAL CELLS IN LYMPHOCYTE-MEDIATED IMMUNE RESPONSES

There has been increasing interest in recent years in other possible effects of lymphocyte–endothelium interaction on immune responses. As noted previously, vascular endothelial cells may bear surface Ia-like (HLA-DR) determinants and stimulate allogeneic mixed lymphocyte reactions (reviewed by Hirschberg *et al.*, 1980). Endothelial cells that bear HLA-DR determinants can effectively substitute for monocytes in the lymphoproliferative response of purified T cells to antigen (Hirschberg *et al.*, 1980; Burger *et al.*, 1981). These interactions are antigen-specific and can be inhibited by the presence of relevant anti-HLA-DR antisera. Similar to the situation seen with monocyte–lymphocyte interactions, shared HLA-DR determinants in the endothelium and lymphocyte donors are required for optimal proliferative responses. Cultured endothelial cells can also act as accessory cells for mitogen-stimulated proliferation of purified T cells (Ashida *et al.*, 1981). Viable endothelial cells and close contact with T cells were required for this response. Cultured fibroblasts or HeLa cells could not serve a similar accessory role.

The *in vivo* biological significance of these interactions is uncertain. However, it is conceivable that lymphocytes are in close proximity to the endothelium of small vessels for a sufficient period of time to react to antigens or other activators presented by the endothelial cells. It is of interest that injected lymphocytes appear to leave the blood into lymphoid tissues of allogeneic and syngeneic recipients at about the same rate. However, Curtis and Renshaw

(1982) have reported increased *in vitro* adherence of allogeneic lymphocytes to pulmonary capillary endothelium.

7. THE VASCULAR ENDOTHELIUM IN INFLAMMATION: GENERAL CONCEPTS

As noted earlier, most inflammatory responses involve increased passage of fluid and/or cellular components from blood to the tissue reaction site. Most investigators feel that the main site of increased vascular permeability in acute inflammation is the postcapillary venule (reviewed by Ryan and Majno, 1977). This is likely related to the relatively loose organization of the intercellular junctions of venules (Simionescu *et al.*, 1976). Several of the vasoactive mediators released in some inflammatory responses cause formation of interendothelial gaps in venules, possibly by contracting the endothelial cells themselves (Ryan and Majno, 1977). Some of the agents may also cause localized or generalized vasodilation. This, along with increased permeability, may lead to prominent loss of fluid into the tissues. A more prominent and persistent vascular leak may result from actual damage to the endothelium by certain noxious processes or agents (Cotran, 1967) or leukocyte products (Marchesi, 1964; Weiss *et al.*, 1981).

8. LYMPHOCYTE–ENDOTHELIUM INTERACTIONS IN INFLAMMATION

As noted above, relatively few lymphocytes appear to pass through vascular endothelium into non-lymphoid tissues. However, the picture is quite different in certain experimental and clinical situations. Vessels lined by high endothelial cells have been reported in certain antigen-induced granulomas (Smith *et al.*, 1970) and experimental chronic arthritis (Graham and Shannon, 1972). Polverini *et al.* (1977) found increased proliferation of vascular endothelial cells in vessels within developing delayed hypersensitivity reactions.

Nightingale and Hurley (1978) investigated further the relationship between vascular endothelial hypertrophy and lymphocyte emigration during inflammatory responses. They found such endothelial hypertrophy in lymphocyte-rich granulomas induced by Freund's complete adjuvant or tubercle bacilli. Such changes were not seen in vessels within inflammatory reactions containing relatively few lymphocytes such as granulomas induced by talc or carrageenan or turpentine-induced neutrophilic abscesses. The vascular patterns in the lymphocyte-rich granulomas were similar to those in the developing Peyer's patch with relatively few venules containing hypertrophied endothelial cells. Also of interest was their finding that vascular permeability for carbon particles was actually less in these vessels than in the lymphocyte-poor granulomas. Thus, the passage of the much larger lymphocytes through these vessel walls was not simply due to a nonspecific increase in permeability. A similar conclusion was reached by Rose and Parrott (1977) in studies of inflamed skin.

The mechanism(s) underlying the selective passage of lymphocytes through vascular endothelium in such granulomas is unclear. However, following sequential histologic analyses of these granulomas, Nightingale and Hurley (1978) concluded that the vascular endothelial hypertrophy followed (and may be a consequence of) the increased lymphocyte passage through the vessel walls. Several investigators (Sholley *et al.*, 1974; Fromer and Klintworth, 1975) have found neovascularization with endothelial cell hypertrophy in inflammatory sites with prior prominent leukocyte infiltration. However, other studies (Sholley and Cotran, 1978) suggest that infiltration of bloodborne mononuclear cells is not a necessary stimulus for vascular regeneration in such sites. Thus, it is possible that leukocytes or their products may facilitate, but are not necessary for, the initiation of endothelial proliferation and vascular growth. The functional significance of this hypertrophy remains unclear. The clinical significance of these findings is underlined by similar findings in lesions of a few patients with rheumatoid arthritis and Hashimoto's thyroiditis.

Freemont (1982) reported that a small percentage of venules in lymphocyte-rich ($> 150/mm^3$) lesions of a variety of human diseases exhibit a structural, ultrastructural, and cytochemical picture very similar to that seen in the HEV of lymphoid tissue.

9. GRANULOCYTE–ENDOTHELIUM INTERACTIONS IN INFLAMMATION

Unlike the situation with lymphocytes, relatively few granulocytes accumulate outside of the bloodstream in lymphoid or nonlymphoid tissue. However, this situation changes considerably in areas of acute inflammation. The leukocytes adhering to and then passing through the vascular endothelium in the early phase of most inflammatory responses are generally neutrophils. It has been postulated that a large efflux of neutrophils may indirectly inhibit or delay the emigration of lymphocytes through the same sites in the vessel wall. This may be because neutrophils appear to pass more readily than lymphocytes between vascular endothelial cells. It has been shown that leukocytes actively migrate between endothelial cells when they leave postcapillary venules during inflammatory responses (Marchesi and Florey, 1960). This emigration has been suggested to be induced by chemotactic factors which enter the interendothelial junctions after being released in the inflammatory reaction site (Ryan and Majno, 1977). As noted previously, vascular endothelial cells infected by certain viruses, or altered by trauma or by circulating immune complexes may express increased receptors for the Fc fragment of gamma globulin (Ryan *et al.*, 1981; Cines *et al.*, 1982). Serum immunoglobulin molecules binding to such sites could possibly act as a ligand, binding neutrophils which also may bear Fc receptors. Factors in the plasma of patients with acute inflammatory disorders have been reported to increase granulocyte adherence; these appear to require the presence of a heat-labile cofactor (Lentnek *et al.*, 1976). Injections of vasoactive agents appear to induce little local neutrophil emigration even when marked increase in vascular permeability occurs; however, injection of such vasoactive substances along with a chemotactic agent increases considerably the local neutrophil response to the latter (Issekutz, 1981).

Immune mechanism can also cause damage to endothelial cells in other ways (Ryan, 1976). Specific antibodies can be formed against antigens on the surface of endothelial cells. Some evidence suggests the presence of cross-reactive antigens between platelets and endothelial cells, possibly responsible for some cases of anaphylactoid purpura and certain other diseases (Lindquist and Osterland, 1971). Some observations suggest that the vascular endothelium may be a primary target in some types of organ graft rejection (Busch *et al.*, 1971). However, the most likely cause of inflammatory damage to vascular endothelium appears to be the deposition of circulating immune complexes.

10. INFLAMMATORY CELL–ENDOTHELIUM INTERACTIONS IN HUMAN ALLERGIC RESPONSES

Because this volume is devoted predominantly to hypersensitivity reactions, discussion of some aspects of inflammatory cell responses in human allergic reactions appears appropriate. It has been generally considered that such IgE-mediated reactions are manifested by vasodilatation, increased permeability, and possibly smooth muscle spasm, depending upon the organ involved. Such events have been thought to be due to the release of histamine and possibly other mediators from mast cells present in perivascular areas. More recently, it has been appreciated that there is an accumulation of inflammatory cells in such sites over a period of hours following antigen challenge (Atkins *et al.*, 1973). In these histologic studies from our laboratory (Atkins *et al.*, 1973; Slott and Zweiman, 1975; Dunsky *et al.*, 1977) it was found that: (1) neutrophils appear in the site earliest, followed by eosinophils; (2) the earliest accumulation of such cells is in the perivascular/periappendigeal areas followed within 1–2 hr by increased accumulation in the interstitium of the dermis; (3) there is suggestive evidence that such inflammatory cells are not necessarily "dead-ended" in such sites. They may leave as well as enter the inflammatory reaction site over a period of hours. At 8 hr after antigenic challenge, the reaction site frequently contains a mixture of granulocytes and mononuclear inflammatory cells (Solley *et al.*, 1976) and fibrin (deShazo *et al.*, 1979). The pathogenic significance of these findings is uncertain.

It is also still unclear what factors lead to the passage of inflammatory cells through the vascular wall into the dermis, which normally contains very few such cells. Histamine itself is only weakly chemotactic for eosinophils *in vitro* (Clark *et al.*, 1975). Intradermal injection of histamine in doses sufficient to mimic the typical wheal and flare response of an IgE-mediation is not followed by local eosinophil accumulation (Atkins *et al.*, 1973). Our group has found that nonhistamine low-molecular-weight material released *in vivo* in allergic reaction sites will induce a transient accumulation of eosinophils when injected intradermally (Ting *et al.*, 1981). These latter findings further support the thesis that at least eosinophils may leave as well as enter allergic reaction tissue sites. It is likely, but not known definitively, that such cellular emigration occurs via the same route (small venules and possibly capillaries) through which inflammatory cells pass from blood to tissue in other types of inflammation.

It is not known whether there are similar factors responsible for passage of neutrophils and blood mononuclear cells into allergic reaction sites. The leukotrienes, particulary LTB_4, have been shown to attract neutrophils as well as eosinophils *in vitro,* and may lead to local accumulation of such cells when injected intradermally (Soter *et al.*, 1983). The low-molecular-weight tetrapeptide(s) termed ECF-A (Goetzl and Austen, 1975) are chemotactic for neutrophils although not as much as for eosinophils. Recent studies (Dorsch and Ring, 1981) suggest that fluid components released in the "immediate" phase of allergic reactions will induce a delayed-appearing reaction when injected into normal skin.

Thus, it is likely that several factors released in human allergic reactions are responsible for both immediate and delayed-appearing manifestations. The role of the vascular endothelium in such responses is not well defined. As mentioned previously, histamine can cause contraction of endothelial cells *in vivo,* leading to increased permeability for blood fluid components. It is not clear whether histamine induces significantly increased leakage of cells. One would suspect that such increased cellular emigration (regardless of the factor responsible) would be preceded by increased inflammatory cell adherence to endothelium of small vessels. However, there does not appear to be any *in vitro* or *in vivo* investigations of the quantitative effects of any factors on the adherence of leukocytes to vascular endothelium in these allergic reactions. There is also no histologic evidence that the vascular endothelium itself is damaged. Thus, there is still much to be learned about the mechanisms underlying the inflammatory responses in allergic reactions.

However, some indirect evidence about the role of leukocyte–endothelium interactions comes from the investigation of drug effects on such reactions. It is worth noting in preface that the passage of fluid and inflammatory cells through small vessel walls may occur by different mechanisms (reviewed by Simon, 1980). Pretreatment of allergic subjects with antihistamines may inhibit antigen skin test whealing reactions but not the later local inflammatory cell accumulation (Atkins *et al.*, 1973). The converse situation is seen with corticosteroid pretreatment. Antigen-induced whealing responses are not affected (Slott and Zweiman, 1975) whereas inflammatory cell responses are markedly diminished (Slott and Zweiman, 1975; Zweiman *et al.*, 1976). It is true that high steroid concentrations in the skin achieved by topical administration may lead to local vasoconstriction as well.

However, the overall pattern of steroid effects noted here may explain why (1) systemic steroids generally do little to prevent or ameliorate the immediate manifestations of human anaphylactic reactions and (2) the beneficial effect in more chronic allergic reactions is delayed in onset for hours to days.

11. CONCLUSION

The findings reviewed in this chapter certainly suggest that interactions of leukocytes with vascular endothelium play important roles in the normal circulation of lymphoid cells. Alterations in such leukocyte–endothelium relations are likely an important early development which is followed by increased emigration of granulocytes and/or mononuclear leukocytes into inflammatory reaction sites.

However, there are several concerns about extending the conclusions noted here to *in vivo* responses in health and disease. Much of the data come from (1) *in vitro* studies using endothelial monolayers or (2) *in vivo* studies in which the migration of labeled leukocytes into one or another organ is investigated. It has been extremely difficult to study *in vivo* the interaction of leukocytes and endothelium itself in physiologic or pathologic states. Ethical and logistic considerations may preclude such studies in humans even if technical approaches are developed. The same concerns can be raised about investigation of the effects of therapeutic agents on such leukocyte–endothelium interactions.

Notwithstanding these caveats, it does appear that progress is being made, albeit slowly. Increasing knowledge about cell membrane receptors, histocompatibility-dependent cell interactions and immune responses, and improved *in vitro* and *in vivo* probes should provide a basis for expanded investigation. The importance of understanding the circulation of leukocytes and its role in localized immune and inflammatory responses should stimulate continued efforts to enhance our knowledge in this area.

ACKNOWLEDGMENT. Supported in part by a grant from the Lupus Foundation of Delaware Valley.

REFERENCES

Anderson, A. O., and Anderson, N. D., 1976, Lymphocyte emigration from high endothelial venules in rat lymph nodes, *Immunology* **31**:731.

Andrews, B. S., Shadforth, M., Cunningham, P., and Davis, J. S., IV, 1981, Demonstration of a Clq receptor on the surface of human endothelial cells, *J. Immunol.* **127**:1075.

Ashida, E. R., Johnson, A. R., and Lipsky, P. E., 1981, Human endothelial cell–lymphocyte interaction, *J. Clin. Invest.* **67**:1490.

Atherton, A., 1972, Quantitative investigation of adhesiveness of circulating leukocytes to blood vessel walls, *J. Physiol.* **222**:447.

Atkins, P. C., Green, G. R., and Zweiman, B., 1973, Histologic studies of human eosinophil and mast cell responses to ragweed, compound 48/80 and histamine, *J. Allergy Clin. Immunol.* **51**:263.

Beesley, J. E., Pearson, J. D., Carleton, J. S., Hubsbury, A., and Gordon, J. L., 1978, Interaction of leukocytes with vascular cells in culture, *J. Cell Sci.* **33**:85.

Burger, D. R., Ford, D., Vetto, R. M., Hamblin, A., Goldstein, A., Hubbard, M., and Dumonde, D. C., 1981, Endothelial cell presentation of antigen to human T cells, *Hum. Immunol.* **3**:209.

Busch, G. J., Reynolds, E. S., Galuanek, E. G., Braun, W. E., and Dammin, G. J., 1971, Human renal allografts: The role of vascular injury in early grant failure, *Medicine (Baltimore)* **50**:29.

Butcher, E. C., and Weissman, I. L., 1980, Cellular, genetic and evolutionary aspects of lymphocyte interactions with high-endothelial venules, *Ciba Found. Symp.* **71**:265.

Butcher, E. C., Kraal, G., Stevens, S. K., and Weissmen, I. L., 1982, Selective migration of immune lymphoblasts and lymphoblast population and the role of endothelial cell recognition, *Adv. Exp. Biol. Med.* **149**:199.

Cines, D. B., Lyss, A. P., Bina, M., Corkey, R., Kefalides, N. A., and Friedman, H. M., 1982, Fc and C3 receptors induced by herpes simplex virus on cultured human endothelial cells, *J. Clin. Invest.* **69**:123.

Clark, R. A. F., Gallin, J. I., and Kaplan, A. D., 1975, The selective activity of histamine, *J. Exp. Med.* **142**:1462.

Cochrane, C. G., and Aikin, B. S., 1966, Polymorphonuclear leukocytes in immunological reactions, *J. Exp. Med.* **124**:733.

Cotran, R. S., 1967, Studies on inflammation: ultrastructure of the prolonged vascular response induced by clostridium oedematous toxin, *Lab. Invest.* **17**:39.

Curtis, A. S. G., and Renshaw, R. M., 1982, Lymphocyte–endothelial interactions and histocompatibility restriction, *Adv. Exp. Med. Biol.* **149**:193.

DeBono, D., 1976, Endothelial leukocyte interactions in vitro. I. Adherence of non-allergized lymphocytes, *Cell. immunol.* **26**:78.

DeBono, D., 1979, Endothelium–lymphocyte interaction in vitro II. Adherence of allergized lymphocytes, *Cell. Immunol.* **44**:64.

deShazo, R., Levinson, A. I., Dvorak, H. F., and Davis, R. W., 1979, The late phase skin reaction: Evidence for activation of the coagulation system in an IgE-dependent reaction in man, *J. Immunol.* **122**:697.

Dorsch, W., and Ring, J., 1981, Induction of late cutaneous reactions by skin blister fluid from allergen tested and normal skin, *J. Allergy Clin. Immunol.* **67**:117.

Dunsky, E. H., Atkins, P. C., and Zweiman, B., 1977, Histologic responses in human skin test reactions to ragweed. 4. Effects of a single intravenous injection of steroids, *J. Allergy Clin. Immunol.* **59**:142.

Ford, W. L., Sedgley, M., Sparshott, M., and Smith, M. E., 1976, The migration of lymphocytes across specialized vascular endothelium, *Cell Tissue Kinet.* **9**:351.

Ford, W. P., 1975, Lymphocyte migration and immune responses, *Prog. Allergy* **19**:1.

Freemont, A. J., 1982, A possible route for lymphocyte migration into diseased tissue, *J. Clin. Pathol.* **36**:161.

Fromer, C. H., and Klintworth, G. K., 1975, An evaluation of the role of leukocytes in the pathogenesis of experimentally induced corneal vascularization, *Am. J. Pathol.* **79**:537.

Gesner, B. M., and Ginsberg, V., 1964, Effects of glycosidases on the fate of transfused lymphocytes, *Proc. Natl. Acad. Sci. U.S.A.* **52**:750.

Gimbrone, M. A., Jr., 1976, Culture of vascular endothelium, *Prog. Hemostasis Thromb.* **3**:1.

Goetzl, E. J., and Austen, K. F., 1975, Purification and synthesis of eosinophilotactic tetrapeptides of human lung tissue: Identification as eosinophilic chemotactic factor of anaphylaxis, ECF-A, *Proc. Natl. Acad. Sci. U.S.A.* **72**:4123.

Goldschneider, I., and MacGregor, D. D., 1968, Migration of lymphocytes and thymocytes in the rat, *J. Exp. Med.* **127**:155.

Gowans, J. L., and Knight, E. J., 1964, The route of recirculation of lymphocytes in the rat, *Proc. R. Soc. London Ser. B* **159**:257.

Gowans, J. L., and Steer, H. W., 1980, The functions and pathways of lymphocyte recirculation, *Ciba Found. Symp.* **71**:113.

Graham, R. C., and Shannon, S. L., 1972, Peroxidase arthritis, *Am. J. Pathol.* **67**:69.

Hashimoto, Y., and Hurd, E. R., 1981, Human neutrophil aggregation and increased adherence to human endothelial cells induced by heat aggregated IgG and immune complexes, *Clin. Exp. Immunol.* **44**:538.

Hirschberg, H., Bergh, O. J., and Thorsby, E., 1980, Antigen presenting properties of vascular endothelial cells, *J. Exp. Med.* **152**:2495.

Issekutz, A. C., 1981, Effect of vasoactive agents on polymorphonuclear leukocyte emigration in vivo, *Lab. Invest.* **45**:234.

Jaffe, E. A., Hoyer, L. W., and Nachman, R. L., 1973, Synthesis of anti-hemophiliac factor antigen by cultured human endothelial cells, *J. Clin. Invest.* **52**:2737.

Lentnek, A. L., Schreiber, A. D., and MacGregor, R. R., 1976, The induction of augmented granulocyte adherence by inflammation, *J. Clin. Invest.* **57**:1098.

Lindquist, K. S., and Osterland, C. K., 1971, Human antibodies to vascular endothelium, *Clin. Exp. Immunol.* **9**:753.

McIntyre, L. V., Adams, G. A., Eskin, S. G., and Martin, R. R., 1982, Leukocyte and platelet interaction with protein and endothelial cell coated model vessels, *Clin. Hemorheol.* **2**:273.

Marchesi, V. T., 1964, Some electron microscopic observations on interactions between leukocytes, platelets, and endothelial cells in acute inflammation, *Ann. N.Y. Acad. Sci.* **116**:274.

Marchesi, V. T., and Florey, H. W., 1960, Electron microscopic observations on the emigration of leukocytes, *O. J. Exp. Physiol.* **45**:343.
Moncada, S., Higgs, E. A., and Vane, J. R., 1977, Human arterial and venous tissues generate prostacyclin, a potent inhibitor of platelet-aggregation, *Lancet* **1**:18.
Nightingale, G., and Hurley, J. V., 1978, Relationship between lymphocyte emigration and vascular endothelium in chronic inflammation, *Pathology* **10**:27.
Polverini, P. J., Cotran, R. S., and Sholley, M. M., 1977, Endothelial proliferation in the delayed hypersensitivity reaction: An autoradiographic study, *J. Immunol.* **118**:527.
Rose, M. L., and Parrott, D. M. V., 1977, Vascular permeability and lymphoblast extravasation into inflamed skin are not related, *Cell. Immunol.* **33**:62.
Ryan, G. B., and Majno, G., 1977, Acute inflammation: A review, *Am. J. Pathol.* **86**:183.
Ryan, T. J. (ed.), 1976, *Microvascular Injury: Vasculitis, Stasis and Ischemia*, pp. 135–143, Saunders, Philadelphia.
Ryan, V. S., Schults, D. R., and Ryan, J. W., 1981, Fc and C3b receptors on pulmonary endothelial cells: Induction by injury, *Science* **214**:557.
Schoefl, G. I., 1972, The migration of lymphocytes across the vascular endothelium in lymphoid tissue: A re-examination, *J. Exp. Med.* **136**:568.
Sholley, M. M., and Cotran, R. S., 1978, Endothelial proliferation in inflammation. II. Autoradiographic studies in X-irradiated leukopenic rats after thermal injury of the skin, *Am. J. Pathol.* **91**:229.
Sholley, M. M., Cavallo, T., and Cotran, R. S., 1974, Endothelial regeneration in acute inflammation, *Anat. Rec.* **178**:462.
Simionescu, M., Simionescu, N., and Palade, G. E., 1976, Recent studies on vascular endothelium, *Ann. N.Y. Acad. Sci.* **275**:64.
Simon, G. T., 1980, Ultrastructure of acute inflammation, *Ciba Found. Symp.* **71**:113.
Slott, R. I., and Zweiman, B., 1975, Histologic studies of human skin test responses to ragweed and compound 48/80. II. Effect of corticosteroid therapy, *J. Allergy Clin. Immunol.* **55**:232.
Smith, J. B., McIntosh, G. H., and Morris, B., 1970, Migration of cells through chronically inflammed tissues, *J. Pathol.* **100**:21.
Solley, G. O., Gleich, G. J., Jordon, R. E., and Schroeter, A. L., 1976, The late phase of the immediate wheal and flare skin reactions: its dependence upon IgE antibodies, *J. Clin. Invest.* **58**:408.
Soter, W., Lewis, R. A., Corey, E. J., and Austen, K. F., 1983, Local effects of synthetic leukotrienes in human skin, *J. Invest. Dermatol.* **80**:115.
Stamper, H. B., Jr., and Woodruff, J. J., 1976, Lymphocyte homing into lymph nodes: In vitro demonstration of the selective affinity of recirculating lymphocytes for high endothelial venules, *J. Exp. Med.* **144**:838.
Stamper, H. B., Jr., and Woodruff, J. J., 1977, An in vitro model of lymphocyte homing, *J. Immunol.* **119**:772.
Thorgeirsson, G., and Robertson, A. L., 1978, The vascular endothelium—Pathobiologic significance, *Am. J. Pathol.* **93**:803.
Ting, S., Zweiman, B., Lavker, R. M., and Dunsky, E. H., 1981, In vivo release of eosinophil chemoattractant activity in human allergic skin reactions, *J. Immunol.* **127**:557.
Weibel, E. R., and Palade, G. E., 1964, New cytoplasmic components in arterial endothelial cells, *J. Cell Biol.* **23**:101.
Weiss, S. J., Young, J., LoBuglio, A. F., Slivka, A., and Nimeh, N. F., 1981, Role of hydrogen peroxide in neutrophil-mediated destruction of cultured endothelial cells, *J. Clin. Invest.* **68**:714.
Zweiman, B., Moskovitz, A. R., and Lisak, R. P., 1982, Quantitative assessment of the adherence of normal human blood mononuclear cells to vascular endothelial cell monolayers, *Cell. Immunol.* **68**:165.
Zweiman, B., Moskovitz, A. R., Flaschen, S., Atkins, P. C., Bedard, P., and Lisak, R. P., 1983, Patterns of adherence of human leukocyte subpopulation to vascular endothelium, *Clin. Res.* **31**:165A.
Zweiman, B., Slott, R. I., Atkins, P. C., 1976, Histologic responses in human skin test responses to ragweed and compound 48/80 effects of alternate-day steroid therapy, *J. All. Clin. Immunol.* **58**:657.

3

Regional Specialization in Antigen Presentation

J. WAYNE STREILEIN, JOAN STEIN-STREILEIN, and JUDITH HEAD

1. INTRODUCTION

Cells of the reticuloendothelial system were recognized as essential to immunologic functions even before a unitary description of the RES was achieved. Metchnikoff, in the late 1800s, first identified phagocytes as important cells in protective immunity. Since then, an enormous amount of experimental data has confirmed the pivotal role of cells of the RES in immune responses.

RE cells participate in immune reactivities at many different levels. It has been almost 20 years since it was first realized that cells of the RES (macrophages) function in and are required for induction of immune responses (Unanue, 1981). Based chiefly on *in vitro* studies, it was learned that cells adherent to glass or plastic were essential to successful lymphocyte responses to mitogens and antigens. These adherent or accessory cells were found to be capable of phagocytosis; consequently, the idea emerged that macrophages are the accessory cells that function in antigen presentation.

Within the past decade, Steinman and his associates (see Steinman and Nussenzweig, 1980) have conducted an elegant series of studies that describe a novel population of cells of bone marrow origin. These cells are deemed to be distinct from phagocytic macrophages and yet they function exceedingly well in antigen presentation. Morphologically, these cells appear to be dendritic, lack significant capacity to phagocytose particulate material, and can be isolated by physical separation techniques from phagocytic macrophages with which they are found in the spleen.

Detailed description of dendritic cells and their putative distinction from

---

J. WAYNE STREILEIN, JOAN STEIN-STREILEIN, and JUDITH HEAD • Departments of Cell Biology, Pathology, and Internal Medicine, University of Texas Health Science Center, Dallas, Texas 75235.

macrophages is beyond the scope of this chapter but the matter of their existence is raised because of the importance of distinguishing between the process of phagocytosis and the process of antigen presentation. While cells capable of presenting antigens to immunocompetent lymphocytes may possess the property of phagocytosis, it would appear that this capacity may be independent of their ability to process and present nominal antigens in immunogenic form on their surface (Chestnut *et al.*, 1980; Lee *et al.*, 1982). A characteristic of antigen-presenting cells, which seems essential to their function, is surface expression of glycoproteins encoded by *I*-region genes within the major histocompatibility complex (Unanue, 1981). It is now believed that the requirement for expression of cell surface Ia molecules on antigen-presenting cells relates to the fact that immunocompetent helper/inducer T lymphocytes use Ia determinants as restricting elements in recognition of nominal antigens. An equally important characteristic of antigen-presenting cells is their capacity to secrete interleukin-1 (IL-1) in response to T-cell recognition of antigen on their surfaces (Oppenheim *et al.*, 1979). IL-1 is an obligate signal in the activation of helper T cells.

With special reference to the purposes of this chapter, the role of antigen-presenting cells in the inductive phase of immune responsiveness can be described in five sequential steps: (1) Antigen capture. In this phase, antigen which arrives at the surface of the antigen-presenting cell, is bound, internalized, probably through endocytosis, processed intracellularly, and then reexpressed on the surface in a particularly immunogenic form. The precise molecular nature of the immunogenic form of antigen remains elusive although recent data bear on this point (Unanue, 1984). (2) Coexpression with Ia determinants (reviewed in Unanue *et al.*, 1984). In an as yet little understood manner, immunogenic antigen and Ia molecules are displayed on the surface of the antigen-presenting cells. Whether or not this involves intimate molecular interactions is unclear, although anti-Ia antibodies can certainly block the antigen-presenting function *in vitro.* (3) Attraction of T cells. Immunogenic antigen on the presenting cell mysteriously attracts immunocompetent T cells to the immediate environment. The molecular basis for this attraction is unclear, and could involve receptor–ligand interactions and/or chemotactic factors. (4) Antigen recognition. Helper and/or inducer T cells with specific surface receptors for nominal antigen and with receptors that allow them to recognize Ia molecules then experience a recognition event. The act of recognition allows the T cell to deliver a signal which induces the antigen-presenting cell to secrete IL-1 and related factors into the environment. (5) T cell activation. These factors have the capacity to activate T cells engaged in antigen recognition; the cells then undergo a series of proliferative and differentiation events.

This brief, simplistic description of the process of antigen presentation has been deduced from a variety of *in vivo* and *in vitro* studies. For the most part, the antigen-presenting cells that have been studied in laboratory animals have been obtained from the peritoneal cavity and/or the spleen. It is the perspective of this chapter that, while there is a general paradigm of antigen presentation that is common, there are in addition unique aspects of antigen presentation which relate to the tissue location of antigen-presenting cells and to the nature of the site of confrontation with the environment.

If one considers the interactions that occur between an organism and its environment, it is apparent that a variety of unique and different interfaces are involved. For example, the confrontation between the skin and the external environment is fundamentally different from that between the lining of the gastrointestinal (GI) tract and the outside world. Each environmental interface has a unique physiological property(ies) that must be carried out in order for the integrity of the organism to be maintained and for communication with the environment to take place. At the same time, each environmental interface offers an opportunity for pathogenic organisms and environmental toxins to threaten the viability of the host, and it is the responsibility of the immune system to provide appropriate protection.

It is the thesis of this chapter that each unique interface between the environment and the organism has required a specialization of antigen-presenting function so that the essential physiologic function of the interface is preserved while immune protection is also achieved.

2. REGIONAL SPHERES OF IMMUNOLOGIC INFLUENCE

In the early 1960s it was discovered that more than one isotype or class of immunoglobulin exists. One newly described class, IgA, was an immunoglobulin particularly enriched in external bodily secretions such as tears, saliva, and intestinal contents. Shortly thereafter, Gowans and Knight (1964) and their collaborators performed an elegant series of experiments using thoracic duct drainage and found that recirculating lymphocytes passing through the lymph are not homogeneous. They found a subpopulation of IgA-bearing B lymphocytes that did not migrate freely throughout the body, but instead was localized to the GI tract. Based on these and related observations, the idea emerged that a segment of the immune system was dedicated to antigenic challenges that occurred chiefly within the GI tract; the term *gut-associated lymphoid tissues* (GALT) was coined (Guy-Grand *et al.*, 1974). For an interval of time, some confusion arose from the possibility that GALT might be important in the ontogeny of lymphocytes in a manner analogous to the ontogenetic role of the bursa of Fabricius in chickens. More recently, it has become clear that GALT refers to a specialized set of lymphoid cells and tissues that provide the GI tract with local immunity, predominantly of the IgA type. It has also been realized that certain other body surfaces are related to the GALT circuit. Bronchus-associated (BALT) (Bienenstock *et al.*, 1973a), mammary gland-associated (MALT) (Parmeley and Beer, 1977), and conjunctiva-associated (CALT) (Axelrod and Chandler, 1979) lymphoid tissues are thought to represent subsets of the GALT network. This has engendered the notion that a portion of the immune apparatus is dedicated to achieving integrated immunity for all mucosal surfaces.

Within the last 5 years, work from our laboratory, along with that of others, has raised the possibility that a unique and relatively separate set of lymphocytes and related tissues are dedicated to antigen confrontations that take place at the level of the skin. Accordingly, a new acronym, SALT, has been coined to denote these skin-associated lymphoid tissues (Streilein, 1978, 1983). We would propose that there are at least these two (and perhaps more) regional spheres of influence within the immune system, and that each sphere uniquely addresses the special physiologic and antigenic demands of body surfaces that confront the environment.

In the following portions of this chapter we will focus attention on GALT, SALT, lung-associated lymphoid tissues, and reproductive tract tissues in an effort to describe the specialized nature of antigen presentation that takes place within each of these tissues. The following questions will be addressed: (1) What is the nature of the barrier to antigen penetration at each interface? (2) What structural specialization allows antigens to gain access beyond this barrier in order to make an impact on the immune system? (3) What are the important and unique features of potential antigen-presenting cells that exist at or near each interface? (4) At what site does meaningful recognition of antigen by immunocompetent cells take place? (5) Is there specialization of lymphocytes that recognize and respond to antigens presented at each of these sites? (6) What are the local and systemic consequences of immune responses that are initiated by antigen presented uniquely at each specialized region?

3. GASTROINTESTINAL TRACT

3.1. BARRIER PROPERTIES

In performing its primary function of nutrient uptake, the GI tract is naturally exposed to large amounts of foreign material, including food and microorganisms. A variety of physiochemical barriers are found along the tract that serve as nonspecific defenses and thus could potentially influence antigenic exposure (Ottaway *et al.*, 1979). Structural barriers include stratified squamous epithelium in the upper parts of the tract and, in the lower regions, columnar cells with extensive tight junctional complexes at their apical borders. The rapid turnover of these columnar epithelial cells may serve to lessen the chance for uptake of foreign molecules or invading organisms by this cell type.

Numerous enzymes are found throughout the alimentary canal and, along with strong gastric acidity, are chemical barriers that could destroy or damage organisms and molecules. Perhaps the most formidable physical barrier is mucus, several forms of which are found in different regions. Both salivary glycoproteins and gastric mucin have been shown to interfere with the binding of bacteria or their toxins to epithelial cells (Gibbons *et al.*, 1976; Strombeck and Harrold, 1974). It has been hypothesized that intestinal mucus reduces exposure of the intestinal absorptive cells by forming not only a physical but also a biochemical barrier (Edwards, 1978). Since it is most likely immiscible with most macromolecules in solution, mucus may prevent access of such molecules to the cell surface. In contrast, small molecules or larger ones that could interact favorably with mucus components could penetrate, creating a molecular filter. Interestingly, the immune system can, by a variety of means, influence the amount of goblet cell mucus in the intestine. Examples include increased mucus release during intestinal anaphylaxis induced by challenging sensitized or passively immunized animals intraduodenally (Lake *et al.*, 1979), and following such injection of antigen–antibody complexes (Walker *et al.*, 1977). In rats infected with *Nippostrongylus brasiliensis,* expulsion of the worms is accompanied by a significant increase in the proportion of goblet cells, an effect that can be adoptively transferred to naive animals using sensitized thoracic duct lymphocytes (Miller and Nawa, 1979). These findings implicate an important nonspecific protective role for intestinal mucus that amplifies specific responses.

Along with these nonspecific factors, it has long been recognized that the GI tract has major lymphoid components that may be important in the generation and effectuation of specific responses to orally acquired substances. Accumulations of lymphocytes occur throughout the GI tract and consist of solitary nodules, found in all parts of the alimentary canal, or organized groups of nodules and diffuse lymphocytes represented by the tonsils, Peyer's patches (PP), and appendix. In certain species, other similar structures are found, such as the sacculus rotundus in the rabbit and the cecal tonsil in fowl. All of these definitive structures together comprise the GALT, representing a considerable mass of lymphoid cells. In addition, the GI tract lamina propria contains significant numbers of free lymphocytes and their progeny, and a considerable population of scattered intraepithelial lymphocytes can be found among intestinal columnar cells.

3.2. STRUCTURAL SPECIALIZATION FOR ANTIGEN PRESENTATION

There is now substantial evidence that the GALT are the primary sites of meaningful antigenic exposure in the GI tract by virtue of specific structural features allowing "controlled contact" between luminal antigens and reactive lymphocytes (LeFevre *et al.*, 1979). The GALT consist of secondary lymphoid tissues, including B-cell-containing follicles and interfollicular areas predominated by T cells, with high endothelial venules where lymphocytes enter from the peripheral blood. Especially in the intestines, this tissue can occupy the gut wall down to the muscularis externa, but most solitary follicles are limited to the mucosa. Above the follicles, the subepithelial connective tissue contains large numbers of lymphoid cells, and, with the overlying epithelium, constitutes a region referred to as the dome. The dome bulges somewhat into the lumen, and both the lamina propria and the epithelium contain large numbers of small and large lymphocytes and macrophages (Faulk *et al.*, 1971).

Unlike lymph nodes, uptake of materials at mucosal sites occurs through the epithelium instead of via afferent lymphatics. There are numerous modifications of this epithelium, previously termed the "follicle-associated epithelium"

(FAE), which can affect antigenic exposure, including smaller, cuboidal cells with fewer microvilli, absence of goblet cells (less mucus), and the presence of many lymphoid cells. Such changes are even seen over small, isolated lymphoid accumulations (Keren *et al.*, 1978). However, the primary structural feature is the presence of scattered cells with a unique morphology, called microfold or membranous (M) cells, which are evenly distributed in the higher portions of the FAE (Smith and Peacock, 1980). These cells have an apical surface with sparsely distributed microvilli and, in some species, small folds, contrasting sharply with the neighboring cuboidal cells with which they form tight junctions at apical borders (Owen and Nemanic, 1978). The cytoplasm is very attenuated, such that the cells form a reticulum of strands around subjacent lymphocytes and macrophages (Owen and Jones, 1974). The apical cytoplasm is often so thin that lymphocytes are separated from the intestinal lumen by as little as 0.3 μm and with the light microscope give the false impression of entering the lumen. Cells with M-cell morphology have been described in virtually all GALT components (Owen and Nemanic, 1978) and are believed to develop from adjacent cuboidal epithelial cells (Bhalla and Owen, 1982).

The tissues comprising GALT are the primary route of entry for many antigens and organisms from the alimentary tract lumen, and studies with particulate tracers and microorganisms imply a key role for the M cell in this process. Colloidal carbon and particles as large as 2 μm (latex beads) fed to mice appear first in PP and not other intestinal areas, with carbon being detected initially within vacuoles of the epithelial cells (LeFevre *et al.*, 1978; Joel *et al.*, 1978). A smaller tracer, ferritin, injected into ileal loops was seen first in pinocytotic vesicles in the FAE cells, and by 1 hr, in adjacent intercellular spaces (Bockman and Cooper, 1973). More recently, horseradish peroxidase (HRP) in the intestinal lumen was observed to very rapidly (by 1 min) adhere to FAE cells, and by 5 min could be found only in M-cell cytoplasmic vesicles (Owen, 1977). By 1 hr, the HRP was found in the intercellular spaces and even within the associated lymphoid cells, thus directly demonstrating that the M cells performed a transport function for luminal substances to subjacent cells. Such transport may in fact be bidirectional, since HRP given intravenously makes its way into the intestinal lumen, again by going primarily through the PP FAE (Bockman and Stevens, 1977).

Studies on experimental GI infections with various microorganisms have generally confirmed the above tracer experiments. Mice inoculated with pathogenic organisms via gastric tube showed initial colonization in ileal PP (Carter and Collins, 1974; Marneerushapisal and Rowley, 1981; Hohmann *et al.*, 1978). Electron microscopy revealed that pathogenic enteroadherent *E. coli* specifically adhered to the M cells of the PP FAE of rabbits several days before they would adhere to columnar cells (Inman and Cantey, 1983). Viruses have been shown not only to attach specifically to the M cells, but also to be taken up into M-cell cytoplasmic vesicles and transported internally to the intercellular space. Both transmissible gastroenteritis virus in pigs and reovirus type I in mice followed this course (Chu *et al.*, 1982; Wolf *et al.*, 1981). They were then found associated with subepithelial lymphocytes and macrophages, either in coated pits on the cell surface or within cytoplasmic vesicles.

A unique feature of the GI tract that may influence antigen uptake and subsequent processing is the presence of the gut flora, which exists in complex ecosystems within the various portions of the tract (Savage, 1977). These organisms affect not only the establishment of pathogens, but also the structure of the bowel, development of GALT, and distribution of lymphoid cells (see Ottaway *et al.*, 1979). In germfree animals, lymphocyte numbers are dramatically decreased in both lamina propria and epithelium, and immunodeficient individuals often have greatly increased gut bacterial populations (Hersh *et al.*, 1970). In addition, gamma globulin synthesis is 50 times greater in conventional mice than in germfree animals (Sell and Fahey, 1964). These observations suggest that immune responses to endogenous gut flora normally do occur, and that an important function of the GALT may be homeostasis of these organisms. Recent studies have shown that the gut microflora can regulate immune responses to orally acquired antigens, probably due to influences of their cell wall lipopolysaccharide (LPS) on lymphocyte subpopulations (Wannemuehler *et al.*, 1982).

Uptake of antigen within the GI tract may also be influenced by the presence of specific antibody. It has been proposed that a major purpose of antibodies, especially secretory IgA, in the intestinal tract is to protect the mucosal surface and the individual by immune exclusion (Walker and Isselbacher, 1977), i.e., prevention of mucosal penetration by organisms, toxins, and immunogenic dietary substances. Treatment of bacteria such as streptococci and *Vibrio cholerae* with specific antibodies prevented adherence to intestinal epithelial cells, with decreased colonization (Williams and Gibbons, 1972; Fubara and Freter, 1973). Lymphocytes bearing Fc receptors for α chains within GALT and the intestinal epithelium have been shown to mediate antibody-dependent cell-mediated cytotoxicity in the presence of secretory IgA antibodies, reducing the numbers of potentially invading organisms (Tagliabue *et al.*, 1983). Specific antitoxins injected into rat ileal loops similarly interfered with binding of radiolabeled bacterial toxins to receptors on intestinal cells, preventing the initial phases of the disease process (Wu and Walker, 1976). The initial adsorption of macromolecules such as albumin and HRP to the mucosal surface of specifically sensitized rats was increased, but subsequently, there was decreased uptake, then degradation of the antigens (Walker *et al.*, 1975). The intestinal mucus may play a role since it could biochemically retain antibody and thus the antigen–antibody complexes formed so that enzymatic processes are more efficient (Edwards, 1978). By a similar process, mucosal immunity may protect against uptake of improperly digested, immunogenic dietary substances.

3.3. ANTIGEN-PRESENTING CELLS OF GALT

Of considerable importance is the fate of substances taken up through the intestinal mucosa. Numerous histologic studies have identified large numbers of macrophages within PP, especially in the epithelium and subepithelial region of the dome, and these cells seem to be involved in avid phagocytosis (Lause and Bockman, 1981; Sobhon, 1971; MacDonald and Carter, 1982). Substances ranging from small compounds such as ferritin and viruses to large particles such as 2-μm latex beads and the protozoan *Giardia* are phagocytosed by these cells immediately after epithelial penetration (Lause and Bockman, 1981; Chu *et al.*, 1982; LeFevre *et al.*, 1978; Owen *et al.*, 1981). There are several possible subsequent courses for ingested substances: (1) they could be retained by these cells and destroyed or sloughed with the cells into the lumen; (2) they could be presented to antigen-reactive cells locally; (3) they could be taken to draining lymph nodes. Phagocytes have been observed to migrate from the subepithelial region to all areas of the PP and to adjacent villi (Faulk *et al.*, 1971; Joel *et al.*, 1978; LeFevre *et al.*, 1978). In normal villi, they tend to accumulate in the tips and there is some evidence that they may be shed during natural sloughing into the lumen. However, phagocytes with ingested material are found in the mesenteric lymph nodes (Joel *et al.*, 1978). There is also considerable evidence that much ingested material is presented to local lymphocytes in follicles and interfollicular areas deep in the patch. Some particles, especially viruses, may be taken up by subepithelial lymphocytes directly as well as by macrophages (Chu *et al.*, 1982; Wolf *et al.*, 1981). In other cases, macrophages containing ingested material have been observed in intimate association with lymphocytes in this region (Owen *et al.*, 1981). A large population of the lymphocytes seen in the dome of the PP are medium and large lymphocytes, suggesting that effective antigen presentation may be occurring even here (Faulk *et al.*, 1971; Owen and Jones, 1974).

Early studies indicated that cells harvested from PP contained none capable of antigen presentation when tested in *in vitro* assays (Kagnoff and Campbell, 1974; Challacombe and Tomasi, 1980). However, several investigators have more recently shown that cells from murine PP are capable of presenting antigen to lymphocytes *in vitro* as well as similar cells from spleen, successful isolation turning on careful enzymatic disruption. These cells support generation of plaque-forming cells to SRBC (Kiyono *et al.*, 1982a) and stimulation of T cells by SRBC and ovalbumin (Richman *et al.*, 1981; MacDonald and Carter, 1982).

Recent immunocytochemical and histologic studies have revealed that PP are well endowed with potential antigen-presenting cells bearing Ia antigens. Besides macrophages, cells with the morphologic characteristics of dendritic cells have been identified histologically in PP follicles, especially in germinal centers, in the interfollicular areas, and in the subepithelial region of the dome (Sobhon, 1971; Sminia *et al.*, 1982; Mayfhofer *et al.*, 1983). Recently, the distribution of Ia^+ cells has been determined directly on rat intestine, using monoclonal antibodies in an immunoperoxidase procedure (Mayrhofer *et al.*, 1983; Sminia *et al.*, 1982, 1983). Two distinct populations of nonlymphoid cells were identified. Strongly Ia^+ cells with dendritic morphology were found in all areas mentioned above as well as in the FAE. These cells were only weakly positive for acid phosphatase and did not require antigenic stimulation to appear ontogenetically. Other cells were phagocytic and strongly acid phosphatase positive, with only slight Ia staining, and represent macrophages. The Ia^+ dendrite-shaped cells are very similar to Steinman's dendritic cells in lymph nodes and spleen. A recent combined functional and morphologic study of murine PP cells confirms that these cells most likely represent the major antigen-presenting cells of the GALT (Spalding *et al.*, 1983).

Cells providing accessory function for T-cell activation were isolated and characterized. In contrast to accessory cells from murine spleens, these PP cells did not adhere to plastic or glass. However, their surface phenotype included Ia antigens and absence of Thy-1, sIg, and Fc receptors. They were irregularly shaped dendritic cells and reacted with a dendritic cell-specific monoclonal antibody.

Although in most studies PP are apparently the primary and perhaps singular site of antigen uptake, this is not the case for every antigen tested, as revealed by experimental intestinal inoculation in the absence of PP. In some cases, proteins injected into isolated ileal loops are capable of immunizing the host, albeit more slowly, when PP have been removed (Hamilton *et al.*, 1981). However, histologic studies revealed that even in such loops, microscopic aggregates of lymphocytes may be found with surface modifications similar to PP and capable of performing the same functions, thus giving the erroneous impression that other routes are involved. In other cases, however, it is clear that other routes can be involved. For example, some organisms and immunoglobulins enter primarily by binding to the absorptive epithelial cells, often by a receptor-mediated process, and thus can potentially bypass direct entry through the PP FAE (Owen *et al.*, 1979; Borthistle *et al.*, 1977). The primary alternative routes include direct drainage into the mesenteric lymph through the central lacteals and transport to the liver in the portal blood with subsequent access to the circulation and spleen. How and to what extent different substances enter are probably significantly influenced by their size and biochemical nature but one primary route does not exclude the possibility of others being used. Haptens more readily gain access to the circulation than antigens of larger molecular weight (Bernstein and Ovary, 1968), but some ingested proteins can be found later in the circulation intact and antigenic at levels ranging from 0.01 to 2% of the ingested amount (Warshaw *et al.*, 1971, 1974).

In this light, another potential antigen-presenting cell that must be considered is the population of Kupffer cells in the liver. These cells, components of the mononuclear phagocyte system, form part of the sinusoidal endothelial lining and can constitute as much as 30% of the total liver cells. Isolated Kupffer cells are phagocytic and express Ia antigens in proportions seen among phagocytic spleen cells (Richman *et al.*, 1979). The subsequent demonstration that Kupffer cells can serve as accessory cells for mitogen stimulation and primary antibody production and present antigen effectively to primed T cells indicates that these cells may play an active role in responses to antigens entering through the portal circulation, instead of merely removing foreign material (Richman *et al.*, 1979; Nadler *et al.*, 1980).

3.4. SITES OF ANTIGEN RECOGNITION

The various routes by which enteric antigens can gain access to the host's lymphoid system suggest that there will be, correspondingly, multiple places where antigen recognition may occur. These include the PP for antigens entering through the FAE, the mesenteric lymph nodes for antigens coming in through central lacteals, and the liver and/or spleen for antigens taken into the intestine's venous vessels. Ingested protein antigens can induce antibody-forming cells in the spleen (Dolezel and Bienenstock, 1971). The role of the liver in antigen processing remains to be elucidated, although experiments in which intestinal blood is diverted away from the liver showed that orally induced tolerance, at least to certain antigens, may depend on this organ (Cantor and Dumont, 1967).

Despite the potential involvement of these routes, there is substantial evidence that the majority of antigens taken in through the intestinal mucosa are processed initially within the PP or perhaps in their microscopic equivalents. Since the lymphatics leaving the PP drain to the mesenteric lymph nodes, it is not surprising that these nodes might be an important secondary site of processing or differentiation of activated cells. Mesenteric lymphadenectomy results in the appearance of nonlymphoid cells in the thoracic duct lymph that have the features of dendritic cells described above (Pugh *et al.*, 1983), including poor adherence, and Ia^{+} surface phenotype. Similar cells are found in lymph from mesenteric lacteals, suggesting that they arise in the intestine, probably in both lamina propria and PP, and migrate normally to the mesenteric lymph nodes where they could function as accessory cells. Thus, antigen, borne by such cells or by macrophages, or perhaps even free, might enter the mesenteric lymph nodes. Tracers and certain bacteria entering the PP are often found subsequently in these draining lymph nodes (Joel *et al.*, 1978; Carter and Collins, 1975). This is not to imply, however, that these nodes are essential in effective antigen processing within the PP. Mesenteric lymphadenectomy long before intestinal exposure to cholera toxoid, using Thirty-Vella loops, did not impair subsequent development of antibody-producing cells (Husband and Gowans, 1978). On the other hand, extensive studies with this antigen in mice and DNP-KLH in rabbits, using intestinal loops with and without PP have confirmed that the PP are the essential sites of antigen processing in the development of mucosal immunity (Cebra *et al.*, 1979).

3.5. SPECIALIZED LYMPHOCYTES THAT TRAFFIC THROUGH GALT

There is now overwhelming evidence that the PP play the key role in a complex circuitry of cell migration and differentiation which results in expression of local immunity in the gut (Ottaway *et al.*, 1979; Lamm, 1976). Cells stimulated within the PP leave to enter a cycle which carries them sequentially to the mesenteric lymph nodes, the thoracic duct lymph, the bloodstream, and finally the lamina propria of the gut, where they mature into plasma cells secreting IgA antibodies. The frequency of PP B cells stimulated by SRBC was substantially reduced after SRBC ingestion, whereas the number of cells responding to unrelated antigens was unaffected, demonstrating that stimulated B cells leave the PP (Kagnoff, 1977). In elegant adoptive transfer studies, Cebra and his colleagues showed that PP cells could repopulate the IgA plasma cells of the gut lamina propria in irradiated rabbits (Craig and Cebra, 1971) and they have provided irrefutable evidence, using chromosomal markers in adoptive transfer experiments, that PP contain large numbers of the precursor lymphocytes destined to become IgA-producing cells (Cebra *et al.*, 1977). Additionally, for a particular antigen, the precursor frequency of IgA-producing cells in PP was considerably higher than in spleens and peripheral nodes. Subsequent transfer studies with mesenteric lymph node cells and thoracic duct lymphocytes revealed even higher numbers of cells, primarily blasts, destined to populate the lamina propria with IgA plasma cells (Guy-Grand *et al.*, 1974; McWilliams *et al.*, 1977). Determinations of the percentage of cells with surface and/or cytoplasmic IgA from these sources confirmed a progressive maturation of B cells as they migrated through these areas, with the least mature in the PP (Lamm, 1976). Similar results have been obtained in migration studies of T lympoblasts from mesenteric lymph nodes, and both the lamina propria and the epithelium were the recipient areas (Guy-Grand *et al.*, 1974; Rose *et al.*, 1976a). Small lymphocytes, whether B or T, do not show this selective migration to the gut, nor do blasts from other sources such as spleen and peripheral lymph nodes (McWilliams *et al.*, 1977; Rose *et al.*, 1976a). Peripheral T blasts will readily enter the inflamed gut, however (Rose *et al.*, 1976b).

All of the above studies point out a dichotomous distribution of the lymphoid cells of the GI tract. The organized lymphoid tissues of the GALT seem to be the areas where induction occurs, whereas the lamina propria and epithelium are the recipients of the generated effector cells. Thus, it is not surprising that the PP are not target tissues of migrating mesenteric lymph node blasts and specific antibody-producing plasma cells are not generally found here after oral immunization (Bienenstock and Dolezel, 1971). The lamina propria, on the other hand, contains large numbers of plasma cells, the overwhelming majority of which secrete IgA, detected by immunofluorescence (Crabbé *et al.*, 1965).

With regard to T cells, PP contain cells capable of inducing GVH reactions (MacDonald and Carter, 1978) as well as precursors for CTL and T_{DTH} (Kagnoff, 1978) and T helpers and T suppressors (see Section 1.3.5.), but few cells with the phenotypes of differentiated effectors (Lyscom and Brueton, 1982). In the mucosa, there are many T cells within the lamina propria and epithelium. In fact, most intraepithelial lymphocytes are T cells, with the majority bearing the cytotoxic/suppressor phenotype (Lyscom and Brueton, 1982) in contrast to lamina propria T cells which are primarily of the helper phenotype (Cerf-Bensussan *et al.*, 1983). It is difficult to reconcile these characteristics with the finding that cytotoxic T cells were abundant in the lamina propria of tumor-immunized animals, whereas intraepithelial T cells had only low activity (Davies and Parrott, 1981). Interestingly, a significant proportion of intraepithelial mononuclear cells with T-cell markers contain cytoplasmic granules, many with histamine (Guy-Grand *et al.*, 1978). These cells were derived originally from PP, and their precise relationship to functional T cells is currently unknown. Natural killer activity is high among intraepithelial lymphocytes according to some but not all reports (Tagliabue *et al.*, 1981), but PP are deficient in this effector cell population. All of these findings indicate that PP and the gut mucosa differ in the T-cell and other populations present; however, a clear understanding of the cell distribution, especially with regard to function, is lacking at present. Conflicting results may be reconciled as techniques for obtaining purified populations from these complex tissues emerge. There is some evidence that one cell type is shared among both PP and mucosal cells, a population with Fc receptors specific for α chain, probably null cells, that mediates ADCC reactions capable of killing bacteria (Tagliabue *et al.*, 1983). Clearly, much further investigation is needed on T cells within both GALT and the intestinal mucosa.

3.6. LOCAL AND SYSTEMIC CONSEQUENCES OF ANTIGEN PRESENTATION AND RECOGNITION WITHIN THE GUT

An important facet of gut-generated immunity is that the seeding of the lamina propria involves a multitude of other mucosal surfaces, including the salivary glands, mammary glands, respiratory tract, and parts of the reproductive tract. This concept of a common mucosal immune system (Bienenstock *et al.*, 1978) wherein cells stimulated with antigen in the GALT generate progeny that migrate to a variety of mucosal sites, is supported by two major lines of evidence: (1) migration of GALT-generated cells not only to gut lamina propria but also to sites like lactating mammary tissue (Roux *et al.*, 1977) and (2) appearance of effector cells and IgA after GI immunization. Oral immunization with ferritin results in the appearance of plasma cells, containing IgA against ferritin, in the subepithelial connective tissue of the intestinal lamina propria, lactating mammary gland, parotid gland, and upper and lower respiratory tract (Weisz-Carrington *et al.*, 1979). Similarly, oral immunization with killed *Streptococcus mutans* induces specific IgA antibodies in a variety of external secretions (Mestecky *et al.*, 1978).

Numerous investigators using both natural and experimental antigens have shown that oral or intestinal exposure to immunogens results in local immunity, manifested by the appearance of specific secretory IgA in the lumen or specific antibody-producing cells in the lamina propria. Useful antigens for studying local GI immunity include *S. mutans* (Michalek *et al.*, 1976), polio virus (Ogra *et al.*, 1980), *Shigella* (Keren *et al.*, 1978), cholera toxin (Pierce *et al.*, 1980; Lange *et al.*, 1979), and DNP–KLH (Cebra *et al.*, 1977). All of these, given orally, lead to the generation of secretory IgA antibodies in the GI tract which, in the case of infectious organisms and their toxins, are capable of protecting the mucosal surface by a variety of mechanisms (see Lamm, 1976; Bienenstock and Befus, 1980). In most of these systems, the generation of local immunity seems to be a complex process, dependent on the nature of the antigen, the dose, and the number and time course of exposures (Pierce, 1978; Keren *et al.*, 1981). Parenteral exposure prior to oral immunization can similarly have variable effects on the success of oral antigenic exposure, from augmentation to suppression, influenced by these same factors (Pierce and Koster, 1980). These complexities, varying in effect from one antigen to another, have considerably hindered the development of effective vaccines against mucosal pathogens, although progress has recently been made with certain organisms, such as polio virus in humans, and several viruses in animals (Ogra *et al.*, 1980).

There is also evidence that GI antigenic exposure leads to local cell-mediated immune reactions. T lymphocytes are essential for effective responses against a number of gut parasites (Miller and Nawa, 1979; Walls *et al.*, 1973) and more T blasts migrate into the intestine when animals are infected with these organisms (Rose *et al.*, 1976a). Many of the morphologic changes, such as crypt hyperplasia and villous atrophy, are likely mediated by these cells during infections (Ferguson and MacDonald, 1977). Recently, cell-mediated immunity to the dietary antigen ovalbumin has been demonstrated by migration inhibition of mesenteric lymph node cells in the presence of ovalbumin (Mowat and Ferguson, 1981). Cyclophosphamide-treated animals had increased intestinal reactions, with the morphologic effects described above, suggesting that normally local cellular reactions are controlled by suppressors that beneficially prevent harmful hypersensitivity reactions to dietary immunogens. When this regulation is prevented, damaging cellular immune reactions can occur, resulting in pathologic symptoms similar to those seen in a number of human diseases. The recent observation that intestinal lymphocytes from patients with Crohn's disease have increased T-cell proliferation to mitogens, but reduced suppressor cell activity *in vitro*, supports this "disturbed immunoregulation" theory for some immunologically mediated intestinal diseases (Goodacre and Bienenstock, 1982).

Oral exposure, besides generating a local immune response, can induce serum antibodies (Ogra *et al.*, 1980) but it often induces no detectable systemic response (Michalek *et al.*, 1976). Interestingly, studies with a wide range of antigens have demonstrated that GI antigenic exposure actually interferes with the host's capacity to react to a subsequent parenteral challenge. In a now-classic study, Chase (1946) found that oral exposure to dinitrochlorobenzene (DNCB) severely reduced the capacity of the host to develop contact sensitivity to the same agent by the standard epidermal sensitization regimen. In other studies, both diminished antibody responses and suppressed DTH reactivity resulted when animals were parenterally challenged after a period of antigen feeding (Tomasi, 1980). Antigens used to reveal this phenomenon, now called oral tolerance, include contact-sensitizing haptens (Asherson *et al.*, 1977), proteins like ovalbumin and BSA (Richman *et al.*, 1978; Thomas and Parrott, 1974), particulate antigens such as heterologous red blood cells (Kagnoff, 1978; André *et al.*, 1975), and organisms like *S. mutans* (Challacombe and Tomasi, 1980). The numerous descriptions of this phenomenon suggest that sytemic unresponsiveness is a common consequence of oral antigen exposure. In two studies, involving both protein antigens and organisms, it has been shown, as long suspected, that systemic suppression and local (GI) immunity are concomitant results of antigenic exposure by the oral route. Feeding *S. mutans* or ovalbumin to mice induced suppression of serum antibody responses and antigen-induced T-cell proliferation in draining nodes after parenteral challenge (Challacombe and Tomasi, 1980; Swarbrick *et al.*, 1979). However, local immunity was elicited, evidenced by secretory IgA antibodies in saliva and by immune exclusion of antigen absorption in the gut. These effects were antigen specific.

Analyses of GALT in animals orally exposed to various antigens have revealed complex regulatory mechanisms that allow such selective reaction to ingested substances, ultimately resulting in mucosal immunity and systemic suppression of many antigens (Strober *et al.*, 1981). The unique antigenic environment within mucosal tissues apparently demands unique immunoregulatory features. Transfer studies on oral tolerance, which is confined to T-dependent antigens, demonstrated that multiple mechanisms may exist, even when the same antigen is involved. In the tolerance induced by ingestion of contact-sensitizing haptens, Asherson *et al.* (1977) demonstrated two distinct suppressor cell populations in peripheral lymphoid tissue. B cells inhibited the adoptive transfer of immunity, while T cells inhibited the proliferative response following skin painting of the hapten when transferred prior to painting. Suppressor B cells were first seen in the PP and mesenteric lymph nodes, then the spleen. Suppressor T cells were found in the spleen and peripheral lymph nodes. Interestingly, spleens from suppressed animals also contained sensitized T cells capable of transferring contact sensitivity, but which were suppressed *in vivo* by these regulatory cells.

Experiments using SRBC revealed that oral exposure induced IgA PFC in the spleen but subsequent challenges systemically yielded little response (André *et al.*, 1973, 1975). This unresponsiveness could be transferred with serum, and immune complexes containing IgA seemed responsible. When cell populations were analyzed, carrier priming of helper T cells had occurred in animals fed SRBC, shown by the capacity of PP T cells to help generate anti-TNP antibodies to TNP-SRBC *in vitro* (Kagnoff, 1975). Using transfer protocols, Kagnoff (1978) demonstrated that T_{DTH} cells could be found in the spleen along with a population of suppressor cells, the net result being suppression of the DTH reaction *in vivo.* T suppressor cells were found in the host's PP and mesenteric lymph nodes 2 days after feeding, but within 4 days had disappeared from these sites and could be found in spleen and thymus (Mattingly and Waksman, 1978). These cells specifically suppressed DTH reactions and IgM and IgG PFC responses and probably act through factors (Mattingly *et al.*, 1980).

Equally extensive studies with the protein antigen ovalbumin have also revealed suppressor cells. Both mesenteric lymph node and splenic T cells from fed animals were able to transfer hyporesponsiveness to syngeneic recipients (Challacombe and Tomasi, 1980; Richman *et al.*, 1978). These cells appeared in the PP at 1 day, and the mesenteric lymph nodes and spleen at 3 days (Richman *et al.*, 1981). Moreover, a distinct isotype-specific regulation was seen. *In vitro,* the PP T cells augmented IgA PFC but decreased generation of IgG PFC. Only the suppressor was seen in the spleen, indicating a compartmentation of helpers for IgA locally and suppressors for IgG systemically. PP contained abundant T cells which, when stimulated with Con A, supported polyclonal activation of B cells for IgA synthesis, whereas similar cells were rare in peripheral lymph nodes or spleen (Elson *et al.*, 1979), supporting this unique regulatory design. Others have argued that the predominance of IgA B-cell precursors and not T cells controls IgA production by PP cells (Tseng, 1982). However, studies with cloned T cells derived from PP have shown that they support immunoglobulin class switching to IgA-bearing cells, although they do not help in the terminal differentiation which presumably takes place elsewhere (Kawanishi *et al.*, 1982).

Oral exposure to ovalbumin suppressed not only IgG antibody production to systemic challenge, but IgE as well (Ngan and Kind, 1978). Thus, only IgA antibodies appear to escape such dramatic suppression resulting from oral antigenic exposure. However, local regulation may occur. T cells with surface receptors for IgA (Tα cells) in high numbers can suppress IgA production after oral immunization (Hoover and Lynch, 1983) but may have T helper activity in the periphery (Endoh *et al.*, 1981). These cells may be regulatory cells whose function depends on a variety of factors, including the microenvironment. A further complexity of GALT immunoregulation has been revealed using a primary *in vitro* humoral response. A contrasuppressor cell with the phenotype Ly 2^+,I–J^+ was found in PP which may act *in vivo* to prevent suppression of IgA responses in GALT (Green *et al.*, 1982). Furthermore, the suppression of systemic T and B cells occurs with different oral doses of ovalbumin, and suppression of T cells, not B cells, is prevented by cyclophosphamide, suggesting that humoral and cell-mediated oral tolerance are mediated by different mechanisms (Mowat *et al.*, 1982).

Thus, the immune response within GALT to ingested antigens appears to be regulated in a complex way that results in activation of local T helper cells, locally generated B cells committed to IgA synthesis and probably T cells which leave to seed mucosal areas, and suppressor cells which emigrate to peripheral lymphoid tissues to inhibit systemic responses. Recent studies have shown that the normal gut flora may play an important role in these immunoregulatory processes. Using a SRBC feeding regimen that generates systemic unresponsiveness in most mice, Kiyono *et al.* (1982b) found that C3H/HeJ mice became systemically sensitized. The T suppressor cells induced in the PP of other mice were not found in this LPS-unresponsive strain, which are also characterized by heightened IgA responses. Similar results were seen in LPS-responsive mice maintained in germfree conditions to prevent exposure to LPS (Wannemuehler *et al.*, 1982). Oral tolerance could be achieved in such animals by prior intentional exposure to LPS. These results support the hypothesis that LPS is important in the induction of suppressor cells in GALT and, therefore, that the gut microflora impacts greatly on responses to ingested antigens.

In addition to the gut flora, other factors that may influence responses to oral antigen include a number of dietary substances. Protein deficiency severely impairs local immune responses to cholera toxin by interfering with induction in PP (Barry and Pierce, 1979). Vitamin A reportedly enhances mucosal immune responses and zinc generally interferes with induction of T suppressor cells (see Bienenstock and Befus, 1980). These findings highlight the need for further study to elucidate the effects of various dietary elements on immune responses at mucosal surfaces. Interestingly, aging does not decrease mucosal immune responses in a murine system as it does the animals' systemic reactivity, primarily because T helper activity is unhindered (Szewczuk and Campbell, 1981). Such findings further point out the functional separation of the mucosal and systemic immune apparatuses.

3.7. SUMMARY

The GI tract is clearly designed to protect itself nonspecifically by a variety of physical and chemical means and to provide specialized local immunity to organisms and other immunogens that do penetrate. Elaborate structural specializations exist to ensure an appropriate response that is local in nature. These include PP with many antigen-presenting cells overlain by M cells specialized for antigen uptake, as well as subpopulations of lymphocytes geared toward local responses. The differentiation of these cells includes "learning" a migration pattern that will bring them back to mucosal sites. Furthermore, other cells within the PP seem specialized to suppress systemic responses, ensuring that responses to most mucosal onslaughts are confined to those areas. Such an extensive and complex arrangement of immunologic features is undoubtedly related to the tremendous number of potential pathogens and antigens entering by this route, including viruses, bacteria, and parasites as well as dietary substances.

4. SKIN

The skin is a highly specialized tissue designed to provide an effective interface between the organism and its environment. It subserves several related, but distinctive, functions: a physicochemical barrier, an important organ in thermoregulation, and a medium through which the organism communicates with its environment. It is the barrier function of skin which makes necessary the matters addressed in this chapter on specialized antigen presentation.

4.1. BARRIER PROPERTIES

The dermis and epidermis together constitute an impressive physicochemical barrier (Marks, 1983; Scheuplein and Bronauch, 1983). They prevent loss from the body of water and precious chemical constituents, while at the same time offering resistance to physical trauma and chemical intrusion. While both dermis and epidermis contribute to the physical barrier, the chemical barrier resides almost exclusively within the epidermis; in fact, virtually all of the barrier function is attributable to special features of the stratum corneum, the outermost layer of the epidermis. It is this thin, saran wrap-like layer which provides a significant impediment to the passage of both aqueous and lipid-soluble materials. The barrier is more effective for water- than for lipid-soluble materials. Among the former, polar molecules move across the stratum corneum less easily than nonpolar ones; ionic and charged molecules are also preferentially impeded. Finally, small molecules can cross the barrier with greater ease than large molecules.

The barrier consists of 10–20 layers of horny, flat, anucleate keratinocytes filled with keratin and linked together by innumerable desmosomes. The membrane of each cell is greatly thickened and the potential intercellular space is virtually occluded with a gluelike substance that thickens the cells' plasma membranes. It is thought that the responsibility for water impermeability resides chiefly in this intercellular material, while resistance to lipid passage resides in the stacked keratinocytes themselves.

Having discussed briefly the chemical barrier properties of the stratum corneum, it is important to acknowledge that the barrier is relative, not absolute. Consequently, potentially antigenic materials, while generally impeded from penetration through the skin, nonetheless may gain access to the tissues immediately beneath the stratum corneum. One would predict from the information just summarized that chemicals with the highest likelihood of penetrating the barrier and therefore becoming "antigens," would preferentially be soluble in lipid rather than water, uncharged (nonpolar), and of low molecular weight. It is perhaps no accident that highly reactive chemicals such as dinitrofluorobenzene (DNFB), oxazolone, and others have been used experimentally to study antigen presentation through the epidermis. These chemicals fit the above description precisely. It is important to realize that the chemical barrier can be compromised physically, e.g., by abrasion of the stratum corneum, puncture with a needle, an insect's bite, and so on. In these instances, molecules and microbial agents of virtually endless antigenic potential can gain access to the epidermal and dermal compartments of the skin.

4.2. STRUCTURAL SPECIALIZATION FOR ANTIGEN PRESENTATION

For molecules which penetrate (either chemically or physically) the barrier of the stratum corneum, there is a great possibility of confronting cells capable of antigen processing and presentation. In this regard, two cutaneous compartments exist: (1) an intraepidermal space—defined as the sinuous, interconnecting spaces between and among the desmosomally attached keratinocytes, and spanning the distance from the basal lamina immediately beneath the basilar layer of keratinocytes and the stratum corneum. Within this space resides a variety of nonepithelial cells, notably epidermal Langerhans cells, that have been the focus of research because of their potential for effective antigen presentation. (2) An intradermal space, the second compartment, is sharply defined distally by the basal lamina at the dermis–epidermis junction, but poorly defined proximally by the junction between the dermal and the subdermal compartment. Within this space reside macrophages and other cells of mesenchymal origin with potential for antigen presentation. These two compartments—the intraepidermal and intradermal—are in communication with each other, since the basal lamina which separates them is semipermeable and permits relatively free exchange of molecules and mobile cells.

Over the past decade, a great deal of attention has been devoted to the role of epidermal Langerhans cells in antigen presentation within the skin. Much less has been learned about antigen-presenting cells within the dermis, and so this discussion will emphasize the former.

4.3. UNIQUE ANTIGEN-PRESENTING CELLS OF THE EPIDERMIS

Epidermal Langerhans cells have emerged during the past few years from the relative oblivion that had characterized their existence following the original description over 100 years ago by Paul Langerhans (1868). Until the 1960s their functional attributes were essentially unknown. Identification of unique cytoplasmic granules by Birbeck *et al.* (1961) using electron microscopy, followed by the demonstration by Breathnach *et al.* (1968) that Langerhans cells were developmentally unrelated to melanocytes and probably of mesenchymal origin, signaled the dawning of a new awareness and interest (Wolff, 1972). Although the evidence was strictly circumstantial, the microscopic description of lymphocytes clustered around Langerhans cells in the epidermis during expression of contact hypersensitivity (Silberberg, 1973; Silberberg *et al.*, 1976) provided the impetus to consider that Langerhans cells might play a role in immune mechanisms within skin. In relatively rapid order, the following critical observations were made: (1) Langerhans cells, exclusively among epidermal cells, bear unique surface markers usually ascribed to cells of the macrophage/monocyte lineage—receptors for Fc of immunoglobulin and C3b of complement (Stingl *et al.*, 1977), Ia antigens (mice) (Rowden *et al.*, 1978), and D/DR antigens (humans) (Klareskog, 1977); (2) Langerhans cells, when present in epidermal cell suspensions, especially if their numbers are enriched by rosetting techniques that take advantage of their Fc receptors, function *in vitro* to stimulate allogeneic lymphocytes in mixed lymphocyte reactions, and to induce proliferation on the part of the antigen-primed T lymphocytes if the epidermal cell preparation had been pulsed with the relevant antigen (Stingl *et al.*, 1978a,b); (3) Langerhans cells are derived at least in part from a stem cell precursor that resides among bone marrow cells (Katz *et al.*, 1979; Frelinger and Frelinger, 1980). In addition, there are other similarities between Langerhans cells and macrophages/monocytes. Steinman (see Steinman and Nussenzweig, 1980) has described a unique leukocyte cell type which he has termed the dendritic cell. This cell shares with Langerhans cells the capacity to stimulate mixed lymphocyte reactions, present antigens to primed T cells, express large amounts of surface Ia, and fails to take part in phagocytosis at an order of magnitude similar to macrophages. Thus, some controversy exists as to the precise cell lineage that Langerhans cells represent. Nonetheless, their capacity to function as antigen-presenting cells—at least as assayed *in vitro*—seems unquestioned, suggesting an important functional role for these cells *in vivo*.

Our laboratory has attempted to determine the physiologic role of Langerhans cells *in vivo*. Our approach was based on the observations that not all cutaneous surfaces are equally endowed with Langerhans cells: murine tail skin and hamster cheek pouch epithelium contain considerably reduced numbers compared to normal body wall skin; the nonlimbic, central region of corneas of all mammalian species studied to date are totally devoid of Langerhans cells (Bergstresser *et al.*, 1980). Exposure of shaved murine and hamster body wall skin to relatively low doses of ultraviolet B radiation induces a transient but significant reduction in normal-appearing Langerhans cells at the site of exposure. Using contact hypersensitivity to simple haptens such as DNFB as a model system, we demonstrated that cutaneous surfaces depleted of normal numbers of normally functioning Langerhans cells are unable to sustain the induction of contact hypersensitivity (Toews *et al.*, 1980; Lynch *et al.*, 1981; Sauder *et al.*, 1981). This evidence makes for a circumstantial argument that

Langerhans cells are in some way necessary to the process by which epicutaneously applied haptens are rendered immunogenic for the immune system. Supporting data for this idea also came from studies employing allografts of corneas in mice (Streilein *et al.*, 1979). In these experiments, the ability of corneas to induce alloimmunity to the Ia alloantigens of their strain of origin was negligible while they retained their capacity to immunize to class I (murine H-2K/D) alloantigens. On the assumption that presentation of Ia alloantigens by Langerhans cells is an analogue of the mechanism by which nominal antigen plus syngeneic Ia represents the antigen-presenting mode of these cells, it was proposed that both types of evidence strongly support a critical role for Langerhans cells in the inductive phase of immune responses initiated by antigens first making impact upon the skin.

Perhaps of even greater significance was the observation that following exposure of mice to haptens placed on cutaneous surfaces depleted of Langerhans cells, a state of specific unresponsiveness was achieved (Toews *et al.*, 1980). This observation added a new dimension to the proposition that Langerhans cells present antigens to immunocompetent cells: it suggests that in the absence of Langerhans cells, antigen placed epicutaneously gains access to the systemic immune apparatus as a *tolerogenic* rather than an immunogenic signal. It emphasizes even more strongly the crucial role for epidermal Langerhans cells in the induction of cutaneous immune reactions. Moreover, it suggests that in the absence of Langerhans cells, antigens introduced through the skin may have the same propensity for inducing unresponsiveness as they do when introduced by other routes.

Evidence in support of this hypothesis has come from Ptak *et al.* (1980), who demonstrated that suspensions of epidermal cells prepared from tail skin, when derivatized with hapten and inoculated intravenously into mice, induce contact hypersensitivity. This places the Langerhans cell in a specialized class of antigen-presenting cells which when derivatized can induce contact hypersensitivity following i.v. inoculation into mice: Langerhans cells, dendritic cells, and elicited peritoneal exudate cells (Britz *et al.*, 1982). Other derivatized lymphoid cells, uninduced peritoneal exudate cells, erythrocytes, and so on all induce unresponsiveness in this way. However, Tamaki *et al.* (1981) reported dissimilar results using hapten-derivatized epidermal cell preparations which in their hands did not produce sensitization after i.v. inoculation. The dilemma produced by these opposing results has yet to be resolved. Because of severe technical problems that bedevil attempts to produce purified preparations of Langerhans cells from skin, we have attempted to circumvent this need by developing experimental strategies to (1) absolutely deplete intact skin of Langerhans cells and test its ability to allow induction of contact hypersensitivity and to induce alloimmunity and (2) use principles of H-2 genetic restriction to examine the earliest phases of antigen presentation within skin. By employing tape-stripping, which virtually eliminates Langerhans cells from skin, we have determined that the Ia alloantigenic load of skin resides almost exclusively in its resident Langerhans cells; and we have used this approach to document the extraordinary speed with which extracutaneous precursors of Langerhans cells can infiltrate grafted skin and assume a proper anatomic position in the epidermis (Streilein *et al.*, 1982). By using hapten derivatized to skin as surrogate transplantation antigens, we have demonstrated that at threshold doses of hapten, the H-2 determinants of the painted skin promote and restrict the effector cells of contact hypersensitivity that are induced. These data strongly argue that essential antigen presentation to immunocompetent lymphocytes takes place within the epidermis itself and/or that cutaneous cells bearing H-2 antigens are able to restrict the repertoire of activated immunocompetent lymphocytes they confront when they traffic to the draining lymph nodes.

Very recently, Sullivan (reported in Streilein, 1985) has been able to produce (via fluorescence-activated cell sorting) virtually pure populations of Ia antigen-bearing epidermal Langerhans cells. Following hapten derivatization, it can be shown that these cells are *only* capable of delivering an immunogenic signal to the immune system—whether they are intradermally, intravenously, or otherwise inoculated. Thus, it would appear to have been proven that epidermal Langerhans cells are uniquely situated within the epidermis to accept antigens which penetrate the barrier of the stratum corneum and convert them uniformly into immunogenic signals. As a consequence, the immune system is obliged to respond by expressing a vigorous cell-mediated response mediated by antigen-specific T lymphocytes—contact hypersensitivity.

4.4. SITE OF ANTIGEN RECOGNITION

Antigens which immunize through the skin must, following their processing and presentation, be recognized by specific immunocompetent lymphocytes. The locus of this important event in the induction of specific immune responses is of considerable interest. Three distinct (and not mutually exclusive) possibilities have been proposed:

1. Antigen is not processed within the skin, but escapes through the dermal lymphatics, is carried passively to the draining regional lymph node wherein antigen-presenting cells present the antigen to lymphocytes passing through that node.
2. Antigen is processed *in situ*—within the epidermis by Langerhans cells or within the dermis by similar (albeit ill-defined) antigen-presenting cells—and then these antigen-presenting cells migrate from the cutaneous site, enter the lymph drainage, and reach the regional lymph node where antigen recognition by T lymphocytes occurs.
3. Recognition of antigen that has been processed and presented by antigen-presenting cells within the skin (i.e., within the epidermis by Langerhans cells) takes place *in situ*, an occurrence which requires that antigen-specific immunocompetent T cells gain access to the intraepidermal compartment and undertake effective antigen recognition therein.

Ample evidence suggests that the first possibility is a physiologic process that can take place during cutaneous immunization. Circumstantial evidence indicates that the second mechanism may also be operative. During cutaneous immunizations, Langerhans cells have been observed histologically to migrate from the epidermis into the dermis and are found subsequently in the draining lymph node (Hoefsmit *et al.*, 1979; Thorbecke *et al.*, 1980; Spry *et al.*, 1980). Neither of these possibilities is unexpected or particularly provocative. However, the third possibility is the most intriguing. If T cells are able to penetrate the epidermis and recognize *for the first time* antigen on Langerhans cells and as a consequence initiate specific immune responses, then the skin must be able to offer a microenvironment (heretofore unsuspected) that supports immune inductive events in a manner similar to that of organized lymphoid tissues such as lymph nodes and spleen.

4.5. SPECIALIZED LYMPHOCYTES THAT TRAFFIC THROUGH THE SKIN

The previous section documents that effective antigen recognition takes place within skin—perhaps within the epidermis itself—and implies that immunocompetent lymphocytes gain access to these cutaneous compartments. This implication raises the question of whether lymphocytes that accomplish this task are unique, or whether they are drawn from the generic pool of recirculating cells that can percolate through many different kinds of tissues. Evidence in favor of the former possibility is now considerable.

In the late 1950s and 1960s, a common experimental maneuver in transplantation immunobiology was to inoculate monodisperse suspensions of lymphocytes into the skin and observe the appearance of a local inflammatory reaction (Brent *et al.*, 1962). The so-called normal lymphocyte transfer reaction, first examined by Brent and Medawar (1964), was achieved by inoculating parental-strain lymphoid cells intradermally into the body wall of F_1 hybrid guinea pigs. Local delayed inflammatory reactions regularly ensued, indicating that a local GVH reaction had taken place. At its most elementary level, this result documents that recognition of antigen (in this case, alloantigens displayed on constitutive dermal cells) was achieved *in situ* by the inoculated lymphocytes. Moreover, the elicitation of a delayed inflammatory reaction signifies that an effector response was also generated; that is, antigen recognition by lymphocytes proceeded to that stage in the efferent limb of the immune response where recruitment of nonspecific host cells was achieved, and a typical inflammatory reaction ensued. This approach was refined by Ramseier and Streilein (1965), who demonstrated that the skin of lethally irradiated Syrian hamsters could serve as a suitable milieu in which alloreactive lymphoid-cell mixtures could be placed and generate local immune reactions.

To test this idea more directly, a series of painstaking and elegant experiments by Macher and Chase (1969a,b) gave the first indication that meaningful antigen recognition leading to the expression of contact hypersensitivity takes place at the site at which hapten is injected. Using DNCB injected into the ears of guinea pigs, these investigators discovered that the site of application had to remain intact during the first 24 hr after skin painting. Excision of the ear prior to the end of the first 24-hr interval failed to achieve sensitization. These investigators reasoned that this length of time was probably required for lymphocytes, trafficking through the haptenated skin, to recognize hapten and thereby become "activated"—a process termed *peripheral sensitization.*

The approach we have taken to answering this important question has been to make use of the observation that hapten-derivatized skin as a graft can be used as a source of immunogen to induce contact hypersensitivity. In typical experiments, shaved abdominal-wall skin is painted with an immunizing dose of DNFB; grafts are then prepared and placed orthotopically on recipient mice. Six days later the ears of the recipient mice are challenged with dilute DNFB to determine whether the animals have developed contact hypersensitivity. Using this approach, we have learned that very much smaller doses of hapten are required to induce systemic sensitization compared with conventional immunizing protocols. Thus, BALB/c mice can be rendered sensitive to DNFB by grafting them with syngeneic skin that has been painted with as little as 20 ml of 0.02% DNFB—approximately 60 times less than the amount applied during conventional sensitizing maneuvers. Using this limiting dose of hapten applied to skin, we then devised an experiment in which recipient mice received hapten-derivatized skin from syngeneic, semiallogeneic, and fully allogeneic donors. When the recipient mice were ear-challenged subsequently, only those animals which received hapten-derivatized skin bearing the host's own H-2 alloantigens displayed significant ear swelling. Recipients of haptenated H-2-incompatible skin grafts developed ear-swelling responses indistinguishable from negative controls. This observation confirms that at limiting doses of hapten, antigen recognition takes place within the skin graft itself. As a consequence, effector T cells are restricted in their ability to express or recognize the hapten by H-2 determinants expressed on the immunizing graft. In this instance, since the H-2 determinants at the ear-challenge site are of recipient type, only derivatized grafts expressing recipient H-2 determinants evoked DTH effector cells able to produce lesions at the challenge site (Streilein *et al.*, 1984).

In a subsequent experiment, mice grafted in this manner were subjected to a reimmunization procedure. Animals that first experienced hapten in the form of derivatized syngeneic or semiallogeneic grafts gave exaggerated responses indicative of a previously immunized state. In contrast, animals first grafted with haptenated allogeneic skin responded to the second immunization regimen by expressing contact hypersensitivity similar in intensity to that of the primary positive controls. This result indicates that at limiting doses of hapten, presentation on H-2 allogeneic skin is an immunologically "null" event. Since these animals displayed no evidence of unresponsiveness, we would argue that little, if any, antigen reprocessing takes place on the recipient's own antigen-presenting cells.

To our minds, we regard this as the strongest evidence to date that peripheral sensitization does in fact occur; moreover, it strongly suggests that the entire immunity sequence—from antigen presentation through effectuation of the response—takes place within the cutaneous environment itself.

4.6. EPIDERMOTROPISM OF T LYMPHOCYTES

Within the past 5 years, especially now that there are reagents that accurately identify types of blood lymphocytes, it has been determined that certain T-lymphocyte malignancies show a distinct propensity for infiltration of the skin. The term *cutaneous T-cell lymphomas/leukemias* (CTCL) has been used to identify this phenomenon (Lutzner *et al.*, 1975; Edelson, 1980). By implication, non-T-cell neoplastic disorders of lymphocytes show no special affinity for skin. Even among T cells, the affinity for skin is not generic, in that in the vast majority of cases in which malignant T cells involve the epidermis, the surface phenotype of the infiltrating cells coincides with the helper/inducer subpopulation (Janossy *et al.*, 1980a). In lesions examined microscopically, these cells demonstrate a striking tendency to localize around marrow-derived dendritic cells that express Ia-like surface antigens (Janossy *et al.*, 1980b). These observations are consistent with the hypothesis that CTCL represent the neoplastic amplification of T lymphocytes that recirculate normally through the blood and into the skin. Presumably, other types of normal T cells do not possess this trafficking capability; consequently, one might expect their neoplastic representatives to infiltrate other types of tissues.

Based in part on these considerations, several groups of investigators have attempted to collect data using nonneoplastic (i.e., normal) T cells, seeking evidence for physiologic epidermotropism of normal T lymphocytes. Radiolabeled unfractionated normal lymphoid cells from mesenteric and peripheral lymph nodes exhibit slightly different homing patterns. Mesenteric lymph node cells tend to localize in the gut, while peripheral lymph node cells tend to home toward peripheral lymph nodes (McWilliams *et al.*, 1975). This difference in homing patterns has been greatly amplified by employing activated lymphoblasts rather than unfractionated or resting lymphocytes. Griscelli *et al.* (1969) first demonstrated that radiolabeled blast cells obtained from rat mesenteric lymph nodes accumulate after adoptive transfer in highest number in the gut wall and the mesenteric lymph nodes. In contrast, peripheral node lymphoblasts selectively localize in peripheral lymph nodes at the expense of the gut and the mesenteric nodes. Rose *et al.* (1976b,c) have determined that T lymphoblasts generated within peripheral lymph nodes will preferentially infiltrate skin when given the choice of inflammatory sites simultaneously present in skin and gut. As expected, mesenteric T lymphocytes display the opposite homing pattern.

Asherson *et al.* (1977) and Tigelaar (personal communication) have employed oxazolone-stimulated peripheral lymph node blast cells to compare their ability to enter sites of allergic contact dermatitis. Blast cells enter these sites with much greater facility than do unstimulated or small lymphocytes. To determine whether their attraction to the skin test site was related in any way to their ability to recognize haptenic antigen, Tigelaar has recently produced alloantigen-activated T-cell blasts and examined their migratory potential in recipients in whom cutaneous inflammation was induced with hapten. The results clearly indicate that the migratory properties of T lymphoblasts are independent of the antigen specificity of the activated cells. T lymphoblasts activated in peripheral lymph nodes of F_1 hybrids demonstrate a distinct proclivity for returning to peripheral lymph nodes of syngeneic mice, and they preferentially infiltrate cutaneous, as opposed to GI, sites of hapten-induced inflammation. Finally, Liden (1967) has accumulated evidence to suggest that the localization of T cells in the skin leads to preferential infiltration by these cells into the epidermis itself. Thus, once localized within the dermis, subpopulations of T cells display a propensity to migrate across the dermis–epidermis junction and infiltrate the epidermis. While this mass of evidence remains largely circumstantial, it is impressive and strongly supports the contention that subpopulations of T lymphocytes exhibit epidermotropism.

Immunocompetent T Cells in Normal Skin

The presence of lymphocytes within the epidermis of normal human and rat skin was initially reported by Andrew and Andrew (1949). These investigators, looking at routine histologic sections by light microscopy, found that lymphoid cells constituted 1 to 4% of the cells in the basal layer of the epidermis. More recently, Lemmel and Fichtelius (1971), in an examination of normal guinea pig skin, also noted the presence of significant numbers of lymphoid cells within the epidermis. Although electron microscopy and immunologic phenotyping of these cells were not used in these studies, the results strongly suggest, but do not prove, that lymphocytes are present not only in the dermis but also in the normal epidermis. Furthermore, small but significant numbers of lymphocytes are present in the afferent lymphatics draining from normal skin to the peripheral lymph nodes (Smith *et al.*, 1970). The cell surface phenotype of these lymphocytes has been examined. In sheep, only 5 to 10% of the lymphocytes in afferent lymph draining the popliteal lymph nodes were B lymphocytes (Miller and Adams, 1977; Hall, 1967; Scollay *et al.*, 1976). T lymphocytes also predominate in human peripheral lymph (Engeset *et al.*, 1974).

Several investigators have suggested that the T cells in normal skin are immunocompetent. Solomon (1963) grafted skin of the parent genotype to a semiallogeneic F_1 hybrid recipient chicken and observed that the graft was rejected. Since, according to conventional transplantation immunology, the F_1 recipient should be genetically incapable of mounting an effective host-versus-graft response against the parental-strain histocompatibility antigens, he concluded that a GVH reaction initiated by cells within the donor graft must have caused the rejection. Reasoning that since newly hatched chicks are immunologically incompetent, cells of hematogenous or lymphoid origin present within skin would be similarly immunoincompetent and incapable of mounting a GVH response, Solomon then grafted newly hatched chick skin of the parent genotype on adult semiallogeneic F_1 recipients. The grafts survived indefinitely, presumably because the putative passenger cells were unable to recognize histoincompatibility in the host. Solomon hypothesized that adult chicken skin contained a population of immunocompetent cells that was able, when confronted by histocompatibility antigens in the recipient, to act against them. In the skin graft bed, the net effect of this confrontation was to elicit a sufficiently intense inflammatory reaction that, although similar to GVH, was ultimately detrimental to the graft survival.

More recently, Barker and Billingham (1972) utilized local lymph node hypertrophy in response to grafting with allogeneic skin in mice to examine further the question of whether cells with immunocompetence are present in a graft of murine skin. When an $(A \times B)F_1$ hybrid was grafted with skin from an identical hybrid donor, essentially no enlargement of the draining axillary lymph node was observed. Alternatively, when an F_1 hybrid was grafted with skin from one of the parent strains, there was significant swelling of the draining lymph node. Thus, the hypertrophy of the node draining the bed of the parental-strain skin graft must have resulted from the confrontation within that node of passenger lymphocytes transferred within the graft, i.e., cells that leave the graft to enter the draining node and there elicit a local GVH reaction. Since it is known that GVH reactions are initiated by T lymphocytes, both of these experiments strongly suggest that immunocompetent T cells are present in normal skin. However, neither of these studies showed conclusively that the immunocompetent T cells present in the skin grafts were within the substance of the skin itself and not simply trapped within the vascular bed. By using antibodies that can discriminate between subpopulations of lymphocytes and newly developed immunofluorescent or immunoperoxidase-staining techniques for *in situ* identification of cells bearing distinctive cell surface markers, investigators should be able to determine the localization and phenotype of lymphoid cells in the clinically normal skin from both humans and laboratory animals.

The idea that specialized lymphocytes leave the bloodstream and penetrate the infiltrate specifically into the dermis and epidermis is an exciting one, but little experimental data exist to illuminate the mechanism. Quite simply, virtually nothing is known about the cellular and molecular bases of this traffic. Over the next several years, research can be expected to focus on the relative contributions made by the following relevant factors: blood flow to the skin, the permeability of the vascular endothelium within the dermis, the presence and nature of surface molecules of dermal endothelial cells that could potentially interact with ligand on specialized lymphocytes and direct their emigration into the extravascular compartments—in analogy to recent information concerning the movement of lymphocytes through high endothelium of postcapillary venules of lymph nodes (de Sousa, 1981; Ford *et al.*, 1976; Woodruff, 1974; Anderson *et al.*, 1976), and to the preferential migration of certain lymphocytes into the GALT (Butcher *et al.*, 1980; Rose *et al.*, 1976a,b; McWilliams *et al.*, 1975, 1977; Hall *et al.*, 1977). In addition, antigen within the epidermal and dermal compartments, as well as factors elaborated during the process of antigen presentation, may influence positively the outflow of lymphocytes into these compartments.

Virtually nothing is known about unique molecular properties of lymphocytes postulated to have a special affinity for skin and epidermis. Considerable excitement attended the very recent discovery of Thy-1$^+$ dendritic cells within mouse epidermis—raising the possibility that *the* T cell with epidermal affinity had been identified (Bergstresser *et al.*, 1983; Tschachler *et al.*, 1983). However, this curious, "new" epidermal dendritic cell is neither a Langerhans cell nor a typical immunocompetent T lymphocyte. Its precise function remains obscure and the search for and identification of epidermotropic immunocompetent T lymphocytes continues.

4.7. LOCAL AND SYSTEMIC CONSEQUENCES OF ANTIGEN PRESENTATION AND RECOGNITION WITHIN THE SKIN

The data heretofore presented have been selected to make a strong argument in favor of the hypothesis that skin-associated lymphoid tissues (SALT) exist as an integrated system of immune surveillance designed uniquely for the skin. SALT is comprised of (1) a specialized set of antigen-presenting cells within the epidermal Langerhans cells; (2) distinctive populations of recirculating T lymphocytes that preferentially infiltrate the skin, especially the epidermis; (3) keratinocytes that provide an anatomically distinct environment for these lymphoreticular cells and secrete into that environment immunoregulatory molecules that can profoundly affect the consequences of immune recognition and differentiation; and (4) a set of draining peripheral lymph nodes, integrating this multicellular system that contains, along with the dermis, blood vessels with endothelial cells whose surface captures lymphocytes passing through the blood.

In an anatomic sense, skin is uniquely situated, and as a consequence, it faces a unique spectrum of antigenic demands. The skin is bombarded daily with the potentially damaging effects of UV radiation, raising the risk of neoplastic transformation of cells, especially within the epidermis. Neoplastic transformation, especially that induced by UV light, is associated with expression of unique neoantigens on the surfaces of the neoplastic cells (Kripke, 1983). It is within this context that a formulation of SALT takes on special meaning. It is proposed that SALT represents the physiologic mechanism created to deal with this special demand placed on skin. The scenario by which SALT operates might be stated as follows: Neoplastic transformation of (for example) a keratinocyte results in the expression of surface neoantigen(s). When transferred to the ubiquitous Langerhans cells, the neoantigen is processed and presented in immunogenic form. One of two subsequent events may then transpire: either (1) peripatetic T lymphocytes with predetermined affinity for the epidermis and with immunologic specificity for the neoantigen migrate into the epidermis, recognize the neoantigen presented on the Langerhans cell, transform *in situ* into effector cells, and directly destroy the adjacent malignant cells, or (2) the neoantigen-bearing Langerhans cell "drops down" into the dermis, flows with the draining lymph into the regional lymph node, settles into the cortex, and forms a nidus for the attraction of antigen-specific lymphocytes that are activated within the node, proliferate, and then disseminate systemically to return predominantly to the skin in order to effect destruction of the neoplastic keratinocytes. We are beginning to collect relevant data supporting the second of these options. However, to date, essentially no information exists that indicates that T cells can be activated directly *in situ*, can destroy the offending cells, and yet never send this signal to be amplified and disseminated systemically. It is conceivable that this may be the fundamental mechanism by which SALT works.

UV radiation, as an oncogenic agent, oncogenic viruses, and other carcinogenic agents may act to induce cutaneous neoplasia. UV light, however, has other effects on the skin that make it particularly appropriate in this consideration. It induces melanization of skin and alters remarkably the functional properties of Langerhans cells with regard to antigen presentation. We would presume that under normal circumstances the immediate effects of UV light (that impair antigen presentation by Langerhans cells) prove to be transient. This is so because activation of melanogenesis by UV light rapidly provides the epidermis with a protective photoshield that allows Langerhans cells to restore their immune functional properties even in the face of continued exposure to UV light.

Failure of SALT could occur through numerous pathways. Inadequate melanization would permit cutaneous cells, including Langerhans cells, to be exposed chronically to the deleterious effects of UV light; in the case of Langerhans cells, UV-induced inability to process neoantigens effectively can lead to the establishment of specific unresponsiveness rather than specific sensitization. In this manner, chronic impairment of the presentation of neoantigens by Langerhans cells would promote, rather than prevent, the escape of neoplastic cells from detection by the immune system. Reduction in the numbers and varieties of antigen-specific recirculating lymphocytes, as might be achieved by chronic immunosuppression, would similarly disrupt SALT, robbing the organism of the antigen-specific and tissue-destructive components of the surveillance mechanism.

4.8. SUMMARY

While the stratum corneum of the skin provides an excellent barrier to the environment, it is an incomplete one. Antigenic material can penetrate through this barrier (or be introduced pathologically) and encounter strategically placed epidermal Langerhans' cells (as well as dermal antigen-presenting cells) that process and present the antigen in highly immunogenic form. Recognition of antigen by immunocompetent T cells may take place locally as well as in the draining lymph nodes and a vigorous cell-mediated immunity is induced which is capable of expression at any cutaneous site. There is strong evidence to suggest that the cells that participate in these reactivities are partially dedicated to a skin-specific ecotaxis. If the skin is robbed of local antigen-presenting cell function at the time and site that antigen is administered, a state of specific unresponsiveness is induced which also acts systemically and interferes with the capacity of the host to develop a protective immune response to that antigen.

5. LUNG

Before considering the lung as a regional sphere of antigen presentation, it is important to consider the structure of the respiratory tract. The unique aspects of local antigen presentation are largely dictated by the anatomic and physiologic features described below. The respiratory tract is often divided into the upper and lower regions with an arbitrary border being established between the air-conducting and air-exchanging regions of the lung. Therefore, in a structural definition of the lung, regions of the upper and lower respiratory tract would be included in the lung (pulmonary) area. Within the lung there are three distinct regions: the airways, the gas exchange parenchyma, and the vasculature. The parenchyma of the lung directly involved in gas exchange makes up approximately 80% of the total mass of the lung. The major airways and blood vessels comprise the remaining 20% (Whitcomb, 1982).

5.1. THE GAS EXCHANGE PARENCHYMA

The human lung contains approximately 300×10^6 individual alveolar sacs (alveoli) joined by an interconnecting network of thin septal or alveolar walls. The alveolar walls are perfused by an extensive capillary network. Blood moves through the alveolar walls almost as a continuous sheet. Blood in the alveolar capillaries is separated from the airspace by a thin tissue barrier, the alveolar capillary membrane.

5.2. THE AIRWAYS

Each alveolar airspace within the lung is in direct contact with the environment through a branching airway. The tracheobronchial tree consists of a series of branching tubes whose main function is to conduct gases into and out of the lung. In addition to serving as conducting tubes, the airways also provide defense mechanisms that protect both the airway mucosa and the alveolar surface from environmental agents (see Section 5.4.1).

The trachea and mainstream bronchi are mediastinal structures that are outside the lung. The remaining subdivisions of the tracheobronchial tree are intrapulmonary structures. The terminal bronchioles are about the 16th generation of airways out of a possible 20–25 generations, and are the last bronchioles that display a continuous mucosal lining. There are at least 3 generations of respiratory bronchioles beyond the terminal bronchioles. Respiratory bronchioles have discontinuous mucosal lining with alveoli protruding occasionally from the airway wall. Beyond are alveolar ducts that are lined with alveoli; the alveolar ducts terminate in alveolar sacs. It would appear that the major portion of the lung is not lined with mucosa.

The upper respiratory tract has been rightfully designated as a mucosal organ and has been shown to be part of the common mucosal membrane immune system (Bienenstock *et al.*, 1978). The lower respiratory tract may be protected by the systemic immune system but appears to have unique mechanisms for inducing immune responses locally within the lung that are independent of both mucosal and systemic immune systems. Because the lung is composed of upper and lower respiratory tracts, it may exploit all three mechanisms for immune defenses.

5.3. THE VASCULATURE

There are two main arterial systems that supply the lung. Venous blood from the right side of the heart is delivered to the alveolar capillary network by the pulmonary artery, whereas the bronchial arteries supply oxygenated blood to the capillaries of the airways, pleura, and major vessels. In addition, the lung contains an extensive lymphatic network. The pulmonary vasculature is intricately involved in maintaining the physiologic function of the lung. The vasculature is also responsible for the delivery of antigens, immunocompetent cells, and inflammatory cells from the systemic circulation to the lung. This makes it possible that antigenic and cellular interactions and/or immune responses may take place in the lung either within the vessels or extravascularly if relevant molecules and cells can penetrate the vascular endothelium. Alternatively, it is important to appreciate that immunologically important molecules and/or cells could penetrate the lung capillary system from either the airways or the pulmonary interstitium and then initiate or focus an immune response within these vessels.

While the passage of molecules and cells from the vessels into the lung parenchyma is more easily imagined than the reverse, there is a sizable body of data to support direct penetration of vessels by inhaled particles (Stuart, 1973; Lauweryns and Baert, 1978) or intratracheally instilled proteins (Bensch and Dominquez, 1971). Earlier workers demonstrated that various proteins (e.g., serum, hemoglobin, and albumin) placed in the trachea migrated from the alveoli into the pulmonary blood capillaries (Drinker *et al.*, 1937; Courtice and Simmonds, 1949; Schultz *et al.*, 1964). However, these investigators could not discern whether the absorption occurred directly into the pulmonary blood capillaries or via the lymphatic capillaries. While morphologic data clearly support that particles penetrate into lymphatic and pulmonary blood capillaries (Casarett, 1964), there are little or no data addressing the pathway by which cells penetrate nor the mechanisms by which experimental particles get into the lumen of the lymphatic or blood capillaries.

5.4. NATURE OF BARRIERS TO ANTIGEN IN PRESENTATION

5.4.1. Airway Defense Mechanisms

It is apparent that each alveolar airspace within the lung is in direct contact with the external environment through a branching airway. The respiratory system is constantly and persistently bombarded with foreign, and noxious agents from the environment. The lungs are largely protected from the inhaled agents by highly specialized specific and nonspecific defense mechanisms of the upper respiratory tract (reviewed in Green *et al.*, 1978; Kaltreider, 1976).

Nonspecific defense mechanisms include reflexes such as the cough or the sneeze, *aerodynamic filtration,* and the *mucociliary escalator.* While a cough can produce sufficiently rapid airflow for the clearance of foreign material or secretions that are located in the larger airways, the airflow in the bronchioles and alveoli is too feeble to effect significant clearance. Foreign material in these regions must be transported to the larger bronchi by ciliary action before the cough has an effect on its clearance. Transport of such material also can be accomplished by the alveolar macrophage through phagocytosis. Any impairment of coughing, ciliary action, or phagocytosis would permit antigens to reach the lung.

Another important defense mechanism of the respiratory tree is the aerodynamic design of the structure. The airways are constructed in such a way as to restrict the size of the particles that can impinge on the various levels of the respiratory tract. Particles 10 μm or more in diameter are trapped in the nose and upper respiratory tract. Particles 0.3 μm or less remain suspended in the inhaled gases and are exhaled. However, particulate matter that is in the range 0.5 to 3 μm may reach the alveoli, sediment, and be subject to the defense mechanisms of the alveolar membrane. The greatest challenge to the entire respiratory tract defenses are particles 2–10 μm in diameter.

Because air velocity drops rapidly at points of bifurcation, most inspired particles land by vertical impaction, and then sediment on the mucous blanket that coats the surface of the larger airways. A *mucociliary* escalator then transports these particles toward the larynx and mouth.

The mucous blanket consists of two layers. A serous watery sol layer is in direct contact with the epithelium and bathes the cilia. The cilia reach up and penetrate and propel the upper zone of mucus toward the mouth. The upper gel layer is highly viscous and comprised of 95% water. Major defense proteins in the soluble layer are IgA and lysozyme. Antibody is produced locally by plasma cells in the airway wall. There is five times as much IgA as IgG in the large airways. Lower in the respiratory tree, IgG is more abundant.

5.4.2. The Cells of the Airways

The walls of the airway can be divided into mucosal and submucosal layers (reviewed in Breeze and Wheeldon, 1978). The mucosa is made up of epithelium, basement membrane, and lamina propria. Cells of the epithelium are attached to the basement membrane. The cellular components of the lamina propria comprise lymphocytes, plasma cells, polymorphonuclear lymphocytes, and numerous mast cells. Many other cell types have been described morphologically but their functional significance is unknown.

The epithelium in the large airways is of the pseudostratified columnar type. The bronchial epithelium contains at least 13 morphologically different cell types. Ciliated and goblet cells make up the majority; there are at least five ciliated cells for every goblet cell. While the function of nonserous cells is unknown, it is generally thought that these are the cells involved in the synthesis of secretory piece and transport of secretory IgA into the bronchial lumen.

In the distal airways, columnar epithelium becomes cuboidal. Goblet cells are not present in the small bronchioles (less than 2 μm in diameter) where clara cells rich in granules of unknown composition are present. As would be expected, these airways do not have a mucous lining layer, reflecting the absence of goblet cells and mucous glands. Instead of mucus, the walls of the distal airways are bathed in a watery protein-rich layer that appears to have surface tension active properties (surfactant).

Lymphoid nodules are important components of the lamina propria at sites along major airways of the tracheobronchial tree. Overlying the nodules the epithelium is modified and consists of flattened nonciliated cells resembling lymphoepithelium in other areas of the body. This lymphoid tissue has been designated bronchus-associated lymphoid tissue (BALT). Because of its association with a mucosa-lined organ, it is often thought to be analogous to PP and GALT (Bienenstock *et al.*, 1973a). BALT is considered in more detail later.

5.4.3. Bronchial Epithelium versus Alveolar Epithelium

Of the two epithelia, fewer cell types have been identified in the alveolar epithelium. In both, barriers use tight junctions between and among the lining cells present (Szidon *et al.*, 1972; Schneeberger-Keeley and Karnovsky, 1968; Lauweryns and Baert, 1978). The major difference between the two epithelia is the absence of ciliated, goblet, and nonserous cells in the alveolar epithelium. Lymphocytes are regularly found in the tracheobronchial epithelium, especially in the extrapulmonary airways. It is not clear if these lymphocytes pass into the lumen or persist in the epithelium. It is known that a small percentage of the bronchoalveolar cells that are removed during lung lavage are lymphocytes. The origin of these cells is not clear. Mast cells appear to be more prevalent in the bronchial epithelium than in the alveolar epithelium (Breeze and Wheeldon, 1978). Alveolar clearance via the pulmonary lymphatics and/or the blood capillaries implies that there is a transport of particles and molecules through the alveolar epithelium and its basement membrane. Although histologic evidence suggests that the bronchoalveolar epithelium is impermeable [the epithelial cells are connected to each other by tight junctions or zonulae occludentes (Szidon *et al.*, 1972; Schneeberger-Keeley and Karnovsky, 1968)], it is known that particles can be transported intact through the epithelium (Bensch and Dominquez, 1971; Gonzalez-Crussi and Boston, 1972) and proteins may reach the alveolar interstitium antigenically intact (Schultz *et al.*, 1963; Bensch *et al.*, 1964). The mechanisms of bronchoalveolar epithelial transport are unknown despite a variety of hypotheses that have been put forth ranging from free penetration through eroded areas (Gross and Westrick, 1964), specialized sumps through which alveolar fluid flows and penetrates (Morrow, 1973), and epithelial cell phagocytosis followed by cellular disruption that liberates the particles into the interstitium and alveolar lumen (Heppleston, 1963). Most data would support the idea of transport through the small alveolar epithelial cells (Lauweryns and Baert, 1974; Sorokin and Brain, 1975). Thus, the bronchoalveolar epithelium is permeable although the precise conditions and mechanisms by which this occurs are unknown.

Once particles and substances pass through the bronchoalveolar epithelium and its basement membrane, the supporting connective tissue is the major route to the blood and lymphatic capillaries. It is widely accepted that particles migrate through the interstitium under the influence of concentration gradients and an orientation of fluid flow toward the lymphatics (Stoub, 1970). A small number of particles may remain in tissue cells and fibers (Gross and Hatch, 1962). Phagocytosis of material passing the epithelial barrier may be achieved by pulmonary interstitial macrophages, a population distinct from alveolar macrophages (Stoub, 1970).

5.4.4. Vascular Endothelium

Antigens on the vascular side of the lung meet an endothelial barrier resting on its basement membrane (made of type IV collagen). Capillary basement membranes in other organs (such as kidney) serve as a barrier to high-molecular-weight substances. In contrast, the alveolar capillary basement membrane does not retard large solutes as effectively as do basement membranes in other capillary beds. It is not known if this is related to type IV collagen or not. This suggests that the endothelial barrier may be a limited fortress.

Still another avenue of particulate and macromolecular transport operates through the lymphatic system. The *lymphatic vascular network* is perivascular and peribronchiolar and connected by vessels that penetrate the pleural surface of the lung. Flow in each system is separated by the orientation of the valves in the lymphatic vessels. Lymph on the surface makes its way to the hilum by traversing the surface. Intraparenchymal lymph is conveyed to the hilum through the perivascular and peribronchial lymph channels. The pulmonary lymph flows into the hilar nodes and then to the tracheobronchial and paratracheal nodes. Ultimately, lung lymph may drain into either the right lymphatic duct or thoracic duct. Substantial species differences exist in the relative drainage of lung lymph (Whitcomb, 1982; Yoffey and Courtice, 1970).

Pulmonary lymphatic vessels consist of endothelial cells that sit on an incomplete basement membrane. Endothelial cells have numerous luminal cytoplasmic projections. The cell layer is discontinuous with open cell junctions and intercellular gaps. It has been proposed that lymphatic channels communicate directly with the interstitial space of the alveolar wall.

Lymphatics play an important role in the clearance of particles from the alveolar surface and in the removal of fluid from the interstitial space of the alveolar wall. Particulate materials may be ingested by alveolar macrophages and transported up the airways. Alternatively, materials could be transported into the interstitium by (1) type I pneumocytes, (2) alveolar macrophages may enter lymphatics through open cell junctions, (3) endocytosis by lymphatic endothelial cells, or (4) be ingested by interstitial macrophages and enter through intercellular junctions. Certainly the discontinuous basement membrane and the open intercellular endothelial junction suggest that the lymphatics are more permeable than the capillary endothelium with respect to the capacity for direct uptake of particles.

In summary, if antigens reach the lung in spite of the highly specialized nonspecific defense mechanisms of the upper respiratory tract, they may enter into the interstitium and/or lymphatic (blood) vessels with little hindrance from physical barriers (Richardson and Bouchard, 1976; Braley *et al.*, 1978). It also seems apparent that antigens within the blood vasculature are able to enter the pulmonary interstitium through the endothelial basement membrane and substances can enter the vasculature from airways or interstitium (Lauweryns and Baert, 1974).

5.5. STRUCTURAL SPECIALIZATIONS ALLOWING ANTIGENS TO GAIN ACCESS INTO THE BODY

Routes by which substances gain access to the body via the lung are:

1. Bronchoalveolar epithelium; internalization via alveolar epithelial cells and transport into the interstitium
2. Lymph drainage, endothelial cells with incomplete basement membrane
3. Endothelial basement membrane of blood vessels
4. Bronchus-associated lymphoid tissue
5. Alveolar macrophage

While the bronchoalveolar epithelium and the vascular endothelium may be barriers, at the same time their specialized structure may aid the access of antigens into an organism in a passive way. Unlike these epithelial and endothelial barriers that may passively admit antigens, BALT and alveolar macrophages may actively route antigens into the body.

5.5.1. BALT

Multiple nodules of lymphoid tissue are scattered within the bronchial mucosa down to the level of the small bronchioles and are particularly numerous about the points of bifurcation where inhaled antigens may impinge. Numerous investigators have observed that the lymphoid collections that constitute BALT are similar to PP in the intestinal wall. Bienenstock *et al.* (1973a) reported a similarity of lymphoid aggregates in the respiratory tract to PP in the intestinal wall. As with PP, BALT consists of lymphoid follicles associated with an overlying dome region beneath the epithelium and contains high-walled endothelial venules (HEV) in the vessels in the parafollicular regions (Racz *et al.*, 1977). Afferent lymphatics have not been identifed. Lymphoepithelium resembles that of PP (Bienenstock *et al.*, 1973a,b; Racz *et al.*, 1977; Vai *et al.*, 1977; Bockman and Cooper, 1973; Owen and Jones, 1974; Owen, 1977). Lymphoepithelial cells with numerous microfolds, called M cells, may function to direct antigens to BALT as do M cells in PP. Plasma cells may be found in BALT occasionally. Macrophages are also present in BALT; their relationship to alveolar macrophages is unknown.

5.5.1a. Lymphoid Follicles. Beneath the BALT lymphoepithelium are well-organized lymphoid aggregates divisible into three morphologic regions: dome, follicular, parafollicular (Racz *et al.*, 1977). BALT is very difficult to study because of technical difficulties of preparation. In one study, 18% of the cells in BALT were found to be T cells, 50% B cells with sIg and IgA predominant isotype (Bienenstock and Johnston, 1976).

Plesch *et al.* (1983) have concluded from their morphologic studies of BALT in the rat that BALT develops in a similar time frame to other lymphoid tissue and can be considered as normal peripheral lymphoid tissue. Interestingly, they observed that IgM-, IgG-, and IgA-bearing B lymphocytes and plasma cells are present as early as 4 days postpartum. Removal of the thymus did not deplete any particular area in BALT, suggesting that the majority of lymphocytes are not thymus dependent.

It appears that lymphocytes could enter BALT through HEV or through M cells, proliferate following antigenic stimulation, and migrate into the pulmonary lymphatics via lymphoepithelium and on to other sites in the body. Despite considerable knowledge of the morphology of BALT, little is known about its functions.

5.5.1b. Antigen Uptake. Because of BALT's location and access to inhaled antigens, the lymphoepithelium appears to be important in the delivery of antigens to lymphocytes. Since BALT lymphoepithelium is nonciliated (Bienenstock *et al.*, 1973a; Plesch *et al.*, 1983; Tenner-Racz *et al.*, 1979; McDermott *et al.*, 1983) and consists of deep invaginations and folds not only in the surface of epithelial cells but also between these cells, the movement of mucus may be sufficiently retarded over BALT to increase the opportunity for absorption of antigen.

Whereas carbon particles instilled into the respiratory tract were not found in BALT (Bienenstock *et al.*, 1973a,b), BCG introduced into the respiratory tract of rabbits was found in the lymphoepithelium and in underlying BALT follicles (Tenner-Racz *et al.*, 1979). Antigen in BALT lymphoepithelial cells may be transferred to the associated lymph follicle, but it is not known whether cells in BALT lymphoepithelium process antigen as do macrophages.

McDermott *et al.* (1983) suggest that the roles of PP might be both to prime cells for an immune response and to act as a site for rapid amplification of predominantly IgA response through lymphocytes disseminated from the PP. By analogy the BALT would not necessarily be the location where priming occurs but might be the site where upon presentation of antigen to the respiratory tract a population of committed cells perhaps already primed at another site (PP, hilar, or mesenteric lymph node) might locally expand and hence be exported primarily to the lung mucosa. If this is true, then these cells would populate preferentially the larger airways where IgA is dominant. These cells may also migrate to the lower respiratory tract (lung) where they share that environment with IgG-secreting cells.

There is little evidence to support the notion that the hilar lymph node (draining lymph node for the lung and BALT) contains predominantly IgA precursors as does the mesenteric lymph node. In unpublished studies in our laboratory (Stein-Streilein) on the isotypes of antibody-forming cells (AFC) in hilar lymph nodes from hamsters immunized to SRBC, the phenotype of the immunoglobulins reflected the serum immunoglobulins, with IgG1 being predominant and only a small percentage being IgG2 and IgA (antisera to hamster heavy chain of immunoglobulin were generous gifts from John Coe, M.D., of Rocky Mountain Laboratories).

IgA plasma cell precursors and T cells are capable of migrating between intestinal sites and the respiratory tract. Therefore, the hypothesis has been proposed that there may be a common mucosal immunologic system (Bienenstock *et al.*, 1978; McDermott and Bienenstock, 1979; Weisz-Carrington *et al.*, 1979). However, studies by McDermott and Bienenstock (1979) demonstrate that B lymphoblasts from tracheobronchial (hilar) nodes preferentially migrate to the lungs and not to the gut. B lymphoblasts from the gut, however, have been shown to populate the lung (Mirski *et al.*, 1981). Perhaps these data suggest that mucosal traffic may be one way, with all roads leading *from* the gut mucosa to other mucosally lined organs. Lymphocyte traffic studies are extremely tedious and difficult to control. While positive results are reassuring, negative results may be a reflection of the sensitivity of the assay. The question remains: how many cells must migrate and be present in the environment to have a biologic impact?

Thus, the extent to which this concept of a common mucosal immune organ includes the lung is not clear. Most striking are the facts that (1) the lung is not entirely lined by a mucosal layer and (2) lymphoblasts from the tracheobronchial (hilar) lymph nodes preferentially home to the lung.

5.5.2. The Alveolar Macrophage

A major function of the alveolar macrophage is to remove particles from the lung (Green *et al.*, 1978; Hocking and Golde, 1979). How this cell interacts in the immune system is unknown. Whether the alveolar macrophage supports or suppresses an immune response appears to depend on the species and *in vitro* assay used (McCombs *et al.*, 1982; Holt, 1980; Holt and Batty, 1980; Ansfield *et al.*, 1979; Yeager *et al.*, 1982; Toews *et al.*, 1984; Schuyler and Todd, 1981; Warner *et al.*, 1981; Lyons and Lipscomb, 1983).

5.6. UNIQUE ANTIGEN-PRESENTING PROCESS

It is not known if there is a unique antigen-presenting cell in the lung. Several candidates have been proposed: alveolar macrophage, interstitial macrophage, blood monocyte, macrophage in BALT, dendritic cell. The alveolar macrophage is a strong candidate but in order for it to be effective it must be able to return to the lung parenchyma, to BALT, or to the draining lymph nodes to find the appropriate environment in which to act. In a GVH disease model, our (Stein-Streilein) laboratory has demonstrated that viable functional cells can cross the bronchoalveolar membrane and induce local and systemic GVH reactions (Stein-Streilein *et al.*, 1981).

There are reports that a small but definite amount of cell-associated radiolabel reaches the draining lymph node after intratracheal inoculation of radiolabeled (^{51}Cr or ^{111}In) bronchoalveolar cells or purified alveolar macrophages (syngeneic) in the guinea pig (Correy *et al.*, 1984). Assuming that the radiolabel that reaches the hilar node is inside the same inoculated macrophage, it can be concluded that the alveolar macrophage is able to return to the lung to present antigen acquired in airways. It was calculated that in the normal guinea pig as many as 8700 alveolar macrophages might reach the hilar lymph node daily.

In addition to the alveolar macrophage, there are macrophages in BALT as well as interstitial macrophages and dendritic cells (Sertl *et al.*, 1986) within the parenchyma of the lung that could present antigen. No published study has indicated that Langerhans cells are present in the lung. Few studies have evaluated antigen presentation with cells in the lung other than the alveolar macrophage.

Site of Antigen Presentation

It is important to consider at what site meaningful recognition of antigen by immunocompetent cells takes place. While there are T cells, B cells, and macrophages in the airways, it seems unlikely that antigen presentation would take place in such a hostile environment that is more concerned with gas exchange than the comfort of cells and antigens. While there is no evidence to rule out the possibility that antigen processing occurs in the pulmonary interstitium, it seems that optimal immune inductions would take place in BALT or the draining lymph nodes. The observation that these tissues hypertrophy after exposure to antigen would support this concept.

Alternatively, it is well known that pathologic changes in the lung parenchyma can be initiated by immune mechanisms. While such mechanisms are generally secondary immune reactions, antigen presentation to the antigen-reactive lymphocytes (memory cells) may often be involved. Under such circumstances, the antigen-presenting cell could be an alveolar macrophage that has entered the interstitium, an interstitial macrophage or even a blood monocyte that has entered the lung parenchyma as well as the dendritic cell (Sertl *et al.*, 1986).

5.7. LOCAL AND SYSTEMIC CONSEQUENCES OF IMMUNE RESPONSES THAT ARE INITIATED BY ANTIGEN PRESENTED TO (IN) THE LUNG

The local and systemic consequences of immune responses that are initiated in the lung are of major importance to the organism. It now appears that, as for the gut, most antigens introduced into the lung have the potential to induce a local sensitization. However, unlike the gut, intratracheally delivered antigens (high dose) are also capable of inducing a systemic immune response (Nash and Holle, 1973; Thomas *et al.*, 1974; Waldman and Henney, 1971; Stein-Streilein *et al.*, 1979a,b; Bice and Schnizilein, 1980). A systemic response can alter the quality of the local immune response (Stein-Streilen *et al.*, 1979b). In fact, titration of the immunizing antigen permits the induction of immune responses in the local pulmonary environment in the absence of a systemic response. However, dilute concentrations of antigen in the lungs will prime the host for a secondary response to a systemic challenge with the same antigen (Stein-Streilein, unpublished observations). Therefore, the low-dose intratracheal antigenic challenge is neither a null event nor a tolerance-producing event.

We (Stein-Streilein, 1983) and others have observed that haptens presented to the lungs (even when inhaled) will induce local and/or systemic sensitization (Karol *et al.*, 1978, 1980; Karol, 1983; Patterson *et al.*, 1978; Enander *et al.*, 1983, 1984). We have observed that intratracheal administration of the water-soluble hapten 2,4,6-trinitrobenzene sulfate (TNBS) sensitizes experimental animals systemically and contact hypersensitivity can be demonstrated by measurement of ear thickness after painting the ear with 2,4,6-trinitrochlorobenzene (TNCB) (Stein-Streilein, 1983). In addition, hapten-modified bronchoalveolar cells can induce sensitization when given (95% alveolar macrophages) intratracheally or intravenously. These observations are similar to those reported by others (Streilein, 1985; Asherson *et al.*, 1977) using Langerhans cell-enriched populations of epidermal cells as the source of hapten-modified cells but not similar to those reported with hapten-modified spleen or lymphoid cells (Miller and Claman, 1976).

The administration of hapten into the gut classically induces tolerance to that hapten (Chase, 1946; Tomasi, 1980; Asherson *et al.*, 1977). However, when antigen is delivered by inhalation (Holt and Leivers, 1982), the site responsible for inducing unresponsiveness is unknown since as much antigen can reach the gut as the lung. It may be concluded that the lung appears to be more like the skin than the gut in its ability to respond to locally instilled hapten by producing hypersensitivity.

Alternatively, if an individual is sensitized systemically to haptens (antigens) and challenged in the lung, an immune response may ensue which may be related to a disease state and histologic changes in the lung (Enander *et al.*, 1983, 1984). Immune responses resulting from inhalation of microorganisms and other agents have been implicated in the etiology of interstitial lung disorders. In many instances, immune effector mechanisms have been associated with pulmonary fibrosis. For example, antibodies to the offending agent (Lopez and Salvaggio, 1976), autoantibodies (McDermott and Bienenstock, 1979; Turner-Warwick, 1980) as well as deposition of immune complexes (Scherzer and Ward, 1978; Dreisen *et al.*, 1978) have been described in patients with pulmonary interstitial disorders. In addition, accumulation of lymphocytes in the lower respiratory tract has been associated with disorders such as hypersensitivity pneumonitis and sarcoidosis, and neutrophils are known to accumulate in idiopathic pulmonary fibrosis (Reynolds *et al.*, 1977; Yeager *et al.*, 1977; Weinberger *et al.*, 1978).

The lower respiratory tract may be separated from the upper respiratory tract (lung) structurally and functionally. It appears that the upper respiratory tract functions as a barrier and generally prevents substances from reaching the lung. In addition to highly specialized nonspecific defense mechanisms, the upper respiratory tract utilizes IgA and appears to communicate with a common mucosal immune system. Substances that are able to reach past this barrier and penetrate the lung proper appear to be met by and induce immune mechanisms uncommon to the upper respiratory tract and the gut. Systemic sensitization rather than unresponsiveness is the common immune response, and IgG antibody predominates. Systemic immunity may not always occur, however; depending on antigen dose, the intrapulmonary immune response may be detected only locally. The ease with which antigens gain access to the parenchyma of the lung once they are in the alveoli suggests that the lower respiratory tract (the lung proper) is *functionally* and *immunologically* inside the body.

5.8. SUMMARY

The respiratory tract possesses highly specialized nonspecific and structural defense mechanisms designed to prevent foreign matter and antigens from gaining access to the body. However, once within the confines of the lung, both particulate and soluble molecules can gain access to the lung parenchyma. When this takes place, sensitization rather than unresponsiveness occurs and the consequences of the immune response may be local, systemic, or both. Immune mechanisms within the lung may be responsible for mediating certain forms of lung disease. The lung appears to resemble the skin more than the gut as a regional sphere of antigen processing, presentation, and immunity.

6. REPRODUCTIVE TRACTS

In comparison with other mucosal systems, the organs of the reproductive and urinary tracts have received relatively little experimental attention. However, it was studies on an associated organ, the lactating mammary gland, that gave rise to the concept of a common mucosal immune system and generated more interest in the immunology of the reproductive systems. Numerous studies have now verified that, like other mucosal sites, the genitourinary tracts and mammary glands produce secretions with significant amounts of secretory IgA (Tomasi and Bienenstock, 1968; Rebello *et al.*, 1975), and, in most sites, the IgA is produced locally by plasma cells in the mucosal connective tissue (Murdoch *et al.*, 1982; Rebello *et al.*, 1975; Weisz-Carrington *et al.*, 1977). Adoptive transfer experiments have shown that, like the gut mucosa, the mammary gland, cervix, and vagina behave as selective recipients of lymphoblasts from the mesenteric lymph nodes, primarily IgA-containing precursors (McDermott and Bienenstock, 1979; Roux *et al.*, 1977).

The reproductive systems have an important feature which sets them apart from other mucosal sites: they are target organs for steroid hormones that dramatically affect both structure and physiological function, and recent studies have revealed that such hormones can also affect mucosal immune function. The preference of the mammary gland for gut-generated IgA blasts is a function of the lactational state and can be induced in virgin animals by combined treatment with estrogen, progesterone, and prolactin (Weisz-Carrington *et al.*, 1978). Moreover, B-lymphoblast migration to the mouse cervicovaginal mucosa fluctuates with the estrous cycle, being greatest in proestrus and least in diestrus, an effect not seen in nonreproductive mucosal sites (McDermott *et al.*, 1980). Interestingly, the uterus has not proven to be a significant recipient for such cells and usually has few plasma cells (McDermott and Bienenstock, 1979; Rebello *et al.*, 1975), even though its fluid often contains abundant IgA. However, these levels, too, are affected by hormones, the highest levels being in proestrus, with accompanying alterations in secretory component production by the epithelial cells (Wira and Sandoe, 1977; Sullivan and Wira, 1981). Estradiol has the greatest positive influence (Wira and Sandoe, 1980).

6.1. BARRIER PROPERTIES AND ANTIGEN PRESENTATION

Little definitive information is available on antigen exposure in the reproductive tracts and to what extent they can serve as sites for the stimulation of immunoreactive cells and generation of effectors. Certainly, in the female reproductive tract significant barriers to antigen exposure exist, especially in the lower regions. Foremost are the acidic pH of the vagina and the cervical mucus. The latter presents a significant barrier not only to potentially pathogenic organisms, but also to spermatozoa (Vickery and Bennett, 1968). These protective biochemical and physical barriers are supplemented by epithelia composed of either columnar cells with tight junctions, as seen in the uterus and endocervix, or stratified squamous epithelium of the vagina and ectocervix, which is even keratinized in some species.

The structure of the male reproductive tract provides barriers presumably designed to guard against exposure of the adult male to sperm autoantigens. The testis contains a well-defined blood–testis barrier, consisting of extremely extensive tight junctions between Sertoli cells (Dym and Fawcett, 1970), in addition to having a microenvironment that seems to interfere locally with antigen processing (Head *et al.*, 1983). The duct system is lined by cuboidal or columnar cells with apical tight junctions. Freeze-fracture studies have revealed, however, that the junctions in different components of the tract have very different degrees of membrane fusion and depths of continguous strands (Fawcett, 1979). The smallest and thus probably least effective junctions are those in the ductuli efferentes, whereas in the rest of the tract they are quite extensive. Although a modest amount of sperm antigens may normally leave these ducts, the extensive literature on antisperm reactivity generated in vasectomized men and animals compared with controls attests to the protective nature of the intact tract (Alexander and Anderson, 1979). Vasectomy forces dramatically increased degradation of sperm in the tract, primarily by phagocytic cells entering the lumen. The primary routes for sperm antigens to escape the tract are through damaged epithelium, transport within macrophages that may reenter the tissues, and as soluble antigen, escaping through the basement membrane (Tung and Alexander, 1980; Marsh and Alexander, 1982).

There are no discrete lymphoid tissues such as PP associated with the reproductive mucosae, in the male or female, that could serve as local sites for either uptake of antigen or antigen processing in a lymphoid microenvironment. Thus, antigens entering at those surfaces must either gain access to draining lymph nodes or to blood vessels or be processed locally by antigen-presenting cells to impinge on the host's immune system. The lymphatic drainage from most reproductive and urinary tissues is quite good, including the ovaries and testes. The exception is a portion of the uterus, which is considered below. Cells with the morphology of Langerhans cells have been described in the vaginal and ectocervical epithelia (Younes *et al.*, 1968) which presumably can function as antigen-presenting cells in these sites, but presence of these cells in the rest of the female and male tracts is unknown. However, we have recently found that Ia^+ cells, most with dendritic morphology and not part of the avidly phagocytic macrophage population, are scattered throughout the uterine endometrium in considerable numbers and are especially abundant around the glands (Head and Gaede, 1986). The antigen-presenting capabilities of these cells remain to be determined.

When considered with regard to function, it is clear that the female reproductive tract, and in particular the uterus, is immunologically unique because of its potential exposure to foreign cells and tissues. Connected as it is to the outside environment, this organ must of course effectively deal with pathogenic organisms. However, it must also cope with periodic exposures to foreign cells, viz. spermatozoa, and serve as the implantation site for genetically alien fetuses and their placentas, without generating destructive immune responses in either case. Harmful reactions not only could cause serious inflammation in the host's tissues, but even jeopardize survival of the viviparous species.

There is now considerable evidence that females do mount immune responses to implanted conceptuses. However, these are characterized primarily by blocking-type antibodies and suppressor cells, with no generation of cytotoxic T cells (CTL), in contrast to the CTL and cytotoxic antibody production characteristically seen in hosts bearing conventional allografts (Head and Billingham, 1982). Although the fetus-derived trophoblast which confronts maternal tissues seems to have some peculiar antigenic properties, it is reasonable to suspect that the site of engraftment, the uterus, may have some special features affecting antigen presentation that play an important role in determining the nature of maternal immune responses during pregnancy and lack of harmful responses to sperm. As mentioned previously, the cycling uterine endometrium, at least in rodents, is well endowed with Ia^+ cells. However, although the uterine musculature has large lymphatics draining to regional nodes, the lymphatic drainage from the endometrium, where implantation occurs, is generally poor, being confined to sparse, very small vessels, at least in virgin rodents (Head and Seelig, 1984). Interestingly, these enlarge during early pregnancy, and become conspicuous, especially at the edges of the decidual capsule of the implantation chamber (Head and Seelig, unpublished observations). The columnar epithelium of the uterus has tight junctional complexes that may restrict molecules from entering between cells. In addition, the uterine basement membrane seems to restrict entry of colloidal materials and cells from the lumen (Head and Lande, 1983). Both viable lymphocytes and India ink placed in the uterine lumen remain there for considerable periods, with only very small numbers or amounts gaining access to draining lymph nodes. The colloidal carbon that does enter the endometrial stroma is taken up by macrophages which are retained locally for at least several days. The combined effect of an undefined epithelial barrier, retention of phagocytes for long periods, poor endometrial lymphatic drainage, and absence of PP-like structures is most likely to limit exposure of the host's lymphoid system to potentially antigenic material entering through the uterine lumen. These barriers may be especially important in ensuring that large amounts of sperm do not penetrate the tissues to immunize the female. Additionally, the apparent barrier to leukocytes entering from the lumen suggests that the abundant cells emigrating into the lumen after mating and phagocytosing sperm have little chance of carrying immunogens back into the female's tissues.

6.2. SITES OF ANTIGEN RECOGNITION AND SPECIALIZED CELL TYPES

In animals receiving antigenic exposure in the uterine lumen, including allogeneic fetal–placental units during pregnancy, it is clear that immune responses do occur in the draining lymph nodes. The uterus-draining lymph nodes enlarge dramatically during allogeneic pregnancy, and show signs of immune processes, such as increases in pyroninophilic cells and mitotic activity (McLean *et al.*, 1980; Maroni and de Sousa, 1973). SRBC inoculated into the lumina of rat uteri induce PFC within those same nodes, albeit at significantly lower numbers than in nodes draining conventional subcutaneous inoculation sites (Lande *et al.*, 1981). The extent to which local processing may also be involved has not been determined. The restriction of some uterus-induced immunity to the local site (see Section 6.3) suggests it may be important. In addition, a series of elegant studies by Clark and his colleagues (Clark *et al.*, 1982; Slapsys and Clark, 1983) has shown that during pregnancy, significant changes occur very early in the uterine stroma of pregnant mice. Small lymphocytes appear in this area shortly after mating that are capable of nonspecifically suppressing the generation of CTL. Although they can be found later within the draining lymph nodes, their initial appearance at the site of implantation suggests they may be regulators of important immune recognition events in the uterine endometrium. These small, non-T suppressors can also be found in a similar time course in pseudopregnant uteri, suggesting that the hormonal milieu and not the presence of allogeneic cells or tissues is responsible.

Other changes in the distribution of cell populations also occur during pregnancy. The virgin mouse uterus does not seem to receive significant numbers of IgA-containing blasts, but a different situation may exist during pregnancy, although this possibility has not been tested directly by trafficking studies. However, IgA-containing plasma cells are reported to increase in the mouse uterus from days 2 to 6 of pregnancy (Bernard *et al.*, 1981). In the decidual capsule, that transformed endometrial stroma around the implanting conceptus, approximately 50% of the cells are Thy-1$^+$ and 15% have Fc receptors, while being negative for Lyt markers and immunoglobulin (Bernard *et al.*, 1978). Indeed, the decidua is becoming increasingly interesting from the immunologic point of view. Aside from containing the suppressor cells of Clark mentioned above, it contains considerable numbers of Ia$^+$ cells (Searle *et al.*, 1983). Moreover, studies on radiation chimeras suggest that significant numbers of decidual cells are derived from precursors originating in the bone marrow (Kearns and Lala, 1982). All of these findings indicate that the decidua, long considered a buffer tissue between maternal tissues and the conceptus, may also be immunologically important and influence how mothers respond to allogeneic fetuses.

A potential role for hormones in the antigen processing that probably occurs during pregnancy must not be overlooked. Studies on pregnant sheep and goats have revealed that the major lymphatic vessels from the uterus and ovary join, their contents mixing prior to arrival in the regional nodes (Staples *et al.*, 1982). In an animal with a mature corpus luteum, lymphocytes and macrophages in uterine lymph would be exposed to levels of progesterone up to 1000-fold higher than jugular vein plasma. Since high concentrations of progesterone interfere with induction of T-cell responses *in vitro* (Pavia *et al.*, 1979), suppression of responses initiated in the uterus or by uterus-derived cells could result, especially during early pregnancy.

6.3. LOCAL AND SYSTEMIC IMMUNE CONSEQUENCES

Recent work with both experimental antigens and infectious agents has revealed that antigenic exposure in the uterus usually leads to generation of responses within the uterus, but has minimal effects at the systemic level. These findings are consistent with the anatomic and tracing studies cited above which suggested that barriers to large-scale antigen uptake in this site exist. For example, DNP–BGG injected into the rat uterine lumen fails to induce the primary serum antibody response seen after subcutaneous inoculation (Lande *et al.*, 1981). However, subsequent challenge, whether intrauterine or intravenous, engenders an excellent secondary response, so that the animals had been primed by the initial intrauterine exposure, probably by small amounts of antigen slowly gaining access to regional nodes. Studies with certain pathogenic organisms have confirmed that, as expected, specific antibodies to organisms infecting the reproductive organs are primarily secretory IgA, although in the uterus IgG often contributes significantly (Ogra and Ogra, 1973; Chipperfield and Evans, 1972), but responses to infectious organisms and toxoids in the reproductive tract are often locally confined. Moreover, although systemically sensitized animals produce good local responses to antigenic challenge in the uterus (Batty and Warrack, 1955; Targowski and Foley, 1983), the reverse apparently does not normally occur. Polio vaccine inoculated into the human vagina and uterus and *M. tuberculosis* into the guinea pig uterus induce local responses of antibody production and leukocytic infiltration, respectively, elicited also by a second local exposure in the latter case, but they do not produce systemic sensitization.

The reproductive tracts are thus mucosal surfaces sharing some features with other such areas, such as reliance primarily on secretory IgA antibody, but with a number of unique features. These include a lack of definitive lymphoid structures, numerous specialized physicochemical barriers to antigen, and the dramatic influence of steroid hormones on their structure and function. The mechanisms of antigen processing in these sites are unknown but appear to be complex, probably influenced by all of the features listed above. The result seems to be predominantly local responses, without significant systemic responses, that may serve to guard against potentially destructive antifetal or antisperm immunity in the female.

7. SUMMARY

In this chapter we have attempted to assemble a diverse array of data which address the existence of regional specialization in antigen presentation, and we have examined briefly the consequences of such presentation. Awareness that regional specialization in antigen processing exists is not general among immunologists and information is spotty and incomplete. But even on the basis of what is currently known, certain patterns emerge which transcend specific tissue requirements. All surfaces display certain barrier functions which limit the free exchange of antigenic material across that surface. Yet each surface is equipped more or less to deal with antigenic material that breaches the barrier. Tables 1 through 4 attempt to summarize what is currently known. Although much obviously remains to be done, what is already clear is that the immune

TABLE 1. BARRIERS TO ANTIGEN ENTRY[a]

Gastrointestinal tract	Upper: Stratified squamous epithelium; mucus
	Lower: Tight junctions among columnar epithelium; low pH of stomach; enzymes in secretions; mucus
Respiratory tract	Upper: Tight junctions of columnar epithelium; mucociliary escalator; aerodynamics in which bore and angle of tubular elements dictate entry on basis of particle size
	Lower: Alveolar macrophages; role of tight junctions among pulmonary epithelia lining alveoli is unclear; role of surfactant is unclear
Genitourinary tract	Upper: Tight junctions of columnar epithelium lining uterus and fallopian tubes; paucity of draining lymphatics
	Lower: Stratified squamous epithelium; pH of vaginal contents and secretions; cervical mucus
Skin	Stratified, keratinized squamous epithelium; stratum corneum excludes lipid- and water-soluble materials

[a]While all surfaces rest on a basement membrane, the role of the components of this structure in limiting antigen entry is ill-defined. Probably it plays a very minor role as a barrier.

TABLE 2. STRUCTURAL SPECIALIZATIONS THAT PROMOTE ANTIGEN ENTRY

Gastrointestinal tract	Epithelium overlying Peyer's patches contains specialized cells—M cells—which facilitate antigen uptake. Products of gut flora, e.g., LPS, may facilitate antigen uptake.
Respiratory tract	M cells have been found as part of BALT as in GI tract.
	Alveolar macrophages may carry antigen across epithelium into pulmonary interstitium.
Genitourinary tract	None known.
Skin	None.

system does not stand as a unitary structure serving equally well all tissues of the body. Not surprisingly, unique functional and therefore structural attributes of each tissue and organ dictate that the generic qualities of immunity be fashioned so that the special needs of each tissue and organ are met while the primary physiologic function of the tissue or organ is maintained. While much remains to be learned about the regional spheres of influence described in this chapter, it would not be surprising to learn that other regions, such as the brain, the eye, and perhaps even organs such as the kidney, will also be endowed with a unique immune capability which serves individual regional needs. This terminal suggestion is buttressed by recent observations that the kidney contains resident cells that bear Ia and serve as antigen-presenting cells (Schreiner *et al.*, 1981; Schreiner and Cotran, 1982), and that astrocytes (glial elements of the CNS) can present myelin basic protein to encephalitogenic T-cell lines (Fontana *et al.*, 1984).

If our thesis concerning the existence of regional specialization in antigen presentation is correct, then there may be far-reaching implications for our understanding of specific types of immune reactions and for the pathogenesis of certain human diseases. It is not too difficult to imagine that derangements of SALT (for example) could result in serious impairment of cutaneous immunity, although this may not be readily apparent when systemic immunity is examined. Regional spheres of immunologic influence may prove to be important in the mechanisms of certain human inflammatory disorders which have so far escaped our understanding.

TABLE 3. SPECIALIZED ANTIGEN-PRESENTING CELLS

Gastrointestinal tract	Dendritic cells that are Ia antigen positive and adherent to surfaces reside near and beneath M cells.
	Kupffer cells of liver can function as antigen-presenting cells.
Respiratory tract	Upper: Similar arrangement to GI tract.
	Lower: Macrophages in alveoli and smallest bronchioles.
Genitourinary tract	Upper: Ia-antigen-bearing cells reside in stroma of endometrium.
	Lower: Langerhans cells in epithelium of ectocervix and vagina.
Skin	Langerhans cells.
	Dermal macrophages.

TABLE 4. CONSEQUENCES OF REGIONAL ANTIGEN PRESENTATION

Gastrointestinal tract	Entire GI tract becomes equipped to produce and be protected by IgA antibody. There is systemic suppression of IgM and IgG synthesis, as well as impaired capacity to mount systemic delayed hypersensitivity responses.
Respiratory tract	Upper: Resembles GALT and arms most mucosal surfaces with IgA-mediated immunity. Lower: May produce strictly local immunity; if systemic immunity is produced, it is characterized by IgG antibody and expression of delayed-type hypersensitivity.
Genitourinary tract	Local immunity, usually of IgA variety, without a systemic response.
Skin	Contact hypersensitivity and IgG humoral immunity. No evidence of locally confined response.

REFERENCES

Alexander, N. J., and Anderson, D. J., 1979, Vasectomy: Consequences of autoimmunity to sperm antigens, *Fertil. Steril.* **32:**253.

Anderson, N. D., Anderson, A. O., and Wyllie, R. G., 1976, Specialized structure and metabolitic activities of high endothelial venules in rat lymphatic tissue, *Immunology* **31:**455.

André, C., Bazin, H., and Heremans, J. F., 1973, Influence of repeated administration of antigen by the oral route on specific antibody-producing cells in the mouse spleen, *Digestion* **9:**166.

André, C., Heremans, J. F., Vaerman, J. P., and Cambiaso, C., 1975, A mechanism for the induction of immunological tolerance by antigen feeding: Antigen–antibody complexes, *J. Exp. Med.* **142:**1509.

Andrew, W., and Andrew, N. L., 1949, Lymphocytes in the normal epidermis of rats and of man, *Anat. Rec.* **104:**217.

Ansfield, M. J., Kaetreider, H. B., Caldwell, J. L., and Herskowitz, F. N., 1979, Hyporesponsiveness of canine bronchoalveolar lymphocytes to mitogens: Inhibition of lymphocyte proliferation by alveolar macrophages, *J. Immunol.* **122:**542.

Asherson, G. L., Zembala, M., Perera, M. A. C. C., Mayhew, B., and Thomas, W. R., 1977, Production of immunity and unresponsiveness in the mouse by feeding contact sensitizing agents and the role of suppressor cells in the Peyer's patches, mesenteric lymph nodes and other lymphoid tissues, *Cell. Immunol.* **33:**145.

Axelrod, A. J., and Chandler, J. W., 1979, Morphologic characteristics of conjunctival lymphoid tissue in the rabbit, in: *Immunology and Immunopathology of the Eye* (A. M. Silverstein and G. R. O'Conner, eds.), p. 292–301, Masson Publications U.S.A., New York.

Barker, C. F., and Billingham, R. E., 1972, Immunologically competent passenger cells in mouse skin, *Transplantation* **14:**525.

Barry, W. S., and Pierce, N. F., 1979, Protein deprivation causes reversible impairment of mucosal immune response to cholera toxoid/toxin in rat gut, *Nature (London)* **281:**64.

Batty, I., and Warrack, G. H., 1955, Local antibody production in the mammary gland, spleen, uterus, vagina, and appendix of the rabbit, *J. Pathol. Bacteriol.* **70:**355.

Bensch, K. G., and Dominquez, E. A., 1971, Studies on the pulmonary air–tissue barrier. Part IV. Cytochemical tracing of macromolecules during absorption, *Yale J. Biol. Med.* **43:**236.

Bensch, K. G., Dominquez, E., and Liebow, A., 1964, Absorption of intact protein molecules across the pulmonary air–tissue barrier, *Science* **157:**1204.

Bergstresser, P. R., Fletcher, C., and Streilein, J. W., 1980, Surface densities of Langerhans cells in relation to rodent epidermal sites with special immunologic properties, *J. Invest. Dermatol.* **74:**77.

Bergstresser, P. R., Tigelaar, R. E., Dees, J. H., and Streilein, J. W., 1983, Thy 1 antigen-bearing dendritic cells populate murine epidermis, *J. Invest. Dermatol.* **81:**286.

Bernard, O., Scheid, M. P., Ripoche, M., and Bennett, D., 1978, Immunological studies of mouse decidual cells. I. Membrane markers of decidual cells in the days after implantation, *J. Exp. Med.* **148**:580.

Bernard, O., Rachman, F., and Bennett, D., 1981, Immunoglobulins in the mouse before implantation, *J. Reprod. Fertil.* **63**:237.

Bernstein, I. D., and Ovary, Z., 1968, Absorption of antigens from the gastrointestinal tract, *Int. Arch. Allergy Appl. Immunol.* **33**:521.

Bhalla, D. K., and Owen, R. L., 1982, Cell renewal and migration in lymphoid follicles of Peyer's patches and cecum—An autoradiographic study in mice, *Gastroenterology* **82**:232.

Bice, D. E., and Schnizilein, C. T., 1980, Cellular immunity induced by immunization of Fischer 344, *Int. Arch. Allergy Appl. Immunol.* **63**:438.

Bienenstock, J., and Befus, A. D., 1980, Mucosal immunology, *Immunology* **41**:249.

Bienenstock, J., and Dolezel, J., 1971, Peyer's patches: Lack of specific antibody-containing cells after oral and parenteral immunization, *J. Immunol.* **106**:938.

Bienenstock, J., and Johnston, N., 1976, A morphologic study of rabbit bronchial lymphoid aggregates, *Lab. Invest.* **35**:343.

Bienenstock, J., Johnston, N., and Perey, D. Y. E., 1973a, Bronchial lymphoid tissue. I. Morphologic characteristics, *Lab. Invest.* **28**:686.

Bienenstock, J., Johnston, N., and Perey, D. Y. E., 1973b, Bronchial lymphoid tissues. II. Functional characteristics, *Lab. Invest.* **28**:693.

Bienenstock, J., McDermott, M., Befus, A. D., and O'Neill, M., 1978, A common mucosal immunological system involving the bronchus, breast, and bowel, *Adv. Exp. Med. Biol.* **107**:53.

Birbeck, M. S., Breathnach, A. S., and Everall, J. D., 1961, An electron microscopic study of basal melanocytes and high level clear cells (Langerhans cells) in vitiligo, *J. Invest. Dermatol.* **37**:51.

Bockman, D. E., and Cooper, M. D., 1973, Pinocytosis by epithelium associated with lymphoid follicles in the bursa of Fabricius, appendix, and Peyer's patches: An electron microscopic study, *Am. J. Anat.* **136**:455.

Bockman, D. E., and Stevens, W., 1977, Gut-associated lymphoepithelial tissue: Bidirectional transport of tracer by specialized epithelial cells associated with lymphoid follicles, *J. Reticuloendothel. Soc.* **21**:245.

Borthistle, B. K., Kubo, R. T., Brown, W. R., and Grey, H. M., 1977, Studies on receptors for IgG on epithelial cells of the rat intestine, *J. Immunol.* **119**:471.

Braley, J. F., Dawson, C. A., Moore, V. L., and Cozzini, B. O., 1978, Absorption of antigen into the circulation of isolated lungs from normal and immunized rabbits, *J. Clin. Invest.* **61**:1240.

Breathnach, A. S., Silvers, W. K., Smith, J., and Heyner, S., 1968, Langerhans cells in mouse skin experimentally deprived of its neural crest component, *J. Invest. Dermatol.* **50**:147.

Breeze, R. G., and Wheeldon, E. B., 1978, The cells of the pulmonary airways, in: *Lung Disease, State of the Art* (J. F. Murray, ed.), pp. 111–183, American Lung Association, New York.

Brent, L., and Medawar, P. B., 1964, Nature of the normal lymphocyte transfer reaction, *Nature (London)* **204**:90.

Brent, L., Brown, J. B., and Medawar, P. B., 1962, Quantitative studies on tissue transplantation immunity. VI. Hypersensitivity reactions associated with the rejection of homografts, *Proc. R. Soc. London* **156**:87.

Britz, J. S., Askenase, P. W., Ptak, W., Steinman, R. M., and Gershon, R. K., 1982, Specialized antigen-presenting cells: Splenic dendritic and peritoneal-exudate cells induced by mycobacteria activate effector T cells that are resistant to suppression, *J. Exp. Med.* **155**:1344.

Butcher, E. C., Scollay, R. G., and Weissman, I. L., 1980, Organ specificity of lymphocyte migration: Mediation by highly selective lymphocyte interaction with organ-specific determinants on high endothelial venules, *Eur. J. Immunol.* **10**:556.

Cantor, H. M., and Dumont, A. E., 1967, Hepatic suppression of sensitization to antigen absorbed into the portal system, *Nature (London)* **215**:744.

Carter, P. B., and Collins, F. M., 1974, The route of enteric infection in normal mice, *J. Exp. Med.* **139**:1189.

Carter, P. B., and Collins, F. M., 1975, Peyer's patch responsiveness to *Salmonella* in mice, *J. Reticuloendothel. Soc.* **17**:38.

Casarett, L. J., 1964, Distribution and excretion of polonium 210. XII. Autoradiographic observations after inhalation of polonium 210 in rats, *Radiat. Res. (Suppl.)* **5**:187.

Cebra, J. J., Kamat, R., Gearhart, P., Robertson, S. M., and Tseng, J., 1977, Origin and differentiation of lymphocytes involved in the secretory IgA response, *Cold Spring Harbor Symp. Quant. Biol.* **41**:201.

Cebra, J. J., Crandall, C. A., Gearhart, P. J., Robertson, S. M., Tseng, J., and Watson, P. M., 1979, Cellular events concerned with the initiation, expression, and control of the mucosal immune response, in: *Immunology of Breast Milk* (P. Ogra and D. Dayton, eds.), pp. 1–16, Raven Press, New York.

Cerf-Bensussan, N., Schneeberger, E. E., and Bhan, A. K., 1983, Immunohistologic and immunoelectron microscopic characterization of the mucosal lymphocytes of human small intestine by the use of monoclonal antibodies, *J. Immunol.* **130**:2615.

Challacombe, S. J., and Tomasi, T. B., Jr., 1980, Systemic tolerance and secretory immunity after oral immunization, *J. Exp. Med.* **152**:1459.

Chase, M. W., 1946, Inhibition of experimental drug allergy by prior feeding of the sensitizing agent, *Proc. Soc. Exp. Biol. Med.* **61**:257.

Chestnut, R., Endies, R., and Grey, H. M., 1980, Antigen recognition by T cells and B cells: Recognition of cross-reactivity between native and denatured forms of globular antigens, *Clin. Immunol. Immunopathol.* **15**:397.

Chipperfield, E. J., and Evans, B. A., 1972, The influence of local infection on immunoglobulin formation in the human endocervix, *Clin. Exp. Immunol.* **11**:219.

Chu, R. M., Glock, R. D., and Ross, R. F., 1982, Changes in gut-associated lymphoid tissues of the small intestine of eight-week-old pigs infected with transmissible gastroenteritis virus, *Am. J. Vet. Res.* **43**:67.

Clark, D. A., Slapsys, R., Croy, B. A., Rossant, J., and McDermott, M. R., 1982, Regulation of cytotoxic T cells in pregnant mice, in: *Immunology of Reproduction* (T. Wegmann and T. J. Gill, III, eds.), pp. 341–361, Oxford University Press, London.

Correy, D., Kulkarni, P., and Lipscomb, M., 1984, The migration of alveolar macrophages into hilar lymph nodes, *Am. J. Pathol.* **115**:321.

Courtice, F. C., and Simmonds, W. J., 1949, Absorption from the lungs, *J. Physiol. (London)* **109**:103.

Crabbé, P. A., Carbonara, A. O., and Heremans, J. F., 1965, The normal human intestinal mucosa as a major source of plasma cells containing γA-immunoglobulin, *Lab. Invest.* **14**:235.

Craig, S. W., and Cebra, J. J., 1971, Peyer's patches: An enriched source of precursors for IgA-producing immunocytes in the rabbit, *J. Exp. Med.* **134**:188.

Davies, M. D. J., and Parrott, D. M. V., 1981, Cytotoxic T cells in small intestine epithelia, lamina propria, and lung lymphocytes, *Immunology* **44**:367.

de Sousa, M., 1981, Circulation of lymphocytes within the lymphoid system, in: *Lymphocyte Circulation: Experimental and Clinical Aspects,* (de Sousa, ed.), Wiley, New York.

Dolezel, J., and Bienenstock, J., 1971, γA and non-γA immune response after oral and parenteral immunization of the hamster, *Cell. Immunol.* **2**:458.

Dreisen, R. B., Schwarz, M. I., Theofipoulos, A. N., and Stanford, R. E., 1978, Circulating immune complexes in idiopathic interstitial pneumonias, *N. Engl. J. Med.* **298**:353.

Drinker, C. K., Warren, M. F., and Machanaham, M., 1937, The absorption of protein solutions from the pulmonary alveoli, J. Exp. Med. **66**:449.

Dym, M., and Fawcett, D. W., 1970, The blood–testis barrier in the rat and the physiological compartmentation of the seminiferous epithelium, *Biol. Reprod.* **3**:308.

Edelson, R. L., 1980, Cutaneous T cell lymphoma: Mycosis fungoides, Sezary syndrome, and other variants, *J. Am. Acad. Dermatol.* **2**:89.

Edwards, P. A., 1978, Is mucus a selective barrier to macromolecules?, *Br. Med. Bull.* **34**:55.

Elson, C. O., Heck, J. A., and Strober, W., 1979, T-cell regulation of murine IgA synthesis, *J. Exp. Med.* **149**:632.

Enander, I., Ahlstedt, S., Kejgren, F. F., and Bjorkstein, B., 1983, Sensitizing ability of derivatives of picryl chloride after exposure of mice on the skin and in the lungs, *Int. Arch. Allergy Appl. Immunol.* **72**:59.

Enander, I., Ahlstedt, S., and Nygren, H., 1984, Mononuclear cells, mast cells and mucous cells as part of the delayed hypersensitivity response to aerosolized antigen in mice, *Immunology* **51**:661.

Endoh, M., Sakai, H., Nomato, Y., Tomino, Y., and Kaneshige, H., 1981, IgA-specific helper activity of T cells in human peripheral blood, *J. Immunol.* **127**:2612.

Engeset, A., Froland, S. S., and Brewer, K., 1974, Low lymphocyte count and few B-lymphocytes in peripheral lymph of patients with CLL, *Scand. J. Haematol.* **13**:93.

Faulk, W. P., McCormick, J. N., Goodman, J. R., Yoffey, J. M., and Fudenberg, H. H., 1971, Peyer's patches: Morphologic studies, *Cell. Immunol.* **1**:500.

Fawcett, D. W., 1979, Interpretation of the sequelae of vasectomy, in: *Vasectomy: Immunologic and Pathophysiologic Effects in Animals and Man* (I. H. Lepow and R. Crozier, eds.), pp. 3–21, Academic Press, New York.

Ferguson, A., and MacDonald, T. T., 1977, Effects of local delayed hypersensitivity in the small intestine, *Ciba Found. Symp.* **46**:305.

Fontana, A., Fierz, W., and Wekerle, H., 1984, Astrocytes present myelin basic protein to encephalitogenic T-cell lines, *Nature (London)* **307**:273.

Ford, W. L., Sedgley, M., Sparshott, M., and Smith, M. E., 1976, The migration of lymphocytes across specialized vascular endothelium. II. The contrasting consequences of treating lymphocytes with trypsin or neuraminidase, *Cell Tissue Kinet.* **9**:351.

Frelinger, J. A., and Frelinger, J. G., 1980, Bone marrow origin of Ia molecules purified from epidermal cells, *J. Invest. Dermatol.* **75**:68.

Fubara, E. S., and Freter, R., 1973, Protection against enteric bacterial infection by secretory IgA antibodies, *J. Immunol.* **111**:395.

Gibbons, R. J., Spinnell, D. M., and Skobe, Z., 1976, Selective adherence as a determinant of the host tropisms of certain indigenous and pathogenic bacteria, *Infect. Immun.* **13**:238.

Gonzalez-Crussi, F., and Boston, R. W., 1972, The absorptive fraction of the neonatal lung: Ultrastructure study of horseradish peroxidase uptake at the onset of ventilation, *Lab. Invest.* **26**:114.

Goodacre, R. L., and Bienenstock, J., 1982, Reduced suppressor cell activity in intestinal lymphocytes from patients with Crohn's disease, *Gastroenterology* **82**:653.

Gowans, J. L., and Knight, E. J., 1964, The route of recirculation of lymphocytes in the rat, *Proc. R. Soc. London Ser. B* **159**:257.

Green, D. R., Gold, J., St. Martin, S., Gershon, R., and Gershon, R. K., 1982, Microenvironmental immunoregulation: Possible role of contrasuppressor cells in maintaining immune responses in gut-associated lymphoid tissues, *Proc. Natl. Acad. Sci. USA* **79**:889.

Green, G., Jakab, J., Low, R., and Davis, G., 1978, Defense mechanisms of the respiratory membrane, in: *Lung Disease, State of the Art* (J. F. Murray, ed.), p. 245–280, American Lung Association, New York.

Griscelli, C., Vasalli, P., and McCluskey, R. T., 1969, The distribution of large dividing lymph node cells in syngeneic recipients after intravenous injection, *J. Exp. Med.* **130**:1427.

Gross, P., and Hatch, T., 1962, Pneumoconiosis the requirements for lymphatic dust transport, *Int. Arch. Gewerbepathol. Gewerbehyg.* **19**:660.

Gross, P., and Westrick, M. C., 1954, The permeability of lung parenchyma to particulate matter, *Am. J. Pathol.* **30**:195.

Guy-Grand, D., Griscelli, C., and Vassalli, P., 1974, The gut-associated lymphoid system: Nature and properties of the large dividing cells, *Eur. J. Immunol.* **4**:435.

Guy-Grand, D., Griscelli, C., and Vassalli, P., 1978, The mouse gut T lymphocyte, a novel type of T cell, *J. Exp. Med.* **148**:1661.

Hall, J. G., 1967, Studies of the cells in the afferent and efferent lymph of lymph nodes draining the site of skin homografts, *J. Exp. Med.* **125**:737.

Hall, J. G., Hopkins, J., and Orlans, E., 1977, Studies on the lymphocytes of sheep. III. Destination of lymph-borne immunoblasts in relation to their tissue of origin, *Eur. J. Immunol.* **7**:30.

Hamilton, S. R., Keren, D. F., Yardley, J. H., and Brown, G., 1981, No impairment of local intestinal immune response to keyhole limpet haemocyanin in the absence of Peyer's patches, *Immunology* **42**:431.

Head, J. R., and Billingham, R. E., 1982, Immunobiological aspects of the maternal–fetoplacental relationship, in: *Clinical Aspects of Immunology*, 4th ed. (P. J. Lachman and D. K. Peters, eds.), pp. 243–282, Blackwell, Oxford.
Head, J. R., and Gaede, S. D., 1986, Ia antigen expression in the rat uterus. *J. Reprod. Immun.* in press.
Head, J. R., and Lande, I. J. M., 1983, Uterine lymphatics: Passage of ink and lymphoid cells from the rat's uterine wall and lumen, *Biol. Reprod.* **28**:941.
Head, J. R., and Seelig, L. L., Jr., 1984, Lymphatic vessels in the uterine endometrium of virgin rats, *J. Reprod. Immunol.* **6**:157.
Head, J. R., Neaves, W. B., and Billingham, R. E., 1983, Immune privilege in the testis. I. Basic parameters of allograft survival, *Transplantation* **36**:423.
Heppleston, A. G., 1963, The disposal of inhaled particulate matter: A unifying hypothesis, *Am. J. Pathol.* **42**:119.
Hersh, T., Flock, M. H., Binder, H. J., Conn, H. O., Prizont, R., and Spiro, H. M., 1970, Disturbance of the jejunal and colonic bacterial flora in immunoglobulin deficiencies, *Am. J. Clin. Nutr.* **23**:1595.
Hoefsmit, E. C. M., Balfour, B. M., and Kamperdigk, E. W. A., 1979, Cells containing Birbeck granules in the lymph and the lymph node, *Adv. Exp. Med. Biol.* **114**:389.
Hohmann, A. W., Schmidt, G., and Rowley, D., 1978, Intestinal colonization and virulence of *Salmonella* in mice, *Infect. Immun.* **22**:763.
Holt, P. G., 1980, Alveolar macrophages. IV. Interspecies differences in activity in proliferating lymphocyte cultures, *Cell. Immunol.* **50**:210.
Holt, P. G., and Batty, J. E., 1980, Comparative studies in the antigen presentation activity of guinea pig and rat alveolar macrophages, *Immunology* **41**:361.
Holt, P. G., and Leivers, S., 1982, Tolerance induction via antigen inhalation isotype specificity, stability and involvement of suppressor T cells, *Int. Arch. Allergy Appl. Immunol.* **67**:155.
Hoover, R. G., and Lynch, R. G., 1983, Isotype-specific suppression of IgA: Suppression of IgA responses in BALB/c mice by Tα cells, *J. Immunol.* **130**:521.
Husband, A. J., and Gowans, J. L., 1978, The origin and antigen-dependent distribution of IgA-containing cells in the intestine, *J. Exp. Med.* **148**:1146.
Inman, L. R., and Cantey, J. R., 1983, Specific adherence of *Escherichia coli* (strain RDEC-1) to membranous (M) cells of the Peyer's patch in *Escherichia coli* diarrhea in the rabbit, *J. Clin. Invest.* **71**:1.
Janossy, G., Tidman, N., and Papagerogiou, E. S., 1980a, Distribution of T lymphocyte subsets in the human bone marrow and thymus: An analysis with monoclonal antibodies, *J. Immunol.* **126**:1608.
Janossy, G., Tidman, N., and Selby, W. S., 1980b, Human T lymphocytes of inducer and suppressor type occupy different microenvironments, *Nature (London)* **288**:81.
Joel, D. D., Laissue, J. A., and LeFevre, M. E., 1978, Distribution and fate of ingested carbon particles in mice, *J. Reticuloendothel. Soc.* **24**:477.
Kagnoff, M. F., 1975, Functional characteristics of Peyer's patch cells. III. Carrier priming of T cells by antigen feeding, *J. Exp. Med.* **142**:1425.
Kagnoff, M. F., 1977, Functional characteristics of Peyer's patch lymphoid cells. IV. Effect of antigen feeding on the frequency of antigen-specific B cells, *J. Immunol.* **118**:992.
Kagnoff, M. F., 1978, Effects of antigen-feeding on intestinal and systemic immune responses. II. Suppression of delayed-type hypersensitivity reactions, *J. Immunol.* **120**:1509.
Kagnoff, M. F., and Campbell, S., 1974, Functional characteristics of Peyer's patch lymphoid cells. I. Induction of humoral antibody and cell-mediated allograft reactions, *J. Exp. Med.* **139**:398.
Kaltreider, B. H., 1976, State of the art: Expression of increased immune mechanisms in the lung, *Am. Rev. Respir. Dis.* **113**:347.
Karol, M., 1983, Concentration-dependent immunologic response to toluene diisocyanate (TDI) following inhalation exposure, *Toxicol. Appl. Pharmacol.* **68**:229.
Karol, M. H., Ioset, H. H., and Riley, E. J., 1978, Hapten specific respiratory hypersensitivity in guinea pigs, *Am. Ind. Hyg. Assoc. J.* **39**:546.

Karol, M. H., Dixon, C., Brady, M., and Alarier, N., 1980, Immunologic sensitization and pulmonary hypersensitivity by repeated inhalation of aromatic isocyanates, *Toxicol. Appl. Pharmacol.* **53**:260.

Katz, S. I., Tamaki, K., and Sachs, D. H., 1979, Epidermal Langerhans cells are derived from cells originating in bone marrow, *Nature (London)* **282**:324.

Kawanishi, H., Saltzman, L. E., and Strober, W., 1982, Characteristics and regulatory function of murine con A-induced, cloned T cells obtained from Peyer's patches and spleen: Mechanisms regulating isotype-specific immunoglobulin production by Peyer's patch B cells, *J. Immunol.* **129**:475.

Kearns, M., and Lala, P. K., 1982, Bone marrow origin of decidual cell precursors in the pseudopregnant mouse uterus, *J. Exp. Med.* **155**:1537.

Keren, D. F., Holt, P. S., Collins, H. H., Gemski, P., and Formal, S. B., 1978, The role of Peyer's patches in the local immune response of rabbit ileum to live bacteria, *J. Immunol.* **120**:1892.

Keren, D. F., Collins, H. H., Gemski, P., Holt, P. S., and Formal, S. B., 1981, Role of antigen form in development of mucosal immunoglobulin A response to *Shigella flexneri* antigens, *Infect. Immun.* **31**:1193.

Kiyono, H., McGhee, J. R., Wannemuehler, M. J., Frangakis, M. V., Spalding, D. M., Michalek, S. M., and Koopman, W. J., 1982a, *In vitro* immune responses to a T cell-dependent antigen by cultures of disassociated murine Peyer's patch, *Proc. Natl. Acad. Sci. USA* **79**:596.

Kiyono, H., McGhee, J. R., Wannemuehler, M. J., and Michalek, S. M., 1982b, Lack of oral tolerance in C3H/HeJ mice, *J. Exp. Med.* **155**:605.

Klareskog, L., Tjernlund, U. M., Forsum, U., and Peterson, P. A., 1977, Epidermal Langerhans cells express Ia antigen, *Nature (London)* **265**:248.

Kripke, M. L., 1983, Immunobiology of UV-radiation-induced skin cancer, in: *Photoimmunology* (J. A. Parrish, M. L. Kripke, and W. L. Morison, eds.), p. 157–173, Plenum Medical Book Co., New York.

Lake, A. M., Bloch, R. J., Neutra, M. R., and Walker, W. A., 1979, Intestinal goblet cell mucus release. II. *In vivo* stimulation by antigen in the immunized rat, *J. Immunol.* **122**:834.

Lamm, M. E., 1976, Cellular aspects of immunoglobulin A, *Adv. Immunol.* **22**:223.

Lande, I. J. M., Head, J. R., and Billingham, R. E., 1981, Immune reactivity induced by antigenic exposure in the rat uterus, *Transplant. Proc.* **13**:1256.

Lange, S., Hansson, H., Molin, S., and Nygren, H., 1979, Local cholera immunity in mice: Intestinal antitoxin-containing cells and their correlation with protective immunity, *Infect. Immun.* **23**:743.

Langerhans, P., 1876, Uber die Nerven di menschlichen Haut, *Virchows Arch. A* **44**:325.

Lause, D. B., and Bockman, D. E., 1981, Heterogeneity, position, and functional capability of the macrophages in Peyer's patches, *Cell Tissue Res.* **218**:557.

Lauweryns, J., and Baert, J. H., 1974, The role of pulmonary lymphatics in the defenses of the distal lung: Morphological and experimental studies on the transport mechanisms of intratracheally instilled particles, *Ann. N.Y. Acad. Sci.* **221**:244.

Lauweryns, J. M., and Baert, J. H., 1978, Alveolar clearance and the role of the pulmonary lymphatics, in: *Lung Disease, State of the Art* (J. F. Murray, ed.), p. 185–243, American Lung Association, New York.

Lee, K. C., Wong, M., and Spitzer, D., 1982, Chloroquine as a probe of antigen processing by accessory cells, *Transplantation* **34**:150.

LeFevre, M. E., Olivo, R., Vanderhoff, J. W., and Joel, D. D., 1978, Accumulation of latex in Peyer's patches and its subsequent appearance in villi and mesenteric lymph nodes, *Proc. Soc. Exp. Biol. Med.* **159**:298.

LeFevre, M. E., Hammer, R., and Joel, D. D., 1979, Macrophages of the mammalian small intestine: A review, *J. Reticuloendothel. Soc.* **26**:553.

Lemmel, E. M., and Fichtelius, K. E., 1971, Lifespan of lymphocytes within intestinal epithelium, Peyer's patch epithelium, epidermis, and liver of mice, *Int. Arch. Allergy Appl. Immunol.* **41**:716.

Liden, S., 1967, The mononuclear-cell infiltrate in allergic contact dermatitis. 3. Selective accumulation of cells from lymph nodes, *Acta Pathol. Microbiol. Scand.* **70**:363.

Lopez, M., and Salvaggio, J., 1976, Hypersensitivity pneumonitis: Current concepts of etiology and pathogenesis, *Annu. Rev. Med.* **27**:453.

Lutzner, M., Edelson, R., and Schein, P., 1975, Cutaneous T-cell lymphoma: The Sezary syndrome, mycosis fungoides, and related disorders, *Ann. Intern. Med.* **83**:534.

Lynch, D. H., Gurish, M. F., and Daynes, R. A., 1981, Relationship between epidermal Langerhans cell density, ATPase activity, and the induction of contact hypersensitivity, *J. Immunol.* **126**:1891.

Lyons, C. R., and Lipscomb, M. F., 1983, Alveolar macrophages in pulmonary immune responses. I. Role in the initiation of primary immune responses and in the selective recruitment of T lymphocytes to the lung, *J. Immunol.* **130**:1113.

Lyscom, N., and Brueton, M. J., 1982, Intraepithelial, lamina propria, and Peyer's patch lymphocytes of the rat small intestine: Isolation and characterization in terms of immunoglobulin markers and receptors for monoclonal antibodies, *Immunology* **45**:775.

McCombs, C. C., Michalski, J. P., Westerfield, B. T., and Light, R. W., 1982, Human alveolar macrophages suppress the proliferative response of peripheral blood lymphocytes, *Chest* **82**: 266.

McDermott, M. R., and Bienenstock, J., 1979, Evidence for a common mucosal immunologic system. I. Migration of B immunoblasts into intestinal, respiratory, and genital tissues, *J. Immunol.* **122**:1892.

McDermott, M. R., Clark, D. A., and Bienenstock, J., 1980, Evidence for a common mucosal immunologic system. II. Influence of the estrous cycle on B immunoblast migration into genital and intestinal tissues, *J. Immunol.* **124**:2536.

McDermott, M. R., Befus, A. D., and Bienenstock, J., 1983, The structural basis of immunity in the respiratory tract, in: *International Review of Experimental Pathology* (G. W. Ricther and M. A. Epstein, eds.), p. 47–112, Academic Press, New York.

MacDonald, T. T., and Carter, P. B., 1978, Mouse Peyer's patches contain T cells capable of inducing the graft-versus-host reaction (GVHR), *Transplantation* **26**:162.

MacDonald, T. T., and Carter, P. B., 1982, Isolation and functional characteristics of adherent phagocytic cells from mouse Peyer's patches, *Immunology* **45**:769.

Macher, E., and Chase, M. W., 1969a, Studies on the sensitization of animals with simple chemical compounds. XI. Fate of labeled picrylchloride and dinitrochlorobenzene after sensitizing injections, *J. Exp. Med.* **129**:81.

Macher, E., and Chase, M. W., 1969b, Studies on the sensitization of animals with simple chemical compounds. XII. Influence of excision of allergenic depots on onset of delayed hypersensitivity and tolerance, *J. Exp. Med.* **129**:103.

McLean, J. M., Shaya, E. I., and Gibbs, A. C. C., 1980, Immune responses to first mating in the female rat, *J. Reprod. Immunol.* **1**:285.

McWilliams, M., Phillips-Quagliata, J. M., and Lamm, M. E., 1975, Characteristics of mesenteric lymph node cells homing to gut-associated lymphoid tissue in syngeneic mice, *J. Immunol.* **115**:54.

McWilliams, M., Phillips-Quagliata, J. M., and Lamm, M. E., 1977, Mesenteric lymph node B lymphoblasts which home to small intestine are precommitted to IgA synthesis, *J. Exp. Med.* **145**:866.

Marks, R., 1983, Mechanical properties of the skin, in: *Biochemistry and Physiology of the Skin* (L. A. Goldsmith, ed.), p. 1237–1254, Oxford University Press, London.

Marneerushapisal, V., and Rowley, D., 1981, Local cell-associated immunity in the Peyer's patches of mouse intestines, *Infect. Immun.* **33**:338.

Maroni, E. S., and de Sousa, M. A. B., 1973, The lymphoid organs during pregnancy in the mouse: A comparison between a syngeneic and an allogenic mating, *Clin. Exp. Immunol.* **31**:107.

Marsh, L. D., and Alexander, N. J., 1982, Vasectomy: Effects on the efferent ducts in *Macaca mulatta*, *Am. J. Pathol.* **107**:310.

Mattingly, J. A., and Waksman, B. H., 1978, Immunologic suppression after oral administration of antigen. I. Specific suppressor cells formed in rat Payer's patches after oral administration of sheep erythrocytes and their systemic migration, *J. Immunol.* **121**:1878.

Mattingly, J. A., Kaplan, J. M., and Janeway, C. A., Jr., 1980, Two distinct antigen-specific suppressor factors induced by the oral administration of antigen, *J. Exp. Med.* **152:**545.

Mayrhofer, G., Pugh, C. W., and Barclay, A. N., 1983, The distribution, ontogeny, and origin in the rat of Ia positive cells with dendritic morphology and of Ia antigen in epithelia, with special reference to the intestine, *Eur. J. Immunol.* **13:**112.

Mestecky, J., McGhee, J. R., Arnold, R. R., Michalek, S. M., Prince, S. J., and Babb, J. L., 1978, Selective induction of an immune response in human secretions by ingestion of bacterial antigen, *J. Clin. Invest.* **61:**731.

Michalek, S. M., McGhee, J. R., Mestecky, J., Arnold, R. R., and Bozzo, L., 1976, Ingestion of *Streptococcus mutans* induces secretory immunoglobulin A and caries immunity, *Science* **192:**1238.

Miller, H. R. P., and Adams, E. P., 1977, Reassortment of lymphocytes in lymph from normal and allografted sheep, *Am. J. Pathol.* **57:**59.

Miller, H. R. P., and Nawa, Y., 1979, *Nippostrongylus brasiliensis:* Intestinal goblet-cell response in adoptively immunized rats, *Exp. Parasitol.* **47:**81.

Miller, S., and Claman, H., 1976, The induction of hapten-specific T cell tolerance by using hapten-modified lymphoid cells. I. Characteristics of tolerance induction, *J. Immunol.* **117:**1519.

Mirski, S., McDermott, M. R., Befus, A. D., and Bienenstock, J., 1981, Selective localization of mesenteric lymphoblasts in mucosal tissues: Effects of altering the number of donor lymphoblasts, *Immunology* **43:**669.

Morrow, P. E., 1973, Alveolar clearance of aerosols, *Arch. Intern. Med.* **131:**101.

Mowat, A. M., and Ferguson, A., 1981, Hypersensitivity in the small intestinal mucosa. V. Induction of cell-mediated immunity to a dietary antigen, *Clin. Exp. Immunol.* **43:**574.

Mowat, A. M., Strobel, S., Drummond, H. E., and Ferguson, A., 1982, Immunological responses to fed protein antigens in mice. I. Reversal of oral tolerance to ovalbumin by cyclophosphamide, *Immunology* **45:**105.

Murdoch, A. J. M., Buckley, C. H., and Fox, H., 1982, Hormonal control of the secretory immune system of the human uterine cervix, *J. Reprod. Immunol.* **4:**23.

Nadler, P. I., Klingenstein, R. J., Richman, L. K., and Ahmann, G. B., 1980, The murine Kupffer cell. II. Accessory cell function in *in vitro* primary antibody responses, mitogen-induced proliferation, and stimulation of mixed lymphocyte responses, *J. Immunol.* **125:**2521.

Nash, D., and Holle, B., 1973, Local and systemic cellular immune responses in guinea pigs given antigen parenterally or directly into the lower respiratory tract, *Clin. Exp. Immunol.* **13:**573.

Ngan, J., and Kind, L. S., 1978, Suppressor T cells of IgE and IgG in Peyer's patches of mice made tolerant by the oral administration of ovalbumin, *J. Immunol.* **120:**861.

Ogra, P. L., and Ogra, S. S., 1973, Local antibody response to polio vaccine in the human female genital tract, *J. Immunol.* **110:**1307.

Ogra, P. L., Fishaut, M., and Gallagher, M. R., 1980, Viral vaccination via the mucosal routes, *Rev. Infect. Dis.* **2:**352.

Oppenheim, J. J., Mizel, S. B., and Meltzer, M. S., 1979, Biological effects of lymphocyte- and macrophage-derived mitogenic "amplification" factors, in: *Biology of the Lymphokines* (S. Cohen, E. Pick, and J. J. Oppenheim, eds.), p. 79, Academic Press, New York.

Ottaway, C. A., Rose, M. L., and Parrott, D. M. V., 1979, The gut as an immunological system, *Int. Rev. Physiol.* **19:**323.

Owen, R. L., 1977, Sequential uptake of horseradish peroxidase by lymphoid follicle epithelium of Peyer's patches in the normal unobstructed mouse intestine: An ultrastructural study, *Gastroenterology* **72:**440.

Owen, R. L., and Jones, A. L., 1974, Epithelial cell specialization within human Peyer's patches: An ultrastructural study of intestinal lymphoid follicles, *Gastroenterology* **66:**189.

Owen, R. L., and Nemanic, P., 1978, Antigen processing structures of the mammalian intestinal tract: An SEM study of lymphoepithelial organs, *Scanning Electron Microsc.* **2:**367.

Owen, R. L., Nemanic, P. C., and Stevens, D. P., 1979, Ultrastructural observations on giardiasis in a murine model, *Gastroenterology* **76:**757.

Owen, R. L., Allen, C. L., and Stevens, D. P., 1981, Phagocytosis of *Giardia muris* by macrophages in Peyer's patch epithelium in mice, *Infect. Immun.* **33:**591.

Parmely, M. J., and Beer, A. E., 1977, Colustral cell mediated immunity and the concept of a common secretory immune system, *J. Diary Sci.* **60**:655–665.

Patterson, R., Zeiss, C. R., Roberts, R. Pruzansky, J. J., and Wolkonsky, P., 1978, Human antihapten antibodies in trimellitic anhydride inhalation reactions: Immunoglobulin classes of antitrimellitic antibodies and hapten inhibition studies, *J. Clin. Invest.* **62**:97.

Pavia, C., Siiteri, P. K., Perlman, J. D., and Stites, D. P., 1979, Suppression of murine allogeneic cell interactions by sex hormones, *J. Reprod. Immunol.* **1**:33.

Pierce, N. F., 1978, The role of antigen form and function in the primary and secondary intestinal immune responses to cholera toxin and toxoid in rats, *J. Exp. Med.* **148**:195.

Pierce, N. F., and Koster, F. T., 1980, Priming and suppression of the intestinal immune response to cholera toxoid/toxin by parenteral toxoid in rats, *J. Immunol.* **124**:307.

Pierce, N. F., Cray, W. C., Jr., and Engel, P. F., 1980, Antitoxic immunity to cholera in dogs immunized orally with cholera toxin, *Infect. Immun.* **27**:632.

Plesch, B., Gamelkoorn, G. J., and van de Ende, M., 1983, Development of bronchus associated lymphoid tissue (Balt) in the rat with special reference to T and B cells, *Dev. Comp. Immunol.* **7**:179.

Ptak, W., Rozycka, D., and Askenase, P. W., 1980, Role of antigen-presenting cells in the development and persistence of contact hypersensitivity, *J. Exp. Med.* **151**:362.

Pugh, C. W., MacPherson, G. G., and Steer, H. W., 1983, Characterization of nonlymphoid cells derived from rat peripheral lymph, *J. Exp. Med.* **157**:1758.

Racz, P., Tenner-Racz, K., Myrvik, Q. N., and Fainter, L. K., 1977, Functional architecture of bronchial associated lymphoid tissue and lymphoepithelium in pulmonary cell-mediated reactions in the rabbit, *J. Reticuloendothel. Soc.* **22**:59.

Ramseier, H., and Streilen, J. W., 1965, Homograft sensitivity reactions in irradiated hamsters, *Lancet* **1**:622.

Rebello, R., Green, F. H. Y., and Fox, H., 1975, A study of the secretory immune system of the female genital tract, *Br. J. Obstet. Gynaecol.* **82**:812.

Reynolds, M. Y., Fulmer, J. D., and Kazmerowski, J., 1977, Analysis of cellular protein content of bronchoalveolar lavage fluid from patients with idiopathic pulmonary fibrosis and chronic hypersensitivity pneumonitis, *J. Clin. Invest.* **59**:165.

Richardson, J. B., and Bouchard, T., 1976, Demonstration of a nonadrenergic inhibitory nervous system in the trachea of the guinea pig, *J. Allergy Clin. Immunol.* **56**:473.

Richman, L. K., Chiller, J. M., Brown, W. R., Hanson, D. G., and Vaz, N. M., 1978, Enterically induced immunologic tolerance. I. Induction of suppressor T lymphocytes by intragastric administration of soluble proteins, *J. Immunol.* **121**:2429.

Richman, L. K., Klingenstein, R. J., Richman, J. A., Strober, W., and Berzofsky, J. A., 1979, The murine Kupffer cell. I. Characterization of the cell serving accessory function in antigen-specific T cell proliferation, *J. Immunol.* **123**:2602.

Richman, L. K., Graeff, A. S., Yarchoan, R., and Strober, W., 1981, Simultaneous induction of antigen-specific helper T cells and IgG suppressor T cells in the murine Peyer's patch after protein feeding, *J. Immunol.* **126**:2079.

Rose, M. L., Parrott, D. M. V., and Bruce, R. G., 1976a, Migration of lymphoblasts to the small intestine. I. Effect of *Trichinella spiralis* infection on the migration of mesenteric lymphoblasts and mesenteric T lymphoblasts in syngeneic mice, *Immunology* **31**:723.

Rose, M. L., Parrott, D. M. V., and Bruce, R. G., 1976b, Migration of lymphoblasts to the small intestine, *Immunology* **31**:723.

Roux, M. E., McWilliams, M., Phillips-Quagliata, J. M., Weisz-Carrington, P., and Lamm, M. E., 1977, Origin of IgA-secreting plasma cells in the mammary gland, *J. Exp. Med.* **146**:1311.

Rowden, G., Phillips, T. M., and Delovitch, T. L., 1978, Expression of Ia antigens by murine keratinizing epithelial Langerhans cells, *Immunogenetics* **7**:465.

Sauder, D. N., Tamaki, K., and Moshell, A. N., 1981, Induction of tolerance to topically applied TNCB using TNP-conjugated ultraviolet light-irradiated epidermal cells, *J. Immunol.* **127**: 261.

Savage, D. C., 1977, Microbial ecology of the gastrointestinal tract, *Annu. Rev. Microbiol.* **31**:107.

Scherzer, H., and Ward, P. A., 1978, Lung injury produced by immune complexes of varying composition, *J. Immunol.* **121**:947.
Scheuplein, R. J., and Bronaugh, R. L., 1983, Percutaneous absorption, in: *Biochemistry and Physiology of the Skin* (L. A. Goldsmith, ed.), pp. 1255–1295, Oxford University Press, London.
Schneeberger-Keeley, E. E., and Karnovsky, M. J., 1968, The structure basis of alveolar-capillary membrane permeability to peroxidase used as a tracer, *J. Cell Biol.* **37**:781.
Schreiner, G. F., and Cotran, R. S., 1982, Localization of an Ia-bearing glomerular cell in the mesangium, *J. Cell Biol.* **94**:483.
Schreiner, G. F., Kiely, J.-M., Cotran, R. S., and Unanue, E. R., 1981, Characterization of resident glomerular cells expressing Ia determinants and manifesting genetically restricted interactions with lymphocytes, *J. Clin. Invest.* **68**:920.
Schultz, A. C., Grisner, J. T., and Grande, F., 1963, Absorption of radioactive albumin from lungs of normal dogs, *J. Lab. Clin. Med.* **61**:494.
Schultz, A. C., Grisner, J. T., Wada, S., and Grande, F., 1964, Absorption of albumin from alveoli of perfused dog lungs, *Am. J. Physiol.* **207**:1300.
Schuyler, M. R., and Todd, L. S., 1981, Accessory cell function of rabbit alveolar macrophages, *Am. Rev. Respir. Dis.* **123**:53.
Scollay, R., Hall, J. R., and Orlans, E., 1976, Studies of the lymphocytes of sheep. II. Some properties of cells in various compartments of the recirculating lymphocyte pool, *Eur. J. Immunol.* **6**:121.
Searle, R. F., Bell, S. C., and Billington, W. D., 1983, Ia antigen-bearing decidual cells and macrophages in cultures of mouse decidual tissue, *Placenta* **4**:139.
Sell, S., and Fahey, J. L., 1964, Relationship between γ-globulin metabolism and low serum γ-globulin in germ-free mice, *J. Immunol.* **93**:81.
Sertl, K., Takemura, T., Tschachler, E., Ferrans, V., Kaliner, M. O., and Shevach, E., 1986, Dendritic cells with antigen-presenting capability reside in airway epithelium, lung parenchyma, and visceral pleura, *J. Exp. Med.* **163**:436.
Silberberg, I., 1973, Apposition of mononuclear cells to Langerhans cells in contact allergic reactions, *Acta Derm. Venereol.* **53**:1.
Silberberg, I., Baer, R. L., and Rosenthal, S. A., 1976, The role of Langerhans cells in allergic contact hypersensitivity: A review of findings in man and guinea pigs, *J. Invest. Dermatol.* **66**:210.
Slapsys, R., and Clark, D., 1983, Active suppression of host-versus-graft reaction in pregnant mice. V. Kinetics, specificity, and in vivo activity of non-T suppressor cells localized to the genital tract of mice during first pregnancy, *Am. J. Reprod. Immunol.* **3**:65.
Smith, J. B., McIntosh, G. H., and Morris, B., 1970, The traffic of cells through tissues: A study of peripheral lymph in sheep, *J. Anat.* **107**:87.
Smith, M. W., and Peacock, M. A., 1980, "M" cell distribution in follicle-associated epithelium of mouse Peyer's patch, *Am. J. Anat.* **159**:167.
Sminia, T., Janse, E. M., and Wilders, M. M., 1982, Antigen-trapping cells in Peyer's patches of the rat, *Scand. J. Immunol.* **16**:481.
Sminia, T., Wilders, M. M., Janse, E. M., and Hoefsmit, E. C. M., 1983, Characterization of non-lymphoid cells in Peyer's patches of the rat, *Immunobiology* **164**:136.
Sobhon, P., 1971, The light and the electron microscopic studies of Peyer's patches in non germ-free adult mice, *J. Morphol.* **135**:457.
Solomon, J. B., 1963, Ontogenesis of immunological competence of lymphocytes in chicken skin, *Transplantation* **1**:327.
Sorokin, S. P., and Brain, J. D., 1975, Pathways of clearance in mouse lungs exposed to iron oxide aerosols, *Anat. Rec.* **181**:581.
Spalding, D. M., Koopman, W. J., Eldridge, J. H., McGhee, J. R., and Steinman, R. M., 1983, Accessory cells in murine Peyer's patch. I. Identification and enrichment of a functional dendritic cell, *J. Exp. Med.* **157**:1646.
Spry, C. J. F., Pflug, J., and Janossy, G., 1980, "Veiled" cells with "Ia-like" membrane antigens in human afferent lymph, *Clin. Exp. Immunol.* **39**:750.
Staples, L. D., Fleet, I. R., and Heap, R. B., 1982, Anatomy of the utero-ovarian lymphatic network and the composition of afferent lymph in relation to the establishment of pregnancy in the sheep and goat, *J. Reprod. Fertil.* **64**:409.

Steinman, R. M., and Nussenzweig, M. C., 1980, Dendritic cells, features, and function, *Immunol. Rev.* **53**:127.
Stein-Streilein, J., 1983, Allergic contact dermatitis induced by intratracheal administration of hapten, *J. Immunol.* **131**:1748.
Stein-Streilein, J., Frazier, J., Gross, G. N., and Hart, D. A., 1979a, Splenic influence on the development of a local pulmonary immune response, *Infect. Immun.* **24**:139.
Stein-Streilein, J., Gross, G. N., and Hart, D. A., 1979b, Comparison of intratracheal and intravenous inoculation of sheep erythrocytes in the induction of local and systemic immune responses, *Infect. Immun.* **24**:145.
Stein-Streilein, J., Lipscomb, M., Hart, D., and Darden, A., 1981, Graft versus host reactions in the lung, *Transplantation* **32**:38.
Stingl, G., Wolff-Schreiner, E. C., Pichler, W. J., Gschnait, F., Knapp, W., and Wolff, W., 1977, Epidermal Langerhans cells bear Fc and C3 receptors, *Nature (London)* **268**:245.
Stingl, G., Katz, S. I., Shevach, E. M., Rosenthal, A. S., and Green, I., 1978a, Analogous functions of macrophages and Langerhans cells in the initiation of the immune response, *J. Invest. Dermatol.* **71**:59.
Stingl, G., Katz, S. I., Clement, L., Green, I., and Shevach, E. M., 1978b, Immunologic functions of Ia-bearing epidermal Langerhans cells, *J. Immunol.* **121**:2005.
Stoub, N. C., 1970, The pathophysiology of pulmonary edema, *Hum. Pathol.* **1**:419.
Streilein, J. W., 1978, Lymphocyte traffic, T cell malignancies and the skin, *J. Invest. Dermatol.* **71**:167.
Streilein, J. W., 1983, Skin-associated lymphoid tissues (SALT): Origins and functions, *J. Invest. Dermatol.* **80**:012s.
Streilein, J. W., 1985, Circuits and signals of the skin associated lymphoid tissues (SALT), *J. Invest. Dermatol.* **85**:10S.
Streilein, J. W., Toews, G. B., and Bergstresser, P. R., 1979, Corneal allografts fail to express Ia antigens, *Nature (London)* **282**:326.
Streilein, J. W., Lonsberry, L. W., and Bergstresser, P. R., 1982, Depletion of epidermal Langerhans cells and Ia immunogenicity from tape-stripped mouse skin, *J. Exp. Med.* **155**:863.
Streilein, J. W., Wood, P. J., Lonsberry, L. W., and Bergstresser, P. R., 1984, Hapten-derivatized skin grafts induce H-2 restricted contact hypersensitivity: Evidence that the immunogenic signal includes H-2 determinants derived from the skin, *Transplantation* **37**:195.
Strober, W., Richman, L. K., and Elson, C. O., 1981, The regulation of gastrointestinal immune responses, *Immunol. Today* **2**:156.
Strombeck, D. R., and Harrold, D., 1974, Binding of cholera toxin to mucins and inhibition by gastric mucin, *Infect. Immun.* **10**:1266.
Stuart, B. O., 1973, Deposition of inhaled aerosols, *Arch. Intern. Med.* **131**:60.
Sullivan, D. A., and Wira, C. R., 1981, Estradiol regulation of secretory component in the female reproductive tract, *J. Steroid Biochem.* **15**:439.
Swarbrick, E. T., Stokes, C. R., and Soothill, J. F., 1979, Absorption of antigens after oral immunization and the simultaneous induction of specific systemic tolerance, *Gut* **20**:121.
Szewczuk, M. R., and Campbell, R. J., 1981, Differential effect of aging on the heterogeneity of the immune response to a T-dependent antigen in systemic and mucosal-associated lymphoid tissues, *J. Immunol.* **126**:472.
Szidon, J. P., Pietra, G. G., and Fishman, A. P., 1972, The alveolar-capillary membrane and pulmonary edema, *N. Engl. J. Med.* **286**:1200.
Tagliabue, A., Luini, W., Soldateschi, D., and Boraschi, D., 1981, Natural killer activity of gut mucosal lymphoid cells in mice, *Eur. J. Immunol.* **11**:919.
Tagliabue, A., Nencioni, L., Villa, L., Keren, D. F., Howell, G. H., and Boraschi, D., 1983, Antibody-dependent cell-mediated antibacterial activity of intestinal lymphocytes with secretory IgA, *Nature (London)* **306**:184.
Tamaki, K., Fujiwara, H., and Katz, S. I., 1981, The role of epidermal cells in the induction and suppression of contact sensitivity, *J. Invest. Dermatol.* **76**:275.
Targowski, S. P., and Foley, J., 1983, Inflammatory responses of the uterus to infusions with antigens, *Am. J. Reprod. Immunol.* **3**:198a.
Tenner-Racz, K., Racz, P., Myrivik, Q. N., Ockers, J. R., and Geester, R., 1979, Uptake and trans-

port of horseradish peroxidase by lymphoepithelium of the bronchus-associated lymphoid tissue in normal and bacillus Calmette-Guerin-immunized and challenged rabbits, *Lab. Invest.* **41**:106.

Thomas, H. C., and Parrott, D. M. V., 1974, The induction of tolerance to a soluble protein antigen by oral administration, *Immunology* **27**:631.

Thomas, W. R., Holt, P. G., and Keast, D., 1974, Local and systemic response of mice after intratracheal and intravenous inoculation of sheep erythrocytes, *Int. Arch. Allergy Appl. Immunol.* **46**:487.

Thorbecke, G. J., Silberberg-Sinakin, I., and Flotte, T. J., 1980, Langerhans cells as macrophages in skin and lymphoid organs, *J. Invest. Dermatol.* **75**:32.

Toews, G. B., Bergstresser, P. R., and Streilein, J. W., 1980, Epidermal Langerhans cell density determines whether contact hypersensitivity or unresponsiveness follows skin painting with DNFB, *J. Immunol.* **124**:445.

Toews, G. B., Vial, W. C., Dunn, M. M., Guzzetta, P., Nunez, G., Stasny, P., and Lipscomb, M. F., 1984, The accessory cell function of human alveolar macrophages in specific T cell proliferation, *J. Immunol.* **132**:181.

Tomasi, T. B., Jr., 1980, Oral tolerance, *Transplantation* **29**:353.

Tomasi, T. B., and Bienenstock, J., 1968, Secretory immunoglobulins, *Adv. Immunol.* **9**:1.

Tschachler, E., Schuler, G., Hutterer, J., Leibl, H., Wolff, K., and Stingl, G., 1983, Expression of Thy-1 antigen by murine epidermal cells, *J. Invest. Dermatol.* **81**:282.

Tseng, J., 1982, Expression of immunoglobulin heavy chain isotypes by Peyer's patch lymphocytes stimulated with mitogens in culture, *J. Immunol.* **128**:2719.

Tung, K. S. K., and Alexander, N. J., 1980, Monocytic orchitis and aspermatogenesis in normal and vasectomized rhesus macaques (*Macacca mulatta*), *Am. J. Pathol.* **101**:17.

Turner-Warwick, M., 1980, Widespread pulmonary fibrosis, in: *Pulmonary Disease and Disorders* (A. Fishman, ed.), pp. 657–688, McGraw-Hill, New York.

Unanue, E. R., 1981, The regulatory role of macrophages in antigenic stimulation. II. Symbiotic relationship between lymphocytes and macrophages, *Adv. Immunol.* **31**:1.

Unanue, E. R., 1984, Regulatory functions of mononuclear phagocytes, *Prog. Immunol.* **5**:973.

Unanue, E. R., Beller, D. I., Lu, C. Y., and Allen, P. M., 1984, Antigen presentation: Comments on its regulation and mechanism, *J. Immunol.* **132**:1.

Vai, F., Fournier, M., and Pariente, R., 1977, Bronchial lympho-epithelial nodules in the rat: Definition and morphological characteristics in optical and electron microscopy, *Biomedicine* **26**:130.

Vickery, B. H., and Bennett, J. P., 1968, The cervix and its secretions in mammals, *Physiol. Rev.* **48**:135.

Waldman, R. H., and Henney, C. S., 1971, Cell-mediated immunity and antibody responses in the respiratory tract after local and systemic immunization, *J. Exp. Med.* **134**:482.

Walker, W. A., and Isselbacher, K. J., 1977, Intestinal antibodies, *N. Engl. J. Med.* **297**:767–773.

Walker, W. A., Wu, M., Isselbacher, K. J., and Bloch, K. J., 1975, Intestinal uptake of macromolecules. III. Studies on the mechanism by which immunization interferes with antigen uptake, *J. Immunol.* **115**:854.

Walker, W. A., Wu, M., and Bloch, K. J., 1977, Stimulation by immune complexes of mucus release from goblet cells of the rat small intestine, *Science* **197**:370.

Walls, R. S., Carter, R. L., Leuchars, E., and Davis, A. J. S., 1973, The immunopathology of trichiniasis in T cell deficient mice, *Clin. Exp. Immunol.* **13**:231.

Wannemuehler, M. J., Kiyono, H., Babb, J. L., Michalek, S. M., and McGhee, J. R., 1982, Lipopolysaccharide (LPS) regulation of the immune response: LPS converts germfree mice to sensitivity to oral tolerance induction, *J. Immunol.* **129**:959.

Warner, L. A., Holt, P. G., and Mayrhofer, G., 1981, Alveolar macrophages. VI. Regulation of alveolar macrophage mediated suppression of lymphocyte proliferation by a putative T cell, *Immunology* **42**:137.

Warshaw, A. L., Walker, W. A., Cornell, R., and Isselbacher, K. J., 1971, Small intestinal permeability to macromolecules: Transmission of horseradish peroxidase into mesenteric lymph and portal blood, *Lab. Invest.* **25**:675.

Warshaw, A. L., Walker, W. A., and Isselbacher, K. J., 1974, Protein uptake by the intestine: Evidence for absorption of intact macromolecules, *Gastroenterology* **66**:987.

Weinberger, S. E., Kelman, J. A., and Elson, N. A., 1978, Bronchoalveolar lavage in interstitial lung disease, *Ann. Intern. Med.* **89**:459.

Weisz-Carrington, P., Roux, M. E., and Lamm, M. E., 1977, Plasma cells and epithelial immunoglobulins in the mouse mammary gland during pregnancy and lactation, *J. Immunol.* **119**:1306.

Weisz-Carrington, P., Roux, M. E., McWilliams, M., Phillips-Quagliata, J. M., and Lamm, M. E., 1978, Hormonal induction of the secretory immune system in the mammary gland, *Proc. Natl. Acad. Sci. USA* **75**:2928.

Weisz-Carrington, P., Roux, M. E., McWilliams, M., Phillips-Quagliata, J. M., and Lamm, M. E., 1979, Organ and isotype distribution of plasma cells producing specific antibody after oral immunization: Evidence for a generalized secretory immune system, *J. Immunol.* **123**:1705.

Whitcomb, M., 1982, *The Lung, Normal and Diseased*, pp. 3–10, Mosby, St. Louis.

Williams, R. C., and Gibbons, R. J., 1972, Inhibition of bacterial adherence by secretory immunoglobulin A: A mechanism of antigen disposal, *Science* **177**:697.

Wira, C. R., and Sandoe, C. P., 1977, Sex steroid hormone regulation of IgA and IgG in rat uterine secretions, *Nature (London)* **268**:534.

Wira, C. R., and Sandoe, C. P., 1980, Hormonal regulation of immunoglobulins: Influence of estradiol on immunoglobulins A and G in the rat uterus, *Endocrinology* **106**:1020.

Wolf, J. L., Rubin, D. H., Finberg, R., Kauffman, R. S., Sharpe, A. H., Trier, J. S., and Fields, B. N., 1981, Intestinal M cells: A pathway for entry of reovirus into the host, *Science* **212**:471.

Wolff, K., 1972, The Langerhans cell, *Curr. Probl. Dermatol.* **4**:79.

Woodruff, J., 1974, Role of lymphocyte surface determinants in lymph node homing, *Cell Immunol.* **13**:378.

Wu, A., and Walker, W. A., 1976, Immunological control mechanism against cholera toxin: Interference with toxin binding to intestinal receptors, *Infect. Immun.* **14**:1034.

Yeager, H. J., Williams, M. C., and Beckman, J. F., 1977, Sarcoidosis: Analysis of cells obtained by bronchial lavage, *Am. Rev. Respir. Dis.* **116**:951.

Yeager, H., Jr., Sweeney, J. A., Herscowitz, H. B., Barsoum, I. J., and Kagan, E., 1982, Modulation of mitogen-induced proliferation of autologous peripheral blood lymphocytes by human alveolar macrophages, *Infect. Immun.* **38**:260.

Yoffey, M. Y., and Courtice, F. C., 1970, *Lymphatics, Lymph and the Lymphomyeloid Complex*, pp. 283–295, Academic Press, New York.

Younces, M. S., Robertson, E. M., Bencosme, S. A., 1968, Electron microscope observations on Langerhans cells in the cervix, *American J. Ob. and Gyn.* **102**:397.

4

Structure and Expression of IgE Receptors

ANTHONY KULCZYCKI, JR.

1. INTRODUCTION

Cell surface receptors with high affinity for IgE are present only on mast cells, basophils, and neoplastic counterparts of these cells. Monomeric IgE is bound to these receptors and displayed on the mast cell/basophil surface. Subsequently, multivalent antigen may bind to adjacent receptor-bound IgE molecules. The primary role of these IgE receptors appears to be transducing this cross-linking signal in order to initiate mediator release. The specificity of these IgE receptors (or Fc_ϵ receptors) on basophils for IgE was first shown in studies with human peripheral blood leukocytes (Ishizaka *et al.*, 1970; Sullivan *et al.*, 1971). More detailed studies of the structure, function, and expression of the high-affinity Fc_ϵ receptors have been made possible by use of the rat basophilic leukemia (RBL) cell line (Kulczycki *et al.*, 1974).

Many cell types express receptors which bind IgE with lower affinity. Since monomeric IgE dissociates rapidly from such receptors, it is presumed that usually an antigen–IgE complex is formed initially. Subsequently, the complex binds via multipoint attachment to multiple lower-affinity receptors and activates the cell. Lower-affinity Fc_ϵ receptors appear to be involved in cellular cytotoxicity against parasites and in regulation of IgE synthesis (Spiegelberg, 1983).

2. INTERACTION OF IgE WITH HIGH-AFFINITY Fc_ϵ RECEPTORS ON MAST CELLS AND BASOPHILS

IgE binding to the cell surface is a temperature-dependent and reversible reaction (Kulczycki *et al.*, 1974; Kulczycki and Metzger, 1974). The interaction of

ANTHONY KULCZYCKI, JR. • Department of Internal Medicine, Division of Allergy and Immunology, Washington University School of Medicine, St. Louis, Missouri 63110.

rat IgE with the rat Fc_ϵ receptor appears to be consistent with the simple bimolecular forward reaction and the first-order reverse reaction:

$$[\text{IgE}] + [\text{receptor}] \underset{k_{-1}}{\overset{k_1}{\rightleftharpoons}} [\text{IgE–receptor complex}]$$

where the forward rate constant, k_1, is approximately 1×10^5 liters/mole-sec at 37°C (Kulczycki and Metzger, 1974). This rather ordinary k_1 value together with the usually low concentrations of IgE explains why IgE binding requires either very long "latent" periods *in vivo* in the passive cutaneous anaphylaxis (PCA) reaction or incubation *in vitro* of basophils with supraphysiologic concentrations of IgE. The reverse rate constant, k_{-1}, of $< 10^{-5}$/sec indicates an affinity constant, K_a $(= k_1/k_{-1})$, of the order of 10^{10} liters/mole (Kulczycki and Metzger, 1974). The affinity of human IgE for the human Fc_ϵ receptor appears to be of a similar order of magnitude (Ishizaka *et al.*, 1973).

Studies on the rate of sequential digestion of IgE by trypsin in the presence and absence of IgE receptor suggest that the $C_\epsilon 3$ (penultimate) domains of IgE are bound to receptor (Perez-Montfort and Metzger, 1982). Resonance energy transfer studies indicate that the $C_\epsilon 2$ domain of receptor-bound IgE is only 35–52 Å from the membrane surface. This suggests that either the C-terminal domain of IgE is buried deeply in the membrane, that the Fc region of IgE is held at an angle to the cell surface, or (most probably) that IgE is bent near its receptor-binding site (Holowka and Baird, 1983).

The interaction of IgE with its receptor had been considered as a species-specific or "homocytotrophic" reaction, because human IgE does not bind appreciably to the rat Fc_ϵ receptor. However, it was found that rat IgE and mouse IgE bind quite well to human IgE receptors (Hempstead *et al.*, 1979; Conrad *et al.*, 1983); thus, the term "homocytotrophic" is no longer used to describe rodent IgE antibodies.

3. STRUCTURE OF MAST CELL/BASOPHIL Fc_ϵ RECEPTORS

The Fc_ϵ receptor appears to be comprised of two subunits, a surface-exposed glycoprotein α-component which binds to IgE and a nonglycosylated β-component more closely associated with the plasma membrane. Initially, the α-component of the rat Fc_ϵ receptor was labeled with ^{125}I and lactoperoxidase, isolated by immunoprecipitation, and characterized as a glycoprotein of about 50,000 daltons in SDS–polyacrylamide gels (Conrad and Froese, 1976; Kulczycki *et al.*, 1976). A similar ^{125}I-labeled α-component has been isolated from human basophils (Hempstead *et al.*, 1979; Conrad *et al.*, 1983). The α-component appears as a broad band in SDS–polyacrylamide gels and in isoelectric focusing gels and such microheterogeneity appears to be a result of *N*-linked carbohydrate moieties (Hempstead *et al.*, 1981a). The rat α-component has been chromatographed in the presence of 6 M guanidine, which is relatively insensitive to heterogeneity of attached oligosaccharides, and exhibits much less heterogeneity, as one might expect (Kumar and Metzger, 1982). The rat α-component contains about 30% carbohydrate and only a modest proportion of hydrophobic amino acids (Kanellopoulos *et al.*, 1980).

In the isolation of Fc_ϵ receptor from RBL cells biosynthetically labeled with amino acids, a macromolecule of about 33,000 daltons was discovered which copurified with the α-component (Kulczycki and Parker, 1979; Holowka *et al.*, 1980) and which could be cross-linked chemically to the α-component (Holowka *et al.*, 1980). This 33K macromolecule is not a glycoprotein (Kulczycki and Parker, 1979; Hempstead *et al.*, 1981a). Since it shares no tryptic peptides with the α-component, it is not simply a fragment of the latter (Hempstead *et al.*, 1981a). Since it can be labeled with 5-iodonaphthyl-1-nitrene, it is likely to have a segment within the plasma membrane (Holowka *et al.*, 1981). Accordingly, the smaller 33K nonglycosylated protein is referred to as the β-component (or β-subunit) whereas the larger surface-exposed glycoprotein macromolecule is termed the α-component.

Enzymatic cleavage of the rat α-component yields fragments of about 32,000 daltons in SDS–polyacrylamide gels. The claim (Goetze *et al.*, 1981) that two independent digestion fragments of the α-component, a 34K "α_1" and a more heavily glycosylated 30K "α_2," can be resolved is not yet well supported by experimental data.

4. EXPRESSION OF MAST CELL AND BASOPHIL IgE RECEPTORS

Mast cells and analogous RBL cells express large numbers of IgE receptors on their surfaces. It is estimated that there are 300,000 receptors per rat mast cell (Conrad *et al.*, 1975) and 300,000 to 1,400,000 receptors per RBL cell (Kulczycki and Metzger, 1974). Smaller-sized human basophils have approximately 40,000 to 100,000 receptors per cell (Ishizaka *et al.*, 1973). With RBL cells it has been shown that the density of IgE receptors varies according to the cell cycle. Cell cultures in their "stationary" phase of growth (when most cells are in the G_1 phase of the cell cycle) synthesize and express the greatest number of receptors per cell (Isersky *et al.*, 1975). Although nonglycosylated nascent α-components (extracted from cells grown in the presence of tunicamycin) are capable of binding both to IgE and to antireceptor antibodies, the α-component of the IgE receptor must be glycosylated before it can be expressed on the cell surface (Hempstead *et al.*, 1981a).

Membrane proteins can be classified in two groups (Brown *et al.*, 1983). "Migrant" proteins such as the LDL and transferrin receptors recycle from one membrane-containing organelle to another via "coated pits." "Resident" proteins such as histocompatibility and theta antigens do not recycle. Both unoccupied and occupied IgE receptors are diffusely distributed on the plasma membrane in studies with RBL cells (Carson and Metzger, 1974; Isersky *et al.*, 1979). Thus, the rat IgE receptor, whether occupied or unoccupied by IgE, appears to function as a "resident" protein in the plasma membrane. The attachment of plasma cell-derived IgE to the Fc_ϵ receptor can be considered for practical purposes as irreversible so that IgE could be viewed as a "resident alien" protein on the basophil/mast cell surface.

Cross-linking of IgE receptors (under physiological conditions by antigen, under experimental conditions by antibodies to IgE or to receptor, or by IgE aggregates) is required for receptor redistribution (Carson and Metzger, 1974) and for internalization of receptor (Isersky *et al.*, 1983). When receptor-bound IgE molecules bind to a multivalent antigen, they become indirectly cross-linked, initiate histamine release from the cell (see Section 5), and the IgE molecules together with their receptors and their bound antigen are internalized. However, the IgE receptor is not recycled nor is it replaced by active synthesis of new receptor (Isersky *et al.*, 1983). Therefore, the IgE receptor does not appear to fit the description of a "migrant" receptor.

The internalization of multivalent antigen bound to antigen-specific IgE molecules (bound, in turn, to Fc_ϵ receptors) does serve to remove both antigen and IgE antibody from the cell surface. It may account, at least in part, for the "antigen-specific desensitization" phenomenon observed when low concentrations of a specific antigen render a mast cell or basophil unresponsive to further challenge with that antigen and yet responsive to a different antigen.

5. THE ROLE OF THE IgE RECEPTOR IN THE RELEASE AND/OR PRODUCTION OF MEDIATORS FROM MAST CELLS AND BASOPHILS

It is now well established that IgE receptors must be brought into apposition on the surface of basophils and mast cells in order for mediator release and production to occur. Initial studies demonstrated that multivalent antigens were required to elicit immediate hypersensitivity responses (to DNP antigens in humans) (Fig. 1A) whereas univalent antigens were ineffective (Farah *et al.*, 1960). Divalent haptens, but not univalent haptens, initiated mediator release from sensitized rabbit basophils (Fig. 1B) (Siraganian *et al.*, 1975). Immediate hypersensitivity responses in man were produced by antibody to human IgE but not by its Fab′ fragments (Fig. 1C,D) (Ishizaka and Ishizaka, 1969). Rabbit antibodies to rat IgE receptor (Fig. 1E) and $F(ab')_2$ fragments of such antibodies (Fig. 1F) induced histamine release from rat mast cells or from RBL cells (Ishizaka *et al.*, 1978; Isersky *et al.*, 1978). Fab′ fragments of such antireceptor antibodies did not initiate release (Fig. 1G) unless brought together by anti-rabbit IgG (Fig. 1H). It was noted that a limited degree of receptor bridging (or receptor cross-linking) was optimal for histamine release; extensive cross-linkage of IgE molecules (and of their attached receptors) led to "capping" on the cell surface and inhibition of release (Becker *et al.*, 1973).

The precise biochemical mechanisms which follow IgE receptor apposition and lead to activation of mast cells and basophils are not known. Since the mediator release process can be initiated within 30 sec of antigen exposure, activation events involving enzymes have received increasing attention.

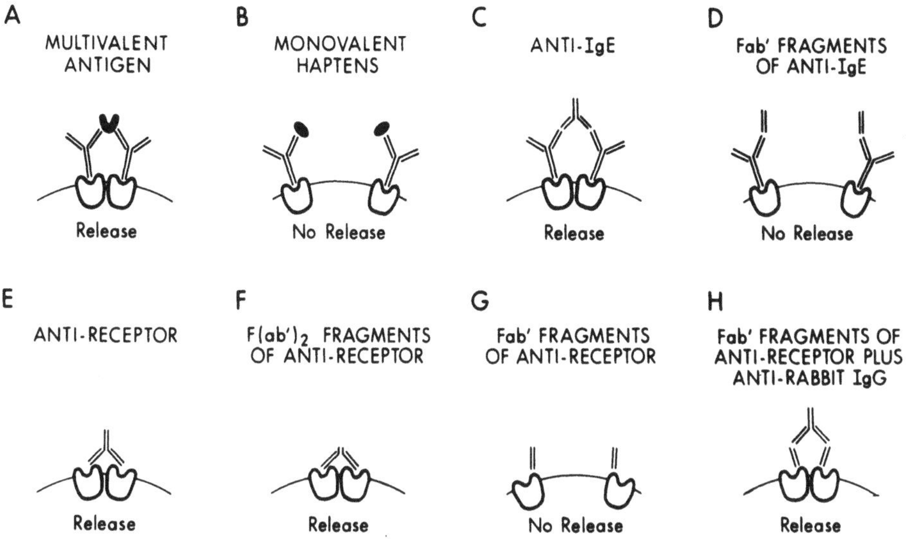

FIGURE 1. Schematic summary of studies leading to establishment that IgE receptor cross-linking is the necessary and sufficient signal for mediator release from mast cells and basophils. (A) Multivalent (including divalent) antigens initiate release; (B) monovalent haptens do not. (C) Anti-IgE initiates mediator release, but (D) its monovalent Fab′ fragments do not. (E) Antibodies to IgE receptor initiate mediator release, as do its divalent $F(ab')_2$ fragments (F). (G) Monovalent Fab′ fragments of rabbit antireceptor antibody do not initiate release unless (H) they become cross-linked by anti-rabbit IgG antibody. (From Kulczycki, 1981, with permission.)

The IgE receptor α-component is phosphorylated by a protein kinase (Hempstead *et al.*, 1981b) and its β-component is also a phosphoprotein (Fewtrell *et al.*, 1982). Antigenic stimulation of intact rat mast cells produces a rapid and marked increase in the phosphorylation of the α-component but no significant change in the phosphorylation of the β-component (Hempstead *et al.*, 1983). Thus, phosphorylation of the IgE receptor α-component (which alters the charge and probably the conformation of the receptor) is attractive as a potential mechanism for initiating or modulating the mediator release process.

6. Fc_ϵ RECEPTORS WITH LOWER AFFINITY AND/OR LESS SPECIFICITY FOR IgE

The human macrophage cell line expresses Fc_ϵ receptors with a K_a of $\sim 4 \times 10^7$ liters/mole for monomeric human IgE (Anderson and Spiegelberg, 1981). Rodent macrophages possess Fc_ϵ receptors with a K_a on the order of 10^7 liters/mole for monomeric rodent IgE (Boltz-Nitulescu *et al.*, 1982; Finbloom and Metzger, 1982). The rat macrophage Fc_ϵ receptor appears to be structurally similar to the high-affinity Fc_ϵ receptor of rat mast cells and RBL cells (Finbloom and Metzger, 1983). One important function of monocyte-macrophage Fc_ϵ receptors involves facilitating the destruction of schistosomules (Capron *et al.*, 1975).

Human and rat B and T lymphocytes express Fc_ϵ receptors (Gonzalez-Molina and Spiegelberg, 1977; Yodoi and Ishizaka, 1979). Lymphocyte Fc_ϵ receptors appear to be similar to monocyte Fc_ϵ receptors in affinity for IgE ($K_a \sim 6 \times 10^7$ liters/mole), in size (both have 47,000 dalton macromolecules), and in antigenic determinants (Melewicz *et al.*, 1982). Since infection of rats with *Nippostrongylus brasiliensis* and exposure of patients to appropriate allergens both induce increases in Fc_ϵ receptor-bearing lymphocyte populations, it is presumed that such populations are involved in regulating IgE synthesis (Yodoi and Ishizaka, 1979; Spiegelberg and Simon, 1981). Fc_ϵ-bearing T lymphocytes are capable of releasing IgE-binding factors which appear to regulate IgE responses *in vitro* (Suemura *et al.*, 1980; Hirashima *et al.*, 1980). The factor which potentiates IgE synthesis is glycosylated (by another soluble T-cell factor) whereas the factor which suppresses IgE production is not glycosylated (Yodoi *et al.*, 1981, 1982). Lymphocyte Fc_ϵ receptors themselves can be induced *in vitro* by incubation with IgE (Yodoi *et al.*, 1979) and, in humans, with cross-linked IgE (Parker *et al.*, 1983).

Rat and human eosinophils also express receptors for IgE. Since they are demonstrated by use of IgE-coated erythrocytes (Capron *et al.*, 1981a), it is presumed that they have a low affinity (compared with basophils) for monomeric IgE. Eosinophil Fc_ϵ receptors appear to play an important role in mediating eosinophil cytotoxicity against parasites (Capron *et al.*, 1981b).

Studies with RBL cells indicate that the high-affinity Fc_ϵ receptor and another cell surface protein (designated H) each are capable of binding to both IgE and IgG. Even though the H protein has considerable affinity for IgE, its substantial affinity for IgG and the enormous excess of IgG present *in vivo* may make H effectively a receptor for IgG (Kepron *et al.*, 1982).

NOTE ADDED IN PROOF

Two additional ~9,000 M_r molecules termed γ-chains are noncovalently associated with each α and β component of the high-affinity IgE receptor (Metzger *et al.*, 1986).

REFERENCES

Anderson, C. L., and Spiegelberg, H. L., 1981, Macrophage receptors for IgE: Binding of IgE to specific IgE Fc receptors on a human macrophage cell line, U937, *J. Immunol.* **126**:2470.

Becker, K. E., Ishizaka, T., Metzger, H., Ishizaka, K., and Grimley, P., 1973, Surface IgE on human basophils during histamine release, *J. Exp. Med.* **138**:394.

Boltz-Nitulescu, G., Plummer, J. M., and Spiegelberg, H. L., 1982, Fc receptors for IgE on mouse macrophages and macrophage-like cell lines, *J. Immunol.* **128**:2265.

Brown, M. S., Anderson, R. G. W., and Goldstein, J. L., 1983, Recycling receptors: The round-trip itinerary of migrant membrane proteins, *Cell* **32**:663.

Capron, A., Dessaint, J.-P., Capron, M., and Bazin, H., 1975, Specific IgE antibodies in immune adherence of normal macrophages to *Schistosoma mansoni* schistosomules, *Nature (London)* **253**:474.

Capron, M., Capron, A., Dessaint, J.-P., Torpier, G., Johansson, S. G. O., and Prin, L., 1981a, Fc receptors for IgE on human and rat eosinophils, *J. Immunol.* **126**:2087.

Capron, M., Bazin, H., Joseph, M., and Capron, A., 1981b, Evidence for IgE-dependent cytotoxicity by rat eosinophils, *J. Immunol.* **126**:1764.
Carson, D. A., and Metzger, H., 1974, Interaction of IgE with rat basophilic leukemia cells. IV. Antibody-induced redistribution of IgE receptors, *J. Immunol.* **113**:1271.
Conrad, D. H., and Froese, A., 1976, Characterization of the target cell receptor for IgE. II. Polyacrylamide gel analysis of the surface IgE receptor from normal rat mast cells and from rat basophilic leukemia cells, *J. Immunol.* **116**:319.
Conrad, D. H., Bazin, H., Sehon, A. H., and Froese, A., 1975, Binding parameters of the interaction between rat IgE and rat mast cell receptors, *J. Immunol.* **114**:1688.
Conrad, D. H., Wingard, J. R., and Ishizaka, T., 1983, The interaction of human and rodent IgE with the human basophil IgE receptor, *J. Immunol.* **130**:327.
Farah, F. S., Kern, M., and Eisen, H., 1960, Specific inhibition of wheal-and-erythema responses with univalent haptens and univalent antibody fragments, *J. Exp. Med.* **112**:1211.
Fewtrell, C., Goetze, A., and Metzger, H., 1982, Phosphorylation of the receptor for IgE, *Biochemistry* **21**:2004.
Finbloom, D. S., and Metzger, H., 1982, Binding of immunoglobulin E to the receptor on rat peritoneal macrophages, *J. Immunol.* **129**:2004.
Finbloom, D. S., and Metzger, H., 1983, Isolation of cross-linked IgE–receptor complexes from rat macrophages, *J. Immunol.* **130**:1489.
Goetze, A., Kanellopoulos, J., Rice, D., and Metzger, H., 1981, Enzymatic cleavage products of the α subunit of the receptor for IgE, *Biochemistry* **20**:6341.
Gonzalez-Molina, A., and Spiegelberg, H. L., 1977, A subpopulation of normal human peripheral B lymphocytes that bind IgE, *J. Clin. Invest.* **59**:616.
Hempstead, B. L., Parker, C. W., and Kulczycki, A., Jr., 1979, Characterization of the IgE receptor isolated from human basophils, *J. Immunol.* **123**:2283.
Hempstead, B. L., Parker, C. W., and Kulczycki, A., Jr., 1981a, The cell surface receptor for IgE: Effect of tunicamycin on molecular properties of receptor from rat basophilic leukemia cells, *J. Biol. Chem.* **256**:10717.
Hempstead, B. L., Kulczycki, A., Jr., and Parker, C. W., 1981b, Phosphorylation of the IgE receptor from ionophore A23187 stimulated intact rat mast cells, *Biochem. Biophys. Res. Commun.* **98**:815.
Hempstead, B. L., Parker, C. W., and Kulczycki, A., Jr., 1983, Selective phosphorylation of IgE receptor in antigen-stimulated rat mast cells, *Proc. Natl. Acad. Sci. USA* **80**:3050.
Hirashima, M., Yodoi, J., and Ishizaka, K., 1980, Regulatory role of IgE-binding factors from rat T lymphocytes. III. IgE-specific suppressive factor with IgE-binding activity, *J. Immunol.* **125**:1442.
Holowka, D., and Baird, B., 1983, Structural studies on the membrane-bound IgE–receptor complex. 2. Mapping of distances between sites on IgE and the membrane surface, *Biochemistry* **22**:3475.
Holowka, D., Hartmann, H., Kanellopoulos, J., and Metzger, H., 1980, Association of the receptor for IgE with an endogenous polypeptide on rat basophilic leukemia cells, *J. Receptor Res.* **1**:41.
Holowka, D., Gitler, C., Bertovici, T., and Metzger, H., 1981, Reaction of 5-iodonaphthyl-1-nitrene with the IgE receptor on normal and tumor mast cells, *Nature (London)* **289**:806.
Isersky, C., Metzger, H., and Buell, D. N., 1975, Cell cycle-associated changes in receptors for IgE during growth and differentiation of a rat basophilic leukemia cell line, *J. Exp. Med.* **141**:1147.
Isersky, C., Taurog, J. D., Poy, G., and Metzger, H., 1978, Triggering of cultured neoplastic mast cells by antibodies to the receptor for IgE, *J. Immunol.* **121**:549.
Isersky, C., Rivera, J., Mims, S., and Triche, T. J., 1979, The fate of IgE bound to rat basophilic leukemia cells, *J. Immunol.* **122**:1926.
Isersky, C., Rivera, J., Segal, D. M., and Triche, T., 1983, The fate of IgE bound to rat basophilic leukemia cells. II. Endocytosis of IgE oligomers and effect on receptor turnover, *J. Immunol.* **131**:388.
Ishizaka, K., and Ishizaka, T., 1969, Immune mechanisms of reversed type reaginic hypersensitivity, *J. Immunol.* **103**:588.
Ishizaka, K., Tomioka, H., and Ishizaka, T., 1970, Mechanisms of passive sensitization. I. Presence of IgE and IgG molecules of human leukocytes, *J. Immunol.* **105**:1459.
Ishizaka, T., Soto, C., and Ishizaka, K., 1973, Mechanisms of passive sensitization. III. Number of IgE molecules and its receptor sites on human basophil granulocytes, *J. Immunol.* **111**:500.

Ishizaka, T., Ishizaka, K., Conrad, D. H., and Froese, A., 1978, A new concept of triggering mechanisms of IgE-mediated histamine release, *J. Allergy Clin. Immunol.* **61**:320.

Kanellopoulos, J. M., Liu, T.-Y., Poy, G., and Metzger, H., 1980, Composition and subunit structure of the cell receptor for IgE, *J. Biol. Chem.* **255**:9060.

Kepron, M. R., Conrad, D. H., and Froese, A., 1982, The cross-reactivity of rat IgE and IgG with solubilized receptors of rat basophilic leukemia cells, *Mol. Immunol.* **19**:1631.

Kulczycki, A., Jr., 1981, Role of IgE and IgE receptors in bronchial asthma, *J. Allergy Clin. Immunol.* **68**:5.

Kulczycki, A., and Metzger, H., 1974, The interaction of IgE with rat basophilic leukemia cells. II. Quantitative aspects of the binding reaction, *J. Exp. Med.* **140**:1676.

Kulczycki, A., and Parker, C. W., 1979, The cell-surface receptor for IgE. I. The use of repetitive affinity chromatography for the purification of a mammalian receptor, *J. Biol. Chem.* **254**:3187.

Kulczycki, A., Isersky, C., and Metzger, H., 1974, The interaction of IgE with rat basophilic leukemia cells. I. Evidence for specific binding of IgE, *J. Exp. Med.* **139**:600.

Kulczycki, A., McNearney, T. A., and Parker, C. W., 1976, The rat basophilic leukemia cell receptor for IgE. I. Characterization as a glycoprotein, *J. Immunol.* **117**:661.

Kumar, N., and Metzger, H., 1982, Gel filtration in 6 M guanidine HCl of the α-subunit of the receptor for IgE, *Mol. Immunol.* **19**:1561.

Melewicz, F. M., Plummer, J. M., and Spiegelberg, H. L., 1982, Comparison of the Fc receptors for IgE on human lymphocytes and monocytes, *J. Immunol.* **129**:563.

Metzger, H., Alcaraz, G., Hohman, R., Kinet, J.-P., Pribluda, V., and Quarto, R., 1986, The receptor with high affinity for IgE, *Ann. Rev. Immunol.* **4**:419.

Parker, C. W., Schechtel, T., Falkenhein, S., and Huber, M., 1983, Induction of IgE receptors on human lymphocytes, *Immunol. Commun.* **12**:1.

Perez-Montfort, R., and Metzger, H., 1982, Proteolysis of soluble IgE receptor complexes: Localization of sites on IgE which interact with the Fc receptor, *Mol. Immunol.* **19**:1113.

Siraganian, R. P., Hook, W. A., and Levine, B. B., 1975, Histamine release from basophils: Activation by bivalent haptens, *Immunochemistry* **12**:149.

Spiegelberg, H. L., 1983, Fc receptors for IgE on macrophages and lymphocytes: IgE synthesis, *Fed. Proc.* **42**:122.

Spiegelberg, H. L., and Simon, R. A., 1981, Increase of lymphocytes with Fc receptors for IgE in patients with allergic rhinitis during the grass pollen season, *J. Clin. Invest.* **68**:845.

Suemura, J., Yodoi, J., Hirashima, M., and Ishizaka, K., 1980, Regulatory role of IgE-binding factors from rat T lymphocytes. I. Mechanisms of enhancement of IgE response by IgE-potentiating factor, *J. Immunol.* **125**:148.

Sullivan, A. L., Grimley, P. M., and Metzger, H., 1971, Electron microscopic localization of IgE on the surface membrane of human basophils, *J. Exp. Med.* **134**:1403.

Yodoi, J., and Ishizaka, K., 1979, Lymphocytes bearing Fc receptors for IgE. I. Presence of human and rat T lymphocytes with Fc_ϵ receptors, *J. Immunol.* **122**:2577.

Yodoi, J., Ishizaka, T., and Ishizaka, K., 1979, Lymphocytes bearing Fc receptors for IgE. II. Induction of Fc_ϵ-receptor bearing rat lymphocytes by IgE, *J. Immunol.* **123**:455.

Yodoi, J., Hirashima, M., and Ishizaka, K., 1981, Lymphocytes bearing Fc receptors for IgE. V. Effect of tunicamycin on the formation of IgE-potentiating factor and IgE-suppressive factor by Con A-activated lymphocytes, *J. Immunol.* **126**:877.

Yodoi, J., Hirashima, M., and Ishizaka, K., 1982, Regulatory role of IgE-binding factors from rat T lymphocytes. V. The carbohydrate moieties in IgE-potentiating factors and IgE-suppressive factors, *J. Immunol.* **128**:289.

5

Interaction of Phagocytic Cells with Immune Complexes of Anaphylactic Antibodies

JEAN-PAUL DESSAINT and ANDRÉ CAPRON

1. INTRODUCTION

The chief mechanism of clearance of antigen–antibody complexes by cells of the reticuloendothelial system is for complexes containing intact IgG molecules to bind the various Fc receptors on the surface of mononuclear and polynuclear phagocytes. Similarly, phagocytosis of IgG antibody-opsonized particles by such cells requires binding to Fc receptors, and macrophages bind immune IgG complexes with up to four orders of magnitude greater affinity than for monomeric IgG (reviewed by McKeever and Spicer, 1980). In response to their interaction with opsonized particles or immune complexes, macrophages either release lysosomal granule enzymes to the surrounding medium (Cardella *et al.*, 1974) or undergo selective phagocytic paralysis (Rabinovitch *et al.*, 1975).

While the role of phagocytic cells in the development of adverse reactions such as serum sickness has been extensively explored (Chapter 9), the most impressive hypersensitivity reaction certainly is of the immediate type. The clustering of the surface receptors for IgE triggers the anaphylactic discharge of the pharmacologically active mediators of mast cells or basophils, and it was shown that small oligomers of IgE, at least dimers, can provide a signal for mast cell degranulation (Segal *et al.*, 1977; Fewtrell and Metzger, 1980; Kagey-Sobotka *et al.*, 1981). Nonreaginic antibodies belonging to one of the IgG subclasses can also be involved in immediate-type reactions. In this respect, rat IgG2a was shown to bind when aggregated to rat mast cells (Halper and Metzger, 1976) and to induce the subsequent release of vasoactive amines (Bach *et al.*, 1971).

Human allergy develops usually in two steps, with the initial binding of

JEAN-PAUL DESSAINT and ANDRÉ CAPRON • Centre d'Immunologie et de Biologie Parasitaire, INSERM U-167, CNRS 624, Institut Pasteur, 59019 Lille Cedex, France.

free, monomeric IgE antibody to Fc_ϵ receptors on mast cells or basophils and the subsequent triggering of the cell on the interaction of a multivalent antigen with the surface-bound IgE. Circulating immune IgE complexes have, however, been characterized in allergic patients, which might induce mast cell degranulation directly (Brostoff *et al.*, 1977). The demonstration that many mediators of anaphylaxis are also produced by phagocytic cells and the characterization on their surface of receptors for anaphylactic antibodies or immune complexes suggest that phagocytic cells can participate in immediate-type reactions not only as recruited inflammatory cells, but also as direct effectors of allergy. But it is certainly in infections by helminths that the sustained release of circulating antigens and the massive production of antiparasite IgE (reviewed by Dessaint, 1982) lead to the formation of the largest amount of circulating immune complexes of anaphylactic antibodies (Santoro *et al.*, 1978; Stevens *et al.*, 1983). Interestingly, this does not result in dramatic anaphylactic reactions (Dessaint, 1982) but instead immune complexes of anaphylactic antibodies were shown to be involved in immune defense against helminths (Capron *et al.*, 1982a,b).

This chapter concerns the interaction of phagocytic cells, like mononuclear phagocytes and eosinophils, with immune complexes through their recently characterized receptors for anaphylactic antibodies and the subsequent activation of the cell, evidenced either in protective immunity against helminth parasites or in allergic diseases.

2. RECEPTORS FOR ANAPHYLACTIC ANTIBODIES ON PHAGOCYTIC CELLS

2.1. RECEPTORS FOR IgE ON MONONUCLEAR PHAGOCYTES

The initial demonstration in our laboratory that rat IgE antibody induced the killing of *Schistosoma mansoni* larvae (schistosomula) by rat peritoneal macrophages (Capron *et al.*, 1975) prompted studies on the binding of IgE to mononuclear phagocytes (Table 1). By using peroxidase-labeled myeloma IgE protein, the specific binding of IgE to the surface of rat peritoneal macrophages could be demonstrated (Dessaint *et al.*, 1979b) (Fig. 1).

A rosette assay using fixed erythrocytes coated with either rat or human myeloma IgE protein under optimal conditions showed that mononuclear phagocytes or macrophage-monocyte cell lines formed IgE rosettes in rats, mice, and humans. The rosetting was specific for IgE since heated or reduced and alkylated IgE did not bind and IgG did not inhibit the rosetting (Melewicz and Spiegelberg, 1980; Boltz-Nitulescu and Spiegelberg, 1981; Joseph *et al.*, 1981; Melewicz *et al.*, 1982a,b; Boltz-Nitulescu *et al.*, 1982).

The percentages of Fc_ϵ receptor-positive cells varied in different species and in cells from different anatomical locations, monocytes showing lower percentages than peritoneal or alveolar macrophages. Interestingly, monocytes or alveolar macrophages from healthy donors formed a lower percentage of IgE rosettes than cells from patients with severe allergic disorders (Melewicz *et al.*,

TABLE 1. CHARACTERIZATION OF RECEPTORS FOR IGE ON MONONUCLEAR PHAGOCYTES AND EOSINOPHILS

Cell studied	Experimental method	Reference
Rat peritoneal macrophage	Anti-IgE rosettes	Capron *et al.* (1975)
	Binding of ^{125}I- or peroxidase-labeled IgE (monomeric or aggregated)	Capron *et al.* (1977b), Dessaint *et al.* (1979b)
	Binding of ^{125}I-labeled IgE (monomeric or aggregated)	Finbloom and Metzger (1982)
Rat peritoneal or alveolar macrophage	IgE rosettes	Boltz-Nitulescu and Spiegelberg (1981)
Mouse macrophages and macrophagelike cell lines	IgE rosettes Binding of ^{125}I-labeled monomeric IgE	Boltz-Nitulescu *et al.* (1982)
Human monocytes	Binding of ^{125}I-labeled aggregated IgE	Lawrence *et al.* (1975)
	IgE rosettes	Melewicz and Spiegelberg (1980)
	Binding of ^{125}I-labeled monomeric IgE	Melewicz *et al.* (1982b)
Human alveolar macrophages	IgE rosettes	Joseph *et al.* (1980), Melewicz *et al.* (1982a)
	Binding of peroxidase-labeled anti-IgE	Joseph *et al.* (1983)
Human macrophage cell line (U937)	IgE rosettes Binding of ^{125}I-labeled monomeric IgE	Anderson and Spiegelberg (1981)
Rat peritoneal eosinophil	Binding of ^{125}I-labeled aggregated IgE	Capron *et al.* (1977b)
	IgE rosettes	Capron *et al.* (1981b)
	IgE + ferritin-labeled anti-IgE	Hubscher (1975)
Human blood eosinophil	IgE rosettes Binding of ^{125}I-labeled monomeric IgE	M. Capron *et al.* (1983, 1984b)

1981a; Joseph *et al.*, 1981, 1983). Likewise, in the rat, infection by helminth parasites was associated with elevated percentages of Fc_ϵ receptor-positive macrophages and the role of IgE in inducing this increase was demonstrated *in vivo* as well as *in vitro* (Spiegelberg *et al.*, 1983).

Saturable binding of radiolabeled monomeric IgE to human, rat, or mouse mononuclear phagocytes can be demonstrated and the fixation is specifically inhibited by native IgE or its Fc_ϵ fragment but not by IgG (Dessaint *et al.*, 1979b; Anderson and Spiegelberg, 1981). About 40,000–500,000 receptors per cell were detected and this number was increased in *Nippostrongylus brasiliensis*-infected rats, in rats bearing an IgE-producing myeloma, or in rats injected with 5 mg IgE (Spiegelberg *et al.*, 1983). An average association constant of $0.8–6 \times 10^7$ M^{-1} for unaggregated IgE was measured, and monomeric IgE dissociated rapidly from the cells ($t_{1/2}$ ~ 2–3 min) (Anderson and Spiegelberg, 1981; Finbloom and

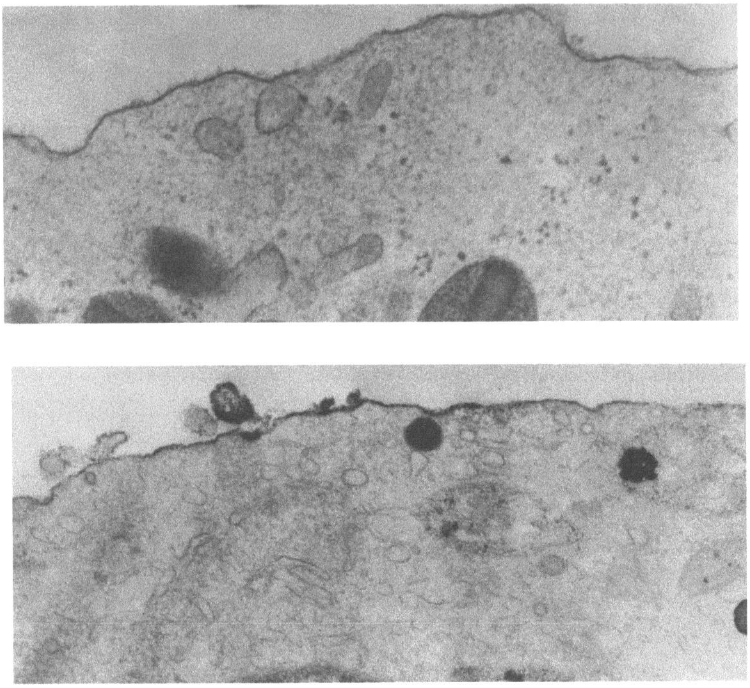

FIGURE 1. Binding of peroxidase-labeled rat IgE to eosinophils and macrophages. (Upper) Human peritoneal eosinophil showing surface-bound IgE, × 59,000; (lower) rat peritoneal macrophage, ×24,800.

Metzger, 1982; Boltz-Nitulescu *et al.*, 1982). Thus, in physiologic concentrations of IgE, the Fc_ϵ receptors of macrophages would be far from saturated with monomeric IgE, although the amount of bound IgE predictably increases with the serum level of IgE, as shown by the rosetting with anti-IgE-coated erythrocytes (Joseph *et al.*, 1981, 1983). Therefore, aggregated IgE or IgE-opsonized particles would not be hindered by monomeric IgE bound to macrophages. Indeed, dimeric IgE binds to rat macrophages with a higher association constant ($2.3 \times 10^8\ M^{-1}$) than monomers, and IgE dimers have a much slower rate of dissociation ($t_{1/2}$ 94 min) (Finbloom and Metzger, 1982). When the Fc_ϵ receptor-positive human monocytic cell line U937 is incubated with radiolabeled IgE monomers or dimers, the binding of each label is competitively inhibited by a moderate (× 50) excess of unlabeled dimers when monomers do not inhibit significantly in these conditions (Dessaint *et al.*, 1983).

An antiserum prepared against lymphocyte Fc_ϵ receptors was shown to inhibit the binding of IgE to human monocytes or to U937 macrophagelike cells. This antiserum precipitated two surface components from U937 cells of approximately 47K and 23K. The comparison made between human mononuclear phagocytes and lymphocytes indicates that both cells bear trypsin-sensitive Fc_ϵ receptors of similar structure and antigenicity, but which differ from the high-affinity monovalent receptor for IgE on basophils or mast cells (Melewicz *et al.*, 1982b).

2.2. RECEPTORS FOR IgE ON EOSINOPHILS

The involvement of IgE antibody in the cytotoxicity of eosinophils against parasites (Capron *et al.*, 1981a) led to the investigation of IgE-binding sites on the surface of eosinophils. By using homologous IgE-coated erythrocytes, the rosetting of a significant proportion of rat or human eosinophils was obtained and the specificity of the binding was shown by inhibition experiments using aggregated IgE or IgG. While no cross-inhibition was observed between the two immunoglobulin classes, suggesting no cross-reactivity of the binding sites for IgE and for IgG, mixed rosettes were seen by using both IgG-coated and IgE-coated indicator cells, indicating that the same eosinophil subpopulation can express both Fc_γ and Fc_ϵ receptors (Capron *et al.*, 1981b). The binding of IgE by eosinophils was also shown by ultrastructural studies (Hubscher, 1975; Capron *et al.*, 1977b) (Fig. 1), by fixation of ^{125}I-labeled IgE (Capron *et al.*, 1981b, 1984b), and by time-course studies in experimental rat schistosomiasis, where surface IgE could be detected on eosinophils from 5 weeks of infection onward by using anti-IgE-coated indicator cells (M. Capron *et al.*, 1983, 1984a).

Both in hypereosinophilic patients and in *S. mansoni*-infected rats, an increase in the proportion of Fc_ϵ receptor-positive eosinophils was observed, an indication of a possible *in vivo* modulation of the expression by eosinophils of their receptor for the IgE. Incubation of normal rat eosinophils either with purified myeloma IgE protein or with the heat-labile antibodies in infected rat serum indeed led to a dose-dependent increase in the proportion of Fc_ϵ receptor-positive eosinophils (M. Capron *et al.*, 1983).

Besides IgE itself, other factors seem to participate in the modulation of eosinophil receptors for IgE. Through the release of their mediators, mast cells, which function as accessory cells in the IgE antibody-dependent cytotoxicity by eosinophils (Capron *et al.*, 1981a), can influence the expression of the Fc_ϵ receptors by eosinophils. ECF-A tetrapeptides in particular were shown to induce *in vitro* a significant increase in the proportion of eosinophils forming IgE rosettes (M. Capron *et al.*, 1983).

Another factor accounting for variations in the expression of Fc_ϵ receptors by eosinophils might be related to the heterogeneity of these cells mainly in hypereosinophilic patients. In such patients, and particularly in those with the hypereosinophilic syndrome, blood eosinophils were mainly formed of abnormal low-density cells with morphological signs of degranulation ("hypodense" eosinophils), which bear an increased number of cell surface receptors including the Fc_ϵ receptor (Capron *et al.*, 1984b) and exhibit higher cytotoxic capacity than eosinophils with normal density (Prin *et al.*, 1983).

The antiserum against the Fc_ϵ receptor of human lymphocytes was recently shown to also inhibit the formation of IgE rosettes by eosinophils from hypereosinophilic patients at the same dilution that previously was reported to block

IgE-dependent rosetting by lymphocytes, monocytes, or macrophagelike cell lines (Capron *et al.*, 1984b).

Thus, the properties of the various receptors for IgE identified so far allow two groups of cell populations bearing two distinct species of Fc_ϵ receptors to be distinguished. Mast cells and basophils are characterized by high-affinity monovalent receptors. In contrast, subpopulations of mononuclear phagocytes, eosinophils, lymphocytes, and platelets bear low-affinity Fc_ϵ receptors showing preferential interaction with aggregated IgE, which appear to have common structural and antigenic properties different from those of the mast cell or basophil receptors (Capron *et al.*, 1983a). However, these differences do not necessarily imply an entirely distinct molecular structure, and a chain with similarities to the β chain of the receptor of rat basophilic leukemia cells has indeed been identified in the Fc_ϵ receptor of rat peritoneal macrophages (Finbloom and Metzger, 1983). The similarities of the low-affinity, multivalent receptors for IgE were recently extended by the demonstration that the B-lymphocyte receptor is also for aggregated IgE, at least as dimer (Lee and Conrad, 1984).

2.3. RECEPTORS FOR ANAPHYLACTIC IgG

Phagocytic cells bear an array of receptors for IgG (McKeever and Spicer, 1980). Among these, one can bind the anaphylactic IgG subclass of the animal species. In the rat for instance, IgG2a was shown to bind to alveolar or peritoneal macrophages through a trypsin-resistant, subclass-specific Fc receptor distinct from the trypsin-sensitive receptor for other IgG subclasses or for IgE (Boltz-Nitulescu *et al.*, 1981). Experimental evidence reported below indicates, however, that rat IgG2c can compete with IgG2a for the eosinophil Fc_γ receptor (Grzych *et al.*, 1984). Interestingly, the binding of IgG2a to eosinophils is increased by ECF-A tetrapeptides (Capron *et al.*, 1981c) as is the case for IgE.

The expression by at least subpopulations of phagocytic cells, of receptors which can bind aggregated anaphylactic antibodies of the IgG or IgE classes and which undergo direct or indirect modulation by IgE (and by ECF-As in the case of eosinophils), suggests therefore that besides the classical interaction of anaphylactic antibodies with mast cells or basophils—a paradigm for the study of cell activation by antibody (Metzger, 1978)—phagocytic cells may also be triggered via their specific receptors by anaphylactic antibodies.

3. FUNCTION OF THE RECEPTORS FOR ANAPHYLACTIC ANTIBODIES

Studies of the interaction of mononuclear phagocytes and eosinophils with anaphylactic antibodies, either cross-linked or complexed with the antigen, have indeed shown that these cells can be activated and release pharmacologically active materials, leading in some instances to the specific killing of foreign targets.

3.1. MACROPHAGE ACTIVATION BY IgE

Activation of mononuclear phagocytes is thought to proceed by successive steps which ultimately lead to the cytotoxic or cytostatic activity of the cells (Cohn, 1978).

The use of different molecular forms of myeloma IgE protein, from purified monomers to various sizes of dimethylsuberimidate-aggregated IgE polymers, together with IgE–anti-IgE or IgE antibody–allergen reactions, has led to the study of some of the biochemical events associated with macrophage "activation" (Table 2).

Incubation of normal (uninduced) rat macrophages with the serum from rats immune to *S. mansoni* induced a marked increase in the lysosomal β-glucuronidase activity of the adherent cells, which was not observed when the immune serum was heated at 56°C or IgE-depleted (Joseph *et al.*, 1977). When normal rat macrophages were successively exposed to myeloma IgE protein and to ε-specific anti-rat IgE or the $F(ab')_2$ fragments therefrom, a significant increase (+755%) in the selective β-glucuronidase release was observed, which proved to be time- and dose-dependent, increasing with both the amount of the IgE added during the first step and that of anti-IgE added afterwards. Unaggregated IgE alone failed to produce such enzyme release, and heating the IgE prevented enzyme release also (Dessaint *et al.*, 1979b). IgE and anti-IgE on macrophages induced an increase in glucosamine incorporation as well as the secretion of plasminogen activator and the generation of superoxide anion (Capron *et al.*, 1980).

TABLE 2. FUNCTIONAL CONSEQUENCES OF (AGGREGATED) IGE–MACROPHAGE INTERACTION

Cell studied	Finding	Reference
Rat macrophages	Increased lysosomal enzyme release	Joseph *et al.* (1977), Dessaint *et al.* (1979a)
	Increased secretion of plasminogen activator	Capron *et al.* (1980)
	Increased generation of superoxide	Capron *et al.* (1980)
	Production of cGMP and increased Ca^{2+} uptake	Dessaint *et al.* (1980)
	Production of SRS-A	Spiegelberg *et al.* (1983)
	Release of leukotriene C_4 by alveolar macrophages	Rankin *et al.* (1982)
	IgE-dependent cytotoxicity	Capron *et al.* (1975)
Mouse macrophages	Release of arachidonic acid	Rouzer *et al.* (1982a)
	Production of PGE_2 and leukotriene C	Rouzer *et al.* (1982b)
Human macrophages or monocytes	Increased lysosomal enzyme release	Joseph *et al.* (1980)
	Increased generation of superoxide	Joseph *et al.* (1980)
	Increased release of PAF-acether	Arnoux *et al.* in Joseph *et al.* (1983)
	IgE-dependent cytotoxicity	Joseph *et al.* (1978)

Similarly, human alveolar macrophages successively incubated with human myeloma IgE protein or the serum from allergic patients and with anti-human IgE, specifically released β-glucuronidase, excreted neutral proteases, and produced the superoxide anion O_2^-. The release was dependent on IgE antibody in patients' sera, as shown by the disappearance of the anti-IgE-induced exocytosis after heating at 56°C or after depletion of IgE of the sensitizing serum. Interestingly, alveolar macrophages collected from allergic patients or cultured with the serum from patients allergic to house mites or to grass pollen released lysosomal enzymes selectively on addition of the specific allergen (+830% increase) and the depletion of IgE but not of IgG prevented enzyme release by the corresponding allergen (Joseph *et al.*, 1980, 1983).

In addition to inducing lysosomal enzyme release by mononuclear phagocytes, IgE could also trigger when cross-linked a sharp increase in cellular cGMP and in calcium uptake, coincident with the release of β-glucuronidase (Dessaint *et al.*, 1980, 1983). Most importantly with respect to allergy, macrophages stimulated by aggregated IgE were shown to release several mediators of immediate-type hypersensitivity, including prostaglandins, prostacyclin, leukotriene C, arachidonic acid, and platelet-activating factor (Rankin *et al.*, 1982, 1984; Rouzer *et al.*, 1982a,b; Spiegelberg *et al.*, 1983; Joseph *et al.*, 1983).

3.2. DIMERIC IgE AS THE MINIMAL DEGREE OF AGGREGATION FOR MACROPHAGE TRIGGERING

Purified rat macrophages were incubated with rat myeloma E protein, which had been nonspecifically cross-linked by dimethylsuberimidate, to compare the activity of various degrees of aggregation of the IgE molecule on macrophage function. Dimeric IgE and larger oligomers induced a selective lysosomal enzyme release together with increasing macrophage intracellular levels of the enzyme, while IgE monomers did not trigger the macrophages. Likewise, the peak in cGMP was induced by dimeric IgE but not by unaggregated IgE (Dessaint *et al.*, 1980). Thus, as for mast cells, dimers or larger oligomers of IgE suffice to induce macrophage activation. As for mast cells also (Fewtrell and Metzger, 1980), the smallest unit signal was the cross-linking of two IgE molecules on the surface of the cell, although larger oligomers proved to be slightly more efficient.

The direct effect of IgE aggregates on macrophages could be demonstrated, owing to the higher affinity of these cells for aggregated IgE when mast cells strongly bind IgE monomers. In a competitive inhibition system where rat mast cells or macrophages were exposed to dimeric IgE in the presence of increasing quantities of IgE monomers, mast cell degranulation was 50% inhibited with a monomer/dimer ratio of 20 : 1 and totally abolished with a 100 : 1 ratio. In contrast, macrophage release of β-glucuronidase was not inhibited at such ratios, and a 50% inhibition was obtained only with a 1000-fold excess of monomers. Thus, with over a 100-fold excess of monomers, which prevented mast cell degranulation, macrophage triggering by IgE was still obtained, indicating a direct activating effect of IgE aggregates on macrophages (Dessaint *et al.*, 1980). Accordingly, stimulation of arachidonic acid release by IgE immune complexes was found not to be inhibited by free IgE antibodies (at 25-fold excess) (Rouzer *et al.*, 1982b).

What is not known is whether IgE aggregates can activate mononuclear phagocytes after their interiorization, in addition to their direct effect through the surface Fc_ϵ receptors. However, the secretion of leukotriene C and of other arachidonic acid metabolites by murine macrophages exposed to particulate immune complexes of IgE can be obtained using IgE-coated particles too large to be ingested by the cells (Rouzer *et al.*, 1982a).

3.3. IgE ANTIBODY-DEPENDENT MACROPHAGE-MEDIATED CYTOTOXICITY

The ultimate step of macrophage activation admittedly is the cytotoxic activity of the cell against foreign targets. The first evidence of an IgE antibody-dependent cell-mediated cytotoxicity mechanism was the observation that normal rat macrophages incubated with the serum from rats immune to *S. mansoni* were able to kill schistosome larvae specifically *in vitro* (Fig. 2). The dependence on IgE of this cytotoxicity was indicated by absorption of the immune serum: only the depletion of IgE was able to totally inhibit the killing reaction (Capron *et al.*, 1975). This finding was further substantiated by various observations concerning rat and human schistosomiasis and filariasis (reviewed by Capron *et al.*, 1982a,b). The involvement of antiparasite IgE antibody as complexed with parasite antigen was indicated by the following experimental evidence: (1) inhibition experiments using rat myeloma IgE added to the immune rat serum during the incubation with the macrophages showed that deaggregated IgE protein exerted little inhibitory activity, in contrast to chemically cross-linked IgE, which prevented schistosomula killing (Joseph *et al.*, 1977); (2) the inducing factor in the immune rat serum was absorbed out not only by *S. mansoni* antigen but also by anti-*S. mansoni* immunosorbent (Capron *et al.*, 1977a); (3) immune complexes containing both antiparasite IgE antibody and schistosome antigen were characterized in the immune rat serum (Santoro *et al.*, 1978); (4) ultracentrifugation of the immune serum and separate testing of the supernatant and the pellet showed that only the latter induced macrophage killing. Almost all antiparasite IgE antibody and 69% of total circulating *S. mansoni* antigen were characterized in this pellet, and the role of IgE in inducing macrophage killing was confirmed by demonstrating that heating the pellet at 56°C abolished the cytotoxic reaction. Thus, IgE antibody triggers normal macrophages to kill the homologous target predominantly when complexed with circulating parasite antigen, in the form of IgE antibody-excess immune complexes (Capron *et al.*, 1977a).

Theoretically, IgE immune complexes could induce macrophage cytotoxicity either by binding primarily to the target (i.e., opsonization) or to the effector cell (i.e., cytophily). The third possibility of nonspecific activation of the macrophages following endocytosis of the complexes can be ruled out because of the

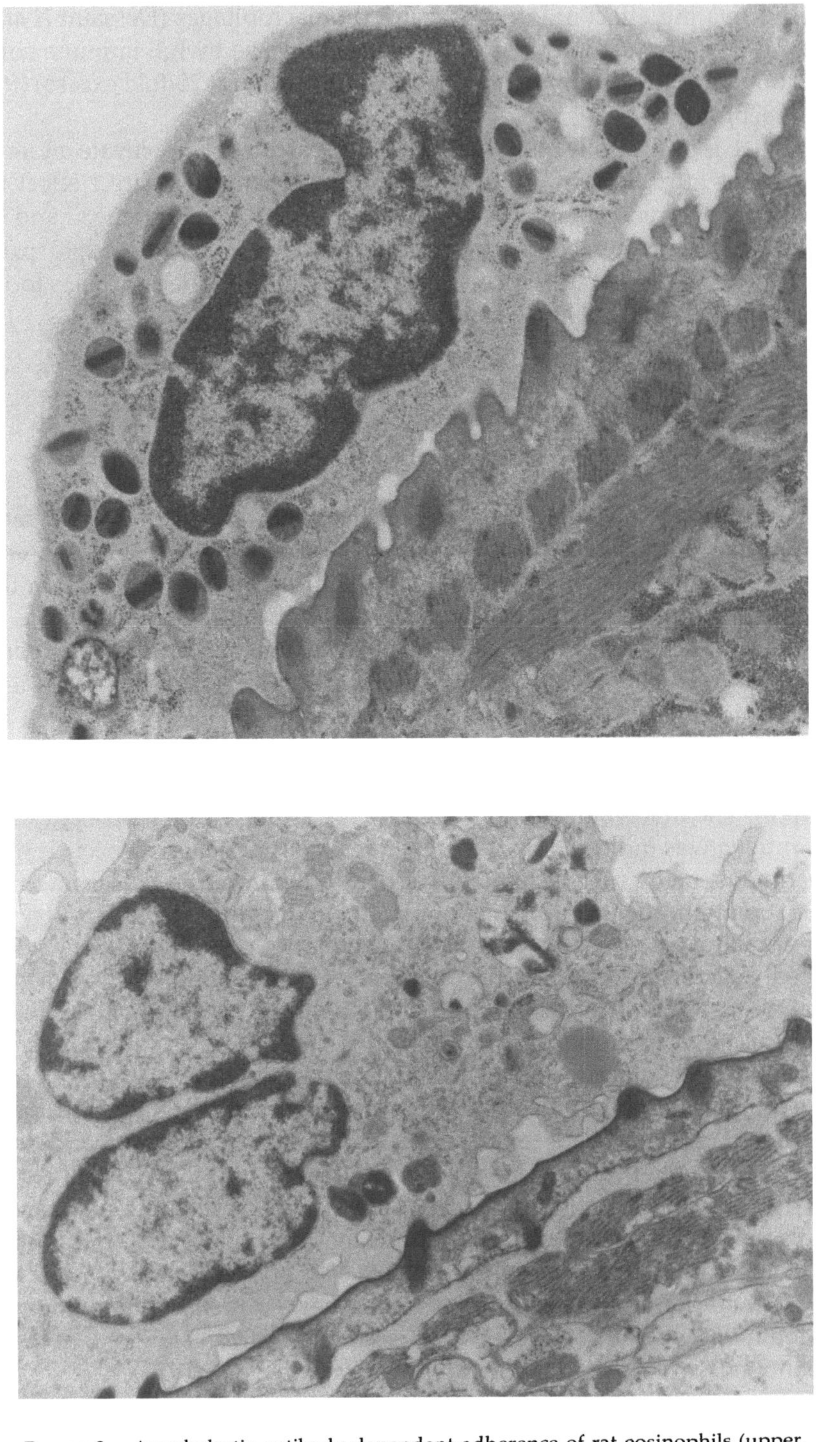

FIGURE 2. Anaphylactic antibody-dependent adherence of rat eosinophils (upper, ×16,600) and macrophages (lower, ×12,000) to *S. mansoni* schistosomula.

strict specificity of the cellular cytotoxicity for the target. By preincubating the target schistosomula with the immune serum before adding normal macrophages, it was shown that no significant cytotoxicity was elicited. Conversely, preincubation of macrophage monolayers with the immune serum was more efficient in inducing the killing than the simultaneous addition of the serum and the parasites to the macrophages (Joseph *et al.*, 1977).

By using the serum of individuals infected by *S. mansoni* and either human monocytes or baboon peritoneal macrophages, IgE immune complexes were shown similarly to induce mononuclear phagocyte-mediated killing of *S. mansoni* schistosomula (Joseph *et al.*, 1978).

A study by Melewicz and Spiegelberg (1980) showed that human monocytes had an enhanced phagocytic activity and displayed specific *in vitro* cytotoxicity for IgE-coated erythrocytes, this cytotoxicity being highly correlated to the percentage of monocytes bearing Fc_ϵ receptors. This IgE-dependent cytotoxicity of blood monocytes was increased in patients with severe atopic disease, whose monocytes bear more Fc_ϵ receptors (Melewicz *et al.*, 1981b).

The role of Fc_ϵ receptors in triggering macrophage cytotoxicity was confirmed by the use of the antireceptor antiserum which inhibits the binding of IgE to mononuclear phagocytes (Melewicz *et al.*, 1982b). This antireceptor antibody induced a dramatic decrease in IgE-dependent macrophage-mediated killing of schistosomula (Capron *et al.*, 1983a,b).

From the induction of biochemical changes such as release of lysosomal enzymes or generation of oxygen metabolites to the triggering of cytotoxicity by IgE immune complexes, the accepted characteristics of macrophage activation have been demonstrated to be induced by the interaction of aggregated or complexed IgE molecules with their specific receptors on the surface of mononuclear phagocytes. While most of these metabolic changes can also be triggered by other immunoglobulin classes via their respective Fc receptors, it should be pointed out that in the antiparasite cytotoxicity system, no other antibody isotype appears to participate, although non-IgE antischistosome antibodies or immune complexes can be detected in the serum of infected humans and rats (Santoro *et al.*, 1978; Stevens *et al.*, 1983). A possible explanation of the exclusive role of IgE in triggering macrophage-mediated cytotoxicity might be found in the delayed activation induced by aggregated IgG: when lysosomal enzyme release or production of cGMP by rat macrophages were triggered by aggregated IgE during the first 15–30 min of incubation, β-glucuronidase release started only after 6 hr of exposure to aggregated IgG. Moreover, macrophages first incubated with IgG aggregates could be additively stimulated by cross-linked IgE, but in contrast IgE-activated macrophages did not respond further to aggregated IgG. In fact, killing of schistosomula by macrophages exposed to IgG oligomers or complexes could only be induced in the presence of complement (Pestel *et al.*, 1984).

At variance with the macrophage-mediated killing of helminth parasites which depends on IgE immune complexes only, eosinophils at least in the rat can be induced to kill parasite targets by anaphylactic antibodies of either isotype (Table 3).

TABLE 3. ANTIBODY-DEPENDENT CELL-MEDIATED CYTOTOXICITY SYSTEMS IN SCHISTOSOMIASIS

	Macrophage-mediated cytotoxicity	Eosinophil-mediated cytotoxicity
Host species	Man, rat	Man, rat, mouse
Nonanaphylactic antibody-dependent cytotoxicity	No[a]	With IgG in humans with complement[b]
Anaphylactic IgG antibody-dependent cytotoxicity	No	In rats
IgE antibody-dependent cytotoxicity	Yes	Yes
Accessory cell	No	Mast cells in rats
Stage of rat infection	⩾ 6 weeks	4–6 weeks (IgG_{2a}-dependent) ⩾ 6 weeks (IgE-dependent)
Blocking by other antibody (sub)classes	No	Yes (IgG_{2a}-dependent cytotoxicity)
Modulation by schistosome-derived factors	Peptides from IgG cleaved by schistosome proteases[d]	Schistosome collagenase[c] Schistosome-derived inhibitor of mast cell degranulation[e]
Passive transfer of protection by armed cells from immune rats	Yes	Yes

[a]IgG-dependent macrophage adherence to schistosomula has been reported by Perez and Smithers (1977), but found poorly efficient in killing (Anwar *et al.* (1979).
[b]Ramalho-Pinto *et al.* (1978), McLaren and Ramalho-Pinto (1979), Anwar *et al.* (1979), McKean and Anwar (1981).
[c]Auriault *et al.* (1983).
[d]Auriault *et al.* (1980, 1984).
[e]Capron *et al.* (1978b).

3.4. EOSINOPHIL CYTOTOXICITY BY COMPLEXES OF ANAPHYLACTIC IgG ANTIBODIES

By using experimental procedures similar to those used in the macrophage-mediated cytotoxicity system, it was shown in rat schistosomiasis that: (1) *S. mansoni* schistosomula opsonized with the serum from rats bled 4 weeks after their infection, i.e., at a period when immunity begins to be detected, were specifically killed by normal rat eosinophils; (2) identical results were obtained with heat-inactivated serum from 4-week-infected rats, ruling out the participation of complement and of heat-labile antibody in the cytotoxicity reaction; (3) immunoaffinity-purified IgG2a antibody from 4-week-infected rat serum added to schistosomula induced their killing by normal eosinophils; in contrast, (4) depletion of IgG2a of 4-week-infected rat serum abrogated its opsonizing activity in eosinophil-mediated killing; (5) rat myeloma IgG2a protein or its Fc fragment competitively inhibited eosinophil cytotoxicity by 4-week-infected rat serum (Capron *et al.*, 1978a,b); (6) a rat monoclonal antibody to the schistosomulum surface of the IgG2a subclass (IPLSm1) was shown to specifically induce killing of the larvae by normal rat eosinophils (Grzych *et al.*, 1982). Interestingly, antischistosome antibodies of the mouse anaphylactic IgG1 subclass were shown to induce antischistosomula cytotoxicity by eosinophils (Ramalho-Pinto *et al.*, 1979), thus confirming the involvement of the anaphylactic IgG subclass of rat and mouse respectively to trigger eosinophil-mediated killing. Although IgG antibody-dependent eosinophil-mediated damage to schistosomula has been demonstrated in humans also (Butterworth *et al.*, 1975, 1977), no information is available on the subclass involved in human schistosomiasis.

Returning to the rat model, while the *in vitro* cytotoxicity reaction was shown to proceed by opsonization of the target schistosomula by IgG2a antibodies, cytophilic arming of eosinophils was also efficient in inducing the cytotoxicity. Indeed, eosinophils from 4-week-infected rat could directly kill nonopsonized schistosomula and IgG2a could be detected on the surface of these cells, at the very period when serum IgG2a antibody induces damage by normal eosinophils. *S. mansoni*-specific IgG2a antibody was detected together with parasite antigens in the contemporary immune serum, and isolation of these IgG2a immune complexes allowed the demonstration that they could induce killing by normal eosinophils. However, by using similar experimental conditions at later periods of the infection (from 6 weeks onward), not only were eosinophils from infected rats found not to be directly cytotoxic for nonopsonized schistosomula, but they were also significantly less efficient than normal cells against antibody-opsonized targets. At these late periods, heat-inactivated ultracentrifugation pellets of the immune rat serum were unable to arm normal eosinophils against nonopsonized targets, and these pellets could inhibit the cytotoxicity of normal cells for schistosomula opsonized with IgG2a antibody from 4-week-infected rat serum, revealing that blocking of the cytotoxic capacity of rat eosinophils could be related to the heat-stable immune complexes. Moreover, rosette assays confirmed that eosinophil Fc_γ receptors were blocked at these periods of the infection (Capron *et al.*, 1979).

In experimental schistosomiasis in rats, important variations in the cytotoxic activity of eosinophils from infected animals seem thus to be conditioned by the regulatory role of immune complexes, which, depending on their composition, can either activate or inhibit the cytotoxic capability of the effector cell. These observations from an experimental model are in agreement with the report by Butterworth *et al.* (1977) on human schistosomiasis that this process of blockade can also occur in human infection.

The blockade of eosinophil cytotoxicity was recently confirmed by the use of a rat IgG2c monoclonal antibody (IPLSm3) which recognizes the same surface antigen of schistosomula as the rat IgG2a monoclonal (IPLSm1) that induces eosinophil-mediated killing. The IPLSm3 IgG2c monoclonal did not exhibit any killing activity *in vitro* against schistosomula in the presence of complement, macrophages, or eosinophils, but strongly inhibited in a dose-dependent manner eosinophil-mediated cytotoxicity by IPLSm1 IgG2a antibody. The blocking activity of IPLSm3 IgG2c was observed also when its $F(ab')_2$ fragment was incubated with schistosomula before the addition of IPLSm1 IgG2a (intact) monoclonal and of normal rat eosinophils, indicating a competition or steric hindrance of the two antibodies for the surface antigen of schistosomula. Preincubation of eosinophils with IgG2c immunoglobulins, specific or not for schistosomes, or with the Fc fragment of IgG2c similarly inhibited killing of schistosomula by IPLSm1 IgG2a monoclonal. Taken together, these experiments indicate that non-IgG2a antischistosome antibody could block IgG2a-dependent eosinophil-mediated cytotoxicity both at the target level and by competition for the eosinophil Fc_γ receptor (Grzych *et al.*, 1984).

While one explanation for the involvement of anaphylactic IgG2a antibodies or complexes in the early stage of immunity to schistosomes in the rat can be drawn from their activating effect on eosinophils via their Fc_γ receptors when other IgG subclasses do not seem to be efficient, another function can be attributed to IgG2a in eosinophil-mediated cytotoxicity to schistosomes. While eosinophils were unequivocally identified as the effector cells responsible for killing of parasitic targets (Butterworth *et al.*, 1975; Capron *et al.*, 1978a,b), it was indeed shown in the rat model that highly purified eosinophils depleted of mast cells were less efficient in achieving schistosomulum killing than mixtures of eosinophils and mast cells (up to 20% of the cell suspension) (Capron *et al.*, 1978a,b). Ultrastructural studies could identify, besides eosinophils, mast cells in close contact with the schistosomula. However, mast cells could be replaced by their supernatants after their degranulation either by anaphylactic IgG2a or by chemical agents, and inhibition of mast cell degranulation reduced the cytotoxic potential of eosinophils in the cell mixture. Among mast cell mediators, ECF-A tetrapeptides were shown to enhance IgG2a-dependent eosinophil cytotoxicity (Capron *et al.*, 1981c), part of this activity being related with the increased expression of eosinophil Fc_γ receptors by ECF-As as mentioned previously. It appears therefore that eosinophil cytotoxic activity depends in the rat system on two signals: one specifically provided by the bridging of the effector eosinophils to their target by eosinophil-activating antibodies of the IgG2a subclass, and the other delivered through the anaphylactic degranulation of mast cells, also induced by IgG2a antibodies.

It must be emphasized that these observations on the selective role of IgG2a antibodies in eosinophil-mediated cytotoxicity to schistosomes were made at the early stage of immunity in the rat, while later in the infection not only was IgG2a-dependent killing not demonstrable, but eosinophils appeared blocked in their ability to be triggered by IgG2a antibodies. It is precisely at this later stage of immunity in the rat that IgE-dependent eosinophil-mediated cytotoxicity could be detected.

3.5. EOSINOPHIL ACTIVATION BY IgE

Experiments using serum from rats infected from 6 weeks onward indeed showed the possibility to induce eosinophil-mediated killing only with unheated immune serum, while addition of fresh complement to heat-inactivated immune serum was inefficient (Fig. 2). Absorption of 8-week-infected rat serum by various anti-rat isotypes or addition of various rat myeloma proteins clearly identified the involvement of IgE antibody in eosinophil cytotoxicity. Again, the accessory function of mast cells in enhancing IgE-dependent killing by eosinophils was demonstrated (Capron *et al.*, 1981a, 1982a). Thus, in the rat, anaphylactic antibodies of either isotype are required both to activate eosinophils via their corresponding Fc_γ or Fc_ϵ receptor and to trigger the release of mediators of anaphylaxis by mast cells which provide an additional signal.

Observations made in human schistosomiasis indicated that IgE antibody-dependent eosinophil-mediated killing of *S. mansoni* schistosomula represented an alternative to the cytotoxicity by IgG (A. Capron *et al.*, 1981b; M. Capron *et al.*, 1983). The prominent role played by the Fc_ϵ receptor of eosinophils was indicated by the marked decrease in the cytotoxicity when eosinophils were preincubated with the antiserum to the cross-reactive Fc_ϵ receptor of lymphocytes, when anti-C3b receptor antibody had no such inhibitory activity (Capron *et al.*, 1984b).

Confirming the observations made on the heterogeneity of human eosinophils for their ability to bind IgE, it was shown that the IgE-dependent cytotoxic capacity was apparently restricted to eosinophils with an abnormally low density (hypodense cells) from highly hypereosinophilic patients (blood counts > $3000/mm^3$) (Capron *et al.*, 1984b). Interestingly, it is in these patients that eosinophils seem to participate in immunopathological lesions in inducing tissue lysis (Durack *et al.*, 1976; Parillo and Fauci, 1978; Spry, 1980; Frigas *et al.*, 1980), and tissue eosinophils such as pleural or pulmonary cells are predominantly found as hypodense cells compared to blood eosinophils (Butterworth *et al.*, 1977; Capron *et al.*, 1984b). Moreover, IgE-mediated eosinophil cytotoxicity was incriminated in chronic eosinophilic pneumonia (McEvoy *et al.*, 1978), and when eosinophils from such patients were examined, the cells were found to bear IgE but not IgG on their surface (Prin *et al.*, 1986).

Since markers of eosinophil activation are poorly defined, most of the information on the consequences of their interaction with anaphylactic antibodies or complexes is drawn from cytotoxicity systems. Extracellular release of rat eosinophil peroxidase was recently shown (Khalife *et al.*, 1985) to be induced by anaphylactic antibodies (IgG2a or IgE) either cross-linked by the corresponding anti-immunoglobulin or in antigen–antibody complexes immobilized on non-phagocytizable surfaces. The triggering of eosinophils by IgE immune complexes can account also for their preferential phagocytosis by the cell (Ishikawa *et al.*, 1974; Fujita *et al.*, 1975; Takenaka *et al.*, 1977).

4. THE CASE OF NEUTROPHILS

No clear-cut evidence has been provided that neutrophils can bind IgE. Using IgE-coated erythrocytes, no significant rosetting was observed with neutrophils (Capron *et al.*, 1984b), although it has been reported that in the mouse, neutrophils could interact with IgE-coated particles (Lopez *et al.*, 1983).

In the case of filarial parasites, complex cell–antibody interactions have been identified, and a synergistic effect of IgG and IgE antibodies with complement and neutrophils has been suggested (Mehta *et al.*, 1980; Aime *et al.*, 1984).

5. SUMMARY AND CONCLUSIONS

This review on the interaction of phagocytic cells with immune complexes containing anaphylactic antibodies reveals several important functional consequences in the areas of parasite immunology and of allergy.

The demonstration that mononuclear phagocytes and eosinophils bear low-affinity receptors for IgE with higher affinity for aggregated IgE molecules suggests that when immune complexes of anaphylactic antibodies are formed, these can interact with phagocytic cells in addition to mast cells or basophils.

As particularly shown with macrophage-mediated cytotoxicity, the participation of IgE in triggering the effector function of phagocytic cells has to be considered at two levels: IgE in aggregated or complexed form, whichever its antibody specificity is, can activate the cell as shown above, but the killing of foreign targets, i.e., the helminth parasite itself, is strictly dependent on the presence of specific IgE which recognizes the corresponding antigen on the target. This could be confirmed in the two cellular models by the use of monoclonal antischistosome IgE antibody. The same is true for IgG2a-dependent killing by eosinophils (Capron *et al.*, 1982a, 1983b).

Confirming the observation made on the molecular form of IgE required to trigger the cell, it was found that only IgE immune complexes, which are a striking feature of parasite infection in humans and experimental animals, were responsible for the induction of cellular cytotoxicity. Likewise, complexes of IgG2a can induce the killing activity of rat eosinophils.

In the context of protective immunity to parasites, if we now try to replace these *in vitro* observations, which were made not only in human and experimental schistosomiasis but also in several models of experimental filariasis (Capron *et al.*, 1982b), several comments can be made on the role played by anaphylactic antibodies complexed to parasite antigens.

Direct evidence of their *in vivo* relevance has recently been provided by passive transfer experiments using either polyclonal immune serum selectively depleted of one or the other anaphylactic isotype or monoclonal antibodies belonging to one of these isotypes. In serum transfer experiments, immune rat serum taken at 6 weeks postinfection, i.e., when both IgG2a and IgE cytotoxicity systems are detected *in vitro,* showed after depletion of either IgG2a or IgE a marked decrease in its capacity to transfer immunity (Capron *et al.*, 1982a). Conversely, IgG2a or IgE monoclonal antibody to *S. mansoni* injected into naive rats induced a high degree of protection to a challenge infection (Grzych *et al.*, 1982; Capron *et al.*, 1983b). It is noteworthy that a protective monoclonal antibody in mice belongs to the mouse anaphylactic IgG1 subclass (Zodda and Phillips, 1982).

While eosinophils and macrophages were shown to accumulate around dead schistosomula in immune animals (Hsü *et al.*, 1971; Von Lichtenberg *et al.*, 1976; Capron *et al.*, 1979), transfer of cells from immune donors could protect naive recipient rats against a challenge schistosome infection. Nonadherent, eosinophil-enriched and adherent, macrophage-rich peritoneal cell preparations from immune rats, injected via intradermal and subcutaneous routes at the precise site of exposure to *S. mansoni* cercariae, were indeed able to confer significant protection. Eosinophils appeared to be the major cell population that transferred immunity when collected at the early stage of infection (4–5 weeks) of the donors, and a direct correlation was found between the logarithm of absolute number of eosinophils injected and the percentage of protection of the recipients. These eosinophils were shown by rosette assays using various anti-rat immunoglobulin-coated erythrocytes to bear IgG2a on their surface. Cytophilic antibodies seemed indeed to participate in arming eosinophils from immune donors at that stage of the infection, since eosinophil-enriched cells from normal rats, sensitized *in vitro* with immune complexes from 4- to 6-week-infected rat serum, could also confer significant protection. When cells were collected from immune rats at 6 or 12 weeks of infection, both eosinophils and macrophages could transfer the immunity, and IgE was identified on the surface of these cells, while arming normal eosinophils or macrophages with immune complexes from contemporary immune rat serum could also confer protection passively (Capron *et al.*, 1984a). Therefore, cell transfer experiments confirmed the relationship established between the kinetics of *in vitro* anaphylactic antibody-dependent eosinophil or macrophage cytotoxicity, of serum IgG2a and IgE antibodies and immune complexes, and of immunity in rat schistosomiasis (Capron *et al.*, 1980, 1982a, 1983b), all the more than blocking of eosinophils could also be demonstrated by the cell transfer experiment using 9-week-infected eosinophil donors (Capron *et al.*, 1984a). This raises the important problem of the kinetics of the production of various anaphylactic antibody isotypes and of their successive or combined participation in protective immune responses against parasites and particularly schistosomes, in collaboration with various effector cell populations bearing the corresponding Fc receptors.

These various findings together with results of IgE suppression in other parasitic systems (Dessein *et al.*, 1981), appear convincing enough to validate the role played by anaphylactic immune complexes and particularly those of IgE antibody, through the Fc_ϵ receptors of various cell populations, in the development of protective immunity against metazoan parasites. This does not by any means confer to IgE an exclusive role in these protective mechanisms and does not preclude the synergistic effect that this antibody class can play together with other anaphylactic and nonanaphylactic antibodies. The interaction of phagocytic cells with anaphylactic antibodies in immune defense against helminth parasites, however, reveals that, besides their well-known function in immediate-type hypersensitivity, anaphylactic antibodies in the form of immune complexes can also have an alternate function in protective immunity, as also suggested by some other circumstantial evidence reviewed elsewhere (Capron and Dessaint, 1975; Dessaint, 1982). It is certainly the merit of parasitic models to have disclosed the beneficial function of IgE in immunity.

The demonstration of circulating IgE immune complexes in atopic diseases (Brostoff *et al.*, 1977; Stevens and Bridts, 1984) raises the question of their participation in the pathogenesis of allergic reactions, all the more since local formation of such complexes may be assumed, resulting from local exposition to the allergen when (free) anaphylactic antibodies circulate.

The recent reports that macrophages, after their interaction with IgE immune complexes, can release leukotriene C and other arachidonic acid metabolites (Rouzer *et al.*, 1982a,b; Rankin *et al.*, 1982, 1984; Spiegelberg *et al.*, 1983) and PAF-acether (Arnoux *et al.*, in Joseph *et al.*, 1983) suggest that the release of many mediators of immediate-type hypersensitivity can in fact be the result of interactions between phagocytic cells and IgE. The demonstration that alveolar macrophages from asthmatic patients bear surface IgE and can be triggered specifically via their surface IgE is also strongly suggestive (Joseph *et al.*, 1981, 1983). Interestingly, the IgE-dependent release of β-glucuronidase was inhibited by 15-min treatment of the macrophages with anti-inflammatory drugs, such as disodium cromoglycate, methylprednisolone, or β_2-mimetics each at a concentration of 100 ng/ml *in vitro*. The inhibition was more effective for allergen-induced exocytosis, which probably involves a smaller number of Fc_ϵ receptors, than for anti-IgE-induced release, which concerns the whole of IgE molecules on the macrophage surface (Joseph *et al.*, 1981).

Finally, supportive evidence of the participation of alveolar macrophages in allergic reactions has been reported—the decrease in macrophage concentrations of lysosomal enzymes found after a local provocation test, while the enzyme levels in bronchoalveolar lavage fluids were significantly higher than in the control lung (Tonnel *et al.*, 1983).

While the exact role of eosinophils in allergic reactions through their activation by anaphylactic antibodies remains to be evaluated, it should be pointed out that the same eosinophil subpopulation, i.e., hypodense cells, is identified in the tissue lesions of the hypereosinophilic syndrome and demonstrates higher cytotoxic capability in the presence of IgE.

Besides the specific message directly related to immunity to parasites, it is obvious that the discovery of the IgE-dependent cellular cytotoxicity mechanisms, subsequently followed by the identification of the corresponding Fc_ϵ receptors on the various effector cells, and the characterization of the mechanisms of cell activation as a consequence of binding of IgE to its specific receptor, has opened an entirely new field of knowledge, which will certainly have in the near future an important input in several areas of human pathology and especially in allergic diseases.

REFERENCES

Aime, N., Haque, A., Bonnel, B., Torpier, G., and Capron, A., 1984, Neutrophil-mediated killing of *Dipetalonema viteae* microfilariae: Simultaneous presence of IgE, IgG antibodies and complement is required, *Immunology* **51**:585.

Anderson, C. L., and Spiegelberg, H. L., 1981, Macrophage receptors for IgE: Binding of IgE to specific IgE Fc receptors on a human macrophage cell line, U937, *J. Immunol.* **126**:2470.

Anwar, A. R. E., Smithers, S. R., and Kay, A. B., 1979, Killing of schistosomula of *Schistosoma mansoni* coated with antibody and/or complement by human leukocytes: *In vitro* requirement for complement in preferential killing by eosinophils, *J. Immunol.* **122**:628.

Auriault, C., Joseph, M., Dessaint, J.-P., and Capron, A., 1980, Inactivation of rat macrophages by peptides resulting from cleavage of IgG by schistosome larvae proteases, *Immunol. Lett.* **2**:135.

Auriault, C., Capron, M., Cesari, I., and Capron, A., 1983, Enhancement of eosinophil effector function by soluble factors released by *S. mansoni:* Role of proteases, *J. Immunol.* **131**:464.

Auriault, C., Joseph, M., Tartar, A., Bout, D., Tonnel, A. B., and Capron, A., 1984, Regulatory role of a tripeptide (TKP) from the second constant domain of immunoglobulin G, *Int. J. Immunopharmacol.* **7**:73.

Bach, M. K., Block, K. J., and Austen, K. F., 1971, IgE and IgG_a antibody-mediated release of histamine from rat peritoneal cells. II. Interaction of IgG_a and IgE at the target cell, *J. Exp. Med.* **133**:772.

Boltz-Nitulescu, G., and Spiegelberg, H. L., 1981, Receptors specific for IgE on rat alveolar and peritoneal macrophages, *Cell. Immunol.* **59**:106.

Boltz-Nitulescu, G., Bazin, H., and Spiegelberg, H. L., 1981, The specificity of Fc receptors for IgG_{2a}, IgG_1/IgG_{2b} and IgE on rat macrophages, *J. Exp. Med.* **154**:374.

Boltz-Nitulescu, G., Plummer, J. M., and Spiegelberg, H. L., 1982, Fc receptors for IgE on mouse macrophages and macrophage-like cell lines, *J. Immunol.* **128**:2265.

Brostoff, J., Stanworth, D. R., and Johns, P., 1977, Immune complexes in atopy, *Lancet* **1**:741.

Butterworth, A. E., Sturrock, R. F., Houba, V., Mahmoud, A. A. F., Sher, A., and Rees, P. H., 1975, Eosinophils as mediators of antibody-dependent damage to schistosomula, *Nature (London)* **256**:727.

Butterworth, A. E., Remold, H. G., Houba, V., David, J. R., Franks, D., David, P. H., and Sturrock, R. F., 1977, Antibody-dependent eosinophil-mediated damage to ^{51}Cr-labeled schistosomula of *Schistosoma mansoni:* Mediation by IgG and inhibition by antigen–antibody complexes, *J. Immunol.* **118**:2230.

Capron, A., and Dessaint, J.-P., 1975, A role for IgE in protective immunity, *IRCS J. Med. Sci.* **3**:477.

Capron, A., Dessaint, J.-P., Capron, M., and Bazin, H., 1975, Specific IgE antibodies in immune adherence of normal macrophages to *Schistosoma mansoni* schistosomules, *Nature (London)* **253**:474.

Capron, A., Dessaint, J.-P., Joseph, M., Rousseaux, R., Capron, M., and Bazin, H., 1977a, Interaction between IgE complexes and macrophages in the rat: A new mechanism of macrophage activation. *Eur. J. Immunol.* **7**:315.

Capron, A., Dessaint, J.-P., Joseph, M., Torpier, G., Capron, M., Rousseaux, R., Santoro, F., and Bazin, H., 1977b, IgE and cells in schistosomiasis, *Am. J. Trop. Med. Hyg.* **26**:39.

Capron, A., Dessaint, J.-P., Capron, M., Joseph, M., and Pestel, J., 1980, Role of anaphylactic antibodies in immunity to schistosomes, *Am. J. Trop. Med. Hyg.* **29**:849.

Capron, A., Dessaint, J.-P., Capron, M., Joseph, M., and Torpier, G., 1982a, Effector mechanisms of immunity to schistosomes and their regulation, *Immunol. Rev.* **61**:41.

Capron, A., Dessaint, J.-P., Haque, A., and Capron, M., 1982b, Antibody-dependent cell-mediated cytotoxicity against parasites, *Prog. Allergy* **31**:234.

Capron, A., Dessaint, J.-P., Capron, M., and Joseph, M., 1983a, IgE receptors on different cell lines and their role in triggering different immunological mechanisms, in: *Proc. XII Congr. Eur. Acad. Allergy Clin. Immunol.* (U. Serafini and E. Errigo, eds.), pp. 17–24, O.I.C. Medical Press, Firenze.

Capron, A., Dessaint, J.-P., Capron, M., and Joseph, M., 1983b, Effector mechanisms against schistosomes, in: *Progress in Immunology V* (T. Tada, ed.), pp. 1305–1307, Academic Press, New York.

Capron, M., Capron, A., Torpier, G., Bazin, H., Bout, D., and Joseph, M., 1978a, Eosinophil-dependent cytotoxicity in rat schistosomiasis: Involvement of IgG_{2a} antibody and role of mast cells, *Eur. J. Immunol.* **8**:127.

Capron, M., Rousseaux, R., Mazingue, C., Bazin, H., and Capron, A., 1978b, Rat mast cell–eosinophil interaction in antibody-dependent eosinophil cytotoxicity to *Schistosoma mansoni* schistosomula, *J. Immunol.* **121**:2518.

Capron, M., Torpier, G., and Capron, A., 1979, *In vitro* killing of *S. mansoni* schistosomula by eosinophils from infected rats: Role of cytophilic antibodies, *J. Immunol.* **123**:2220.

Capron, M., Bazin, H., Joseph, M., and Capron, A., 1981a, Evidence for IgE-dependent cytotoxicity by rat eosinophils, *J. Immunol.* **126**:1764.

Capron, M., Capron, A., Dessaint, J.-P., Johansson, S. G. O., and Prin, L., 1981b, Fc receptors for IgE on human and rat eosinophils, *J. Immunol.* **126**:2087.

Capron, M., Capron, A., Goetzl, E. J., and Austen, K. F., 1981c, Tetrapeptides of the eosinophil chemotactic factor of anaphylaxis (ECF-A) enhance eosinophil Fc receptor, *Nature (London)* **289:**71.

Capron, M., Capron, A., Joseph, M., and Verwaerde, C., 1983, IgE receptors on phagocytic cells and immune response to schistosoma infection, *Monogr. Allergy* **18:**33.

Capron, M., Nogueira-Queiroz, J. A., Papin, J. P., and Capron, A., 1984a, Interaction between eosinophils and antibodies: *In vivo* protective role against rat schistosomes, *Cell. Immunol.* **83:**60.

Capron, M., Spiegelberg, H. L., Prin, L., Bennich, H., Butterworth, A. E., Pierce, R. J., Ouaissi, M. A., and Capron, A., 1984b, Role of IgE receptors in effector function of human eosinophils, *J. Immunol.* **132:**462.

Cardella, C. J., Davies, P., and Allison, A. C., 1974, Immune complexes induce selective release of lysosomal hydrolases from macrophages, *Nature (London)* **247:**46.

Cohn, Z., 1978, The activation of mononuclear phagocytes: Fact, fancy and function, *J. Immunol.* **121:**813.

Dessaint, J.-P., 1982, Anaphylactic antibodies and their significance, *Clin. Immunol. Allergy* **2:**621.

Dessaint, J.-P., Capron, A., Joseph, M., and Bazin, H., 1979a, Cytophilic binding of IgE to the macrophage. II. Immunologic release of lysosomal enzyme from macrophages by IgE in the rat: A new mechanism of macrophage activation, *Cell. Immunol.* **46:**24.

Dessaint, J.-P., Torpier, G., Capron, M., Bazin, H., and Capron, A., 1979b, Cytophilic binding of IgE to the macrophage. I. Binding characteristics of IgE on the surface of macrophages in the rat, *Cell. Immunol.* **46:**12.

Dessaint, J.-P., Waksman, B. H., Metzger, H., and Capron, A., 1980, Cytophilic binding of IgE to the macrophage. III. Involvement of cyclic GMP and calcium in macrophage activation by dimeric or aggregated rat myeloma IgE, *Cell. Immunol.* **51:**280.

Dessaint, J.-P., Capron, A., Joseph, M., Auriault, C., and Pestel, J., 1983, Macrophage-mediated IgE ADCC to helminth parasites and IgE-dependent macrophage activation, in: *Macrophage Mediated Antibody-Dependent Cellular Cytotoxicity* (H. S. Koren, ed.), pp. 315–338, Dekker, New York.

Dessein, A. J., Parker, W. L., James, S. L., and David, J. R., 1981, IgE antibody and resistance to infection. I. Selective suppression of the IgE antibody response in rats diminishes the resistance and the eosinophil response to *T. spiralis* infection, *J. Exp. Med.* **153:**423.

Durack, D. T., Sumi, S. M., and Klebanoff, S. J., 1976, Neurocytotoxicity of human eosinophils, *Proc. Natl. Acad. Sci. USA* **73:**1443.

Fewtrell, C., and Metzger, H., 1980, Larger oligomers of IgE are more effective than dimers in stimulating rat basophil leukemia cells, *J. Immunol.* **125:**701.

Finbloom, D. S., and Metzger, H., 1982, Binding of immunoglobulin E to the receptor on rat peritoneal macrophages, *J. Immunol.* **129:**2004.

Finbloom, D. S., and Metzger, H., 1983, Isolation of cross-linked IgE–receptor complexes from rat macrophages, *J. Immunol.* **130:**1489.

Frigas, E., Loegzring, D. A., and Gleich, G. J., 1980, Cytotoxic effects of the guinea pig eosinophil major basic protein on tracheal epithelium, *Lab. Invest.* **42:**35.

Fujita, Y., Rubinstein, E., Greco, D. B., Reisman, R. E., and Arbesman, C. E., 1975, Antigen–antibody complexes in or on eosinophils in nasal secretions, *Int. Arch. Allergy Appl. Immunol.* **48:**577.

Grzych, J. M., Capron, M., Bazin, H., and Capron, A., 1982, *In vitro* and *in vivo* effector function of rat IgG_{2a} monoclonal anti-*S. mansoni* antibodies, *J. Immunol.* **129:**2739.

Grzych, J. M., Capron, M., Dissous, C., and Capron, A., 1984, Blocking activity of rat monoclonal antibodies in experimental schistosomiasis, *J. Immunol.* **133:**998.

Halper, J., and Metzger, H., 1976, The interaction of IgE with rat basophilic leukemia cells. VI. Inhibition by IgG_a immune complexes, *Immunochemistry* **13:**907.

Hsü, J. Y. L., Lust, G. L., and Hsü, H. F., 1971, The fate of challenge schistosome cercariae in a monkey immunized by cercariae exposed to high doses of X-irradiation, *Proc. Soc. Exp. Biol. Med.* **136:**727.

Hubscher, T., 1975, Role of the eosinophil in the allergic reactions. I. EDI, an eosinophil derived inhibitor of histamine release, *J. Immunol.* **114:**1379.

Ishikawa, T., Wicher, K., and Arbesman, C. E., 1974, *In vitro* and *in vivo* studies on uptake of antigen–antibody complexes by eosinophils, *Int. Arch. Allergy Appl. immunol.* **46:**230.

Joseph, M., Dessaint, J.-P., and Capron, A., 1977, Characteristics of macrophage cytotoxicity induced by IgE immune complexes, *Cell. Immunol.* **34**:247.
Joseph, M., Capron, A., Butterworth, A. E., Sturrock, R. F., and Houba, V., 1978, Cytotoxicity of human and baboon mononuclear phagocytes against schistosomula in *vitro:* Induction by immune complexes containing IgE and *Schistosoma mansoni* antigens, *Clin. Exp. Immunol.* **33**:48.
Joseph, M., Tonnel, A., Capron, A., and Voisin, C., 1980, Enzyme release and superoxide anion production by human alveolar macrophages stimulated with immunoglobulin E, *Clin. Exp. Immunol.* **40**:416.
Joseph, M., Tonnel, A. B., Capron, A., and Dessaint, J.-P., 1981, The interaction of IgE antibody with human alveolar macrophages and its participation in the inflammatory processes of lung allergy, *Agents Actions* **11**:619.
Joseph, M., Tonnel, A. B. T., Torpier, G., Capron, A., Arnoux, B., and Benveniste, J., 1983, Involvement of immunoglobulin E in the secretory processes of alveolar macrophages from asthmatic patients, *J. Clin. Invest.* **71**:221.
Kagey-Sobotka, A., Dembo, M., Goldstein, B., Metzger, H., and Lichtenstein, L. M., 1981, Qualitative characteristics of histamine release from human basophils by covalently cross-linked IgE dimers and trimers, *J. Immunol.* **127**:2285.
Khalife, J., Capron, M., Grzych, J. M., Bazin, H., and Capron, A., 1985, Extracellular release of rat eosinophil peroxidase (EPO). I. Role of anaphylactic immunoglobulins, *J. Immunol.* **134**:1968.
Lawrence, D. A., Weigle, W. D., and Spiegelberg, H. L., 1975, Immunoglobulins cytophilic for human lymphocytes, monocytes, and neutrophils, *J. Clin. Invest.* **55**:368.
Lee, W. T., and Conrad, D. H., 1984, The murine lymphocyte receptor for IgE. II. Characterization of the multivalent nature of the B lymphocyte receptor for IgE, *J. Exp. Med.* **152**:1790.
Lopez, A. F., Strath, M., and Sanderson, C. J., 1983, Mouse immunoglobulin isotypes mediating cytotoxicity of target cells by eosinophils and neutrophils, *Immunology* **48**:503.
McEvoy, J. D. S., Donald, K. J., and Edwards, P. L., 1978, Immunoglobulin level and electron microscopy in eosinophil pneumonia, *Am. J. Med.* **64**:529.
McKean, J. R., and Anwar, A. R. E., 1981, *Schistosoma mansoni:* Complement and antibody damage mediated by human eosinophils and neutrophils, in killing schistosomula *in vitro, Exp. Parasitol.* **51**:307.
McKeever, P. E., and Spicer, S. S., 1980, Surface receptors of mononuclear phagocytes, in: *The Reticuloendothelial System,* Volume 1, *Morphology* (I. Carr and W. T. Daems, eds.), pp. 161–258, Plenum Press, New York.
McLaren, D. J., and Ramalho-Pinto, F. J., 1979, Eosinophil-mediated killing of schistosomula of *S. mansoni in vitro:* Synergistic effect of antibody and complement, *J. Immunol.* **123**:327.
Mehta, K., Sindhu, R. K., Subrahmanyam, D., and Nelson, D. S., 1980, IgE-dependent adherence and cytotoxicity of rat spleen and peritoneal cells to *Litomosoides carinii* microfilaria, *Clin. Exp. Immunol.* **41**:107.
Melewicz, F. M., and Spiegelberg, H. L., 1980, Fc receptors for IgE on a subpopulation of human peripheral blood monocytes, *J. Immunol.* **125**:1026.
Melewicz, F. M., Zeiger, R. S., Mellon, M. H., O'Connor, R. D., and Spiegelberg, H. L., 1981a, Increased peripheral blood monocytes with Fc receptors for IgE in patients with severe allergic disorders, *J. Immunol.* **126**:1592.
Melewicz, F. M., Zeiger, R. S., Mellon, M. H., O'Connor, R. D., and Soiegelberg, H. L., 1981b, Increased IgE-dependent cytotoxicity by blood mononuclear cells of allergic patients, *Clin. Exp. Immunol.* **43**:526.
Melewicz, F. M., Kline, L. E., Cohen, A. B., and Spiegelberg, H. L., 1982a, Characterization of Fc receptors for IgE on human alveolar macrophages, *Clin. Exp. Immunol.* **49**:364.
Melewicz, F. M., Plummer, J. M., and Spiegelberg, H. L., 1982b, Comparison of the Fc receptors for IgE on human lymphocytes and monocytes, *J. Immunol.* **129**:563.
Metzger, H., 1978, The IgE–mast cell system as a paradigm for the study of antibody mechanisms, *Immunol. Rev.* **41**:187.
Parillo, J. E., and Fauci, A. S., 1978, Human eosinophils: Purification and cytotoxic capability of eosinophils from patients with the hypereosinophilic syndrome, *Blood* **51**:457.
Perez, H., and Smithers, S. R., 1977, *Schistosoma mansoni* in the rat: The adherence of macrophages to schistosomula *in vitro* after sensitization with immune serum, *Int. J. Parasitol.* **7**:315.

Pestel, J., Dessaint, J.-P., Joseph, M., Bazin, H., and Capron, A., 1984a, Macrophage triggering by aggregated immunoglobulins. II. Comparison of IgE and IgG aggregates or immune complexes, *Clin. Exp. Immunol.* **57**:404.

Prin, L., Capron, M., Tonnel, A. B., Bletry, O., and Capron, A., 1983, Heterogeneity of human peripheral blood eosinophils: Variability in cell density and cytotoxic ability in relation to the level and origin of hypereosinophilia, *Int. Arch. Allergy Appl. Immunol.* **72**:336.

Prin, L., Capron, M., Gosset, P., Kustniez, J. P., Bletry, O., Tonnel, A. B., and Capron, A., 1986, Eosinophilic lung disease: Immunological studies of blood and alveolar eosinophils, *Clin. Exp. Immunol.* **63**:249–257.

Rabinovitch, M., Manejias, R. E., and Nussenzweig, V., 1975, Selective phagocytic paralysis induced by immobilized immune complexes, *J. Exp. Med.* **142**:827.

Ramalho-Pinto, F. J. D., McLaren, D. J., and Smithers, S. R., 1978, Complement-mediated killing of schistosomula of *Schistosoma mansoni* by rat eosinophils in *vitro*, *J. Exp. Med.* **147**:147.

Ramalho-Pinto, F. J. D., De Rossi, R., and Smithers, S. R., 1979, Murine *Schistosoma mansoni* anti-schistosomula antibodies and the IgG subclasses involved in the complement and eosinophil-mediated killing of schistosomula *in vitro*, *Parasite Immunol.* **1**:295.

Rankin, J. A., Hitchcock, M., Merrill, W. W., Bach, M. K., Brashler, J. R., and Askenase, P. W., 1982, IgE-dependent release of leukotriene C_4 from alveolar macrophages, *Nature (London)* **297**:329.

Rankin, J. A., Hitchcock, M., Merrill, W. W., Huang, S. S., Brashler, J. R., Bach, M. K., and Askenase, P. W., 1984, IgE immune complexes induce immediate and prolonged release of leukotriene C_4 (LTC_4) from rat alveolar macrophages, *J. Immunol.* **132**:1993.

Rouzer, C. A., Scott, W. A., Hamill, A. L., Liu, F.-T., Katz, D. H., and Cohn, Z. A., 1982a, IgE immune complexes stimulate arachidonic acid release by mouse peritoneal macrophages, *Proc. Natl. Acad. Sci. USA* **79**:5656.

Rouzer, C. A., Scott, W. A., Hamill, A. L., Liu, F.-T., Katz, D. H., and Cohn, Z. A., 1982b, Secretion of leukotriene C and other arachidonic acid metabolites by macrophages challenged with immunoglobulin E immune complexes, *J. Exp. Med.* **156**:1077.

Santoro, F., Capron, M., Joseph, M., Rousseaux-Prevost, R., and Capron, A., 1978, Circulating antigens and immune complexes in *Schistosoma mansoni*-infected rats: Characterization by radioimmunoprecipitation-PEG assay (RIPEGA), *Clin. Exp. Immunol.* **32**:435.

Segal, D. M., Taurog, J. D., and Metzger, H., 1977, Dimeric immunoglobulin E serves as a unit signal for mast cell degranulation, *Proc. Natl. Acad. Sci. USA* **74**:2993.

Spiegelberg, H. L., Boltz-Nitulescu, G., Plummer, J. M., and Melewicz, F. M., 1983, Characterization of the IgE Fc receptors on monocytes and macrophages, *Fed. Proc.* **42**:124.

Spry, C. J. F., 1980, Eosinophilia and hypereosinophilic syndrome, *Trans. R. Soc. Trop. Med. Hyg.* **74**(Suppl.):3.

Stevens, W. J., and Bridts, C. H., 1984, IgG-containing and IgE-containing circulating immune complexes in patients with asthma and rhinitis, *J. Allergy Clin. Immunol.* **73**:276.

Stevens, W. J., Feldmeier, H., Bridts, C. H., and Daffalla, A. A., 1983, IgG and IgE circulating immune complexes, total serum IgE and parasite related IgE in patients with mono- or mixed infection with *Schistosoma mansoni* and/or *S. haematobium:* Influence of therapy, *Clin. Exp. Immunol.* **52**:142.

Takenaka, T., Okuda, M., Kawabori, S., and Kubo, K., 1977, Extracellular release of peroxidase from eosinophils by interaction with immune complexes, *Clin. Exp. Immunol.* **28**:56.

Tonnel, A. B., Joseph, M., Gosset, P., and Fournier, E., 1983, Stimulation of alveolar macrophages in asthmatic patients after local provocation test, *Lancet* **1**:1406.

Von Lichtenberg, F., Sher, A., Gibbons, N., and Doughty, B. L., 1976, Eosinophil-enriched inflammatory response to schistosomula in the skin of mice immune to *Schistosoma mansoni* infection in mice, *Am. J. Path.* **84**:479.

Zodda, R. M., and Phillips, S. M., 1982, Monoclonal antibody-mediated protection against *Schistosoma mansoni* infection in mice, *J. Immunol.* **129**:2326.

6

Leukocyte Function in Human Allergic Disease

MICHAEL R. SIMON

1. INTRODUCTION

Human allergic disease is mediated by antigen-specific IgE which fixes to a high-affinity receptor on the surfaces of mast cells and basophils. Antigen bridging of adjacent IgE molecules results in the release of mediators of Type I hypersensitivity. These include histamine, neutrophil chemotactic factors, eosinophil chemotactic factor of anaphylaxis (ECF-A), and platelet-activating factor (PAF) (Plaut and Lichtenstein, 1978). In turn, these substances affect various other leukocytes.

The presence of histamine receptors has been established on human leukocytes (Melmon *et al.*, 1972), including monocytes, T lymphocytes, B lymphocytes, and neutrophils (Saxon *et al.*, 1977; Smart and Kay, 1981). There are two known types of histamine receptors on leukocytes: an H_1 high-affinity receptor and an H_2 low-affinity receptor. When occupied by an agonist, the H_1 receptor causes an augmentation of cellular reactivity, while the H_2 receptor agonist causes a diminution of cellular reactivity (Ogden *et al.*, 1979; Cavagnaro and Osband, 1983). The H_2 receptor effects are mediated by an increase in intracellular cAMP (Bourne *et al.*, 1972) while the H_1 effects are secondary to increased cGMP (Marom and Casale, 1983). The histamine antagonists also alter cellular function, although the mechanism by which this occurs is not established.

Human allergic disease is associated with a multiplicity of immunological abnormalities. In this chapter, after the immunological effects in humans of the components of the Type I hypersensitivity reaction are outlined, the abnormalities associated with specific diseases will be reviewed. Then the observed

MICHAEL R. SIMON • Medical and Research Services, Veterans Administration Medical Center, Allen Park, Michigan 48101; and Department of Medicine, Wayne State University School of Medicine, Detroit, Michigan 48201.

abnormalities will be related to the allergic mediators when possible. Except where otherwise stated, only work in humans is discussed.

2. EXPRESSION OF HISTAMINE RECEPTORS ON HUMAN MONONUCLEAR AND POLYMORPHONUCLEAR LEUKOCYTES (TABLE 1)

Histamine receptors are present on 8% of B cells which bear complement receptors, on 20% of those with surface IgM, and on 50% of those with surface IgG. These latter cells are able to mature into IgG-secreting cells (Ballet and Merler, 1976). Saxon *et al.* (1977) report that 40% of B cells are histamine receptor-positive. In addition, 34% of tonsillar T cells bear histamine receptors (Ballet and Merler, 1976), as do 10% of peripheral blood T cells (Saxon *et al.*, 1977). One-half of $OKT8^+$ T suppressor/cytotoxic cells have histamine receptors and function as suppressor cells (Lima and Rocklin, 1981; Damle and Gupta, 1981b; Cavagnaro and Osband, 1983). $OKT8^+$ lymphocytes which are histamine receptor-negative have cytotoxic activity (Osband and Hamilton, 1982; Cavagnaro and Osband, 1983). Whether histamine receptor-bearing lymphocytes are the proliferating cells in one-way mixed lymphocyte cultures and participate in mixed lymphocyte culture-induced cytotoxicity (Ballet and Merler, 1976) remains controversial (Tartakovsky *et al.*, 1979). T lymphocytes with Fc receptors for IgG (T gamma cells) (Rocklin *et al.*, 1980a), concanavalin A (Con A)-inducible suppressor cells (Fineman *et al.*, 1979; Rocklin and Haberek-Davidson, 1981), and antigen-induced suppressor cells (Rocklin *et al.*, 1980a,b) have also been shown to possess histamine receptors. Circulating suppressor cells in patients with lymphoma also possess histamine receptors (Zarling *et al.*, 1980). Depletion of histamine receptor-positive lymphocytes results in diminished pokeweed mitogen (PWM)- and Con A-induced thymidine uptake with no change in phytohemagglutinin (PHA)- and antigen-induced thymidine uptake (Tartakovsky *et al.*, 1979), providing further evidence that Con A-responsive lymphocytes carry histamine receptors.

TABLE 1. HISTAMINE RECEPTOR-BEARING MONONUCLEAR CELLS

Cell type	% of cells bearing histamine receptors
B cells with complement receptors	8
B cells with surface IgM	20
B cells with surface IgG	50
Tonsillar T cells	34
Peripheral blood T cells	10
$OKT8^+$ T suppressor/cytotoxic cells	≃50
Con A-inducible suppressor cells	present
Antigen-inducible suppressor cells	present
Monocytes	≃33–52

Histamine receptors are found on 33 to 52% of monocytes (Saxon *et al.*, 1977; Smart and Kay, 1981) and on 25% of neutrophils (Smart and Kay, 1981). The monocyte receptors are of the H_1 (Smart and Kay, 1981) and H_2 classes (Lappin and Whaley, 1980).

Histamine receptor-positive T cells are able to produce a lymphokine (lymphocyte mitogenic factor), and in turn can be activated in response to that lymphokine (Ballet and Merler, 1976). Such cells also participate in the production of leukocyte migration inhibition factor (LIF) (Reichman *et al.*, 1979). In addition, histamine receptor-positive B cells are able to secrete IgG (Ballet and Merler, 1976). Histamine receptor-bearing leukocytes also carry receptors for epinephrine and prostaglandin (PG) E_1 (Weinstein and Melmon, 1976).

Studies with murine leukocytes (reviewed by Plaut and Lichtenstein, 1982) have raised some question as to whether murine leukocytes adsorbed onto histamine–albumin–Sepharose conjugates are actually histamine receptor-bearing cells. It is thought that the histamine structure may have been modified, changing the specificity of the pharmacoabsorbent. Human studies using histamine–albumin–Sepharose conjugates include the following: Melmon *et al.* (1972), Ballet and Merler (1976), Weinstein and Melmon (1976), Reichman *et al.* (1979), Tartakovsky *et al.* (1979), Zarling *et al.* (1980), Rocklin *et al.* (1980a), Rocklin and Haberek-Davidson (1981). In addition, again using murine lymphocytes, it has been reported that membrane binding and cellular uptake of histamine occur simultaneously, making it difficult to determine the specificity, affinity, and number of binding sites per cell (Wang *et al.*, 1983).

3. HISTAMINE EFFECTS ON CELL SURFACE RECEPTORS (TABLE 2)

Histamine modulates the expression of many cell surface receptors. Surface receptors for antigens which are present on sheep erythrocytes as measured by sheep red blood cell (SRBC) rosette (T rosette, E rosette) formation have long been recognized as a marker of T lymphocytes. The process of SRBC rosette formation is inhibited by agents which elevate cAMP (Chisari and Edgington, 1974; Lang *et al.*, 1978), including histamine. Concentrations of histamine as low as 10^{-6} M result in decreased numbers of lymphocytes forming SRBC rosettes

TABLE 2. EFFECTS OF HISTAMINE ON CELL SURFACE RECEPTOR EXPRESSION

Cell	Receptor	Effect	Histamine receptor
Lymphocyte	SRBC	Decrease	H_2
T mu	Fc portion of IgG	Increase	H_2 agonist, impromidine
T mu	OKT8	Increase	H_2 agonist, impromidine
T mu	OKT4	Decrease	H_2 agonist, impromidine
T mu	β_2-Microglobulin	Increase	H_2 agonist, impromidine
T gamma	Fc portion of IgG	Decrease	H_2 antagonist, cimetidine
T gamma	Fc portion of IgG	Decrease	H_1 agonist, pyridylethylamine

(DeCock *et al.*, 1977; Galli *et al.*, 1979). Since this phenomenon is prevented by the H_2 blocker cimetidine, it is probably mediated via H_2 receptors (Malolepszy *et al.*, 1980). Cimetidine alone results in decreased proportions of T lymphocytes forming active SRBC rosettes (Lang *et al.*, 1981b).

Histamine effects on the expression of peripheral blood T-cell receptors have been studied using theophylline-sensitive and -resistant SRBC rosettes. In those T cells which form theophylline-resistant E rosettes and predominantly possess receptors for IgM (T helper cell enriched), the H_2 agonist impromidine causes increased expression of β_2-microglobulin, of receptors for the Fc segment of IgG (Fc gamma), and of the OKT8 differentiation antigen. Impromidine decreases the expression of the OKT4 differentiation antigen on these cells and causes them to act as suppressor cells. As expected, cimetidine antagonizes the increase in Fc gamma on Fc mu receptor T cells which is caused by impromidine. Cimetidine also acts alone to cause decreased expression of Fc gamma on cells which are already Fc gamma$^+$. Furthermore, the H_1 agonist 2-(2-pyridyl)ethylamine causes decreased expression of Fc gamma receptors (Birch and Polmar, 1981; Birch *et al.*, 1982). Work by Nair and Schwartz (1983) has complemented these findings by showing that histamine exposure for 24 hr results in increased percentages of T cells bearing the OKT8 differentiation antigen. In other work, cimetidine caused a decrease in the percentage of OKT8$^+$ cells (White and Ballow, 1985). Other agents such as theophylline and isoproterenol, which raise cAMP, can cause decreased expression of Fc mu, suggesting that cyclic nucleotide changes may be the mechanism by which H_2 agonists modulate T-cell receptors (Klein *et al.*, 1979; Gupta *et al.*, 1979b).

4. EFFECTS OF HISTAMINE, HISTAMINE AGONISTS, AND ANTAGONISTS ON CELLULAR FUNCTIONS *IN VITRO* AND *IN VIVO*

4.1. HISTAMINE SUPPRESSION OF LYMPHOCYTE FUNCTION (TABLES 3 AND 4)

Histamine stimulates the intracellular accumulation of cAMP in human lymphocytes (Marone *et al.*, 1980; Wang and Zweiman, 1981). However, elevations of leukocyte intracellular cAMP produced by histamine can diminish the stimulatory effects of histamine itself and of β-adrenergic agonists which are also stimulators of cAMP. This effect is termed *heterologous desensitization* and lasts for 24 hr (Safko *et al.*, 1981). Other effects of cAMP are the inhibition of unstimulated and PHA-stimulated lymphocyte thymidine uptake and PHA-induced morphological blast transformation (Smith *et al.*, 1971; Mendelsohn *et al.*, 1973). Con A-stimulated thymidine incorporation is also inhibited by cAMP (Ogden *et al.*, 1979). Conversely, PHA and Con A stimulation of Leu-3$^+$ (OKT4$^+$ helper) cells and Leu-2$^+$,9.3$^+$ (OKT8$^+$ cytotoxic) cells, but not Leu-2$^+$,9.3$^-$ (OKT8$^+$ suppressor) cells, result in a marked increase in cAMP following incubation with histamine. This effect is blocked by the H_2 antagonist cimetidine. The presence

TABLE 3. HISTAMINE EFFECTS ON LYMPHOCYTE FUNCTION

Function	Effect	Cell type	Agonist/receptor type
Thymidine incorporation			
PHA	Inhibition	$OKT8^+$ lymphocyte (9.3^-)	Histamine/H_2
Con A	"	"	"
PWM	None		
Alloantigen	Inhibition	Lymphocyte	Histamine
Tetanus toxoid	"	"	"
Lymphokine release			
LIF	"	"	"
Lymphocyte chemotactic factor	Augmentation	"	Histamine/H_2
Lymphocyte migration inhibition factor	"	"	Histamine/H_1
HSF	"	$OKT8^+$ lymphocyte	Histamine/H_2, and other
Immunoglobulin production			
Spontaneous	None		
PWM-induced	Inhibition	$OKT8^+$ lymphocyte (9.3^-)	Histamine
Cytotoxicity			
Spontaneous	None		
Alloantigen-induced	Inhibition	Lymphocyte	"
Lymphocyte mobility	"	T lymphocyte	Histamine/H_1 and H_2

of Leu-2^+,9.3^- suppressor cells prevents the mitogen-induced increased histamine responsiveness of Leu-2^+,9.3^+ cytotoxic cells (Khan *et al.*, 1985).

In one of the first comprehensive studies of histamine effects on cellular immune function (Ballet and Merler, 1976), 10^{-3} M histamine was added to mononuclear cell cultures containing histamine receptor-bearing cells. This resulted in marked inhibition of alloantigen-induced lymphocyte cytotoxicity. Alloantigen- and tetanus toxoid-induced thymidine incorporation were also markedly diminished. Subsequent work has confirmed that histamine inhibits lymphocyte thymidine incorporation induced by Con A (Wang and Zweiman, 1978; Ogden and Hill, 1980; Thomas *et al.*, 1981). Ogden *et al.* (1979) found that relatively high concentrations of histamine (10^{-5} to 10^{-3} M) suppress Con A-induced thymidine uptake while lower concentrations (10^{-7} M) are stimulatory. These concentration-dependent histamine effects are attributed to the different affinities of the H_1 and H_2 receptors. Lymphocyte cultures stimulated with suboptimal Con A concentrations respond more effectively to histamine-induced inhibition than with optimal concentrations (Hebert *et al.*, 1980). PHA-induced thymidine uptake is also inhibited by histamine (Wang and Zweiman, 1978; Staszak *et al.*, 1980; Gordon *et al.*, 1981; Thomas *et al.*, 1981). This effect is also best seen at suboptimal stimulatory concentrations of PHA (Goodwin *et al.*, 1979; Martinez *et al.*, 1979; Hebert *et al.*, 1980; Meretey *et al.*, 1981) and is medi-

TABLE 4. HISTAMINE AGONIST/ANTAGONIST EFFECTS ON LYMPHOCYTE FUNCTION

Function	Effect	Cell type	Agent tested
Thymidine incorporation			
PMA	Inhibition	Lymphocyte	2-Pyridylethylamine (H_1 agonist)
PWM	"	"	"
	"	Theophylline-sensitive T cell	"
	Inhibition of helper function	Theophylline-resistant T cell	Impromidine (H_2 agonist)
	Augmentation	Lymphocyte	Cimetidine (H_2 antagonist)
Con A	Inhibition	"	Diphenhydramine, clemastine (H_1 antagonists)
	Augmentation	"	Cimetidine
PHA	"	"	"
Streptokinase/streptodornase	"	"	"
Alloantigen	"	"	"
B-cell maturation	Decreased suppression when theophylline-resistant T cells are present	Theophylline-sensitive T cell	2-Pyridylethylamine
	Augmentation when theophylline-resistant T cells are absent	"	"
PWM-induced IgG production	Inhibition	$OKT8^+$ cell	"
	"	"	Dimaprit (H_2 agonist)
	"	"	Diphenhydramine
	Augmentation	Lymphocyte	Cimetidine
Spontaneous lymphocyte-mediated cytotoxicity	Inhibition	"	Diphenhydramine, clemastine
	"	"	Cimetidine
Suppressor cell activity			
Histamine-induced	"	T suppressor cell	Cimetidine, metiamide (H_2 antagonists)
Con A-induced	"	"	Cimetidine

ated by H_2 receptors (Hebert *et al.*, 1980; Thomas *et al.*, 1981) on $OKT8^+$ T cells (Damle and Gupta, 1981b). These H_2 receptor-mediated histamine-induced T suppressor cells are included in the noncytotoxic (defined as negative with murine monoclonal antibody 9.3) subpopulation of Leu-2^+ ($OKT8^+$) cells (Sansoni *et al.*, 1985). Thomas *et al.* (1981) showed that the suppressor cell stimulated by histamine is radiosensitive. Staszak and Goodwin (1980) established a correlation between histamine-induced and PGE_2-induced inhibition of lymphocyte thymidine incorporation even though histamine inhibited PGE_2 production by mixed mononuclear cells from peripheral blood. Histamine activation of an intermediary adherent suppressor cell in the inhibition of mixed lymphocyte- and Con A-induced reactivity has been suggested (Hebert *et al.*, 1980; Hall *et al.*, 1983). These cells maximize histamine-induced suppression. Histamine preincubation is as effective an inhibiting influence as coculture with histamine (Thomas *et al.*, 1981). Lymphocytes precultured for 24 hr with histamine suppress autologous natural killer and antibody-dependent cellular cytotoxicity (Nair and Schwartz, 1983). Histamine also inhibits T-cell release of LIF (Rigal *et al.*, 1979). It has been suggested that histamine mediates Con A-induced suppressor cell activity (Osband *et al.*, 1980). Histamine-induced acceleration of T-cell mobility is blocked by both H_1 and H_2 antagonists (Smogorzewska *et al.*, 1981). Recently, histamine has been reported to cause release of a lymphocyte chemotactic factor and a lymphocyte migration inhibition factor when acting via the H_2 and H_1 receptors, respectively (Center *et al.*, 1983). The chemotactic factor is released by $OKT8^+$ lymphocytes and attracts $OKT4^+$ cells, while the lymphocyte migration inhibition factor is released by $OKT4^+$ cells. This latter factor is composed of a 35,000-dalton fraction which acts on $OKT4^+$ lymphocytes and a 75,000-dalton fraction which acts on both $OKT4^+$ and $OKT8^+$ lymphocytes (Berman *et al.*, 1984).

One of the mechanisms by which histamine suppresses cellular function is by causing the elaboration of soluble suppressor factor(s). Rocklin *et al.* (1980a) have reported, and Meretey *et al.* (1981) have confirmed, that *in vitro* histamine activation of T suppressor cells results in the elaboration of a soluble histamine-induced suppressor factor (HSF). This activation is inhibited by H_2 blockers. However, data presented by Beer *et al.* (1982a) suggest that diphenhydramine, an H_1 antagonist, and pyridylethylamine, an H_1 agonist, may also augment HSF production or its effects on monocytes. Consequently, it cannot be concluded that histamine is acting solely via the H_2 receptor. HSF inhibits PHA-induced (Rocklin and Melmon, 1978), and autologous and allogeneic mixed lymphocyte culture (MLC)-induced (Damle and Gupta, 1981a), thymidine incorporation (Table 5). The cells which produce this factor are $OKT8^+$ T cells (Damle and Gupta, 1981a). Monocytes, or a monocyte-derived factor (possibly interleukin-1), are required for the generation of HSF (Beer *et al.*, 1982b). HSF acts in part by augmenting monocyte production of PGE_2. However, one fraction of HSF which causes suppression does not increase PGE_2 production (Beer *et al.*, 1982a; Rocklin *et al.*, 1983). Prostaglandins of the E series inhibit lymphocyte thymidine uptake (Ogden *et al.*, 1979; Staszak and Goodwin, 1980; Goodwin and Webb, 1980).

4.2. HISTAMINE AGONIST EFFECTS ON B-CELL FUNCTION (TABLES 3 AND 4)

Many of the effects of histamine, its agonists and antagonists—described below—cannot be explained using the concept of two histamine receptors. The H_1 receptor agonist 2-(2-pyridyl)ethylamine has been reported to inhibit PWM-

and phorbol myristate acetate (PMA)-induced thymidine incorporation (Gordon *et al.*, 1981). The finding of 2-(2-pyridyl)ethylamine inhibition of PWM-induced thymidine incorporation was confirmed using cultures of theophylline-sensitive T cells (suspensions enriched in suppressor cells) (Birch and Polmar, 1982). These two findings appear consistent with the hypothesis that the H_1 agonist is augmenting T suppressor cell activity since low-dose histamine has been reported to augment cellular function via the H_1 receptor (Ogden and Hill, 1980; Cavagnaro and Osband, 1983). Birch and Polmar (1982) preincubated theophylline-sensitive (suppressor enriched) T cells with 2-(2-pyridyl)ethylamine. These cells provided decreased suppressor activity in PWM-driven B-cell maturation to plasma cells when helper-enriched (theophylline resistant) T cells were added to the B cells. Preincubation of theophylline-sensitive (suppressor enriched) T lymphocytes with 2-(2-pyridyl)ethylamine also resulted in augmentation of PWM-driven B-cell maturation to plasma cells when a theophylline-resistant helper cell-enriched population was not added to the B-cell cultures. This is consistent with their earlier finding that 2-(2-pyridyl)ethylamine caused decreased expression of Fc gamma receptors on suppressor cells. However, as documented above, suppressor cells have been found to have H_2 and not H_1 receptors. A third histamine receptor could explain similar effects by H_1 and H_2 agonists. H_1 receptor augmentation of T helper cell function of the theophylline-sensitive cells is also theoretically possible. Birch and Polmar (1982) also studied the H_2 receptor agonist impromidine. Impromidine pretreatment of helper cell-enriched theophylline-resistant T cells resulted in decreased PWM-induced thymidine incorporation. Untreated theophylline-resistant T cells markedly augmented PWM-driven B-cell maturation, and that augmentation was lost when impromidine-pretreated theophylline-resistant T cells were used instead. These findings were interpreted by Birch and Polmar (1982) to represent induction of suppressor cell activity among the theophylline-resistant helper-enriched T cells since addition of helper cells could not overcome the decreased reactivity caused by the impromidine pretreatment. The presence of H_2 receptors on theophylline-resistant helper T cells is implied by the ability of impromidine to alter such cells and induce more Fc gamma expression and by the abolition or decrease of PHA-, Con A-, and PWM-induced lymphocyte reactivity by the depletion of histamine receptor-bearing lymphocytes as reported by Ballet and Merler (1976).

Although histamine has been found to have no effect on PWM-induced thymidine uptake (Gordon *et al.*, 1981; Meretey *et al.*, 1981; Thomas *et al.*, 1981) or spontaneous IgG production, it does suppress PWM-induced IgG production via a radiosensitive $OKT8^+$ cell. The H_1 and H_2 agonists 2-pyridylethylamine and dimaprit and the H_1 antagonist diphenhydramine all cause inhibition of PWM-induced IgG production (Lima and Rocklin, 1981). The H_2 histamine-induced Leu-2^+ ($OKT8^+$) suppressor cell which suppresses PWM-induced immunoglobulin production is a noncytotoxic (9.3^-) cell (Sansoni *et al.*, 1985). The ability of activated T cells to trigger B-cell immunoglobulin secretion is inhibited by HSF probably by direct inhibition of the action of MLC-derived helper factors on resting B cells. This factor also inhibits activation of T helper cells in a plaque-forming cell assay and inhibits the expression of HLA-DR antigens on activated

TABLE 5. EFFECTS OF HISTAMINE-INDUCED SUPPRESSOR FACTOR

Cell affected	Function
Lymphocyte	Inhibition of PHA-induced thymidine incorporation
OKT8+ T cell (by inference)	Inhibition of allogeneic mixed lymphocyte culture-induced thymidine incorporation
OKT4+ T cell (by inference)	Inhibition of autologous mixed lymphocyte culture-induced thymidine incorporation
OKT4+ T cell	Inhibition of expression of HLA-DR antigens following activation
B cell–plasma cell	Inhibition of immunoglobulin secretion probably via T helper cell effects
Monocyte	Induction of prostaglandin E production

T helper cells (Garovoy *et al.*, 1983) (Table 5). Since H_1 and H_2 agonists and/or antagonists all cause similar metabolic effects in some circumstances, it may be necessary to postulate a third histamine receptor (see Section 4.3).

4.3. HISTAMINE ANTAGONIST EFFECTS ON LYMPHOCYTE FUNCTION (TABLE 4)

The *in vitro* effects of H_1 and H_2 receptor antagonists on leukocyte functions in the absence of added histamine have been studied. The H_1 blockers diphenhydramine and clemastine cause decreased Con A-induced lymphocyte reactivity (Ogden and Hill, 1980) and decreased spontaneous lymphocyte-mediated cytotoxicity, respectively. Cimetidine (an H_2 blocker) has also been reported to decrease spontaneous lymphocyte-mediated cytotoxicity (Lang *et al.*, 1981a). As noted above, Lima and Rocklin's (1981) data suggest that diphenhydramine inhibits PWM-induced IgG production. These effects on lymphocyte function are not readily explained by current concepts of H_1 or H_2 receptor activity. It is conceivable that binding of a histamine antagonist to the receptor might have functional consequences even without the presence of histamine.

The H_2 blockers have been more extensively studied. The addition of cimetidine to PHA-stimulated lymphocyte cultures has been reported to cause an increase in lymphocyte thymidine uptake (Gifford *et al.*, 1980; Palacios and Alarcon-Segovia, 1981; Meretey *et al.*, 1981; Thomas *et al.*, 1981), an increase in the proportion of responding lymphocytes, and an increase in the average incremental lymphocyte growth rate as determined by change in cell volume distribution (Robertson *et al.*, 1979). Other examples of cimetidine-augmented lymphocyte function include increased PWM-induced thymidine uptake (Gifford *et al.*, 1980; Palacios and Alarcon-Segovia, 1981) and immunoglobulin synthesis (Palacios and Alarcon-Segovia, 1981), increased streptokinase/streptodornase- and alloantigen-induced thymidine incorporation (Gifford *et al.*, 1980), and increased

Con A-induced lymphocyte reactivity (Ogden and Hill, 1980). The H_2 receptor antagonists metiamide and burimamide augment guinea pig lymphocyte reactivity (Beets and Dale, 1979).

Metiamide blocks histamine-induced suppressor T-cell activity (Martinez *et al.*, 1979). Cimetidine abrogates Con A-induced and histamine-induced suppressor T-cell function *in vitro* (Palacios and Alarcon-Segovia, 1981; Meretey *et al.*, 1981). These findings are consistent with the report that Con A-inducible suppressor cells carry histamine receptors (Fineman *et al.*, 1979; Rocklin and Haberek-Davidson, 1981). Cimetidine preincubation of mononuclear cells from patients with disseminated cancer restored and/or augmented the local graft-versus-host (GVH) reaction caused when those cells were injected into partially immunologically suppressed rats (Mavligit *et al.*, 1981). Finally, cimetidine in the absence of histamine was found to have no intrinsic action on histamine-inducible suppressor cells (Rocklin and Haberek-Davidson, 1981). Cimetidine also does not reverse *in vitro* suppressor effects of cord blood T cells, or of lymphocytes from patients with common variable hypogammaglobulinemia, on PWM-induced B-cell secretion of immunoglobulin (Morito *et al.*, 1980; White *et al.*, 1984). Effects like those of cimetidine were not seen when ranitidine, another H_2 antihistamine, was added to PHA-, Con A-, or PWM-stimulated lymphocyte cultures (Puppo *et al.*, 1982).

Histamine itself was found to have no effect on spontaneous lymphocyte-mediated cytotoxicity (Lang *et al.*, 1981a). However, when added directly to the assay system, clemastine and cimetidine inhibit antibody-dependent cellular (Lang *et al.*, 1981b) and natural killer (Nair and Schwartz, 1983) cytotoxicity, the latter effect by decreasing the binding ability of the cytotoxic lymphocytes. This effect is reversible by histamine (Nair and Schwartz, 1983). Cimetidine, at 10^{-7} M, directly inhibits NK cell function. At 10^{-6} to 10^{-4} M, cimetidine-induced inhibition of monocyte suppressor cells results in recovery of NK activity (Ruiz-Arguelles *et al.*, 1982). When lymphocytes are preincubated with cimetidine for 48 hr, NK activity is enhanced (Allen *et al.*, 1984). This contrasts with the effect of histamine preincubation, which suppresses such activity (Nair and Schwartz, 1983).

Cimetidine administered *in vivo* has also been shown to have effects on *in vitro* immunological assays. Serum taken from patients receiving cimetidine causes augmentation of *in vitro* PHA-induced reactivity of lymphocytes from persons not receiving cimetidine (Smith *et al.*, 1979). Lymphocytes from subjects receiving cimetidine which are cultured with autologous serum supplements may also exhibit enhanced PHA-induced thymidine incorporation as compared with pre-cimetidine treatment values (Simon *et al.*, 1981). Lymphocytes from three of four patients with chronic mucocutaneous candidiasis were able to produce LIF *in vitro* in response to candidal antigen only during periods of *in vivo* administration of cimetidine to the patients (Jorizzo *et al.*, 1980). Group data from an earlier study of 12 dyspeptic cimetidine-treated patients compared with similar data from 5 healthy untreated subjects revealed no significant overall differences in BCG-induced production of LIF (McGregor *et al.*, 1977). One patient with common variable hypogammaglobulinemia and marked T-cell suppressor activity exhibited increased *in vitro* IgG secretion and decreased suppressor function in coculture experiments while receiving cimetidine treatment. Simultaneously, the patient's serum IgG concentration also increased (White and Ballow, 1985). Mononuclear cells from subjects receiving cimetidine mediate a more vigorous GVH reaction when injected into partially immunosuppressed rats than do cells from untreated subjects (Talpaz *et al.*, 1982). Therefore, reconstituted *in vivo* effector function is also enhanced.

Cimetidine-induced effects in the absence of added histamine might conceivably be explained by the unexplored possibility that cimetidine acts as an H_1 agonist producing an increase in cGMP and thereby augmenting cellular function. Gordon *et al.* (1981) suggested that some histamine actions are not mediated by currently defined histamine receptors. One possibility is that cimetidine occupies a third type of histamine receptor which is not coupled to adenylate cylase. A third class of histamine receptor (H_3), which is pharmacologically distinguishable from H_1 and H_2 receptors, has recently been described in experiments on rat cerebral cortex (Arrang *et al.*, 1983). This receptor operates at histamine concentrations 100 times smaller than those needed to stimulate H_1 and H_2 receptors in this system. *N*-Methylhistamine and *N,N'*-dimethylhistamine were found to be H_3 receptor agonists while impromidine and burimamide were found to be competitive H_3 antagonists.

4.4. *IN VIVO* EFFECTS OF H_2 BLOCKADE

In vivo immunological effects of cimetidine were first reported by Avella *et al.* (1978), who found that delayed hypersensitivity skin test erythema and induration were significantly increased as compared to the pre-cimetidine test results. These results were mainly secondary to augmentation of responsiveness to streptokinase/streptodornase (Jones, 1978), although an increased erythematous response while on cimetidine was not restricted to that specific antigen. Conversion of a negative delayed hypersensitivity response to streptokinase/streptodornase into a positive response occurred in 7 of 8 patients with Crohn's disease while they were receiving cimetidine. Two of these individuals also developed positive delayed reactivity to histoplasmin (Bicks and Rosenberg, 1980). Similarly, all of 4 patients with chronic mucocutaneous candidiasis converted previously negative delayed candidal skin tests to positive tests while receiving cimetidine (Jorizzo *et al.*, 1980). Finally, cimetidine has been reported to augment dinitrochlorobenzene-induced contact dermatitis (Breuillard and Szapiro, 1978; Daman *et al.*, 1978). Enhancement of delayed hypersensitivity was not confirmed in a double-blind study of patients with Crohn's disease who received cimetidine (Hamid *et al.*, 1983). However, it was not reported whether these patients were simultaneously receiving adrenal corticosteroid treatment. Cimetidine augmentation of streptokinase/streptodornase delayed reactivity could not be confirmed in one study (Wolfe *et al.*, 1979). As opposed to the effects observed with cimetidine, ranitidine *in vivo* caused no enhancement of delayed hypersensitivity skin reactions in 20 patients (Bovo *et al.*, 1981). The absence of ranitidine effects *in vitro* and *in vivo* may result from the possibility that ranitidine is highly specific for gastric H_2 receptors. Ranitidine differs from cimetidine by lacking an imidazole ring and by replacement of the cyanoguanidine group by 1,2-diaminonitroethene.

4.5. *IN VIVO* INTERACTION OF IMMEDIATE AND DELAYED HYPERSENSITIVITY

The first observations of a relationship between immediate and delayed hypersensitivity were reported by Jillson and Huppert (1949). They noted that in 10 of 13 patients with immediate allergy to trichophytin, delayed reactivity was absent, while delayed reactivity was present in all of 10 patients without an immediate wheal and flare. More recently, similar results using tuberculin (Tarlo *et al.*, 1977), trichophytin (Kaaman, 1978; Rajka and Barlinn, 1979), and *Penicillium*, and *Cladosporium* antigen skin tests (Rajka and Barlinn, 1979) have been reported. Also, most subjects with immunological skin reactivity to *Staphylococcus aureus* manifest either an immediate or a delayed-type hypersensitivity to the antigen, but not both types (Rudzki *et al.*, 1964; Sipos, 1965). Slavin *et al.* (1963) showed that the delayed skin reaction was decreased by the immediate wheal and flare reaction induced by either a chemical histamine releaser or by immunological histamine release due to an unrelated antigen. If the immediate wheal and flare were caused by a given antigen, the delayed-type hypersensitivity reaction to the same antigen was completely suppressed. Pepys (1953), in earlier work, had shown that simultaneously administered intradermal histamine and tuberculin resulted in diminution or abolition of the delayed tuberculin reactivity.

4.6. HISTAMINE EFFECTS ON NEUTROPHIL AND MONOCYTE FUNCTION (TABLE 6)

Neutrophils are affected by increases in intracellular cAMP with resulting decreased chemotaxis and random migration (Tse *et al.*, 1972; Hatch *et al.*, 1977). Neutrophil antibody-dependent cellular cytotoxicity is also inhibited by increased cAMP (Gale and Zighelboim, 1974). Polymorphonuclear leukocytes respond to histamine with a rise in intracellular cAMP (Zurier *et al.*, 1974; Busse and Sosman, 1977; Anderson *et al.*, 1977). Nevertheless, neutrophils are less responsive to adenylate cyclase agonists, including histamine, than are lymphocytes (Marone *et al.*, 1980).

Histamine has been shown to cause inhibition of neutrophil chemotaxis in several studies (Hill *et al.*, 1975; Anderson *et al.*, 1977; Radermecker and Maldague, 1981). Histamine preincubation followed by washing of the cells still produces this effect (Hill and Quie, 1975). This inhibition, as well as histamine-induced stimulation of chemokinesis (random movement) and inhibition of lysosomal enzyme release, have been related to histamine-induced increases in

TABLE 6. HISTAMINE AND ANTAGONIST EFFECTS ON NEUTROPHIL AND MONOCYTE FUNCTION

	Function	Effect	Agent tested/receptor type
Neutrophils	Chemotaxis	Inhibition	Histamine/H_2
		"	4-Methylhistamine (H_2 agonist)
		"	Imidazole acetic acid
		Augmentation	2-Pyridylethylamine (H_1 agonist)
		"	Cimetidine (H_2 antagonist)
	Chemokinesis	"	Histamine
	Oxidative metabolism	Inhibition	Histamine
	Lysosomal enzyme release	"	Histamine/H_2
	Histaminase release	"	Imidazole acetic acid
Monocytes	Antibody-dependent cytotoxicity	"	Histamine/H_2
	C_2 production	"	Histamine/H_2
		"	Dimaprit and 4-methylhistamine (H_2 agonists)

cAMP (Zurier *et al.*, 1974; Busse and Sosman, 1976; Anderson *et al.*, 1977). Antigen-induced suppression of chemotaxis in atopic persons may also be secondary to allergic mediator release (Rubin *et al.*, 1978; Radermecker and Maldague, 1981). Imidazole acetic acid, a product of histamine catabolism, inhibits C3b-induced neutrophil histaminase release (Herman *et al.*, 1979) and zymosan-induced chemotaxis (Herman and Colten, 1980).

Histamine-induced inhibition of neutrophil chemotaxis is mimicked by the H_2 agonist 4-methylhistamine, while the H_1 agonist pyridylethylamine enhances the response (Goetzl, 1978). Histamine-induced decreases in neutrophil chemotaxis and lysosomal enzyme release have been shown to be blocked by the H_2 antihistamines burimamide and metiamide, respectively (Hill *et al.*, 1975; Busse and Sosman, 1976, 1977). As in the other systems described previously, cimetidine in the absence of added histamine augments chemotaxis (Radermecker and Maldague, 1981).

Recently, histamine effects on *N*-formyl-methionyl-leucyl-phenylalanine stimulation of neutrophil oxidative metabolism, degranulation, membrane potential changes, and locomotion were extensively studied by Seligmann *et al.* (1983). It was found that histamine, the H_2 agonist dimaprit, the H_1 antagonist pyrilamine, and the H_1 agonists thiazole, 2-methylhistamine, and 2-pyridylethylamine all inhibited superoxide anion production. Neutrophil degranulation as measured by the release of lysozyme and β-glucuronidase was inhibited by histamine, dimaprit, and thiazole. Histamine and dimaprit also inhibited changes in membrane potential stimulated by *N*-formyl-methionyl-leucyl-phenylalanine. Histamine and both H_1 and H_2 agonists enhanced chemokinesis. In each of the above assays, cimetidine had no intrinsic activity, but blocked the effects of histamine. Seligmann *et al.* (1983) postulate that the neutrophil possesses a single histamine receptor which differs from other H_2 receptors with respect to H_1 and H_2 agonist specificity. They also suggest that the histamine effects may be mediated through very early effects on cyclic nucleotide metabolism or through direct changes in ion fluxes.

In 90% monocyte preparations, histamine causes increased intracellular cAMP concentrations and inhibits antibody-dependent cell-mediated cytotoxicity (Herlin and Kragballe, 1980). This effect is blocked by cimetidine. Elevated monocyte cAMP levels are also associated with decreased chemotaxis (Gallin *et al.*, 1978). In addition, histamine has been shown to inhibit chemoattractant-induced monocyte polarization which precedes actual chemotactic movement (Stephens and Snyderman, 1982). Histamine also inhibits monocyte production of the second component of complement. This effect is reproduced by the H_2 agonists dimaprit and 4-methylhistamine, but not by the H_1 agonist 2-(2-aminoethyl)thiazide dihydrochloride. It is blocked by cimetidine, an H_2 antagonist, but not by chlorpheniramine, an H_1 antagonist. Neither antagonist alone had any effect (Lappin and Whaley, 1980).

4.7. SUMMARY

In summary, H_2 agonist activity increases the expression of the surface markers of T suppressor cells, while H_2 antagonists and H_1 agonists have the opposite effect. Histamine has an inhibitory action on most lymphocyte functions, usually via H_2 receptor stimulation. H_2 receptor antagonists, in the absence of added histamine, have the opposite effect, possibly by inhibiting T suppressor cell activity. HSF inhibits both T and B cell functions. Histamine, acting via the H_2 receptor, also inhibits neutrophil and monocyte function. *In vivo,* histamine inhibits the expression of delayed-type hypersensitivity.

5. EFFECTS OF OTHER MEDIATORS OF IMMEDIATE HYPERSENSITIVITY

5.1. NEUTROPHIL CHEMOTACTIC FACTOR AND EOSINOPHIL CHEMOTACTIC FACTOR OF ANAPHYLAXIS

High-molecular-weight chemotactic factor is released during immediate hypersensitivity reactions in humans (Atkins *et al.*, 1977). It has been found in the blood of patients during both the early and late asthmatic responses (Nagy *et al.*, 1982) and following exercise-induced asthma (Nagy, 1983; Lee *et al.*, 1983). Exposure of neutrophils in the absence of a chemotactic gradient to low concentrations of chemotaxins results in specific chemotactic deactivation, while exposure to high concentrations may cause nonspecific deactivation (Nelson *et al.*, 1978). Cold-challenged subjects with cold urticaria produce a chemotactic factor which is capable of causing chemotactic deactivation (Wasserman *et al.*, 1977), and *in vivo* chemotactic deactivation has also been demonstrated (Center *et al.*, 1979).

As already noted, ECF-A is released from basophils and mast cells during immediate hypersensitivity reactions. Some chemotactic deactivation of neutrophils may occur after contact with ECF-A (Wasserman *et al.*, 1975).

Chemotactic factors cause elevated intracellular cAMP in human neutrophils, and chemotactic factor deactivation of this response is stimulus specific (Simchowitz *et al.*, 1980).

5.2. PLATELET-ACTIVATING FACTOR, SEROTONIN HYDROXYEICOSATETRAENOIC ACID, AND THE LEUKOTRIENES

Platelet activation probably occurs during antigen-induced immediate hypersensitivity airway reactions in humans (Knauer *et al.*, 1981). Serotonin, which is released from human platelets during activation by PAF (McManus *et al.*, 1982), induces production of a monocyte chemotactic factor by peripheral blood leukocytes (Foon *et al.*, 1976). This serotonin-induced monocyte chemotactic factor may cause chemotactic deactivation of monocytes and contribute to the impaired monocyte chemotactic response seen in some allergic conditions.

PAF, which causes platelet aggregation, may be released from basophils (Plaut and Lichtenstein, 1978) and from monocytes (Sanchez-Crespo *et al.*, 1980) during allergic reactions. Aggregation of platelets is associated with production of 12-L-hydroxy-5,8,10,14-eicosatetraenoic acid (HETE) (Hamberg *et al.*, 1974; Nugteren, 1975), which in turn can cause chemotactic deactivation of neutrophils (Goetzl *et al.*, 1977; Goetzl and Gorman, 1978). The physiological relevance of this sequence is supported by the IgE-mediated immunological release from human lung fragments of lipid factors which cause neutrophil chemotactic deactivation (Valone *et al.*, 1977). In addition, hydroperoxyeicosatetraenoic acid, which is the metabolic precursor of HETE, inhibits PHA- and Con A-induced T-cell thymidine incorporation while augmenting thymidine uptake by $OKT8^+$ T cells (Gualde *et al.*, 1985).

Leukotrienes D_4 and E_4 are released by mast cells following activation (MacGlashan *et al.*, 1982). These compounds have been shown to inhibit murine (Webb *et al.*, 1982) and human (Rola-Pleszczynski *et al.*, 1982) mitogen-induced mononuclear cell thymidine incorporation.

6. IgE AND THE MACROPHAGE

Consequences of human allergy are understood to derive from effects of mediators of immediate hypersensitivity. These mediators are released from mast cells and basophils following antigen-mediated bridging of adjacent IgE molecules on their cell surfaces. In recent years, IgE-mediated cellular effects on macrophages have been documented in humans. It was first shown that human peripheral blood monocyte cytotoxicity to *S. mansoni* schistosomula can be mediated by antigen-specific IgE (Joseph *et al.*, 1978). Because of the sedimentation characteristics of the IgE, it was concluded that IgE–immune complexes are involved. These same workers (Joseph *et al.*, 1980, 1983) extended their experiments and showed that human alveolar macrophages can be induced to release neutral proteases and lysosomal β-glucuronidase following sequential incubations with IgE and anti-human IgE. Superoxide anion production parallels the enzyme release. They also established that fresh macrophages possess adherent cell surface IgE. Preincubation of macrophages in sera from allergic subjects followed by addition of the relevant allergen also resulted in enzyme release (Joseph *et al.*, 1980, 1981, 1983). This work was subsequently extended (Joseph *et al.*, 1981, 1983) to show that fresh alveolar macrophages from subjects with asthma can be activated by addition of the appropriate allergen and that enzyme release *in vitro* can be inhibited by methylprednisolone, by the selective β_2 agonist albuterol, and by sodium cromoglycate. Finally, Melewicz and Spiegelberg (1980) showed that between 11 and 32% of human peripheral blood monocytes possess Fc receptors for IgE and phagocytose the antigen to which the IgE is bound. The presence of human macrophage Fc receptors for IgE was confirmed by Joseph *et al.* (1981, 1983).

IgE-induced activation of monocytes/macrophages has been found to be mediated via increases in cGMP in guinea pig macrophages, which is the only system studied to date (Dessaint *et al.*, 1980; Rouveix *et al.*, 1980, 1981). It is known that cGMP decreases neutrophil membrane deformability, and that this latter effect is associated with decreased chemotaxis (although random movement is not affected) (Miller, 1975). Furthermore, similar membrane events in neutrophils and macrophages following chemotactic stimuli are thought to occur (Wilkinson, 1976). Monocyte membrane deformability might therefore be affected in a similar manner. Although decreased cGMP has been associated with decreased chemotaxis (Gallin *et al.*, 1978), their precise relationship is unclear (Sandler *et al.*, 1975). Nevertheless, IgE-related membrane events may account for the finding of decreased monocyte chemotaxis in some patients with chronic mucocutaneous candidiasis (Snyderman *et al.*, 1973) and atopic dermatitis (Snyderman *et al.*, 1977). Interestingly, Snyderman *et al.* (1977) found that chemotaxis of monocytes from normal subjects could be suppressed by serum from atopics and to a minimal extent by IgE myeloma protein, while Furukawa and Altman (1978) found depressed monocyte chemotaxis only in persons whose total serum IgE was at least 1000 IU/ml.

In addition to possible direct effects on macrophage mobility by IgE, the production of arachidonic acid metabolites and PAF by such activated macrophages is also possible (Sanchez-Crespo *et al.*, 1980; Joseph *et al.*, 1983). These, in turn, may have the effects alluded to in Section 5.2. It has also been demonstrated that human alveolar macrophages produce leukotriene B_4 when stimulated with the calcium ionophore A23187 (Fels *et al.*, 1982). It is possible that IgE-induced macrophage activation has a similar effect. Leukotriene B_4 inhibits PHA-induced lymphocyte thymidine incorporation and LIF production (Payan and Goetzl, 1983). These effects may occur as a result of induction of suppressor lymphocyte activity (Rola-Pleszczynski *et al.*, 1982). This concept is supported by the work of Gualde *et al.* (1985), who found that leukotriene B_4 augments thymidine incorporation by mitogen-stimulated OKT8$^+$ cells while inhibiting such incorporation by mixed T cells.

7. IMMUNOLOGICAL ABNORMALITIES WHICH MAY RELATE TO IMMEDIATE HYPERSENSITIVITY

7.1. RESPIRATORY ALLERGY (TABLE 7)

A variety of immunological parameters have been studied in patients with atopic diseases. Allergic rhinitis and allergic asthma are two of the most common atopic diseases and are initiated by unusually small amounts of antigen. Nevertheless, such patients have been found to have higher than normal plasma levels of histamine and neutrophil chemotactic factor of anaphylaxis (Allen *et al.*, 1978; Martinez *et al.*, 1979; Deal *et al.*, 1980). Patients with some physical urticarias have also been shown to have elevated plasma histamine levels (Kaplan *et al.*, 1975).

To a certain extent, cell function is reflected by the proportions of E rosette-forming cells and T gamma and T mu cells, since expression of these surface receptors depends upon active metabolic processes (Chisari and Edgington, 1974; Lang *et al.*, 1978; Klein *et al.*, 1979; Gupta *et al.*, 1979a). One of the most widely studied parameters is the total number of T cells in atopic patients. Total T-cell numbers and proportions in these studies were determined by SRBC rosetting. Numerous studies report normal percentage of total E rosettes and normal or increased numbers of active E rosettes in atopic patients (Yocum *et al.*, 1976; Hsieh, 1976; Ghazanshahi *et al.*, 1976; Saraclar *et al.*, 1977; Verhaegen *et al.*, 1977; Thomson *et al.*, 1977; Neiburger *et al.*, 1978; Lang *et al.*, 1978; Pryjma *et al.*, 1978;

TABLE 7. LEUKOCYTE ABNORMALITIES IN PATIENTS WITH RESPIRATORY ALLERGY

Abnormality	Possible cause
Decreased SRBC rosettes in severe disease	*In vivo* histamine exposure, H_2 effect
Decreased T gamma cells	*In vivo* histamine exposure, H_1 effect
Decreased sensitivity to histamine suppression of PHA-induced thymidine uptake	Histamine desensitization following *in vivo* exposure
Decreased generation of histamine-induced T suppressor cells	*In vivo* histamine desensitization or genetic predisposition
Increased short-lived T suppressor cell activity	? HSF effect
Decreased suppression of spontaneous IgE production	*In vivo* histamine desensitization of histamine receptor-bearing T suppressor cells
Increased autologous mixed lymphocyte culture thymidine incorporation	Decreased HSF secondary to histamine desensitization of histamine-inducible HSF-producing T suppressor cell
Generation of helper cells rather than suppressor cells in autologous mixed lymphocyte culture	Genetic predisposition
Decreased neutrophil chemotaxis	*In vivo* histamine exposure, H_2 effect; *in vivo* chemotactic deactivation by ECF-A, NCF-A, HETE
Phagocytosis	?
Decreased monocyte response to HSF	?

Canonica *et al.*, 1979; Cernelc *et al.*, 1979; Harper *et al.*, 1980; Gupta *et al.*, 1980; Perelmutter and Potvin, 1980; Anfosso *et al.*, 1982). Six reports show decreased percentage of total rosettes in some subgroups of patients (Gupta *et al.*, 1975; Strannegard *et al.*, 1976; Byrom *et al.*, 1978; Lang *et al.*, 1978; Canonica *et al.*, 1979; Anfosso *et al.*, 1982) although Canonica *et al.* (1979) report that only the more severely ill patients had decreased T cells. A few studies have found decreased absolute and relative numbers of T rosettes (Strannegard and Strannegard, 1977b; Stanciu *et al.*, 1983). Decreased T rosettes were found in infants of atopic parents, some of whom were fed cow's milk and some of whom had an increased eosinophil count (Juto and Strannegard, 1979) and in asthmatic subjects with low serum IgA concentrations (Clot *et al.*, 1979).

The proportions of T cells with receptors for the Fc portion of IgM (T mu) and IgG (T gamma) have also been studied (Canonica *et al.*, 1979; Gupta *et al.*, 1980; Perelmutter and Potvin, 1980; Anfosso *et al.*, 1982; Zielinski and Oehling, 1982). In all five studies the proportion of T mu in atopic patients was found to be normal. In two of the reports, T gamma proportions were decreased, although in one of the two, the proportion was only decreased in more severely ill patients (Canonica *et al.*, 1979; Anfosso *et al.*, 1982). One study showed an increased T mu/T gamma ratio with proportions of each type of T cell in the allergic patients still within normal limits (Perelmutter and Potvin, 1980). Numbers of suppressor T cells and the helper/suppressor T-cell ratio as determined by monoclonal antibodies to the T-cell differentiation antigens are also normal (Matloff *et al.*, 1983). Histamine receptor-bearing mononuclear cells are present in equal numbers in allergic and normal subjects (Martinez *et al.*, 1979). Beer *et al.* (1982c) reported that normal and atopic subjects have significant overlap in the numbers of lymphocytes which carry H_1 and H_2 receptors, although the mean number of H_2 receptors on the atopics' lymphocytes is lower.

The lability of E rosettes in response to agents which elevate cAMP is unclear. Verhaegen *et al.* (1977) report that E rosette proportions from atopic patients are decreased by histamine preincubation, while those from normal subjects are unchanged. A seemingly contradictory report by Lang *et al.* (1978) showed that total T rosettes of atopic patients and healthy subjects were unchanged, but that active E rosettes from asthmatics were more inhibited by isoproterenol than were active rosettes from normal subjects.

More direct studies of cellular function in allergic patients have been undertaken. Allergic patients have been found to have normal responses to delayed hypersensitivity skin testing (Yocum *et al.*, 1976; Harper *et al.*, 1980), and to *in vitro* mononuclear cell stimulation with Con A (Yocum *et al.*, 1976; Harper *et al.*, 1980; Lydyard *et al.*, 1981), PHA (Yocum *et al.*, 1976; Strannegard, 1979; Harper *et al.*, 1980; Hsieh, 1983), PWM (Yocum *et al.*, 1976; Strannegard *et al.*, 1976), and PMA (Strannegard, 1979). Only one study reported lower than normal thymidine uptake in response to PHA and Con A in patients with asthma and/or eczema (Strannegard *et al.*, 1976). These same workers showed that sera from atopic patients with high IgE levels suppressed Con A-induced reactivity (Strannegard and Strannegard, 1977b). Further work by Strannegard (1981) revealed data suggesting that monocytes from atopic patients functioned less well than monocytes from controls in terms of enhancement of PMA-induced lymphocyte stimulation. Monocytes from atopic persons also produce less PGE_2 in response to HSF than do monocytes from normal persons (Matloff *et al.*, 1983). However, antibody-dependent monocyte-mediated and lymphocyte-mediated cytotoxicity is normal in patients with respiratory allergy (Thomson *et al.*, 1977; Kragballe *et al.*, 1980).

Abnormally decreased phagocytosis of yeast by neutrophils from atopic patients has also been reported (Turner *et al.*, 1978). Depressed neutrophil chemotaxis in patients with respiratory allergy (Hill *et al.*, 1976a; Martinez Cairo-Cueto *et al.*, 1981; Faulkner *et al.*, 1982) has been confirmed in all but one study (Furukawa and Altman, 1978). In addition, Radermecker and Maldague (1981) reported that the depressed neutrophil chemotaxis is correctable with cimetidine. Finally, IgE-containing circulating immune complexes have been described in patients with asthma and rhinitis (Stevens and Bridts, 1984). These immune complexes may play a causal role in the decreased neutrophil chemotaxis reported in atopic patients (Ito *et al.*, 1983).

Histamine-induced suppression of immunological function has also been studied in allergic patients. While Wang and Zweiman (1978) found atopics to have normal susceptibility to the suppressive effects of histamine on PHA-induced mixed mononuclear cell thymidine uptake, others found that mononuclear cells from atopics were less sensitive to such suppression than cells from normal people (Martinez *et al.*, 1979). This latter finding is consistent with the report that mononuclear cells from allergic subjects were less able to generate histamine-induced T suppressor cell activity (Beer *et al.*, 1982c) and less HSF (Matloff *et al.*, 1983) than were cells from healthy control subjects. However, mononuclear cells from atopics were found to be more sensitive to histamine-induced suppression of suboptimal Con A stimulation of thymidine uptake, possibly because of increased activity of atopics' prostaglandin-producing adherent suppressor cells in this system (Hall *et al.*, 1983).

Various other suppressor cell assays have been studied. Patients with hay fever as compared to healthy subjects were found to have more short-lived suppressor cell activity following a 24-hr preincubation prior to Con A stimulation (Lydyard *et al.*, 1981). Saxon *et al.* (1980) found that mononuclear cells from allergic patients had less T-cell inhibition of spontaneous IgE production than would have been predicted in normal subjects. Con A-induced inhibition of PWM-induced IgE production was found to be present in patients with ragweed allergy (Too, 1979). In addition, patients with ragweed allergy have normal levels of streptokinase/streptodornase and ragweed antigen-specific suppressor cells generated by preculture activation (Rocklin *et al.*, 1980b). These authors also reported that hyposensitization injections with ragweed pollen extract caused increased ragweed antigen-specific suppressor activity. This finding was confirmed by Dorval *et al.* (1980). Finally, Con A-induced T-cell suppression of PHA-induced thymidine incorporation has been variously reported to be normal (Martinez *et al.*, 1979; Delafuente *et al.*, 1980; Beer *et al.*, 1982c; Church and Nye, 1982) or decreased (Harper *et al.*, 1980). Dorval *et al.* (1980) reported Con A-induced T-cell suppression of PHA-stimulated thymidine incorporation to be normal in immune complex-negative patients and abnormally increased in immune complex-positive subjects. They also found that Con A-induced T-cell suppression of MLC reactivity and PWM-induced immunoglobulin production is normal. This latter finding confirms results of Martinez *et al.* (1979).

Rocklin and Beer (1983) argue that diminished generation of histamine-induced T suppressor cell activity is constitutional rather than acquired, since patients with systemic mastocytosis with high levels of circulating histamine have normal histamine-induced suppressor function (Beer *et al.*, 1982c). The report that persons who carry the HLA-B12 antigen are less responsive to histamine inhibition of Con A-induced thymidine uptake (Staszak *et al.*, 1980) also supports this concept. This contention, however, does not address the findings of normal Con A-induced suppressor activity in atopics (Martinez *et al.*, 1979; Beer *et al.*, 1982c) since Con A-inducible suppressor cells are thought to encompass the histamine-inducible cell population (Fineman *et al.*, 1979; Zarling *et al.*, 1980; Rocklin and Haberek-Davidson, 1981). It has also been suggested that Con A-induced suppressor activity is actually mediated via histamine (Osband *et al.*, 1980). In addition, atopics do not predominantly express the HLA-B12 antigen. Consequently, the question remains unanswered.

The autologous mixed lymphocyte reaction has been studied in asthmatic subjects by two laboratories. Hsieh (1983) found autologous mixed lymphocyte reaction-induced thymidine incorporation to be normal, while Hwang *et al.* (1983) found it to be normal in five subjects and increased in seven. Interestingly, five of these latter seven had received hyposensitization with parenteral allergen for more than 1 year. Those with normal responses had not received hyposensitization. In addition, Hsieh (1983) found that normal cells activated in the autologous mixed lymphocyte reaction resulted in suppression of PHA-induced thymidine incorporation and PWM-induced IgE production. However, similarly activated cells from asthmatics resulted in augmentation of PHA reactivity and PWM-induced IgE secretion.

7.2. ATOPIC DERMATITIS (TABLE 8)

Atopic dermatitis (atopic eczema) is strongly associated with atopy although its etiology and pathogenesis remain a matter of contention. The vast majority of patients with atopic dermatitis are colonized, if not actively infected, with *S. aureus* (Selwyn and Chalmers, 1965; Leyden *et al.*, 1974; Rogge and Hanifin, 1976; Aly *et al.*, 1977). *S. aureus* protein A causes basophil histamine release without the intermediary of antistaphylococcal IgE (Marone *et al.*, 1982). In addition, most patients with atopic dermatitis are allergic to *S. aureus* antigens as shown by immediate hypersensitivity skin testing (Welbourn *et al.*, 1976) and by RAST testing (Walsh *et al.*, 1981). This has been confirmed in seven patients with atopic dermatitis who were found to have an average of 33 IU/ml of *S. aureus*-specific IgE (Geha *et al.*, 1981). This allergy to *Staphylococcus* results in an ongoing allergic eczematous reaction on the skin wherever the antigen has access to sensitized cutaneous mast cells (Mitchell *et al.*, 1982). Such access would be

TABLE 8. LEUKOCYTE ABNORMALITIES IN PATIENTS WITH ATOPIC DERMATITIS

Abnormality	Possible cause
Decreased SRBC rosettes in many studies	*In vivo* histamine exposure, H_2 effect
Decreased T gamma cells	*In vivo* histamine exposure, H_1 effect
Decreased $OKT8^+$ cells	?
Decreased theophylline-sensitive T cells	*In vivo* histamine exposure with increased cAMP
Decreased delayed-type hypersensitivity skin reactivity	*In vivo* histamine-induced suppression
Decreased thymidine uptake induced by suboptimal Con A and PHA concentrations, and by antigens	*In vivo* histamine-induced suppression
Inhibition of mitogen-induced thymidine uptake by autologous plasma	HSF effect in plasma
Decreased LIF production	*In vivo* histamine-induced suppression
Decreased mitogen- and antigen-induced suppressor cell activity	? *In vivo* histamine-induced desensitization of histamine receptor-bearing T suppressor cells; decreased T gamma cells secondary to *in vivo* histamine exposure, H_1 effect
Decreased alloantigen-induced cytotoxicity	*In vivo* histamine exposure
Defective $OKT4^+$ helper function	HSF effect *in vivo*
Decreased leukocyte responsiveness to agents which elevated cAMP	Heterologous desensitization secondary to *in vivo* histamine exposure
Decreased neutrophil chemotaxis	*In vivo* histamine exposure, H_2 effect; chemotactic deactivation by NCF-A, ECF-A, HETE
Decreased neutrophil antibody-dependent cytotoxicity, phagocytosis, lysozyme release, hexose monophosphate shunt activation, decreased β-glucuronidase release	*In vivo* histamine exposure
Decreased monocyte chemotaxis	Membrane effect of cell surface IgE bridging
Decreased monocyte antibody-dependent cytotoxicity	*In vivo* histamine exposure, H_2 effect

facilitated by associated allergies with cutaneous manifestations. In fact, 54% of children with atopic eczema developed such symptoms in response to food antigens during double-blind placebo-controlled challenges (Sampson, 1983). Such patients also exhibit increased plasma histamine concentrations after ingestion of food allergens (Sampson and Jolie, 1984). Other less direct evidence for the association of immediate hypersensitivity with atopic dermatitis is the finding of elevated serum IgE concentrations in affected persons (Juhlin *et al.*, 1969; Johnson *et al.*, 1974). Fifty to eighty percent of those affected with atopic dermatitis eventually develop allergic asthma and/or rhinitis (Pasternack, 1965; Stifler, 1965; Hoffman *et al.*, 1975; O'Loughlin *et al.*, 1977). Most patients with atopic dermatitis possess positive immediate hypersensitivity skin tests and much of the elevated circulating IgE is also specific for various dietary and environmental antigens (Hoffman *et al.*, 1975; Mitchell *et al.*, 1982; Chapman *et al.*, 1983; Zachary and MacDonald, 1983).

Lymphocytes bearing cell surface receptors for SRBC have been studied extensively in patients with atopic dermatitis. Some workers have reported that absolute numbers and/or proportions of such lymphocytes are normal in patients with atopic dermatitis (Dupree *et al.*, 1975; Grove *et al.*, 1975; McGeady *et al.*, 1976; Rachelefsky *et al.*, 1976; Secher *et al.*, 1977; Thestrup-Pedersen *et al.*, 1977; Bousquet *et al.*, 1978; Byrom *et al.*, 1979; Amblard *et al.*, 1980; Schuster *et al.*, 1980; Jensen *et al.*, 1981; Soppi *et al.*, 1982) while others have found lower than normal values(Gottlieb and Hanifin, 1974; Luckasen *et al.*, 1974; McGeady and Buckley, 1975; Carapeto *et al.*, 1976; Hovmark, 1977; Rola-Pleszczynski and Blanchard, 1981; Stingl *et al.*, 1981; Valverde *et al.*, 1983). The data presented by Andersen and Hjorth (1975) show decreased numbers and proportions of SRBC-rosetting lymphocytes for a subgroup of patients, with normal values in the remainder. However, the group values become lower than normal as a result of the abnormal subgroup. Active E rosetting cells have been reported to be normal (Bousquet *et al.*, 1978; Valverde *et al.*, 1983).

More recent studies have evaluated the proportions of lymphocytes bearing the pan-T-cell differentiation antigen, OKT3. Leung *et al.* (1981) and Faure *et al.* (1982) report decreased group mean percentages of OKT3$^+$ cells, although the results of the former study are heavily influenced by a subgroup of patients with low values. A subsequent study (Leung *et al.*, 1983) found 6 of 8 subjects to have normal proportions of OKT3$^+$ cells. Others have also reported normal proportions of OKT3$^+$ lymphocytes in a total of 38 patients with atopic dermatitis (Butler *et al.*, 1982; Businco *et al.*, 1983; Armitstead and Ewan, 1983). Surface markers for T suppressor cells have usually been reported to be decreased, whether T cells bearing Fc receptors for IgG (T gamma) (Cooper *et al.*, 1980; Schuster *et al.*, 1980; Stingl *et al.*, 1981), or OKT8$^+$ cells are enumerated (Leung *et al.*, 1981, 1983; Butler *et al.*, 1982; Faure *et al.*, 1982; Zachary and MacDonald, 1983). These studies all reported that T helper cells as denoted by Fc mu or by OKT4 are normal. However, three studies of a total of 34 patients have reported normal numbers of OKT8$^+$ lymphocytes (Joffe *et al.*, 1983; Businco *et al.*, 1983; Armitstead and Ewan, 1983). There are reports of low or normal proportions of theophylline-sensitive (suppressor enriched) T rosettes in patients with atopic dermatitis (Rola-Pleszczynski and Blanchard, 1981; Soppi *et al.*, 1982).

Studies of mononuclear cell function have also been done. Several reports document that patients with atopic dermatitis have fewer positive delayed hypersensitivity skin tests to recall antigens (Rajka, 1967, 1968; McGeady and Buckley, 1975; Hovmark, 1975; Grove *et al.*, 1975; Rogge and Hanifin, 1976; Rajka and Barlinn, 1979; Amblard *et al.*, 1980), although some show normal delayed skin reactivity to candidal and streptococcal antigens (Palacios *et al.*, 1966; Dupree *et al.*, 1975). Rajka and Barlinn (1979) document that the absence of delayed reactivity to fungal antigens is usually associated with the presence of immediate hypersensitivity to that antigen. Transient suppression of tuberculin reactivity has been correlated with the activity of eczematous lesions (Uehara, 1977). There is also resistance to contact sensitization with *Rhus* antigen (Jones *et al.*, 1973) and dinitrochlorobenzene (Palacios *et al.*, 1966; Lobitz *et al.*, 1972; Rogge and Hanifin, 1976; Pauly *et al.*, 1978). However, contact dermatitis to other antigens apparently occurs with normal frequency (Epstein and Mohajerin, 1964; Cronin *et al.*, 1970). *In vitro* studies of lymphocyte function have all revealed normal thymidine incorporation in response to PHA stimulation (Schoepf and Boehringer, 1974; Gottlieb and Hanifin, 1974; McGeady and Buckley, 1975; Dupree *et al.*, 1975; Grove *et al.*, 1975; Hovmark, 1975, 1977; Rogge and Hanifin, 1976; Thestrup-Pedersen *et al.*, 1977; Horsmanheimo *et al.*, 1979; Stingl *et al.*, 1981; Soppi *et al.*, 1982; Valverde *et al.*, 1983). Rachelefsky *et al.* (1976) report abnormally decreased responses at lower stimulatory concentrations of PHA. Andersen and Hjorth (1975) found that PHA responsiveness became abnormally low in the presence of autologous serum in a subgroup of their patients. Mixed lymphocyte reactivity has also been found to be normal (Gottlieb and Hanifin, 1974; Leung *et al.*, 1983) as has PWM-induced thymidine incorporation (Thestrup-Pedersen *et al.*, 1977; Bousquet *et al.*, 1978; Valverde *et al.*, 1983). Con A-induced reactivity has been found to be normal by some (Horsmanheimo *et al.*, 1979; Stingl *et al.*, 1981; Soppi *et al.*, 1982) while others report normal values only in patients with mild eczema (Ogden *et al.*, 1979) or can demonstrate subtle decreases in reactivity by using suboptimal stimulatory doses (Schopf *et al.*, 1978). Spontaneous uptake of thymidine has been variously found to be normal (Grove *et al.*, 1975; Hovmark, 1977; Horsmanheimo *et al.*, 1979) or increased (Rachelefsky *et al.*, 1976; Bousquet *et al.*, 1978; Schopf *et al.*, 1978; Soppi *et al.*, 1982). Candida-induced, but not tetanus toxoid-induced, thymidine uptake was lower than normal (McGeady and Buckley, 1975). Thestrup-Pedersen *et al.* (1977) found that autologous plasma inhibited Con A-, PWM-, and PPD-induced reactivity. Another normal cellular function has been PHA-induced production of lymphotoxin (Dupree *et al.*, 1975). Lymphocyte production of LIF has been reported to be normal following stimulation with PPD (Thestrup-Pedersen *et al.*, 1977; Soppi *et al.*, 1982), PHA, PWM, and Con A (Soppi *et al.*, 1982). However, Horsmanheimo *et al.* (1979) found abnormally low PHA-stimulated production of LIF. These same workers also reported that lymphocytes from only about one-third of atopic dermatitis patients produce LIF following stimulation with Con A (Horsmanheimo and Horsmanheimo, 1979). Functional assays have revealed increased (Ogden *et al.*, 1979; Elliott and Hanifin, 1979), normal (Schuster *et al.*, 1979), or decreased (Rola-Pleszczynski and Blanchard, 1981) short-lived suppressor cell activity. Con A-induced suppression of PHA-stimulated thymidine incorporation has also been found to be normal (Stingl *et al.*, 1981; Armitstead and Ewan, 1983) or decreased (Jensen *et al.*, 1981; Rola-Pleszczynski and Blanchard, 1981; Butler *et al.*, 1982; Valverde *et al.*, 1983). Ogden *et al.* (1979) and Armitstead and Ewan (1983) found Con A-induced suppression of Con A reactivity to be normal, while Stingl *et al.* (1981) found it to be decreased. This latter finding correlated with disease activity. Rola-Pleszczynski and Blanchard (1981) and Butler *et al.* (1982) also found that Con A-induced suppression of PWM-induced immunoglobulin synthesis was lower than normal and that the decreased suppressor function that they studied was correlated with serum IgE concentrations and decreased numbers of $OKT8^{+}$ T lymphocytes. Soppi *et al.* (1982) reported that Con A-, PHA-, and PPD-induced suppressor cell activity on PHA-, Con A-, and PWM-stimulated deoxyuridine incorporation are all normal.

Prostaglandin-producing suppressor cell function was also within the range of that of control subjects (Soppi *et al.*, 1982; Armitstead and Ewan, 1983). Contrary to the findings of Stingl *et al.* (1981), none of the parameters studied by Soppi *et al.* (1982) correlated with the severity of the disease. Extensive analysis of abnormally low cell-mediated lympholysis in 12 of 19 patients with atopic dermatitis (Leung *et al.*, 1983) revealed that cytotoxic T cells were decreased within the $OKT8^+$ T-cell subset and that $OKT4^+$ helper cell function was also defective.

Mixed mononuclear cell responsiveness to the suppressive effects of theophylline and isoproterenol on mitogenic stimulation was more pronounced in patients with atopic dermatitis than in normal people (Rola-Pleszczynski and Blanchard, 1981). These latter findings are difficult to explain in view of the results of most workers that atopic dermatitis leukocytes are less rather than more responsive to agents which elevate cAMP (see below). However, it might be postulated that these results could be accounted for by increased suppressor activity of theophylline-sensitive T cells secondary to ongoing exposure to endogenous histamine. Histamine-induced suppression of antigen- and mitogen-stimulated thymidine uptake is reported to be normal (Armitstead and Ewan, 1983).

Other leukocyte abnormalities have been described in patients with atopic dermatitis. Busse and Lee (1976) found that, while basal levels of cAMP were normal in leukocytes from patients with atopic eczema, the increase in levels following stimulation with β-adrenergic agonists was less than that found in lymphocytes and granulocytes from normal subjects. Similar work by Parker *et al.* (1977) revealed low basal cAMP levels with a subnormal response to theophylline and PGE_1, as well as adrenergic agents. This abnormality correlated with the patients' clinical condition. Herlin and Kragballe (1980) and Safko *et al.* (1981) respectively demonstrated a similarly deficient histamine-induced increase in cAMP levels in monocytes and mixed mononuclear cells from patients with atopic dermatitis. Augmentation of cAMP by isoproterenol or PGE_1 was also subnormal (Galant *et al.*, 1979; Herlin and Kragballe, 1980). Safko *et al.* (1981) found that a 15-min preincubation with 10^{-6} M histamine resulted in refractoriness to subsequent effects of histamine, isoproterenol, and PGE_1 for 24 hr in mononuclear cells from normal persons. This is called *heterologous desensitization.* Busse and Sosman (1976) and Busse and Lantis (1979) respectively found that there was subnormal isoproterenol-induced, and H_2 receptor-mediated, histamine-induced, inhibition of zymosan-stimulated neutrophil lysosomal enzyme release.

Rogge and Hanifin (1976) found atopic dermatitis patients to have a defect in neutrophil chemotaxis which remitted as the disease severity decreased. Studies by Snyderman *et al.* (1977) confirmed that a subgroup of patients who were not free of infection had neutrophils which responded subnormally to a chemotactic stimulus. Subsequent work by Dahl *et al.* (1978) showed chemotaxis to be normal in most patients with atopic dermatitis. Other studies of neutrophil function have shown decreased antibody-dependent cytotoxicity associated with decreased phagocytosis, lysozyme release, and hexose monophosphate shunt activation (Kragballe and Borregaard, 1981) and decreased β-glucuronidase release following stimulation with aggregated immunoglobulin, activated complement (zymosan), yeast, or anti-human leukocyte globulin (Ring and Lutz, 1983). Decreased phagocytosis has been associated with increased disease severity (Michaelsson, 1973).

Monocyte studies have revealed depression of antibody-dependent cytotoxicity (Kragballe, 1979; Herlin and Kragballe, 1980; Kragballe *et al.*, 1980; Kragballe and Borregaard, 1981; Kragballe and Herlin, 1981) and chemotaxis (Rogge and Hanifin, 1976; Fischer *et al.*, 1977; Snyderman *et al.*, 1977; Furukawa and Altman, 1978). Histamine preincubation can suppress antibody-dependent monocyte cytotoxicity. The prevention of histamine-induced depression of monocyte antibody-dependent cytotoxicity by cimetidine, but not by diphenhydramine, suggests that histamine acts via the H_2 receptor (Herlin and Kragballe, 1980). These same workers found that the incremental cAMP response of monocytes from patients with atopic dermatitis was less than that of monocytes from healthy subjects (Kragballe and Herlin, 1981). The data presented reveal that the atopic dermatitis patients fall into a normal and a subnormal group with regard to their monocyte cAMP responses. It appears that there is a relationship between maximum increase in cAMP and cytotoxicity in the group which is abnormal [although Kragballe and Herlin (1981) do not draw this conclusion]. There is no such relationship in the atopic dermatitis group as a whole. The chemotactic abnormality could be reproduced in normal monocytes by preincubation in sera from patients with atopic dermatitis but not by preincubation with histamine. In addition, Snyderman *et al.* (1977) showed that myeloma IgE induced a small inhibition of monocyte chemotaxis. Furukawa and Altman (1978) found an association of elevated IgE and decreased monocyte chemotaxis. The question as to whether antigen-specific IgE plus antigen is responsible for the antichemotactic effect remains unanswered.

Johnson *et al.* (1960), Ring *et al.* (1979), and Safko and Hanifin (1980) have found elevated histamine levels in skin and plasma of patients with atopic dermatitis. It is tempting to postulate that such elevated histamine concentrations, especially locally increased tissue concentrations in proximity to mast cells, may ultimately be shown to account for: (1) the finding of decreased SRBC-rosetting lymphocytes in some of these patients (DeCock *et al.*, 1977; Galli *et al.*, 1979; Malolepszy *et al.*, 1980); (2) the finding of abnormally low numbers of T gamma cells since an H_1 histamine agonist has been shown to decrease the expression of Fc gamma receptors (Birch and Polmar, 1981); (3) fewer positive delayed hypersensitivity skin tests since intradermal histamine diminishes delayed reactivity (Pepys, 1953); (4) decreased lymphocyte proliferation in response to lower stimulatory concentrations of PHA since histamine-induced inhibition of PHA-induced reactivity is only seen with suboptimal PHA concentrations (Martinez *et al.*, 1979; Ogden and Hill, 1980; Meretey *et al.*, 1981); (5) decreased neutrophil chemotaxis (Hill *et al.*, 1975; Anderson *et al.*, 1977; Radermecker and Maldague, 1981); (6) decreased histamine-induced inhibition of neutrophil lysosomal enzyme release (Busse and Sosman, 1976, 1977) resulting from receptor desensitization; (7) abnormal T helper cell function (Birch and Polmar, 1982); (8) abnormally depressed monocyte-mediated antibody-dependent cytotoxicity (Herlin and Kragballe, 1980). It is not known whether histamine activation of suppressor cells renders such cells less responsive in *in vitro* suppressor cell assays.

7.3. SYNDROME OF RECURRENT SEVERE BACTERIAL INFECTIONS, ATOPIC DERMATITIS, HYPERIMMUNOGLOBULINEMIA E, AND DEFECTIVE NEUTROPHIL CHEMOTAXIS

Beginning with the report of Davis *et al.* (1966), a large series of papers has appeared documenting a syndrome which encompasses recurrent severe bacterial infections, atopic dermatitis, defective neutrophil chemotaxis, and hyperimmunoglobulinemia E (Buckley *et al.*, 1972; Kikkawa *et al.*, 1973; Hill and Quie, 1974; Hill *et al.*, 1974; Pincus *et al.*, 1975; Church *et al.*, 1976; Jacobs and Norman, 1977; Paslin and Norman, 1977; Blum *et al.*, 1977; Komiyama *et al.*, 1977; Issekutz *et al.*, 1978; DeCree *et al.*, 1978; Schopfer *et al.*, 1979; Businco *et al.*, 1981). Classical respiratory allergy or urticaria may also be present in some of these patients (Hill and Quie, 1975; Foster and Goetzl, 1978).

Most patients with this syndrome have been reported to have normal numbers of T lymphocytes as shown by receptors for SRBC (Katona *et al.*, 1980) or by the presence of the OKT3 differentiation antigen (Kraemer *et al.*, 1982), although some have decreased $OKT3^+$ lymphocytes (Geha *et al.*, 1981). Two reports indicate that T suppressor cells are decreased, as evidenced by T cells with Fc receptors for IgG (Katona *et al.*, 1980) or by decreased $OKT8^+$ cells (Geha *et al.*, 1981). These findings are not uniform, since Kraemer *et al.* (1982) found the $OKT8^+$ cells to be present in normal numbers in seven such patients whose infections and/or eczematous inflammation were in remission. All three reports agree that lymphocytes with Fc receptors for IgM or with the OKT4 differentiation antigen are present in normal numbers. Some of the patients described also had abnormal *in vivo* or *in vitro* cell-mediated immunity (Buckley *et al.*, 1972; Kikkawa *et al.*, 1973; Church *et al.*, 1976; Komiyama *et al.*, 1977; Issekutz *et al.*, 1978; DeCree *et al.*, 1978; Katona *et al.*, 1980; Businco *et al.*, 1981; Soderberg-Warner *et al.*, 1983). Detailed functional studies have revealed decreased Con A-induced suppression of PHA-, PWM-, and MLR-induced thymidine incorporation (Geha *et al.*, 1981; Soderberg-Warner *et al.*, 1983). However, suppressor T cells from patients were able to normally suppress spontaneous (*in vivo* stimulated) *in vitro* IgE production (Kraemer *et al.*, 1982). Impaired monocyte chemotaxis has also been reported (Foster and Goetzl, 1978). *S. aureus*-stimulated mononuclear cells from these patients produced unusually large amounts of two inhibitors of leukocyte chemotaxis. One factor inhibits only neutrophil chemotaxis while the other inhibits chemotaxis of both neutrophils and monocytes (Donabedian and Gallin, 1983).

Of special interest is the finding that, in some subjects, the defect in neutrophil chemotaxis is intermittent and therefore an acquired defect (Issekutz *et al.*, 1978; Mawhinney *et al.*, 1980). One study reports remission of both clinical symptoms and abnormal neutrophil chemotaxis when food allergens were avoided (Fontan *et al.*, 1976). Another study showed delayed *in vivo* neutrophil migration when patients were free of infection and transiently abnormal *in vitro* neutrophil chemotaxis in one patient during an episode of staphyloccocal pneumonia (Soderberg-Warner *et al.*, 1983). Others have noted a correlation of the severity of the dermatitis with the severity of the defect in neutrophil chemotaxis (Hill *et al.*, 1974; Dahl *et al.*, 1976). A similar seasonally recurrent syndrome, associated with allergic rhinitis rather than eczema, also occurs (Hill *et al.*, 1976a). Incubation of patients' leukocytes with ragweed pollen antigens resulted in decreased neutrophil chemotaxis, suggesting that mediators of immediate hypersensitivity may be playing a role.

Buckley *et al.* (1972) were the first to report the presence of immediate hypersensitivity to *S. aureus* in two such patients. The documentation of the presence of antistaphylococcal IgE antibodies in 27 patients with this syndrome (Schopfer *et al.*, 1979, 1980; Berger *et al.*, 1980; Geha *et al.*, 1981) led to the suspicion that histamine release caused by staphylococci might be responsible for inhibiting polymorphonuclear leukocyte chemotaxis and for predisposing these patients to recurrent staphylococcal infections. This hypothesis has been given added support by the observation of Mawhinney *et al.* (1980) that administration of the H_2 antagonist cimetidine to a patient with this syndrome resulted in clinical remission and correction of the defect in neutrophil chemotaxis.

7.4. CHRONIC MUCOCUTANEOUS CANDIDIASIS

Chronic mucocutaneous candidiasis is a syndrome in which there is ongoing infection of the skin and mucous membranes with *Candida*. The syndrome includes a multiplicity of immunological abnormalities. These include the absence of delayed hypersensitivity to recall antigens, especially to candidal antigen, and an associated absence or diminution of *in vitro* antigen- and mitogen-induced lymphocyte thymidine incorporation or lymphokine production (Hermans *et al.*, 1969; Valdimarsson *et al.*, 1970; Kirkpatrick *et al.*, 1971; Kirkpatrick and Smith, 1974). In some patients, delayed-type hypersensitivity to candidal antigen is absent but *in vitro Candida*-stimulated lymphocyte thymidine uptake may be present (Frisk *et al.*, 1974). Decreased Con A-induced suppression of PHA-stimulated thymidine uptake has also been reported (Arulanantham *et al.*, 1979). Additional immunological abnormalities in a minority of patients with this syndrome include decreased monocyte chemotaxis (Snyderman *et al.*, 1973), phagocytosis (Sams *et al.*, 1979), and chemiluminescence (Bortolussi *et al.*, 1981). The most frequently reported defect in neutrophil function is decreased chemotaxis (Clark *et al.*, 1973; Van Scoy *et al.*, 1975; Hill and Quie, 1975; Lawton *et al.*, 1976; Cates *et al.*, 1980; Businco *et al.*, 1981), although abnormal phagocytosis, killing, and nitroblue tetrazolium reduction have also been reported (Djawari *et al.*, 1977; Walker and Urbaniak, 1980). In some patients the abnormality in cellular immunity or neutrophil function was related to serum factors (Twomey *et al.*, 1975; Walker and Urbaniak, 1980). Defective cellular immunity, in turn, has been associated with defective catabolism of mannan by monocytes from some patients with this disease (Fischer *et al.*, 1982).

Chronic mucocutaneous candidiasis occurs as well in the context of the hyperimmunoglobulinemia E syndrome (Buckley *et al.*, 1972; Clark *et al.*, 1973; Kirkpatrick and Smith, 1974; Van Scoy *et al.*, 1975; Hill and Quie, 1975; Dahl *et al.*, 1976; Quie and Cates, 1978; Gupta *et al.*, 1979a; Katona *et al.*, 1980; Berger *et al.*, 1980; Businco *et al.*, 1981). Some patients suffer with concomitant typical recurrent infection with *S. aureus* (Kirkpatrick *et al.*, 1971; Kirkpatrick and Smith, 1974; Hill and Quie, 1975; Dahl *et al.*, 1976; Businco *et al.*, 1981), or dermatophytosis (Hermans *et al.*, 1969; Kirkpatrick *et al.*, 1971; Kirkpatrick and Smith, 1974; Mobacken and Lindholm, 1974; Gupta *et al.*, 1979a; Sams *et al.*, 1979). Some have eczema (Van Scoy *et al.*, 1975; Dahl *et al.*, 1976; Businco *et al.*, 1981).

Because of the presence of high titers of anticandidal IgE in patients with systemic or mucosal candidiasis (Mathur *et al.*, 1977) or with the hyperimmunoglobulinemia E syndrome (Berger *et al.*, 1980), and the finding of type I hypersensitivity to candidal antigen in others (Pluss and Kadas, 1954; Buckley *et al.*, 1972; Kirkpatrick and Smith, 1974; Van Scoy *et al.*, 1975; Rothschild *et al.*, 1976; Kirkpatrick *et al.*, 1976; Takeya *et al.*, 1976; Simon *et al.*, 1979), it is conceivable that allergy to *Candida* may be a complicating factor in a subset of patients with chronic mucocutaneous candidiasis. In fact, in 13 of 20 of the patients studied in the references just cited, the immediate wheal and flare reaction was associated with an absent delayed-type hypersensitivity response. Of additional interest in this respect is the report of Jorizzo *et al.* (1980), in which cimetidine treatment of 4 patients with chronic mucocutaneous candidiasis resulted in the positive conversion of delayed hypersensitivity skin tests to candidal antigen in all 4 patients, and of *Candida*-induced production of LIF in 3. These 3 patients in whom *Candida*-induced lymphokine production was restored had cutaneous allergy to *Candida* (Simon and Sams, 1980).

It is possible that some of the defects in cell-mediated immunity in these patients is related to histamine release, especially at the site of the infection (Simon *et al.*, 1979). In some, the defect in neutrophil chemotaxis may also be attributed to the inhibitory effect of histamine (Clark, 1978) or desensitization resulting from exposure to other mediators of immediate hypersensitivity. In at least some patients, the serum inhibitor of chemotaxis is not histamine (Cates *et al.*, 1980). The defect in monocyte chemotaxis, which is occasionally found, may conceivably be related to monocyte cell surface IgE. However, it is unlikely that the reported decreases in monocyte phagocytosis or chemiluminescence could be explained in this way.

7.5. CHRONIC DERMATOPHYTOSIS

Chronic dermatophyte infections have been found to be associated with type I hypersensitivity (Jones *et al.*, 1974b,c; Hanifin *et al.*, 1974; Hay, 1979; Hunziker and Brun, 1980). Patients with chronic dermatophytosis frequently have elevated serum IgE (Brahmi *et al.*, 1980; Hunziker and Brun, 1980; Balogh *et al.*, 1981) and positive wheal and flare skin responses to inhalant allergens (Jones *et al.*, 1974b) or concomitant allergic disease (Hay and Brostoff, 1977; Hunziker and Brun, 1980). Immediate skin allergy to trichophytin is especially common among such patients and is often associated with the diminution or absence of delayed-type hypersensitivity skin reactions to the fungal antigen (Jones *et al.*, 1974b,c; Hanifin *et al.*, 1974; Sorensen and Jones, 1976; Hay and Brostoff, 1977; Kaaman, 1978; Rajka and Barlinn, 1979; Hay, 1979; Hunziker and Brun, 1980; Balogh *et al.*, 1981; Stahl and Svejgaard, 1982). There are also reports of diminished frequency of delayed-type hypersensitivity in patients in whom immediate reactivity was not studied (Tagami *et al.*, 1977; Kaaman *et al.*, 1979; Brahmi *et al.*, 1980). Absent delayed reactivity in the context of type I allergy can be overcome by repeated injection of antigen (Hunziker and Brun, 1980). Diminished delayed hypersensitivity to *Rhus* has also been reported, but reactivity to other antigens is normal (Sorensen and Jones, 1976), except in severely infected patients (Brahmi *et al.*, 1980). In one study in which cimetidine was administered prior to, but not concurrent with, the delayed hypersensitivity skin tests, delayed hypersensitivity was not augmented (Hay, 1979).

In vitro parameters of immunological function have also been evaluated in patients with dermatophytosis. Total proportions of E-rosetting T lymphocytes have been reported to be normal in most patients (Brahmi *et al.*, 1980; Petrini and Kaaman, 1981; Hay and Shennan, 1982), although Balogh *et al.* (1981) found decreased active and total E rosettes in patients with elevated serum IgE and Brahmi *et al.* (1980) found decreased T cells in patients with visceral dissemination of their infection. The only study of T-cell Fc receptors and helper and suppressor cell differentiation antigens revealed inconclusive and essentially normal results (Petrini and Kaaman, 1981).

Functional *in vitro* studies of cellular immunity have found variably decreased (Hanifin *et al.*, 1974; Brahmi *et al.*, 1980; Hay and Shennan, 1980; Sohnle and Collins-Lech, 1982) or normal (Kaaman *et al.*, 1979; Stahl and Svejgaard, 1982; Hay and Shennan, 1982) lymphocyte thymidine incorporation in response to the antigenic extract of the infecting fungus. Normal macrophage migration inhibition factor production has also been reported (Brahmi *et al.*, 1980). However, only 8 of 19 patients were found to produce sufficient LIF to yield a migration index of 0.8 or less (Hay and Brostoff, 1977). Normal thymidine uptake in response to stimulation with PHA (Miller *et al.*, 1973; Hunziker and Brun, 1980; Brahmi *et al.*, 1980; Stahl and Svejgaard, 1982), Con A (Hunziker and Brun, 1980), and PWM (Brahmi *et al.*, 1980; Stahl and Svejgaard, 1982) has also been found. Only with visceral dissemination does mitogenic lymphocyte reactivity decrease (Brahmi *et al.*, 1980). Sohnle and Collins-Lech (1982) found that *in vitro* cimetidine did not enhance the decreased antigen-induced thymidine uptake in their subjects as a group, although the broad standard error of the group's percent change indicated that lymphocytes from some subjects did show increased reactivity. Finally, 3 patients with defective neutrophil chemotaxis but normal random migration, phagocytosis, and bactericidal activity have been described (Miller *et al.*, 1973).

It is of interest to note that following experimental dermatophyte infection in humans, the emergence of cell-mediated immunity to the dermatophyte antigen is followed by resolution of the dermatophytic lesions (Jones *et al.*, 1974a). Among patients who are treatment failures, 58% had a wheal and flare reaction, but only 12% had delayed-type reactivity to trichophytin (Hay, 1979).

There is a high frequency of immediate hypersensitivity skin reactions to dermatophytic antigens in atopic persons (Wood and Cruickshank, 1962). This, in turn, would predispose to chronic dermatophytosis because of the immunosuppressive effects of histamine released at the site of infection as previously discussed.

7.6. ALLERGIC BRONCHOPULMONARY ASPERGILLOSIS, COCCIDIOIDOMYCOSIS, AND HISTOPLASMOSIS

Allergic bronchopulmonary aspergillosis is characterized in part by type I hypersensitivity to *Aspergillus fumigatus*, intractable asthma, bronchopulmonary *Aspergillus* infection, serum precipitins to *Aspergillus* antigen, and elevated serum IgE concentrations (Rosenberg *et al.*, 1977). Allergy to *Aspergillus* is the *sine qua non* of this disease. Information on delayed-type hypersensitivity skin reactions to *Aspergillus* antigen is available in the literature on 13 patients. All had positive wheal and flare reactions but only 3 had positive *Aspergillus* delayed skin reactivity (Chan-Yeung *et al.*, 1971; Hanson *et al.*, 1976; Haslam *et al.*, 1976; Geha, 1977; Rosenberg *et al.*, 1977; Laham *et al.*, 1979). Skin test anergy to other antigens is not present (Haslam *et al.*, 1976; Hanson *et al.*, 1976; Geha, 1977; Forman *et al.*, 1978).

In vitro parameters of cellular immunity in patients with this disease have not been extensively studied. A preliminary report of five patients by Flaherty *et al.* (1979) indicates decreased total numbers of peripheral blood T lymphocytes, and decreased T cells with Fc receptors for IgM. Monocyte-mediated antibody-dependent cellular cytotoxicity was also reduced. One additional patient was found to have a normal proportion of circulating T cells (Geha, 1977). *In vitro* lymphocyte thymidine uptake has revealed the expected reactivity to *Aspergillus* antigen in only a portion of the patients (Haslam *et al.*, 1976; Rosenberg *et al.*, 1977; Geha, 1977; Forman *et al.*, 1978; Laham *et al.*, 1979). *Aspergillus, Candida,* and tuberculin antigen-induced production of LIF and thymidine incorporation is also only variably positive (Haslam *et al.*, 1976). PHA-induced (Chan-Yeung *et al.*, 1971; Haslam *et al.*, 1976; Hanson *et al.*, 1976; Rosenberg *et al.*, 1977; Geha, 1977; Forman *et al.*, 1978), and PWM-induced thymidine uptake (Rosenberg *et al.*, 1977; Geha, 1977) are usually normal.

Infection with *Coccidioides immitis* has been associated with defective cellular immunity (Catanzaro *et al.*, 1975). However, the presence of type I hypersensitivity has just recently been appreciated. Many patients with coccidioidomycosis have been found to have elevated total serum IgE concentrations which correlate with increased coccidioidin-specific IgE (Cox and Arnold, 1979; Cox *et al.*, 1982). These patients also have multiple positive immediate hypersensitivity skin tests and circulating antigen-specific IgE to common inhalant allergens, suggesting the presence of atopy. Of 11 patients with active disease, 10 exhibited a wheal and flare response to coccidioidin. Seven of these ten lacked a delayed hypersensitivity response to that antigen. Similarly, 6 of 7 patients with inactive coccidioidomycosis exhibited a type I skin response and 4 of these 6 lacked delayed skin reactivity to coccidioidin (Cox and Arnold, 1979).

Histoplasmosis is similarly associated with depressed cellular immunity (Cox, 1979). Type I hypersensitivity to histoplasmin was found in five of eight patients tested (Cox and Arnold, 1980). Two of these five had negative delayed reactivity to histoplasmin. Many patients with this disease also were found to have elevated total serum IgE (Cox and Arnold, 1980).

Since delayed hypersensitivity is important in controlling fungal infection, and since histamine may inhibit cellular reactivity, it is possible that allergy to *Aspergillus*, *Coccidioides*, and *Histoplasma* may contribute to the pathogenicity of these fungi.

8. CONCLUSION

In recent years, many have written that allergic disease is caused by abnormal lymphocyte numbers and function. As can be seen from this review, it is more probable that it is the atopic condition that causes the leukocyte abnormalities. It is a genetic predisposition which allows the development of the atopic disease. The capacity to produce IgE antibodies to specific antigens is inherited in association with specific transplantation alloantigens. HLA-Dw2 is an almost perfect genetic marker for the IgE and IgG immune responses to short ragweed pollen antigen Ra5 (Marsh *et al.*, 1982). HLA-MB3 is found in 100% of Japanese subjects with an IgE response to *Dermatophagoides farinae* (house dust mite) antigen (Nakai *et al.*, 1982). There is an interaction between HLA-linked immune response genes and a non-HLA IgE-regulating gene. Allergy to an antigen is the result of the HLA-D-determined genetic propensity for an antigen-specific IgE response. This is regulated by antigen-specific suppressor T cells (Rocklin *et al.*, 1980b). The magnitude of the response is determined in part by the IgE class, antigen-nonspecific regulating gene (Marsh *et al.*, 1981). Regulatory aspects of this IgE antibody response may indeed result in measurable abnormalities in $OKT4^+$ helper cell function (Leung *et al.*, 1983) or in the types of cellular effects seen after activation in the autologous mixed lymphocyte reaction (Hsieh, 1983). In contrast, it seems very unlikely that the vast number and variety of abnormalities associated with allergic disease represent primary or basic immunological defects. Most can be attributed to allergic mediator-induced deactivation, desensitization, or direct agonist effects. Certainly, observations of decreased numbers of total T cells or suppressor T cells cannot be explanations for a very selective elevation of the concentration of one class of antibodies to one or a few antigens. It is hoped that this review of such a complex and confusing area will be helpful in the formulation of new hypotheses to explain the existing observations and in the treatment of patients suffering the immunological consequences of the type I hypersensitivity reaction.

ACKNOWLEDGMENTS. The author wishes to thank Ms. Mary Euth for making the preparation of the manuscript possible, and Mrs. Caroline Desmond for her excellent skills and tireless efforts in using the word processor for this work. Her work was invaluable in the completion of this chapter. This work was supported by the Veterans Administration.

REFERENCES

Allen, D. H., Mathison, D. A., Wagner, P. D., Arroyave, C. M., Plow, E., and Tan, E. M., 1978, Mediator release during methacholine-induced bronchoconstriction in asthmatic patients, *J. Allergy Clin. Immunol.* **61:**140.

Allen, J. I., Grant, B. W., Syropoulus, M., Eagon, J. C., and Kay, M. E., 1984, Cimetidine augments NK function of normals and CLL patients, *Clin. Res.* **32:**342A.

Aly, R., Maibach, H. I., and Schinefield, H. R., 1977, Microbial flora of atopic dermatitis, *Arch. Dermatol.* **113:**780.

Amblard, P., Reymond, J. L., Pin, J., Bruttman, G., Fayolle, G., and Volatier, E., 1980, Exploration immunoallergologique de 30 sujets atopiques, *Ann. Dermatol. Venereol.* **107:**1007.

Andersen, E., and Hjorth, N., 1975, B lymphocytes, T lymphocytes and phytohaemagglutinin responsiveness in atopic dermatitis, *Acta Dermatovenereol.* **55:**345.

Anderson, R., Glover, A., and Rabson, A. R., 1977, The *in vitro* effects of histamine and metiamide on neutrophil motility and their relationship to intracellular cyclic nucleotide levels, *J. Immunol.* **118:**1690.

Anfosso, F., Alcaraz, G., Vervloet, D., and Charpin, J., 1982, Variations in T cell subsets during hyposensitization of grass-sensitive patients with formaldehyde modified extracts, *Allergy* **37:**613.

Armitstead, J. G., and Ewan, P. W., 1983, Suppressor T cell numbers and function in atopic patients with high serum IgE levels, *J. Clin. Lab. Immunol.* **12:**179.

Arrang, J.-M., Garbarg, M., and Schwartz, J.-C., 1983, Auto-inhibition of brain histamine release mediated by a novel class (H_3) of histamine receptor. *Nature (London)* **302:**832.

Arulanantham, K., Dwyer, J. M., and Genel, M., 1979, Evidence for defective immunoregulation in the syndrome of familial candidiasis endocrinopathy, *N. Engl. J. Med.* **300:**164.

Atkins, P. C., Norman, M., Weiner, H., and Zweiman, B., 1977, Release of neutrophil chemotactic activity during immediate hypersensitivity reactions in humans, *Ann. Intern. Med.* **86:**415.

Avella, J., Binder, H. J., Madsen, J. E., and Askenase, P. W., 1978, Effect of histamine H2-receptor antagonists on delayed hypersensitivity, *Lancet* **1:**624.

Ballet, J. J., and Merler, E., 1976, The separation and reactivity *in vitro* of a subpopulation of human lymphocytes which bind histamine: Correlation of histamine reactivity with cellular maturation, *Cell. Immunol.* **24:**250.

Balogh, E., Forizs, E., Debreczeni, M., and Szabolcsy, M., 1981, Serum IgE level and T-cell count in chronic dermatophytosis, *Mykosen* **24:**84.

Beer, D. J., Rosenwasser, L. J., Dinarello, C. A., and Rocklin, R. E., 1982a, Cellular interactions in the generation and expression of histamine-induced suppressor activity, *Cell. Immunol.* **69:**101.

Beer, D. J., Dinarello, C. A., Rosenwasser, L. J., and Rocklin, R. E., 1982b, Human monocyte-derived soluble product(s) has an accessory function in the generation of histamine- and concanavalin A-induced suppressor T cells, *J. Clin. Invest.* **70:**393.

Beer, D. J., Osband, M. E., McCaffrey, R. P., Soter, N. A., and Rocklin, R. E., 1982c, Abnormal histamine-induced suppressor-cell function in atopic subjects, *N. Engl. J. Med.* **306:**454.

Beets, J. L., and Dale, M. M., 1979, Inhibition of guinea pig lymphocyte activation by histamine and histamine analogues, *Br. J. Pharmacol.* **66**:365.

Berger, M., Kirkpatrick, C. H., Goldsmith, P. K., and Gallin, J. I., 1980, IgE antibodies to *Staphylococcus aureus* and *Candida albicans* in patients with the syndrome of hyperimmunoglobulin E and recurrent infections, *J. Immunol.* **124**:2437.

Berman, J. S., Cruickshank, W. W., Center, D. M., and Beer, D. J., 1984, Monoclonal antibody definition of the cell of origin and target cell specificity of histamine-induced human lymphokines modulating T cell motility, *Clin. Res.* **32**:343A.

Bicks, R. O., and Rosenberg, E. W., 1980, Reversal of anergy in Crohn's disease by cimetidine, *Lancet* **1**:552.

Birch, R. E., and Polmar, S. H., 1981, Induction of Fc gamma receptors on a subpopulation of human T lymphocytes by adenosine and impromidine, an H_2 histamine agonist, *Cell. Immunol.* **57**:455.

Birch, R. E., and Polmar, S. H., 1982, Pharmacological modification of immunoregulatory T lymphocytes. I. Effect of adenosine, H_1 and H_2 histamine agonists upon T lymphocyte regulation of B lymphocyte differentiation *in vitro, Clin. Exp. Immunol.* **48**:218.

Birch, R. E., Rosenthal, A. K., and Polmar, S. H., 1982, Pharmacological modification of immunoregulatory T lymphocytes. II. Modulation of T lymphocyte cell surface characteristics, *Clin. Exp. Immunol.* **48**:231.

Blum, R., Geller, G., and Fish, L. A., 1977, Recurrent severe staphylococcal infections, eczematoid rash, extreme elevations of IgE, eosinophilia, and divergent chemotactic responses in two generations, *J. Pediatr.* **90**:607.

Bortolussi, R., Faulkner, G., Lee, S. H. S., and Ozere, R., 1981, Phagocytosis of *Candida albicans* in chronic mucocutaneous candidiasis, *Pediatr. Res.* **15**:1287.

Bourne, H. R., Lichtenstein, L. M., and Melmon, K. L., 1972, Pharmacologic control of allergic histamine release in vitro: Evidence for an inhibitory role in 3′,5′-adenosine monophosphate in human leukocytes, *J. Immunol.* **108**:695.

Bousquet, J., Andary, M., Clot, J., Guilhou, J. J., and Meynadier, J., 1978, *In vitro* immunological studies in atopic and contact dermatitis, *Ann. Allergy* **41**:170.

Bovo, P., Vaona, B., Angelini, G., Cavallini, G., Montesi, G., Rosa, B., and Scuro, L. A., 1981, Ranitidine and delayed hypersensitivity, *Ital. J. Gastroenterol.* **13**:49.

Brahmi, Z., Liautaud, B., and Marill, F., 1980, Depressed cell-mediated immunity in chronic dermatophytic infections, *Ann. Immunol. (Inst. Pasteur)* **131C**:143.

Breuillard, F., and Szapiro, E., 1978, Cimetidine in a acquired tolerance to dinitrochlorobenzene, *Lancet* **1**:726.

Buckley, R. H., Wray, B. B., and Belmaker, E. Z., 1972, Extreme hyperimmunoglobulinemia E and undue susceptibility to infection, *Pediatrics* **49**:59.

Businco, L., Laurenti, F., Rossi, P., Galli, E., and Aiuti, F., 1981, A child with atopic features, raised serum IgE, and recurrent infection treated with levamisole, *Arch. Dis. Child.* **56**:60.

Businco, L., Benincori, N., and Cantani, A., 1983, Clinical and immunological aspects of atopic dermatitis: A review and update, in: *International Allergy Workshop Advances in Pediatric Allergy* (L. Businco, ed.), pp. 195–236, Excerpta Medica/Elsevier, Amsterdam.

Busse, W. W., and Lantis, S. D. H., 1979, Impaired H2 histamine granulocyte response in active atopic eczema, *J. Invest. Dermatol.* **73**:184.

Busse, W. W., and Lee, T.-P., 1976, Decreased adrenergic responses in lymphocytes and granulocytes in atopic eczema, *J. Allergy Clin. Immunol.* **58**:586.

Busse, W. W., and Sosman, J., 1976, Histamine inhibition of neutrophil lysosomal enzyme release: An H2 histamine receptor response, *Science* **194**:737.

Busse, W. W., and Sosman, J., 1977, Decreased H2 histamine response of granulocytes of asthmatic patients, *J. Clin. Invest.* **59**:1080.

Butler, M., Atherton, D., and Levinsky, R. J., 1982, Quantitative and functional deficit of suppressor T cells in children with atopic eczema, *Clin. Exp. Immunol.* **50**:92.

Byrom, N. A., Caballero, F., Campbell, M.-A., Chooi, M., Lane, A. M., Hugh-Jones, K., Timlin, D. M., and Hobbs, J. R., 1978, T-cell depletion and in vitro thymosin inducibility in asthmatic children, *Clin. Exp. Immunol.* **31**:490.

Byrom, N. A., Staughton, R. C. D., Campbell, M.-A., Timlin, D. M., Chooi, M., Lane, A. M.,

Copeman, P. W. M., and Hobbs, J. R., 1979, Thymosin-inducible 'null' cells in atopic eczema, *Br. J. Dermatol.* **100**:499.

Canonica, G. W., Mingari, M. C., Melioli, G., Colombatti, M., and Moretta, L., 1979, Imbalances of T cell subpopulations in patients with atopic diseases and effect of specific immunotherapy, *J. Immunol.* **123**:2669.

Carapeto, F. J., Winkelmann, R. K., and Jordon, R. E., 1976, T and B lymphocytes in contact and atopic dermatitis, *Arch. Dermatol.* **112**:1095.

Catanzaro, A., Spitler, L. E., and Moser, K. M., 1975, Cellular immune response in coccidioidomycosis, *Cell. Immunol.* **15**:360.

Cates, K. L., Grady, P. G., Shapira, E., and Davis, A. T., 1980, Cell-directed inhibition of polymorphonuclear leukocyte chemotaxis in a patient with mucocutaneous candidiasis, *J. Allergy Clin. Immunol.* **65**:431.

Cavagnaro, J., and Osband, M. E., 1983, Successful *in vitro* primary immunization of human peripheral blood mononuclear cells and its role in the development of human-derived monoclonal antibodies, *Biotechniques* **1**:30.

Center, D. M., Soter, N. A., Wasserman, S. I., and Austen, K. F., 1979, Inhibition of neutrophil chemotaxis in association with experimental angioedema in patients with cold urticaria: A model of chemotactic deactivation in vivo, *Clin. Exp. Immunol.* **35**:112.

Center, D. M., Cruikshank, W. W., Berman, J. S., and Beer, D. J., 1983, Functional characteristics of histamine receptor-bearing mononuclear cells. I. Selective production of lymphocyte chemoattractant lymphokines with histamine used as a ligand, *J. Immunol.* **131**:1854.

Cernelc, P., Kralj, J., and Cernelc, D., 1979, T and B lymphocyte rosettes in tonsils of atopic and nonatopic children, *Int. J. Pediatr. Otorhinolaryngol.* **1**:157.

Chan-Yeung, M., Chase, W. H., Trapp, W., and Grzybowski, S., 1971, Allergic bronchopulmonary aspergillosis: Clinical and pathological study of three cases, *Chest* **59**:33.

Chapman, M. D., Rowntree, S., Mitchell, E. B., DiPrisco de Fuenmajor, M. C., and Platts-Mills, T. A. E., 1983, Quantitative assessments of IgG and IgE antibodies to inhalant allergens in patients with atopic dermatitis, *J. Allergy Clin. Immunol.* **72**:27.

Chisari, F. V., and Edgington, T. S., 1974, Human T lymphocyte E rosette function. I. A process modulated by intracellular cyclic AMP, *J. Exp. Med.* **140**:1122.

Church, J. A., and Nye, C., 1982, The effects of *in vitro* antigen exposure upon suppressor cell function in atopic disease, *Ann. Allergy* **48**:199.

Church, J. A., Frenkel, L. D., Wrights, D. G., and Bellanti, J. A., 1976, T lymphocyte dysfunction, hyperimmunoglobulinemia E, recurrent bacterial infections, and defective neutrophil chemotaxis in a Negro child, *J. Pediatr.* **88**:982.

Clark, R. A., 1978, Disorders of granulocyte chemotaxis, in: *Leukocyte Chemotaxis: Methods, Physiology, and Clinical Implications* (J. I. Gallin and P. G. Quie, eds.), pp. 329–356, Raven Press, New York.

Clark, R. A., Root, R. K., Kimball, H. R., and Kirkpatrick, C. H., 1973, Defective neutrophil chemotaxis and cellular immunity in a child with recurrent infections, *Ann. Intern. Med.* **78**:515.

Clot, J., Bousquet, J., Guendon, R., and Michel, F. B., 1979, T-, B- and Fc-gamma-receptor-bearing lymphocytes in asthma, *Clin. Allergy* **9**:245.

Cooper, K. D., Wuepper, K. D., and Hanifin, J. M., 1980, T cell subset enumeration and functional analysis in atopic dermatitis, *Clin. Res.* **28**:566A.

Cox, R. A., 1979, Immunological studies of patients with histoplasmosis, *Am. Rev. Respir. Dis.* **120**:193.

Cox, R. A., and Arnold, D. R., 1979, Immunoglobulin E in coccidioidomycosis, *J. Immunol.* **123**:194.

Cox, R. A., and Arnold, D. R., 1980, Immunoglobulin E in histoplasmosis, *Infect. Immun.* **29**:290.

Cox, R. A., Baker, B. S., and Stevens, D. A., 1982, Specificity of immunoglobulin E in coccidioidomycosis and correlation with disease involvement, *Infect. Immun.* **37**:609.

Cronin, E., Bandmann, H. J., Calnan, C. D., Fregert, S., Hjorth, N., Magnusson, B., Maibach, H. I., Malten, K., Meneghini, C. L., Pirila, V., and Wilkinson, D. S., 1970, Contact dermatitis in the atopic, *Acta Dermatovenereol.* **50**:183.

Dahl, M. V., Greene, W. H., Jr., and Quie, P. G., 1976, Infection, dermatitis, increased IgE, and impaired neutrophil chemotaxis: A possible relationship, *Arch. Dermatol.* **112**:1387.

Dahl, M. V., Cates, K. L., and Quie, P. G., 1978, Neutrophil chemotaxis in patients with atopic dermatitis without infection, *Arch. Dermatol.* **114**:544.
Daman, L. A., Rosenberg, E. W., and Drake, L., 1978, Treatment of alopecia areata with dinitrochlorobenzene, *Arch. Dermatol.* **114**:1036.
Damle, N. K., and Gupta, S., 1981a, Autologous mixed lymphocyte reaction in man. II. Histamine-induced suppression of the autologous mixed lymphocyte reaction by T-cell subsets defined with monoclonal antibodies, *J. Clin. Immunol.* **1**:241.
Damle, N. K., and Gupta, S., 1981b, $OKT8^+$ T cells mediate histamine-induced suppressor T cell function, *IRCS Med. Sci.* **9**:911.
Davis, S. D., Schaller, J., and Wedgewood, R. J., 1966, Job's syndrome, recurrent 'cold' staphylococcal abscesses, *Lancet* **1**:1013.
Deal, E. C., Jr., Wasserman, S. I., Soter, N. A., Ingram, R. H., and McFadden, E. R., Jr., 1980, Evaluation of role played by mediators of immediate hypersensitivity in exercise-induced asthma, *J. Clin. Invest.* **65**:659.
DeCock, W., DeCree, J., and Verhaegen, H., 1977, Restoration by levamisole of histamine inhibited E rosette formation of T lymphocytes of patients with allergies, *Int. Arch. Allergy Appl. Immunol.* **54**:176–182.
DeCree, J., Emmery, L., Timmermans, J., Eeckels, R., DeCock, W., and Verhaegen, H., 1978, Defective neutrophil chemotaxis and raised serum IgE levels in a child with recurrent bacterial infections and eczema, *Arch. Dis. Child.* **53**:144.
Delafuente, J. C., Bethea, L. N., and Panush, R. S., 1980, Suppressor cell function in atopic diseases, *Ann. Allergy* **44**:331.
Dessaint, J.-P., Waksman, B. H., Metzger, H., and Capron, A., 1980, Cytophilic binding of IgE to the macrophage. III. Involvement of cyclic GMP and calcium in macrophage activation by dimeric or aggregated rat myeloma IgE, *Cell. Immunol.* **51**:280.
Djawari, D., Hornstein, O. P., Gross, J., and Meinhof, W., 1977, Defect of phagocytosis and intracellular killing of *Candida albicans* by granulocytes in patients with familial and non-familial chronic mucocutaneous candidiasis, *Arch. Dermatol. Res.* **260**:159.
Donabedian, H., and Gallin, J. I., 1983, Two inhibitors of neutrophil chemotaxis are produced by hyperimmunoglobulin E recurrent infection syndrome mononuclear cells exposed to heat-killed staphylococci, *Infect. Immun.* **40**:1030.
Dorval, G., Yang, W. H., Goodfriend, L., Roy, R., Espinoza, L. R., and Hebert, J., 1980, Circulating immune complexes may be associated with increased suppressor T-cell activity in atopic allergy, *Clin. Immunol. Immunopathol.* **16**:245.
Dupree, E., Friedman, J. M., Land, R. N., and Goldman, A. S., 1975, Cell-mediated immunity in atopic dermatitis, *J. Allergy Clin. Immunol.* **55**:102.
Elliott, S. T., and Hanifin, J. M., 1979, Lymphocyte response to phytohemagglutinin in atopic dermatitis: Enhancement after in vitro culture, *Arch. Dermatol.* **115**:1424.
Epstein, S., and Mohajerin, A. H., 1964, Incidence of contact sensitivity in atopic dermatitis, *Arch. Dermatol.* **90**:284.
Faulkner, J., Thompson, R. A., and Morrison-Smith, J., 1982, Polymorph function in asthma, *Clin. Allergy* **12**:137.
Faure, M. R., Nicholas, J. F., Thivolet, J., Gaucherand, M. A., and Czernielewski, J. M., 1982, Studies on T-cell subsets in atopic dermatitis: Human T-cell subpopulations defined by specific monoclonal antibodies, *Clin. Immunol. Immunopathol.* **22**:139.
Fels, A. O. S., Pawlowski, N. A., Cramer, E. B., King, T. K. C., Cohn, Z. A., and Scott, W. A., 1982, Human alveolar macrophages produce leukotriene B_4, *Proc. Natl. Acad. Sci. USA* **79**:7866.
Fineman, S. M., Mudawwar, F. B., and Geha, R. S., 1979, Characteristics and mechanisms of action of the concanavalin A-activated suppressor cell in man, *Cell. Immunol.* **45**:120.
Fischer, A., Pichat, L., Audinot, M., and Griscelli, C., 1982, Defective handling of mannan by monocytes in patients with chronic mucocutaneous candidiasis resulting in a specific cellular unresponsiveness, *Clin. Exp. Immunol.* **47**:653.
Fischer, T. J., Rachelefsky, G. S., Gard, S. E., and Stiehm, E. R., 1977, Defective monocyte chemotaxis in atopic dermatitis, *Ann. Allergy* **38**:308.

Flaherty, D. K., Imbeau, S., Nichols, D., and Reed, C. E., 1979, Immunologic parameters in asymptomatic subjects with allergic bronchopulmonary aspergillosis, *J. Allergy Clin. Immunol.* **63**:204.

Fontan, G., Lorente, F., Garcia Rodriguez, M. C., and Ojeda, J. A., 1976, Defective neutrophil chemotaxis and hyperimmunoglobulinemia E: A reversible defect? *Acta Paediatr. Scand.* **65**:509.

Foon, K. A., Wahl, S. M., Oppenheim, J. J., and Rosenstreich, D. L., 1976, Serotonin-induced production of a monocyte chemotactic factor by human peripheral blood leukocytes, *J. Immunol.* **117**:1545.

Forman, S. R., Fink, J. N., Moore, V. L., Wang, J., and Patterson, R., 1978, Humoral and cellular immune responses in *Aspergillus fumigatus* pulmonary disease, *J. Allergy Clin. Immunol.* **62**:131.

Foster, S. C., and Goetzl, E. J., 1978, Ascorbate therapy in impaired neutrophil and monocyte chemotaxis, *Arch. Ophthalmol.* **96**:2069.

Frisk, A., Von Stedingk, L.-V., and Wasserman, J., 1974, Lymphocyte stimulations in *Candida albicans* infections, *Sabouraudia* **12**:87.

Furukawa, C. T., and Altman, L. C., 1978, Defective monocyte and polymorphonuclear leukocyte chemotaxis in atopic disease, *J. Allergy Clin. Immunol.* **61**:288.

Galant, S. P., Underwood, S., Allred, S., and Hanifin, J. M., 1979, Beta adrenergic receptor binding on polymorphonuclear leucocytes in atopic dermatitis, *J. Invest. Dermatol.* **72**:330.

Gale, R. P., and Zighelboim, J., 1974, Modulation of polymorphonuclear leukocyte-mediated antibody-dependent cellular cytotoxicity, *J. Immunol.* **113**:1793.

Galli, E., Rossi, P., Businco, E., Perlini, R., and Businco, L., 1979, Rosette E ed istamina. Comportamento delle rosette E dopo incubazione con istamina e durante la stagione di pollinazione in soggetti pollinosici. Effetto del levamisolo, *Folia Allergol. Immunol. Clin.* **26**:173.

Gallin, J. I., Sandler, J. A., Clyman, R. I., Manganiello, V. C., and Vaughn, M., 1978, Agents that increase cyclic AMP inhibit accumulation of cGMP and depress human monocyte locomotion, *J. Immunol.* **120**:492.

Garovoy, M. R., Reddish, M. A., and Rocklin, R. E., 1983, Histamine-induced suppressor factor (HSF): Inhibition of helper T cell generation and function, *J. Immunol.* **130**:357.

Geha, R. S., 1977, Circulating immune complexes and activation of the complement sequence in acute allergic bronchopulmonary aspergillosis, *J. Allergy Clin. Immunol.* **60**:357.

Geha, R. S., Reinherz, E., Leung, D., McKee, K. T., Jr., Schlossman, S., and Rosen, F. S., 1981, Deficiency of suppressor T cells in the hyperimmunoglobulin E syndrome, *J. Clin. Invest.* **68**:783.

Ghazanshahi, S., Townley, R., Chaperone, E., and Villacorte, G., 1976, T and B lymphocyte rosettes in bronchial asthma. *Ann. Allergy* **36**:324.

Gifford, R. R. M., Hatfield, S. M., and Schmidtke, J. R., 1980, Cimetidine-induced augmentation of human lymphocyte blastogenesis by mitogen, bacterial antigen, and alloantigen, *Transplantation* **29**:143.

Goetzl, E. J., 1978, Regulation of the polymorphonuclear leukocyte chemotactic response by immunological reactions, in: *Leukocyte Chemotaxis* (J. I. Gallin and P. G. Quie, eds.), pp. 161–177, Raven Press, New York.

Goetzl, E. J., and Gorman, R. R., 1978, Chemotactic and chemokinetic stimulation of human eosinophil and neutrophil polymorphonuclear leukocytes by 12-L-hydroxy-5,8,10-heptadecatrienoic acid (HHT), *J. Immunol.* **120**:526.

Goetzl, E. J., Woods, J. M., and Gorman, R. R., 1977, Stimulation of human eosinophil and neutrophil polymorphonuclear leukocyte chemotaxis and random migration by 12-L-hydroxy-5,8,10,14-eicosatetraenoic acid, *J. Clin. Invest.* **59**:179.

Goodwin, J. S., Messner, R. P., and Williams, R. C., Jr., 1979, Inhibitors of T-cell mitogenesis: Effect of mitogen dose, *Cell. Immunol.* **45**:303.

Goodwin, J. S., and Webb, D. R., 1980, Regulation of the immune response by postaglandins, *Clin. Immunol. Immunopathol.* **15**:106.

Gordon, D., Lewis, G. P., and Nouri, A. M. E., 1981, Inhibition of lymphocyte proliferation by histamine and related compounds not mediated via H_1- or H_2-receptors, *Br. J. Pharmacol.* **74**:137.

Gottlieb, B. R., and Hanifin, J. M., 1974, Circulating T-cell deficiency in atopic dermatitis, *Clin. Res.* **22**:159A.

Grove, D. I., Reid, J. G., and Forbes, I. J., 1975, Humoral and cellular immunity in atopic eczema, *Br. J. Dermatol.* **92**:611.

Gualde, N., Atluru, D., and Goodwin, J. S., 1985, Effect of lipoxygenase metabolites of arachidonic acid on proliferation of human T cells and T cell subsets, *J. Immunol.* **134:**1125.

Gupta, S., Frenkel, R., Rosenstein, M., and Grieco, M. H., 1975, Lymphocyte subpopulations, serum IgE and total eosinophil counts in patients with bronchial asthma, *Clin. Exp. Immunol.* **22:**438.

Gupta, S., Fikrig, S., and Good, R., 1979a, Effects of agents modulating cyclic nucleotides on T cell subsets and locomotion of T cells, *J. Allergy Clin. Immunol.* **63:**143.

Gupta, S., Kirkpatrick, C. H., and Good, R. A., 1979b, Subpopulations of human T lymphocytes. XI. T cells with receptors for IgM and IgA and locomotion of T and non-T cells in peripheral blood from patients with chronic mucocutaneous candidiasis, *Clin. Immunol. Immunopathol.* **14:**86.

Gupta, S., Fikrig, S., and Good, R. A., 1980, Subpopulations of human T lymphocytes. XIII. T cell subpopulations (T mu and T gamma) in children with bronchial asthma, *Int. Arch. Allergy Appl. Immunol.* **61:**293.

Hall, T. J., Hudspith, B. N., and Brostoff, J., 1983, The role of prostaglandins in the modulation of histamine suppression of mitogen responses in atopic and normal subjects, *Int. J. Immunopharmacol.* **5:**107.

Hamberg, M., Svensson, J., and Samuelsson, B., 1974, Prostaglandin endoperoxides: A new concept concerning the mode of action and release of prostaglandins, *Proc. Natl. Acad. Sci. USA* **71:**3824.

Hamid, J., Ross, I. N., and Ward, C., 1983, The effect of cimetidine on immunological parameters in Crohn's disease: A double blind trial, *Immunol. Lett.* **6:**293.

Hanifin, J. M., Ray, L. F., and Lobitz, W. C., Jr., 1974, Immunological reactivity in dermatophytosis, *Br. J. Dermatol.* **90:**1.

Hanson, G., Flod, N., Wells, I., Novey, H., and Galant, S., 1977, Bronchocentric granulomatosis: A complication of allergic bronchopulmonary aspergillosis, *J. Allergy Clin. Immunol.* **59:**83.

Harper, T. B., Gaumer, H. R., Waring, W., Brannon, R. B., and Salvaggio, J. E., 1980, A comparison of cell-mediated immunity and suppressor T-cell function in asthmatic and normal children, *Clin. Allergy* **10:**555.

Haslam, P., Lukoszek, A., Longbottom, J. L., and Turner-Warwick, M., 1976, Lymphocyte sensitization to *Aspergillus fumigatus* antigens in pulmonary disease in man, *Clin. Allergy* **6:**277.

Hatch, G. E., Nichols, W. K., and Hill, H. R., 1977, Cyclic nucleotide changes in human neutrophils induced by chemoattractants and chemotactic modulators, *J. Immunol.* **119:**450.

Hay, R. J., 1979, Failure of treatment in chronic dermatophyte infections, *Postgrad. Med. J.* **55:**608.

Hay, R. J., and Brostoff, J., 1977, Immune responses in patients with chronic Trichophyton rubrum infections, *Clin. Exp. Dermatol.* **2:**373.

Hay, R. J., and Shennan, G., 1982, Chronic dermatophyte infections. II. Antibody and cell-mediated immune responses, *Br. J. Dermatol.* **106:**191.

Hebert, J., Beaudoin, R., Aubin, M., and Fontaine, M., 1980, The regulator effect of histamine on the immune response: Characterization of the cells involved, *Cell. Immunol.* **54:**49.

Herlin, T., and Kragballe, K., 1980, Impaired monocyte cyclic AMP responses and monocyte cytotoxicity in atopic dermatitis, *Allergy* **35:**647.

Herman, J. J., and Colten, H. R., 1980, Specific modulation of complement-dependent human granulocyte function by imidazole acetic acid, *J. Allergy Clin. Immunol.* **66:**274.

Herman, J. J., Brenner, J. K., and Colten, H. R., 1979, Inhibition of histaminase release from human granulocytes by products of histaminase activity, *Science* **206:**77.

Hermans, P. E., Ulrich, J. A., and Markowitz, H., 1969, Chronic mucocutaneous candidiasis as a surface expression of deep seated abnormalities: Report of a syndrome of superficial candidiasis, absence of delayed hypersensitivity and aminoaciduria, *Am. J. Med.* **47:**503.

Hill, H. R., and Quie, P. G., 1974, Raised serum-IgE levels and defective neutrophil chemotaxis in three children with eczema and recurrent bacterial infections, *Lancet* **1:**183.

Hill, H. R., and Quie, P. G., 1975, Defective neutrophil chemotaxis associated with hyperimmunoglobulinemia E, in: *The Phagocytic Cell in Host Resistance* (J. A. Bellanti and D. H. Dayton, eds.), pp. 249–266, Raven Press, New York.

Hill, H. R., Quie, P. G., Pabst, H. F., Ochs, H. D., Clark, R. A., Klebanoff, S. J., and Wedgwood, R. J., 1974, Defect in neutrophil granulocyte chemotaxis in Job's syndrome of recurrent "cold" staphylococcal abscesses, *Lancet* **2:**617.

Hill, H. R., Estensen, R. D., Quie, P. G., Hogan, N. A., and Goldberg, N. D., 1975, Modulation of human neutrophil chemotactic responses by cyclic 3′,5′-guanosine monophosphate and cyclic 3′,5′-adenosine monophosphate, *Metabolism* **24:**447.

Hill, H. R., Williams, P. B., Krueger, G. G., and Janis, B., 1976a, Recurrent staphylococcal abscesses associated with defective neutrophil chemotaxis and allergic rhinitis, *Ann. Intern. Med.* **85:**39.

Hill, H. R., Estensen, R. D., Hogan, N. A., and Quie, P. G., 1976b, Severe staphylococcal disease associated with allergic manifestations, hyperimmunoglobulinemia E, and defective neutrophil chemotaxis, *J. Lab. Clin. Med.* **88:**796.

Hoffman, D. R., Yamamoto, F. Y., Geller, B., and Haddad, Z., 1975, Specific IgE antibodies in atopic eczema, *J. Allergy Clin. Immunol.* **55:**256.

Horsmanheimo, M., and Horsmanheimo, A., 1979, Leukocyte migration inhibition factor in atopic dermatitis: Induction by concanavalin A *in vitro*, *J. Invest. Dermatol.* **72:**128.

Horsmanheimo, M., Horsmanheimo, A., Banov, C. H., Ainsworth, S. K., and Fudenberg, H. H., 1979, Cell-mediated immunity *in vitro* in atopic dermatitis, *Arch. Dermatol.* **115:**161.

Hovmark, A., 1975, An *in vivo* and *in vitro* study of cell-mediated immunity in atopic dermatitis, *Acta Dermatovenereol.* **55:**181.

Hovmark, A., 1977, An *in vitro* study of depressed cell-mediated immunity and of T and B lymphocytes in atopic dermatitis, *Acta Dermatovenereol.* **57:**237.

Hsieh, K.-H., 1976, Study of E rosettes, serum IgE and eosinophilia in asthmatic children, *Ann. Allergy* **37:**383.

Hsieh, K.-H., 1983, Different immunoregulatory functions of T lymphocytes responding in autologous mixed lymphocyte reaction between allergic patients and normals, *Ann. Allergy* **50:**110.

Hunziker, N., and Brun, R., 1980, Lack of delayed reaction in presence of cell-mediated immunity in trichophytin hypersensitivity, *Arch. Dermatol.* **116:**1266.

Hwang, K.-C., Fikrig, S. M., Friedman, H. M., and Gupta, S., 1983, Autologous mixed lymphocyte reaction in man. VII. Autologous mixed lymphocyte reaction in patients with bronchial asthma, allergic rhinitis and atopic dermatitis, *Allergy* **38:**113.

Issekutz, A. C., Lee, K. Y., and Biggar, W. D., 1978, Neutrophil chemotaxis in two patients with recurrent staphylococcal skin infections and hyperimmunoglobulin E, *J. Lab. Clin. Med.* **92:**640.

Ito, S., Shinomiya, K., and Mikawa, H., 1983, Suppressive effect of IgE soluble immune complex on neutrophil chemotaxis, *Clin. Exp. Immunol.* **51:**407.

Jacobs, J. C., and Norman, M. E., 1977, A familial defect of neutrophil chemotaxis with asthma, eczema and recurrent skin infections, *Pediatr. Res.* **11:**732.

Jensen, J. R., Cramers, M., and Thestrup-Pedersen, K., 1981, Subpopulations of T lymphocytes and non-specific suppressor cell activity in patients with atopic dermatitis, *Clin. Exp. Immunol.* **45:**118.

Jillson, O. F., and Huppert, M., 1949, The immediate wheal and the 24–48 hour tuberculin type, edematous reactions to trichophytin, *J. Invest. Dermatol.* **12:**179.

Joffe, M. I., Kew, M., and Rabson, A. R., 1983, Lymphocyte subtypes in patients with atopic eczema, protein calorie malnutrition, SLE and liver disease, *J. Clin. Lab. Immunol.* **10:**97.

Johnson, E. E., Irons, J. S., Patterson, R., and Roberts, M., 1974, Serum IgE concentration in atopic dermatitis—Relationship to severity of disease and presence of atopic respiratory disease, *J. Allergy Clin. Immunol.* **54:**94.

Johnson, H. H., DeOreo, G. A., Lascheid, W. P., and Mitchell, F., 1960, Skin histamine levels in chronic atopic dermatitis, *J. Invest. Dermatol.* **34:**237.

Jones, F. G. C., 1978, Cimetidine and the delayed-hypersensitivity response, *Lancet* **1:**880.

Jones, H. E., Lewis, C. W., and McMarlin, S. L., 1973, Allergic contact sensitivity in atopic dermatitis, *Arch. Dermatol.* **107:**217.

Jones, H. E., Reinhardt, J. H., and Rinaldi, M. G., 1974a, Acquired immunity to dermatophytes, *Arch. Dermatol.* **109:**840.

Jones, H. E., Reinhardt, J. H., and Rinaldi, M. G., 1974b, Immunologic susceptibility to chronic dermatophytosis, *Arch. Dermatol.* **110:**213.

Jones, H. E., Reinhardt, J. H., and Rinaldi, M. G., 1974c, Model dermatophytosis in naturally infected subjects, *Arch. Dermatol.* **110:**369.

Jorizzo, J. L., Sams, W. M., Jr., Jegasothy, B. V., and Olansky, A. J., 1980, Cimetidine as an immunomodulator: Chronic mucocutaneous candidiasis as a model, *Ann. Intern. Med.* **92**:192.

Joseph, M., Capron, A., Butterworth, A. E., Sturrock, R. F., and Houba, V.,1978, Cytotoxicity of human and baboon mononuclear phagocytes against schistosomula *in vitro:* Induction by immune complexes containing IgE and *Schistosoma mansoni* antigens, *Clin. Exp. Immunol.* **33**:48.

Joseph, M., Tonnel, A. B., Capron, A., and Voisin, C., 1980, Enzyme release and superoxide anion production by human alveolar macrophages stimulated with immunoglobulin E, *Clin. Exp. Immunol.* **40**:416.

Joseph, M., Tonnel, A. B., Capron, A., and Dessaint, J.-P., 1981, The interaction of IgE antibody with human alveolar macrophages and its participation in the inflammatory processes of lung allergy, *Agents Actions* **11**:619.

Joseph, M., Tonnel, A. B., Torpier, G., Capron, A., Arnoux, B., and Benveniste, J., 1983, Involvement of immunoglobulin E in the secretory processes of alveolar macrophage from asthmatic patients, *J. Clin. Invest.* **71**:221.

Juhlin, L., Johansson, G. O., Bennich, H., Hogman, C., and Thyresson, N., 1969, Immunoglobulin E in dermatoses: Levels in atopic dermatitis and urticaria, *Arch. Dermatol.* **100**:12.

Juto, P., and Strannegard, O., 1979, T lymphocytes and blood eosinophils in early infancy in relation to heredity for allergy and type of feeding, *J. Allergy Clin. Immunol.* **64**:38.

Kaaman, T., 1978, The clinical significance of cutaneous reactions to trichophytin in dermatophytosis, *Acta Dermatovenereol.* **58**:139.

Kaaman, T., Petrini, B., and Wasserman, J., 1979, *In vivo* and *in vitro* immune responses to trichophytin in dermatophytosis, *Acta. Dermatovenereol.* **59**:229.

Kaplan, A. P., Gray, L., Shaff, R. E., Horakova, Z., and Beaven, M. A., 1975, *In vivo* studies of mediator release in cold urticaria and cholinergic urticaria, *J. Allergy Clin. Immunol.* **55**:394.

Katona, I. M., Tata, G., Scanlon, R. T., and Bellanti, J. A., 1980, Hyper IgE syndrome: A disease with suppressor T cell deficiency, *Ann. Allergy* **45**:295.

Khan, M. M., Sansoni, P., Engelman, E. G., and Melmon, K. L., 1985, Pharmacologic effects of autocoids on subsets of T cells: Regulation of expression/function of histamine-2 receptors by a subset of suppressor cells, *J. Clin. Invest.* **75**:1578.

Kikkawa, Y., Kamimura, K., Hamajima, T., Sekiguchi, T., Kawai, T., Takenaka, M., and Tada, T., 1973, Thymic alymphoplasia with hyper-IgE-globulinemia, *Pediatrics* **51**:690.

Kirkpatrick, C. H., and Smith, T. K., 1974, Chronic mucocutaneous candidiasis: Immunologic and antibiotic therapy, *Ann. Intern. Med.* **80**:310.

Kirkpatrick, C. H., Rich, P. R., and Bennett, J. E., 1971, Chronic mucocutaneous candidiasis: Model-building in cellular immunity, *Ann. Intern. Med.* **74**:955.

Kirkpatrick, C. H., Ottenson, E. A., Smith, T. K., Wells, S. A., Burdick, J. F., 1976, Reconstitution of defective cellular immunity with foetal thymus and dialysable transfer factor. Long-term studies in a patient with chronic musocutaneous candidiasis, *Clin. Exp. Immunol.* **23**:414.

Klein, G. L., Ong, K. S., and Underwood, S. B., 1979, Theophylline suppression of T lymphocytes bearing Fc receptors for IgM, *J. Allergy Clin. Immunol.* **63**:144.

Knauer, K. A., Lichenstein, L. M., Adkinson, N. F., and Fish, J. E., 1981, Platelet activation during antigen-induced airway reactions in asthmatic subjects, *N. Engl. J. Med.* **304**:1404.

Komiyama, A., Morosawa, H., Hanamura, K., Miyagawa, T., and Akabane, T., 1977, Defective neutrophil chemotaxis, hyperimmunoglobulinemia E, and impaired cellular immunity in a child with recurrent pyogenic infections and chronic eczema, *Acta Haematol. Jpn.* **40**:513.

Kraemer, M. J., Ochs, H. D., Furukawa, C. T., and Wedgwood, R. J., 1982, *In vitro* studies of the hyper-IgE disorders: Suppression of spontaneous IgE synthesis by allogeneic suppressor T lymphocytes, *Clin. Immunol. Immunopathol.* **25**:157.

Kragballe, K., 1979, Antibody-dependent monocyte-mediated cytotoxicity in severe atopic dermatitis, *Allergy* **34**:35.

Kragballe, K., and Borregaard, N., 1981, Mechanisms of decreased antibody-dependent cytotoxicity mediated by monocytes and neutrophils in atopic dermatitis, *Acta Dermatovenereol.* **61**:11.

Kragballe, K., and Herlin, T., 1981, Antibody-dependent monocyte-mediated cytotoxicity in atopic dermatitis: Relation to clinical state and cyclic AMP response, *Allergy* **36**:27.

Kragballe, K., Herlin, T., and Jensen, J. R., 1980, Impaired monocyte-mediated cytotoxicity in atopic dermatitis, *Arch. Dermatol. Res.* **269**:21.

Laham, M. N., Allen, R. C., and Green, J. C., 1979, Allergic bronchopulmonary aspergillosis caused by *Aspergillus terreus:* Specific lymphocyte sensitization and antigen-directed serum opsonic activity, *J. Allergy Clin. Immunol.* **63**:205.

Lang, I., Gergely, P., and Petranyi, G., 1981a, Effect of histamine-receptor blocking on human spontaneous lymphocyte-mediated cytotoxicity, *Scand. J. Immunol.* **14**:573.

Lang, I., Torok, K., Gergely, P., Nekam, K., and Petranyi, G., 1981b, Effect of histamine receptor blocking on human antibody-dependent cell-mediated cytotoxicity, *Scand. J. Immunol.* **13**:361.

Lang, P., Goel, Z., and Grieco, M. H., 1978, Subsensitivity of T lymphocytes to sympathomimetic and cholinergic stimulation in bronchial asthma, *J. Allergy Clin. Immunol.* **61**:248.

Lappin, D., and Whaley, K., 1980, Effects of histamine on monocyte complement production. 1. Inhibition of C2 production mediated by its action on H2 receptors, *Clin. Exp. Immunol.* **41**:497.

Lawton, J. W. M., Costello, C., Barclay, G. R., Urbaniak, S. J., Darg, C., Raeburn, J. A., Uttley, W. S., and Kay, A. B., 1976, The effect of transfer factor on neutrophil function in chronic mucocutaneous candidiasis, *Br. J. Haematol.* **33**:137.

Lee, T. H., Nagakura, T., Papageorgiou, N., Iikura, Y., and Kay, A. B., 1983, Exercise-induced late asthmatic reactions with neutrophil chemotactic activity, *N. Engl. J. Med.* **308**:1502.

Leung, D. Y. M., Rhodes, A. R., and Geha, R. S., 1981, Enumeration of T cell subsets in atopic dermatitis using monoclonal antibodies, *J. Allergy Clin. Immunol.* **67**:450.

Leung, D. Y. M., Wood, N., Dubey, D., Rhodes, A. R., and Geha, R. S., 1983, Cellular basis of defective cell-mediated lympholysis in atopic dermatitis, *J. Immunol.* **130**:1678.

Leyden, J. J., Marples, R. R., and Klingman, A. M., 1974, *Staphylococcus aureus* in the lesions of atopic dermatitis, *Br. J. Dermatol.* **90**:525.

Lima, M., and Rocklin, R. E., 1981, Histamine modulates *in vitro* IgG production by pokeweed mitogen-stimulated human mononuclear cells, *Cell. Immunol.* **64**:324.

Lobitz, W. C., Jr., Honeyman, J. F., and Winkler, N. W., 1972, Suppressed cell-mediated immunity in two adults with atopic dermatitis, *Br. J. Dermatol.* **86**:317.

Luckasen, J. R., Sabad, A., Goltz, R. W., and Kersey, J. H., 1974, T and B lymphocytes in atopic eczema, *Arch. Dermatol.* **110**:375.

Lydyard, P. M., Brostoff, J., Pack, S., and Parry, H. F., 1981, Evidence of diminished suppressor T cell activity in patients with atopy and SLE, *Immunobiology* **158**:173.

McGeady, S. J., and Buckley, R. H., 1975, Depression of cell-mediated immunity in atopic eczema, *J. Allergy Clin. Immunol.* **56**:393.

McGeady, S., Saraclar, Y., and Mansmann, H. C., Jr., 1976, Normal T-cell numbers found in atopic children, *J. Allergy Clin. Immunol.* **57**:194.

MacGlashan, D. W., Jr., Schleimer, R. P., Peters, S. P., Schulman, E. S., Adams, G. K., III, Newball, H. H., and Lichtenstein, L. M., 1982, Generation of leukotrienes by purified human lung mast cells, *J. Clin. Invest.* **70**:747.

McGregor, C. G. A., Cochran, A. J., Ogg, L. J., Gray, G. R., Smith, I. S., and Gillespie, G., 1977, Immunological and other laboratory studies of patients receiving short-term cimetidine therapy, *Lancet* **1**:122.

McManus, L. M., Hanahan, D. J., and Pickard, R. N., 1981, Human platelet stimulation by acetyl glyceryl ether phosphorylcholine, *J. Clin. Invest.* **67**:903.

Malolepszy, J., Kraus-Filarska, M., Szeliga, W., Litwa, M., and Panaszek, B., 1980, Effect of histamine on the spontaneous E rosette formation after blockade of histamine H_2 receptors with cimetidine, *Arch. Immunol. Ther. Exp.* **28**:891.

Marom, Z., and Casale, T. B., 1983, Mast cells and their mediators, *Ann. Allergy* **50**:367.

Marone, G., Thomas, L. L., and Lichtenstein, L. M., 1980, The role of agonists that activate adenylate cyclase in the control of cAMP metabolism and enzyme release by human polymorphonuclear leukocytes, *J. Immunol.* **125**:2277.

Marone, G., Poto, S., Petracca, R., Triggiani, M., DeLutio Di Castelguidone, E., and Condorelli, M., 1982, Activation of human basophils by staphylococcal protein A. I. The role of cyclic AMP, arachidonic acid metabolites, microtubules and microfilaments, *Clin. Exp. Immunol.* **50**:661.

Marsh, D. G., Meyers, D. A., and Bias, W. B., 1981, The epidemiology and genetics of atopic allergy, *N. Engl. J. Med.* **305**:1551.

Marsh, D. G., Hsu, S. H., Roebber, M., Ehrlich-Kautzky, E., Freidhoff, L. R., Meyers, D. A., Pollard, M. K., and Bias, W. B., 1982, HLA-Dw2: A genetic marker for human immune response to short ragweed pollen allergen Ra5. I. Response resulting primarily from natural antigenic exposure, *J. Exp. Med.* **155**:1439.

Martinez, J. D., Santos, J., Stechschulte, D. J., and Abdou, N. I., 1979, Nonspecific suppressor cell function in atopic subjects, *J. Allergy Clin. Immunol.* **64**:485.

Martinez Cairo-Cueto, S., Espinosa-Ayala, J., Lopez-Roman, M., and Lugo, J. C., 1981, Defective chemotaxis of PMN cells of patients with bronchial asthma, *Arch. Invest. Med.* **12**:505.

Mathur, S., Goust, J-M., Horger, E. O., Fudenberg, H. H., 1977, Immunoglobulin E anti-*Candida* antibodies and candidiasis, *Infect. Immun.* **18**:257.

Matloff, S. M., Kiselis, I. K., and Rocklin, R. E., 1983, Reduced production of histamine-induced suppressor factor (HSF) by atopic mononuclear cells and decreased prostaglandin E_2 output by HSF-stimulated atopic monocytes, *J. Allergy Clin. Immunol.* **72**:359.

Mavligit, G. M., Calvo, D. B., III, Patt, Y. Z., and Hersh, E. M., 1981, Immune restoration and/or augmentation of local xenogeneic graft versus host reaction by cimetidine *in vitro*, *J. Immunol.* **126**:2272.

Mawhinney, H., Killen, M., Fleming, W. A., and Roy, A. D., 1980, The hyperimmunoglobulin E syndrome—A neutrophil chemotactic defect reversible by histamine H_2 receptor blockade?, *Clin. Immunol. Immunopathol.* **17**:483.

Melewicz, F. M., and Spiegelberg, H. L., 1980, Fc receptors for IgE on a subpopulation of human peripheral blood monocytes, *J. Immunol.* **125**:1026.

Melmon, K. L., Bourne, H. R., Weinstein, J., and Sela, M., 1972, Receptors for histamine can be detected on the surface of selected leukocytes, *Science* **177**:707.

Mendelsohn, J., Multer, M. M., and Boone, R. F., 1973, Enhanced effects of prostaglandin E_1 and dibutyryl cyclic AMP upon human lymphocytes in the presence of cortisol, *J. Clin. Invest.* **52**:2129.

Meretey, K., Room, G., and Maini, R. N., 1981, Effects of histamine on the mitogenic response of human lymphocytes and its modification by cimetidine and levamisole, *Agents Actions* **11**:84.

Michaelsson, G., 1973, Decreased phagocytic capacity of the neutrophil leucocytes in patients with atopic dermatitis, *Acta Dermatovenereol.* **53**:279.

Miller, M. E., 1975, Pathology of chemotaxis and random mobility, *Semin. Hematol.* **12**:59.

Miller, M. E., Norman, M. E., Koblenzer, P. J., and Schonauer, T., 1973, A new familial defect of neutrophil movement. *J. Lab. Clin. Med.* **82**:1.

Mitchell, E. B., Chapman, M. D., Pope, F. M., Crow, J., Jouhal, S. S., and Platts-Mills, T. A. E., 1982, Basophils in allergen-induced patch test sites in atopic dermatitis, *Lancet* **1**:127.

Mobacken, H., and Lindholm, L., 1974, Chronic mucocutaneous candidiasis, dermatophytosis and defective cellular immunity in monozygotic twins, *Acta Dermatovenereol.* **54**:203.

Morito, T., Bankhurst, A. D., and Williams, R. C., 1980, Studies on the pharmacologic manipulation of suppressor cells associated with impaired immunoglobulin production, *J. Lab. Clin. Med.* **96**:232.

Nagy, L., 1983, Serum neutrophil chemotactic activity and exercise induced asthma, *Eur. J. Respir. Dis.* **64**:161.

Nagy, L., Lee, T. H., and Kay, A. B., 1982, Neutrophil chemotactic activity in antigen-induced late asthmatic reactions. *N. Engl. J. Med.* **306**:497.

Nair, M. P. N., and Schwartz, S. A., 1983, Effect of histamine and histamine antagonists on natural and antibody-dependent cellular cytotoxicity of human lymphocytes *in vitro*, *Cell. Immunol.* **81**:45.

Nakai, Y., Kano, T., Moriuchi, J., Wakisaka, A., Aizawa, M., Tsuneta, H., Togashi, K., Sakiyama, Y., Wagatsuma, Y., and Matsumoto, S., 1982, Association between the MB system and asthma in the Japanese, *Hum. Immunol.* **4**:265.

Neiburger, R. G., Neiburger, J. B., and Dockhorn, R. J., 1978, Distribution of peripheral blood T and B lymphocyte markers in atopic children and changes during immunotherapy, *J. Allergy Clin. Immunol.* **61**:88.

Nelson, R. D., McCormack, R. T., Fiegel, V. D., and Simmons, R. L., 1978, Chemotactic deactivation of human neutrophils: Evidence for nonspecific and specific components, *Infect. Immun.* **22**:441.

Nugteren, D. H., 1975, Arachidonate lipoxygenase in blood platelets, *Biochim. Biophys. Acta* **380**:299.

Ogden, B. E., and Hill, H. R., 1980, Histamine regulates lymphocyte mitogenic responses through activation of specific H_1 and H_2 histamine receptors, *Immunology* **41**:107.

Ogden, B. E., Krueger, G. G., and Hill, H. R., 1979, Lymphocyte suppressor activity in atopic eczema, *Clin. Exp. Immunol.* **35**:269.

O'Loughlin, S., Diaz-Perez, J. L., Gleich, G. J., and Winkelmann, R. K., 1977, Serum IgE in dermatitis and dermatosis: An analysis of 497 cases, *Arch. Dermatol.* **113**:309.

Osband, M., and Hamilton, D., 1982, $OKT8^+$ cells with histamine H_2 receptors are enriched for suppressor cell activity while those without H_2 receptors are enriched for cytotoxic activity, *Clin. Res.* **30**:354A.

Osband, M., Gallison, D., Miller, B., Agarwal, R. P., and McCaffrey, R., 1980, Concanavalin A activation of suppressor cells is mediated by the release of histamine and is blocked by cimetidine, *Clin. Res.* **28**:365A.

Palacios, J., Fuller, E. W., and Blaylock, W. K., 1966, Immunological capabilities of patients with atopic dermatitis, *J. Invest. Dermatol.* **47**:484.

Palacios, R., and Alarcon-Segovia, D., 1981, Cimetidine abrogates suppressor T cell function *in vitro*, *Immunol. Lett.* **3**:33.

Parker, C. W., Kennedy, S., and Eisen, A. Z., 1977, Leukocyte and lymphocyte cyclic AMP responses in atopic eczema, *J. Invest. Dermatol.* **68**:302.

Paslin, D., and Norman, M. E., 1977, Atopic dermatitis and impaired neutrophil chemotaxis in Job's syndrome, *Arch. Dermatol.* **113**:801.

Pasternack, B., 1965, The prediction of asthma in infantile eczema: A statistical approach, *J. Pediatr.* **66**:164.

Pauly, C. R., Artis, W. M., and Jones, H. E., 1978, Atopic dermatitis, impaired cellular immunity, and molluscum contagiosum, *Arch. Dermatol.* **114**:391.

Payan, D. G., and Goetzl, E. J., 1983, Specific suppression of human T lymphocyte function by leukotriene B_4, *J. Immunol.* **131**:551.

Pepys, J., 1953, Effect of local adrenaline and histamine on tuberculin reactions, *Acta Allergol.* **6**:265.

Perelmutter, L., and Potvin, L., 1980, Role of Tgamma and Tmu sub-set lymphocytes in allergic disease, *Ann. Allergy* **45**:18.

Petrini, B., and Kaaman, T., 1981, T-lymphocyte subpopulations in patients with chronic dermatophytosis, *Int. Arch. Allergy Appl. Immunol.* **66**:105.

Pincus, S. H., Thomas, I. T., Clark, R. A., Ochs, H. D., 1975, Defective neutrophil chemotaxis with variant ichthyosis, hyperimmunoglobulinemia E, and recurrent infections, *J. Ped.* **87** (6, part 1):908.

Plaut, M., and Lichtenstein, L. M., 1978, Cellular and chemical basis of the allergic inflammatory response: Component parts and control mechanisms, in: *Allergy: Principles and Practice* (E. Middleton, Jr., C. E. Reed, and E. F. Ellis, eds.), pp. 115–138, Mosby, St. Louis.

Plaut, M., and Lichtenstein, L. M., 1982, Histamine and immune responses, in: *Pharmacology of Histamine Receptors* (C. R. Ganellin and M. E. Parsons, eds.), pp. 392–435, John Wright, PSG Inc., Littleton, Mass.

Pluss, J. M., and Kadas, L., 1954, A case of monilial granulomas associated with widespread superficial moniliasis and epidermophytosis, *J. Invest. Dermatol.* **23**:317.

Pryjma, J., Scislicki, A., Rusinek, B., and Pryjma, K., 1978, T and B lymphocytes, immunoglobulins and complement components in bronchial asthma, *Allergol. Immunopathol.* **6**:209.

Puppo, F., Corsini, G., and Mangini, P., 1982, Ranitidine does not modify lymphocyte *in vitro* reactivity to the mitogens PHA, Con A and PWM, *IRCS Med. Sci.* **10**:538.

Quie, P. G., and Cates, K. L., 1978, Clinical manifestations of disorders of neutrophil chemotaxis, in: *Leukocyte Chemotaxis: Methods, Physiology, and Clinical Implications* (J. I. Gallin and P. G. Quie, eds.), pp. 307–328, Raven Press, New York.

Rachelefsky, G. S., Opelz, G., Mickey, M. R., Kiuchi, M., Terasaki, P. I., Siegel, S. C., and Stiehm, E. R., 1976, Defective T cell function in atopic dermatitis, *J. Allergy Clin. Immunol.* **57**:569.

Radermecker, M., and Maldague, M.-P., 1981, Depression of neutrophil chemotaxis in atopic individuals, *Int. Arch. Allergy Appl. Immunol.* **65:**144.
Rajka, G., 1967, Delayed dermal and epidermal reactivity in atopic dermatitis (Prurigo Besnier). I. Delayed reactivity to bacterial and mold allergens, *Acta Derm. Venereol.* **47:**158.
Rajka, G., 1968, Delayed dermal and epidermal reactivity in atopic dermatitis (Prurigo Besnier). III. Further studies with bacterial and viral allergens, *Acta Derm. Venereol.* **48:**186.
Rajka, G., and Barlinn, C., 1979, On the significance of the trichophytin reactivity in atopic dermatitis, *Acta Dermatovenereol.* **59:**45.
Reichman, B. L., Handzel, Z. T., Segal, S., Weinstein, Y., and Levin, S., 1979, Participation of a histamine-Sepharose-adherent subpopulation of human mononuclear cells in the production of leucocyte migration inhibition factor (LIF) in healthy children, *Clin. Exp. Immunol.* **37:**562.
Rigal, D., Monier, J. C., and Souweine, G., 1979, The effect of histamine on leukocyte migration test in man. I. Demonstration of a LIF production inhibitor (LIF-PI), *Cell. Immunol.* **46:**360.
Ring, J., and Lutz, J., 1983, Decreased release of lysosomal enzymes from peripheral leukocytes of patients with atopic dermatitis, *J. Am. Acad. Dermatol.* **8:**378.
Ring, J., Senter, T., Cornell, R. C., Arroyave, C. M., and Tan, E. M., 1979, Plasma complement and histamine changes in atopic dermatitis, *Br. J. Dermatol.* **100:**521.
Robertson, A. J., Peden, N. R., Saunders, J. H. B., Gibbs, J. H., Potts, R. C., Brown, R. A., Wormsley, K. G., and Swanson-Beck, J., 1979, Cimetidine and the immune response, *Lancet* **2:**420.
Rocklin, R. E., and Beer, D. J., 1983, Histamine and immune modulation, *Adv. Intern. Med.* **28:**225.
Rocklin, R. E., and Haberek-Davidson, A., 1981, Histamine activates suppressor cells *in vitro* using a coculture technique, *J. Clin. Immunol.* **1:**73.
Rocklin, R. E., and Melmon, K. L., 1978, Production and assay of a histamine-induced suppressor factor (HSF) by human lymphocytes, *Clin. Res.* **26:**520A.
Rocklin, R. E., Breard, J., Gupta, S., Good, R. A., and Melmon, K. L., 1980a, Characterization of the human blood lymphocytes that produce a histamine-induced suppressor factor (HSF), *Cell. Immunol.* **51:**226.
Rocklin, R. E., Sheffer, A. L., Greineder, D. K., and Melmon, K. L., 1980b, Generation of antigen-specific suppressor cells during allergy desensitization, *N. Engl. J. Med.* **302:**1213.
Rocklin, R. E., Kiselis, I., Beer, D. J., Rossi, P., Maggi, F., and Bellanti, J. A., 1983, Augmentation of prostaglandin and thromboxane production *in vitro* by monocytes exposed to histamine-induced suppressor factor (HSF), *Cell. Immunol.* **77:**92.
Rogge, J. L., and Hanifin, J. M., 1976, Immunodeficiencies in severe atopic dermatitis: Depressed chemotaxis and lymphocyte transformation, *Arch. Dermatol.* **112:**1391.
Rola-Pleszczynski, M., and Blanchard, R., 1981, Abnormal suppressor cell function in atopic dermatitis, *J. Invest. Dermatol.* **76:**279.
Rola-Pleszczynski, M., Borgeat, P., and Sirois, P., 1982, Leukotriene B_4 induces human suppressor lymphocytes, *Biochem. Biophys. Res. Commun.* **108:**1531.
Rosenberg, M., Patterson, R., Mitzer, R., Cooper, B. J., Roberts, M., and Harris, K. E., 1977, Clinical and immunologic criteria for the diagnosis of allergic bronchopulmonary aspergillosis, *Ann. Intern. Med.* **86:**405.
Rothschild, H., Wilson, M., Lopez, M., Salvaggio, J., and Bice, D., 1976, An immunological investigation of a family with chronic mucocutaneous candidiasis, *Int. Arch. Allergy Appl. Immunol.* **52:**291.
Rouveix, B., Badenoch-Jones, P., Sarno, S., and Turk, J. L., 1980, Lymphokine-induced macrophage aggregation: The possible role of cyclic nucleotides, *Immunopharmacology* **2:**319.
Rouveix, B., Larno, S., Badenoch-Jones, P., and Lechat, P., 1981, Lymphokine-induced macrophage aggregation: Involvement of cyclic-GMP and microtubules, *Agents Actions* **11:**622.
Rubin, J. L., Griffiths, R. W., and Hill, H. R., 1978, Allergen-induced depression of neutrophil chemotaxis in allergic individuals, *J. Allergy Clin. Immunol.* **62:**301.
Rudzki, E., Moskalewska, K., Maciejowska, E., and Blaszczyk, M., 1964, Fruhallergie auf Streptokokken und Staphylokokken bei Menschen und experimentell infizierten Kaninchen, *Arch. Klin. Exp. Dermatol.* **220:**535.

Ruiz-Arguelles, A., Seroogy, K. B., and Ritts, R. E., Jr., 1982, *In vitro* effect of cimetidine on human cell-mediated cytotoxicity. I. Inhibition of natural killer cell activity, *Cell. Immunol.* **69**:1.

Safko, M. J., and Hanifin, J. M., 1980, Receptor desensitization as a possible basis for the beta adrenergic theory of atopic disease, *Clin. Res.* **28**:22A.

Safko, M. J., Chan, S.-C., Cooper, K. D., and Hanifin, J. M., 1981, Heterologous desensitization of leukocytes: A possible mechanism of beta adrenergic blockade in atopic dermatitis, *J. Allergy Clin. Immunol.* **68**:218.

Sampson, H. A., 1983, Role of immediate food hypersensitivity in the pathogenesis of atopic dermatitis, *J. Allergy Clin. Immunol.* **71**:473.

Sampson, H. A., and Jolie, P. L., 1984, Increased plasma histamine concentrations after food challenges in children with atopic dermatitis, *N. Engl. J. Med.* **311**:372.

Sams, W. M., Jr., Jorizzo, J. L., Snyderman, R., Jegasothy, B. V., Ward, F. E., Weiner, M., Wilson, J. G., Yount, W. J., and Dillard, S. B., 1979, Chronic mucocutaneous candidiasis immunologic studies of three generations of a single family, *Am. J. Med.* **67**:948.

Sanchez-Crespo, M., Alonso, F., and Egido, J., 1980, Platelet-activating factor in anaphylaxis and phagocytosis. I. Release from human peripheral polymorphonuclears and monocytes during the stimulation by ionophore A23187 and phagocytosis but not from degranulating basophils, *Immunology* **40**:645.

Sandler, J. A., Gallin, J. I., and Vaughan, M., 1975, Effects of serotonin, carbamylcholine, and ascorbic acid on leukocyte cyclic GMP and chemotaxis, *J. Cell Biol.* **67**:480.

Sansoni, P., Silverman, E. D., Khan, M. M., Melmon, K. L., and Engelman, E. G., 1985, Immunoregulatory T cells in man: Histamine-induced suppressor T cells are derived from a Leu 2^+ ($T8^+$) subpopulation distinct from that which gives rise to cycotoxic T cells, *J. Clin. Invest.* **75**:650.

Saraclar, Y., McGeady, S. J., and Mansmann, H. C., Jr., 1977, Lymphocyte subpopulations of atopic children and the effect of therapy upon them, *J. Allergy Clin. Immunol.* **60**:301.

Saxon, A., Morledge, V. D., and Bonavida, B., 1977, Histamine-receptor leucocytes (HRL): Organ and lymphoid subpopulation distribution in man, *Clin. Exp. Immunol.* **28**:394.

Saxon, A., Morrow, C., and Stevens, R. H., 1980, Subpopulations of circulating B cells and regulatory T cells involved in *in vitro* immunoglobulin E production in atopic patients with elevated serum immunoglobulin E, *J. Clin. Invest.* **65**:1457.

Schoepf, E., and Boehringer, D., 1974, IgE and cell-mediated immunity in atopic dermatitis, *J. Dermatol.* **1**:133.

Schopf, E., Kapp, A., and Kim, C. W., 1978, T-cell function in atopic dermatitis: Controlled examination of concanavalin A dose–response relations in cultured lymphocytes, *Arch. Dermatol. Res.* **262**:37.

Schopfer, K., Baerlocher, K., Price, P., Krech, U., Quie, P. G., and Douglas, S. D., 1979, Staphylococcal IgE antibodies, hyperimmunoglobulinemia E and *Staphylococcus aureus* infections, *N. Engl. J. Med.* **300**:835.

Schopfer, K., Douglas, S. D., and Wilkinson, B. J., 1980, Immunoglobulin E antibodies against *Staphylococcus aureus* cell walls in the sera of patients with hyperimmunoglobulinemia E and recurrent staphylococcal infection, *Infect. Immun.* **27**:563.

Schuster, D. L., Pierson, D., Bongiovanni, B., and Levinson, A. I., 1979, Suppressor cell function in atopic dermatitis associated with elevated immunoglobulin E, *J. Allergy Clin. Immunol.* **64**:139.

Schuster, D. L., Bongiovanni, B. A., Pierson, D. L., Barbaro, J. F., Wong, D. T. O., and Levinson, A. I., 1980, Selective deficiency of a T cell subpopulation in active atopic dermatitis, *J. Immunol.* **124**:1662.

Secher, L., Svejgaard, E., and Sonderstrup Hansen, G., 1977, T and B lymphocytes in contact and atopic dermatitis, *Br. J. Dermatol.* **97**:537.

Seligmann, B. E., Fletcher, M. P., and Gallin, J. I., 1983, Histamine modulation of human neutrophil oxidative metabolism, locomotion, degranulation, and membrane potential changes, *J. Immunol.* **130**:1902.

Selwyn, S., and Chalmers, D., 1965, Dispersal of bacteria from skin lesions: A hospital hazard, *Br. J. Dermatol.* **77**:349.

Simchowitz, L., Atkinson, J. P., and Spilberg, I., 1980, Stimulus-specific deactivation of chemotactic

factor-induced cyclic AMP response and superoxide generation by human neutrophils, *J. Clin. Invest.* **66**:736.
Simon, M. R., and Sams, W. M., Jr., 1980, Cimetidine and immunologic response (letter by Simon and response by Sams), *Ann. Intern. Med.* **93**:152.
Simon, M. R., Tubergen, D., Cassidy, J., Silva, J., Jr., and Magilavy, D., 1979, Chronic mucocutaneous candidiasis clinically exacerbated by type I hypersensitivity, *Clin. Immunol. Immunopathol.* **14**:56.
Simon, M. R., Salberg, D. J., and Crane, S. A., 1981, *In vivo* cimetidine augmentation of phytohemagglutinin induced human lymphocyte thymidine uptake, *Transplantation* **31**:400.
Sipos, K., 1965, Analyse der intracutanen Reaktionen von Pyokokkenantigenen, *Arch. Klin. Exp. Dermatol.* **222**:430.
Slavin, R. G., Tennenbaum, J. I., Becker, R. J., Feinberg, A. R., and Feinberg, S. M., 1963, Cell transfer of delayed hypersensitivity to ragweed from atopic subjects treated with emulsified ragweed extracts, *J. Allergy* **34**:368.
Smart, L. M., and Kay, A. B., 1981, Histamine receptors on human peripheral blood leucocytes, *Clin. Exp. Immunol.* **44**:581.
Smith, J. W., Steiner, A. L., and Parker, C. W., 1971, Human lymphocyte metabolism: Effects of cyclic and noncyclic nucleotides on stimulation by phytohemagglutinin, *J. Clin. Invest.* **50**:442.
Smith, M. D., Couhig, E., Miller, J. J., and Salaman, J. R., 1979, Cimetidine and the immune response, *Lancet* **1**:1406.
Smogorzewska, E. M., Layward, L., and Soothill, J. F., 1981, T lymphocyte mobility defects and effects of ascorbic acid, histamine and complexed IgG, *Clin. Exp. Immunol.* **43**:174.
Snyderman, R., Altman, L. C., Frankel, A., and Blaese, R. M., 1973, Defective mononuclear leukocyte chemotaxis: A previously unrecognized immune dysfunction. Studies in a patient with chronic mucocutaneous candidiasis, *Ann. Intern. Med.* **78**:509.
Snyderman, R., Rogers, E., and Buckley, R. H., 1977, Abnormalities of leukotaxis in atopic dermatitis, *J. Allergy Clin. Immunol.* **60**:121.
Soderberg-Warner, M., Rice-Mendoza, C. A., Mendoza, G. R., and Stiehm, E. R., 1983, Neutrophil and T lymphocyte characteristics of two patients with hyper-IgE syndrome, *Pediatr. Res.* **17**:820.
Sohnle, P. G., and Collins-Lech, C., 1982, Analysis of the lymphocyte transformation response to *Pityrosporum orbiculare* in patients with tinea versicolor, *Clin. Exp. Immunol.* **49**:559.
Soppi, E., Viander, M., Soppi, A.-M., and Jansen, C. T., 1982, Cell-mediated immunity in untreated and PUVA treated atopic dermatitis, *J. Invest. Dermatol.* **79**:213.
Sorensen, G. W., and Jones, H. E., 1976, Immediate and delayed hypersensitivity in chronic dermatophytosis, *Arch. Dermatol.* **112**:40.
Stahl, D., and Svejgaard, E., 1982, Lymphocyte transformation *in vitro* in acute dermatophytosis: A follow-up study, *Acta Dermatovenereol.* **62**:289.
Stanciu, L., Urcan, S., Parau, N., Radu, D., Itu, I., and Georoceanu, A., 1983, Some aspects of cellular and humoral immunity in bronchial asthma and allergic rhinitis, *Rev. Roum. Med. Interne* **21**:19.
Staszak, C., and Goodwin, J. S., 1980, Is prostaglandin a mediator for the inhibitory action of histamine, hydrocortisone, and isoproterenol?, *Cell. Immunol.* **54**:351.
Staszak, C., Goodwin, J. S., Troup, G. M., Pathak, D. R., and Williams, R. C., Jr., 1980, Decreased sensitivity to prostaglandin and histamine in lymphocytes from normal HLA-B12 individuals: A possible role in autoimmunity, *J. Immunol.* **125**:181.
Stephens, C. G., and Snyderman, R., 1982, Cyclic nucleotides regulate the morphologic alterations required for chemotaxis in monocytes, *J. Immunol.* **128**:1192.
Stevens, W. J., and Bridts, C. H., 1984, IgG-containing and IgE-containing circulating immune complexes in patients with asthma and rhinitis, *J. Allergy Clin. Immunol.* **73**:276.
Stifler, W. C., Jr., 1965, A 21 year follow-up of infantile eczema, *J. Pediatr.* **66**(Suppl.):166.
Stingl, G., Gazze, L. A., Czarnecki, N., and Wolff, K., 1981, T cell abnormalities in atopic dermatitis patients: Imbalances in T cell subpopulations and impaired generation of Con A-induced suppressor cells, *J. Invest. Dermatol.* **76**:468.
Strannegard, I.-L., 1979, Lymphocyte stimulation with phorbol myristate acetate in atopic and nonatopic individuals, *Int. Arch. Allergy Appl. Immunol.* **58**:175.

Strannegard, I.-L., 1981, Adherent suppressor cells in atopic disease, *Int. Arch. Allergy Appl. Immunol.* **64**:308.
Strannegard, I.-L., and Strannegard, O., 1977a, Increased sensitivity of lymphocytes from atopic individuals to histamine-induced suppression, *Scand. J. Immunol.* **6**:1226.
Strannegard, I.-L., and Strannegard, O., 1977b, Influence of serum from atopic children on T lymphocytes, *Int. Arch. Allergy Appl. Immunol.* **55**:217.
Strannegard, I.-L., Lindholm, L., and Strannegard, O., 1976, T lymphocytes in atopic children, *Int. Arch. Allergy Appl. Immunol.* **50**:684.
Tagami, H., Watanabe, S., Ofuji, S., and Minami, K., 1977, Trichophytin contact sensitivity in patients with dermatophytosis, *Arch. Dermatol.* **113**:1409.
Takeya, K., Nomoto, K., Matsumoto, T., Miyake, T., and Himeno, K., 1976, Chronic mucocutaneous candidiasis accompanied by enhanced antibody production, *Clin. Exp. Immunol.* **25**:497.
Talpaz, M., Medina, J. E., Patt, Y. Z., Goepfert, H., Guillamondegui, O. M., Wong, W., and Mavligit, G. M., 1982, The immune restorative effect of cimetidine administration *in vivo* on the local graft-versus-host reaction of cancer patients, *Clin. Immunol. Immunopathol.* **24**:155.
Tarlo, S. M., Day, J. H., Mann, P., and Day, M. P., 1977, Immediate hypersensitivity to tuberculin: *In vivo* and *in vitro* studies, *Chest* **71**:33.
Tartakovsky, B., Segal, S., Shani, A., Hellerstein, S., Weinstein, Y., and Bentwich, Z., 1979, Segregation of human peripheral blood lymphocytes according to their affinity for insolubilized histamine: Principal differences between males and females, *Clin. Exp. Immunol.* **38**:166.
Thestrup-Pedersen, K., Ellegaard, J., Thulin, H., and Zachariae, H., 1977, PPD and mitogen responsiveness of lymphocytes from patients with atopic dermatitis, *Clin. Exp. Immunol.* **27**:118.
Thomas, Y., Huchet, R., and Granjon, D., 1981, Histamine-induced suppressor cells of lymphocyte mitogenic response, *Cell. Immunol.* **59**:268.
Thomson, N. C., Sandilands, G. P., Gray, K., and Reid, F., 1977, Increase in peripheral blood "null" cells in extrinsic bronchial asthma, *Clin. Exp. Immunol.* **30**:429.
Too, T. J., 1979, Concanavalin A-induced suppressor cells in ragweed allergy, *J. Allergy Clin. Immunol.* **63**:143.
Tse, R. L., Phelps, P., and Urban, D., 1972, Polymorphonuclear leukocyte motility in vitro. VI. Effect of purine and pyrmidine analogues: Possible role of cyclic AMP, *J. Lab. Clin. Med.* **80**:264.
Turner, M. W., Mowbray, J. F., Harvey, B. A. M., Brostoff, J., Wells, R. S., and Soothill, J. F., 1978, Defective yeast opsonization and C2 deficiency in atopic patients, *Clin. Exp. Immunol.* **34**:253.
Twomey, J. J., Waddell, C. C., Krantz, S., O'Reilly, R., L'Esperance, P., and Good, R. A., 1975, Chronic mucocutaneous candidiasis with macrophage dysfunction, a plasma inhibitor, and coexistent aplastic anemia, *J. Lab. Clin. Med.* **85**:968.
Uehara, M., 1977, Atopic dermatitis and tuberculin reactivity, *Arch. Dermatol.* **113**:1226.
Valdimarsson, H., Riches, H. R. C., Holt, L., and Hobbs, J. R., 1970, Lymphocyte abnormality in chronic mucocutaneous candidiasis, *Lancet* **1**:1259.
Valone, F. H., Leid, R. W., Jr., and Goetzl, E. J., 1977, Immunologic release of lipid factors chemotactic for human eosinophil and neutrophil polymorphonuclear leukocytes (PMNL), *Fed. Proc.* **36**:1328A.
Valverde, E., Vich, J. M., Huguet, J., Garcia-Calderon, J. V., and Garcia-Calderon, P. A., 1983, An *in vitro* study of lymphocytes in patients with atopic dermatitis, *Clin. Allergy* **13**:81.
Van Scoy, R. E., Hill, H. R., Ritts, R. E., Jr., and Quie, P. G., 1975, Familial neutrophil chemotaxis defect, recurrent bacterial infections, mucocutaneous candidiasis, and hyperimmunoglobulinemia E, *Ann. Intern. Med.* **82**:766.
Verhaegen, H., DeCock, W., and DeCree, J., 1977, Histamine receptor-bearing peripheral T lymphocytes in patients with allergies, *J. Allergy Clin. Immunol.* **59**:266.
Walker, S. M., and Urbaniak, S. J., 1980, A serum-dependent defect of neutrophil function in chronic mucocutaneous candidiasis, *J. Clin. Pathol.* **33**:370.
Walsh, G. A., Richards, K. L., Douglas, S. D., and Blumenthal, M. N., 1981, Immunoglobulin E anti-*Staphylococcus aureus* antibodies in atopic patients, *J. Clin. Microbiol.* **13**:1046.
Wang, S. R., and Zweiman, B., 1978, Histamine suppression of human lymphocyte responses to mitogens, *Cell. Immunol.* **36**:28.

Wang, S. R., and Zweiman, B., 1981, Inhibitory effects of corticosteroids and histamine on human lymphocytes, *J. Allergy Clin. Immunol.* **67**:39.
Wang, Y., Kristensen, F., Joncourt, F., Slauson, D. O., and De Weck, A. L., 1983, Analysis of ^{3}H-histamine interaction with lymphocytes: Receptor binding or uptake?, *Clin. Exp. Immunol.* **54**:501.
Wasserman, S. I., Whitmer, D., Goetzl, E. J., and Austen, K. F., 1975, Chemotactic deactivation of human eosinophils by the eosinophil chemotactic factor of anaphylaxis, *Proc. Soc. Exp. Biol. Med.* **148**:301.
Wasserman, S. I., Soter, N. A., Center, D. M., and Austen, K. F., 1977, Cold urticaria recognition and characterization of a neutrophil chemotactic factor which appears in serum during experimental cold challenge, *J. Clin. Invest.* **60**:189.
Webb, D. R., Nowowiejski, I., Healy, C., and Rogers, T. J., 1982, Immunosuppressive properties of leukotriene D_4 and E_4 *in vitro, Biochem. Biophys. Res. Commun.* **104**:1617.
Weinstein, Y., and Melmon, K. L., 1976, Control of immune responses by cyclic AMP and lymphocytes that adhere to histamine columns, *Immunol. Commun.* **5**:401.
Welbourn, E., Champion, R. H., and Parish, W. E., 1976, Hypersensitivity to bacteria in eczema. I. Bacterial culture, skin tests and immunofluorescent detection of immunoglobulins and bacterial antigens, *Br. J. Dermatol.* **94**:619.
White, W. B., and Ballow, M., 1985, Modulation of suppressor-cell activity by cimetidine in patients with common variable hypogammaglobulinemia, *N. Engl. J. Med.* **312**:198.
White, W. B., Ballow, M., and Desbonnet, C. R., 1984, Cimetidine modulates suppressor activity in common variable hypogammaglobulinemia, *Clin. Res.* **32**:362A.
Wilkinson, P. C., 1976, Cellular and molecular aspects of chemotaxis of macrophages and monocytes, in: *Immunobiology of the Macrophage* (D. S. Nelson, ed.), pp. 349–365, Academic Press, New York.
Wolfe, J. D., Plaut, M., Norman, P. S., and Lichtenstein, L. M., 1979, The effect of an H_2 receptor antagonist on immediate and delayed skin test reactivity in man, *J. Allergy Clin. Immunol.* **63**:208.
Wood, S. R., and Cruickshank, C. N. D., 1962, The relation between trichophytin sensitivity and fungal infection, *Br. J. Dermatol.* **74**:329.
Yocum, M. W., Strong, D. M., and Larkin, J. D., 1976, Competent cellular immunity in allergic rhinitis patients with elevated IgE, *J. Allergy Clin. Immunol.* **57**:384.
Zachary, C. B., and MacDonald, D. M., 1983, Quantitative analysis of T-lymphocyte subsets in atopic eczema, using monoclonal antibodies and flow cytofluorimetry, *Br. J. Dermatol.* **108**:411.
Zarling, J. M., Berman, C., and Raich, P. C., 1980, Depressed cytotoxic T-cell responses in previously treated Hodgkin's and non-Hodgkin's lymphoma patients. Evidence for histamine receptor-bearing suppressor cells, *Cancer Immunol. Immunother.* **7**:243.
Zielinski, A., and Oehling, A., 1982, Cytochemical markers for T mu and T gamma: A comparative study, *Allergol. Immunopathol.* **10**:115.
Zurier, R. B., Weissmann, G., Hoffstein, S., Kammerman, S., and Tai, H. H., 1974, Mechanisms of lysosomal enzyme release from human leukocytes. II. Effects of cAMP and cGMP, autonomic agonists, and agents which affect microtubule function, *J. Clin. Invest.* **53**:297.

7

Complement-Mediated Injury

MICHEL D. KAZATCHKINE and URS E. NYDEGGER

1. INTRODUCTION

The complement system functions as part of the immune system to protect the individual from microbial infection by mediating biological reactions that in many instances are synergistic with the function of cells from the reticuloendothelial system. Both systems are found in primitive animals and have evolved to serve both specific (immune) and natural (non-antibody-dependent) host defense. Complement activation can be triggered by antigen–antibody complexes and by surfaces of microorganisms or cells that express distinct biochemical characteristics. Activation of the complement system results in opsonization of the target and its interaction with cells of the RES bearing specific receptors for bound complement components, in generation of biologically active diffusible peptides that promote the inflammatory reaction, and may result in lysis of the activating target. These finely regulated reactions also carry the potential for complement-mediated tissue injury in the host (Austen, 1978).

2. PROTEINS OF THE COMPLEMENT SYSTEM AND PATHWAYS OF COMPLEMENT ACTIVATION

2.1. PROTEINS OF THE COMPLEMENT SYSTEM

The complement system consists of 19 glycoproteins (Table 1) present in plasma where they represent approximately 5% of the total protein content and, at lesser concentrations, in extravascular fluids. Upon complement activation, complement components interact within distinct and finely regulated functional units (Fig. 1): two pathways of activation, the classical and the alternative pathway which both form specific enzymatic complexes termed "C3 convertases"

MICHEL D. KAZATCHKINE • INSERM U-28 and Unité d'Immunopathologie, Hôpital Broussais, 75014 Paris, France. URS E. NYDEGGER • Central Hematology Laboratory Inselspital, Bern, Switzerland.

TABLE 1. PROTEINS OF THE COMPLEMENT SYSTEM

	Protein	Molecular weight	Serum concentration (μg/ml)	Cleavage fragments and their generation	Relevant function in complement activation
	1. Classical pathway C3 and C5 convertases				
Generation	C1q	410,000	70	None	Recognition of classical pathway activators
	C1r	85,000	35	C1r	C1r in the C1r2s2 subunit of C1 cleaves C1s into $C\bar{1}s$ following binding of C1q to activators
	C1s	85,000	70	C1s	$C\bar{1}s$ cleaves C4 and C2
	C2	117,000	25		Substrate for $C\bar{1}s$
				C2a: cleavage of C2 by $C\bar{1}s$	C2a in the C4b,2a complex cleaves C3; C2a in the (C3b)nC4b,2a complex cleaves C5
				C2ai: dissociation of C2a from C4b,2a	
				C2b: cleavage of C2 by $C\bar{1}s$	
	C4	180,000	400		Substrate for C1s
				C4a: cleavage of C4 by $C\bar{1}s$	
				C4b: cleavage of C4 by $C\bar{1}s$	C4b binds to targets; bound C4b complexes with C2a to form the classical pathway convertase C4b,2a
				C4c and C4d: cleavage of C4b by I	
Regulation	C1 inh	105,000	180	None	C1 inh inhibits activation of C1; disassembles activated C1, inhibits C4 cleavage by C1s
	C4bp	1.2–1.5 × 10^6	250	None	C4bp dissociates C4b,2a and serves as cofactor of I in cleavage of C4b
	I	90,000	50	None	Endopeptidase: cleaves C4b into C4c and C4d
	2. Alternative pathway C3 and C5 convertases				
ation	C3	185,000	1400		Participates in the initial alternative pathway C3 convertase; substrate for the classical and alternative pathway C3 convertases
				C3a: cleavage of C3 by C3 convertases	Anaphylatoxin
				C3b: cleavage of C3 by C3 convertases	C3 binds to targets; bound C3b binds B to form the amplification C3 convertase; participates in assembly of the classical and alternative pathway C5 convertases; ligand for the C3b receptor (CR1)

Gener				C3bi: cleavage of C3b by I	Ligand for the C3bi receptor (CR3)
				C3c and C3dg: cleavage of C3bi by I	C3dg is a ligand for the C3d receptor (CR2)
				C3d and C3g: cleavage of C3dg by tissue proteases	
	B	95,000	250		Binds to bound C3b to form the alternative pathway amplification convertase
				Ba: cleavage of B by D	
				Bb: cleavage of B by D	Bb in the C3b,Bb complex cleaves C3 Bb in the (C3b)nBb complex cleaves C5
				Bbi: dissociation of Bb from C3b,Bb	
	D	25,000	2	None	Cleaves B in the C3b,B complex
Regulation	P	220,000	25	None	Stabilizes the C3b,Bb and (C3b)nBb complexes
	H	150,000	400	None	H dissociates C3b,Bb; prevents formation of C3b,Bb and (C3b)nBb; cofactor of I in cleavage of C3b
	I	90,000	50	None	Endopeptidase: cleaves bound C3b into C3bi, and C3bi into C3c and C3dg
3. C5b–9 effector sequence					
Generation	C5	200,000	160		Substrate for the classical and alternative pathway C5 convertases
				C5a: cleavage of C5 by C5 convertases	Anaphylatoxin
				C5b: cleavage of C5 by C5 convertases	C5 binds to C6
	C6	130,000	75	None?	Binds to C5; protease with as yet unidentified complement substrate
	C7	120,000	65	None	Participates in the C5b–7 complex that may insert in cell membranes
	C8	155,000	80	None	Binds to C5b–7
	C9	79,000	200	None	Several molecules of C9 can bind to one C5b–8 complex
Regulation	S protein	88,000	?	None	Binds to C5b–7 in the fluid phase and prevents its insertion in membranes

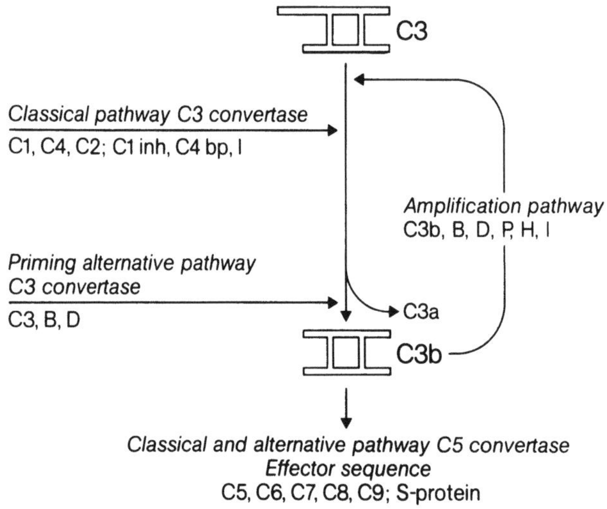

FIGURE 1. Complement activation. Cleavage of C3 may result from the assembly of the classical pathway C3 convertase, the "priming" alternative pathway convertase, or the amplification C3 convertase. Generation of C3b triggers formation of C5 convertases and assembly of the C5b–9 effector sequence.

that cleave the third complement component, C3, and generate the major cleavage fragment C3b; a single amplification (feedback) pathway comprised of alternative pathway proteins that augments C3 cleavage once initial C3b has been generated; a common effector sequence comprised of components C3b and C5 to C9 that generates the opsonizing, vasoactive, leukocyte-attracting, immune regulatory, and cytolytic activities of activated complement. The classical pathway of activation comprises the proteins C1q, C1r, C1s, C2, and C4 and the regulatory proteins C1 inhibitor (C1 inh), C4-binding protein (C4bp), and I (previously termed C3b inactivator). The alternative and amplification pathways are constituted of proteins C3, B, D and regulatory proteins P, H, I. When not engaged in the assembly of a membrane-bound C5b–9 complex, components C5, C6, and C7 bind the serum S protein to form a fluid-phase cytolytically inactive S C5b–9 complex. Complement proteins are synthesized by the liver and by monocytes/macrophages in their inactive precursor forms. During complement activation, up to the C5 cleavage step, multimolecular proteolytic complexes become assembled that cleave native complement proteins into a large fragment which contributes to participate in the reaction cascade and small cleavage fragments that are freely diffusible in the fluid phase. Cells of the immune system and of the RES including polymorphonuclear leukocytes, monocytes/macrophages, and lymphocytes express specific receptors for some complement proteins and their cleavage fragments (see Section 3.3) which upon ligand–receptor interaction elicit the various cellular responses that result in the inflammatory process.

2.2. THE CLASSICAL PATHWAY OF COMPLEMENT ACTIVATION (FIG. 2)

The classical pathway C3 convertase C4b,2a is formed by the interaction of three components: C1 which acquires enzymatic activity ($C\bar{1}$) upon recognition of and binding to activators (Table 2) (Ziccardi, 1983); C4 which is cleaved by $C\bar{1}$

and serves as a receptor for C2; C2 which upon binding to C4 and cleavage by $C\bar{1}$, serves as the catalytic unit of the C3 convertase.

C1 is a macromolecular complex of 750,000 daltons that consists in serum of two subunits, C1q and a calcium-dependent tetramer of C1r and C1s, denoted C1r2s2. C1q, the recognition subunit of the C1 complex, is an 11 S protein that consists of six globular heads capable of binding to a polypeptide sequence within the CY2 and $C\mu 4$ domains of immunoglobulin, and of a collagenlike stem. Binding of C1q to complexed IgM or to closely positioned IgG molecules enhances the affinity of C1 for Fc's and triggers C1 activation. The role of antigen in the antigen–antibody complex is merely to aggregate IgG molecules in order to facilitate multivalent binding of C1q (Jaton, 1982). Binding of C1q to an immune activator increases the affinity between C1q and C1r2s2 and induces a conformational change in C1q resulting in cleavage and conversion of C1r to an active serine esterase $C\bar{1}r$ that activates the other C1r subunit in the C1r2 dimer. The substrate of C1r is $C\bar{1}s$ which, upon activation, is cleaved into two disulfide-linked chains ($C\bar{1}s$) with the smaller chain carrying a proteolytic site for C4 and C2. C1s cleaves the α chain of C4 to release the anaphylatoxin C4a and yield the C4b fragment which expresses a labile binding site that mediates covalent attachment of the molecular to cell membranes or to the Fd portion of complexed IgG in the vicinity of the $C\bar{1}$ enzyme (Law *et al.*, 1980; Campbell *et al.*, 1980). C2

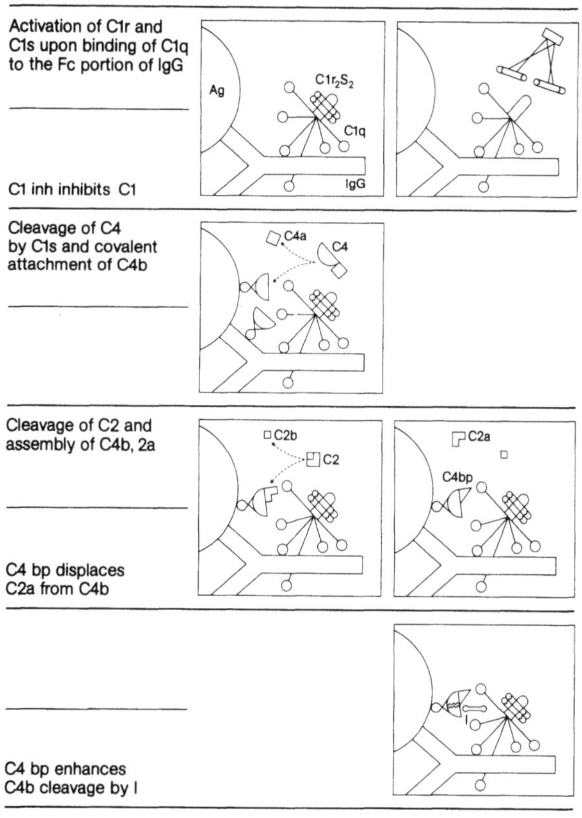

FIGURE 2. Molecular events underlying activation of the classical pathway by antigen–antibody complexes. Formation of the classical pathway C3 convertase is depicted in the diagrams on the left and in the upper parts of the legends. Regulation of convertase formation is depicted in the diagrams on the right and in the lower parts of the legends.

TABLE 2. C1 ACTIVATORS

Fc portion of complexed immunoglobulins
IgG1, IgG2, IgG3
IgM
Isolated peptides from the CH_2 domain of immunoglobulins
Nonimmune activators
RNA viruses
LPS
Some strains of salmonella, *E. coli*, and *Neisseria*
DNA, polynucleotides
Heparin–protamine complexes
Cytoskeletal filaments
Heart mitochondrial membranes
Myelin membranes
Ant venom polysaccharides
Monosodium urate crystals

then binds reversibly to surface-bound C4b in a magnesium-dependent reaction. Cleavage of C2 by C1s releases C2b and generates the bimolecular classical pathway C3 convertase C4b,2a in which the C2a subunit carries the C3-cleaving site (Müller-Eberhard *et al.*, 1967). In addition to antigen–antibody complexes, a number of nonimmune activators including viruses and gram-negative bacteria may activate C1 (Table 2) and provide the classical complement pathway with a protective role against infection in a nonimmune host. Antibody-dependent activation of C1 by murine leukemia virus was shown to require the binding of C1q and C1s to a specific membrane protein of the virus (Bartholomew and Esser, 1980) but such interactions are not involved in C1 activation by other nonimmune activators. Furthermore, the capacity of C1 inh to block the spontaneous activation of C1 and the activation of C1 by DNA and heparin (Ziccardi, 1982) is likely to limit the potential physiopathological relevance of C1 activation by nonimmune substances. Control of C1 function by C1 inh involves at least three different types of interaction between C1 and the control protein: (1) by interacting with a transitional form between native and activated C1, C1 inh prevents activation of C1r and C1s; (2) by strongly binding to the active sites of $C\bar{1}r$ and $C\bar{1}s$, C1 inch blocks the proteolytic activity of activated $C\bar{1}$ and prevents cleavage of C4 and C2; (3) C1 inh actively disassembles immune complex-bound $C\bar{1}$ to release two [C1rC1s(C1 inh)$_2$] complexes from each inactivated C1 molecule. C1q then remains bound to the activators in a form that favors its interaction with specific receptors on leukocytes. The important role of C1 inh in the control of classical pathway activation is illustrated by the hypercatabolism of C4 and C2 which is observed in patients with hereditary angioedema who are genetically deficient in C1 inh function.

The function of the C4b,2a convertase, once it has been formed, is regulated by spontaneous disassembly of the convertase with release of an inactive fragment C2ai and by the accelerated decay of C2a from C4b by the control protein C4bp. By binding to C4b (Gigli *et al.*, 1979), C4bp prevents the interaction between C4b and C2 and facilitates the proteolytic cleavage-inactivation of C4b by

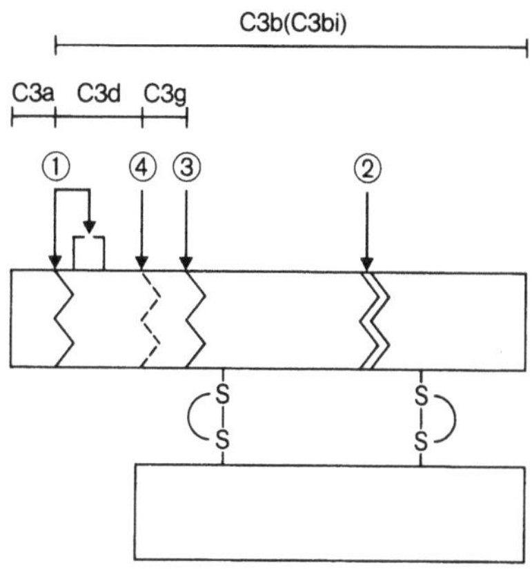

FIGURE 3. Schematic representation of the C3 molecule and localization of its cleavage sites. (1) Cleavage of the α chain in native C3 by the classical and alternative pathway C3 convertases. Upon generation, C3b expresses a labile binding site in the C3d region of the molecule that mediates covalent binding of C3b to bystander surfaces. (2) Cleavage of C3b by I generates C3bi. (3) Cleavage of C3bi by I generates C3c and C3dg. (4) Cleavage of C3dg by trypsin generates C3d and C3g.

factor I. I cleaves the α chain of C4b at two sites yielding a large C4c fragment that is released in the fluid phase and a small fragment, C4d, which remains bound to the activator. On the surface of most human peripheral blood cells, formation and function of C4b,2a is also controlled by two membrane-associated proteins—the C3b receptor (CR1) and decay-accelerating factor (DAF)—which bind to C4b and accelerate decay-dissociation of the classical C3 convertase (Iida and Nussenzweig, 1983; Nicholson-Weller *et al.*, 1982).

Cleavage of C3 by C4b,2a occurs near the N-terminus of the α-polypeptide chain of C3 and results in release of the C3a anaphylatoxin and generation of C3b (Fig. 3). For an extremely short time after its generation, C3b, like C4b, can form a covalent bond with cell surfaces or immune complexes. Formation of this bond depends on exposure of a reactive glutamyl-cysteinyl thioester group in the C3d region of C3b which is capable by transacylation of forming a stable covalent bond with hydroxyl groups or amino groups on a bystander surface (Law and Levine, 1977; Tack *et al.*, 1980; Tack, 1983). If a suitable surface "acceptor" group is not immediately available, hydrolysis of the internal thioester bond of C3b results in a form of the molecule that can no longer attach to target surfaces. Covalently bound C3b may recruit the alternative pathway proteins to form an amplification C3 convertase (Fig. 1) that will result in amplified C3 cleavage and C3b deposition. Binding of multiple C3b molecules to the target surface will then change the specificity of the C3 convertase C4b,2a to a C5 convertase [denoted C4b,2a(C3b)n] and trigger activation of the effector sequence C5–C9. Surface-bound C3b also serves as the natural ligand for C3b receptors on cells of the RES.

2.3. THE ALTERNATIVE PATHWAY OF COMPLEMENT ACTIVATION

Whereas the classical pathway is predominantly activated by antigen–antibody complexes, triggering of the alternative pathway (reviewed in Kazatchkine and Nydegger, 1982) may occur in the absence of immunoglobulin as demonstrated by *in vitro* restoration of alternative pathway activity with purified complement components (Fearon and Austen, 1977a; Schreiber and Müller-Eberhard, 1978). Thus, the alternative pathway represents a natural system for resistance to infection in the nonimmunized host. The main characteristics and the functions of the alternative pathway proteins C3, B, D and regulatory proteins P, H, I are listed in Table 1. Two convertases are formed in the reaction sequence of the alternative pathway: the "priming" C3 convertase which generates initial C3b molecules in plasma, and the amplification C3 convertase which utilizes C3b that is bound to an "activating"surface of the alternative pathway and results in amplified C3 cleavage and C3b generation on that surface (Fig. 1). The priming convertase involves a form of C3 that results from spontaneous hydrolysis of the internal thioester bond in the native molecule (Pangburn *et al.*, 1981). This C3b-like C3 binds factor B in plasma to form a C3-cleaving enzyme in the presence of factors D and P. Continuous C3 turnover in plasma occurs at a slow rate because it is controlled by the inhibitory proteins H and I. Upon generation in the fluid phase, C3b randomly attaches to bystander surfaces following exposure of the reactive group in the internal thioester bond of the molecule. When bound to an activating surface of the alternative pathway, C3b interacts with factors B and D to form the amplification C3 convertase. Surface-fixed C3b binds B in a magnesium-dependent reaction (Götze and Müller-Eberhard, 1971), exposing a site in complexed B that is sensitive to cleavage by D. D is a serine protease circulating as an active enzyme that releases the Ba fragment from C3b,B and uncovers the C3-cleaving site on the Bb fragment which remains bound to C3b. Once formed, the amplification convertase C3b,Bb cleaves C3 and generates C3b molecules which bind to the target surface. As occurs with the classical pathway convertase, binding of multiple C3b molecules in the vicinity of the C3b,Bb enzyme changes the specificity of the alternative pathway convertase from a C3 convertase (C3b,Bb) to a C5 convertase [C3b,Bb(C3b)n] and initiates activation of the effector sequence C5–C9. The C3b,Bb complex rapidly loses enzymatic activity by spontaneous intrinsic decay from Bb from C3b,Bb. By binding to C3b,Bb, properdin (P) decreases the rate of dissociation of the convertase and increases the stability of the enzymatic complex (Fearon and Austen, 1975).

Binding of B to C3b is unaffected by the nature of the surface to which C3b is bound. Discrimination between activating and nonactivating surfaces of the alternative pathway depends on the relative capacity of the regulatory protein H to bind to C3b, allow inactivation of C3b by I, and prevent formation of the C3b,Bb enzyme (Fearon and Austen, 1977a,b; Kazatchkine *et al.*, 1979a; Horstmann *et al.*, 1985) (Fig. 4). When C3b is bound to a nonactivating surface, H binds to most C3b sites with a high affinity that is 20- to 50-fold greater than that of B for C3b, thus preventing formation of the C3b,B complex, accelerating decay-dissociation of the C3b,Bb convertase, and serving as a cofactor for inactivation of C3b by I. In contrast, C3b that is bound to an activator of the alternative pathway, is protected from inactivation and activator-bound C3b,Bb sites are protected from dissociation by the regulatory proteins, because the affinity of H for bound C3b on such surfaces is relatively low and similar to that of B for C3b. Two cell surface characteristics, the relative content in sialic acid and in

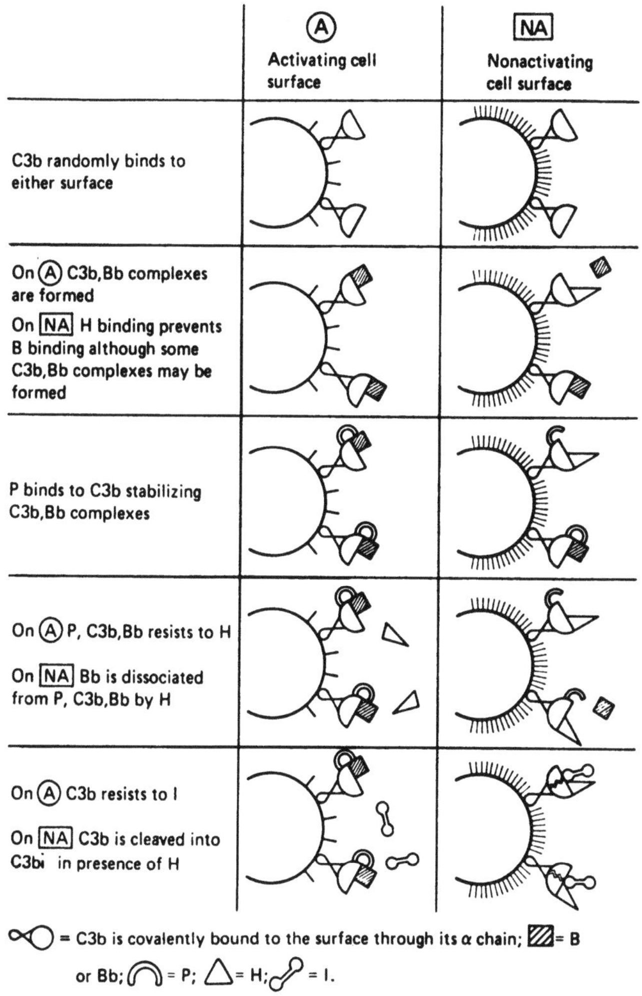

FIGURE 4. Molecular events underlying formation and regulation of the amplification C3 convertase on activating (A) and nonactivating (NA) surfaces of the alternative pathway. Degree of surface hatching reflects the relative amount of surface-associated sialic acid or sulfated mucopolysaccharide. (From Kazatchkine and Nydegger, 1982, with permission.)

sulfated mucopolysaccharides, modulate the affinity of H for surface-bound C3b on heterologous and homologous cells. Activators of the human alternative pathway such as erythrocytes from rabbits (Platts-Mills and Ishizaka, 1974) and from some strains of mice (Nydegger *et al.*, 1978) are relatively poor in cell surface sialic acid as compared with nonactivating sheep erythrocytes; furthermore, sheep erythrocytes from which sialic acid residues have been removed or chemically modified (Fearon, 1978) and guinea pig erythrocytes that have been treated with influenza virus neuraminidase (Lambré *et al.*, 1982) acquire the capacity to activate the alternative pathway. The extent of activation is related to the relative number of C3b sites on the cells that bind H with low affinity and proportional to the amount of sialic acid removed. Capsular sialic acid was shown to prevent activation of the alternative pathway by type III group B streptococci; other bacterial species having capsular sialic acid such as group C *Neisseria meningitidis* and K1 *E. coli* also escape complement activation and are pathogenic for man. In an analogous fashion to sialic acid, heparin bound to zymosan was found to inhibit the alternative pathway-activating capacity of the particles by increasing the affinity of surface-bound C3b for factor H (Kazatchkine *et al.*, 1979b). Alternatively, incorporation of LPS from *E. coli* into the membrane of sheep erythrocytes decreased the affinity of H for cell-bound C3b and induced lysis of the cells in human serum through activation of the alternative pathway (Pangburn *et al.*, 1980).

Additional membrane-associated factors prevent formation of the amplification and classical C3 convertases on the surface of cells, protecting cells in the circulation from the deleterious effects of activation of autologous complement on their surface. The C3b receptor (CR1) either in purified form or in the cell membrane (Iida and Nussenzweig, 1983) inhibits classical and alternative pathway activation. It resembles H and C4bp in that it strongly binds to surface-fixed C3b and C4b, decay-dissociates preformed C3b,Bb and C4b,2a complexes, and serves as a cofactor for cleavage of C3b and C4b by I. Decay-accelerating factor (DAF) is a 70,000-dalton glycoprotein present on erythrocytes, lymphocytes, and platelets that only acts within the cell membrane (Medof *et al.*, 1984) to prevent assembly of the classical and alternative pathway C3 convertases (Nicholson-Weller *et al.*, 1982). The essential role of DAF in controlling complement activation in cell membranes is suggested by the finding that complement-sensitive red cells of patients with paroxysmal nocturnal hemoglobinuria are deficient in DAF but express normal numbers of CR1 (Pangburn *et al.*, 1983).

2.4. INTERRELATIONS BETWEEN THE CLASSICAL AND ALTERNATIVE PATHWAYS

Although the alternative pathway phylogenetically precedes the classical pathway and can function in C2- or C4-deficient serum, several biochemical and functional features are shared between the classical and alternative pathways. C3b that has covalently bound to a surface or an immune complex following classical pathway activation may interact with alternative pathway proteins to form amplification C3 convertase sites (Fig. 1), indicating that the amplification pathway has a modulating role on classical pathway activation. There are structural and/or functional analogies between C3 and C4, C2 and B, C1s and D. For example, cleavage of C3 and of C4 yields a small anaphylatoxic fragment and a major fragment that covalently attaches to surfaces. As mentioned above, function of cell-bound C3b and C4b is regulated by the same membrane-associated regulatory factors CR1 and DAF. Unlike the genes coding for C4, C2, and B, C3 genes are not localized in the major histocompatibility complex (Lachmann and Hobart, 1979; Whitehead *et al.*, 1982) but the regulatory proteins CR1, C4bp, and H that bind C3b and C4b have recently been found to be encoded for within the same cluster of genes (Lublin *et al.*, 1985). Although the alternative pathway can be activated in the absence of specific antibody, IgG antibody has been shown to initiate and/or enhance activation in several experimental models. Thus, alternative pathway activation by rabbit erythrocytes, mouse erythrocytes, and zymosan is decreased in hypogammaglobulinemic and in absorbed human serum (Scheinkein and Ruddy, 1981). Human anti-measles virus IgG antibodies enhance alternative pathway activation by virus-infected cultured cells and are required for their lysis (Sissons *et al.*, 1979). Enhancement is independent of the immunoglobulin Fc fragment. It occurs in some systems through an IgG-mediated increase in the rate of C3 cleavage by cell-bound C3b,Bb and of C3b deposition (Moore *et al.*, 1981). Alternatively, specific antibody can cover up sites that are down-regulators of alternative pathway activation; e.g., antibody against group B streptococci exerts its alternative pathway-activating effect by masking capsular sialic acid (Edwards *et al.*, 1982). Aggregated human myeloma proteins of all classes activate the alternative pathway. Activation of the alternative pathway by immune complexes leads to covalent binding of C3b to IgG and to solubilization of the immune aggregates (Fujita *et al.*, 1981) (see Section 3.4).

3. EFFECTOR FUNCTIONS OF THE COMPLEMENT SYSTEM

3.1. THE TERMINAL C5b–9 COMPLEX

Interaction of the terminal complement components C5–C9 on the surface of susceptible cells results in the formation of protein-lined hydrophilic transmembrane channels which cause osmotic lysis of the cells (reviewed in Mayer *et al.*, 1981; Bhakdi and Tranum-Jensen, 1983). Formation of the C5b–9 complex on target cells is initiated by the cleavage of C5 into C5a and C5b by cell-bound C5 convertases. As mentioned in Sections 2.2 and 2.3, C5 convertase activity is acquired by the classical and alternative C3 convertases after C3b molecules have attached on the target membrane to a site that is adjacent to the enzyme; this C3b binds C5 in a form that renders it susceptible to cleavage by the convertases (Vogt *et al.*, 1978). Freshly generated C5b is then rapidly released in an inactive form in the fluid phase unless it forms a stable bimolecular C5b,6 complex with C6. The C5b,6 complex can reversibly bind to cell membranes; with the addition of C7–C9, it can lyse unsensitized bystander cells, a process known as reactive lysis (Lachmann and Thompson, 1970). Binding of one molecule of C7 to C5b,6 creates a trimolecular C5b–7 complex that expresses exposed apolar sites which permit the firm insertion of the complex into the lipid bilayer of the target membrane. If activation occurs in the absence of a lipid bilayer, the serum S protein binds to C5b–7, masking its hydrophobic sites and suppressing the capacity of the complex to insert into cell membranes (Podack *et al.*, 1977). C8 (one molecule) and C9 (several molecules) then sequentially bind to C5b–7 whether it is inserted in a membrane or complexed with the S protein in the fluid phase so as to form the membrane-bound m(C5b–9) complex or the cytolytically inactive S C5b–9 complex in serum. Uptake of C8 by membrane-bound C5b–7 leads to further insertion of the complex in the target membrane and formation of unstable transmembrane "ion" channels; binding of several (six?) molecules of C9 per m(C5b–8) monomer leads to formation of the stable m(C5b–9) complex which can be isolated from detergent-solubilized membranes (Fig. 5). The large transmembrane channel does not, however, appear to be essential for cell lysis since one molecule of C9 is probably sufficient to form a functional C5b–9 lesion (Müller-Eberhard, 1984). Upon electron microscopic analysis, m(C5b–9) has the appearance of a thin-walled cylinder rimmed by an annulus that projects above the membrane (Bhakdi and Tranum-Jensen, 1978). The cylindrical appearance of the complex is partly or exclusively due to polymerized C9 molecules (Tschopp *et al.*, 1982). Several self-protective mechanisms contribute to the relative inefficiency of human complement in inducing autologous cell lysis. These include the membrane-associated factors which control function of C3 and C5 convertases on autologous cell surfaces (i.e., CR1 and DAF), plasma high-density lipoproteins which have been shown to inhibit full insertion of C9 after binding of C5b–8 to human erythrocyte membranes (Rosenfeld *et al.*, 1983), and the relative inefficiency of homologous C9 as compared with C9 from other species to participate in channel formation with human C5b–8 (Hänsch *et al.*, 1981). Nucleated cells are more resistant to membrane attack by complement because of cell membrane renewal and of the shorter life span of complement channels in membranes of nucleated cells (Ramm *et al.*, 1983). Removal of phospholipids and disruption of the lipid bilayer which occurs during insertion of C5b–9 in target membranes are essential for lysis of viruses whose envelopes do not function as osmotic barriers (Nemerow and Cooper, 1981; Welsh *et al.*, 1976). On bacterial surfaces, specific biochemical characteristics determine the relative susceptibility of the bacteria to the cytotoxic action of human complement by modulating the efficiency of penetration, the molecule configuration, and the state of aggregation of C5b–9 on serum-sensitive and serum-resistant strains (reviewed

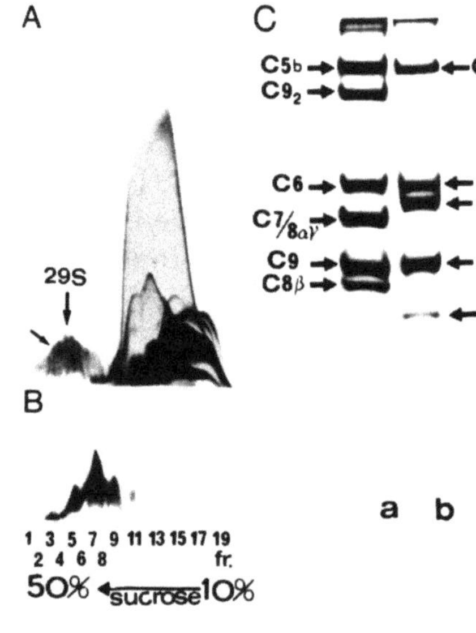

FIGURE 5. Isolated m(C5b–9) complexes from erythrocyte membranes. (A, B) Rocket electrophoreses of proteins solubilized from erythrocyte membrane, developed with antibodies against human serum proteins (A) and with specific anti-(C5b–9) antibodies (B). (C) SDS–polyacrylamide gel electrophoresis of human (a) and rabbit (b) m(C5b–9) complexes. (From Bhakdi and Tranum-Jensen, 1983, with permission.)

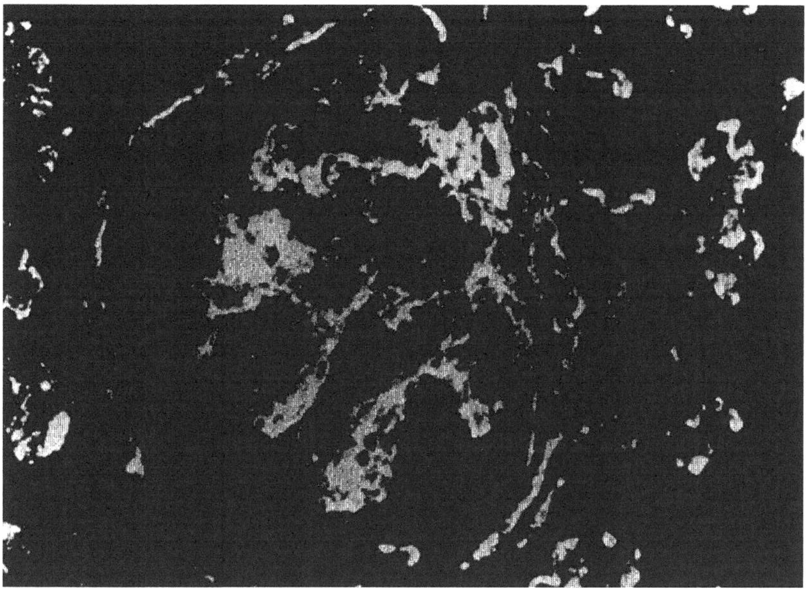

FIGURE 6. Staining with anti-C5b–9 neoantigen antibodies of a glomerulus in a renal biopsy specimen from a patient with diabetic glomerulosclerosis.

in Frank *et al.*, 1984). Thus, binding of m(C5b–9) to targets does not necessarily explain *per se* the cause of cell death and tissue lesions. This should be critically considered when interpreting the immunohistochemical finding of neoantigens expressed by the C5b–9 complex (Kolb and Müller-Eberhard, 1975) in pathological tissues. C5b–9 neoantigens have been found at the site of the lesions in biopsies from patients with systemic lupus erythematosus (Biesecker *et al.*, 1981, 1982), other forms of glomerulonephritis with immune deposits (Hinglais *et al.*, 1986), and myasthenia gravis (Sahashi *et al.*, 1980); the antigens are only found in glomerular subepithelial deposits at a late phase during the course of membraneous nephritis, indicating that the occurrence of proteinuria in the disease is independent of the terminal complement sequence (Cosyns *et al.*, 1986). C5b–9 neoantigens have also been found on shedded membrane remnants in connective matrices in renal biopsies from patients with non-immune-complex-mediated nephritis such as diabetic glomerulosclerosis (Fig. 6) and in normal kidneys (Falk *et al.*. 1983; Hinglais *et al.*, 1986).

3.2. BIOLOGICALLY ACTIVE PEPTIDES DERIVED FROM COMPLEMENT ACTIVATION

The anaphylatoxins C3a, C4a, and C5a are generated as a result of the cleavage of the α chains of C3, C4, and C5 near their N-termini. The peptides do not remain at the site of activation but are released into plasma and extracellular fluids where their binding to specific cellular receptors will induce various inflammatory responses and immunoregulatory properties that are mediated in a hormonelike fashion. The polypeptide chains of human C3a, C4a, and C5a contain 74–78 amino acid residues of which a third are homologous, indicating common genetic ancestry. Extensive structure–function studies using synthetic analogue peptides have localized the active sites of C3a and C4a in a linear C-terminal portion of the molecules (Hugli and Erickson, 1977; Hugli, 1981). C3a and C4a probably bind to the same receptor on target cells, although C4a interacts less effectively with the receptor, making it a 100-fold less potent anaphylatoxin than is C3a. In serum the C-terminal arginines of C3a and C4a are rapidly removed by carboxypeptidase N, only allowing the active anaphylatoxins to diffuse for a short time before they are converted to stable, biologically inactive C3a des Arg and C3a des Arg derivatives. Alterations in cellular functions in the presence of purified C3a and/or binding experiments using the purified peptide suggest the presence of C3a receptors on mast cells, basophils, smooth muscle cells, monocytes, lymphocytes, and platelets. Binding of C3a to mast cells and basophils induces degranulation with secretion of histamine and other mediators of anaphylaxis (Glovsky *et al.*, 1979); the release of histamine is likely to mediate the contractile response of guinea pig ileum to C3a which can be blocked with antihistamine agents. It also participates with other mediators such as PGE_2 in C3a-induced alterations in vascular permeability. More recently, human C3a and synthetic analogues of the C-terminal sequence of C3a, were shown to suppress both specific and polyclonal antibody responses of murine splenic lymphocytes without affecting T- or B-cell proliferative responses to antigen or mitogens (Morgan *et al.*, 1982).

In addition to being the most potent anaphylatoxin, C5a induces a variety of biological effects that are critical for the initiation of the inflammatory reaction, by interacting with specific receptors on leukocytes. Approximately $1–3 \times 10^5$ receptors have been demonstrated on human neutrophils that bind C5a with a high affinity of $0.3–1 \times 10^9\ M^{-1}$ (Chenoweth and Hugli, 1978). Other cell types which respond to C5a are mast cells, basophils, eosinophils, and monocytes. The interaction between C5a and its receptor involves both the C-terminal part of the molecule and a separate site that is close to the N-terminal portion. Thus, the C-terminal arginine is important for the induction of biological activities of C5a. However, C5a des Arg, the stable derivative released in plasma by carboxypeptidase N, although it is unable to cause smooth muscle cell contraction, may bind with low affinity to cause smooth muscle cell contraction, may bind with low affinity to leukocytes and induce chemotaxis, enhanced adhesiveness and enzyme secretion. C5a is chemotactic at nanomolar concentrations, causing directed migration of neutrophils, eosinophils, basophils, and monocytes against the gradient of concentration that occurs when the peptide diffuses away from the site of complement activation (Shin *et al.*, 1968; Hausman *et al.*, 1972). Neutrophil activation by C5a leads to stimulation of oxidative metabolism in the cells (Goldstein *et al.*, 1975), release of granular enzymes from cells that have been pretreated with cytochalasin B or that have adhered to a surface (Henson *et al.*, 1978; Dahinden *et al.*, 1983), and increased membrane expression of C3b receptors (Fearon and Collins, 1983) and of the Mo1 antigen (Berger *et al.*, 1984). C5a and C5a des Arg also enhance the adhesiveness of neutrophils to foreign surfaces and endothelial cells (Tonnesen *et al.*, 1984) and reversibly aggregate the cells *in vitro* and *in vivo* (Craddock *et al.*, 1977). The latter effect is secondary to enhanced expression of the adhesion-promoting molecule Mo1 (Anderson *et al.*, 1984); it mediates intravascular aggregation of leukocytes and leukopenia observed in individuals undergoing hemodialysis with membranes that activate complement (Craddock *et al.*, 1977; Arnaout *et al.*, 1985) or cardiopulmonary bypass during cardiac surgery (Chenoweth *et al.*, 1981; Van Oeveren *et al.*, 1985). In contrast to C3a, human C5a was found to enhance humoral immune responses, antigen-induced T-cell proliferation, and IL1 secretion by monocytes (Morgan *et al.*, 1983).

Complement activation also causes leukocytosis following leukopenia. Leukocytosis is mediated by the soluble C3e fragment derived from C3c (Ghebrehiwet and Müller-Eberhard, 1979) that induces mobilization of neutrophils from the bone marrow. Neutrophils express specific receptors for C3e.

3.3. INTERACTIONS BETWEEN COMPLEMENT PROTEINS BOUND TO TARGETS OF COMPLEMENT ACTIVATION AND SPECIFIC CELLULAR RECEPTORS

Complement fragments generated during complement activation and some complement proteins that are bound to the targets of complement activation, may interact with receptors on effector cells to trigger phagocytosis and/or other specific cellular responses which contribute to the inflammatory process. These fragments include the C3b, C3bi, C3dg, and C3d cleavage fragments of C3 (Fig. 7), C1q, and H. C3b that is covalently bound to cell surfaces or immune complexes may be converted to C3bi by I which cleaves the α chain of C3b at two sites to release the 3000-dalton C3f peptide. This cleavage requires H as a cofactor. In the presence of CR1, I can then catalyze cleavage of an additional peptide bond in the α chain of C3bi, to give rise to C3dg which remains bound to the target

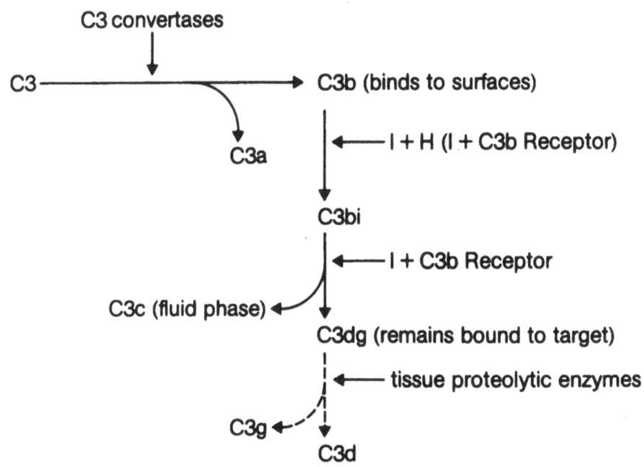

FIGURE 7. Cleavage of C3.

surface and to the C3c fragment that is released in the fluid phase (Lachmann *et al.*, 1982; Ross *et al.*, 1982). C3dg probably represents the final cleavage product of C3 in plasma, although it can be further degraded into C3d and C3g by trypsin, plasmin, or elastase, a reaction that may only occur at the site of the tissue lesion. The binding specificities and cell distribution of the three types of receptors, CR1, CR2, and CR3, which bind the C3b, C3bi, and C3dg/C3d fragments covalently attached to surfaces (reviewed in Fearon and Wong, 1983) are summarized in Table 3.

The C3b receptor, CR1, is a polymorphic glycoprotein (Dykman *et al.*, 1983) of approximately 200,000 daltons that was first purified from human erythrocyte membranes (Fearon, 1979). Electrophoretic analysis of immunoprecipitates with anti-CR1 antibodies from erythrocytes, neutrophils, monocytes, B and T lymphocytes, and podocytes demonstrated no major differences in CR1 on these cell types. There are 200–900 CR1 sites on each erythrocyte with an average of 700 sites/cell and about 5000–10,000 sites/cell on resting neutrophils and monocytes. CR1 expression on the latter cells can be enhanced up to eightfold in the presence of C5a and other chemoattractants (Fearon and Collins, 1983). The number of CR1 sites on erythrocytes is inherited as an autosomal codominant trait in normal individuals (Wilson *et al.*, 1982); it is decreased in patients with systemic lupus erythematosus (Miyakawa *et al.*, 1981; Iida *et al.*, 1982) and glomerular CR1 is absent in patients with severe proliferative SLE nephritis (Kazatchkine *et al.*, 1982). Decrease in CR1 expression in SLE may be the consequence of both genetic (Wilson *et al.*, 1982) and acquired factors related to disease activity (Walport *et al.*, 1985). The function of CR1 on erythrocytes could be the transport of immune complexes and their delivery to the RES (Cornacoff *et al.*, 1983); thus,

TABLE 3. HUMAN C3 RECEPTORS

Receptor	Ligand[a]	Molecular weight	Monoclonal antibodies specific for epitopes on receptors	Cell distribution
CR1	C3b C4b, C3bi	200,000 (polymorphic)	Monoclonal antibodies raised against the purified molecule	Erythrocytes, neutrophils, eosinophils, monocytes, B lymphocytes, some T lymphocytes, follicular dendritic cells, glomerular podocytes
CR3	C3bi C3dg?	α chain: 165,000 β chain: 90,000	Anti-Mac1, anti-Mo1, OKM1, OKM10, Leu-15	Neutrophils, eosinophils, monocytes, large granular lymphocytes, follicular dendritic cells
CR2	C3dg, C3d C3bi, C3b	145,000	Anti-B2, HB-5, OKB7, BL13	B lymphocytes, follicular dendritic cells

[a]The underlined ligand is that which binds to the receptor with highest affinity.

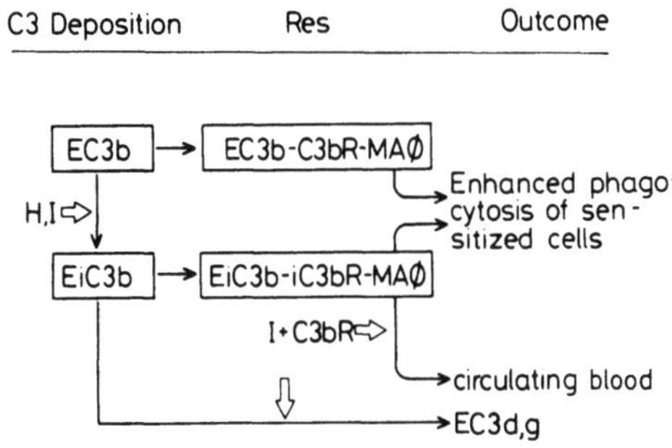

FIGURE 8. The role of regulatory proteins and of the C3b receptor (C3bR) for *in vivo* processing of C3b-coated erythrocytes. Res, reticuloendothelial system. (From Nydegger and Kazatchkine, 1983, with permission.)

C3b-coated immune complexes that have bound to CR1 on erythrocytes are released upon exposure to I that cleaves C3b into C3bi and C3dg, which could lead to interaction of the complexes with CR3 and CR2 (Medof *et al.*, 1982, 1983) (Fig. 8). The main function of CR1 on neutrophils and monocytes is to enhance phagocytosis of IgG-coated particles (Ehlenberger and Nussenzweig, 1977) and mediate internalization of small ligands bearing C3b (Fearon *et al.*, 1981). Binding of fibronectin or other connective tissue components to C3b-coated particles confers the ability to ingest the opsonized targets in the absence of antibody on both monocytes and C5a-stimulated neutrophils. Interaction of bound C3b with CR1 has also been shown to induce enzyme release from neutrophils and trigger the oxidative metabolism in neutrophils and monocytes (Schreiber *et al.*, 1982; Schopf *et al.*, 1982). A role for CR1 on B lymphocytes in B-cell differentiation has been suggested by the observations that $F(ab')_2$ anti-CR1 enhanced polyclonal antibody secretion by peripheral blood mononuclear cells in the presence of pokeweed mitogen (Daha *et al.*, 1984) and specific anti-TNP immune response *in vitro* (L. Weis, J. F. Delfraissy, and M. D. Kazatchkine, unpublished observations).

The predominant or exclusive ligand of CR3 is C3bi. The receptor is present on neutrophils, monocytes, large granular lymphocytes, and follicular dendritic cells. The expression of CR3 is up-regulated by chemoattractants and decreased by interferons. CR3 consists of two polypeptide chains, an α chain of 165,000 daltons that expresses the Mac1/OKM1 and Mo1 determinants and a β chain of 90,000 daltons that is shared with two other leukocyte membrane proteins, LFA1 and p 150,95 (Arnaout *et al.*, 1983; Wright *et al.*, 1983; Sanchez Madrid *et al.*, 1983). Binding of C3bi-coated particles to CR3 on phagocytes is considerably enhanced in the presence of divalent cations. CR3 is probably the most important C3 receptor for phagocytosis of opsonized bacteria at least for the following reasons: C3b is rapidly converted in serum into C3bi; CR3 binds with high affinity to surface-bound C3bi; it also binds to unopsonized yeast and to some bacterial surfaces; it can trigger phagocytosis of C3bi-coated particles in the presence or absence of IgG (in the absence of antibody, phagocytosis was shown to be enhanced by a T-derived soluble factor); finally, neutrophil responses to serum-opsonized zymosan, such as induction of chemiluminescence by the cells, are inhibited by anti-CR3 but not by anti-CR1. On human K cells and neutrophils, CR3 may enhance antibody-mediated cellular cytotoxicity of targets coated with C3bi (Perlmann *et al.*, 1981). The role of CR3 in phagocytosis and that of the Mo1 α chain in promoting adherence of neutrophils to surfaces explain the complex defects in neutrophil functions reported in patients with recurrent bacterial infections whose cells are deficient in the Mac1/Mo1 antigen and/or in the common β chain of the Mac1–LFA1–p 150,95 family of molecules (Anderson *et al.*, 1984). Binding of C3dg and C3d to neutrophils, which was first thought to be mediated by CR3, has recently been shown to depend on a distinct fourth type of C3 receptor (Vik and Fearon, 1985); how this receptor is related to the membrane structure that mediates C3d binding to cultured human monocytes (Inada *et al.*, 1983) is unclear.

CR2 is a glycoprotein of 140,000–145,000 daltons expressed on B lymphocytes at discrete stages of differentiation and on some human B-cell lines (Iida *et al.*, 1983; Weiss *et al.*, 1984; Barel *et al.*, 1981). CR2 binds surface-fixed C3d and the C3d region of C3dg, C3bi, and C3b and represents the B-lymphocyte receptor for Epstein–Barr virus (EBV) (Fingeroth *et al.*, 1984). Monoclonal antibodies which were originally raised against B cells and subsequently found to react with CR2, recognize at least two distinct antigenic subclusters on B lymphocytes, one comprised of the B2 (Nadler *et al.*, 1981; Iida *et al.*, 1983) and BL13 epitopes, the other defined by HB-5 antibody (Weiss *et al.*, 1984). Functional characterization of these clusters could be of physiological relevance since the epitopes are differentially expressed during B-cell maturation. A third epitopic region defined by monoclonal antibody OKB7 may represent the EBV-binding site. Preliminary evidence suggests that CR2 is involved in B-cell proliferation. In lymph node follicles, the receptor which is present on B cells and on follicular dendritic cells could be of importance for the trapping of immune complexes bearing C3dg, a complement-dependent process that is essential for the generation of B memory cells (Klaus and Humphrey, 1977).

The biological relevance of cellular receptors for H and C1q is poorly understood. The presence of an H receptor has been reported on B lymphocytes, neutrophils, and monocytes. Treatment of these cells with purified H triggered the release of endogenous I and stimulated a respiratory burst in human monocytes. Lymphocytes, neutrophils, and platelets express a C1q receptor (Tenner and Cooper, 1981) which binds to C1q that remains fixed to classical pathway activators following dissociation of macromolecular C1 by C1 inh. Particle-bound C1q has been shown to enhance cell-mediated cytotoxic reactions against targets that were not sensitized with antibody and to stimulate the oxidative metabolism in neutrophils (Tenner and Cooper, 1982).

3.4. COMPLEMENT-MEDIATED PROCESSING OF IMMUNE COMPLEXES

The role of complement in handling immune complexes in the circulation and at the site of a tissue lesion is decisive in that it may favor removal of the immune aggregates or mediate a clinically overt inflammatory process (reviewed in Nydegger and Kazatchkine, 1986). Activation of the classical or the alternative pathway by immune complexes results in covalent binding to the antigen and/or to the immunoglobulin of cleavage fragments of activated complement that allow an optimal interaction of the complexes with cellular receptors. Complement activation is also involved in maintaining immune complexes in solution and in the transport of C3-bearing complexes to the RES. Interaction of classical pathway components with immune complexes at the stage of their formation prevents immune precipitation (Schifferli *et al.*, 1982) whereas incubation of preformed complexes with serum leads to their disaggregation and solubilization through an alternative pathway-mediated mechanism (Fujita *et al.*, 1981). The multiple roles of complement in determining the physicochemical state of immune complexes, their rate of removal from the circulation and degradation in cells from the RES and in modulating the secondary immune response, contribute to the adverse consequences with regard to humoral immune defense which occur in hypocomplementemic patients. This is relevant both to the occurrence of immune complex disease in patients with inherited complement deficiencies (Schifferli and Peters, 1983) and to the self-perpetuating process during immune complex disease in patients with acquired abnormalities of the complement system.

REFERENCES

Anderson, D. C., Schmalsteig, F. C., Arnaout, M. A., Kohl, S., Tosi, M. F., Dana, N., Buffone, G. J., Hughes, B. J., Brinkley, B. R., Dickey, W. D., Abramson, J. S., Springer, T. A., Boxer, L. A., Hollers, J. M., and Smith, C. W., 1984, Abnormalities of polymorphonuclear leucocyte function associated with a heritable deficiency of high molecular weight glycoproteins (gp 138): Common relationship to diminished cell adherence, *J. Clin. Invest.* **74**:536.

Arnaout, M. A., Todd, R. F., III, Dana, N., Melamed, J., Schlossman, S. F., and Colten, H. R., 1983, Inhibition of phagocytosis of complement C3 or immunoglobulin G coated particles and of C3bi binding by monoclonal antibodies to a monocyte-granulocyte membrane glycoprotein (Mo1), *J. Clin. Invest.* **72**:171.

Arnaout, M. A., Hakim, R. M., Todd, R. F., III, Dana, N., and Colten, H. R., 1985, Increased expression of an adhesion-promoting surface glycoprotein in the granulocytopenia of hemodialysis, *N. Engl. J. Med.* **312**:457.

Austen, K. F., 1978, Homeostasis of effector systems which can also be recruited for immunologic reactions, *J. Immunol.* **121**:793.

Barel, M., Charriaut, C., and Frade, R., 1981, Isolation and characterization of a C3b receptor-like molecule from membranes of a human B lymphoblastoid cell line (Raji), *FEBS Lett.* **136**:111.

Bartholomew, R. M., and Esser, A. F., 1980, Mechanism of antibody-independent activation of the first component of complement (C1) on retrovirus membranes, *Biochemistry* **19**:2847.

Berger, M., O'Shea, J., Cross, A. S., Folks, T. M., Chused, T. M., Brown, E. J., and Frank, M. M., 1984, Human neutrophils increase expression of C3bi as well as C3b receptors upon activation, *J. Clin. Invest.* **74**:1566.

Bhakdi, S., and Tranum-Jensen, J., 1978, Molecular nature of the complement lesion, *Proc. Natl. Acad. Sci. USA* **75**:5655.

Bhakdi, S., and Tranum-Jensen, J., 1983, Membrane damage by complement, *Biochim. Biophys. Acta* **737**:343.

Biesecker, G., Katz, S., and Koffler, D., 1981, Renal localization of the membrane attack complex in systemic lupus erythematosus nephritis, *J. Exp. Med.* **154**:1779.

Biesecker, G., Lavin, L., Ziskind, M., and Koffler, D., 1982, Cutaneous localization of the membrane attack complex in discoid and systemic lupus erythematosus, *N. Engl. J. Med.* **306**:264.

Campbell, R. D., Dodds, A. W., and Porter, R. R., 1980, The binding of human complement component C4 to antibody–antigen aggregates, *Biochem. J.* **189:**67.
Chenoweth, D. E., and Hugli, T. E., 1978, Demonstration of specific C5a receptor on intact human polymorphonuclear leucocytes, *Proc. Natl. Acad. Sci. USA* **75:**3943.
Chenoweth, D. E., Cooper, S. W., Hugli, T. E., Steward, R. W., Blakstone, E. H., and Kirklin, J. W., 1981, Complement activation during cardiopulmonary bypass: Evidence for generation of C3a and C5a anaphylatoxins, *N. Engl. J. Med.* **304:**497.
Cornacoff, J. B., Hebert, L. A., Smead, W. L., Van Aman, M. E., Birmingham, D. J., and Waxman, F. J., 1983, Primate erythrocyte–immune complex clearing mechanism, *J. Clin. Invest.* **71:**236.
Cosyns, J. P., Kazatchkine, M. D., Bhakdi, S., Mandet, C., Grossetete, J., Hinglais, N., and Bariety, J., 1986, Immunohistochemical analysis of C3 cleavage fragments, factor H, and C5b–9 terminal complex of complement in de novo membranous glomerulonephritis occurring in patients with renal transplant, *Clin. Nephrol.* in press.
Craddock, P. R., Hammerschmidt, D. E., White, J. G., Dalmasso, A. P., and Jacob, H. S., 1977, Complement (C5a)-induced granulocyte aggregation in vitro: A possible mechanism for complement-mediated leucostasis and leucopenia, *J. Clin. Invest.* **60:**260.
Daha, M. R., Bloem, A. C., Bast, E. J., and Ballieux, R. E., 1984, Immunoglobulin production by human peripheral lymphocytes induced by anti-C3 receptor antibodies, *J. Immunol.* **132:**1197.
Dahinden, C. A., Fehr, J., and Hugli, T. E., 1983, Role of cell surface contact on the kinetics of superoxide production by granulocytes, *J. Clin. Invest.* **72:**113.
Dykman, T. R., Cole, J. L., Iida, K., and Atkinson, J. P., 1983, Polymorphism of human erythrocyte C3b/C4b receptor, *Proc. Natl. Acad. Sci. USA* **80:**1698.
Edwards, M. S., Kasper, D. L., Jennings, H. F., Baker, C. J., and Nicholson-Weller, A., 1982, Capsular sialic acid prevents activation of the alternative complement pathway by type III group B streptococci, *J. Immunol.* **128:**1278.
Ehlenberger, A. G., and Nussenzweig, V., 1977, The role of membrane receptors for C3b and C3d in phagocytosis, *J. Exp. Med.* **145:**357.
Falk, R. J., Dalmasso, A. P., Kim, Y., Tsai, C. H., Scheinman, J. I., Gewurz, H., and Michael, A. F., 1983, Neoantigen of the polymerized ninth component of complement: Characterization of a monoclonal antibody and immunohistochemical localization in renal disease, *J. Clin. Invest.* **72:**560.
Fearon, D. T., 1978, Regulation by membrane sialic acid of beta-1H-dependent decay-dissociation of amplification C3 convertase of the alternative complement pathway, *Proc. Natl. Acad. Sci. USA* **75:**1971.
Fearon, D. T., 1979, Regulation of the amplification C3 convertase of human complement by an inhibitory protein isolated from human erythrocyte membrane, *Proc. Natl. Acad. Sci. USA* **76:**5867.
Fearon, D. T., and Austen, K. F., 1975, Properdin binding to C3b and stabilization of the C3b-dependent C3 convertase, *J. Exp. Med.* **142:**856.
Fearon, D. T., and Austen, K. F., 1977a, Activation of the alternative complement pathway due to resistance of zymosan-bound amplification convertase to endogenous regulatory mechanisms, *Proc. Natl. Acad. Sci. USA* **74:**1683.
Fearon, D. T., and Austen, K. F., 1977b, Activation of the alternative complement pathway with rabbit erythrocytes by circumvention of the regulatory action of endogenous control proteins, *J. Exp. Med.* **146:**22.
Fearon, D. T., and Collins, L. A., 1983, Increased expression of C3b receptors on polymorphonuclear leucocytes induced by chemotactic factors and by purification procedures, *J. Immunol.* **130:**370.
Fearon, D. T., and Wong, W. W., 1983, Complement ligand–receptor interactions that mediate biological responses, *Annu. Rev. Immunol.* **1:**243.
Fearon, D. T., Kaneko, I., and Thomson, G., 1981, Membrane distribution and adsorptive endocytosis by C3b receptors on human polymorphonuclear leucocytes, *J. Exp. Med.* **153:**1615.
Fingeroth, J. D., Weis, J. J., Tedder, T. F., Strominger, J. L., Biro, A., and Fearon, D. T., 1984, Epstein Barr virus receptors of human B lymphocytes is the C3d receptor, *Proc. Natl. Acad. Sci. USA* **81:**4510.
Frank, M. M., Joiner, K. A., and Brown, E. J., 1984, Parry–thrust: Host defense and bacterial

resistance to complement action, in: *Progress in Immunology V* (Y. Yamamura and T. Tada, eds.), pp. 387–396, Academic Press, New York.

Fujita, T., Takata, Y., and Tamura, N., 1981, Solubilization of immune precipitates by six isolated alternative pathway proteins, *J. Exp. Med.* **154:**1743.

Ghebrehiwet, B., and Müller-Eberhard, H. J., 1979, C3e: An acidic fragment of human C3 with leucocytosis-inducing activity, *J. Immunol.* **123:**616.

Gigli, I., Fujita, T., and Nussenzweig, V., 1979, Modulation of the classical pathway C3 convertase by the plasma proteins C4 binding protein and C3b inactivator, *Proc. Natl. Acad. Sci. USA* **76:**6596.

Glovsky, M. M., Hugli, T. E., Ishizaka, T., Lichtenstein, E. M., and Erickson, B. W., 1979, Anaphylatoxin-induced histamine release with human leucocytes: Studies of C3a leucocyte binding and histamine release, *J. Clin. Invest.* **64:**804.

Goldstein, I. M., Feit, F., and Weissmann, G., 1975, Enhancement of nitroblue tetrazolium dye reduction by leucocytes exposed to a component of complement in the absence of phagocytosis, *J. Immunol.* **114:**516.

Götze, O., and Müller-Eberhard, H. J., 1971, The C3-activator system: An alternative pathway of complement activation, *J. Exp. Med.* **134:**90s.

Hänsch, G. M., Hammer, C. H., Vanguri, P., and Shin, M. L., 1981, Homologous species restriction in lysis of erythrocytes by terminal complement proteins, *Proc. Natl. Acad. Sci. USA* **78:**5118.

Hausman, M. S., Snyderman, R., and Mergenhagen, S. E., 1972, Humoral mediators of chemotaxis of mononuclear leucocytes, *J. Infect. Dis.* **125:**6.

Hensen, P. M., Zanolari, B., Schwartzman, N. A., and Hong, S. R., 1978, Intracellular control of human neutrophil secretion. I. C5a-induced stimulus-specific desensitization and the effects of cytochalasin B, *J. Immunol.* **121:**851.

Hinglais, N., Kazatchkine, M. D., Bhakdi, S., Appay, M. D., Mandet, C., Grossetete, J., and Bariety, J., 1986, Immunohistochemical study of the C5b–9 complex in normal and diseased human kidneys, *Kidney Inter.*, in press.

Horstmann, R. D., Pangburn, M. K., and Müller-Eberhard, H. J., 1985, Active participation of factor H in the recognition function of the alternative pathway of complement, *J. Immunol.* **134:**1101.

Hugli, T. E., 1981, The structural basis for anaphylatoxin in chemotactic functions of C3a and C5a, *CRC Crit. Rev. Immunol.* **1:**321.

Hugli, T. E., and Erickson, B. W., 1977, Synthetic peptides with the biological activities and specificity of human C3a anaphylatoxin, *Proc. Natl. Acad. Sci. USA* **74:**1826.

Iida, K., and Nussenzweig, V., 1983, Functional properties of membrane-associated complement receptor CR1, *J. Immunol.* **130:**1876.

Iida, K., Mornaghi, R., and Nussenzweig, V., 1982, Complement receptor (CR1) deficiency in erythrocytes from patients with systemic lupus erythematosus, *J. Exp. Med.* **153:**1427.

Iida, K., Nadler, L., and Nussenzweig, V., 1983, Identification of the membrane receptor for the complement fragment C3d by means of a monoclonal antibody, *J. Exp. Med.* **158:**1021.

Inada, S., Brown, E. J., Gaither, T. A., Hammer, C. H., Takahashi, T., and Frank, M. M., 1983, C3d receptors are expressed on human monocytes after in vitro cultivation, *Proc. Natl. Acad. Sci. USA* **80:**2351.

Jaton, J. J., 1982, Structural and biological effects of antigen binding to antibodies: A reevaluation of the data, in: *Immunohemotherapy* (U. E. Nydegger, ed.), pp. 7–11, Academic Press, New York.

Kazatchkine, M. D., and Nydegger, U. E., 1982, The human alternative complement pathway: Biology and immunopathology of activation and regulation, *Prog. Allergy* **30:**193.

Kazatchkine, M. D., Fearon, D. T., and Austen, K. F., 1979a, Human alternative complement pathway: Membrane-associated sialic acid regulates the competition between B and beta-1H for cell-bound C3b, *J. Immunol.* **122:**75.

Kazatchkine, M. D., Fearon, D. T., Silbert, J. E., and Austen, K. F., 1979b, Surface-associated heparin inhibits zymosan-induced activation of the human alternative pathway by augmenting the regulatory action of the control proteins on particle-bound C3b, *J. Exp. Med.* **150:**1202.

Kazatchkine, M. D., Fearon, D. T., Appay, M. D., Mandet, C., and Bariety, J., 1982, Immunohistochemical study of the human glomerular C3b receptor in normal kidney and in 75 cases of renal diseases, *J. Clin. Invest.* **69:**900.

Klaus, G. G. B., and Humphrey, J. H., 1977, The generation of memory cells. I. The role of C3 in the generation of B memory cells, *Immunology* **33**:31.
Kolb, W. P., and Müller-Eberhard, H. J., 1975, Neoantigens of the membrane attack complex of human complement, *Proc. Natl. Acad. Sci. USA* **72**:1687.
Lachmann, P. J., and Hobart, M. J., 1979, The genetics of the complement system, 231–260, in: *Human Genetics: Possibilities and Realities,* Excerpta Medica, Amsterdam.
Lachmann, P. J., and Thompson, R. A., 1970, Reactive lysis: The complement-mediated lysis of unsensitized cells. II. The characterization of activated reactor as C5,6 and the participation of C8 and C9, *J. Exp. Med.* **131**:643.
Lachmann, P. J., Pangburn, M. K., and Olroyd, R. G., 1982, Breakdown of C3 after complement activation: Identification of a new fragment, C3g, using monoclonal antibodies, *J. Exp. Med.* **156**:205.
Lambré, C., Kazatchkine, M. D., Maillet, F., and Thibon, M., 1982, Guinea pig erythrocytes after their contact with influenza virus acquire the ability to activate the human alternative complement pathway through virus-induced desialation of the cells, *J. Immunol.* **128**:629.
Law, S. K., and Levine, R. P., 1977, Interaction between the third complement protein and cell surface macromolecules, *Proc. Natl. Acad. Sci. USA* **74**:2701.
Law, S. K., Lichtenberg, N. A., Holcombe, F. H., and Levine, R. P., 1980, Interaction between the labile binding sites of the fourth (C4) and fifth (C5) human complement proteins and erythrocyte cell membranes, *J. Immunol.* **125**:634.
Lublin, D. M., Rodriguez de Cordoba, S., Dykman, T. R., Rubinstein, P., and Atkinson, J. P., 1985, Loci for CR1, C4 binding protein and factor H are linked in man, *Complement* **1**:161a.
Mayer, M. M., Michaels, D. W., Ramm, L. E., Shin, M. L., Whitlow, M. B., and Willoughby, J. B., 1981, Membrane damage by complement, *CRC Crit. Rev. Immunol.* **2**:133.
Medof, M. E., Iida, K., Mold, C., and Nussenzweig, V., 1982, Unique role of the complement receptor CR1 in the degradation of C3b associated with immune complexes, *J. Exp. Med.* **156**:1739.
Medof, M. E., Lam, T., Prince, G. M., and Mold, C., 1983, Requirement for human red blood cells in inactivation of C3b in immune complexes and enhancement of binding to spleen cells, *J. Immunol.* **130**:1336.
Medof, M. E., Kinoshita, T., and Nussenzweig, V., 1984, Inhibition of complement activation on the surface cells after incorporation of decay-accelerating factor (DAF) into their membranes, *J. Exp. Med.* **160**:1558.
Miyakawa, Y., Yamada, A., Kosaka, K., Tsuda, F., and Mayumi, M., 1981, Defective immune adherence (C3b) receptor on erythrocytes from patients with systemic lupus erythematosus, *Lancet* **2**:493.
Moore, F. D., Jr., Austen, K. F., and Fearon, D. T., 1981, IgG on mouse erythrocytes augments activation of the human alternative complement pathway by enhancing deposition of C3b, *J. Immunol.* **126**:1805.
Morgan, E. L., Weigle, W. O., and Hugli, T. E., 1982, Anaphylatoxin-mediated regulation of the immune response. I. C3a-mediated suppression of human and murine humoral immune response, *J. Exp. Med.* **155**:1412.
Morgan, E. L., Thoman, W. L., Weigle, W. O., and Hugli, T. E., 1983, Anaphylatoxin-mediated regulation of the immune response. II. C5a-mediated enhancement of human humoral and T cell-mediated immune responses, *J. Immunol.* **130**:1257
Müller-Eberhard, H. J., 1984, The membrane attack complex, *Springer Semin. Immunopathol.* **7**:93.
Müller-Eberhard, H. J., Polley, M., and Calcott, M. A., 1967, Formation and functional significance of a molecular complex derived from the second and fourth components of human complement, *J. Exp. Med.* **125**:359.
Nadler, L. M., Stashenko, P., Hardy, R., Von Agthoven, A., Terhorst, C., and Schlossman, S. M., 1981, Characterization of a human B cell specific antigen (B2) distinct from B1, *J. Immunol.* **126**:1941.
Nemerow, G. R., and Cooper, N. R., 1981, Isolation of Epstein Barr virus and studies of its neutralization by human IgG and complement, *J. Immunol.* **127**:272.
Nicholson-Weller, A., Burge, J., Fearon, D. T., Weller, P. F., and Austen, K. F., 1982, Isolation of a

human erythrocyte membrane glycoprotein with decay-accelerating activity for C3 convertases of the complement system, *J. Immunol.* **129**:184.

Nydegger, U. E., and Kazatchkine, M. D., 1983, The role of complement in immune clearance of blood cells, *Springer Semin. Immunopathol.* **6**:373.

Nydegger, U. E., and Kazatchkine, M. D., 1986, Modulation by complement of immune complex processing in health and disease in man, *Prog. Allergy* in press.

Nydegger, U. E., Fearon, D. T., and Austen, K. F., 1978, Autosomal locus regulates inverse relationship between sialic acid content and capacity of mouse erythrocytes to activate human alternative complement pathway, *Proc. Natl. Acad. Sci. USA* **75**:6078.

Pangburn, M. K., Morrison, D. C., Schreiber, R. D., and Müller-Eberhard, H. J., 1980, Activation of the alternative complement pathway: Recognition of surface structures on activators by bound C3b, *J. Immunol.* **124**:977.

Pangburn, M. K., Schreiber, R. D., and Müller-Eberhard, H. J., 1981, Formation of the initial C3 convertase of the alternative complement pathway: Acquisition of C3b-like activities by spontaneous hydrolysis of the putative thioester in native C3, *J. Exp. Med.* **154**:856.

Pangburn, M. K., Schreiber, R. D., and Müller-Eberhard, H. J., 1983, Deficiency of an erythrocyte membrane protein with complement regulatory activity in paroxysmal nocturnal hemoglobinuria, *Proc. Natl. Acad. Sci. USA* **80**:5430.

Perlmann, H. P., Perlmann, R. D., Schreiber, R. D., and Müller-Eberhard, H. J., 1981, Interaction of target cell-bound C3bi and C3d with human lymphocyte receptors: Enhancement of antibody-mediated cellular cytotoxicity, *J. Exp. Med.* **153**:1592.

Platts-Mills, T. A. E., and Ishizaka, K., 1974, Activation of the alternative pathway of human complement by rabbit cells, *J. Immunol.* **113**:348.

Podack, E. R., Kolb, W. P., and Müller-Eberhard, H. J., 1977, The C5B–7 complex: Formation, isolation, properties, and subunit composition, *J. Immunol.* **119**:2024.

Ramm, L. E., Whitlow, M. B., Koski, C. L., Shin, M. L., and Mayer, M. M., 1983, Elimination of complement channels from the plasma membrane of U937, a nucleated cell line: Temperature dependence of the elimination rate, *J. Immunol.* **131**:1411.

Rosenfeld, S. I., Packman, C. H., and Leddy, J. P., 1983, Inhibition of the lytic action of cell-bound terminal complement components by human high density lipoproteins and apoproteins, *J. Clin. Invest.* **71**:795.

Ross, G. D., Lambris, J. D., Cain, J. A., and Newman, S. L., 1982, Generation of three different fragments of bound C3 with purified I or serum. I. Requirements for factor H versus CR1 cofactor activity, *J. Immunol.* **129**:2051.

Sahashi, K., Engle, A. G., and Lambert, E. H., 1980, Ultrastructural localization of the terminal and lytic ninth complement component (C9) at the motor end-plate in myasthenia gravis, *J. Neuropathol. Exp. Neurol.* **39**:160.

Sanchez Madrid, F., Nagy, J. A., Robbins, E., Simon, P., and Springer, T. A., 1983, A human leucocyte differentiation antigen family with distinct subunits and a common subunit: The lymphocyte function associated (LFA1), the C3bi complement receptor (OKM1/Mac1) and the p 150,95 molecule, *J. Exp. Med.* **158**:1785.

Scheinkein, H. A., and Ruddy, S., 1981, The role of immunoglobulins in alternative pathway activation by zymosan, *J. Immunol.* **126**:7.

Schifferli, J. A., and Peters, D. K., 1983, Complement, the immune complex lattice and the pathophysiology of complement deficiency syndromes, *Lancet* **2**:957.

Schifferli, J. A., Woo, P., and Peters, D. K., 1982, Complement-mediated inhibition of immune precipitation. I. Role of the classical and alternative pathways, *Clin. Exp. Immunol.* **47**:555.

Schopf, R. E., Hammann, K. P., Scheiner, O., Lemmel, E. M., and Dierich, M. P., 1982, Activation of human monocytes by both human beta-1H and C3b, *Immunology* **46**:307.

Schreiber, R. D., and Müller-Eberhard, H. J., 1978, Assembly of the cytolytic alternative pathway of complement from 11 isolated plasma proteins, *J. Exp. Med.* **148**:1722.

Schreiber, R. D., Pangburn, M. K., Bjornson, A. B., Brothers, M. A., and Müller-Eberhard, H. J., 1982, The role of C3 fragments in endocytosis and extracellular cytotoxic reactions by polymorphonuclear leucocytes, *Clin. Immunol. Immunopathol.* **23**:335.

Shin, H. S., Snyderman, R., Friedman, E., Mellors, A., and Mayer, M. M., 1968, Chemotactic and anaphylatoxic fragment cleaved from the fifth component of guinea pig complement, *Science* **162**:361.

Sissons, J. G. P., Cooper, N. R., and Oldstone, M. B. A., 1979, Alternative complement pathway mediated lysis of measles virus-infected cells: Induction by IgG antibody bound to individual viral glycoproteins and comparative efficacy of $F(ab')_2$ and F(ab') fragments, *J. Immunol.* **123**:2144.

Tack, B. F., 1983, The γ-Cys-β-Glu thiolester bond in human C3, C4 and β_2 macroglobulin, *Springer Semin. Immunopathol.* **6**:259.

Tack, B. F., Harrison, R. A., Janatova, J., Thomas, M. L., and Prahl, J. W., 1980, Evidence for presence of an internal thioester bond in third component of human complement, *Proc. Natl. Acad. Sci. USA* **77**:5764.

Tenner, A. J., and Cooper, N. R., 1981, Identification of types of cells in peripheral blood that bind C1q, *J. Immunol.* **125**:1658.

Tenner, A. J.,and Cooper, N. R., 1982, Stimulation of a human polymorphonuclear leucocyte oxidative response by the C1q subunit of the first complement component, *J. Immunol.* **128**:2547.

Tonnesen, M. G., Smedly, L. A., and Henson, P. M., 1984, Neutrophil–endothelial cell interactions: Modulation of neutrophil adhesiveness induced by complement fragments C5a and C5a des Arg and formyl-methionyl-leucyl-phenylalanine in vitro, *J. Clin. Invest.* **74**:1581.

Tschopp, J., Müller-Eberhard, H. J., and Podack, E. R., 1982, Formation of transmembrane tubules by spontaneous polymerization of hydrophilic complement protein C9, *Nature (London)* **298**:534.

Van Oeveren, W., Kazatchkine, M. D., Descamps Latscha, B., Maillet, F., Fischer, E., Carpentier, A., and Wildewuur, C. R. H., 1985, Deleterious effects of cardiopulmonary bypass: A prospective study of bubble versus membrane oxygenation, *J. Thorac. Cardiovasc. Surg.* **89**:888.

Vik, D. P., and Fearon, D. T., 1985, Neutrophils express a receptor for iC3b, C3dg and C3d that is distinct from CR1, CR2 and CR3, *J. Immunol.* **134**:2571.

Vogt, W., Schmidt, G., Von Buttlar, B., and Dieminger, L., 1978, A new function of the activated third component of complement: Binding to C5, an essential step for C5 activation, *Immunology* **34**:29.

Walport, M. J., Ross, G. D., Mackworth-Young, C., Watson, J. V., Lida, J., and Lachmann, P. J., 1985, Family studies of erythrocyte complement receptor type 1 levels: Reduced levels in patients with SLE are acquired, not inherited, *Complement* **1**:167a.

Weiss, J. J., Tedder, T. F., and Fearon, D. T., 1984, Identification of a 145,000 Mr membrane protein as the C3d receptor (CR2) of human B lymphocytes, *Proc. Natl. Acad. Sci. USA* **81**:885.

Welsh, R. M., Lampert, P. W., Burner, P. A., and Oldstone, M. B. A., 1976, Antibody complement interactions with purified lymphocytic choriomeningitis virus, *Virology* **73**:59.

Whitehead, A. S., Solomon, E., Chambers, S., Bodmer, W. F., Povey, S., and Fey, G., 1982, Assignment of the structural gene for the third component of human complement to chromosome 19, *Proc. Natl. Acad. Sci. USA* **79**:5021.

Wilson, J. G., Wong, W. W., Schur, P. H., and Fearon, D. T., 1982, Mode of inheritance of decreased C3b receptors on erythrocytes of patients with systemic lupus erythematosus, *N. Engl. J. Med.* **397**:981.

Wright, S. D., Rao, P. E., Von Voorhis, W. C., Craigmyle, L. S., Iida, K., Talle, M. A., Westberg, E. F., Goldstein, G., and Silverstein, S. C., 1983, Identification of the C3bi receptor of human monocytes and macrophages by using monoclonal antibodies, *Proc. Natl. Acad. Sci. USA* **80**:5699.

Ziccardi, R. J., 1982, A new role for C1 inhibitor in homeostasis: Control of activation of the first component of human complement, *J. Immunol.* **128**:2505.

Ziccardi, R. J., 1983, The first component of human complement (C1): Activation and control, *Springer Semin. Immunopathol.* **6**:213.

8

Antibody-Mediated Cytotoxicity

JAMES W. KAZURA

1. INTRODUCTION

Contact of an Fc receptor (FcR)-bearing effector cell and a potential biologic target may occur in the presence of antibodies directed against the surface of the target. When this interaction culminates in damage or killing of the biologic target, it is referred to as antibody-dependent cell-mediated cytotoxicity (ADCC). A multitude of studies indicate that ADCC reactions are involved in immune functions which may be beneficial or deleterious to the whole animal. For example, allograft and tumor rejection, viral resistance (Cerottini and Brunner, 1974; Kohl *et al.*, 1977), and host defense against multicellular helminths (Ellner and Mahmoud, 1982) are in large part effected by ADCC reactions. On the other hand, life-threatening disease states such as immune thrombocytopenic purpura and immune hemolytic anemia also involve ADCC reactions (Kurlander *et al.*, 1978).

In vitro studies of ADCC have employed effector and target cells of multiple types and sources. Monocytes, macrophages, various lymphoid cell populations, granulocytes, and platelets have all been shown to be capable of mediating toxic reactions against antibody-sensitized targets, such as red blood cells, tumor cells, allogeneic lymphocytes, bacteria, viruses, worms, and so on. The diversity of these effector cell–target interactions indicates that conclusions based on results in one experimental system may not necessarily be extrapolatable to another. The mechanism by which a neutrophil lyses a phagocytosed red cell, for example, may be quite different from how a macrophage inhibits DNA synthesis of an antibody-coated tumor cell. In this regard, the purpose of the present review is not to describe the details or unique attributes of the multiple types of ADCC; rather, attention will be focused on the general characteristics of ADCC which results in its important functional role in a wide variety of immunologic reactions. In the first portion of the review, the various effector cells and their possible modes of cytotoxic activity will be emphasized. The role of differ-

JAMES W. KAZURA • Division of Geographic Medicine, Department of Medicine, Case Western Reserve University and University Hospitals, Cleveland, Ohio 44106.

ent classes of immunoglobulin antibodies as ligands for effector cell–target contact and cytoxicity will then be discussed.

2. EFFECTOR CELLS

The common feature of effector cells which allows them to mediate ADCC is the presence of FcR on their surface. In some situations, these FcR-bearing cells may also participate in cytotoxic reactions that do not depend on antibody as a ligand for attachment to target cells. Certain lymphoid cell populations, for example, may serve as killer (K) cells in ADCC and as the effectors of natural killing (NK) activity. The major classes of effector cells important in ADCC are described below.

2.1. LYMPHOID CELLS

Investigations utilizing mouse and human cells have shown that mononuclear cells with morphologic and surface characteristics of lymphocytes may act as effector cells in ADCC reactions. It was initially believed that ADCC was mediated by null cells, which did not possess surface markers for B or T lymphocytes (Perlmann and Cerottini, 1979). Further investigations, however, have shown that T cells as well as non-T cells may participate in ADCC (Lamon *et al.*, 1975). Zoller *et al.*, (1983) have extensively characterized the murine splenic lymphoid cells that mediate IgG and IgM ADCC against SRBC. Utilizing a discontinuous Percoll gradient to separate lymphoid cells on the basis of density, these investigators found three distinct populations of cells with cytolytic activity. One was a low-density population which lysed IgM- and IgG-coated SRBC. These lymphoid cells did not have detectable receptors for Fc or C3 (C3R); their cytolytic capacity was reduced by preincubation with anti-Thy-1 antibodies and complement. A second population of cells of medium density preferentially effected cytolysis of IgG- versus IgM-coated red cells and was both FcR and C3R positive. Only a small proportion of these cells were reduced in ADCC activity by anti-Thy-1 and complement. A third fraction of lymphoid cells exhibited relatively high density. These cells were not as active in ADCC as the former two populations and were FcR positive. All three lymphoid cell populations were nonadherent, nonphagocytic, and sIg negative.

Analyses of human peripheral blood lymphocyte populations which are active in ADCC have in general indicated that T cells, defined by their ability to form rosettes with SRBC, are not the primary cells responsible for ADCC (Brier *et al.*, 1975). Rather, "unactivated" or "resting" peripheral blood mononuclear cells which mediate ADCC appear to have the surface phenotype of a monocyte which is Ia negative (Kay and Horwitz, 1980). These cells are FcR positive (Pichler *et al.*, 1979), possess surface antigens identified by the monoclonal antibody OKMl (which reacts with cells of myelomonocytic origin) (Zarling and Kung, 1980), and are large granular mononuclear cells by light microscopy (Timonen *et al.*, 1981). Since they bind a monoclonal antibody which reacts with monocytes and myeloid cells but fail to mature to monocytes during *in vitro* cultivation, it has been suggested that most of these cytotoxic effector cells are of myeloid origin (Kay and Horwitz, 1980).

In contrast to studies which utilized "resting" lymphoid cells, experiments with "activated" lymphocytes have shown that leukocytes with T-cell surface markers are able to mediate ADCC. Rimm *et al.* (1981) obtained an SRBC rosette-positive population of human peripheral blood mononuclear cells and incubated it with allogeneic Laz 156 cells to effect activation. This activation led to a marked increase in ADCC activity compared to unstimulated cells. Lymphoid cells effecting ADCC against autologous lymphocytes had the surface phenotype of inducer or helper cells (T4) as well as that associated with suppressor cells (T8).

Several investigations have suggested that lymphocytes active in ADCC may also mediate NK activity. Rimm *et al.* (1981) noted that activated lymphocytes with NK activity were T8 positive, similar to a subset of ADCC effectors described above. A detailed examination of the relationship of NK and ADCC activity has been reported by Bradley and Bonavida (1982). Large granular lymphocytes were separated from blood by Percoll density gradient centrifugation. These cells were then mixed with NK (K562 or MOLT 4 cells) or ADCC targets (RAJI or P815 cells), dispersed as a single-cell suspension, and embedded in agarose. Individual effector cells bound both NK and ADCC targets and killed each of the attached K562 (NK target) and RAJI (ADCC target) or K562 and P815 (ADCC target) cells. These findings indicated that a single cell effected killing of both NK and ADCC targets. A portion of effector lymphocytes thus possesses a recognition system for both NK and ADCC activities.

2.2. MONOCYTES AND MACROPHAGES

ADCC mediated by monocytes or macrophages is, like that mediated by lymphocytes, dependent on the presence of FcR-positive cells that can bind and subsequently release toxic molecules. In assay systems employing human red blood cells as targets, phagocytosis and intracellular lysis are generally thought to be the primary events in ADCC effected by these cells (MacDonald *et al.*, 1975). When larger nucleated targets such as tumor cells are used, the effector molecules which damage the targets are presumably delivered to extracellular sites (Shaw *et al.*, 1978). The specificity of monocyte/macrophage ADCC is, like other types of ADCC, dependent on the presence of surface FcR. Monocytes and macrophages have been shown to bear FcR for IgG and IgE (Urbaniak, 1979; Anderson and Spiegelberg, 1981; Melewicz and Spiegelberg, 1980).

Cells of monocyte origin undergo marked alterations during their life span in the body. In humans, they originate in the bone marrow as pluripotent stem cells, circulate in the bloodstream, and eventually become tissue macrophages; alterations in cellular functions and surface characteristics occur during this time and are influenced by local immunologic and nonimmunologic reactions (van Furth and Cohn, 1968). In this section, attention will be focused on the possible role of these factors in altering the capacity of monocytes and macrophages to effect ADCC.

Garagiola *et al.* (1981) compared the ability of rabbit peripheral blood monocytes and alveolar macrophages in ADCC against ox red blood cells and CEM-T lymphoblast tumor cells. With ox red cells as the target and rabbit IgG antibodies as opsonins, monocytes were significantly more effective than alveolar macrophages in lysing these targets. In contrast, alveolar macrophages exhibited a greater ability than monocytes to damage lymphoblastoid tumor cells. The differences in the efficiency of target killing by these two effector cells were associated with differences in their capacity to bind the antibody-sensitized targets. These results imply that the transition from a monocyte to tissue macrophage and changes in ADCC efficiency against specific targets may be related to alterations of FcR accessibility, mobility, and/or density.

Monocyte/macrophage activity may be influenced by exogenous inflammatory stimuli as well as by differentiation in tissues (Nathan and Cohn, 1980). Ralph *et al.* (1980) have shown that ADCC activities of mouse peritoneal macrophages and a macrophage cell line against several tumor targets were enhanced by preincubation of the effector cells for several days with LPS and the tumor promoter PMA. In addition, lymphokine activity (derived from supernatants of BCG-immune spleen cells stimulated with PPD) significantly increased macrophage ADCC activity. Thus, soluble mediators derived from antigen-stimulated lymphocytes can modify this important macrophage function (Leonard *et al.*, 1978).

2.3. GRANULOCYTES

The three types of granulocytes or polymorphonuclear leukocytes (PMN) present in human peripheral blood (neutrophils, eosinophils, and basophils) have been shown to participate in ADCC reactions against both phagocytosable and nonphagocytosable target cells (Lovchik and Hong, 1977; Butterworth *et al.*, 1975; Kazura *et al.*, 1981). PMN possess FcR for IgG, IgA (Fanger *et al.*, 1980), and IgE (Capron *et al.*, 1981). These cells are thus capable of binding and causing damage to targets coated with specific antibodies of several types (see Section 3). Investigations employing helminths suggest that eosinophils may be particularly well-suited as effector cells against extracellular nonphagocytosable targets (Ellner and Mahmoud, 1982).

2.4. MECHANISMS OF EFFECTOR CELL-MEDIATED CYTOTOXICITY

The diversity of cell types capable of acting as effectors of ADCC underscores the complexity and multiplicity of mechanisms that may eventually damage or kill antibody-sensitized targets. Granulocytes and monocytes, for example, are characterized by the presence of metabolic pathways and enzymes that result in the reduction of atmospheric oxygen to potentially toxic molecules such as hydrogen peroxide (Babior, 1978). T lymphocytes, on the other hand, are not believed to be capable of reducing oxygen in a similar manner. An understanding of the biochemical basis of effector cell-mediated damage is further complicated by the different targets employed in various studies, such as nonnucleated erythrocytes, nucleated lymphoblastoid or other tumor cells, and multicellular helminths. The objectives of this section will be to provide a general framework for classifying the biochemical processes involved in ADCC and describing the alterations in the surfaces of target cells associated with this cytotoxic event.

2.4.1. Oxidative versus Nonoxidative Damage

Because of the importance of oxidative metabolism in PMN- and monocyte-mediated host defense against bacteria and other microorganisms (Klebanoff, 1980), attention has appropriately been focused on the possible role of these effector cell products as mediators of ADCC. Using red cells as targets, Fleer *et al.* (1979) noted that monocytes from subjects with chronic granulomatous disease (CGD), which are not capable of reducing oxygen, were as effective as cells of normal individuals in mediating ADCC. It was thus suggested that monocyte ADCC was related to effector cell degranulation and not generation of reduced oxygen products. In contrast, other groups have concluded from studies conducted under anaerobic conditions as well as experiments with CGD effector cells that neutrophil and monocyte ADCC against sensitized erythrocytes are dependent on the production of toxic oxygen radicals (Clark and Klebanoff, 1977; Borregaard and Kragballe, 1982). The possible reasons for these discrepancies may be related to differences in the experimental systems. In this regard, Katz *et al.* (1980) studied the capacity of effector cells (neutrophils, monocytes, and lymphocytes) to mediate ADCC against antibody-sensitized red cells in suspension and against identical targets in a monolayer. Neutrophils and monocytes obtained from subjects with CGD were less effective than cells of normal donors in their capacity to lyse red cells in suspension. When cytotoxicity against red cells in a monolayer was compared for CGD and normal leukocytes, however, there were no differences in the ability of the two to effect hemolysis. In addition, CGD leukocytes were found to have an increased capacity to damage red cells in suspension when phagocytosis was prevented by addition of colchicine. These findings suggested that granulocyte ADCC against red cells is dependent on oxidative mechanisms when the targets are phagocytosed (i.e., in suspension). Extracellular lysis of red cells, on the other hand, appears to be mediated by a process that does not require the generation of oxidants.

Conkling *et al.* (1982) have performed experiments which indicate that the importance of oxidative versus nonoxidative processes in ADCC against red cell targets may also be related to the type of effector leukocyte utilized in the assay system. The ability of human PMN, blood monocytes, and alveolar macrophages to lyse chicken red cells were compared. When each cell type was incubated with targets under hypoxic conditions, effector cell lysis of red cells was decreased but not totally absent, suggesting that both oxidative and nonoxidative mechanisms were involved in the cytotoxic reactions. The relative efficacy of effector cells in mediating red cell lysis was greatest for alveolar macrophages, followed by PMN and finally monocytes.

Investigations employing nonphagocytosable multicellular helminths have been especially helpful in elucidating the biochemical basis of neutrophil- and eosinophil-mediated ADCC against these targets. With regard to the role of nonoxidative mechanisms, Butterworth *et al.* (1979) showed that eosinophils released a granular material, major basic protein, upon interaction with opsonized schistosomula of *Schistosoma mansoni*. When purified, this material was able to damage the organisms in a cell-free assay. Pincus *et al.* (1981) found that eosinophils were capable of damaging these organisms under anaerobic conditions. Results of our studies employing eosinophils of subjects with CGD indicated that eosinophils may employ nonoxidative as well as oxidative mechanisms in damaging schistosomula (Kazura *et al.*, 1981). CGD eosinophils demonstrated a diminished but not complete loss of capacity to kill the parasites compared to cells of control individuals. These results contrasted those obtained with CGD neutrophils, which totally lacked the ability to kill schistosomula. These and other studies (Venge *et al.*, 1980; Bass and Szejda, 1979) indicate that lysosomal contents and oxidative products are important in effecting damage against extracellular targets in ADCC reactions.

2.4.2. Alterations in Target Cell Surfaces Associated with ADCC

The interaction of FcR-positive effector cells and antibody-sensitized targets results in a close physical association of the two. Because the surface of the target is readily accessible to toxic effector molecules generated extracellularly or intracellularly in the phagolysosome, alterations of this structure have been studied to determine the membrane pertubations which lead to cell death. Simone and Henkart (1980) utilized resealed erythrocyte ghosts packaged with various dyes and substances of known molecular weights as targets of ADCC. With lymphocytes or neutrophils as effector cells, proteins and dextran with molecular weights of less than 500,000 were released from resealed erythrocyte ghosts. Materials with a molecular weight of less than approximately 300,000 were released by monocytes. These results contrasted those of studies of antibody- and complement-induced lysis (no effector cells present). In this cytotoxic system, the molecular weight cutoff of molecules released from resealed erythrocytes was one-tenth that found using lymphocytes or neutrophils. Based on these findings and on electron microscopic examination of membrane lesions in resealed erythrocyte ghosts (Dourmashkin *et al.*, 1980), it may be inferred that ADCC target cell damage is initiated by the development of membrane pores of a specific size. These breaches in the integrity of the target cell membrane may then allow escape of intracellular solutes and other materials required for maintenance of viability (Mayer, 1977).

3. ANTIBODY CLASSES ACTIVE IN ADCC

The specificity of the ADCC reaction is conferred by the interaction of the Fc portion of the immunoglobulin molecule bound to the target with the FcR on the surface of the effector cell. A large number of investigations have confirmed that antibodies of the IgG class are important in this process (Perlmann and Cerottini, 1979). Over the past several years, antibodies of other classes as well as various IgG subclasses have been found to play a role in ADCC.

3.1. IgG SUBCLASSES

Detailed examination of the role of various IgG subclasses in ADCC has been hampered by the technical difficulties of obtaining these materials from polyvalent antiserum. Relatively small amounts of each IgG isotype can be purified. In addition, there is often variability in the antibody activity of each isotype. The development of monoclonal antibodies of various IgG isotypes has provided larger quantities of material of greater purity. Ralph and Nakoinz (1983) assessed the activity of a series of monoclonal antibodies directed against surface antigens of murine T lymphoma targets. Murine macrophages and human nonadherent mononuclear cells (K cells) were used as effectors. Peritoneal macrophages effected a 20 to 30% release of ^{125}I-labeled deoxyuridine from targets in the presence of 1/10 to 1/100 dilutions of monoclonal antibodies of the IgG1, IgG2a, IgG2b, and IgG3 subclasses. Heterologous polyvalent antisera were significantly more active with a macrophage cell line or human K cells as effectors (40 to 50% tumor cell damage). These findings indicate that murine antibodies of each IgG subclass can act as ligands in ADCC against tumor cells. Similar conclusions were made with erythrocytes as target cells (Ralph *et al.*, 1980). With regard to human immunoglobulin, it has been noted that IgG1, IgG2, and IgG3 antibodies participate in monocyte-mediated ADCC against red cells (Holm *et al.*, 1974; Larsson *et al.*, 1975). The importance of antigen density on the target cell surface and antigen–antibody affinity in determining the efficiency of binding and triggering of effector cells by each of these IgG subclasses needs further study.

3.2. IgM

The ability of IgM to act as an antibody in ADCC reactions has been controversial, largely because of the possibility that small amounts of IgG may be present in many of these preparations (Andersson *et al.*, 1981; Greenberg and Lydyard, 1979). To circumvent the technical difficulties of preparing purified IgM from polyvalent serum, Zoller *et al.* (1982) utilized four murine monoclonal IgM antibodies against SRBC and assessed their ability to cooperate in ADCC with mouse spleen cells and human peripheral blood lymphocytes as effectors.

It was found that each of the IgM monoclonal antibodies induced target lysis by effector cells; however, relatively higher amounts of these antibodies compared to those of the IgG class were required. There were several findings which suggested that the activity of the IgM preparation was not due to small amounts of contaminating IgG. First, the physicochemical properties of the antibody as determined by Sephadex G-200 chromatography were those of IgM and not IgG. Second, passage of the IgM preparations over protein A–Sepharose did not affect antibody activity. Finally, trypsin treatment of effector cells failed to alter IgM-dependent cytotoxicity whereas it decreased the capacity of effector cells to mediate IgG-dependent target cell lysis. These findings strongly support the concept that IgM may act as a ligand in ADCC. This class of immunoglobulin, however, appears to be less effective than IgG in this regard.

3.3. IgA

Conflicting results regarding the possible role of IgA antibodies in ADCC have been obtained in studies using bacterial and nonbacterial targets (Quie *et al.*, 1968; Wilson, 1972). However, the development of better methods of purifying this class of immunoglobulin and the clear demonstration of FcR on the surface of potential effector cells (Strober *et al.*, 1978; Fanger *et al.*, 1980) indicate that IgA antibody may be involved in ADCC independently or synergistically with other immunoglobulin classes. Lowell *et al.* (1980a) showed that human monocytes and lymphocytes killed group C meningococci in the presence of serum IgA obtained from persons immunized with meningococcal polysaccharide. IgA was as effective as postimmunization IgG in ADCC and was not dependent on the presence of complement for its activity. In related studies, it was demonstrated that IgA obtained from individuals convalescing from disseminated meningococcal disease was also active in monocyte-mediated ADCC (Lowell *et al.*, 1980b). A possible role for secretory IgA (sIgA) in ADCC against bacterial targets was suggested by the experiments of Shen and Fanger (1981). PMN- and monocyte-mediated lysis of ^{51}Cr-labeled ox red cells was demonstrable when anti-ox cell IgG but not anti-ox sIgA antibodies were added. When anti-ox cell sIgA was mixed with IgG antibodies, however, a marked increase in the degree of red cell lysis was observed; sIgA from control unimmunized animals did not enhance the cytotoxic ability of PMN or monocytes. These findings suggest that sIgA may act in a synergistic manner with IgG antibodies by increasing the binding of target erythrocytes to effector cells and/or by stimulating greater production of toxic effector molecules.

Since sIgA is present in large amounts at mucosal surfaces and secretions, the possible role of these immunoglobulins in ADCC occurring at these sites has been investigated. Tagliabue *et al.* (1983) have assessed the capacity of murine sIgA (obtained from intestinal secretions) to participate in lymphocyte-mediated ADCC against the enteric pathogen *Shigella* X16. When lymphocytes from the spleen, popliteal lymph nodes, or thymus were utilized as effector cells, their antibacterial activity against *Shigella* was not affected by addition of antibacterial sIgA. In contrast, the antibacterial activity of gut-associated lymphoid tissue (lymphocytes derived from Peyer's patches) was markedly enhanced by sIgA antibodies. Thus, sIgA at mucosal surfaces may be important in ADCC reactions occurring at mucosal surfaces, such as those of the gastrointestinal or respiratory tract. In addition, these results suggest that effector cells localized in mucosal sites may be especially well-suited to participate in these specialized forms of ADCC.

3.4. IgE

A functional role for antibodies of the IgE class has been implicated in ADCC reactions associated with helminthic infections and allergic disorders. Joseph *et al.* (1978) noted that monocytes incubated with IgE-containing immune complexes were able to kill schistosomula of *S. mansoni*. More recently, Capron *et al.* (1984) examined the roles of IgE antibodies and FcR for IgE in eosinophil-mediated killing of schistosomula. It was found that selected immune sera contained high levels of antiparasite IgE antibodies. These sera led to eosinophil-mediated parasite toxicity that was inhibitable by preincubation of effector cells with aggregated IgE or anti-FcR IgE antibodies. ADCC by these cells therefore appeared to be dependent on the interaction of opsonic IgE antibodies with FcR on the eosinophil surface. With regard to the possible importance of IgE-dependent cell-mediated cytotoxicity in the pathogenesis of allergic disorders, Melewicz *et al.* (1981) have shown that mononuclear cells from patients with severe atopy effected a greater degree of cytotoxicity against IgE-coated red cells compared to cells from normal individuals or those with mild allergies. The degree of target cell lysis correlated with the proportion of monocytes having detectable FcR for IgE.

4. FUTURE CONSIDERATIONS

In vitro models of ADCC reactions have provided valuable insight into the mechanisms of how mammalian cells interact with and eventually eliminate particulate targets not recognized as self. Comparisons of cytotoxic mechanisms involved in eliminating ingested red cells or bacteria versus nonphagocytosable targets such as helminthic parasites have led to a better understanding of the different factors that trigger effector cells to deliver toxic molecules into intra- versus extracellular sites. However, there remain several major areas in need of further study. First, although ADCC has clearly been demonstrated to eliminate targets *in vitro*, the possible role of this reaction in the intact host needs to be addressed. This will require *in vivo* experiments, such as determination of the ability of IgE antibodies to confer resistance against helminthic infections. Second, the possible therapeutic application of ADCC reactions warrants further exploration. In this regard, recent studies suggest that effector cells precoated or "franked" with antitumor antibodies may be effective in eliminating neoplastic cells in an antigen-specific manner (Simone, 1982). Finally, the factors which modulate the efficiency of cytotoxicity, such as antigen density on the target and FcR density and mobility on the effector cell surface, require further delineation.

REFERENCES

Anderson, C. L., and Spiegelberg, H. L., 1981, Macrophage receptors for IgE: Binding of IgE to specific IgE Fc receptors on a human macrophage cell line, U937, *J. Immunol.* **126**:2470.

Andersson, B., Skogland, A.-C., Ronnholm, M., Lindsten, T., Lamon, E. W., Collison, E. W., and Wahia, A. S., 1981, Functional aspects of IgM and IgG Fc receptors on murine T-lymphocytes, *Immunol. Rev.* **56**:21.

Babior, B. M., 1978, Oxygen-dependent microbial killing of phagocytes, *N. Engl. J. Med.* **298**:659.

Bass, D. A., and Szejda, P., 1979, Mechanisms of killing of newborn larvae of *Trichinella spiralis* by neutrophil and eosinophil leukocytes: Killing by generators of hydrogen peroxide in vitro, *J. Clin. Invest.* **65**:1558.

Borregaard, N., and Kragballe, K., 1982, The oxygen-dependent cell-mediated cytotoxicity of human monocytes and neutrophils, *Adv. Exp. Med. Biol.* **141**:71.

Bradley, T. P., and Bonavida, B., 1982, Mechanisms of cell-mediated cytotoxicity at the single cell level. IV. Natural killing and antibody-dependent cellular cytotoxicity can be mediated by the same human effector cell as determined by the two-target conjugate assay, *J. Immunol.* **129**:2260.

Brier, A. M., Chess, L., and Schlossman, S. F., 1975, Human antibody-dependent cellular cytotoxicity, *J. Clin. Invest.* **56**:1580.

Butterworth, A. E., Sturrock, R. F., Houba, V., Mahmoud, A. A. F., Sher, A., and Rees, P. H., 1975, Eosinophils as mediators of antibody-dependent damage to schistosomula, *Nature (London)* **256**:727.

Butterworth, A. E., Wassom, D. L., Gleich, G. J., Loegering, D. A., and David, J. R., 1979, Damage to schistosomula of *Schistosoma mansoni* induced directly by eosinophil major basic protein, *J. Immunol.* **122**:221.

Capron, M., Capron, A., Dessaint, J.-P., Torpier, G., Johanson, S. G. O., and Prin, L., 1981, Fc receptors for IgE on human and rat eosinophils, *J. Immunol.* **126**:2087.

Capron, M., Spiegelberg, H. L., Prin, L., Bennich, H., Butterworth, A. E., Pierce, R. J., Ouassi, M. A., and Capron, A., 1984, Role of IgE receptors in effector function of human eosinophils, *J. Immunol.* **132**:462.

Cerottini, J. C., and Brunner, K. T., 1974, Cell-mediated cytotoxicity, allograft rejection and tumor immunity, *Adv. Immunol.* **18**:67.

Clark, R. A., and Klebanoff, S. J., 1977, Studies on the mechanism of antibody-dependent polymorphonuclear leukocyte-mediated cytotoxicity, *J. Immunol.* **119**:1413.

Conkling, P., Klassen, D. K., and Sagone, A. L., Jr., 1982, Comparison of antibody-dependent cytotoxicity mediated by human polymorphonuclear cells, monocytes, and alveolar macrophages, *Blood* **60**:1290.

Dourmashkin, R. R., Deteix, D., Simone, C. B., and Henkart, P., 1980, Electron microscopic demonstration of lesions in target cell membranes associated with antibody-dependent cellular cytotoxicity, *Clin. Exp. Immunol.* **42**:554.

Ellner, J. J., and Mahmoud, A. A. F., 1982, Phagocytes and worms: David and Goliath, revisited, *Rev. Infect. Dis.* **4**:698.

Fanger, M. W., Shen, L., Pugh, J., and Bernier, G. M., 1980, Subpopulations of human peripheral granulocytes and monocytes express receptors for IgA, *Proc. Natl. Acad. Sci. USA* **77**:3640.

Fleer, A., Roos, P., van den Borne, A. E. G., and Engelfrist, C. P., 1979, Cytotoxic activity of human monocytes towards sensitized red cells is not dependent on the generation of reactive oxygen species, *Blood* **54**:407.

Garagiola, D. M., Huard, T. K., and LoBuglio, A. F., 1981, Comparison of monocyte and alveolar macrophage antibody-dependent cellular cytotoxicity and Fc-receptor activity, *Cell. Immunol.* **64**:359.

Greenberg, A. H., and Lydyard, P. M., 1979, Observation of IgG1 anti-DNP hybridoma-mediated ADCC and the failure of three IgM anti-DNP hybridomas to mediate ADCC, *J. Immunol.* **123:**861.

Holm, G., Engvall, E., Hammarstrom, S., and Natvig, J. B., 1974, Antibody-induced hemolytic activity of human blood monocytes, *Scand. J. Immunol.* **3:**173.

Joseph, M., Capron, A., Butterworth, A. E., Sturrock, R. F., and Houba, V., 1978, Cytotoxicity of human and baboon mononuclear phagocytes against schistosomula in vitro: Induction by immune complexes containing IgE and *Schistosoma mansoni* antigens, *Clin. Exp. Immunol.* **33:**48.

Katz, P., Simone, C. B., Henkart, P. A., and Fauci, A. S., 1980, Mechanism of antibody-dependent cellular cytotoxicity, *J. Clin. Invest.* **65:**55.

Kay, H. D., and Horwitz, D. A., 1980, Evidence by reactivity with hybridoma antibodies for a probable myeloid origin of peripheral blood cells active in natural cytotoxicity and antibody-dependent cell-mediated cytotoxicity, *J. Clin. Invest.* **66:**847.

Kazura, J. W., Fanning, M. M., Blumer, J. T., and Mahmoud, A. A. F., 1981, Role of cell-generated H_2O_2 in granulocyte-mediated killing of schistosomula of *Schistosoma mansoni*, *J. Clin. Invest.* **67:**93.

Klebanoff, S. J., 1980, Oxygen metabolism and the toxic properties of phagocytes, *Ann. Intern. Med.* **93:**480.

Kohl, S., Starr, S. E., Oleske, J. M., Shore, S. L., Ashman, R. B., and Nakmias, A. J., 1977, Human monocyte-macrophage-mediated antibody-dependent cytotoxicity to herpes simplex virus infected cells, *J. Immunol.* **118:**729.

Kurlander, R. J., Rosse, W. F., and Logue, G. L., 1978, Quantitative influence of antibody and complement coating of red cells on monocyte-mediated lysis, *J. Clin. Invest.* **61:**3109.

Lamon, E. W., Whitten, H. D., Skurzak, H. M., Andersson, B., and Lindin, B., 1975, IgM antibody-dependent cell-mediated cytotoxicity in the Moloney sarcoma virus system: The involvement of T and B lymphocytes as effector cells, *J. Immunol.* **115:**1288.

Larsson, A., Pisarri-Salsano, S., Ohlander, C., Natvig, J. B., and Perlmann, P., 1975, Destruction of dextran-coated target cells by normal human lymphocytes and monocytes, *Scand. J. Immunol.* **4:**421.

Leonard, E. J., Ruco, L. P., and Meltzer, M. S., 1978, Characterization of macrophage activation factor, a lymphokine that causes macrophages to become cytotoxic for tumor cells, *Cell. Immunol.* **41:**347.

Lovchik, J. C., and Hong, R., 1977, Antibody-dependent cell-mediated cytolysis (ADCC): Analyses and projections, *Prog. Allergy* **22:**1.

Lowell, G. H., Smith, L. F., Griffis, J. M., and Brandt, B. M., 1980a, Antibody-dependent mononuclear cell-mediated anti-meningococcal activity, *J. Clin. Invest.* **66:**260.

Lowell, G. H., Smith, L. F., Griffis, J. M., and Brandt, B. M., 1980b, IgA-dependent, monocyte-mediated, antibacterial activity, *J. Exp. Med.* **152:**452.

MacDonald, H. R., Bonnard, G. D., Sordat, B., and Zawodnik, S. A., 1975, Antibody-dependent cell-mediated cytotoxicity: Heterogeneity of effector cells in human peripheral blood, *Scand. J. Immunol.* **4:**487.

Mayer, M., 1977, Mechanism of cytolysis by lymphocytes: A comparison with complement, *J. Immunol.* **119:**1195.

Melewicz, F. M., and Spiegelberg, H. L., 1980, Fc receptors for IgE on a subpopulation of human peripheral blood monocytes, *J. Immunol.* **125:**1026.

Melewicz, F. M., Zeiger, R. S., Mellon, M. H., O'Connor, R. D., and Spiegelberg, H. L., 1981, Increased IgE-dependent cytotoxicity by blood mononuclear cells of allergic patients, *Clin. Exp. Immunol.* **43:**526.

Nathan, C. F., and Cohn, Z. A., 1980, Role of oxygen-dependent mechanisms in antibody-dependent lysis of tumor cells, *J. Exp. Med.* **152:**198.

Perlmann, H., and Cerottini, J. C., 1979, Cytotoxic lymphocytes, in: *The Antigens* (M. Sela, ed.), pp. 173–283, Academic Press, New York.

Pichler, W., Gendelman, F., and Nelson, D., 1979, Fc receptors on human T lymphocytes. II. Cytotoxic capabilities of T gamma, T mu, B, and L cells, *Cell. Immunol.* **42:**410.

Pincus, S. H., Butterworth, A. E., David, J. R., Robbins, M., and Vadas, M. A., 1981, Antibody-

dependent eosinophil-mediated damage to schistosomula of *Schistosoma mansoni:* Lack of requirement for oxidative metabolism, *J. Immunol.* **126:**1794.
Quie, P. G., Messner, R. P., and Williams, R. C. Jr., 1968, Phagocytes in subacute bacterial endocarditis. Localization of the primary opsonic site to Fc fragment, *J. Exp. Med.* **128:**553.
Ralph, P., and Nakoinz, I., 1983, Cell-mediated lysis of tumor targets directed by murine monoclonal antibodies of IgM and all IgG isotypes, *J. Immunol.* **131:**1028.
Ralph, P., Nakoinz, I., Diamond, B., and Yelton, D., 1980, All major classes of murine IgG antibody mediate macrophage phagocytosis and lysis of erythrocytes, *J. Immunol.* **125:**1885.
Rimm, I. J., Schlossman, S. F., and Reinherz, E. L., 1981, Antibody-dependent cellular cytotoxicity and natural killer-like activity are mediated by subsets of activated T cells, *Clin. Immunol. Immunopathol.* **21:**134.
Shaw, G. M., Levy, P. C., and LoBuglio, A. F., 1978, Human monocyte antibody-dependent cell-mediated cytotoxicity to tumor cells, *J. Clin. Invest.* **62:**1172.
Shen, L., and Fanger, M. W., 1981, Secretory IgA antibodies synergize with IgG in promoting ADCC by human polymorphonuclear cells, monocytes, and lymphocytes, *Cell. Immunol.* **59:**75.
Simone, C. B., 1982, Directed effector cells selectively lyse human tumour cells, *Nature (London)* **297:**234.
Simone, C. B., and Henkart, P. A., 1980, Permeability changes induced in erythrocyte ghost targets by antibody-dependent cytotoxic effector cells: Evidence for membrane pores, *J. Immunol.* **124:**954.
Strober, W., Hague, N. E., Lum, L. G., and Henkart, P. A., 1978, IgA-Fc receptors on mouse lymphoid cells, *J. Immunol.* **121:**2440.
Tagliabue, A., Nencioni, L., Villa, L., Keren, D. F., Lowell, G. H., and Boraschi, D., 1983, Antibody-dependent cell-mediated anti-bacterial activity of intestinal lymphocytes with secretory IgA, *Nature (London)* **306:**184.
Timonen, T., Ortaldo, J. R., and Herberman, R., 1981, Characteristics of human large granular lymphocytes and relationship to natural killer and K cells, *J. Exp. Med.* **153:**569.
Urbaniak, S. J., 1979, ADCC (K-cell) lysis of human erythrocytes sensitized with rhesus alloantibodies: Investigation of in vitro culture variables, *Br. J. Haematol.* **42:**315.
van Furth, R., and Cohn, Z. A., 1968, The origin and kinetics of mononuclear phagocytes, *J. Exp. Med.* **128:**415.
Venge, P., Dahl, R., Hallgre, R., and Olsson, I., 1980, Cationic proteins of human eosinophils and their role in the inflammatory reaction, in: *The Eosinophil in Health and Disease* (A. A. F. Mahmoud and K. F. Austen, eds.), pp. 131–142, Grune & Stratton, New York.
Wilson, I., 1972, Studies on the opsonic activity of human secretory IgA using an in vitro phagocytosis system, *J. Immunol.* **108:**726.
Zarling, J., and Kung, P. C., 1980, Monoclonal antibodies which distinguish between human NK cells and cytotoxic T lymphocytes, *Nature (London)* **288:**394.
Zoller, M., Heyman, B., Andrighetto, G., and Wigzell, H., 1982, IgG- and IgM-induced cellular cytotoxicity, *Scand. J. Immunol.* **16:**379.
Zoller, M., Andrighetto, G. C., Heyman, B., Lamon, E. W., and Wigzell, H., 1983, Characterization of effector cells mediating IgG and IgM antibody-dependent cellular cytotoxicity, *Scand. J. Immunol.* **17:**19.

9

Inflammatory Lymphokines in Hypersensitivity Reactions

MARION C. COHEN, JANICE K. GUTOWSKI, and STANLEY COHEN

1. INTRODUCTION

The immune system is responsible for both the detection of foreign substances and the response to them. Thus, immunologic function requires both recognition and reaction. Recognition involves the sequence of events by which the immune system is made aware of a foreign substance (antigen). It involves the interaction of the antigen with an appropriate cell surface receptor and leads to the activation of lymphocytes for both proliferation and differentiation. The products of differentiation include not only antibodies, but also memory cells, specialized effector cells, and a variety of nonantibody, hormonelike mediators. These latter agents, which have been shown to play a role in both afferent (inductive stages) and efferent (effector responses) manifestations of immunity, are known as lymphokines.

The actual discovery of the lymphokines grew out of two unrelated areas of immunologic investigation, delayed hypersensitivity and mechanisms of antibody formation. Historically, the term "lymphokine" was invoked to describe certain biologic mediators of the delayed hypersensitive state. Subsequently, certain cell–cell interactions involved in antibody production were shown to be dependent on soluble lymphocyte-derived factors, and those agents were recognized to be lymphokines as well.

"Delayed hypersensitivity" is the name given to the cutaneous manifestation of cell-mediated immunity. It is also called "tuberculin"-type hypersensitivity, especially in the older literature. The delayed reaction is a slowly evolving inflammatory reaction (24–48 hr to reach maximal size) which develops at the site of injection of antigen into a previously sensitized individual. It appears

MARION C. COHEN, JANICE K. GUTOWSKI, and STANLEY COHEN • Department of Pathology, University of Connecticut Health Center, Farmington, Connecticut 06032.

grossly as an indurated lump and microscopically, at least in humans and guinea pigs, as a collection of lymphocytes and macrophages which are usually collectively referred to as "mononuclear" cells. The delayed hypersensitivity reaction can be transferred to a normal, nonimmune individual by sensitized lymphocytes but not by serum.

Although the skin reaction is the best studied form of cellular hypersensitivity, this kind of immunologic response is of much broader biologic significance; it appears to play an essential role in infectious immunity, tumor immunity, and the response to transplanted organs and tissues. In these various manifestations, cellular hypersensitivity is a protean process. It need not be very delayed; in locations other than the skin, it can reach maximal intensity in as little as 4 hr. Moreover, there are a number of biologically important situations in which the major infiltrating cell population is granulocytic rather than mononuclear. Finally, this class of immunologic reaction is no more or less cell-mediated than antibody-dependent reactions which are ultimately due to the participation of a lymphocyte or plasma cell. With the exception of those reactions which involve direct cell killing, most reactions of cell-mediated immunity are dependent not on the lymphocyte as effector, but as stated above, on the action of soluble lymphocyte products known as lymphokines.

2. HISTOLOGIC MANIFESTATIONS OF CELLULAR IMMUNITY

Classically, cellular hypersensitivity reactions are said to be composed predominantly of mononuclear cells. In these situations, only a few of the lymphocytes present are specifically sensitized and there seems to be no evidence of preferential accumulation of such sensitized cells at the test site. Evidence for this comes mainly from two types of experiments: (1) transfer studies with radiolabeled sensitized donor lymphocytes, and (2) immunization of animals with two unrelated antigens and injection of [^{3}H]thymidine so as to label only proliferating cells of one specificity (Cohen *et al.*, 1967b; McCluskey and Leber, 1974). In these kinds of experiments, the animals were sensitized to two unrelated antigens but possessed radiolabeled cells of only one specificity. Autoradiography of the skin sites demonstrated no significant difference in the number of labeled cells at each reaction site.

The indurated nature of the delayed hypersensitivity reaction has been thought to be due to the vast numbers of infiltrating inflammatory cells. However, Dvorak and his co-workers have shown that fibrin deposits are a feature of delayed hypersensitivity reactions in humans and guinea pigs and this results in the induration which is observed (Colvin and Dvorak, 1979). It has also been shown that delayed reactions could be visibly suppressed by anticoagulants (Nelson, 1965; Cohen *et al.*, 1967a). This suppression did not result in a decrease in the cellular infiltrate, but rather in the gross appearance, which is an indication of induration. Thus, the induration observed in delayed hypersensitivity reactions does not seem to be due to the infiltrating cells but rather to the deposition of fibrin, possibly in association with bound extravascular water.

The histologic features of delayed hypersensitivity reactions in guinea pigs are frequently used as the reference standard in identifying similar reactions in other species. Thus, a mononuclear cell infiltrate is the hallmark of a delayed reaction. However, in cellular hypersensitivity in the mouse, neutrophil infiltration is common and precedes a mononuclear cell response (Crowle, 1975).

Furthermore, when guinea pigs are immunized by procedures that avoid the use of mycobacterial adjuvants, a delayed reaction is produced which is relatively nonindurated and is characterized by extensive infiltration of basophils. This was originally recognized as a response to soluble protein antigens (so-called "Jones–Mote" reactivity) and is now also referred to as "cutaneous basophil hypersensitivity (CBH)." CBH reactions are seen in response to a variety of immunogens including contact allergens, vaccinia virus, allogeneic tumor cells, insect bites, and skin allografts (Colvin and Dvorak, 1979).

In certain circumstances, the eosinophil may play a major role. Accelerated reactivity on reinjection of antigen at skin sites of prior (healed) delayed hypersensitivity reactions has been observed. This second, or "retest" reaction, is more intense and is characterized by the presence of large numbers of eosinophils (Arnason and Waksman, 1963).

These various histologic patterns demonstrate that the typical mononuclear cell infiltrate is only one manifestation of delayed hypersensitivity reactions. Common to all of these reactions is the presence of a small number of specifically sensitized lymphocytes that interact with antigen locally and release a variety of lymphokines which are responsible for the effects observed.

3. GENERAL PROPERTIES OF EFFECTOR LYMPHOKINES

The production of nonantibody mediators by antigen-activated lymphocytes was first demonstrated in 1966 by Bloom and Bennett (1966) and by David (1966) in studies involving the effector mechanisms of delayed hypersensitivity. Since that time, it has become evident that lymphocyte-derived mediators, collectively known as lymphokines (Dumonde *et al.*, 1969), play a role not only in delayed hypersensitivity reactions, but also in most other cell-mediated immune reactions as well (Cohen, 1977). As originally described, lymphokine activities were involved in the effector arm of the immune response. However, the lymphokine concept has been expanded to include various afferent molecules which mediate cell interactions involved in the induction of immune responses as well but which will not be discussed in this chapter.

Lymphokines can be produced by specific antigenic stimulation of sensitized lymphocytes or by nonspecific activation of nonimmune cells. Both B and T lymphocytes can synthesize lymphokines, if activated by an appropriate stimulus. In addition, many long-term lymphoid cell lines spontaneously release lymphokines in the absence of exogenous stimulation. The cellular sources of lymphokines will be described in greater detail in a later section. In general, lymphokines do not represent preformed cellular components. They appear to be synthesized and released as a result of the interaction between the inducing agent and membrane receptors. In this respect, lymphokines are quite different from the mediators of anaphylaxis and from the enzymes found in phagocytic cells, which are preformed, stored in intracellular granules, and merely released as a result of activation. With some exceptions, lymphokine production requires RNA and protein synthesis, but not DNA synthesis. The exact mechanism of lymphokine release is not known.

It should be noted that a variety of nonlymphoid cells are also capable of producing soluble mediators. Cohen *et al.* (1974a) have suggested the more general term "cytokines" for the biologically active factors produced by both lymphoid and nonlymphoid cells. Many of the non-lymphocyte-derived factors share biochemical and functional characteristics with the lymphokines and certain of them share antigen determinants, suggesting structural similarities (Yoshida *et al.*, 1975a).

Each lymphokine was discovered and defined on the basis of its biologic activity and target cell *in vitro.* Approximately 100 lymphokine activities were described as of 1979 (Waksman, 1979); and we have stopped counting since then. It should be noted, however, that under most circumstances, multiple activities are produced simultaneously and are probably not due to 100 different molecules. The biologic effects of the lymphokines are varied, including alteration of cell motility, activation of inflammatory cells, induction of proliferation, regulation of cell cooperation, and expression of cell surface phenotype. The range of target cells for these effects is equally broad including inflammatory, immunologic, reparative, neoplastic, and stem cells.

All lymphokines are proteins and most are glycoproteins. With the exception of leukocyte migration inhibition factor (LIF), which demonstrates esterase activity (Bendtzen, 1977a; Rocklin and Rosenthal, 1977), lymphokines do not appear to be enzymatic in nature. Lymphokines generally act in an antigen-nonspecific fashion and their effects often cross species lines. Although specific membrane receptors have not been demonstrated for most lymphokines, the mechanism of lymphokine action probably involves binding to such receptors, disturbance of some complex cellular equilibrium, and a sequence of intracellular biochemical events which ultimately cause the specific lymphokine effect. This hypothesis is supported by the ability of low-molecular-weight substances to interact with known membrane receptors and mimic chemotactic, mitogenic, cytotoxic, and migration inhibitory effects of lymphokines (Rocklin and Rosenthal, 1977; Schiffmann *et at.*, 1975; Clark *et al.*, 1975; Kay *et al.*, 1971).

Although lymphokine activities have all been described in *in vitro* systems, there is little doubt as to the *in vivo* significance of these soluble mediators. Evidence for this relationship includes: (1) the ability of lymphokines produced *in vivo* to exert biologic effects when introduced into an intact animal; (2) the recovery of lymphokine activity from tissues and fluids derived from sites of ongoing immunologically mediated inflammatory reactions; and (3) the ability of antilymphokine antibodies to suppress manifestations of cell-mediated immunity *in vivo.* Furthermore, systemic administration of lymphokine-containing supernatants to experimental animals results in desensitization and cutaneous anergy which mimics that seen in granulomatous and lymphoproliferative diseases in which serum lymphokine activity is high (Yoshida *et al.*, 1975c).

4. LYMPHOKINES AFFECTING INFLAMMATORY CELLS

The best studied of the lymphokines involved in inflammatory responses are those that affect macrophages. Studies on *in vitro* correlates of delayed hypersensitivity by Rich and Lewis (1932) first demonstrated that tuberculin preparations inhibited the migration of cells from spleen fragments of peripheral blood taken from actively immunized animals. This assay was simplified by George and Vaughan (1962) who developed the capillary tube method. Bloom and Bennett (1966) and David (1966) independently found that the migration of peritoneal exudate macrophages from such tubes could be inhibited by a soluble factor, which was named migration inhibition factor (MIF). MIF was released by sensitized lymphocytes in the exudate suspension reacting with specific antigen. If a source of MIF was included in the incubating medium, the migration inhibition reaction could be demonstrated using peritoneal exudate cells from nonimmunized animals. The best source of MIF proved to be supernatant fluids from cultures of sensitized lymphocytes incubated with specific antigen. MIF is of historical importance because it was the first lymphokine to be described.

MIF is produced not only by specific antigenic stimulation of sensitized cells, but also by nonspecific activation of lymphoid cells. Thus, activation of lymphocytes from a variety of animal species with mitogens such as Con A or PHA or with antigen–antibody complexes, results in the production of MIF, which shares many common properties with the antigen-induced factor (Kotkes and Pick, 1975; Geczy *et al.*, 1975; Remold, *et al.*, 1972). Some investigators have detected MIF activity in allogeneic mixed leukocyte cultures (Bartfield and Atoynatan, 1971) whereas others have not (Geczy *et al.*, 1976b). MIF activity has also been detected in culture supernatants from a variety of long-term cell lines (Remold *et al.*, 1983; Yoshida *et al.*, 1976; Poste, 1975; Tubergen *et al.*, 1972; Papageorgiou *et al.*, 1972).

Evidence from a variety of experimental systems clearly indicates that T cells make MIF (Woody *et al.*, 1975: Rocklin *et al.*, 1974; Clinton *et al.*, 1974; Yoshida *et al.*, 1973b). Murine MIF is produced by T cells bearing the Ly-1^+,2^-,3^- phenotype (Adelman *et al.*, 1980; Kühner *et al.*, 1980). The overwhelming evidence indicates a strict requirement of syngeneic macrophages for T-cell MIF production (Landolfo *et al.*, 1977; Nelson and Leu, 1975; Ohisi and Onoue, 1975). However, in one study, macrophages could be replaced by a soluble factor produced by LPS-stimulated adherent cells (Ohisi and Onoue, 1976).

MIF is not exclusively a T-cell product; purified populations of B cells have also been shown to make MIF in human, mouse, and guinea pig systems (Adelman *et al.*, 1980; Bloom *et al.*, 1975; Rocklin *et al.*, 1974; Yoshida *et al.*, 1973b). Although in one report soluble antigen was capable of serving as an inducing agent for B-cell MIF production (Rocklin *et al.*, 1974), most of these studies demonstrated a requirement for B-cell activation by mitogens. Interestingly, T cells have been shown to produce a soluble factor which can suppress the production of MIF by B cells (Cohen and Yoshida, 1977). This observation may explain why B cells do not appear to play a role in cell-mediated immune reactions *in vivo*.

Different strains of mice exhibit different levels of MIF production in response to mitogenic or antigenic stimulation. However, these strain variations do not directly correlate with H-2 haplotype (Adelman *et al.*, 1978). MIF produced in one strain of mouse is capable of inhibiting the migration of macrophages from unrelated strains, suggesting a lack of genetic restrictions on MIF activity (Adelman *et al.*, 1978). The interaction of MIF with its target cells can even transcend species barriers (Gorski, 1974; Gadol and Waldman, 1973; Svejcar *et al.*, 1971; Thor *et al.*, 1968).

Several investigators have shown that MIF production is independent of cell proliferation (Visakorpi, 1974; Rocklin, 1973; Bloom *et al.*, 1972). MIF production by stimulated lymphocytes is blocked by inhibitors of protein synthesis (Bloom, 1971) and there appears to be a requirement for macromolecular synthesis during the first 2 hr after stimulation (Varesio *et al.*, 1980).

Generally, MIF from all species appear to be proteins with molecular weights roughly between 12,000 and 82,000, as estimated by gel filtration (Remold and Mednis, 1977; Kühner and David, 1976; Remold, 1973; Dumonde *et al.*, 1972; Rocklin *et al.*, 1972; Yoshida and Reisfeld, 1970). The differences in reported molecular weights may be due to inconsistencies inherent in gel filtration or to the heterogeneity of MIF molecules. Thus, reports in the literature point to the existence of more than one molecular species for guinea pig, mouse, and human MIF (Boult *et al.*, 1981; Weiser *et al.*, 1981; Sorg, 1980; Remold and Mednis, 1977; Yoshida and Reisfeld, 1970). Human MIF is postulated to contain disulfide bonds and MIF activity was irreversibly blocked by the reducing agent L-cysteine (Bartfield and Atoynatan, 1971). In a more recent report, however, the biologic activity of human MIF was shown to be independent of disulfide bonding (Ewan *et al.*, 1980). Evidence for the subunit structure of human MIF is drawn from the results of Possanza *et al.* (1979), who showed that human cell line MIF loses activity upon dialysis. Although activity was not recovered in either the retentate or dialysate alone, the two components could reassociate and be biologically active when the two fractions were mixed.

The best known manifestation of MIF activity is the inhibition of macrophage migration *in vitro*. MIF activity is also associated with increased target cell aggregation and clumping (Salvin *et al.*, 1971; Pick *et al.*, 1969; David *et al.*, 1964), inhibition of macrophage spreading and increased macrophage adherence (stickiness) (Mooney and Waksman, 1970; Fauve and Dekaris, 1968; Weiss and Glaves, 1968), alterations in electrophoretic mobility and transmembrane potential of target cells (Jenssen *et al.*, 1975), alterations in phagocytic and pinocytotic activity (Meade *et al.*, 1974; Nathan *et al.*, 1971), and alterations in cell surface tension (Thrasher *et al.*, 1973). Ultrastructurally, MIF-treated target cells exhibit a retraction of cell processes and extensive cytoplasmic vacuolation (Nath *et al.*,

1973), enhanced microvilli formation (Thorne *et al.*, 1975), and loss of a cell surface glycoprotein at areas of cell contact (Dvorak *et al.*, 1972).

MIF appears to act by binding to receptors on the macrophage cell surface. Receptor models for the MIF–target cell interaction are supported by the fact that MIF activity can be removed from supernatants by incubation with macrophages (Leu *et al.*, 1972). Furthermore, trypsin treatment destroys responsiveness to MIF, but this effect is reversed by removal of the enzyme and further culture for 24 hr, suggesting the regeneration of MIF receptor (Manheimer and Pick, 1973). Evidence suggests that α-L-fucose may be a biologically active part of the MIF receptor in both humans and guinea pigs (Rocklin, 1976; Remold, 1973). More recent data suggest that the MIF receptor on macrophages may actually be a fucose-containing glycolipid (Poste *et al.*, 1979a, b; Higgins *et al.*, 1978).

Although the exact mechanism(s) by which MIF causes inhibition of cell migration is not known, various theories have been put forward: (1) cessation of movement may be due to direct or indirect effects of MIF on cell membranes; (2) migration inhibition may result from the interference of MIF with some metabolic process necessary for motility; (3) migration may be blocked by an effect on cytoskeletal elements of the cell, such as microtubules and microfilaments. The evidence for and against each of these theories has been reviewed by Pick (1979). In addition, Geczy and Hopper (1981) have postulated that surface-bound fibrin, which appears as a consequence of the MIF–macrophage interaction, forms a network which surrounds the cell and thereby restricts migration *in vivo*.

Early studies of MIF resulted in the formulation of three important methodologic principles. First, control supernatants, which were obtained from the sensitized cells incubated in the absence of antigen, were always reconstituted with antigen prior to use. Second, controls for viability were also included. Third, reversibility of effect was demonstrated. This last point means that if the inhibited preparations are allowed to remain in culture, the cells escape from inhibition and ultimately begin to migrate. The last two principles are of great practical importance since the distinction between toxicity and migration inhibition remains a very important one. A surprisingly large number of experiments in the literature are difficult to interpret because this distinction was not always appreciated.

Another important lymphokine that affects migration properties of macrophages is macrophage chemotactic factor (MCF). By definition, chemotaxis is movement in the direction of an increasing concentration gradient of the factor. The usual assay system involves the Boyden double chamber technique (Boyden, 1962) in which the test substance is placed in the lower compartment and the target cell suspension in the upper compartment. Cells migrate through a micropore filter toward the test substance. After a given period of time, the number of cells migrating through the filter in the test preparation is counted and compared to the number migrating in a control preparation. Increased numbers of cells in the presence of the gradient are taken as evidence of a positive chemotactic response. More recently, new techniques involving the use of migration in agarose have been developed which allow for the quantitation of chemotactic responses (discussed in Orr and Ward, 1978). Guinea pig MCF has been separated from MIF using acrylamide gel disc electrophoresis and shown to be heat stable (56°C for 30 min) with a molecular weight between 35,000 to 55,000 (Ward *et al.*, 1969, 1970). Further, it is important to distinguish between factors that are chemotactic for cells and those that merely enhance random movement. When the Boyden chamber is used, this distinction is easy to make. If the test substance is placed in both compartments, rather than the bottom only, as in the usual assay, the gradient is abolished but the cells are in contact with the mediator. If this procedure abolishes enhanced migration, then the substance is chemotactic; if increased migration occurs, then the substance is nonspecifically potentiating cell movement. Studies in the guinea pig by Wahl *et al.* (1974a) revealed that splenic, peripheral blood, thymic, and peritoneal exudate lymphocytes could also produce MCF. MCF could be detected 6–8 hr after stimulation, preceding lymphocyte proliferation. Furthermore, guinea pig MCF was sensitive to proteolytic enzymes, resistant to neuraminidase, and antigenically distinct from C2 and C5, which are precursors of complement-derived chemotactic peptides. These investigators reported that guinea pig MCF had an approximate molecular weight of 12,500, in contrast to the findings of Ward *et al.* (1969). It may be that there is more than one MCF in the guinea pig or that aggregated forms of the mediator exist.

Studies on human MCF showed that human blood, spleen, and tonsil lymphocytes could elaborate a factor that was chemotactic for rabbit macrophages (Rocklin and David, 1971). Snyderman *et al.* (1972) showed that mitogen- or antigen-stimulated human lymphocytes elaborate a factor that could attract homologous mononuclear cells. Like guinea pig MCF, human MCF is heat stable (56°C for 30 min), distinct from homologous C3 and C5, and has a molecular weight of approximately 12,500 (Altman *et al.*, 1973). MCF production in other species has also been described (Meltzer, 1976; Ward and Volkman, 1975; Altman and Kirchner, 1972, 1974).

Both T and B cells produce MCF in response to mitogenic stimulation (Mackler *et al.*, 1974; Wahl *et al.*, 1974b). B cells are also induced to produce MCF by agents that bind to C3 and Fc receptors (Mackler *et al.*, 1974; Wahl *et al.*, 1974b). Antigen-stimulated guinea pig lymphocytes require viable macrophages to produce MCF, whereas mitogen-stimulated cells are capable of producing MCF without macrophage cooperation (Wahl *et al.*, 1975b). MCF production by both human and guinea pig lymphocytes is inhibited by pharmacologic doses of glucocorticosteroids (Wahl *et al.*, 1975a; Ruhl *et al.*, 1974). Guinea pig MCF activity is inhibited by incubation with α-L-fucose or α-L-rhamnose, suggesting that these monosaccharides may be part of the MCF receptor binding site (Amsden *et al.*, 1978).

Activated lymphocyte supernatants can alter many macrophage functions in addition to migration properties. These functions include increased adherence, enhanced phagocytic and pinocytotic activity, increased glucosamine uptake, increased glucose oxidation via the hexose monophosphate shunt, and increased bacteriostatic or bactericidal activity (Pick, 1979). These activities have also been associated with macrophage activation factor (MAF) activity. MIF and MAF have not been effectively separated by the usual methods of protein chemistry and it has been suggested that they may actually be the same (Nathan *et al.*, 1973). Macrophages preincubated in MIF-rich supernatants were found to be capable of killing tumor cells but not normal cells (Churchill *et al.*, 1975). Further, trypsinization of the activated macrophages did not diminish their cytotoxic potential, suggesting that the effect was not due merely to adsorbed materials on the cell surface.

Stimulation of macrophage migration in conventional MIF assays has also been observed. This activity has been called macrophage enhancement factor (MEF) (Weisbart *et al.*, 1974) or migration stimulation factor (MStF) (Aaskov and Anthony, 1976). It is quite possible the two factors (enhancing and inhibiting) are chemically identical but exert paradoxical effects depending on the state of the target cells, concentration of the mediator itself, or presence or absence of serum cofactors. The fact that the enhancing activity can appear in preparations presumed to contain MIF reemphasizes the difficulty of dealing with crude lymphokine preparations and bioassays.

Lymphokines that affect the behavior of granulocytes are fewer in number than those that affect macrophages. A lymphocyte-derived chemotactic factor for each granulocyte type (basophil, eosinophil, and neutrophil) has been found as well as an inhibitory factor for neutrophils.

The lymphocytes of animals primed for CBH are capable of secreting mediators on subsequent exposure to specific antigen *in vitro*. A basophil chemotactic factor (BCF) has been detected in supernatants derived from these animals (Ward *et al.*, 1975; Boetcher and Leonard, 1973). BCF is a nondialyzable macromolecule which is moderately heat-sensitive and bands on sucrose density gradients in a biphasic pattern similar to that of MCF. Absorption of lymphokine fluids with monocytes results in a loss of detectable activity for basophils but not monocytes; however, absorption with basophils has no effect on either activity, thus suggesting that the chemotactic factors for basophils and monocytes are distinct. BCF has also been described in culture fluids of stimulated human peripheral blood lymphocytes (Lett-Brown *et al.*, 1976; Kay and Austen, 1972).

Eosinophil chemotactic factor (ECF) derived from lymphocyte cultures has unusual properties. Eosinophils have been observed in inflammatory infiltrates at sites of various kinds of immunologic reactions, some of which appear to be dependent on a state of cellular hypersensitivity. However, while it was thus suspected that there must be a lymphokine with chemotactic activity for eosinophils, no such factor was detected in antigen-stimulated lymphocyte cultures. However, investigation of activated supernatants found them to contain an inactive precursor (ECF_p) which could be activated by reaction with specific preformed immune complexes *in vitro* to generate a potent eosinophil chemotactic factor (Torisu *et al.*, 1973; Cohen and Ward, 1971). Studies using antigen- and antibody-coupled immunosorbent columns have demonstrated that ECF_p behaves as if associated with a fragment of antigen and that its activation to ECF is associated with a loss of that antigenic determinant. To explain these data, a model for activation has been developed in which ECF_p combines with a free antibody site in the immune complex by coprecipitation. As a consequence of this interaction, a modified, active, antigen-free mediator molecule (ECF) is then released. It should be noted that ECF, on injection into guinea pig skin, produces an inflammatory reaction in which the predominant cell type is the eosinophil, attesting to its effectiveness *in vivo*. Furthermore, there is indirect evidence that ECF may play a role in the cutaneous "retest" reaction (Leber *et al.*, 1973) and in experimental autoimmune thyroiditis (Cohen *et al.*, 1974c).

Indications that neutrophils may have an integral role in cell-mediated immune reactions have been strengthened by findings that lymphocytes produce a neutrophil chemotactic factor (Ward *et al.*, 1969) and a leukocyte migration inhibition factor (LIF) (Bendtzen, 1976; Rocklin, 1974). Ward *et al.* (1969) described a lymphokine from guinea pig lymphocytes which is chemotactic for neutrophils. This factor is separable from MCF and MIF by polyacrylamide gel electrophoresis. Using the Boyden double chamber assay, it was found to be less potent than MCF (Ward *et al.*, 1970).

LIF is produced by incubation of human blood mononuclear cells with specific antigen (Rocklin, 1974) or Con A (Rocklin, 1975). Its effect can be demonstrated using conventional capillary tube assays and in the agarose assay described by Clausen (1971). LIF can be separated from MIF on the basis of physicochemical properties and its monosaccharide inhibition profile (Rocklin, 1976). LIF has been shown to have serine esterase activity (Bendtzen, 1976; Rocklin, 1976) with a preference for arginine esters and amides as substrates (Bendtzen, 1977b). Additional data suggest that LIF may be a multifunctional or an allosteric enzyme regulated by cGMP (Bendtzen, 1977c). The relationship between this enzyme activity and its effect on neutrophils is suggestive, but not yet established. A neutrophil migration inhibition factor derived from T cells (NIF-T) which can be distinguished from LIF has also been described (Weisbart *et al.*, 1980). The presence of B cells or macrophages is required for its production.

5. OTHER MIGRATION-MODIFYING LYMPHOKINES

Fractions of supernatants capable of inhibiting macrophage migration have also been found to affect endothelial cell migration. Recently, Cohen *et al.* (1982) reported that MIF-rich fractions of culture supernatants are also capable of inhibiting the migration of endothelial cells from agarose droplets. Whether inhibition of macrophage and endothelial cell migration is due to the same factor, however, remains to be determined.

Stimulation of guinea pig lymphocytes with antigen, Con A, alloantigen, or anti-immunoglobulin has been shown to induce the production of factors which are chemotactic for rat lymphocytes (Ward *et al.*, 1971, 1977). These lymphocyte chemotactic factors are selective in that the MLC-derived factor only attracts T cells, whereas the anti-immunoglobulin-induced factor only attracts B cells.

Human lymphocytes stimulated with PPD or PHA produce a factor which is chemotactic for human fibroblasts (Postlethwaite *et al.*, 1976). This factor was shown to have a molecular weight of approximately 22,000 and to be distinct from human MCF. More recently, mitogen- and antigen-activated guinea pig lymphocytes were shown to produce a fibroblast chemotactic factor (Postlethwaite and Kang, 1980). This factor has a molecular weight of 80,000, is heat labile, trypsin sensitive, and distinct from guinea pig MCF.

6. LYMPHOKINES AFFECTING VASCULAR PERMEABILITY

The first indication of the existence of *in vivo* effects of lymphokines came from finding that there are lymphocyte-derived mediators which affect vascular permeability. These mediators are evaluated by their ability to induce permeability changes at an intradermal injection site. Macroscopically, erythema and induration are seen and on histologic sections, cellular infiltrates and fibrin deposition may be found.

The first lymphocyte-derived factors shown to affect vascular permeability were not strictly lymphokines because they were found in tissue extracts of lymph nodes and thymus rather than being secreted into the medium by intact viable lymphocytes. Both lymph node permeability factor (LNPF) and thymic permeability factor (TPF) (Lykke *et al.*, 1967; Willoughby *et al.*, 1962, 1964) cause increased permeability of both venules and capillaries and stimulate a massive emigration of leukocytes into injection sites so that by 24 hr, the resulting lesions resemble classic tuberculin reactions. It was found that vasoactive amines or pharmacologic mediators of immediate hypersensitivity were not involved (Schild and Willoughby, 1967).

Several groups of investigators demonstrated that intradermal injection of culture supernatants from stimulated lymphocytes resulted in skin reactions that resembled delayed hypersensitivity reactions, while injection of control material resulted in a transient erythema and neutrophil infiltrate (Yoshida *et al.*, 1973a; Valdimarsson and Gross, 1973; Zschiesche *et al.*, 1972; Maillard *et al.*, 1972; Schwartz *et al.*, 1970; Pick *et al.*, 1969; Bennett and Bloom, 1968). The mediator responsible for this activity was called skin reactive factor (SRF). Its effect is probably the result of a combination of monocyte and neutrophil chemotactic factors, a substance that increases vascular permeability, and possibly MIF.

Fibrin accumulation and clotting are known to be regular features of delayed hypersensitivity reactions (Colvin and Dvorak, 1975). Anticoagulants have been found that inhibit the evolution of such reactions (Cohen *et al.*, 1967a; Nelson, 1965). These *in vivo* observations may be explained by the finding of a platelet-aggregating factor (PAF) in culture supernatants of antigen-stimulated rabbit lymphocytes (Lavelle *et al.*, 1975) and Con A-stimulated human lymphocyte culture supernatants (Coeugniet and Bendixen, 1976a). In addition to the induction of the platelet release reactions, these latter investigators found that the supernatants could decrease clotting time, accelerate clot retraction, and directly coagulate fibrinogen (Coeugniet and Bendixen, 1976a, b). While the role these lymphokines play *in vivo* has not been defined, it is most likely related to the fibrin deposits seen in delayed hypersensitivity reactions.

7. *IN VIVO* EFFECTS OF LYMPHOKINES

Much of the data on lymphokines have been accumulated using *in vitro* assay studies. Only recently has it been possible to obtain direct evidence for *in vivo* effects as well. Several different approaches have been taken which provide these data. These include (1) injection of exogenous lymphokines, (2) detection of lymphokine production *in vivo*, and (3) suppression of *in vivo* reactions by agents which interfere with lymphokine production and/or effect.

A nonspecific peritoneal inflammatory exudate can be induced in delayed hypersensitive guinea pigs by the intraperitoneal injection of glycogen. Three or four days following injection of this irritant, the exudates are comprised mainly of mononuclear cells, of which macrophages predominate. If specific antigen is then injected intraperitoneally, within 5 hr there is a drop of 90% in the absolute number of cells recovered in the peritoneal fluid. Over the next 24 to 48 hr the macrophage number will then increase. This reaction is known as the macrophage disappearance reaction (MDR) and has been shown to be a manifestation of cell-mediated immunity (Sonozaki and Cohen, 1971a; Nelson and Boyden, 1963).

The decrease in the number of macrophages is apparently due to increased cell adhesiveness induced by specific antigen, resulting in the clumping and sticking of macrophages to peritoneal surfaces. Available evidence shows the MDR to be an *in vivo* analogue of migration inhibition *in vitro*. Thus, it provides a model system to explore lymphokine activity *in vivo*.

The MDR can be transferred to nonimmune animals by intraperitoneal injections of sensitized lymphocytes and specific antigen (Sonozaki and Cohen, 1971b). If the lymphocytes are enclosed in micropore chambers so that only soluble factors can escape, the transfer is still successful. Furthermore, the MDR can also be transferred using cell-free, MIF-rich supernatants from antigen-stimulated lymphocyte cultures. In these experiments the possibility of contaminating immunoglobulins was eliminated by passage of the supernatants through Sephadex columns. Thus, there is strong evidence that the MDR is mediated by a soluble, lymphocyte-derived factor.

Attempts to find MIF activity in extracts of skin reaction sites in delayed hypersensitive animals have not met with success. While it might be expected that inhibition factors should be present at such sites, it is possible that these mediators play no role in these reactions or that the amount released is so small that it remains undetectable.

Another approach has been to inject skin extract fluids into nonimmunized guinea pigs bearing peritoneal exudates. This resulted in a macrophage appearance reaction with a two- to three-fold increase in macrophage content observed. It seemed likely that this accumulation of macrophages might be due to MCF activity. This was confirmed (Cohen *et al.*, 1973) when the extracts of sites of delayed reactions to various protein antigens, or of sites of contact reactions to *o*-chlorobenzoyl chloride, a potent sensitizing agent, were examined by means of the Boyden double chamber assay system for the presence of *in vitro* chemotactic activity. Chemotactic activity toward macrophages and lymphocytes, but not neutrophils, could be found in these extract fluids. Extracts from normal skin or from sites of nonspecific inflammatory reactions did not show such activity. Physicochemical properties of these chemotactic factors were similar to those of chemotactic factors derived from antigen-stimulated lymphocyte culture supernatants. Thus, lymphokine activity could be recovered from the site of a cell-mediated reaction.

Mediator activity has been found in the serum of experimental animals under a variety of conditions. MIF activity has been found in the serum of sensitized guinea pigs and mice following intravenous challenge with antigen (Yoshida and Cohen, 1974; Salvin *et al.*, 1973; Yamamoto and Takahashi, 1971; Krejci *et al.*, 1968). Salvin *et al.* (1973) also found that acid-labile interferon was liberated into the blood of BCG-infected mice challenged intravenously with Old Tuberculin.

Yoshida *et al.* (1969) reported that rats or guinea pigs immunized with various antigens in Freund's complete adjuvant showed increases in peripheral blood monocytes for at least 2 weeks following immunization. However, in analogy with the MDR in the peritoneal cavity, when specific antigen was injected intravenously into such immunized animals, there was a prompt reduction in the number of circulating monocytes. The maximal disappearance from the circulation occurred at 6 hr following antigen administration and there was a slow return to normal over the next 24 hr. The effect of antigen on blood monocytes was shown to be a function of the state of delayed hypersensitivity of the animals and was therefore a manifestation of cellular immunity.

Yoshida and Cohen (1974) were able to detect MIF in the serum of guinea pigs immunized by intramuscular injection of the BCG strain of tubercle bacilli and subsequently challenged intravenously with BCG. This procedure induces an intense state of delayed hypersensitivity associated with massive splenomegaly as well as peripheral lymphadenopathy. MIF activity was maximal at 6 to 12 hr which corresponds approximately to the period of greatest reduction in the number of circulating monocytes.

These experiments prompted studies of the effects of administering preformed exogenous lymphokines to experimental animals (Yoshida and Cohen, 1974). It was found that intravenous injection of MIF-containing, but not control, supernatants into normal animals resulted in a drop in circulating monocytes comparable to that observed in immunized animals challenged with specific antigen. In addition, it was noted that animals treated in this way developed anergy, i.e., they were transiently unable to mount a skin reaction to the test antigen.

Passive transfer of the desensitized state can be accomplished by systemic administration of serum from a desensitized animal. Physicochemical characterization of the responsible factor in the serum showed that it was nondialyzable, 10,000–45,000 molecular weight, and stable at 56°C. Recent studies on the mechanism of MIF-induced desensitization and anergy suggest that desensitization is a two-stage process (Yoshida *et al.*, 1983). Initially, high levels of circulating lymphokines block skin reactivity by preempting potential target cells and abolishing local chemotactic gradients. Later, when systemic lymphokine levels drop, desensitization is maintained by macrophage-derived soluble suppressor factors.

The detection of MIF in the serum of experimental animals raised the possibility that MIF could be found in human sera in the course of certain diseases. Diseases associated with anergy were first considered since exogenously administered MIF was known to suppress cutaneous anergy. Detectable serum MIF was found in patients with various lymphoproliferative disorders such as chronic lymphatic leukemia, multiple myeloma, Hodgkin's disease, non-Hodgkin's lymphoma, and the Sezary syndrome (Yoshida *et al.*, 1975d; Cohen *et al.*, 1974b). MIF was thus found in B-cell as well as T-cell disease in accordance with the observations already discussed on the ability of B cells to make lymphokines. Serum MIF has also been found with great frequency in patients with sarcoidosis (Yoshida *et al.*, 1978). It is tempting to speculate that one explanation for the cutaneous anergy observed in these diseases involves the presence of circulating lymphokines in analogy with studies in experimental animals.

Serum MIF is also detectable in the posttransplantation hepatic dysfunction syndrome (Torisu *et al.*, 1975). These patients have derangements in liver function that are often reflected in episodic variations in chemical parameters such as serum glutamic-oxaloacetic transaminase (SGOT) levels. For reasons that are unclear, such patients have transient elevations in serum MIF activity which precede elevations in SGOT. While the significance of this observation is unknown, the data document another situation is which a lymphokine activity may be detected *in vivo.*

Granulomatous inflammation is characteristic of a small group of diseases including tuberculosis, sarcoidosis, fungal infections, syphilis, cat-scratch fever, and actinomycosis. It is reasonable to assume that lymphokines may be involved in some, if not all, of these disease processes. Attempts have been made to recover soluble mediators from animals with experimental granulomas since the development of this kind of inflammation appears to be dependent on the particulate nature of the inducing agent and its ability to induce a state of cell-mediated immunity.

Fluids recovered from the granulomatous lungs of BCG-sensitized rabbits following intratracheal administration of antigen demonstrate a lymphokinelike activity (Galindo *et al.*, 1975). The lavage fluid contained a soluble factor that aggregated normal alveolar macrophages and had physical properties similar to macrophage aggregation factor (MAgF).

Masih and Yoshida (1978) have found that in guinea pigs injected with Freund's complete adjuvant, most of the lung parenchyma was replaced by multiple granulomas by 5 or 6 days following immunization. MIF and mitogenic activity were found in aqueous extracts of the lungs of these animals.

In another model of disease, Boros *et al.* (1973) reported that cultures of intact schistosome egg granulomas from mouse livers exposed to soluble antigen obtained from homogenized parasite eggs *in vitro* resulted in MIF production. James and Colley (1975) showed that intact schistosome egg granulomas maintained in culture without exogenous antigen released eosinophil stimulation promoter (ESP). It is unclear why exogenous antigen was required for MIF but not ESP production.

Antisera with specificity for lymphokines have been difficult to develop. They have, however, provided evidence for the *in vivo* role of soluble mediators. A major problem confronting the development of antilymphokine antibodies is the fact that culture fluids used as the mediator source contain multiple lymphokine activities as well as products of cell metabolism. Yoshida *et al.* (1975b) were able to circumvent this difficulty by utilizing a two-step immunization procedure. Partially purified lymphokine-containing supernatants were initially adsorbed with rabbit antiserum raised against corresponding control supernatant in order to reduce the level of contaminants common to both preparations. The lymphokine preparations purified in this manner were then used as immunogens to obtain antilymphokine serum (ALKS). These antiserum preparations were shown to remove MIF, MCF, and SRF (Yoshida *et al.*, 1975b), but not a number of other activities (Kuratsuji *et al.*, 1976) from antigen-activated guinea pig lymphocyte culture fluids. In addition, local injection of ALKS resulted in a 70% suppression of delayed hypersensitivity reactions in sensitized and immunized animals (Yoshida *et al.*, 1975b). The anticontrol antiserum had no effect on the evolution of skin lesions. It should be noted that ALKS exhibited no cytotoxic activity for lymphocytes.

Taking another approach, Geczy *et al.* (1975, 1976a) also prepared antisera with antilymphokine activity. These investigators partially purified lymphokines by performing a series of physicochemical separation procedures, including Sephadex G-200 and G-100 chromatography, Pevikon block electrophoresis, and polyacrylamide gel electrophoresis. The resulting fraction containing MIF, mitogenic factors, and SRF activities was used to produce ALKS in rabbits. This antiserum could completely absorb MIF from antigen-stimulated cultures, but had no antimitogenic or anti-SRF activity. As was the case in the study by Yoshida *et al.* (1975b), local injection of ALKS, but not control rabbit serum, around a skin test site completely ablated the delayed hypersensitivity response of an appropriately sensitized animal to specific antigenic challenge. Again, the ALKS had no toxic effects on lymphocytes. Subsequent studies using the immunoglobulin fraction of a "second-generation" antiserum (raised in goats by immunizing with purified mediators which eluted from ALKS immunosorbent column) showed that this immunoglobulin (ALyG) also could inhibit delayed hypersensitivity reactions *in vivo*. However, ALyG had no effect on nonspecific inflammatory responses. In addition, ALyG was shown to partially inhibit the initial inflammation associated with the intradermal transfer of allogeneic lymphocytes, and totally block the subsequent flare-up reaction (Hentges *et al.*, 1977).

In addition to being biologic probes for lymphokine activity *in vivo*, antilymphokine reagents have proved useful for investigating the molecular similarities and differences among mediators. Thus, Yoshida *et al.* (1975) have shown that a virus-induced MIF-like factor obtained from a nonlymphoid source (cytokine) could be adsorbed by ALKS and therefore shared antigenic determinants with conventional MIF. In contrast, MIF obtained from a continuously proliferating human cell line was recognized by this antiserum as antigenically distinct from guinea pig, antigen-induced MIF (Yoshida *et al.*, 1976).

Antilymphokine antibodies can also be used as tools to study the nature of the molecules that bind to them. Sorg and Geczy (1976), using radiolabeled lymphokines, reported that ALKS recognized three lymphocyte products with molecular weights of 60,000, 45,000, and 30,000. More recently, Onozaki *et al.* (1980) used radiolabeled lymphokines to study the material which reacted with an antibody preparation previously shown to bind specifically to MIF. Using this technique, MIF was identified as a 35,000-molecular-weight protein with a pI of 5–5.5.

8. CONCLUSION

There is little doubt that most of the experimental manifestations of cell-mediated immunity are mediated by lymphokines. Even in those instances where direct cell effects are observed (cytotoxic T cells, NK cells, cytotoxic macrophages), lymphokines are involved in inducing or expanding these populations. The general significance of lymphokines in the overall biology of the organism, however, remains to be determined. Those lymphokines involved in inflammation are the best defined and are the ones which have been examined in this chapter. It should be noted, however, that many of the properties of these lymphokines are those which would be required in growth and development. For example, factors affecting cell migration, cell adhesion, cell differentiation, and cell proliferation must play a role in embryogenesis. Thus, those regulatory factors that at present are only poorly understood are likely to be very similar to lymphokines functionally, and possibly chemically as well. The existence of cytokines, as discussed in this chapter, further attests to the general biologic significance of these phenomena.

The common denominator of the effector lymphokines is that they can mobilize, attract, and activate a variety of circulating cells that can participate in local inflammation. In addition, they can influence vascular permeability and the clotting system. Most importantly, lymphokines can amplify the effect of the interaction between sensitized lymphocytes and specific antigen and thus generate a biologically useful response.

On the basis of *in vitro* effects, the number of activities attributed to lymphokines is extremely large. However, not all of them have been correlated with *in vivo* activity and thus may not be biologically relevant. It is apparent that inflammatory lymphokines play a role *in vivo*. This comes from observations of cutaneous reactions including the detection of chemotactic lymphokines in skin extracts, the induction of inflammatory skin reactions by local injection of lymphokines, and the suppression of skin reactions by systemic administration of lymphokines or by the local administration of antilymphokine antibodies. It is not clear which lymphokine(s) is responsible for these effects. While many immunologists have assumed that these effects are due to the presence of MIF, there is no direct evidence for this. In fact, as has been noted, extracts of skin reactions have chemotactic but not migration inhibitory activity (Cohen *et al.*, (1973). On the other hand, MIF probably plays a role in other manifestations of cellular hypersensitivity. Direct evidence for this is seen in the macrophage disappearance reaction (Sonozaki and Cohen, 1971b).

With the exception of γ-interferon, we have not reached the stage where sources of purified effector lymphokines are available, but with the application of modern biotechnologic procedures, this should not be far off. Procedures utilized for afferent mediators such as interleukin-2 should have general applicability for all categories of lymphokines. When these reagents are prepared, it should be possible to define which lymphokines play a role *in vivo* and how they can be manipulated in host defense.

ACKNOWLEDGMENTS. Supported by NIH Grants AI-12477, AI-16706, and CA-32319.

REFERENCES

Aaskov, J. G., and Anthony, H. M., 1976, Assay of human lymphokines *in vitro*. Evidence for a migration stimulation factor (MStF) which interferes with the macrophage migration inhibition assay, *Aust. J. Exp. Biol. Med.* **54**:527.

Adelman, N., Cohen, S., and Yoshida, T., 1978, Strain variations in murine MIF production, *J. Immunol.* **121**:209.

Adelman, N. E., Ksiazek, J., Yoshida, T., and Cohen, S., 1980, Lymphoid sources of murine migration inhibition factor, *J. Immunol.* **124**:825.

Altman, L. C., and Kirchner, H., 1972, The production of a monocyte chemotactic factor by agammaglobulinemic chicken spleen cells, *J. Immunol.* **109**:1149.

Altman, L. C., and Kirchner, H., 1974, Mononuclear leukocyte chemotaxis in the chicken: Definition of a phylogenetically specific lymphokine, *Immunology* **26**:393.

Altman, L. C., Snyderman, R., Oppenheim, J. J., and Mergenhagen, S., 1973, A human mononuclear leukocyte chemotactic factor: Characterization, specificity, and kinetics of production by homologous leukocytes, *J. Immunol.* **110**:801.

Amsden, A., Ewan, V., Yoshida, T., and Cohen, S., 1978, Studies on cellular receptors for lymphokines. I. Interaction of chemotactic factors with monosaccharides, *J. Immunol.* **120**:542.

Arnason, B. G., and Waksman, B. H., 1963, The retest reaction in delayed sensitivity, *Lab. Invest.* **12**:737.

Bartfield, H., and Atoynatan, T., 1971, Activity and properties of macrophage migration inhibitory factor produced by mixed lymphocyte culture, *Nature New Biol.* **230**:246.

Bendtzen, K., 1976, Some physiochemical properties of human leucocyte migration inhibitory factor (LIF), *Acta Pathol. Microbiol. Scand.* **C84**:471.

Bendtzen, K., 1977a, Human leukocyte migration inhibition factor (LIF). I. Effect of synthetic and naturally occurring esterase and protease inhibitors, *Scand. J. Immunol.* **6**:126.

Bendtzen, K., 1977b, Human leukocyte migration inhibition factor (LIF). III. Further investigations on the serine protease nature of this lymphokine and its preference for arginine amides, *Scand. J. Immunol.* **6**:1055.

Bendtzen, K., 1977c, Human leukocyte migration inhibitory factor (LIF). IV. 3′,5′-cGMP protects LIF against inactivation by the esterase inhibitor phenylmethylsulfonyl fluoride, *Scand. J. Immunol.* **6**:1357.

Bennett, B., and Bloom, B. R., 1968, Reactions *in vivo* and *in vitro* produced by a soluble substance associated with delayed-type hypersensitivity, *Proc. Natl. Acad. Sci. USA* **59**:756.

Bloom, B. R., 1971, *In vitro* approaches to the mechanism of cell-mediated immune reactions, *Adv. Immunol.* **13**:101.
Bloom, B. R., and Bennett, B., 1966, Mechanisms of a reaction *in vitro* associated with delayed-type hypersensitivity, *Science* **153**:80.
Bloom, B. R., Gaffney, J., and Jimenez, L., 1972, Dissociation of MIF production and cell proliferation, *J. Immunol.* **109**:1395.
Bloom, B. R., Stoner, G., Gaffney, J., Shevach, E., and Green, I., 1975, Production of migration inhibitory factor and lymphotoxin by non-T cells, *Eur. J. Immunol.* **5**:218.
Boetcher, D. A., and Leonard, E. J., 1973, Basophil chemotaxis: Augmentation by a factor from stimulated lymphocyte cultures, *Immunol. Commun.* **2**:421.
Boros, D. L., Warren, K. S., and Pelley, R. P., 1973, The secretion of migration inhibitory factor by intact schistosome egg granulomas maintained *in vitro*, *Nature (London)* **246**:224.
Boult, D., Kühner, A. L., Klimetzek, V., Remold, H. G., and David, J. R., 1981, Fractionation of murine macrophage inhibitory factor into two distinct species, *Cell. Immunol.* **63**:198.
Boyden, S. V., 1962, The chemotactic effect of mixtures of antibody and antigen on polymorphonuclear leucocytes, *J. Exp. Med.* **115**:453.
Churchill, W. H., Jr., Piessens, W. F., Sulis, C. A., and David, J. R., 1975, Macrophages activated as suspension cultures with lymphocyte mediators devoid of antigen become cytotoxic for tumor cells, *J. Immunol.* **115**:781.
Clark, R. A. F., Gallin, J. I., and Kaplan, A. P., 1975, The selective eosinophil chemotactic activity of histamine, *J. Exp. Med.* **142**:1462.
Clausen, J. E., 1971, Tuberculin-induced migration inhibition of human peripheral leucocytes in agarose medium, *Acta Allergol.* **26**:56.
Clinton, B. A., Majoc, T. J., and Aspinall, R. L., 1974, The abrogation of macrophage migration inhibition by pre-treatment of immune exudate cells with anti-theta antibody and complement, *J. Immunol.* **112**:1741.
Coeugniet, E., and Bendixen, G., 1976a, Lymphokines and thrombosis. I. Thrombocyte aggregating activity released by human lymphocytes stimulated with concanavalin-A, *Acta Allergol.* **31**:94.
Coeugniet, E., and Bendixen, G., 1976b, Lymphokines and thrombosis. II. Procoagulant activity produced by human lymphocytes stimulated with concanavalin A (Con A), *Acta Allergol.* **31**:107.
Cohen, M. C., Picciano, P. T., Douglas, W. J., Yoshida, T., Kreutzer, D. L., and Cohen, S., 1982, Migration inhibition of endothelial cells by lymphokine-containing supernatants, *Science* **215**:301.
Cohen, S., 1977, The role of cell-mediated immunity in the induction of inflammatory responses, *Am. J. Pathol.* **88**:501.
Cohen, S., and Ward, P. A., 1971, *In vitro* and *in vivo* activity of a lymphocyte and immune complex-dependent chemotactic factor for eosinophils, *J. Exp. Med.* **133**:133.
Cohen, S., and Yoshida, T., 1977, Suppression of B cell MIF production by T cells and soluble T cell-derived factors, *J. Immunol.* **119**:719.
Cohen, S., Benacerraf, B., McCluskey, R. T., and Ovary, Z., 1967a, Effect of anticoagulants on delayed hypersensitivity reactions, *J. Immunol.* **98**:351.
Cohen, S., McCluskey, R. T., and Benacerraf, B., 1967b, Studies on the specificity of cellular infiltrates of delayed hypersensitivity reactions, *J. Immunol.* **98**:269.
Cohen, S., Ward, P. A., Yoshida, T., and Burek, C. L., 1973, Biologic activity of extracts of delayed hypersensitivity skin reaction sites, *Cell. Immunol.* **9**:363.
Cohen, S., Bigazzi, P. E., and Yoshida, T., 1974a, Commentary: Similarities of T-cell function in cell-mediated immunity and antibody production, *Cell. Immunol.* **12**:150.
Cohen, S., Fisher, B., Yoshida, T., and Bettigole, R. E., 1974b, Serum migration inhibitory activity in patients with lymphoproliferative diseases, *N. Engl. J. Med.* **290**:882.
Cohen, S., Rose, N. R., and Brown, R. C., 1974c, The appearance of eosinophils during the development of experimental autoimmune thyroiditis in the guinea pig, *Clin. Immunol. Immunopathol.* **2**:256.
Colvin, R. B., and Dvorak, H. F., 1975, Role of the clotting system in cell-mediated hypersensitivity. II. Kinetics of fibrinogen/fibrin accumulation and vascular permeability changes in tuberculin and cutaneous basophil hypersensitivity reactions, *J. Immunol.* **114**:377.
Colvin, R. B., and Dvorak, H. F., 1979, Role of granulocytes in cell-mediated immunity, in: *Mecha-*

nisms of Immunopathology (S. Cohen, P. A. Ward, and R. T. McCluskey, eds.), pp. 69–89, Wiley, New York.

Crowle, A. J., 1975, Delayed hypersensitivity in the mouse, *Adv. Immunol.* **20**:197.

David, J. R., 1966, Delayed hypersensitivity *in vitro:* Its mediation by cell-free substances formed by lymphoid cell–antigen interaction, *Proc. Natl. Acad. Sci. USA* **56**:72.

David, J. R., Al-Askari, S., Lawrence, H. S., and Thomas, L., 1964, Delayed hypersensitivity *in vitro*. I. The specificity of inhibition of cell migration antigens, *J. Immunol.* **93**:264.

Dumonde, D. C., Wolstencroft, R. A., Panayi, G. S., Matthew, M., Morely, J., and Howson, W. T., 1969, Lymphokines: Non-antibody mediators of cellular immunity generated by lymphocyte activation, *Nature (London)* **224**:38.

Dumonde, D. C., Page, D. A., Matthew, M., and Wolstencroft, R. A., 1972, Role of lymphocyte activation products (LAP) in cell-mediated immunity. I. Preparation and partial purification of guinea pig LAP, *Clin. Exp. Immunol.* **10**:25.

Dvorak, A. M., Hammond, M. E., Dvorak, H. F., and Karnovsky, M. J., 1972, Loss of cell surface material from peritoneal exudate cells associated with lymphocyte-mediated inhibition of macrophage migration from capillary tubes, *Lab. Invest.* **27**:561.

Ewan, V., Yoshida, T., and Cohen, S., 1980, Differential susceptibility of human and guinea pig lymphokines to reduction and alkylation, *Mol. Immunol.* **17**:1585.

Fauve, R. M., and Dekaris, D., 1968, Macrophage spreading: Inhibition in delayed hypersensitivity, *Science* **160**:795.

Gadol, N., and Waldman, R. H., 1973, Migration inhibitory factor induction by non-specific mitogenic stimulation of mouse lymphocytes, *J. Reticuloendothel. Soc.* **14**:398.

Galindo, B., Myrvik, Q. M., and Love, S. H., 1975, A macrophage agglutinating factor produced during a pulmonary delayed hypersensitivity reaction, *J. Reticuloendothel. Soc.* **18**:295.

Geczy, C. L., and Hopper, K. E., 1981, A mechanism of migration inhibition in delayed-type hypersensitivity reactions. II. Lymphokines promote pro-coagulant activity of macrophages *in vitro*, *J. Immunol.* **126**:1059.

Geczy, C. L., Friedrich, W., and de Weck, A. L., 1975, Production and *in vivo* effect of antibodies against guinea pig lymphokines, *Cell. Immunol.* **19**:65.

Geczy, C. L., Geczy, A. F., and de Weck, A. L., 1976a, Antibodies to guinea pig lymphokines. II. Suppression of delayed hypersensitivity reactions by a "second generation" goat antibody against guinea pig lymphokines, *J. Immunol.* **117**:66.

Geczy, C. L., Geczy, A. F., and de Weck, A. L., 1976b, Antibodies to guinea pig lymphokines. IV. Suppression of the mixed leukocyte culture reaction by anti-lymphokine globulin, *J. Immunol.* **117**:1824.

George, M., and Vaughan, J. H., 1962, *In vitro* cell migration as a model for delayed hypersensitivity, *Proc. Soc. Exp. Biol. Med.* **111**:514.

Gorski, A. J., 1974, Murine spleen leukocytes as an inhibitable target system for human MIF, *Cell Immunol.* **12**:315.

Hentges, F., Geczy, C. L., Geczy, A. F., and de Weck, A. L., 1977, Antibodies to guinea pig lymphokines. V. Suppression of normal lymphocyte transfer, immune lymphocyte transfer, and direct reactions in the guinea pig by anti-lymphokine globulin, *Immunology* **32**:905.

Higgins, T. J., Sabatino, A. P., Remold, H. G., and David, J. R., 1978, Possible role of macrophage glycolipids as receptors for migration inhibitory factor (MIF), *J. Immunol.* **121**:880.

James, S. L., and Colley, D. G., 1975, Eosinophils and immune mechanisms: Production of the lymphokine eosinophil stimulation promoter (ESP) *in vitro* by isolated intact granulomas, *J. Reticuloendothel. Soc.* **18**:283.

Jenssen, H. L., Redmann, K., and Köhler, H. F., 1975, The effect of a mediator of cellular immunity on the transmembrane potential of macrophages, *Acta Biol. Med. Ger.* **34**:1907.

Kay, A. B., and Austen, K. F., 1972, Chemotaxis of human basophil leukocytes, *Clin. Exp. Immunol.* **11**:557.

Kay, A. B., Stechschulte, D. J., and Austen, K. F., 1971, An eosinophil leukocyte chemotactic factor of anaphylaxis, *J. Exp. Med.* **133**:602.

Kotkes, P., and Pick, E., 1975, Studies on the inhibition of macrophage migration induced by soluble antigen–antibody complexes, *Clin. Exp. Immunol.* **19**:105.

Krejci, J., Svejcar, J., Pekarek, J., and Johanovsky, J., 1968, Demonstration of migration inhibiting

substances in sera of guinea pigs with delayed type hypersensitivity to ovalbumin influenced with systemic antigen administration, *Z. Immunitätsforsch.* **136**:259.
Kühner, A. L., and David, J. R., 1976, Partial characterization of murine migration inhibitory factor (MIF), *J. Immunol.* **116**:140.
Kühner, A. L., Cantor, H., and David, J. R., 1980, Ly phenotype of lymphocytes producing murine migration inhibitory factor (MIF), *J. Immunol.* **125**:1117.
Kuratsuji, T., Yoshida, T., and Cohen, S., 1976, Anti-lymphokine antibody. II. Specificity of biological activity, *J. Immunol.* **117**:1985.
Landolfo, S., Herberman, R. B., and Holden, H. T., 1977, Stimulation of mouse migration inhibition factor (MIF) production from MSV-immune lymphocytes by soluble tumor-associated antigen: Requirement for histocompatible macrophages, *J. Immunol.* **118**:1244.
Lavelle, K. J., Ransdell, B. A., and Trygstad, C. W., 1975, Identification of a new platelet aggregating factor released by sensitized lymphocytes, *Clin. Immunol. Immunopathol.* **3**:492.
Leber, P. D., Milgrom, M., and Cohen, S., 1973, Eosinophils in delayed hypersensitivity skin reaction sites, *Immunol. Commun.* **2**:615.
Lett-Brown, M. A., Boetcher, D. A., and Leonard, E. J., 1976, Chemotactic responses of normal human basophils to C5a and to lymphocyte-derived chemotactic factor, *J. Immunol.* **117**:246.
Leu, R. W., Eddleston, A. L., Hadden, J. W., and Good, R. A., 1972, Mechanism of action of migration inhibitory factors (MIF). I. Evidence for a receptor for MIF present on the peritoneal macrophage but not on the alveolar macrophage, *J. Exp. Med.* **136**:589.
Lykke, A. W. J., Willoughby, D. A., and Kosche, E. R., 1967, Thymic permeability factor: Its relationship to the lymph node permeability factor and its antagonism by pyridinol carbamate agents, *J. Pathol. Bacteriol.* **94**:381.
McCluskey, R. T., and Leber, P. D., 1974, Cell-mediated reactions *in vivo*, in: *Mechanisms of Cell-Mediated Immunity* (R. T. McCluskey and S. Cohen, eds.), pp. 1–24, Wiley, New York.
Mackler, B. F., Altman, L., Rosenstreich, D. L., and Oppenheim, J. J., 1974,Induction of lymphokine production by EAC and of blastogenesis by soluble mitogens during human B cell activation, *Nature (London)* **249**:834.
Maillard, J. L., Pick, E., and Turk, J. L., 1972, Interactions between "sensitized lymphocytes" and antigen *in vitro*. V. Vascular permeability induced by skin-reactive factor, *Int. Arch. Allergy Appl. Immunol.* **42**:50.
Manheimer, S., and Pick, E., 1973, The mechanism of action of soluble lymphocytic mediators. I. A pulse exposure test for the measurement of macrophage migration inhibitory factor, *Immunology* **24**:1027.
Masih, N., and Yoshida, T., 1978, Lymphokines in pulmonary granuloma, *Fed. Proc.* **37**:486.
Meade, C. J., Lachmann, P. J., and Brenner, S., 1974, A sensitive assay for cellular hypersensitivity based on the uptake of radioactive colloid gold, *Immunology* **27**:227.
Meltzer, M. S., 1976, Chemotactic response of mouse macrophages to culture fluids from mitogen-stimulated spleen cells, *Clin. Immunol. Immunopathol.* **6**:238.
Mooney, J. J., and Waksman, B. H., 1970, Activation of normal rabbit macrophage monolayers by supernatants of antigen-stimulated lymphocytes, *J. Immunol.* **105**:1138.
Nath, I., Poulter, L. W., and Turk, J. L., 1973, Effect of lymphocyte mediators on macrophages *in vitro:* A correlation of morphological and cytochemical changes, *Clin. Exp. Immunol.* **13**:455.
Nathan, C. F., Karnovsky, M. L., and David, J. R., 1971, Alterations of macrophage functions by mediators from lymphocytes, *J. Exp. Med.* **133**:1356.
Nathan, C. F., Remold, H. G., and David, J. R., 1973, Characterization of a lymphocite factor which alters macrophage functions, *J. Exp. Med.* **137**:275.
Nelson, D. S., 1965, The effect of anticoagulants and other drugs on cellular and cutaneous reactions to antigen in guinea pigs with delayed-type hypersensitivity, *Immunology* **9**:219.
Nelson, D. S., and Boyden, S. V., 1963, The loss of macrophages from peritoneal exudates following the injection of antigens into guinea pigs with delayed-type hypersensitivity, *Immunology* **6**:264.
Nelson, R. D., and Leu, R. W., 1975, Macrophage requirement for production of guinea pig migration inhibitory factor (MIF) *in vitro*, *J. Immunol.* **114**:606.
Ohisi, M., and Onoue, K., 1975, Functional activation of immune lymphocytes by antigenic stimula-

tion in cell-mediated immunity. I. Requirement for macrophages in antigen-induced MIF production by guinea pig immune lymphocytes *in vitro*, *Cell. Immunol.* **18**:220.

Ohisi, M., and Onoue, K., 1976, Functional activation of immune lymphocytes by antigenic stimulation in cell-mediated immunity. II. Analysis of the macrophage-replacing activity of the LPS-stimulated macrophage culture supernatant in the antigenic activation of purified immune lymphocytes, *Cell. Immunol.* **26**:295.

Onozaki, K., Haga, S., Miura, K., Homma, Y., and Hashimoto, T., 1980, Production of an antibody against guinea pig MIF. II. Analysis of the antibody-reacting material using radiolabelled lymphokines, *Cell. Immunol.* **55**:465.

Orr, W., and Ward, P. A., 1978, Quantitation of leukotaxis in agarose by three different methods, *J. Immunol. Methods* **20**:95.

Papageorgiou, T. S., Henley, W. L., and Glade, P. R., 1972, Production and characterization of migration inhibitory factor(s) (MIF) of established lymphoid and non-lymphoid cell lines, *J. Immunol.* **108**:494.

Pick, E., 1979, Mechanism of action of migration inhibitory lymphokines, in *The Biology of the Lymphokines* (S. Cohen, E. Pick, and J. J. Oppenheim, eds.), pp. 60–119, Academic Press, New York.

Pick, E., Krejci, J., Cech, K., and Turk, J. L., 1969, Interaction between sensitized lymphocytes and antigen release *in vitro*. I. The release of a skin reactive factor, *Immunology* **17**:741.

Possanza, G., Cohen, M. C., Yoshida, T., and Cohen, S., 1979, Human macrophage migration inhibition factor: Evidence for subunit structure, *Science* **205**:300.

Poste, G., 1975, Production of a serine protease with macrophage migration inhibitory factor activity by virus-transformed cells and human tumor cell lines, *Cancer Res.* **35**:2558.

Poste, G., Allen, H., and Matta, K. L., 1979a, Cell surface receptors for lymphokines. II. Studies on the carbohydrate composition of the MIF receptor on macrophages using synthetic saccharides and plant lectins, *Cell. Immunol.* **44**:89.

Poste, G., Kirsh, R., and Fidler, I. J., 1979b, Cell surface receptors for lymphokines. I. The possible role for macrophage migration inhibitory factor (MIF) and macrophage activation factor (MAF), *Cell. Immunol.* **44**:71.

Postlethwaite, A. E., and Kang, A. H., 1980, Characterization of guinea pig lymphocyte-derived chemotactic factor for fibroblasts, *J. Immunol.* **124**:1462.

Postlethwaite, A. E., Snyderman, R., and Kang, A. H., 1976, The chemotactic attraction of human fibroblasts to a lymphocyte-derived factor, *J. Exp. Med.* **144**:1188.

Remold, H. G., 1973, Requirements for alpha-L-fucose on the macrophage membrane receptor for MIF, *J. Exp. Med.* **138**:1065.

Remold, H. G., and Mednis, A. D., 1977, Two migration inhibitory factors with different chromatographic behavior and isoelectric point, *J. Immunol.* **118**:2015.

Remold, H. G., David, R. A., and David, J. R., 1972, Characterization of migration inhibitory factor (MIF) from guinea pig lymphocytes stimulated with concanavalin A, *J. Immunol.* **109**:578.

Remold, H. G., Mednis, A., Kawaguchi, T., Bersch, N., and Golde, D. W., 1983, Production of migration inhibitory factor by a human T lymphoblast cell line, *Cell. Immunol.* **78**:305.

Rich, A. R., and Lewis, M. R., 1932, The nature of allergy in tuberculosis as revealed by tissue culture studies, *Bull. Johns Hopkins Hosp.* **50**:115.

Rocklin, R. E., 1973, Production of migration inhibitory factor by non-dividing lymphocytes, *J. Immunol.* **110**:674.

Rocklin, R. E., 1974, Products of activated lymphocytes: Leukocyte inhibitory factor (LIF) distinct from migration inhibitory factor (MIF), *J. Immunol.* **112**:1461.

Rocklin, R. E., 1975, Partial characterization of leukocyte inhibitory factor by concanavalin A-stimulated human lymphocytes ($LIF_{con\ A}$), *J. Immunol.* **114**:1161.

Rocklin, R. E., 1976, Role of monosaccharides in the interaction of two lymphocyte mediators with their target cells, *J. Immunol.* **116**:816.

Rocklin, R. E., and David, J. R., 1971, Detection *in vitro* of cellular hypersensitivity to drugs, *J. Allergy Clin. Immunol.* **48**:276.

Rocklin, R. E., and Rosenthal, A. S., 1977, Evidence that human leukocyte inhibitory factor (LIF) is an esterase, *J. Immunol.* **119**:249.

Rocklin, R. E., Remold, H. G., and David, J. R., 1972, Characterization of human migration inhibitory factor (MIF) from antigen-stimulated lymphocytes, *Cell. Immunol.* **5**:436.

Rocklin, R. E., MacDermott, R. P., Chess, L., Schlossman, S. F., and David, J. R., 1974, Studies on mediator production by highly purified human T and B lymphocytes, *J. Exp. Med.* **140**:1303.

Ruhl, H., Vogt, W., Bochert, G., Schmidt, S., Moelle, R., and Schaoua, H., 1974, Effect of L-asparaginase and hydrocortisone on human lymphocyte transformation and production of a mononuclear leucocyte chemotactic factor *in vitro, Immunology* **26**:989.

Salvin, S. B., Sell, S., and Nichio, J., 1971, Activity *in vitro* of lymphocytes and macrophages in delayed hypersensitivity, *J. Immunol.* **104**:655.

Salvin, S., Younger, J. S., and Lederer, W. H., 1973, Migration inhibitory factor and interferon in the circulation of mice with delayed hypersensitivity, *Infect. Immun.* **7**:68.

Schiffmann, E., Corcoran, B. A., and Wahl, S. M., 1975, N-Formylmethionyl peptides as chemoattractants for leukocytes, *Proc. Natl. Acad. Sci. USA* **72**:1059.

Schild, H. O., and Willoughby, D. A., 1967, Possible pharmacological mediators of delayed hypersensitivity, *Br. Med. Bull.* **23**:46.

Schwartz, H. J., Leon, M. A., and Pelley, R. P., 1970, Concanavalin A-induced release of skin reactive factor from lymphoid cells, *J. Immunol.* **104**:265.

Snyderman, R., Altman, L. C., Hausman, M. S., and Mergenhagen, S. E., 1972, Human mononuclear leukocyte chemotaxis: A quantitative assay for humoral and cellular chemotactic factors, *J. Immunol.* **108**:857.

Sonozaki, H., and Cohen, S., 1971a, The effect of sensitized lymphocytes on peritoneal exudate macrophages in the guinea pig, *J. Immunol.* **106**:1404.

Sonozaki, H., and Cohen, S., 1971b, The macrophage disappearance reaction: Mediation by a soluble lymphocyte-derived factor, *Cell. Immunol.* **2**:341.

Sorg, C., 1980, Characterization of murine macrophage migration inhibitory activities (MIF), in *Biochemical Characterization of Lymphokines* (A. L. de Weck, F. Kristensen, and M. Landy, eds.), pp. 37–39, Academic Press, New York.

Sorg, C., and Geczy, C. L., 1976, Antibodies to guinea pig lymphokines. III. Reactions with radiolabelled lymphocyte activation products, *Eur. J. Immunol.* **6**:688.

Svejcar, J., Pekarek, J., and Johanovsky, J., 1971, Studies on the mechanism of delayed type hypersensitivity in tissue cultures. XIV. Species non-specificity of biologically active substances (migration inhibiting factor) released during the interaction of hypersensitive lymphocytes with antigen *in vitro, Z. Immunitaetsforsch. Allerg. Klin. Immunol.* **141**:119.

Thor, D. E., Jurezig, R. D., Veach, S. R., Miller, E., and Dray, S., 1968, Cell migration inhibition factor released by antigen from human peripheral lymphocytes, *Nature (London)* **219**:755.

Thorne, E. G., Kolsch, I., Schwartzmann, N., and Weston, W. L., 1975, *In vivo* effects of lymphokines: An ultrastructural study, *Clin. Res.* **23**:455a.

Thrasher, S. G., Yoshida, T., van Oss, C. J., Cohen, S., and Rose, N. R., 1973, Alteration of macrophage interfacial tension by supernatants of antigen-activated lymphocyte cultures, *J. Immunol.* **110**:321.

Torisu, M., Yoshida, T., Ward, P. A., and Cohen, S., 1973, Lymphocyte-derived eosinophil chemotactic factor. II. Studies on the mechanism of activation of the precursor substance by immune complexes, *J. Immunol.* **111**:1450.

Torisu, M., Yoshida, T., and Cohen, S., 1975, Serum migration inhibitory activity in patients with post-transplantation hepatic dysfunction, *Clin. Immunol. Immunopathol.* **3**:369.

Tubergen, D. G., Feldman, J. D., Pollock, E. M., and Lerner, R. A., 1972, Production of macrophage migration inhibition factor by continuous cell lines, *J. Exp. Med.* **135**:255.

Valdimarsson, H., and Gross, N. J., 1973, Human skin responses to products of concanavalin A-activated lymphocytes, *J. Immunol.* **111**:485.

Varesio, L., Holden, H., and Taramelli, D., 1980, Mechanism of lymphocyte activation. II. Requirements for macromolecular synthesis in the production of lymphokines, *J. Immunol.* **125**:2810.

Visakorpi, R., 1974, The effect of irradiation on macrophage migration inhibition, *Immunology* **27**:145.

Wahl, S. M., Altman, L. C., and Oppenheim, J. J., 1974a, *In vitro* studies of a chemotactic lymphokine in the guinea pig, *Int. Arch, Allergy Appl. Immunol.* **46**:1974.

Wahl, S. M., Iverson, G. M., and Oppenheim, J. J., 1974b, Induction of guinea pig B cell lymphokine synthesis by mitogenic and non-mitogenic signals to Fc, Ig and C3 receptors, *J. Exp. Med.* **140**:1631.

Wahl, S. M., Altman, L. C., and Rosenstreich, D. L., 1975a, Inhibition of *in vitro* lymphokine synthesis by glucocorticosteroids, *J. Immunol.* **115**:476.

Wahl, S. M., Wilton, J. M., Rosenstreich, D. L., and Oppenheim, J. J., 1975b, The role of macrophages in the production of lymphokines by T and B lymphocytes, *J. Immunol.* **114**:1296.

Waksman, B. H., 1979, Overview: Biology of the lymphokines, in: *The Biology of the Lymphokines* (S. Cohen, E. Pick, and J. J. Oppenheim, eds.), pp. 585–616, Wiley, New York.

Ward, P. A., and Volkman, A., 1975, The elaboration of leukotactic mediators during the interaction between parental-type lymphocytes and F_1 hybrid cells, *J. Immunol.* **115**:1394.

Ward, P. A., Remold, H. G., and David, J. R., 1969, Leukotactic factor produced by sensitized lymphocytes, *Science* **163**:1079.

Ward, P. A., Remold, H. G., and David, J. R., 1970, The production by antigen-stimulated lymphocytes of a leukotactic factor distinct from migration inhibitory factor, *Cell. Immunol.* **1**:162.

Ward, P. A., Offen, C. D., and Montgomery, J. R., 1971, Chemoattractants of leukocytes with special reference to lymphocytes, *Fed. Proc.* **30**:1721.

Ward, P. A., Dvorak, H. F., Cohen, S., Yoshida, T., Data, R., and Selvaggio, S., 1975, Chemotaxis of basophils by lymphocyte-dependent and lymphocyte-independent mechanisms, *J. Immunol.* **114**:1523.

Ward, P. A., Unanue, E. R., Goralnick, S. J., and Schreiner, G. F., 1977, Chemotaxis of rat lymphocytes, *J. Immunol.* **119**:416.

Weisbart, R. H., Bluestone, R., Goldberg, L. S., and Pearson, C. M., 1974, Migration enhancement factor: A new lymphokine, *Proc. Natl. Acad. Sci. USA* **71**:875.

Weisbart, R. H., Golde, D. W., Spolter, L., Eggena, P., and Rinderknecht, H., 1980, Neutrophil migration inhibition factor from T lymphocytes (NIF-T): A new lymphokine, in: *Biochemical Characterization of Lymphokines* (A. L. de Weck, F. Kristensen, and M. Landy, eds.), pp. 105–107, Academic Press, New York.

Weiser, W. Y., Greineder, D. K., Remold, H. G., and David, J. R., 1981, Studies on human migration inhibitory factor: Characterization of three molecular species, *J. Immunol.* **126**:1958.

Weiss, L., and Glaves, D., 1975, Effects of migration inhibiting factor(s) on the *in vitro* detachment of macrophages, *J. Immunol.* **115**:1362.

Willoughby, D. A., Boughton, B., Spector, W. G., and Schild, H. O., 1962, A vascular permeability factor extracted from normal and sensitized guinea pig lymph node cells, *Life Sci.* **1**:347.

Willoughby, D. A., Spector, W. G., and Boughton, B., 1964, A lymph node permeability factor in the tuberculin reaction, *J. Pathol. Bacteriol.* **87**:353.

Woody, J. N., Ahmed, A., Knudsen, R. C., Strong, D. M., and Sell, K. W., 1975, Human T cell heterogeneity as delineated with a specific human thymus lymphocyte antiserum: *In vitro* effects on mitogen responses, mixed leukocyte culture, cell-mediated lymphocytotoxicity, and lymphokine production, *J. Clin. Invest.* **55**:956.

Yamamoto, K., and Takahashi, Y., 1971, Macrophage migration inhibition by serum from desensitized animals previously sensitized with tubercle bacilli, *Nature New Biol.* **233**:261.

Yoshida, T., and Cohen, S., 1974, Lymphokine activity *in vivo* in relation to circulating monocyte levels and delayed skin reactivity, *J. Immunol.* **112**:1540.

Yoshida, T., and Reisfeld, R. A., 1970, Two fractions with macrophage inhibitory activity from sensitized lymphocyte cultures, *Nature (London)* **226**:856.

Yoshida, T., Benacerraf, B., McCluskey, R. T., and Vassalli, P., 1969, The effect of intravenous antigen on circulating monocytes in animals with delayed hypersensitivity, *J. Immunol.* **102**:804.

Yoshida, T., Nagai, R., and Hashimoto, T., 1973a, Lack of species specificity of a skin-reactive factor released from sensitized guinea pig spleen cells, *Lab. Invest.* **29**:329.

Yoshida, T., Sonozaki, H., and Cohen, S., 1973b. The production of migration inhibition factor by B and T cells of the guinea pig, *J. Exp. Med.* **138**:784.

Yoshida, T., Bigazzi, P. E., and Cohen, S., 1975a, Biologic and antigenic similarity of virus-induced migration inhibition factor to conventional, lymphocyte-derived migration inhibition factor, *Proc. Natl. Acad. Sci. USA* **72**:1641.

Yoshida, T., Bigazzi, P. E., and Cohen, S., 1975b. The production of anti-guinea pig lymphokine antibody, *J. Immunol.* **114**:688.

Yoshida, T., Cohen, S., Bigazzi, P. E., Kuratsuji, T., and Amsden, A., 1975c, Inflammatory mediators in culture filtrates of *Escherichia coli*, *Am. J. Pathol.* **81**:389.

Yoshida, T., Edelson, R., Cohen, S., and Green, I., 1975d, Migration inhibitory activity in serum and cell supernatants in patients with Sezary syndrome, *J. Immunol.* **114**:915.

Yoshida, T., Kuratsuji, T., Takada, A., Takada, Y., Minowada, J., and Cohen, S., 1976, Lymphokine-like factors produced by human lymphoid cell lines with B or T cell surface markers, *J. Immunol.* **117**:548.

Yoshida, T., Siltzbach, L. E., and Cohen, S., 1978, Migration inhibitory activity in serum of patients with sarcoidosis, in: *8th Int. Sarcoidosis Conf. (Abstr.)*, Cardiff, Wales.

Yoshida, T., Suko, M., and Cohen, S., 1983, Antigen-specific and non-specific regulation of lymphokine production, in: *Interleukins, Lymphokines and Cytokines* (J. J. Oppenheim and S. Cohen, eds.), pp. 391–394, Academic Press, New York.

Zschiesche, W., Eisenhut, I., and Küchler, W., 1972, Identification and characterization of skin reactive factor (SRF) in supernatants of human lymphocyte cultures stimulated by PHA, *Z. Immunitätsforsch.* **143**:377.

10

Macrophage Procoagulants, Fibrin Deposition, and the Inflammatory Response

RICHARD L. EDWARDS, VEETA A. EWAN, and FREDERICK R. RICKLES

1. INTRODUCTION: THE INFLAMMATORY RESPONSE

In 1938, Menkin provided a comprehensive view of the inflammatory response: "Inflammation may be broadly defined as the complex vascular, lymphatic, and local tissue reaction elicited in higher animals by the presence of microorganisms or of nonviable irritants. It represents a basic reaction to injury, in which the deleterious agent tends to be localized and ultimately destroyed." Menkin characterized several phases in the development of the inflammatory lesion beginning with increased fluid passage through the capillary endothelial wall and the formation of a fibrinous network which immobilizes the irritant. The lesion is then completed by the development of local thrombi, occlusion of draining lymphatics, and subsequent infiltration of leukocytes. Menkin proposed that the early fixation of the irritant allowed an interval during which leukocytes could assemble for the purpose of phagocytosis. Furthermore, these three phases—local fluid influx, fixation in a focal fibrinous meshwork, and progressive leukocyte infiltration—represent a necessary prelude to the development of immunity.

The presence of fibrin within sites of inflammation was recognized over a century ago (Addison, 1841, 1842). Recent studies have supported historical observations that the deposition of fibrin is a common consequence of the inflammatory response. Furthermore, it has become clear that the cellular component of the reaction is central to both the deposition of extravascular fibrin within inflammatory lesions and the subsequent resolution of fibrin by fibrinoly-

RICHARD L. EDWARDS, VEETA A. EWAN, and FREDERICK R. RICKLES • Department of Medicine, University of Connecticut School of Medicine, Farmington, Connecticut 06032; and the Veterans Administration Medical Center, Newington, Connecticut 06111.

sis. While the role of the neutrophil in this sequence of events remains poorly understood, it is now generally accepted that cells of the monocyte-macrophage system generate cell-focused and cell-derived procoagulants which contribute to fibrin deposition in the inflammatory response. This review will examine the evidence for the participation of immune cells and mediators in the evolution, expression, and regulation of coagulation within an inflammatory reaction, with a particular emphasis on the role of the monocyte. Finally, we will examine the evidence which supports the importance of monocyte-focused coagulation in the development of a variety of inflammatory disorders. The reader is referred to other reviews of the interrelationships among coagulation, fibrinolysis, complement activation, kinins, and the cellular components of the inflammatory response (Ratnoff, 1969; Zimmerman, 1976; Dunn and Willoughby, 1981; Sundsmo and Fair, 1983; Geczy, 1983).

2. FIBRIN AND INFLAMMATION

Fibrin deposition is a common feature of inflammatory reactions to nonspecific stimuli such as burns (Glenn *et al.*, 1942; Menkin, 1933), infection (Menkin, 1938), and injury inflicted by trauma, chemicals, or radiation (Movat *et al.*, 1960). Fibrin deposition is also a common feature of the antigen-specific, immune-mediated response of delayed-type hypersensitivity (DTH) (Colvin and Dvorak, 1975b; Colvin *et al.*, 1973). Models of each of these types of inflammation will be considered below.

One of the most extensively studied experimental models of "nonspecific" fibrin deposition is the generalized Shwartzman reaction (GSR), a thrombohemorrhagic reaction produced by two sequential intravenous injections of endotoxin. In the classic Shwartzman reaction, the provoking dose of endotoxin is separated from the preparative dose by 18–24 hr. Glomerular thrombosis, capillary thrombosis of other organs, and secondary renal cortical necrosis are most profound in rabbits (Apitz, 1935), but features of intravascular coagulation common to the GSR are also observed in other experimental animals. Endotoxin [bacterial lipopolysaccharide (LPS)] is known to activate many of the components of the inflammatory response including complement, kinins, platelets, and coagulation factors (McKay *et al.*, 1959; Horowitz *et al.*, 1962; Morrison and Cochrane, 1974; Morrison and Ulevitch, 1978; Sundsmo and Fair, 1983). Moreover, disseminated intravascular blood coagulation can occur *in vivo* as a component of the response to bacterial sepsis in man (McGehee *et al.*, 1967; Evans *et al.*, 1969).

Studies in experimental animals undergoing the GSR, utilizing intravenous infusions of radiolabeled fibrinogen to quantify fibrin deposition *in vivo*, have demonstrated accumulation of isotope in the spleen, lungs, kidney, and liver. Anticoagulation with warfarin or heparin inhibits fibrin deposition and also impairs the accumulation of isotope in this experimental model (Lipinski *et al.*, 1974). The typical histological findings of glomerular fibrin deposition and capillary thrombosis of other organs (McKay *et al.*, 1959) can also be prevented by anticoagulation with heparin (Good and Thomas, 1953) or warfarin (Shapiro and McKay, 1958) before inducing the GSR.

Fibrin deposition, consequent to specific antigen stimulation, has been rigorously examined in DTH reactions in guinea pigs and humans. Biopsies of contact dermatitis in humans and intradermal delayed-type skin reactions in normal guinea pigs and humans are also characterized by fibrin deposition (Colvin and Dvorak, 1975b; Colvin *et al.*, 1973). Radiolabeled fibrinogen was found to be deposited as urea-insoluble fibrinogen derivatives at the site of developing tuberculin reactions over the interval from 3 to 18 hr after intradermal injection of PPD in sensitized animals (Colvin and Dvorak, 1975b). Local fibrin deposition at skin test sites was inhibited by prior treatment with anticoagulants (Colvin and Dvorak, 1975b). Skin test reactions in afibrinogenemic individuals (Colvin *et al.*, 1979), and in individuals anticoagulated with warfarin (Edwards and Rickles, 1978), exhibit erythema and mononuclear cell infiltrates, but lack characteristic induration. These findings suggest that fibrin deposition is directly related to and possibly responsible for the induration of delayed-type cutaneous hypersensitivity reactions.

Several other immune-related reactions are characterized by fibrin deposition, and a subset of these can be modified by anticoagulants. The renal allograft rejection reaction, for example, is accompanied by evidence of intravascular coagulation, and fibrin deposition in the rejected organ (Kincaid-Smith, 1967; Starzl *et al.*, 1968). Immunoglobulin, complement, and fibrinogen have been immunochemically localized in the lesions of acute, subacute, and chronic glomerulonephritis (Koffler and Paronetto, 1965). Anticoagulants may have a beneficial effect in some renal disorders characterized by glomerular and/or vascular lesions (Kincaid-Smith *et al.*, 1970). Other immune-related diseases associated with fibrin deposition include experimental allergic encephalomyelitis (Paterson, 1982), rheumatoid arthritis (Zvaifler, 1973), systemic lupus erythematosus (Paronetto and Koffler, 1965), immediate hypersensitivity of the IgE type (De Shazo *et al.*, 1979), as well as the Arthus reaction (Stetson, 1951).

3. CELLS, FIBRIN, AND INFLAMMATION

Several coagulation pathways have been implicated in the pathogenesis of extravascular fibrin deposition. Increased capillary permeability observed in inflammatory lesions may expose the contact factors of plasma to activators such as collagen in the extravascular space (Ratnoff, 1969; Zimmerman, 1976). Alternatively, inflammatory cells associated with the lesion may supply one or more procoagulants capable of activating clotting further down the "cascade." Early studies, primarily utilizing the GSR as the inflammatory model, implicated the granulocyte as the source of cellular procoagulant activity (PCA). However, more recent studies have suggested that the primary cell involved in supplying PCA in this and other cellular immune reactions is the monocyte/macrophage. In this section we will review the data supporting this concept.

In vivo studies of the Shwartzman reaction have supported a role for white blood cells in the development of the GSR. Animals rendered neutrophenic by treatment with nitrogen mustard (Stetson and Good, 1951) or busulfan (Müller-Berghaus *et al.*, 1976) failed to mount a GSR in response to LPS. Moreover, radiolabeled fibrin deposition did not occur in such neutropenic animals following *in vivo* exposure to LPS (Lipinski *et al.*, 1974). It is of interest that the GSR can be induced by a single dose of endotoxin in rabbits rendered neutropenic with nitrogen mustard (Forman *et al.*, 1969), if the endotoxin injection is accompanied by the passive transfer of leukocytes. Further, it appears that *in vitro* exposure of leukocytes to LPS prior to intravenous injection produces the same results (Horn and Collins, 1968; Niemetz and Fani, 1971; Kociba and Griesemer, 1972).

Several authors have suggested that the elaboration of cellular procoagulants *in vivo* or *in vitro*, which are discussed in detail in Section 4, is the mechanism by which "LPS-activated leukocytes" elicit the Shwartzman reaction (Niemetz and Fani, 1971; Lerner *et al.*, 1971; Kociba and Giesemer, 1972; Thiagarjan and Niemetz, 1980; Rothberger *et al.*, 1983). The earlier studies suggested that the neutrophil (or granulocyte) was the cell responsible for the generation of PCA. Indeed, Lerner *et al.* (1977) reported that peripheral blood cells from a patient with agranulocytosis failed to produce PCA on exposure to LPS *in vitro.* These results supported the previous studies on experimental animals. More recent experiments have provided an alternative cellular pathway for the mediation of LPS-induced PCA generation (Rickles and Rick, 1977; Edwards *et al.*, 1979; Rothberger *et al.*, 1983; Østerud and Flaegstad, 1983). These studies have demonstrated both *in vitro* and *in vivo* expression of *monocyte* PCA in the presence of trace concentrations of LPS. Mononuclear cells recovered from the bone marrow, spleen, or blood of animals undergoing the GSR exhibit potent PCA (Rothberger *et al.*, 1983; Thiagarjan and Niemetz, 1980). It is conceivable that LPS activation of monocytes rather than neutrophils may have been responsible for the coagulation abnormalities observed in the earlier studies of the GSR. This view is supported by the failure of a heterologous antineutrophil antibody to inhibit the GSR (Margaretten and McKay, 1969) and by the failure to demonstrate PCA in purified populations of LPS-activated neutrophils (Rickles *et al.*, 1983).

It is of interest that the granulocyte-enriched populations used by Kociba and Griesemer (1972) to induce a GSR-like lesion in rabbits may have been contaminated with as many as 4.6×10^6 mononuclear cells, more than enough to account for the observed leukocyte PCA. It is difficult to interpret older studies which utilized chemotherapeutic drugs or heteroantisera to produce granulocytopenia since these agents may also impair the production and/or function of mononuclear cells. Similarly, the case study reported by Lerner *et al.* (1977), in which cells from a patient with agranulocytosis were assayed for PCA, did not include an assessment of mononuclear cell function. The absence of demonstrable leukocyte PCA in that patient may have reflected a more generalized abnormality affecting cells of both the granulocyte and monocyte series, which share a common stem cell. Thus, the studies performed to date have not completely answered the question of whether the cellular PCA associated with the GSR is primarily generated by granulocytes, monocytes, or both.

Other experimental models have provided additional support for the role of leukocyte procoagulants in the inflammatory response. The macrophage disappearance reaction (MDR) (Nelson and Boyden, 1963) constitutes one of the best *in vivo* models for DTH. In this model, sterile glycogen or paraffin oil is injected intraperitoneally and induces an infiltration of macrophages, which peaks between 3 and 4 days after injection (Cohn and Benson, 1965). When antigen is then injected into a hypersensitive animal, peritoneal macrophages adhere to one another and rapidly disappear from the peritoneal cavity. The MDR is inhibited in animals pretreated with the anticoagulant drugs heparin or warfarin (Nelson, 1965) and can be mimicked by the intraperitoneal injection of thrombin (Jokay and Karczag, 1973). These experimental results suggest that activation of the coagulation system may play a central role in the process of macrophage adherence.

Using immunofluorescent microscopy and an antiserum reactive with fibrinogen, fibrin, and fibrin degradation products (Fib), Colvin and Dvorak (1975a) examined the antibody-binding patterns of peritoneal exudate cells from guinea pigs. Exudate cells from normal guinea pigs exhibited distinctive fine-speckled or fibrillar netlike staining patterns. By alternating between the use of fluorescence and phase microscopy, most of the Fib-positive cells could be identified as macrophages. Notably, small lymphocytes, granulocytes, and erythrocytes never exhibited Fib staining, although occasional cells intermediate in size between typical macrophages and small lymphocytes were found.

When peritoneal exudate cells were recovered from warfarin-anticoagulated guinea pigs and assessed for Fib-binding, the netlike pattern was absent, while the frequency of the speckled pattern almost doubled. These results suggested that the netlike pattern of Fib-binding represented cell-associated fibrin generation while the speckled pattern represented the binding of fibrinogen to cells without the formation of fibrin.

The MDR has also been studied using radiolabeled fibrinogen *in vivo*. Hopper *et al.* (1981) demonstrated by SDS–PAGE that eluates of macrophages from immunized animals contained radiolabeled fibrin, indicating the ability of these cells to participate in the proteolytic cleavage of the injected fibrinogen. Cells from nonimmune animals failed to convert fibrinogen to fibrin. The immune specificity of macrophage-associated fibrin deposition has been examined more closely. Peritoneal macrophages from unimmunized animals, or from immune animals challenged with an irrelevant antigen *in vivo* bound little labeled fibrinogen. Moreover, macrophages from these same animals exhibited patchy, granular patterns of fibrin(ogen) binding when exposed to an irrelevant antigen *in vitro*. Peritoneal macrophages recovered from animals sensitized with BCG and boosted with PPD, or animals sensitized and boosted with ovalbumin, bound radiolabeled fibrinogen *in vivo* and demonstrated a typical netlike pattern of cell surface fibrin-related immunofluorescence when exposed to appropriate antigen *in vitro* (Hopper *et al.*, 1981). In further studies of this model, Geczy and Hopper (1981) showed that a quantitative relationship existed between the amount of PCA elaborated by the peritoneal exudate cells and the binding of [^{125}I]fibrinogen. They observed a direct correlation between macrophage migration inhibition in response to specific antigens and macrophage PCA generation.

4. LEUKOCYTE PROCOAGULANT ACTIVITY, THE GENERATION OF FIBRIN, AND INFLAMMATION

In the previous section we examined evidence that the generation of fibrin is integral to the expression of both nonspecific inflammatory reactions (e.g., response to LPS) and specific inflammatory reactions (e.g., DTH) and suggested a potential role for monocytes in the mediation of these events. This section will examine characteristics of leukocyte-associated procoagulants and their role in the pathogenesis of fibrin deposition.

4.1. CELL OF ORIGIN OF LEUKOCYTE PROCOAGULANT ACTIVITY

While most authors now agree that the monocyte is the cell responsible for the *in vitro* generation of PCA observed in studies of mixed leukocyte populations, there remains some controversy regarding the potential for neutrophil expression of PCA *in vivo.* The *in vivo* studies of the GSR (Section 2) fail to provide clear evidence supporting the ability of neutrophils to generate PCA. Studies of patients with acute progranulocytic leukemia (APL), however, have added some support to the hypothesis that cells of the granulocyte lineage may contain PCA. Disseminated intravascular coagulation (DIC) occurs with a high frequency in patients with APL, particularly following the initiation of therapy (Drapkin *et al.*, 1978). This increased incidence of DIC has been attributed to the release of PCA from the granular fraction of the leukemic promyelocytes (Gralnick *et al.*, 1972; Gralnick and Abrell, 1973; Gralnick and Sultan, 1975; Gouault-Heilman *et al.*, 1975). Extracts of blood cells from patients with APL contain a PCA which is factor VII-dependent, heat-labile, and partially neutralized by a heterologous antiserum raised against a crude preparation of human brain tissue factor (TF) (Gouault-Heilman *et al.*, 1975). While these studies suggest that cells of the granulocyte series contain a granule-associated PCA with the properties of TF, it must be noted that purified populations of leukemic promyelocytes were not utilized and therefore one cannot exclude the possibility that the PCA might have been produced by monocytes contaminating the leukemic population. Alternatively, malignant promyelocytes, defined as cells of the neutrophil lineage by morphology and cytochemistry, may express phenotypic characteristics of monocytes (Kornberg *et al.*, 1982). The specificity of the heterologous anti-TF antibody used in these studies was unclear and further complicates the interpretation of the PCA data in these experiments.

Studies of the properties of cell lines derived from malignant neoplasms have provided some further insight into the origins of cell-derived procoagulants. The generation of TF in cultures of the inducible cell line HL-60 supports the conclusion that the monocyte/macrophage is the cell responsible for leukocyte PCA generation. The HL-60 cell line was derived from human progranulocytic leukemia cells and HL-60 cells generate TF PCA *only* when induced to develop monocyte or macrophage characteristics (Kornberg *et al.*, 1982, 1983; Rickles *et al.*, 1983). Thus, although monocytes share a common stem cell with neutrophils (Cline, 1977), only the monocyte appears to develop the capacity to generate PCA.

Several studies utilized purified populations of peripheral blood mononuclear cells in an attempt to better define the nature of the leukocytes which generate PCA. Rivers *et al.* (1975), Hiller *et al.* (1977), and Edwards *et al.* (1979) showed that LPS induces the elaboration of a TF-like PCA in partially purified populations of human and rabbit (Edwards *et al.*, 1985) monocytes *in vitro.* Single-cell PCA assays utilizing cells dispersed in plasma-containing media have shown that macrophages serve as a focus for fibrin deposition *in vitro.* These studies utilized histochemical staining (Levy and Edgington, 1980) or a fibrin-specific monoclonal antibody (Hogg *et al.*, 1981; Hogg, 1983a) (Fig. 1) to demonstrate fibrin strands deposited around both viable and fixed monocytes. The single-cell studies demonstrated that the macrophage could serve as a focus of fibrinogen binding and fibrin generation *in vitro.*

Lymphocytes and neutrophils failed to generate measurable PCA in the studies utilizing purified subpopulations of mononuclear cells and failed to bind fibrin in the single-cell assays. We have confirmed these earlier studies utilizing more highly purified populations of human monocytes (> 99%) and neutrophils (> 99%) (Rickles *et al.*, 1983). Monocytes expressed TF PCA following exposure to endotoxin *in vitro* whereas neutrophils failed to develop any significant PCA following stimulation with endotoxin (Fig. 2). Although we cannot say that neutrophils are unrelated to the activation of coagulation and fibrinolysis in the inflammatory response, there is at present no convincing evidence that they supply PCA.

Recent data, however, provide circumstantial evidence for the participation of one or more neutrophil-derived neutral proteases in the pathological activation of blood coagulation *in vivo.* Neutral proteases from neutrophilic leukocyte granules can cleave (and potentially activate) a variety of coagulation substrates, including factor XIII, fibrinogen, prothrombin, factors V, X, VII, IX, VIII, and XII (Schmidt *et al.*, 1974; Plow and Edgington, 1975; Henriksson and Nilsson, 1979; Karpaki *et al.*, 1982). Conversely, physiologic concentrations of coagulation factors, formed during the *in vitro* activation of coagulation, have been shown to stimulate the release of these same leukocyte proteases (Plow, 1982; Wachtfogel *et al.*, 1983). Leukocyte proteases have also been demonstrated in the circulation of patients with leukemia and/or septicemia in association with DIC (Egbring *et al.*, 1977; Brower and Harpel, 1983), although the pathophysiologic significance of these observations remains uncertain. Early studies utilizing various crude plasma assay systems *in vitro* suggested that these enzymes may act as procoagulants (Kingdon *et al.*, 1978). However, recent data utilizing purified proteins in more specific assays suggest that the procoagulant function reported for at least one enzyme, granulocyte elastase, might have been an artifact. The interaction of leukocyte elastase with purified human factor IX does *not* lead to

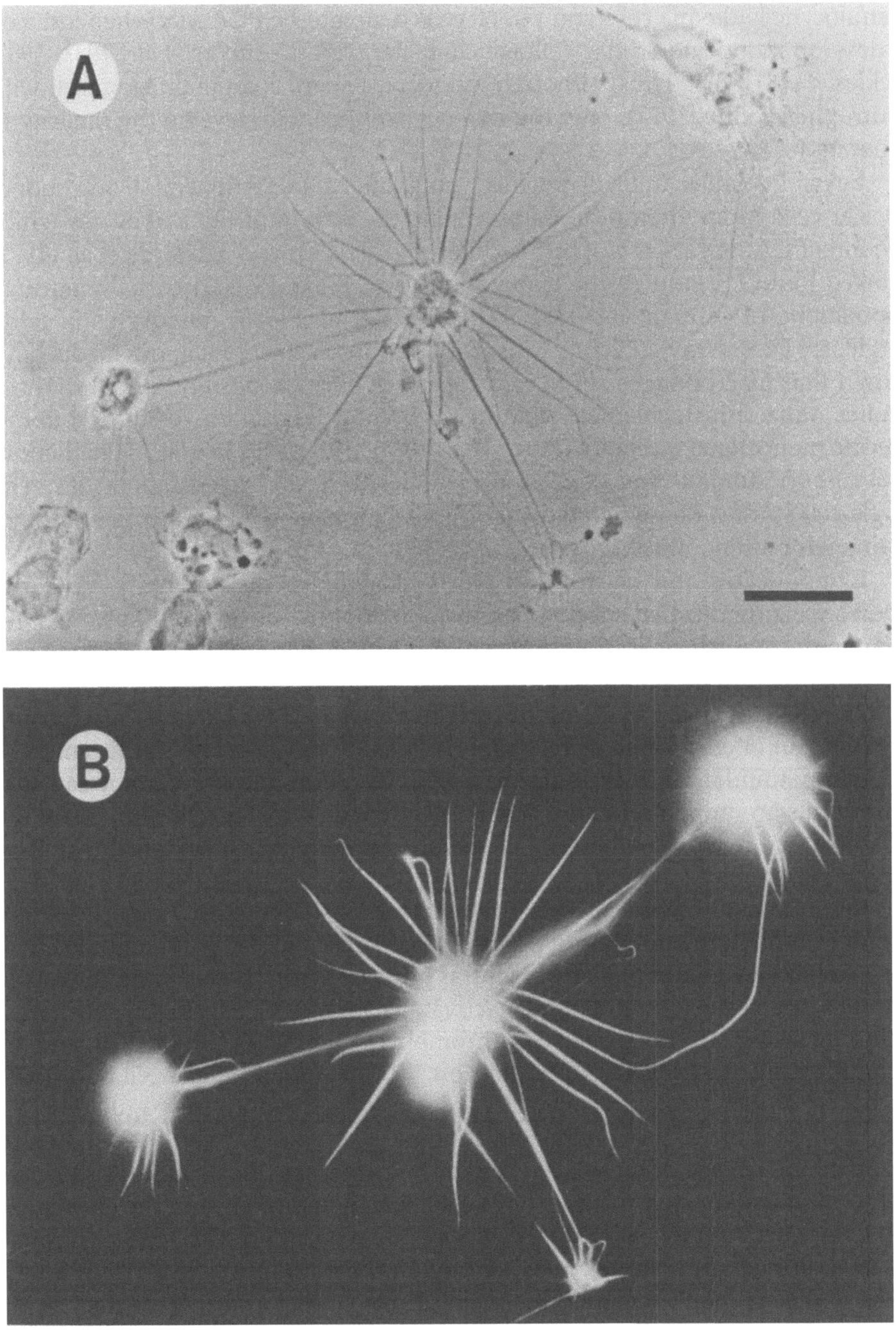

FIGURE 1. Human monocytes are associated with the formation of fibrin. A human monocyte displaying needlelike projections formed after a 10-min incubation with fresh human plasma shown by phase microscopy (A) and fluorescence microscopy (B) after reaction with antifibrin monoclonal antibody. (From Hogg, 1983a, with permission.)

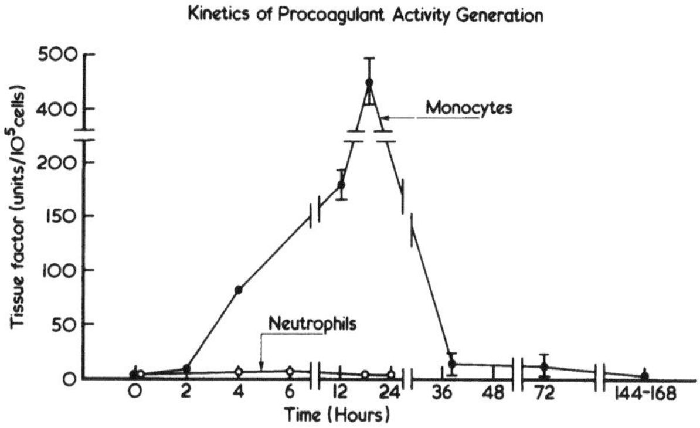

FIGURE 2. Expression of tissue factor procoagulant activity (PCA) in human blood cells. Enriched populations of peripheral blood monocytes (70–95% pure) and neutrophils (> 99% pure) were grown in tissue culture and exposed to LPS (10 μg/ml). PCA was determined on intact, viable cells at each of the time points indicated on the horizontal axis. PCA is expressed in units per 10^5 cells as compared to a rabbit brain TF standard, where a clotting time of 60 sec is equivalent to 100 units of activity. The height of the error bars represent the standard error of the mean of at least three cell cultures. (From Edwards and Rickles, 1984, with permission.)

the release of an activation peptide and subsequent activation of factor X as had been suggested previously. Under a variety of conditions *in vitro*, elastase failed to activate factor X but instead rapidly cleaved the protein, rendering it inactive in clotting assays (Takaki *et al.*, 1983). It is likely, therefore, that many of the observations of PCA in studies of neutrophil populations will need to be reexamined using purified systems—homogeneous cell populations, enzymes, and clotting substrates. At this time, therefore, insufficient evidence exists to implicate any blood cell other than the monocyte in the generation of leukocyte PCA!

4.2. CHARACTERIZATION OF LEUKOCYTE PROCOAGULANTS

It is now clear from the fibrin-specific immunofluorescence studies of Hogg (1983a, b) and the single-cell fibrin-plaque assay results of Levy and Edgington (1980) that PCA generation leads directly to the formation of fibrin on the surface of monocyte/macrophages. However, the nature of the PCA(s) generated by monocytes is less clear. Apparently different procoagulants have been described in cells from different species, different anatomic locations, stimulated by different agonists under different conditions either *in vitro* or *in vivo*.

Procoagulants may initiate blood coagulation at one of several steps because of the multiple enzymatic reactions involved in the coagulation cascade. In brief, blood coagulation may be activated physiologically either by the intrinsic contact activation sequence or by the extrinsic (TF-dependent) sequence. The intrinsic cascade is initiated by exposure of Hageman factor (factor XII) to a foreign surface with the subsequent cleavage and activation of the zymogen factors XI, and IX. In the presence of factor VIII (a nonenzymatic cofactor), activated factor IX can then cleave factor X and initiate the final common pathway, resulting in the conversion of prothrombin to thrombin and the ultimate conversion of fibrinogen to fibrin. The extrinsic cascade is initiated by the interaction between TF and factor VII (Fig. 3). The TF–factor VII complex can then participate in the

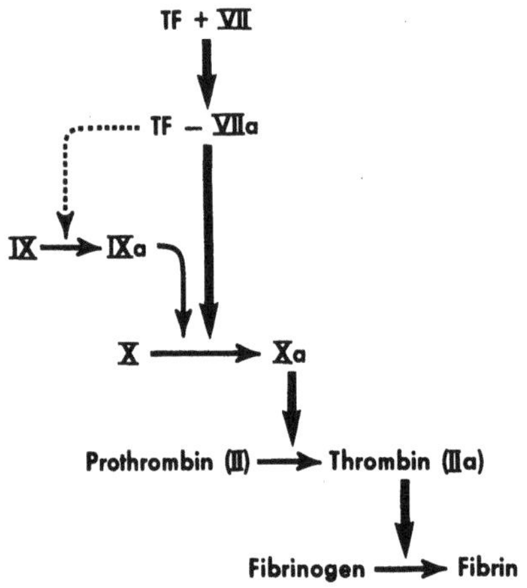

FIGURE 3. Tissue factor pathway of blood coagulation. The dotted arrow represents the proposed mechanism for tissue factor-induced activation of the intrinsic coagulation cascade. (From Edwards and Rickles, 1980b, with permission.)

activation of factor X and lead to the initiation of the final pathway of clotting described above. Recently, activation of factor IX by the TF–factor VII complex has been described and may play an important role in the activation of coagulation *in vivo* (Zur *et al.*, 1982; Nemerson and Bach, 1982; Bajaj *et al.*, 1983). Activated factor X, as well as thrombin, can cleave single-chain factor VII to the more active two-chain factor VIIa and thus greatly accelerate (about 85-fold) the process of factor X activation and, ultimately, fibrin generation (reviewed in Nemerson and Bach, 1982).

Cell-generated procoagulants could function by initiating cleavage and activation of any of the coagulation factors mentioned. The identity of a particular PCA is generally determined in assays using plasma substrates from subjects congenitally deficient in specific coagulation factors. Thus, inability of a PCA to induce clotting of a factor-deficient substrate suggests that the PCA is acting before (above) the site of the deficient factor in the clotting sequence. The endpoints utilized in these assays can be detected by the formation of a fibrin clot or by cleavage of luminogenic or chromogenic substrates. Thus, assays using specific factor-deficient plasmas can often confirm the identity of a PCA. It must be noted, however, that activated clotting factors may show striking species specificity for their substrates and therefore clotting assays using factor-deficient substrate plasmas obtained from different species may be unreliable.

Although not all-inclusive, Tables 1 and 2 attempt to summarize much of the available information regarding PCA generation by mononuclear cell populations. Table 1 emphasizes that the type of PCA is dependent on the anatomic location and species from which the cells are isolated. Table 2 places emphasis on the inducing agent (agonist) utilized and the species of the monocyte/macrophage donor.

In certain species, the identity of the PCA has proven difficult to determine. This is particularly true of cellular PCA from species like mouse and guinea pig, which exhibit poor reactivity with human plasma, and hence cannot be reliably characterized with human deficient plasmas (Kadish *et al.*, 1983). Use of chromogenic or luminogenic peptide substrates with limited specificity (Østerud *et al.*,

TABLE 1. CLASSIFICATION OF MONOCYTE/MACROPHAGE PROCOAGULANT ACTIVITIES

Procoagulant activity	Species	Cell source (monocyte purity, %)	Reference
Tissue factor	Human	Blood	
		(85)	Rivers *et al.* (1975)
		(>97)	Hiller *et al.* (1977)
		(>85)	Edwards *et al.* (1979)
		(90–95)	Prydz *et al.* (1979)
		(>95)	Schwartz and Edgington (1981b)
		(>99)	Semeraro *et al.* (1983)
		Pulmonary alveoli	
		(90–98)	Rickles *et al.* (1984)
		Peritoneal cavity	
		(>99)	Semeraro *et al.* (1983)
		Milk	
		(>99)	Semeraro *et al.* (1983)
		Cell lines	
		U937	Hudig *et al.* (1981), Lyberg *et al.* (1982c), Ewan *et al.* (1983)
		HL60	Young *et al.* (1981), Kornberg *et al.* (1982), Rickles *et al.* (1983)
	Rabbit	Blood	
		(?)	Lorenzet *et al.* (1983), Edwards *et al.*
		(?)	(1983)
		Pulmonary alveoli	
		(98)	Rothberger *et al.* (1983)
		(>96)	Sitrin *et al.* (1983)
		Peritoneal cavity	
		(>90)	Garg and Niemetz (1973)
		Thymus	
		(14)	Rothberger *et al.* (1983)
		Spleen	
		(36)	Rothberger *et al.* (1983)
		Bone marrow	
		(11)	Rothberger *et al.* (1983)
	Dog	Renal parenchyma	
		(31)	Rothberger *et al.* (1984)
	Mouse	Macrophage cell line	Lyberg *et al.* (1982b)
Factor X activator	Rabbit	Liver	
		(>95)	Maier and Ulevitch (1981)
	Mouse	Peritoneal cavity	
		(90–95)	Shands (1983)
Prothrombin activator	Human	Blood	
		(7–25)	Hogg (1983b)
	Mouse	Blood	
		(>96)	Schwartz *et al.* (1982)
Other PCAs (II, VII, IX, X)	Human	Blood	
		(?)	Østerud and Bjørklid (1982), Tsao *et al.* (1984)
	Mouse	Peritoneal cavity	
		(90–95)	Shands (1983)
		(?)	Østerud *et al.* (1980,1981), Lindahl *et al.* (1982)

TABLE 2. AGENTS WHICH INDUCE PCA GENERATION BY MONOCYTES/MACROPHAGES

Agent	Type of PCA	Species	Reference
LPS[a]	Tissue factor	Human	Rivers *et al.* (1975), Rickles *et al.* (1977), Levy *et al.* (1981b)
		Rabbit	Niemetz (1972b), Rothberger *et al.* (1983)
		Mouse	Amlie *et al.* (1981)
	Factor VII	Human	Tsao *et al.* (1984)
		Mouse	Shands (1983)
	X activator	Rabbit	Maier and Ulevitch (1981)
		Mouse	Chapman *et al.* (1983)
	Prothrombinase	Mouse	Schwartz *et al.* (1982)
	?	Guinea pig	Geczy and Hopper (1981)
Antigens[b]	Tissue factor	Human	Rickles and Edwards (1981), Van Ginkel *et al.* (1981), Geczy and Meyer (1982)
	?	Mouse	Geczy *et al.* (1983)
	?	Guinea pig	Geczy and Hopper (1981)
Lectins	Tissue factor	Human	Rickles *et al.* (1973), Edwards *et al.* (1979), Lyberg and Prydz (1980)
		Mouse	Amlie *et al.* (1981)
Complement	Tissue factor	Human	Muhlfelder *et al.* (1979), Prydz *et al.* (1977)
Immune complexes	Tissue factor	Human	Prydz *et al.* (1979), Rothberger *et al.* (1977)
		Mouse	Amlie *et al.* (1981)
	?	Mouse	Schwartz and Edgington (1981a)
Aggregated IgG	Tissue factor	Human	Rothberger *et al.* (1977)
	?	Mouse	Schwartz and Edgington (1981a)
Allogeneic cells	Tissue factor	Human	Rothberger *et al.* (1978), Van Ginkel *et al.* (1981), Helin and Edgington (1983)
Lymphokines (uncharacterized)	Tissue factor	Human	Edwards and Rickles (1980a), Helin
	?	Mouse	and Edgington (1984)
	?	Guinea pig	Geczy (1983), Farram *et al.* (1983)
	?	Rat	Geczy and Hopper (1981)
			Geczy (1983)
γ-Interferon	?	Mouse	de Moerloose *et al.* (1985)
Virus	?	Mouse	Levy *et al.* (1981a)
Lipoproteins[c]			
VLDL	Tissue factor	Human	Schwartz *et al.* (1981)
HDL, IDL	Prothrombinase	Human	Schwartz *et al.* (1981)
Phorbol esters	Tissue factor	Human	Lyberg and Prydz (1981)
FMLP[d]	Tissue factor	Human	Janco and Morris (1985)

[a]LPS, lipopolysaccharide from gram-negative bacteria (usually *E. coli* LPS, strain 026-B6 or 011-B4).
[b]Antigens studied have included PPD, SK-SD, mumps, fungal antigens, ovalbumin, and human gamma globulin.
[c]VLDL, very-low-density lipoproteins; HDL, high-density lipoproteins; IDL, intermediate-density lipoproteins.
[d]FMLP, *N*-formyl-methionyl-leucyl-phenylalanine (a chemotactic peptide).

1981; Shands, 1983; Lindahl *et al.*, 1982) or the examination of the cleavage patterns of purified human clotting factors (Schwartz *et al.*, 1982), though not definitive, serves to shed some light on these questions. In spite of the confusion which is present in the literature, substantial evidence exists for the production of two or three types of PCA by cells of the monocyte/macrophage system (summarized in Table 1).

4.2.1. Tissue Factor

TF, defined minimally in all of the quoted studies by its dependence on factor VII for expression of PCA, is also characterized by sensitivity to phospholipase C, binding to Con A, resistance to serine protease inhibitors (e.g., diisopropyfluorophosphate), and sedimentability (Nemerson and Bach, 1982). This extremely hydrophobic, lipophilic membrane protein has been purified to homogeneity from bovine brain and has an apparent molecular weight of 43,000 (Bach *et al.*, 1981). Preliminary characterization of the protein from human brain has been published (Liu and McCoy, 1975; Bjørklid and Storm, 1977), and purification of TF to apparent homogeneity from human tissue has just recently been reported by two different laboratories (Broze *et al.*, 1985; Guha *et al.*, 1986). The human protein appears to have physical and chemical properties similar to the bovine protein. Delipidated TF apoprotein has little biological activity; however, the addition of phospholipid restores abundant procoagulant function (Nemerson, 1968). Carbohydrate comprises at least 6% of the glycoprotein (Bjørklid and Storm, 1977) and is important for PCA (Pitlick, 1976). The active procoagulant material, in the presence of adequate phospholipid, greatly enhances the ability of factor VII to activate the extrinsic cascade of clotting. Although the TF–factor VIIa complex is most effective in the activation of factor X, there is now substantial evidence that the complex can also participate in the activation of factor IX (Østerud and Rapaport, 1977; Zur *et al.*, 1982; Bajaj *et al.*, 1983). Therefore, TF, which is a minor constituent of cell membranes, is an extremely potent stimulus capable of triggering both coagulation pathways (Fig. 3).

TF activity has been described in cultures of a variety of mammalian cells including human fibroblasts (Green *et al.*, 1971; Maynard *et al.*, 1975; Korn *et al.*, 1982), granulocytes (Lerner *et al.*, 1971; Niemetz, 1972b), endothelial and smooth muscle cells (Zeldis *et al.*, 1972; Maynard *et al.*, 1975), and mononuclear cells (Rickles *et al.*, 1973; Hattler *et al.*, 1973; Edwards and Rickles, 1978; Edwards *et al.*, 1979). TF may exist in cells in an inactive or precursor form since unstimulated cells manifest little if any PCA, and, after stimulation, PCA is expressed within 2 hr (Rickles *et al.*, 1977). TF activity is expressed on the cell surface and some activity can be shed into the culture media (Zacharski and McIntyre, 1973; Bona *et al.*, 1986). The amount of TF PCA in stimulated mononuclear cell cultures increases until peak activity is reached approximately 6–18 hr following stimulation (Rickles *et al.*, 1977). Most of the TF activity remains associated with cells or portions of cell membranes as demonstrated by the sedimentability of TF activity when culture supernatants are centrifuged at 100,000*g* (Rickles *et al.*, unpublished observations).

Cell membrane perturbation appears to be an important step in the sequence of events leading to TF activation. TF generation may be stimulated by a variety of agonists that affect membrane function (see Section 5). Following exposure to an appropriate stimulus, cell-associated TF PCA is generated and may be measured using any of the assays outlined above. Cell disruption and/or gentle protease digestion of the cell surface results in an increase in apparent TF PCA measured in these cell culture systems.

Various pharmacologic agents are capable of inhibiting the elaboration of monocyte TF activity. Puromycin and actinomycin D are effective inhibitors of PCA generation in human and rabbit leukocyte cultures whereas cytosine arabinoside, mitomycin, and sublethal irradiation are ineffective (Rickles *et al.*, 1977; Niemetz, 1972a). These findings suggest that RNA and protein synthesis are necessary for full elaboration of PCA but that DNA synthesis and cell division are not. More recent studies have suggested that the antimicrotubular agents colchicine and vinblastine also inhibit the generation and release of PCA from human monocytes, possibly by interfering with stimulus-mediated activation of the monocyte (de Prost *et al.*, 1979).

4.2.2. Factor X Activator

Macrophage-associated activators of factor X have also been described in rabbits and mice. Studies of these activators have used chromogenic substrates to circumvent the species specificity problems posed in assays of blood coagulation. Maier and Ulevitch (1981) characterized the procoagulant induced by endotoxin in rabbit hepatic macrophages. Utilizing human plasma substrates, the cell surface procoagulant was independent of factors VII, VIII, IX, and XI but was absolutely dependent on the presence of factor X. Activation of factor X by the PCA was measured using S-2222, a chromogenic substrate specifically cleaved by factor Xa. It is not yet possible to determine the identity of the factor X-activating species they described. However, certain features suggest that the PCA may be composed of a TF–factor VII complex. Like activated VII, the rabbit X activator was highly sensitive to the serine protease inhibitor diisopropylfluorophosphate (DFP). Mixing TF (from rabbit brain tissue) and the macrophage PCA produced a tenfold increase in apparent PCA, an effect similar to that observed when TF is reacted with factor VII (Nemerson and Bach, 1982). Therefore, the factor X activator described by Maier and Ulevitch may be a complex of TF and factor VII. The source of the factor VII-like activity remains unclear but several recent studies suggest that factor VII may be generated by macrophages (Østerud *et al.*, 1980; Tsao *et al.*, 1984; Shands, 1983, 1985; van Dam-Miers, 1985).

Shands (1983, 1985) has provided preliminary data characterizing a macrophage-associated PCA arising in mouse peritoneal exudate macrophages stimulated with LPS *in vitro*. Washed mouse macrophages demonstrated TF-like PCA in an assay system utilizing a source of factor VII and a chromogenic substrate. No factor X-cleaving activity was observed in sonicates of washed cells. However, when the macrophages were lysed in their spent culture medium, significant factor X-cleaving activity was observed, suggesting that the culture medium contained a substance necessary for expression of this PCA. The

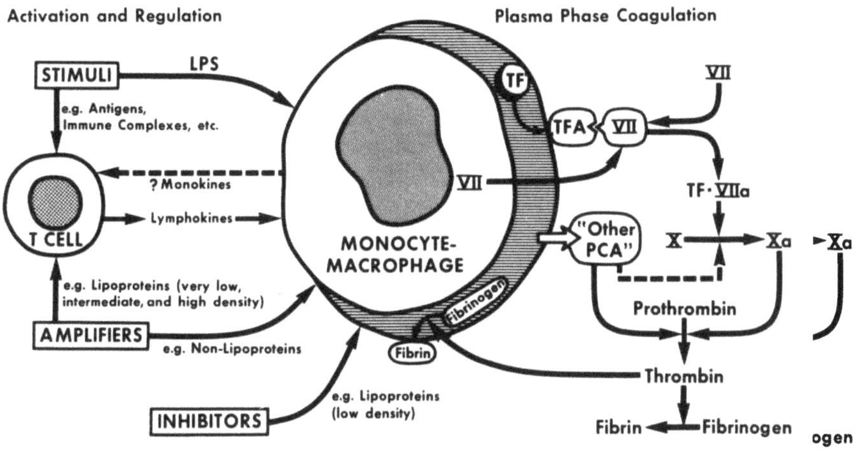

FIGURE 4. Activation and regulation of monocyte-mediated blood coagulation. The putative pathways involved in the stimulation and regulation of monocyte-derived procoagulant activity are represented on the left side. Individual procoagulants are represented on the right side along with an indication of their primary site of action in the coagulation cascade. TF, tissue factor antigen; TFA, tissue factor procoagulant activity. The pathways represented by the dotted arrows are less well-characterized than those represented by solid arrows (see text for more detailed discussion of the individual pathways). The net result of these regulatory and effector pathways is the generation of fibrin on the cell surface or in the plasma phase. (From Edwards and Rickles, 1984, with permission.)

factor X activator was DFP-sensitive and was evident in assay systems deficient in exogenous factor VII activity. This PCA was increased when the washed macrophages were suspended in cell-free supernatants from cultures supplemented with vitamin K. If macrophages were suspended in the supernatant from warfarin-treated cells, no factor X activator was observed. However, mixtures of warfarin-treated cells and control supernatants demonstrated full activity. These studies suggest that a vitamin K-dependent protein present in macrophage culture supernatants is integral to the formation of the serine protease factor X activator developed in this culture system. These data support the hypothesis that cell-associated TF and cell-derived factor VII are combined to produce a proteolytically active TF–factor VIIa complex capable of activating factor X.

Recently, more direct evidence has been presented for the synthesis of factor VII by human monocytes (Østerud *et al.*, 1980; Tsao *et al.*, 1984). This finding, taken along with that of Broze (1982), substantiating the binding of human factor VII to human monocytes, supports the concept that cell-bound TF is the receptor for factor VII, be it plasma- or cell-derived (Fig. 4).

4.2.3. Prothrombin Activator

A third PCA, a prothrombin activator, has been described in human and mouse peripheral blood monocytes grown in tissue culture under certain conditions. Under physiologic conditions in human or bovine plasma, a complex composed of activated factor X (Xa), thrombin-activated factor V, phospholipid, and calcium is required for the proteolytic cleavage of prothrombin which results in the formation of thrombin (Nesheim *et al.*, 1979, 1981). This complex is usually referred to as "prothrombinase." In the human monocyte system, the identification of the cellular PCA as a prothrombin activator has been based on its ability to induce the clotting of plasmas deficient in both extrinsic (VII) and intrinsic (XII and IX) coagulation factors. No PCA was expressed in the absence of prothrombin. Coagulation was demonstrated by one-stage clotting assays in the studies of Schwartz *et al.* (1981), while Hogg (1983b) demonstrated the generation around macrophages of needles with the microscopic appearance of fibrin. In both cases, cleavage of isotopically labeled prothrombin into fragments was demonstrated. The fragments resembled those generated when purified factor Xa was added to the same substrates (Schwartz *et al.*, 1981), although Hogg (1983b) described the ability of LPS-activated macrophages to produce a larger intermediate cleavage product of prothrombin. The identity of the prothrombin-cleaving species have not been determined with certainty although the similarity of cleavage products suggests analogy with factor Xa (or a XA–V complex).

In a murine system utilizing one-stage clotting assays, Schwartz *et al.* (1982) have also partially characterized a prothrombin activator. This murine prothrombinase was able to clot human plasmas deficient, respectively, in factors VII, XII, VIII, X, and V. The generation of thrombin was demonstrated directly as follows. The combination of cell homogenate (prothrombinase) with purified human prothrombin, calcium, and human fibrinogen caused clot formation, which did not occur if any of the components was omitted. As in the experiments with human cells, cleavage products of ^{125}I-labeled human prothrombin were generated by incubation with activated murine monocytes which were similar to those generated by human factor Xa. Even if the products could be shown to be the same as those generated by human factor Xa, purification of the monocyte prothrombinase will be required to establish homology with that protease.

4.2.4. Other Procoagulants

As noted in Table 1, small but significant concentrations of other procoagulants appear to be generated by human and mouse monocytes/macrophages under certain conditions. As noted in Section 4.2.2, mouse and human peripheral blood monocytes, under appropriate culture conditions, stimulated by LPS, generate sufficient factor VII-like activity (in a serum-free culture system) to shorten the clotting time of human factor VII-deficient plasma or activate factor X in the presence of TF (Shands, 1983, 1985; Tsao *et al.*, 1984). In addition to factor VII, mouse peritoneal exudate cells also generate the other vitamin K-dependent clotting factors II, IX, and X following LPS stimulation *in vitro* (Østerud *et al.*, 1980, 1981; Lindahl *et al.*, 1982; van Dam-Miers, 1985), although this finding must be interpreted with caution in view of the cross-species assay systems utilized.

Even less is known about the nature of the cellular PCA(s) in species such as the guinea pig, where cross-species clotting assays are even more problematic. For example, guinea pig PCA(s) have little activity in assays which utilize human plasmas as substrates (Quick, 1936; Kadish *et al.*, 1983). Therefore, in the absence of purification data, it is difficult to draw specific conclusions regarding the characteristics of cellular PCA when homologous, deficient substrate plasmas are not available.

5. REGULATION OF MONOCYTE/MACROPHAGE PROCOAGULANT GENERATION

5.1. ACTIVATION OF MONOCYTE PROCOAGULANT(S)

Monocyte PCA generation may be induced by exposure to a wide range of stimuli *in vitro*. Many of these stimuli are capable of activating other "immune responses" or are themselves products of immune reactions. Table 2 lists some of the known agents which can induce monocyte PCA generation along with the identity of the PCA produced and the species studied. Figure 4 presents a summary schema of our current understanding of the interconnecting pathways for the activation and regulation of monocyte procoagulants.

The most potent stimulus for monocyte PCA generation is LPS, a common contaminant of most laboratory reagents. LPS can stimulate the generation of significant human monocyte TF PCA at concentrations of as little as 0.1 ng/ml (Rickles *et al.*, 1977). Induction of TF generation in human monocytes is dependent on the integrity of the lipid A portion of the LPS molecule. Polymyxin B, an agent which binds to the lipid A region of LPS, inhibits its ability to induce TF generation (Rickles and Rick, 1977). Although most studies have demonstrated that TF is the PCA generated by LPS-stimulated human monocytes, others have shown that LPS is capable of inducing the production of different PCAs in monocyte cultures derived from other species. As shown in Table 2, LPS can induce mouse monocytes to generate the PCAs described in the previous section: TF, factor VII, factor X activator, prothrombinase. Similarly, rabbit monocytes can generate TF or factor X activator following LPS exposure. Recent studies by Shands (1985), however, suggest that factor VII is generated constitutively in mouse macrophage cultures and is not enhanced by LPS stimulation. As noted previously, the PCA generated by guinea pig monocytes has not been fully characterized.

It is now clear that monocyte PCA generation in several species can be stimulated in response to specific antigen recognition. PPD, streptokinase–streptodornase (SK–SD), mumps, fungal antigens, ovalbumin, and human gamma globulin stimulate monocyte PCA generation in an immune-specific manner (see Table 2). In both human and animal studies, antigens have been shown to induce PCA generation in mononuclear cell cultures only when the monocytes are obtained from a donor who has been sensitized by previous exposure to the same antigen. Cellular collaboration between antigen-presenting cells and HLA-DR-identical T cells plays a prominent role in antigen-induced PCA generation (Helin and Edington, 1983). As with LPS-induced PCA generation, the type of PCA produced appears to be dependent on the species of the donor, anatomic source of the monocyte/macrophage, culture conditions, and PCA assay system utilized. Assessment of antigen-induced monocyte PCA generation has been reported to be a highly sensitive and reproducible method for evaluating cell-mediated immunity which is technically less complex than the standard methods currently being utilized for such evaluations (Geczy and Meyer, 1982).

Other stimuli capable of inducing monocyte PCA and listed in Table 2 include the lectins PHA and Con A, complement components (C5 chemotactic fragments and C3b), immune complexes, chemotactic peptides, and aggregated IgG. In addition, recent studies have demonstrated monocyte PCA generation associated with the mixed lymphocyte reaction in both human and murine systems. In the human, PCA generation in allogeneic cell cultures appears to be dependent on the presence of different HLA-DR (Ia-like) antigens on the stimulatory and responding cells. Moreover, PCA generation has been shown to be temporally discordant from the proliferative responses characteristic of allogeneic reactions, the PCA being generated as early as 18 hr after initiation of the allogeneic cell culture (Helin and Edgington, 1983).

The close relationship between immune reactivity and monocyte PCA generation is further supported by the ability of lymphokines to stimulate monocyte PCA generation. Edwards and Rickles (1980a) demonstrated that a lymphokine-rich supernatant from PHA-stimulated human lymphocyte cultures, which contained < 0.1 ng/ml of LPS, induced TF generation by monocyte-enriched cultures. Addition of LPS-free PHA directly to monocyte cultures failed to stimulate monocyte PCA generation in the absence of either lymphocytes or lymphokines. Lymphokine production appears to be essential in the induction of monocyte PCA generation in human cell cultures following allogeneic stimulation (Helin and Edgington, 1984). Lymphocyte–monocyte contact was not required but lymphocyte–monocyte collaboration was necessary and could be replaced by spent media from lymphoblastoid cell lines or mixed lymphocyte cultures. Lymphokine-induced monocyte PCA generation in other species has been demonstrated (Farram *et al.*, 1983; Geczy, 1983; de Moerloose *et al.*, 1985). However, the nature of the PCA(s) generated in the nonhuman cells used in these experiments has (have) not been determined. Furthermore, with the exception of the study by de Moerloose *et al.* (1985), in which PCA was generated by cultured WEHI 256 cells in response to recombinant γ-interferon (devoid of measurable LPS), the nature of the lymphokine(s) responsible for the induction of monocyte (macrophage) PCA has also not been determined with certainty.

Several components of normal plasma have been shown to induce monocyte PCA generation. Schwartz *et al.*, (1981) demonstrated that some plasma lipoproteins have the ability to stimulate monocyte PCA generation. When lipoproteins were separated into different classes on the basis of density, the very-low-density lipoproteins (VLDL) induced TF generation by human monocytes whereas high-density (HDL) and intermediate-density (IDL) lipoproteins induced the generation of a monocyte prothrombinase. T cells proved to be the target of lipoprotein stimulation since exposure of T cells alone (but not monocytes alone) prior to coculture was sufficient to transfer the stimulatory effect. It is of interest that low-density lipoproteins (LDL) were not stimulatory and actually inhibited the stimulatory effects of the other lipoprotein classes. LDL appear to interfere with the monocyte response to HDL, IDL, and/or VLDL stimulation (Schwartz *et al.*, 1981). Edwards and Perla (1984) have reported that low concentrations (< 1%) of heat-inactivated serum from some normal human subjects induced monocyte PCA generation in unstimulated cell cultures and enhanced monocyte TF (MTF) generation in LPS-stimulated cultures. Although the serum stimulatory factor has not been fully characterized, activity was present in LPS-free lipoprotein-free serum and plasma and did not require lymphocytes for its full stimulatory activity.

It is important to note that many of the effects of a variety of biologically active agonists can be mimicked by trace concentrations of LPS (Westphal, 1975), including the activation of monocyte PCA by as little as 0.1 ng/ml of LPS (Rickles and Rick, 1977). This exquisite sensitivity to LPS along with the ubiquitous contamination of biological reagents and tissue culture media (Rickles *et al.*, 1977), makes it difficult to interpret many of the published studies regarding activation and regulation of monocyte PCA generation. It is imperative that all such reports include a careful assessment of the level of LPS contamination in the culture systems utilized. Statements such as "an assay utilizing *Limulus polyphemus* amoebocyte lysate failed to demonstrate measurable LPS contamination" are not sufficient since the sensitivity of various commercial preparations of *Limulus* lysates varies considerably. The presence of as little as 50 pg/ml of LPS can stimulate monocyte PCA generation and can, therefore, obscure the activation of PCA generation induced by other suboptimal stimuli. Moreover, small amounts of contaminating LPS may potentiate the effect of otherwise poor stimuli and induce the generation of PCA in amounts inappropriately great for the potency of the stimulus under consideration. Similar confounding effects of LPS can be observed in studies attempting to clarify the role of lymphocyte populations in the regulation of MTF generation, as noted in the following section.

5.2. MODULATION OF MONOCYTE PROCOAGULANT GENERATION

5.2.1. Amplification

A number of factors have been shown to modify the generation of PCAs following *in vitro* monocyte stimulation. Several authors have reported that monocyte PCA generation is dependent on the presence of T lymphocytes. Edwards and Rickles (1980a) first described the absolute T-lymphocyte dependence of human MTF generation following stimulation by the T-cell mitogen PHA and the relative T-lymphocyte dependence for LPS-induced MTF generation. In subsequent studies, Edwards *et al.* (1984) demonstrated that small but significant TF generation occurred in LPS-stimulated human monocyte cultures which were essentially free of lymphocytes (> 97% viable monocytes as assessed by latex particle ingestion). When purified T lymphocytes were added back to monocytes in these experiments and stimulated with LPS, the amount of TF generation increased and reached a peak at a T lymphocyte–monocyte ratio of 4 : 1 (Fig. 5).

There remains disagreement, however, on whether LPS-stimulated monocyte PCA generation can occur in the complete absence of lymphocytes. Lyberg and Prydz (1983) have reported that freshly isolated human monocytes can generate PCA following stimulation by LPS, PHA, or antigens in the absence of lymphocytes, whereas others claim an absolute lymphocyte requirement for monocyte PCA generation in response to LPS in cells obtained from humans (Helin *et al.*, 1983) or mouse (Levy and Edgington, 1982). Interpretation of these studies is difficult because of the use of different assay systems for the detection of LPS contamination and the amount of PCA generated. High "background" levels of PCA in unstimulated cultures of lymphocyte-free monocytes may represent direct stimulation of monocyte PCA generation by small amounts of LPS contaminating the culture reagents. It is interesting that several of the studies which claim to demonstrate a requirement for lymphocyte help for LPS-induced MTF generation have reported measurable PCA in the "LPS-free" control cultures.

More recently, Semeraro *et al.* (1983) have utilized prolonged *in vitro* culture (72 hr) of adherent monocytes in order to study TF generation by highly purified (> 99% pure) human macrophage populations. In their study, lymphocyte-free macrophages derived from peripheral blood, peritoneum, and milk produced no PCA prior to stimulation. Each of these macrophage populations demonstrated the ability to generate TF PCA following LPS stimulation. Although this study supports the original conclusion of Edwards and Rickles (1980a) that monocytes can be activated directly, one must be careful in drawing conclusions about normal monocyte regulation based on studies of macrophages prepared by prolonged periods of adherence in tissue culture. Such macrophage preparations may contain specific subpopulations of cells which are capable of surviving such manipulations but which may not represent the characteristics of unseparated

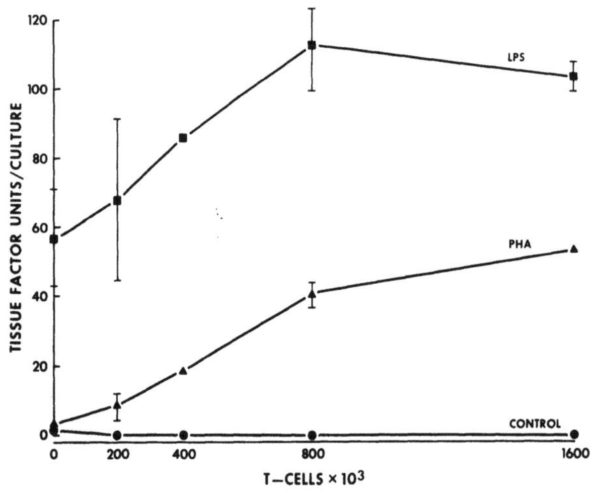

FIGURE 5. Effect of added T cells on monocyte tissue factor generation. Monocyte tissue factor generation by unstimulated (control), PHA-, or LPS-stimulated cultures is plotted as a function of the number of T cells added. All cultures contained 250 × 10³ monocytes. (Reproduced from Edwards and Rickles, 1980a, with permission.)

populations of peripheral blood monocytes. Nevertheless, it is generally agreed that T lymphocytes are capable of enhancing the generation of PCA by monocyte cultures stimulated by LPS, lectins, and immune complexes as well as several other stimuli. More recent studies have demonstrated distinct subpopulations of T cells which are capable of enhancing or suppressing human MTF generation (Helin *et al.*, 1983).

5.2.2. Inhibition

Several pharmacologic agents inhibit the generation of monocyte PCA. Edwards and Rickles (1978) first reported the inhibition of MTF PCA expression by *in vivo* treatment with the anticoagulant sodium warfarin. In that study of normal volunteers and in another study of cancer patients (Edwards and Rickles, 1981), monocytes obtained from anticoagulated subjects generated significantly less PCA following LPS stimulation than monocytes obtained from the same subjects before anticoagulation. Similar results have been reported in rabbits anticoagulated with warfarin (Edwards *et al.*, 1986). Shands (1985) has reported that warfarin (5 μM) inhibits the *in vitro* generation of PCA by mouse peritoneal macrophages. Similarly, Maier (1982) has shown that warfarin inhibits the generation of DFP-sensitive factor X activator by LPS-stimulated rabbit hepatic macrophages.

Østerud *et al.* (1980), Shands (1983), and Tsao *et al.* (1984) have demonstrated factor VII synthesis by stimulated monocytes, which can then be expressed on the surface of these cells. Since warfarin alters the functional properties of factor VII by inhibiting the vitamin K-dependent carboxylation of glutamic acid residues of the synthesized protein (Suttie and Stenflo, 1977), the warfarin-induced reduction in monocyte PCA could represent a specific effect of warfarin on monocyte factor VII. That is, the PCA of TF, the putative receptor for factor VII on the cell surface (Broze, 1982), could be regulated by the posttranslational modification of the ligand (factor VII) that binds to it.

Recent studies in our laboratory (Edwards *et al.*, 1986) have demonstrated decreased rabbit MTF expression by cells grown in plasma from warfarin-treated animals. Since the TF assay is performed in the presence of adequate amounts of factor VII, these studies suggest that dysfunctional factor VII in warfarin-treated plasma may bind to MTF at the time of its generation in culture and block the subsequent expression of PCA. Further studies of the interaction of decarboxylated factor VII and MTF are under way in our laboratory and may shed some light on this interesting pathway for PCA generation.

A variety of pharmacologic agents have been investigated for stimulatory or inhibitory effects on monocyte PCA generation *in vitro*. Dexamethasone (Prydz and Lyberg, 1980) and methylprednisone (Muhlfelder *et al.*, 1982), agents known to interfere with monocyte-mediated inflammatory responses, inhibit human monocyte PCA generation following exposure to a variety of stimuli. Other anti-inflammatory and cytotoxic agents have produced variable degrees of inhibition *in vitro* depending on the dose of the agent and the source of the target cells (Geczy, 1983; Prydz and Lyberg, 1980). Sitrin *et al.* (1984) have reported that indomethacin (10^{-6} M) inhibits LPS-stimulated TF generation by rabbit alveolar macrophages. Moreover, prostaglandins PGE_2 and PGE_1 (10^{-9} M) reversed the indomethacin-induced suppression and did not directly induce PCA generation in the absence of LPS. Crutchley (1984) has reported that phospholipase inhibitors and lipoxygenase inhibitors block LPS-induced MTF generation whereas the cyclooxygenase inhibitors aspirin and indomethacin showed no ability to impair TF generation in their culture system. Similar interactions between prostaglandins and prostaglandin inhibitors have been demonstrated for TF synthesis in other cell systems (Korn *et al.*, 1982). Other investigators, however, have failed to demonstrate either suppression of PCA by indomethacin or enhancement by prostaglandins (Prydz and Lyberg, 1980; Edgington and Pizzolato, 1983) in cultures of human monocytes. Thus, as in many other studies of monocyte/macrophage PCA generation, the effects of prostaglandins may vary depending on the anatomic source of the phagocyte and the species being examined.

6. THE POTENTIAL RELATIONSHIP BETWEEN MONOCYTE/MACROPHAGE PROCOAGULANT GENERATION AND THROMBOEMBOLIC COMPLICATIONS OF DISEASE

6.1. CONDITIONS ASSOCIATED WITH ENHANCED MONOCYTE PROCOAGULANT GENERATION AND THROMBOEMBOLIC DISEASE

6.1.1. Immune Disorders

A number of inflammatory and immune disorders including the renal allograft rejection reaction, systemic lupus erythematosus, and glomerulonephritis are associated with laboratory evidence of activation of blood coagulation and histologic evidence of fibrin deposition *in vivo* (see Section 2). Several pathologic states have also been associated with increased monocyte PCA either *in vitro* or *in vivo*. Hattler *et al.* (1973) reported that mononuclear cells obtained from a renal allograft undergoing rejection and cultured *in vitro* for 24 hr expressed increased PCA when compared to peripheral blood mononuclear cells or kidney parenchymal cells from the same patient. The PCA was inhibited by a heterologous anti-TF antibody. More recently, Halloran *et al.* (1982) examined peripheral blood mononuclear cell PCA in cells obtained from 35 renal allograft recipients. They found a short-lived increase in PCA in mononuclear cell suspensions obtained from patients in the immediate postoperative period and also found a more sustained increase in PCA at the time of allograft rejection. In a dog model of renal allograft rejection, Rothberger and colleagues demonstrated a similar increase in PCA, which was TF-like, in monocytes isolated from the graft itself (Rothberger *et al.*, 1984). The enhanced PCA described in these studies may represent activation of monocyte procoagulant as a part of the inflammatory response to the transplant. More recently, Cole *et al.* (1985) have reported enhanced monocyte PCA of the prothrombinase type in peripheral blood cells obtained from patients with lupus erythematosus and focal or diffuse proliferative lupus nephritis. Monocyte PCA may be responsible for initiating at least some of the blood coagulation events commonly associated with the allograft rejection response and the glomerulonephritis associated with lupus erythematosus.

In view of the ability of immune complexes to stimulate monocyte PCA generation *in vitro,* it is possible that patients with increased circulating immune complex levels would have enhanced *in vivo* monocyte PCA. Lyberg *et al.* (1982d) examined the PCA of peripheral blood monocytes from patients with rheumatologic diseases and increased circulating immune complex levels. They reported that patient monocytes had enhanced PCA after *in vitro* incubation in unstimulated cultures but these monocytes showed no increase in PCA generation in cultures stimulated *in vitro* by PHA or immune complexes. Moreover, sera from the patients with rheumatologic disorders induced increased PCA in cultures of normal monocytes when added to the culture media at a concentration of 20%. These findings suggest that circulating immune complexes may induce enhanced monocyte PCA generation *in vivo* and may contribute to the microvascular thrombosis and fibrin deposition common in inflammatory disorders.

Other chronic inflammatory disorders have been associated with increased monocyte PCA. Levy *et al.* (1981a) reported increased PCA generation by murine peripheral blood mononuclear cells in association with murine hepatitis virus infection in genetically susceptible BALB/c mice. Cells from mice of the hepatitis virus-resistant strain, A/J, failed to generate PCA following viral exposure whereas cells from the partially resistant strain of C3H mice generated intermediate levels of PCA. In view of the close relationship between the immune response to the hepatitis virus and the development of the pathologic sequelae of hepatitis (Edgington and Chisari, 1975), the development of monocyte PCA may serve as a marker for pathologic activation of the cellular immune response in these animals. More recent studies (Levy *et al.*, 1984) have suggested that monocyte PCA may serve as a marker for the development of chronic hepatitis in man. In a series of 225 patients with liver disease, 93% of those with chronic active hepatitis demonstrated a 75- to 90-fold increase in total monocyte PCA as compared to normal control subjects or patients with other liver diseases. Of 12 patients studied during the course of acute hepatitis B infection, 4 had elevated monocyte PCA and all 4 developed chronic active hepatitis within 1 year.

Edwards *et al.* (1984a) have demonstrated increased TF generation by peripheral blood mononuclear cells obtained from patients with active Crohn's disease. Moreover, spontaneous generation of MTF in an LPS-free culture system correlated with simultaneous plasma levels of fibrinopeptide A, a peptide generated during *in vivo* activation of blood coagulation. In a prospective study of 15 patients with inflammatory bowel disease, this relationship proved highly significant ($r = 0.89$, $p > 0.001$), suggesting that *in vivo* activation of monocyte PCA may contribute to the activation of coagulation *in vivo*. In another recent study, Ottaway *et al.* (1984) demonstrated spontaneous expression of a PCA with the characteristics of prothrombinase by peripheral blood mononuclear cells from two patients with Whipple's disease. In each of these chronic inflammatory disorders, endotoxin, viruses, immune complexes, or other products of the immune response may be responsible for *in vivo* stimulation of monocyte PCA generation.

6.1.2. Infectious Disease

As described in Section 2, LPS activation of leukocyte PCA(s) has been implicated in the pathogenesis of the coagulation abnormalities associated with gram-negative sepsis. Enhanced blood monocyte PCA generation *in vivo* has been observed in rabbits undergoing the GSR following intravenous injection of LPS. Increased PCA has been described in monocytes obtained from peripheral blood (Niemetz, 1972b) as well as in cells from the bone marrow, spleen, peritoneal cavity, and lung (Rothberger *et al.*, 1983). Similar studies in our laboratory (Edwards and Brande, 1985) have demonstrated increased peripheral blood MTF generation following the intravenous injection of as little as 1 μg of LPS into a 4-kg rabbit. Østerud and Flaegstad (1983) have reported that monocytes obtained from the blood of patients with meningococcal infection have markedly increased PCA as compared to monocytes from normal volunteers. In those experiments, the monocytes were rapidly isolated from peripheral blood, immediately disrupted and assayed for PCA, thereby minimizing the potential for *in vitro* generation of PCA. These experiments suggest that monocyte procoagulants can be elaborated *in vivo* in response to endotoxins and, therefore, may participate in the activation of blood coagulation events associated with bacterial sepsis.

6.1.3. Cancer

Blood coagulation may play a central role in the growth and spread of malignant tumors (reviewed in Rickles and Edwards, 1983). Although there are several potential mechanisms for the activation of blood coagulation in neoplasia, evidence from our laboratory indicates that peripheral blood monocytes from patients with malignant disease have enhanced PCA (Edwards *et al.*, 1981). Purified monocyte suspensions obtained from cancer patients generate increased MTF activity in both unstimulated and LPS-stimulated tissue cultures as compared to similar monocyte preparations from normal individuals. This enhanced PCA generation appears to represent *in vivo* "preactivation" of the monocytes and is not further enhanced by *in vitro* exposure to T cells (Morgan *et al.*, 1985). Moreover, as in inflammatory bowel disease, a strong correlation has been observed between PCA generation by unstimulated peripheral blood mononuclear cells and *in vivo* blood coagulation as measured by plasma fibrinopeptide A levels.

Although some of the enhancement in MTF activity may be explained by the increased number of monocytes found in cancer patients, those patients with increased fibrinopeptide A levels (indicating increased rates of *in vivo* blood coagulation) demonstrated increased PCA generation even when the data were expressed per monocyte. Since macrophage-associated fibrin deposition is an early finding in a mouse model of mammary carcinoma (Dvorak *et al.*, 1978), macrophage procoagulants might be generated soon after implantation of the tumor. Indeed, Lorenzet *et al.* (1983) have reported increased monocyte PCA in peritoneal exudate cells and tumor-associated macrophages obtained from rabbits bearing a malignant carcinoma. Each of these studies supports the hypothesis that monocyte procoagulants may be generated *in vivo* in animals with neoplastic disease. Increased availability of monocyte PCA may play a role in the abnormal activation of blood coagulation commonly associated with neoplasia.

6.2. CONDITIONS ASSOCIATED WITH IMPAIRED MONOCYTE PROCOAGULANT GENERATION

In contrast to the increasing number of disorders associated with enhanced expression of monocyte PCA, impaired generation of monocyte PCA is rare. However, defective monocyte PCA expression, which has been described in patients with immunodeficiency diseases (Ballow and Rickles, 1980) and in subjects with leukemia (Cortellazzo *et al.*, 1981, 1983), provides further support for several of the *in vitro* observations reviewed previously. For example, two distinct patterns of defective MTF expression have been reported in patients with severe combined immunodeficiency disease (Table 3). Those patients with disease characterized primarily by impaired T-cell function (patients 1 and 2) failed to express normal levels of MTF following stimulation with the lectin PHA (a T-cell-dependent stimulus) but were able to generate MTF in response to LPS. Patients with evidence of a more global immune dysfunction (patients 3 and 4), presumably due to defects of a stem cell common to both T cells and monocytes (Pahwa *et al.*, 1978), failed to generate normal MTF activity in response to either stimulus. We interpret these data to indicate that normal monocytes can elaborate PCA in response to LPS directly in the absence of functional T cells. Studies

TABLE 3. MONOCYTE TISSUE FACTOR GENERATION IN RESPONSE TO PHA AND LPS IN SEVERE COMBINED IMMUNODEFICIENCY DISEASE (SCID)[a]

Patient description	Monocytes (%)	Tissue factor (units × 10^{-4}/monocyte) PHA	LPS
Patient 1—T-cell deficiency	16	15.9	34.6
Patient 2—Thymic hypoplasia	38	0.5	36.3
Patient 3—SCID with circulating B cells	35–50[b]	6.2 ± 1.2[c]	11.7 ± 5.2
Patient 4—Adenosine deaminase deficiency	82–90	2.1 ± 0.1	3.8 ± 0.6
Normal controls	8–20	40.3 ± 9.3	53.1 ± 4.7

[a]Modified from Ballow and Rickles (1980).
[b]Range of percentage of monocytes in cell culture.
[c]Mean monocyte tissue factor generation ± 1 S.E.M.

of cells from patients with either chronic myelogenous or chronic lymphocytic leukemia have demonstrated impaired MTF generation in response to LPS (Cortellazzo *et al.*, 1981, 1983). The latter studies provide additional evidence that optimal expression of monocyte PCA requires both normal monocytes and normal T lymphocytes. Monocyte PCA, therefore, can be considered a sensitive marker for both monocyte function and cell-mediated immunity.

It should be noted that none of these patients, whose cells had impaired MTF generation *in vitro*, had evidence of either a congenital or an acquired hemorrhagic disorder. Thus, while these "relative" MTF deficiency states provide important clues to the regulation of MTF and *extravascular* fibrin deposition, they also suggest that MTF expression has little or no role in the maintenance of normal *intravascular* hemostasis.

7. SUMMARY

Thus, the regulation of monocyte/macrophage PCA is dependent on a complex network of interacting pathways, some of which amplify and others of which inhibit the response of the monocyte/macrophage. In all probability many more pathways will emerge. The construct illustrated in Fig. 4, therefore, is a simplified view of the two major stimulatory pathways: (1) the T-cell-dependent pathway, activated by immune recognition and mediated by lymphokine(s); (2) the T-cell-independent pathway, activated by direct perturbation of monocytes by such stimuli as LPS. At least two or three different PCAs can be expressed by monocytes/macrophages from different species, depending on the anatomic site of origin of the cell and the types of stimuli imposed. Inhibition of PCA expression is accomplished by at least one set of regulatory lipoproteins and other inhibitory loops may be found.

The ultimate result of these multiple interactions is the deposition of fibrin on the cell surface or in the surrounding milieu. It is our belief that this close relationship between coagulation reactions and inflammatory reactions, resulting in fibrin deposition, represents a fundamental host defense designed to delimit the inflammatory response. Nevertheless, the precise role of monocyte procoagulants *in vivo* remains unclear. A number of potential mechanisms exist for activation of coagulation in both inflammatory and neoplastic disorders and the finding of enhanced monocyte PCA by no means established its importance in physiologic or pathophysiologic responses *in vivo*. Further studies, possible with agents capable of specific inhibition of monocyte procoagulants *in vivo*, will be necessary to define the precise importance of these procoagulants in clinical disorders.

ACKNOWLEDGMENTS. This work was supported by research grants CA 30651, CA 22202, and HL 27722 from the National Institutes of Health, research grant 83-957 from the American Heart Association, research grant CH321 from the American Cancer Society, and the research service of the Veterans Administration.

REFERENCES

Addison, W., 1841, Colourless globules in the buffy coat of the blood, *London Med. Gaz.* **27:**477.

Addison, W., 1842, On the colourless corpuscles, and on the molecules and cytoblasts in the blood, *London Med. Gaz.* **30:**144.

Amlie, E., Lyberg, T., Kaplun, A., Hetland, Ø., and Prydz, H., 1981, Thromboplastin activity of mouse peritoneal macrophages, *Thromb. Res.* **24:**61.

Apitz, K., 1935, A study of the generalized Shwartzman phenomenon, *J. Immunol.* **29:**255.

Bach, R., Nemerson, Y., and Konigsberg, W., 1981, Purification and characterization of bovine tissue factor, *J. Biol. Chem.* **256:**8324.

Bajaj, S. P., Rapaport, S. I., and Russell, W. A., 1983, Redetermination of the limiting step in the activation of factor IX by factor VIIa/tissue factor: Explanation for different electrophoretic radioactivity profiles obtained on activation of ^{3}H and ^{125}I-labelled factor IX, *Biochemistry* **22:**4047.

Ballow, M., and Rickles, F. R., 1980, Impaired mononuclear cell tissue factor generation in patients with immunodeficiency diseases, *Blood* **56:**402.

Bjørklid, E., and Storm, E., 1977, Purification and some properties of the protein component of tissue thromboplastin from human brain, *Biochem. J.* **165:**89.

Bona, R., Lee, E., and Rickles, F. R., 1986, Transport of tissue factor to membrane vesicles: role of N-linked glycosylation, *Clin. Res.* **34:**655.

Brower, M. S., and Harpel, P. C., 1983, Alpha-1-antitrypsin–human leukocyte elastase complexes in blood: Quantification by an enzyme-linked differential antibody immunosorbent assay and comparison with alpha-2-plasmin inhibitor–plasmin complexes, *Blood* **61:**842.

Broze, G. J., Jr., 1982, Binding of human factor VII and VIIa to monocytes, *J. Clin. Invest.* **70:**526.

Broze, G., Jr., Leykam, J. E., Schwartz, B. D., and Milletichs, J. P., 1985, Purification of human brain tissue factor, *J. Biol. Chem.* **260:**10917.

Chapman, H. A., Jr., Zdenek, V., and Hibbs, J. B., Jr., 1983, Coordinate expression of macrophage procoagulant and fibrinolytic activity in vitro and in vivo, *J. Immunol.* **130:**261.

Cline, M. J., 1977, Cellular kinetics of monocytes and macrophages, in: *Hematology* (W. J. Williams, E. Beutler, A. J. Erslev, and R. W. Rundles, eds.), pp. 869–874, McGraw–Hill, New York.

Cohn, Z. A., and Benson, B., 1965, The differentiation of mononuclear phagocytes: Morphology, cytochemistry, and biochemistry, *J. Exp. Med.* **121:**153.

Cole, E. H., Schulman, J., Urowitz, M., Keystone, E., Williams, C., and Levy, G. A., 1985, Monocyte procoagulant activity in glomerulonephritis associated with systemic lupus erythematosus, *J. Clin. Invest.* **75:**861.

Colvin, R. B., and Dvorak, H. F., 1975a, Fibrinogen/fibrin on the surface of macrophages: Detection, distribution, binding requirements, and possible role in macrophage adherence phenomena, *J. Exp. Med.* **142:**1377.

Colvin, R. B., and Dvorak, H. F., 1975b, Role of the clotting system in cell-mediated hypersensitivity. II. Kinetics of fibrinogen/fibrin accumulation and vascular permeability changes in tuberculin and cutaneous basophil hypersensitivity reactions, *J. Immunol.* **114:**377.

Colvin, R. B., Johnson, R. A., Mihm, M. C., Jr., and Dvorak, H. F., 1973, Role of the clotting system in cell-mediated hypersensitivity. I. Fibrin deposition in delayed skin reactions in man, *J. Exp. Med.* **138:**686.

Colvin, R. B., Mosesson, M. W., and Dvorak, H. F., 1979, Delayed-type hypersensitivity skin reactions in congenital afibrinogenemia lack fibrin deposition and induration, *J. Clin. Invest.* **63:**1302.

Cortellazzo, S., Viero, P., Barbui, T., Colucci, M., and Semeraro, N., 1981, Reduced generation of procoagulant activity by endotoxin-stimulated mononuclear cells from patients with chronic myeloid leukemia, *Br. J. Haematol.* **48:**501.

Cortellazzo, S., Viero, P., Barbui, T., Colucci, M., and Semeraro, N., 1983, Impaired production of mononuclear cell procoagulant activity in chronic lymphocytic leukaemia, *J. Clin. Pathol.* **36:**37.

Crutchley, D. J., 1984, Effects of inhibitors of arachidonic acid metabolism on thromboplastin activity in human monocytes, *Biochem. Biophys. Res. Commun.* **119:**179.

de Moerloose, P. A., Reber, G. M., and Rickles, F. R., 1985, Interferon-gamma induces procoagulant activity in macrophage cell lines. *Thromb. Haem.* **54**(Suppl.):1173.
de Prost, D., Cramer, E., and Hakim, J., 1979, Inhibition by antimicrotubular agents of the procoagulant activity generated by the blood monocytes, *Am. J. Hematol.* **6**:229.
De Shazo, R. D., Levinson, A. I., Dvorak, H. F., and Davis, R. W., 1979, The late phase skin reaction: Evidence for activation of the coagulation system in a IgE-dependent reaction in man, *J. Immunol.* **122**:692.
Drapkin, R. L., Gee, T. S., Dowling, M. D., Arlin, Z., McKenzie, S., Kempin, S., and Clarkson, B., 1978, Prophylactic heparin in acute promyelocytic leukemia, *Cancer* **41**:2484.
Dunn, C. J., and Willoughby, D. A., 1981, Leukocyte and macrophage migration inhibitory activities in inflammatory exudates—Involvement of the coagulation system, *Lymphokines* **4**:231.
Dvorak, A. M., Connell, A. B., Proppe, K., and Dvorak, H. F., 1978, Immunologic rejection of mammary adenocarcinoma (TA-3-St) in C57BL/6 mice: Participation of neutrophils and activated macrophages with fibrin formation, *J. Immunol.* **120**:1240.
Edgington, T. S., and Chisari, F. V., 1975, Immunological aspects of hepatitis B infection, *Am. J. Med. Sci.* **270**:213.
Edgington, T. S., and Pizzolato, M. C. 1983, Regulatory effects of prostaglandins in the T cell instructed monocyte procoagulant response, *Thromb. Diath. Haemorrh. Suppl.* **50**:148.
Edwards, R. L., and Brande, W., 1985, Effect of bacterial lipopolysaccharide on monocyte tissue factor expression, *Thromb. Haemostasis* **54**:263.
Edwards, R. L., and Perla, D., 1984, Effect of serum on monocyte tissue factor generation, *Blood* **64**:707.
Edwards, R. L., and Rickles, F. R., 1978, Delayed hypersensitivity: Effect of systemic anticoagulation, *Science* **200**:541.
Edwards, R. L., and Rickles, F. R., 1980a, The role of human T cells (and T cell products) for monocyte tissue factor generation, *J. Immunol.* **125**:606.
Edwards, R. L., and Rickles, F. R., 1980b, The role of monocyte tissue factor in the immune response, *Lymphokine Rep.* **1**:181.
Edwards, R. L., and Rickles, F. R. 1981, Abnormalities of blood coagulation in patients with cancer: Mononuclear cell tissue factor generation, *J. Lab. Clin. Med.* **98**:917.
Edwards, R. L., and Rickles, F. R., 1984, Macrophage procoagulants, *Prog. Hemostasis Thromb.* **7**:183
Edwards, R. L., Rickles, F. R., and Bobrove, A. M., 1979, Mononuclear cell tissue factor: Cell of origin and requirements for activation, *Blood* **54**:359.
Edwards, R. L., Rickles, F. R., and Cronlund, M., 1981, Abnormalities of blood coagulation in patients with cancer: Mononuclear cell tissue factor generation, *J. Lab. Clin. Med.* **98**:917.
Edwards, R. L., Rickles, F. R., and Ewan, V. A., 1983, Monocyte procoagulant activity: Recognition by monoclonal anti-tissue factor antibody and evidence for in vivo regulation, *Thromb. Haemostasis* **50**(Suppl.):175.
Edwards, R. L., Cosgrove, J., and Rickles, F. R., 1984, Monocyte tissue factor generation does not require lymphocyte cooperation, *Blood* **64**(Suppl.):262a.
Edwards, R. L., Levine, J. B., Brande, W., and Rickles, F. R., 1984a, Monocyte tissue factor generation and plasma fibrinopeptide A levels are increased in inflammatory bowel disease, *Blood Suppl.* **64**:67a.
Edwards, R. L., Scheiber, E., and Brande, W., 1986, The effect of sodium warfarin on rabbit monocyte tissue factor expression, *Thromb. Res.* **42**:125.
Egbring, R., Schmidt, W., Fuchs, G., and Havemann, K., 1977, Demonstration of granulocyte proteases in plasma of patients with acute leukemia and septicemia with coagulation defects, *Blood* **49**:219.
Evans, R. W., Glick, B., Kimball, F., and Lobell, M., 1969, Fatal intravascular consumption coagulopathy in meningococcal sepsis, *Am. J. Med.* **46**:910.
Ewan, V. A., Edwards, R. L., and Rickles, F. R., 1983, Expression of procoagulant activity in a human monocyte-like cell line. *J. Lab. Clin. Med.* **101**:401.
Farram, E., Geczy, C. L., Moon, D. K., and Hopper, K., 1983, The ability of lymphokine and

lipopolysaccharide to induce procoagulant activity in mouse macrophage cell lines, *J. Immunol.* **130**:2750.

Forman, E. N., Abilgaard, C. F., Bolger, J. F., Johnson, C. A., and Schulman, I., 1969, Generalized Shwartzman reaction: Role of the granulocyte in intravascular coagulation and renal cortical necrosis, *Br. J. Haematol.* **16**:507.

Garg, S. K., and Niemetz, J., 1973, Tissue factor activity of normal and leukemic cells, *Blood* **42**:729.

Geczy, C. L., 1983, The role of clotting processes in the action of lymphokines on macrophages, *Lymphokines* **8**:201.

Geczy, C. L., and Hopper, K. E., 1981, A mechanism of migration inhibition in delayed-type hypersensitivity reactions. II. Lymphokines promote procoagulant activity of macrophages *in vitro*, *J. Immunol.* **126**:1059.

Geczy, C. L., and Meyer, P. A., 1982, Leukocyte procoagulant activity in man: An *in vitro* correlate of delayed-type hypersensitivity, *J. Immunol.* **128**:331.

Geczy, C. L., Farram, E., Moon, D. K., Meyer, P. A., and McKenzie, I. F. C., 1983, Macrophage procoagulant activity as a measure of cell-mediated immunity in the mouse, *J. Immunol.* **130**:2743.

Glenn, W. W. L., Peterson, D. K., and Drinker, C. K., 1942, The flow of lymph from burned tissue, with particular reference to the effects of fibrin formation upon lymph drainage and composition, *Surgery* **12**:685.

Good, R. A., and Thomas, L., 1953, Studies on the generalized Shwartzman reaction. IV. Prevention of the local and generalized Shwartzman reactions with heparin, *J. Exp. Med.* **97**:871.

Gouault-Heilman, M., Chardon, E., Sultan, C., and Josso, F., 1975, The procoagulant factor of leukaemic promyelocytes: Demonstration of immunologic cross-reactivity with human brain tissue factor. *Br. J. Haematol.* **30**:151.

Gralnick, H. R., and Abrell, E., 1973, Studies of the procoagulant and fibrinolytic activity of promyelocytes in acute promyelocytic leukaemia, *Br. J. Haematol.* **24**:89.

Gralnick, H. R., and Sultan, C., 1975, Acute promyelocytic leukemia: Haemorrhagic manifestation and morphologic criteria, *Br. J. Haematol.* **29**:373.

Gralnick, H. R., Bagley, J., and Abrell, E., 1972, Heparin treatment for the hemorrhagic diathesis of acute promyelocytic leukemia, *Am. J. Med.* **52**:167.

Green, D., Ryan, C., Malandruccolo, N., and Nadler, H. L., 1971, Characterization of the coagulant activity of cultured fibroblasts, *Blood* **37**:47.

Gula, A., Bach, R., Konigsberg, W., and Nemerson, V., 1986, Affinity purification of human tissue factor: interaction of factor VII and tissue factor in detergent micelles, *Proc. Natl. Acad. Sci. (USA).* **83**:299.

Halloran, P., Aprile, M., Haddad, G., and Robinette, M., 1982, The significance of elevated procoagulant activity in the monocytes of renal transplant recipients, *Transplant. Proc.* **14**:669.

Hattler, B. G., Jr., Rocklin, R. E., Ward, P. A., and Rickles, F. R., 1973, Functional features of lymphocytes recovered from a human renal allograft, *Cell. Immunol.* **9**:289.

Helin, H. J., and Edgington, T. S., 1983, Allogeneic induction of the human T cell-instructed monocyte procoagulant response is rapid and is elicited by HLA-DR, *J. Exp. Med.* **158**:962.

Helin, H., and Edgington, T. S., 1984, A distinct "slow" cellular pathway involving soluble mediators for the T cell instructed induction of monocyte tissue factor activity in an allogeneic immune response, *J. Immunol.* **132**:2457.

Helin, H. J., Fox, R. I., and Edgington, T. S., 1983, The instructor cell for the human procoagulant monocyte response to bacterial lipopolysaccharide is a Leu-3a$^+$ T cell by fluorescence-activated cell sorting, *J. Immunol.* **131**:749.

Henriksson, P., and Nilsson, I. M., 1979, Effects of leukocytes, plasmin, and thrombin on clotting factors: A comparative *in vitro* study, *Thromb. Res.* **16**:301.

Hiller, E., Saal, J. G., and Riethmüller, G., 1977, Procoagulant activity of activated monocytes, *Haemostasis* **6**:347.

Hogg, N., 1983a, Human monocytes are associated with the formation of fibrin, *J. Exp. Med.* **157**:473.

Hogg, N., 1983b, Human monocytes have prothrombin cleaving activity, *Clin. Exp. Immunol.* **53**:725.

Hogg, N., Slusarenko, M., Cohen, J., and Reiser, J., 1981, Monoclonal antibody with specificity for monocytes and neurons, *Cell* **24:**875.

Hopper, K. E., Geczy, C. L., and Davies, W. A., 1981, A mechanism of migration inhibition in delayed-type hypersensitivity reactions. I. Fibrin deposition on the surface of elicited peritoneal macrophages *in vivo*, *J. Immunol.* **126:**1052.

Horn, R. G., and Collins, R. D., 1968, Studies on the pathogenesis of the generalized Shwartzman reaction, *Lab. Invest.* **18:**101.

Horowitz, H. I., Des Prez, R. M., and Hook, E. W., 1962, Effects of bacterial endotoxin on rabbit platelets. II. Enhancement of platelet factor 3 activity in vitro and in vivo, *J. Exp. Med.* **116:**619.

Hudig, D., and Bajaj, S. P., 1982, Tissue factor-like activity of the human monocytic tumor cell line U937, *Thromb. Res.* **27:**321.

Janco, R. L., and Morris, P. J., 1985, Regulation of monocyte procoagulant by chemoattractants, *Blood* **65:**545.

Jokay, I., and Karczag, E., 1973, Thrombin-induced macrophage disappearance reaction in mice, *Experientia* **29:**334.

Kadish, J. L., Wenc, K. M., and Dvorak, H. F., 1983, Tissue factor activity of normal and neoplastic cells: Quantitation and species specificity, *J. Natl. Cancer Inst.* **70:**551.

Karpaki, J., Varadi, K., and Elodi, S., 1982, Effect of granulocyte proteases on human coagulation factors IX and X: The protective effect of calcium, *Hoppe Seylers Z. Physiol. Chem.* **363:**521.

Kincaid-Smith, P., 1967, Histological diagnosis of rejection of renal homografts in man, *Lancet* **2:**849.

Kincaid-Smith, P., Laver, M. C., and Fairley, K. F., 1970, Dipyridamole and anticoagulants in renal disease due to glomerular and vascular lesions, *Med. J. Aust.* **1:**145.

Kingdon, H. S., Herion, J. C., and Rausch, P. G., 1978, Cellular activation of factor IX (Christmas factor), *Thromb. Res.* **13:**501.

Kociba, G. J., and Griesemer, R. A., 1972, Disseminated intravascular coagulation induced with leukocyte procoagulant, *Am. J. Pathol.* **69:**407.

Koffler, D., and Paronetto, F., 1965, Immunofluorescent localization of immunoglobulins, complement, and fibrinogen in human diseases. II. Acute, subacute, and chronic glomerulonephritis, *J. Clin. Invest.* **44:**1665.

Korn, J. H., Rickles, F. R., Ewan, V. A., Edwards, R. L., Myers, T. J., and Onder, O., 1982, Mononuclear cell modulation of fibroblast procoagulant activity. *J. Lab. Clin. Med.* **99:**657.

Kornberg, A., Treves, A., Rachmilewitz, E. A., and Fibach, E., 1983, Generation of procoagulant activity (PCA) by macrophage-like cells derived from acute and chronic myeloid leukaemia cells in response to phorbol esters, *Scand. J. Haematol.* **31:**102.

Kornberg, A., Treves, A., Rachmilewitz, E. A., and Fibach, E., 1983, Generation of procoagulant activity (PCA) by macrophage-like cells derived from acute and chronic myeoloid leukaemia cells in response to phorbol esters, *Scand. J. Haematol.* **31:**102.

Lerner, R. G., Goldstein, R., and Cummings, G., 1971, Stimulation of human leukocyte thromboplastic activity by endotoxin, *Proc. Soc. Exp. Biol. Med.* **138:**145.

Lerner, R. G., Goldstein, R., and Cummings, G., 1977, Endotoxin induced disseminated intravascular clotting: Evidence that it is mediated by neutrophil production of tissue factor, *Thromb. Res.* **11:**253.

Levy, G. A., and Edgington, T. S., 1980, Lymphocyte cooperation is required for amplification of macrophage procoagulant activity, *J. Exp. Med.* **151:**1232.

Levy, G. A., and Edgington, T. S., 1982, The major histocompatibility complex requirement for cellular collaboration in the murine lymphoid procoagulant response stimulated by bacterial lipopolysaccharide, *J. Immunol.* **128:**1284.

Levy, G. A., Leibowitz, J. L., and Edgington, T. S., 1981a, Induction of monocyte procoagulant activity by murine hepatitis virus type 3 parallels disease susceptibility in mice, *J. Exp. Med.* **154:**1150.

Levy, G. A., Schwartz, B. A., and Edgington, T. S., 1981b, The kinetics and metabolic requirements for direct lymphocyte induction of human procoagulant monokines by bacterial lipopolysaccharide, *J. Immunol.* **127:**357.

Levy, G. A., Helin, H., and Edgington, T. S., 1984, The pathobiology of viral hepatitis and immunologic activation of the coagulation protease network, *Semin. Liver Dis.* **4:**59.

Lindahl, U., Kolset, S. O., Bøgwald, J., Østerud, B., and Seljelid, R., 1982, Studies, with a luminogenic peptide substrate, on blood coagulation factor X/Xa produced by mouse peritoneal macrophages, *Biochem. J.* **206**:231.

Lipinski, B., Nowak, A., and Gurewich, V., 1974, The organ distribution of ^{125}I-fibrin in the generalized Shwartzman reaction and its relation to leucocytes, *Br. J. Haematol.* **28**:221.

Liu, D. T. H., and McCoy, L. E., 1975, Tissue extract thromboplastin: Quantitation, fractionation and characterization of protein components, *Thromb. Res.* **7**:199.

Lorenzet, R., Peri, G., Locati, D., Allavena, P., Colucci, M., Semararo, N., Mantovani, A., and Donati, M. G., 1983, Generation of procoagulant activity by mononuclear phagocytes: A possible mechanism contributing to blood clotting activation within malignant tissues, *Blood* **62**:271.

Lyberg, T., and Prydz, H., 1980, Lectin stimulation of tissue thromboplastin activity in human monocytes in vitro, *Thromb. Haemostasis* **42**:1574.

Lyberg, T., and Prydz, H., 1981, Phorbol esters induce the synthesis of thromboplastin activity in human monocytes, *Biochem. J.* **194**:699.

Lyberg, T., and Prydz, H., 1983, Is lymphocyte co-operation necessary for thromboplastin synthesis by human monocytes?, *Clin. Exp. Immunol.* **53**:731.

Lyberg, T., Hetland, O., and Prydz, H., 1982a, Synthesis of thromboplastin protein by a murine macrophage-like cell line, *Thromb. Haemostasis* **47**:154.

Lyberg, T., Nilsson, K., and Prydz, H., 1982b, Synthesis of thromboplastin by U-937 cells, *Br. J. Haematol.* **51**:631.

Lyberg, T., Prydz, H., Baklien, K., and Høyeraal, H. M., 1982c, Effect of immune-complex containing sera from patients with rheumatic diseases on thromboplastin activity of monocytes, *Thromb. Res.* **25**:193.

McGehee, W. G., Rapaport, S. I., and Hjort, P. F., 1967, Intravascular coagulation in fulminent meningococcemia, *Ann. Int. Med.* **67**:250.

McKay, D. G., Gitlin, D., and Craig, J. M., 1959, Immunochemical demonstration of fibrin in the generalized Shwartzman reaction, *Arch. Pathol.* **67**:270.

Maier, R. V., 1982, Characterization of the procoagulant activity induced in rabbit Kupffer cells by bacterial lipopolysaccharide, *J. Reticuloendothel. Soc.* **32**:64.

Maier, R. V., and Ulevitch, R. J., 1981, The induction of a unique procoagulant activity in rabbit hepatic macrophages by bacterial lipopolysaccharides, *J. Immunol.* **127**:1596.

Margaretten, W., and McKay, D. G., 1969, The effect of leukocyte antiserum on the generalized Shwartzman reaction, *Am. J. Pathol.* **57**:299.

Maynard, J. R., Heckman, C. A., Pitlick, F. A., and Nemerson, Y., 1975, Association of tissue factor activity with the surface of cultured cells, *J. Clin. Invest.* **55**:814.

Maynard, J. R., Drayer, B. E., Stemerman, M. B., and Pitlick, F. A., 1977, Tissue factor coagulant activity of cultured human endothelial and smooth muscle cells and fibroblasts, *Blood* **50**:387.

Menkin, V., 1933, Fixation mechanism in sterile inflammation, *Proc. Soc. Exp. Biol. Med.* **30**:1069.

Menkin, V., 1938, The role of inflammation in immunity, *Physiol. Rev.* **18**:366.

Morgan, D., Edwards, R. L., and Rickles, F. R., 1985, T cell collaboration is not required for the enhanced monocyte tissue factor generation in cancer patients, *Clin. Res.* **33**:350.

Morrison, D. C., and Cochrane, C. G., 1974, Direct evidence for Hageman factor (factor XII) activation by bacterial lipopolysaccharide, *J. Exp. Med.* **140**:797.

Morrison, D. C., and Ulevitch, R. J., 1978, The effects of bacterial endotoxin on host mediation systems, *Am. J. Pathol.* **93**:527.

Movat, H. Z., More, R. H., and Wolochow, D., 1960, Cellular and intercellular changes after mechanical, chemical, or radiation injury of connective tissue, *Br. J. Exp. Pathol.* **41**:97.

Muhlfelder, T. W., Niemetz, J., Kreutzer, D., Beebe, D., Ward, P. A., and Rosenfeld, S. I., 1979, C5 chemotactic fragment induces leukocyte production of tissue factor activity, a link between complement and coagulation, *J. Clin. Invest.* **63**:147.

Muhlfelder, T. W., Niemetz, J., and Kang, S., 1982, Glucocorticoids inhibit the generation of leukocyte procoagulant (tissue factor) activity, *Blood* **60**:1169.

Müller-Berghaus, G., Bohn, E., and Höbel, W., 1976, Activation of intravascular coagulation by endotoxin: The significance of granulocytes and platelets, *Br. J. Haematol.* **33**:213.

Nelson, D. S., 1965, The effects of anticoagulants and other drugs on cellular and cutaneous reactions to antigen in guinea-pigs with delayed-type hypersensitivity, *Immunology* **9**:219.
Nelson, D. S., and Boyden, S. V., 1963, The loss of macrophages from peritoneal exudates following the injection of antigens into guinea-pigs with delayed-type hypersensitivity, *Immunology* **6**:264.
Nemerson, Y., 1968, The phospholipid requirements of tissue factor in blood coagulation, *J. Clin. Invest.* **47**:72.
Nemerson, Y., and Bach, R., 1982, Tissue factor revisited, *Prog. Hemostasis Thromb.* **6**:237.
Nesheim, M. E., Taswell, J. B., and Mann, K. G., 1979, The contribution of bovine factor V and factor Va to the activity of prothrombinase, *J. Biol. Chem.* **254**:10952.
Nesheim, M. E., Eid, S., and Mann, K. G., 1981, Assembly of prothrombinase in the absence of prothrombin, *J. Biol. Chem.* **256**:9874.
Niemetz, J., 1972a, The role of protein synthesis on the generation of tissue factor activity by leukocytes, *Proc. Soc. Exp. Biol. Med.* **139**:1276.
Niemetz, J., 1972b, Coagulant activity of leukocytes: Tissue factor activity, *J. Clin. Invest.* **51**:307.
Niemetz, J., and Fani, K., 1971, The role of leukocytes in blood coagulation and the generalized Shwartzman reaction, *Nature New Biol.* **232**:247.
Østerud, B., and Bjørklid, E., 1982, Human factor VII associated with endotoxin stimulated monocytes in whole blood, *Biochem. Biophys. Res. Commun.* **108**:620.
Østerud, B., and Flaegstad, T., 1983, Increased tissue thromboplastin activity in monocytes of patients with meningococcal infection: Related to an unfavorable prognosis, *Thromb. Haemostasis* **49**:5.
Østerud, B., and Rapaport, S. I., 1977, Activation of factor IX by the reaction product of tissue factor and factor VII: Additional pathway for initiation of blood coagulation, *Proc. Natl. Acad. Sci. USA* **74**:5260.
Østerud, B., Lindahl, U., and Seljelid, R., 1980, Macrophages produce blood coagulation factors, *FEBS Lett.* **120**:41.
Østerud, B., Bögwald, J., Lindahl, U., and Seljelid, R., 1981, Production of blood coagulation factor V and tissue thromboplastin by macrophages in vitro, *FEBS Lett.* **127**:154.
Ottaway, C. A., Warren, R. E., Saibil, F. G., Fung, L. G., Fair, D. S., and Levy, G. A., 1984, Monocyte procoagulant activity in Whipple's disease, *J. Clin. Immunol.* **4**:348.
Pahwa, S. G., Smithwick, E. M., Grimes, E. R., O'Reilly, R. J., Pahwa, R. N., and Good, R. A., 1978, Chemotactic defects in severe combined immunodeficiency, *J. Pediatr.* **92**:43.
Paronetto, F., and Koffler, D., 1965, Immunofluorescent localization of immunoglobulins, complement, and fibrinogen in human diseases. I. Systemic lupus erythematosus, *J. Clin. Invest.* **44**:1657.
Paterson, P. Y., 1982, Molecular and cellular determinants of neuroimmunologic inflammatory disease, *Fed. Proc.* **41**:2569.
Pitlick, F. A., 1976, Binding of bovine brain tissue factor to concanavalin A-Sepharose, *Biochim. Biophys. Acta* **428**:27.
Plow, E., 1982, Leukocyte elastase release during blood coagulation: A potential mechanism for activation of the alternative fibrinolytic pathway, *J. Clin. Invest.* **69**:564.
Plow, E. F., and Edgington, T. S., 1975, An alternate pathway for fibrinolysis. I. The cleavage of fibrinogen by leukocyte proteases at physiologic pH, *J. Clin. Invest.* **56**:30.
Prydz, H., and Lyberg, T., 1980, Effect of some drugs on thromboplastin (factor III) activity of human monocytes in vitro, *Biochem. Pharmacol.* **29**:9.
Prydz, H., Allison, A. C., and Schorlemmer, H. V., 1977, Further link between complement activation and blood coagulation, *Nature (London)* **270**:173.
Prydz, H., Lyberg, T., Deteix, P., and Allison, A. C., 1979, In vitro stimulation of tissue thromboplastin (factor III) activity in human monocytes by immune complexes and lectins, *Thromb. Res.* **15**:465.
Quick, A. J., 1936, On various properties of thromboplastin acqueous tissue extracts, *Am. J. Physiol.* **114**:282.
Ratnoff, O. D., 1969, Some relationships among hemostasis, fibrinolytic phenomena, immunity, and the inflammatory response, *Adv. Immunol.* **10**:148.
Rickles, F. R., and Edwards, R. L., 1981, Activation of monocyte tissue factor (MTF) in response to antigen recognition, *Blood Suppl.* **58**:225a.

Rickles, F. R., and Edwards, R. L., 1983, Activation of blood coagulation in cancer: Trousseau's syndrome revisited, *Blood* **62**:14.
Rickles, F. R., and Rick, P. D., 1977, Structural features of *Salmonella typhimurium* lipopolysaccharide required for activation of tissue factor in human mononuclear cells, *J. Clin. Invest.* **59**:1188.
Rickles, F. R., Hardin, J. A., Pitlick, F. A., Hoyer, L. W., and Conrad, M. E., 1973, Tissue factor activity in lymphocyte cultures from normal individuals and patients with hemophilia A, *J. Clin. Invest.* **52**:1427.
Rickles, F. R., Levin, J., Hardin, J. A., Barr, C. F., and Conrad, M. E., Jr., 1977, Tissue factor generation by human mononuclear cells: Effects of endotoxin and dissociation of tissue factor generation from the mitogenic response, *J. Lab. Clin. Med.* **89**:792.
Rickles, F. R., de Moerloose, P., Hancock, W. W., Ewan, V. A., Boyd, A. W., and Burgess, A. W., 1983, Expression of tissue factor in human blood cells, *Blood Suppl.* **62**:293a.
Rickles, F. R., Hancock, W. W., Boyd, A. W., Stanley, I., Ewan, V. A. and Burgess, A. W., 1984, Tissue factor related antigen(s) (TF:RAg): Preliminary molecular characterization and tissue distribution determined with the monoclonal antibody HTF-1, *Clin. Res.* **32**:320a.
Rivers, R. P. A., Hathaway, W. E., and Weston, W. L., 1975, The endotoxin-induced coagulant activity of human monocytes, *Br. J. Haematol.* **30**:311.
Rothberger, H., Zimmerman, T. S., Spiegelberg, H. L., and Vaughan, J. H., 1977, Leukocyte procoagulant activity: Enhancement of production in vitro by IgG and antigen–antibody complexes, *J. Clin. Invest.* **59**:549.
Rothberger, H., Zimmerman, T. S., and Vaughan, J. H., 1978, Increased production and expression of tissue thromboplastin-like procoagulant activity in vitro by allogeneically stimulated human leukocytes, *J. Clin. Invest.* **62**:649.
Rothberger, H., Dove, F. B., Lee, T.-K., McGee, M. P., and Kardon, B., 1983, Procoagulant activity of lymphocyte-macrophage populations in rabbits: Selective increases in marrow, blood, and spleen cells during Shwartzman reactions, *Blood* **61**:712.
Rothberger, H., Barringer, M., and Meredith, J., 1984, Increased tissue factor activity of monocytes/macrophages isolated from canine renal allografts, *Blood* **63**:623.
Schmidt, W., Egbring, R., and Havemann, K., 1974, Effect of elastase-like and chymotrypsin-like neutral proteases from human granulocytes on isolated clotting factors, *Thromb. Res.* **6**:315.
Schwartz, B. S., and Edgington, T. S., 1981a, Lymphocyte collaboration is required for induction of murine monocyte procoagulant activity by immune complexes, *J. Immunol.* **127**:438.
Schwartz, B. S., and Edgington, T. S., 1981b, Immune complex-induced human monocyte procoagulant activity. I. A rapid unidirectional lymphocyte-instructed pathway, *J. Exp. Med.* **154**:892.
Schwartz, B. S., Levy, G. A., Curtiss, L. K., Fair, D. S., and Edgington, T. S., 1981, Plasma lipoprotein induction and suppression of the generation of cellular procoagulant activity in vitro, *J. Clin. Invest.* **67**:1650.
Schwartz, B. S., Levy, G. A., Fair, D. S., and Edgington, T. S., 1982, Murine lymphoid procoagulant activity induced by bacterial lipopolysaccharide and immune complexes is a monocyte prothrombinase, *J. Exp. Med.* **155**:1464.
Semeraro, N., Biondi, A., Lorenzet, R., Locati, D., Mantovani, A., and Donati, M. B., 1983, Direct induction of tissue factor synthesis by endotoxin in human macrophages from diverse sites, *Immunology* **50**:529.
Shands, J. W., Jr., 1983, The endotoxin-induced procoagulant of mouse exudate macrophages: A factor-X activator, *Blood* **62**:333.
Shands, J. W., Jr., 1985, Macrophage factor X activator formation: Metabolic requirements for synthesis of components, *Blood* **65**:169.
Shapiro, S. S., and McKay, D. G., 1958, The prevention of the generalized Shwartzman reaction with sodium warfarin, *J. Exp. Med.* **107**:377.
Sitrin, R. G., Kaltreider, H. B., Ansfield, M. J., and Webster, R. O., 1983, Procoagulant activity of rabbit alveolar macrophages, *Am. Rev. Respir. Dis.* **128**:282.
Sitrin, R. G., Kaltreider, H. B., and Goldyne, M. E., 1984, Prostaglandin E is required for the augmentation of procoagulant activity of LPS-stimulated rabbit alveolar macrophages, *J. Immunol.* **132**:867.

Starzl, T. E., Lerner, R. A., Dixon, F. J., Groth, C. G., Brettschneider, L., and Terasaki, P. I., 1968, Shwartzman reaction after human renal homotransplantation, *N. Engl. J. Med.* **278**:642.

Stetson, C. A., 1951, Similarities in the mechanisms determining the Arthus and Shwartzman phenomena, *J. Exp. Med.* **94**:347.

Stetson, C. A., and Good, R. A., 1951, Studies on the mechanism of the Shwartzman phenomenon: Evidence for the participation of polymorphonuclear leukocytes in the phenomenon, *J. Exp. Med.* **93**:49.

Sundsmo, J. S., and Fair, D. S., 1983, Relationship among the complement, kinin, coagulation and fibrinolytic systems in the inflammatory reaction, *Clin. Physiol. Biochem.* **1**:225.

Suttie, J. W., and Stenflo J., 1977, Oral anticoagulant therapy: The biosynthetic basis, *Semin. Hematol.* **15**:365.

Takaki, A., Enfield, D. L., and Thompson, A. R., 1983, Cleavage and inactivation of factor IX by granulocyte elastase, *J. Clin. Invest.* **72**:1706.

Thiagarajan, P., and Niemetz, J., 1980, Procoagulant-tissue factor activity of circulating peripheral blood leukocytes: Results of in vivo studies, *Thromb. Res.* **17**:891.

Tsao, B. P., Fair, D. S., Curtiss, L. K., and Edgington, T. S., 1984, Monocytes can be induced by lipopolysaccharide-triggered lymphocytes to express functional VII/VIIa protease activity, *J. Exp. Med.* **159**:1042.

van Dam-Miers, M. C. E., Muller, A. D., van Deijk, W. A., and Hemker, M. C., 1985, Clotting factors secreted by monocytes and macrophages: Analytical considerations, *Thromb. Res.* **37**:9.

Van Ginkel, C. J. W., Zeijlemaker, W. P., Stricker, L. A. M., Oh, J. I. H., and Van Aken, W. G., 1981, Enhancement of monocyte thromboplastin activity by antigenically stimulated lymphocytes—A link between immune reactivity and blood coagulation, *Eur. J. Immunol.* **11**:579.

Wachtfogel, Y. T., Kucick, U., James, H. L., Scott, C. F., Schapira, M., Zimmerman, M., Cohen, A. B., and Colman, R. W., 1983, Human plasma kallikrein releases neutrophil elastase during blood coagulation, *J. Clin. Invest.* **72**:1672.

Westphal, O., 1975, Bacterial endotoxins, *Int. Arch. Allergy Appl. Immunol.* **49**:1.

Young, T. J., Niemetz, J., and Ohnuma, T., 1981, Procoagulant activity in HL-60, ML-2 and KG-1 human myelogenous leukemia cell lines, *Blood,* **58**:81a.

Zacharski, L., and McIntyre, O. R., 1973, Procoagulant synthesis by cultured fibroblasts triggered by cell adhesion, *Nature (London)* **232**:338.

Zeldis, S. M., Nemerson, Y., Pitlick, F. A., and Lentz, T. L., 1972, Tissue factor (thromboplastin): Localization to plasma membranes by peroxidase-conjugated antibodies, *Science* **175**:766.

Zimmerman, T. S., 1976, The coagulation mechanism and the inflammatory response, in: *Textbook of Immunopathology* (P. A. Miescher and H. J. Müller-Eberhard, eds.), Volume 1, pp. 95–115, Grune & Stratton, New York.

Zur, M., Radcliffe, R. D., Oberdick, J., and Nemerson, Y., 1982, The dual role of factor VII in blood coagulation, *J. Biol. Chem.* **257**:5628.

Zvaifler, N. J., 1973, The immunopathology of joint inflammation in rheumatoid arthritis, *Adv. Immunol.* **16**:265.

11

Leukotrienes

MICHAEL K. BACH

1. INTRODUCTION

Slow-reacting substance of anaphylaxis (SRS-A) has been known for over 40 years and has been suspected to play an important role in the symptomatology of asthma for a major part of that time. There has been a virtual explosion of interest in this substance and in some of the chemically related molecules in the 5 years or so which have passed since the structures of the leukotrienes were established and since their chemical synthesis made them more readily available. This has led to detailed studies of the actions of the leukotrienes and to the discovery of their actions on various smooth muscle preparations. In addition, many other activities, both *in vitro* and *in vivo*, which may be linked to asthma and to other inflammatory conditions have been recognized.

Several reviews of work in this area are available (Orange and Austen, 1969a; Samuelsson *et al.*, 1980; Goetzl, 1981; Goetzl *et al.*, 1984; Lewis and Austen, 1984). In the present chapter, I will bring the information up to date as far as possible in this rapidly progressing field, and I will target my comments to a readership which is not primarily involved in research on leukotrienes but which, nonetheless, stands to reap considerable benefits from the knowledge which is being acquired in this area.

2. NOMENCLATURE

The old designation *slow-reacting substance of anaphylaxis* will be used only in the historical context in this review. The modern nomenclature for the slow-reacting substances and for structurally related materials was coined by Samuelsson. The name *leukotriene* (LT), which is the basis for this nomenclature, is derived from the fact that these molecules are produced by leukocytes and that they all have three conjugated double bonds. The letter designations following

MICHAEL K. BACH • Department of Hypersensitivity Diseases Research, The Upjohn Company, Kalamazoo, Michigan 49001.

the abbreviation LT are given in the order in which the materials were identified. Thus, LTA stands for the central 5,6-epoxide intermediate in the formation of all other 5,6 and 5,12 leukotrienes; LTB is the biologically active, 5(*S*),12(*R*)-dihydroxy-6,14-*cis*-8,10-*trans*-eicosatetraenoic acid. The other dihydroxy acids which are derived from the opening of the epoxide in LTA_4 are specified as to the position of the hydroxyls and as to their stereochemistry (e.g., 5*S*,12*S*-LTB_4). This is to be contrasted with other 5,12-dihydroxy acids (designated DHETEs) which are formed by the double lipoxygenation of arachidonate at positions 5 and 12. LTC, D, E, and F are 6-sulfidopeptide derivatives in which the peptide portion is, respectively, glutathione, cysteinyl glycine, cysteine, and γ-glutamyl cysteine (Samuelsson *et al.*, 1979; Samuelsson and Hammarström, 1980). The subscripts following the letter designations indicate the total number of ethylenic double bonds in the molecule. Thus, the materials which are derived from arachidonic acid all have the subscript "4" while analogues which are formed from eicosapentaenoic acid are given the subscript "5," and those derived from dihomo-γ-linolenic acid have the subscript "3."

More recently, it has become apparent that primary lipoxygenolytic attack on arachidonate at positions other than No. 5 can give rise to metabolites which suggest that an epoxide intermediate was again formed. Thus, the formation of a 14,15 epoxide and its conversion to a 15-hydroxy-14*S*-glutathionyl eicosatetraenoic acid have been described (Sok *et al.*, 1982; Rådmark, *et al.*, 1982), while 8,15-dihydroxy acids have also been reported (Maas *et al.*, 1981; Lundberg *et al.*, 1981). The trivial designation of these molecules includes the position of substitution; thus, 14,15-LTA_4, 8,15-LTB_4, and so on.

3. HISTORICAL

3.1. DISCOVERY OF SRS-A

Evidence for the existence of spasmogenic substances in the plasma of patients suffering from asthma dates back to a report by Zeydner in 1922 (cited by Harkavy, 1930). When Harkavy applied aqueous solutions of ethanolic extracts of the sputum of asthmatic patients to smooth muscle preparations of cat jejunum or rabbit duodenum in a chymograph, the muscle strips contracted only in the presence of specimens which had been collected during an asthmatic attack which had been induced by inhalation of antigen and not in the presence of extracts obtained from sputum collected before the challenge or after the attack had subsided. The name *slow-reacting substance* (SRS) was coined some 10 years later by Feldberg and Kellaway (1938), and by Kellaway and Trethewie (1940), who found that perfusion of small intestines or lungs of ovalbumin-immunized guinea pigs with solutions containing antigen resulted in the elaboration of a spasmogenic substance which could be detected in the effluents. The name *slow-reacting substance of anaphylaxis,* which persisted for the next 18 years in the history of these materials, was coined by Brocklehurst (1960), who also suggested that, in the absence of evidence for the identity of the materials which are elicited in tissues from different species, a superscript be employed to designate the species of origin of the material being studied. Much of this early progress was reviewed by Brocklehurst (1962), who suggested, based on indirect arguments, that SRS-A may be responsible for eliciting the symptoms of asthma. The more intensive effort through the 1960s has been summarized by Orange and Austen (1971).

The immunologic basis for the production of SRS-A received considerable attention particularly after the identification of IgE as the immunoglobulin which is responsible for atopy. Thus, it was of interest to demonstrate that rodent IgE could elicit the production of SRS-A in rodents under conditions which were suitable for the release of histamine. Both IgE and the heat-stable homocytotropic antibody of the rat, IgG2a, were shown to be capable of eliciting this response (Orange and Austen, 1969b). However, several factors hinted that there may be a dissociation between the release of preformed mediators of anaphylaxis and the production of SRS-A:

1. It was consistently observed that while IgE was much better at eliciting histamine release, IgG2a caused much more SRS-A to be formed.
2. Polymorphonuclear neutrophils (PMNs) rather than mast cells were primarily responsible for SRS-A production when IgG2a was used (Orange *et al.*, 1967).
3. Pharmacologic dissociation of the production of the two mediators was possible since disodium cromoglycate selectively inhibited histamine release while diethylcarbamazine selectively inhibited SRS-A formation (Orange and Austen, 1968).

3.2. PROOF OF STRUCTURE

Two events helped overcome many of the problems which had hampered progress in the determination of the structure of SRS-A: One was the discovery (Bach and Brashler, 1974) that challenge of rat peritoneal cells both *in vitro* and *in vivo* with the calcium ionophore A23187 in the presence of calcium could induce the production of large amounts of an SRS. The other event was the discovery that the incorporation of large amounts of cysteine in the *in vitro* incubations could markedly increase the amount of SRS or SRS-A produced (Orange and Chang, 1975). These findings were followed by the observation that, in addition to the primary cells which had to be isolated from rats, rat basophil leukemia (RBL-1) cells, an established tumor cell line, also could be induced to produce an SRS under similar conditions (Jakschik *et al.*, 1977b), making it possible to generate large amounts of SRS. However, the demonstration of the identity of the components which constituted SRS from different cell types and SRS-A from different species had to be deferred until the structures of the components were fully defined.

The observation that production of SRS could be enhanced by cysteine immediately suggested that cysteine may be a precursor of the molecule. It was also found that arylsulfatases from many different sources inactivated SRS and SRS-A (Orange *et al.*, 1974). In fact, the ability of these enzymes to inactivate a new slow-reacting substance became one of the accepted criteria for the characterization of that substance as an SRS.

The inactivation of SRSs by arylsulfatases led to the assumption that the SRSs had a sulfate ester linkage in them and that the arylsulfatases were inactivating the active spasmogens by cleaving this linkage. It developed much later that the inactivation of SRSs by arylsulfatases was due to the contamination of these enzyme preparations with a ubiquitous leucine amino peptidase which was able to convert the biologically active LTD_4 to the much less active LTE_4 (Sok *et al.*, 1980). The major role of the cysteine which was used in the incubations to produce leukotrienes, and of the other mercaptocarboxylic acids which had been found to be more active than cysteine in this regard (Bach and Brashler, 1979), was shown to be the inhibition of this enzyme, while at the same time promoting the conversion of LTC_4 to the biologically more active LTD_4. Two groups independently examined the possibility that the ionophore-induced SRSs might be derived from arachidonic acid (Jakschik *et al.*, 1977a; Bach *et al.*, 1977). The approach was both to demonstrate that production of the SRSs could be inhibited by inhibitors of arachidonate metabolism such as 4,8,11,14-eicosatetraynoic acid (ETYA) and at the same time to demonstrate that radioactivity from highly labeled arachidonic acid could be incorporated into the purified biologically active molecules. Only a fraction of one percent of the total radiolabeled arachidonate which was taken up by the cells was excreted in association with SRS (Bach *et al.*, 1977, 1980c). Nonsteroidal anti-inflammatory compounds (NSAICs) at concentrations which are sufficient to inhibit the cyclooxygenase of the cells did not inhibit production of the spasmogens. In fact, their addition appeared to enhance production of SRS (Engineer *et al.*, 1976). This supported the conclusion that the SRSs were not prostaglandins and suggested, in addition, that they may be derived from a previously unrecognized pathway of arachidonate metabolism.

The determination of the structures of the SRSs depended on the evolving technology of high-pressure liquid chromatography, which first yielded evidence that the SRSs are composed of two or more physically separable materials (Yecies *et al.*, 1979a; Bach *et al.*, 1979). Another key observation was the finding that purified preparations of SRSs had a triplet absorption band in the 270- to 290-nm range in the ultraviolet (Morris *et al.*, 1979).

The interpretation of the triplet absorption peaks was arrived at in studies which were initially unrelated to the quest for the structure of SRS-A. Samuelsson's laboratory had been studying the products obtained when arachidonic acid is attacked by various lipoxygenases. They had found that rabbit PMNs metabolized arachidonate to a family of dihydroxy acids which were unique at the time because they had a triplet of absorption peaks in the ultraviolet with λ_{max} at 259, 269, and 279 nm (Borgeat and Samuelsson, 1979a). The triplet peaks suggested the existence of three conjugated double bonds in the molecules. Some of these products had hydroxyl groups at the 5 and 12 positions while others were substituted at the 5 and 6 positions. Studies with radioactive O_2 revealed that only the hydroxyl in position 5 was derived from atmospheric oxygen while the second hydroxyl was derived from water. This led to the suggestion that these dihydroxy acids were derived from a hypothetical 5,6-epoxy-eicosatetraenoic acid intermediate (Borgeat and Samuelsson, 1979b) and that the substitution at position 12 was due to the resonance nature of the intermediates which would be formed when this epoxide is opened. They coined the name *leukotrienes* for the molecules having the absorption in the ultraviolet and designated the hypothetical epoxide intermediate as leukotriene A_4. LTA_4 was subsequently trapped as the methyl ester and purified (Rådmark *et al.*, 1980a). Its conversion both to the biologically active dihydroxy acid, LTB_4, and to one of the components of SRS-A, LTC_4, have been described as well (Rådmark *et al.*, 1980b,c). The stereochemistry of LTA_4 and LTB_4 were confirmed by total synthesis (Rådmark *et al.*, 1980b; Corey *et al.*, 1980). LTB_4 is 5(*S*),12(*R*)-dihydroxy-6-*cis*-8,10-*trans*-14-*cis*-eicosatetraenoic acid.

The resemblance of the ultraviolet absorption of SRS to that of the dihydroxy acids, and the observation that radioactivity from ^{35}S-labeled cysteine was incorporated into SRS in a thiol ether linkage (Parker *et al.*, 1979) led to the suggestion that SRS may be related to the dihydroxy-eicosatetraenoic acids. The observation that the biologic activity of the SRSs was destroyed by incubation with soybean lipoxygenase (Sirois, 1979), and that the ultraviolet spectrum of the inactivated product was shifted to a higher wavelength, was evidence that the fourth ethylenic bond in the molecule which had been out of conjugation was now shifted into conjugation. This defined the location of the thiol ether substituent at position 6 and led to the proposed structure of the first component of SRS (Murphy *et al.*, 1979). The correct structure of the peptide portion and the stereochemistry of the molecule were confirmed by comparison with a series of totally synthetic molecules (Hammarström *et al.*, 1979, 1980).

A second major component of SRS was shown to be the product of γ-glutamyl transpeptidase action on LTC_4, and was designated LTD_4 (Örning *et al.*, 1980). The structure of this substance, which is the major component of SRS-A in both human lung and guinea pig lung, was independently confirmed by a team led by Piper and Morris (Morris *et al.*, 1980a,b,c). A third component of the SRSs was subsequently identified as the 6-cysteinyl derivative, LTE_4, which is formed by cleaving the peptide bond in cysteinyl glycine in LTD_4 (Lewis *et al.*, 1980). LTE_4 can accept a γ-glutamyl residue from glutathione to form LTF_4 (Anderson *et al.*, 1982), although the presence of this compound in SRSs has not been reported.

With the definition of the structures of the leukotrienes, the rigorous characterization of the SRSs from different species became possible. This has been accomplished by comparing the components of SRS from mastocytoma cells to those from rat peritoneal mononuclear cells (Bach *et al.*, 1980a,b), and from anaphylactically or ionophore-challenged guinea pig lung and RBL cells (Morris *et al.*, 1980b). Other comparisons are being added to this list.

4. ASSAY

The standard assay for the thiol ether leukotrienes until very recently had been the guinea pig ileum. In order to achieve the necessary selectivity, the response to histamine has to be blocked with an antihistamine, and spontaneous activity is usually suppressed by the inclusion of atropine or another anticholinergic agent in the muscle bath (Orange and Austen, 1976). It has been useful to pretreat leukotriene-containing preparations with chymotrypsin or to heat them in alkali in order to avoid possible interference in the assay by spasmogenic peptides which may contaminate crude biological preparations (Bach, unpublished results). It is also useful to check responses using the selective end-organ antagonist of the leukotrienes, FPL 55712. When used at the appropriate concentration ($< 1\ \mu g/ml$), inhibition of the spasmogenic response by this compound is highly indicative that the responding material is a leukotriene (Augstein *et al.*, 1973). A problem with the assay is the inability to discriminate between the various thiol ether leukotrienes. In a stop-flow arrangement, the relative sensitivity of the assay to LTD_4, LTC_4, and LTE_4 is in the ratio 1.0 : 0.3 : 0.12. It is quite easy to detect 80 pg of LTD_4 in a 3-ml assay bath. Radioimmunoassays for the leukotrienes are gradually becoming available. One, described by Hayes *et al.* (1983), is some ten times more sensitive than the bioassay but suffers with respect to selectivity within the sulfidopeptide leukotriene family. The same lack of specificity has been experienced by other groups (Levine *et al.*, 1981; Aharony *et al.*, 1983) while others (e.g., Lindgren *et al.*, 1983; Wynalda *et al.*, 1984) have reported extreme specificity even to the point of discriminating between the natural 11-*cis* and the spontaneously arising 11-*trans* isomers of LTC_4. Because of the lack of specificity, and the absence for the moment of a specific radioimmunoassay for LTD_4 and LTE_4, it is commonly recognized that analysis of unknown samples and especially biologic fluids by radioimmunoassay must be coupled with high-pressure liquid chromatographic separation of the agonists (Leitch *et al.*, 1983b; Peters *et al.*, 1983; Ziltener *et al.*, 1983). Methods for the rapid extraction and quantitative isolation of these substances are continuing to be refined (Metz *et al.*, 1982; Clancy and Hugli, 1983; Borgeat, 1984; Verhagen *et al.*, 1984b). The need for this procedure is the more important due to the general finding that unfractionated biologic fluids, and especially serum and plasma, contain unidentified substances which cross-react with the antibodies to sulfidopeptide leukotrienes but which do not cochromatograph with any of the known sulfidopeptide leukotrienes. The apparent abundance of these materials is of the order of 0.5 to 1 ng/ml of LTC_4 equivalents.

The original assay for LTB_4 relied on the potent chemotactic activity of this material for PMNs. However, this response is not selective for LTB_4 even among the metabolites of arachidonate or the products of 5-lipoxygenase. Guinea pig parenchymal tissue was found to contract in response to LTB_4 in a dose-dependent fashion and this response was proposed as a bioassay (Sirois *et al.*, 1981c). The response could be totally inhibited by adding NSAICs to the muscle bath and this afforded a means of differentiating the response from that due to the thiol ether leukotrienes. The preparation of immunogens which can be used to immunize rabbits with LTB_4 has been described (Young *et al.*, 1983), as have radioimmunoassays which are selective for LTB_4 (Salmon *et al.*, 1982; Lewis *et al.*, 1982b). A monoclonal antibody to LTB_4 which, however, only has approximately one-tenth the sensitivity of the polyclonal rabbit antibodies, has recently been described (J. Y. Lee *et al.*, 1984).

5. STRUCTURE–ACTIVITY STUDIES

Studies of structure–activity relationships are useful in defining the properties of the enzymes involved in the metabolism of the leukotrienes and in defining the properties of the receptors for these substances. The 5-lipoxygenase can accept a variety of polyunsaturated long-chain fatty acids as substrate. Thus, acids with 19 to 21 carbons and having three or more unconjugated double bonds have been shown to be converted by the enzyme from RBL-1 cells to metabolites which comigrated with 5-HETE and 5,12-DHETEs on thin-layer chromatography (Jakschik *et al.*, 1980). The double bond at position 5 was essential for activity. Replacement of the 5,6 double bond by an acetylenic bond resulted in a potent inhibitor of the enzyme (Corey and Munroe, 1982). Similarly, 7,7-dimethyl analogues of arachidonate as well as the 7-ethano derivative were potent inhibitors of the 5-lipoxygenase (Nicolaou *et al.*, 1983; Welton *et al.*, 1984).

There is great selectivity in the structure of LTB_4. Stereoisomers of the 5 and 12 hydroxyls, as well as stereoisomers of the double bond configuration, have virtually no biologic activity (Ford-Hutchinson *et al.*, 1981, 1982; Lewis *et al.*, 1981b; Sirois *et al.*, 1981b; Dahinden *et al.*, 1984). Acetylation of the hydroxyls, or amidation of the carboxylic acid group reduced biologic activity and, in the latter case, resulted in the formation of a partial agonist (Goetzl and Pickett, 1981; Showell *et al.*, 1982b). Further metabolism of LTB_4 leads to the formation of 5,12,20-trihydroxy-eicosatetraenoic acid and 5,12-dihydroxy-1,20-eicosatetraendioate, which retain approximately 25% of the activity of LTB_4 (Palmblad *et al.*, 1982). The ability of eicosapentaenoic acid, which is relatively abundant in fish oil, to substitute for arachidonate is of considerable interest at the moment. LTB_5, the product which is derived from this acid via the 5-lipoxygenase, the LTA synthetase, and the LTA epoxide hydrolase, is reported to have 1/8th or 1/30th the biologic activity of LTB_4 (T. H. Lee *et al.*, 1984b; Terano *et al.*, 1984).

LTC_4, LTD_4, and LTE_4 differ considerably, both in their absolute and in their relative potencies in different assay systems. On the other hand, LTC_3, LTD_3, and LTC_5 and LTD_5, which are formed by mastocytoma cells when dihomo-γ-linolenic acid or eicosapentaenoic acid replaces arachidonate, have activities which closely resemble those of LTC_4 and LTD_4, respectively (Hammarström, 1980, 1981). The 11-*trans* isomers of LTC_4 and LTD_4 contaminate most biologic preparations of LTC_4 and LTD_4 and are somewhat less active than the more abundant 11-*cis* isomer. Considerable effort has gone into studying the effect of the stereochemistry at the 5 and 6 positions and of the triene double bonds on the activity of analogues of LTC_4 and LTD_4 (Tsai *et al.*, 1982; Baker *et al.*, 1982) and also into more detailed studies of structure–activity relationships (Drazen *et al.*, 1981; Young *et al.*, 1981; Lewis *et al.*, 1981a; Jones *et al.*, 1982b). The following structural features appear to have a major impact on the biologic activity of the molecules:

1. The carboxylic acid group of glycine, but not that of the fatty acid.
2. A free amino group.

3. The hydroxyl function at position 5.
4. The conformations of positions 5 and 6. Specifically, only the 5*S*,6*R* conformation had high activity.
5. A long aliphatic chain beyond carbon No. 6. However, *cis–trans* changes in the 9 and 11 double bonds, and even saturation of all the double bonds resulted in compounds which retained some activity. Shifting the substituents to 14*S*-peptidyl-15-hydroxy LTC_4 (Sok *et al.*, 1982) resulted in the loss of most of the activity.

On the other hand, oxidation of the thiol ether to the sulfone resulted in compounds which retained most if not all of their activity (Girard *et al.*, 1982).

Docosahexaenoic acid is not a substrate for the 5-lipoxygenase. The synthetic analogue of LTC_4 (LTC_6) has been prepared and has only 0.01% the smooth muscle-contracting activity on the guinea pig ileum of LTC_4 (Corey *et al.*, 1983). A series of sulfidopeptide leukotriene analogues have been prepared in which the ethylenic bond at position 14 is replaced by an acetylenic bond, presumably to render the molecules resistant to attack by the 15-lipoxygenase, and the thiol ether is modified in a variety of manners. While some of these compounds had considerable activity, none were found to be antagonists of the natural leukotrienes (Cho *et al.*, 1983).

Moving the polar substituents along the fatty acid chain has had varying results: On the one hand, 8,9-LTC_3, the product derived from dihomo-γ-linolenic acid (Baker *et al.*, 1983), had essentially the same activity as LTC_4 on the guinea pig lung strip (Dahlén *et al.*, 1983b); on the other hand, 2-nor leukotriene analogues had only 10% of the contractile activity of the natural leukotrienes but were competitive antagonists of the latter compounds (Gleason *et al.*, 1983).

The structural requirements for other actions of the leukotrienes are different from those which may involve a receptor on smooth muscle cells. Thus, the hydrophobic residue was more critical for the competitive inhibition of arylsulfatase by the leukotrienes than it was in the action of these compounds on smooth muscle (Weller *et al.*, 1981). Similarly, the ability to induce mucus production in human lung explants (see Section 10.5) did not require the presence of any of the carbons of the fatty acid chain beyond carbon No. 8 (Coles *et al.*, 1983).

6. RECEPTORS

The structure–activity studies clearly suggested the existence of specific receptors for the leukotrienes and, furthermore, the possibility that the receptors on different organs and tissues may indeed be different. Receptors for LTB_4 were first described by Goldman and Goetzl (1982), by Goldman *et al.* (1983), and by Kreisle and Parker (1983) although the equilibrium binding constant which was reported from these studies differed by as much as 1000-fold. Specificity of binding of structural analogues of LTB_4 to these receptors was very similar to the specificity of the same analogues for replacing LTB_4 in the activation of the cells. Further study by Goldman and Goetzl (1984) revealed the existence of both low- and high-affinity receptors for LTB_4 on PMNs. Interestingly, the presence of the high-affinity receptors (K_D 3.9×10^{-10} M, 4400 receptors/cell) was associated with chemotactic response while the activation of the low-affinity receptors (K_D 6.8×10^{-8} M, 250,000/cell) was associated with induction of degranulation. In addition to PMNs, a subset of T cells has been shown to possess high-affinity receptors for LTB_4. This subset appears to be the $OKT8^+$ population which is ordinarily associated with suppressor function. On the other hand, the $OKT4^+$ subset, which is associated with helper function, appears to have a low-affinity receptor for LTB_4 (Payan *et al.*, 1984). Receptor occupancy on T cells is also reflected in functional changes in that the same group has reported the dose-dependent inhibition of lymphokine generation and mitogen-induced proliferation with submicromolar concentrations of LTB_4 (Payan and Goetzl, 1983).

Krell *et al.* (1983) suggested the existence of two classes of receptors for sulfidopeptide leukotrienes based on carefully constructed Schild plots for the inhibitory activity of the antagonist FPL 55712 (see Section 12.4). This was a more extensive, quantitative treatment of an observation which had initially been made by Drazen *et al.* (1980). Several groups showed that binding of LTC_4 to membrane fractions from homogenized lung was saturable (Pong *et al.*, 1983; Hogaboom *et al.*, 1983). Structure–activity studies revealed that the saturation by LTC_4 occurred at concentrations which were at least 100 times lower than the concentrations of LTD_4 which were required, while FPL 55712 required at least 1000 times the concentration of LTC_4, as compared to LTD_4, to compete for the receptors. Further study has shown that binding of LTC_4 is markedly affected by high concentrations of calcium ion, while guanine nucleotides did not affect binding in contrast to the data with LTD_4 (see Section 7; Bruns *et al.*, 1983). A continuous cell line which is derived from the vas deferens of the Syrian hamster, and which is designated DDT_1 MF-2, was shown to possess specific receptors for LTC_4 with a K_d reported as 5 and 21 nM (Krilis *et al.*, 1983b; Clark *et al.*, 1984). The subcellular distribution of the receptor for LTC_4 on ileal smooth muscle cells has been investigated. A large fraction of these receptors appear to be located intracellularly and are only available for binding once the cells have been homogenized. The functional significance for this localization is not clear (Krilis *et al.*, 1984).

Receptors selective for LTD_4 have been identified in membrane preparations from lung (Pong and DeHaven, 1983; Mong *et al.*, 1984; Cheng and Townley, 1984a,b). FPL 55712 is a competitive inhibitor for the binding of LTD_4 to these receptors (Mong *et al.*, 1984), and both GTP and sodium ion increase the rate of binding while decreasing the maximum extent of binding. On the other hand, calcium and other divalent metals increased the extent of binding (Pong and DeHaven, 1983). Indirect evidence suggests the existence of a third class of receptors which are presumably selective for LTE_4 (Cheng and Townley, 1984c; Krilis *et al.*, 1984), although no direct demonstration of the existence of such a receptor has been reported.

7. CELLS PRODUCING LEUKOTRIENES AND ELICITING STIMULI

7.1. VARIOUS TISSUES, VARIOUS STIMULI

The tissue which has received the most attention as a potential source of leukotrienes is the lung. Thus, production of SRS-A has been shown for chopped human, monkey, and guinea pig lung, sensitized with antibody and challenged with antigen; perfused guinea pig lung; perfused rat lung challenged with antigen (Farmer *et al.*, 1975); and more recently for guinea pig lung challenged with the calcium ionophore (Piper and Seale, 1978). Cat nasal mucosa has been reported to generate an SRS though the characterization of the product(s) is incomplete (Anggård and Strandberg, 1981). Recently, the spasmogen in the perfusate of cat paws which had been exposed to compound 48/80 has been characterized as a mixture of LTD_4 and LTE_4 (Houglum *et al.*, 1980). Various other organs have been reported to generate leukotrienes, including synovial fluid (Lewis *et al.*, 1982a), kidney (Benveniste *et al.*, 1981; Van Praag and Farber, 1981, 1983), and spleen (Malik and Wang, 1981). Various secretions have been found to contain leukotrienes, including tears (Bisgaard *et al.*, 1984), seminal fluid (Saad and Burka, 1983), and the sputum of patients with chronic bronchitis and cystic fibrosis and, to a lesser extent, asthma (O'Driscoll *et al.*, 1984).

Leukotriene generation appears to be generally associated with an inflammatory response, regardless of the tissue being studied. Thus, suitable stimulation (e.g., with A23187) has resulted in the production of sulfidopeptide leukotrienes in a variety of vascular preparations (Piper *et al.*, 1983), the gastrointestinal tract (Wöbling *et al.*, 1983a,b), and brain (Dembinska-Kieć *et al.*, 1984; Moskowitz *et al.*, 1984; Lindgren *et al.*, 1984).

SRS has been reported to be present as a "preformed" mediator in human lung (Turnbull *et al.*, 1976). This may imply that these leukotrienes play a role in the homeostatic control of smooth muscle tone or in mucus production. Anaphylatoxins have been reported to induce the generation of leukotrienes in guinea pig lung (Regal and Pickering, 1981; Stimler *et al.*, 1982), and very recently, it has been reported that the vasoconstriction which can be induced in perfused rat lung preparations by the addition of a platelet-activating factor (PAF) to the perfusing fluid appears to be due to the generation of leukotrienes in these preparations in a nonimmunologic manner (Voelkel *et al.*, 1982).

7.2. POLYMORPHONUCLEAR LEUKOCYTES

The 5-lipoxygenase pathway of PMNs can be activated with A23187 giving rise to LTB and other DHETEs (Borgeat and Samuelsson, 1979a; Ford-Hutchinson *et al.*, 1980). An "eosinophil chemotactic factor" (ECF), which these cells also produce upon stimulation with the ionophore, has now been shown to be largely identical to LTB_4 (Czarnetzki *et al.*, 1983). It is also possible to activate PMNs for the production of leukotrienes by more physiologically relevant stimuli. Thus, activation with PAF resulted in the production of ECF (Czarnetzki and

Benveniste, 1981), and activation with serum-coated zymosan (Claesson *et al.*, 1981), LPS (Bremm *et al.*, 1984), or carrageenan *in vivo* (Siegel *et al.*, 1981) also resulted in formation of products of the 5-lipoxygenase pathway. However, activation of PMNs with the chemotactic peptide formylmethionyl-leucyl-phenylalanine (fMLP) resulted in the predominant formation of ω-oxidation products of LTB_4, 5,12,20-trihydroxy-eicosatetraenoic acid, and the 1,20-dicarboxylic acid (Palmblad *et al.*, 1982). Prostaglandins have been reported to inhibit the production of LTB by PMNs (Ham *et al.*, 1983), a regulatory role which may be interesting in view of the fact that the very same activating signals which cause the production of LTB can also activate other cells in the inflammatory reaction, especially the mononuclear cells, to produce the prostaglandins which might then help to limit the extent of the response.

PMNs can also produce thiol ether leukotrienes. Much of the SRS which was produced by ionophore-challenged human peripheral blood leukocytes could be traced back to the PMNs in the preparation (Conroy *et al.*, 1976). Human PMNs which were incubated in the presence of cysteine or glutathione and the ionophore generated some LTC_4 (Hansson and Rådmark, 1980), and this has been recently confirmed by using the radioimmunoassay to measure the LTC_4 which was produced (Aeringhaus *et al.*, 1982). However, when PMNs are highly activated, such as by the use of high concentrations of phorbol diesters, they rapidly destroy LTC_4 by a peroxidative cleavage of the thiol ether linkage giving rise to a mixture of 5*S*,12*S*- and 5*S*,12*R*-6-*trans*-LTB_4 (Lee *et al.*, 1982a). It is now clear that the extent of LTC production by pure PMNs is only about 10% of the production of LTB by these cells while the bulk of the LTC which has often been reported to have been produced by PMNs may well be due to the presence of contaminating eosinophils in the preparations (see further in Section 7.5; Verhagen *et al.*, 1984a). Nonetheless, there is apparently variability between species in that bovine PMNs have been reported to produce more LTC_4 than LTB_4 (Walstra *et al.*, 1984).

7.3. MACROPHAGES AND MONONUCLEAR CELLS

The initial observation that SRS could be generated with the calcium ionophore by peritoneal exudate cells from rats, and that neither mast cells nor PMNs appeared to be required for this process, eventually led to the demonstration that a weakly phagocytic subpopulation of mononuclear cells was responsible (Bach and Brashler, 1978; Orange *et al.*, 1980). However, all attempts to find a physiologically relevant stimulus with which leukotriene production might be induced in these cells were unsuccessful (Bach *et al.*, 1982a). On the other hand, murine peritoneal and alveolar macrophages could be induced to generate large amounts of LTC_4 upon stimulation with zymosan (Bretz *et al.*, 1980; Rouzer *et al.*, 1980, 1982a), and mouse peritoneal macrophages and rat alveolar macrophages could also be induced to produce LTC_4 upon challenge with immune complexes containing IgE (Rouzer *et al.*, 1982b; Rankin *et al.*, 1982, 1984).

Human peripheral blood mononuclear cells can generate an SRS upon stimulation with the calcium ionophore (Gunter and Stechschulte, 1980; Goldyne *et al.*, 1984; Williams *et al.*, 1984) although the nature of the arachidonate metabolites which human mononuclear cells produce appears to be critically dependent on the presence of contaminating platelets and/or serum in the preparations (Pawlowski *et al.*, 1983). Human peritoneal macrophages have been reported to produce LTB_4 and LTC_4 upon stimulation (Du *et al.*, 1983). Human alveolar macrophages have been reported to generate ECF (Czarnetzki, 1981) or LTB_4 (Bray *et al.*, 1981b). The capacity of cells from bronchoalveolar lavage to generate LTB_4 has been studied in more detail recently, and a capacity to produce as much as 7 to 23 ng LTB/10^6 cells has been reported (MacDermot *et al.*, 1984; Martin *et al.*, 1984). Leukotriene production in cells from healthy volunteers and from patients with pulmonary diseases has been compared, and the production of significant amounts of LTD_4 has been reported (Damon *et al.*, 1983; Godard *et al.*, 1983, 1984).

7.4. MAST CELLS AND BASOPHILS

Rat peritoneal mast cells were shown to be capable of generating SRS upon stimulation with the calcium ionophore (Yecies *et al.*, 1979a,b), but the amount produced per cell was no greater than the amount that was produced by the much more abundant mononuclear cells, suggesting that the contribution of the peritoneal mast cells to the total pool of leukotrienes produced was small. This observation seemed to be in conflict with the notion that mast cells "should be" the primary source of SRS-A by virtue of the belief that asthma was linked to atopy which, in turn, is caused by the activation of mast cells and basophils. A possible explanation can be found in the recent report that so-called E-type (bone marrow-derived) murine mast cells are capable of producing large amounts of LTC_4 (Razin *et al.*, 1982). LTB_4 production by these cells takes place at approximately one-tenth the rate of LTC_4 production (Mencia-Huerta *et al.*, 1983). A recent report also documents that dog mastocytoma cells can produce large amounts of LTB, LTC, and LTD (Phillips *et al.*, 1983).

Turning from rodents to humans, a report several years ago documented that leukemic basophils produced leukotrienes along with three other mediators of anaphylaxis (Lewis *et al.*, 1975). Recently, the ability of enzymatically disaggregated and highly purified human lung mast cells to produce large amounts of LTD_4 was reported (MacGlashan *et al.*, 1982). Dog mastocytoma cells have also been reported to produce large amounts of leukotrienes when they are anaphylactically challenged with canine IgE, ionophore, or a platelet factor IV-derived peptide but not with compound 48/80. The abundance of LTB_4, LTC_4, and LTD_4 in the cell supernatants was roughly comparable (Goetzl *et al.*, 1983).

7.5. EOSINOPHILS

As already noted, the variable production of LTC by PMNs which has been encountered by many investigators has been explained by the contamination of the PMN preparations by variable numbers of eosinophils (Weller *et al.*, 1983; Shaw *et al.*, 1984). Henderson *et al.* (1984) have compared the production of all the products of the 5-lipoxygenase pathway in eosinophils from normal donors and in cells from patients with hypereosinophilic syndrome and found the spectrum of metabolites to be comparable (Henderson *et al.*, 1984).

7.6. OTHER CELLS

Equine eosinophils stimulated with the calcium ionophore generate LTB_4, LTC_4, and LTD_4 (Jörg *et al.*, 1982). On the other hand, human eosinophils apparently predominantly produce products of the 15-lipoxygenase (Turk *et al.*, 1982). The total amount of LTB_4 produced in these cells was only approximately 15% of the amount of 5,15-DHETE produced.

Rabbit platelets have been reported to produce LTC and LTD (Mencia-Huerta *et al.*, 1981) although no evidence of a 5-lipoxygenase has otherwise been found in these cells. Similarly, lipoxygenase products have, reportedly, been found when human T cells were incubated with arachidonate (Goetzl, 1982) although it is not clear if this production is due to the T cells or the 10–14% other cells which contaminated the preparations.

8. BIOSYNTHESIS, INTERCONVERSION, AND METABOLIC BREAKDOWN

The pathway to the formation of the biologically active leukotrienes, beginning with phospholipid-bound arachidonic acid, is summarized in Fig. 1.

8.1. ARACHIDONATE MOBILIZATION

The first step in leukotriene biosynthesis is the mobilization of the precursor polyunsaturated fatty acid from the phospholipids of the cell membrane. This step can be circumvented by supplying either arachidonic acid or a structural analogue of arachidonate exogenously, although there are a growing number of examples which will not be considered here in detail which suggest that the utilization of exogenous arachidonate (or alternative substrate) is not necessarily the same as that of endogenous, phospholipid- or lipid-bound arachidonate. The products from exogenous 5,8,11-eicosatrienoic acid have been especially useful in detailed and elegant studies of the metabolism of the thiol ether leukotrienes by isolated cells and tissues (Ormstad *et al.*, 1982), their distribution in the body, and their elimination. Such studies have been reported for the mouse (Appelgren and Hammarström, 1982), the guinea pig (Hammarström, 1982), and the monkey (Hammarström *et al.*, 1981).

The mobilization of arachidonate from phospholipids is mediated by a phospholipase. Formally, the reaction is that of a phospholipase A_2, but there is evidence to suggest that, at least in some systems, the reaction is actually a

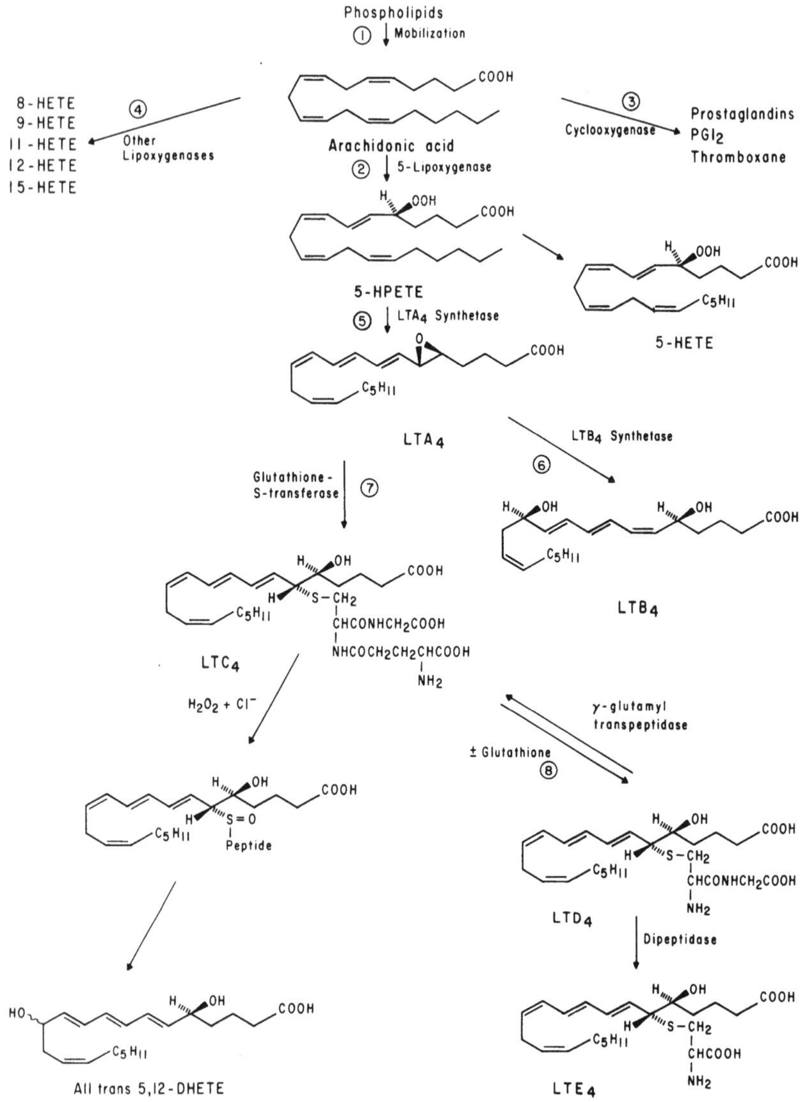

FIGURE 1. Schematic for the biosynthesis and metabolism of the leukotrienes.

combination of the actions of a phospholipase C acting on phosphatidylinositol and the further metabolism of the diacyl glycerol which is formed to either monoacyl glycerol plus free fatty acid or to phosphatidic acid. The phosphatidic acid, in turn, gives rise to lysophosphatidic acid and free fatty acid (Lapetina *et al.*, 1981). The so-called selective inhibitors of phospholipase A_2, *p*-bromophenacylbromide and mepacrine, are actually not sufficiently selective to permit discrimination between the direct, phospholipase A_2-dependent pathway and this indirect pathway (Hofmann *et al.*, 1982).

It is not clear if the mobilization of arachidonate makes this compound equally accessible to all the enzymes which a given cell may have for metabolizing it. There have been repeated suggestions in the literature that mobilization may be selective. Some of these suggestions have been critically reviewed (Irvine, 1982). More recent reports have suggested that different secretagogues activated phospholipid metabolism along different pathways in the same cells (Homma *et al.*, 1982; Clancy *et al.*, 1983; Bonney and Humes, 1984). There have also been suggestions that peritoneal macrophages contain two distinct phospholipase A_2 activities (Wightman *et al.*, 1981a) in addition to a phospholipase C (Wightman *et al.*, 1981b), and that the differential activation of these enzymes may explain why the cells produce prostaglandins in response to certain stimuli and leukotrienes in response to others (Humes *et al.*, 1982). It also appears that mobilized arachidonate from one cell can be excreted by that cell unchanged and can then be utilized by another cell in its milieu (Marcus *et al.*, 1984).

8.2. 5-LIPOXYGENASE

The role of lipoxygenase in the formation of the leukotrienes was initially inferred from the observation that nonselective inhibitors of arachidonate metabolism, such as ETYA, inhibited formation of the leukotrienes whereas selective inhibitors of the cyclooxygenase did not. A supposedly selective inhibitor of the 12-lipoxygenase of platelets, 5,8,11-eicosatriynoic acid (Hammarström, 1977), was reported to also inhibit leukotriene synthesis (Örning and Hammarström, 1980); other analogues of arachidonate in which the double bond at position 5 was replaced by an acetylenic bond have also been reported to be active inhibitors. Exogenous 5-HPETE, the product of the action of the 5-lipoxygenase on arachidonate, has been shown to be incorporated into the thiol ether leukotrienes and LTA_4 has been recognized as an intermediate in this synthesis (Hammarström and Samuelsson, 1980); (Falkenhein *et al.*, 1980). Little progress has been made in the purification of the 5-lipoxygenase. The enzyme is present in the high-speed supernatant fraction from homogenates of RBL-1 cells and is calcium-dependent (Jakschik *et al.*, 1980, 1982). Furukawa *et al.* (1984) achieved further purification by relying on the ability of the enzyme to bind to ATP, which they had previously reported stimulates the activity of the enzyme. The substrate specificity of the 5-lipoxygenase is quite broad and eicosapentaenoic acid is actually a better substrate than is arachidonic acid. Conversion of 5,8,11-eicosatrienoic acid to LTA_3 was reported to occur at approximately half the rate of the conversion of arachidonate to LTA_4 (Jakschik *et al.*, 1983).

8.3. REGULATORY EFFECTS OF 12-HPETE AND OF 15-HPETE

The production of 5-HETE and LTB_4 by peripheral blood leukocytes was recently reported to show a complicated dependence on the extent of platelet contamination of the leukocyte preparations (Maclouf *et al.*, 1982). This was explained by the finding that 12-HPETE, a product of the 12-lipoxygenase of platelets, caused a specific and dose-dependent enhancement of the activity of the 5-lipoxygenase of human leukocytes.

The product of the 15-lipoxygenase, 15-HPETE, has been reported to enhance the production of SRS-A in anaphylactically challenged guinea pig lung (Engineer *et al.*, 1978). This action may be due to the inhibition of PGI_2 formation by 15-HPETE, which would result in lower levels of AMP. Increased levels of cAMP are known to inhibit the release and production of mediators of anaphylaxis. However, an alternative explanation seems possible since 15-HPETE was shown to cause increased production of other hydroxy-eicosatetraenoic acids, and this stimulation could be blocked by ETYA (Koide *et al.*, 1982). Thus, it is possible that 15-HPETE may stimulate the action of 5-lipoxygenase.

15-HPETE is broken down to the more stable hydroxide, 15-HETE, and the two compounds were nearly equipotent as inhibitors of the 5-lipoxygenase (Vanderhoek *et al.*, 1980). This action of 15-HETE has been used, in part, to document a role for products of the 5-lipoxygenase in the action of T lymphocytes (Goetzl, 1982; Bailey *et al.*, 1982). The production of 15-HETE in human leukocytes increases exponentially with the concentration of exogenous arachidonate (Fruteau de Laclos *et al.*, 1984). Thus, the nature of the metabolites which are produced under a given set of conditions is necessarily a complicated function of the specific cells which may be present in the incubation, the activating stimulus, and the exogenous concentration of arachidonate. In this regard, it is interesting that although eicosapentaenoic acid, which is mobilized upon stimulation in the PMNs from donors that had ingested large amounts of this acid, can be converted to LTB_5, it actually causes a profound inhibition of LTB_4 generation in those cells (Prescott, 1984).

8.4. LTA_4 AND LTB_4 SYNTHETASES

The hypotheses concerning the existence of LTA_4, and its role as an intermediate in the biosynthesis of LTB_4 and LTC_4 were subsequently confirmed by the isolation of the methanolysis product of LTA_4 from PMNs (Borgeat and Samuelsson, 1979b; Rådmark *et al.*, 1980a) and the demonstration that synthetic LTA_4 could be converted by the cells to LTB_4 and LTC_4 (Rådmark *et al.*, 1980b,c). Both the LTA_4 synthetase and the LTA_4 hydrolase (i.e., LTB_4 synthetase) are present in the high-speed supernatant fractions of RBL-1 cell homogenates. ETYA (Bokoch and Reed, 1981) as well as several other tri- and tetraacetylenic long-chain fatty acids (Jakschik *et al.*, 1981) have been shown to inhibit the formation of LTA_4 in RBL-1 cell homogenates. More recently, it was suggested that the antifilarial drug diethylcarbamazine selectively inhibits the conversion of 5-HPETE to LTA_4 in murine mastocytoma cells (Mathews and Murphy, 1982) though data from the author's laboratory (Bach *et al.*, 1986) indicate that the latter compound can inhibit the subsequent step in the generation of the sulfidopeptide leukotrienes as well.

Little is known about the LTA_4 synthetase of mammalian cells. However, Shimizu *et al.* (1984) reported that 5-lipoxygenase from potato tubers which had been purified to homogeneity was capable both of generating 5-HPETE and of converting 5-HPETE to LTA_4. Jakschik and Kuo (1983) presented kinetic evidence that the lipoxygenase and the LTA synthetase are inactivated in the course of the synthesis of their respective products, presumably by oxidative action of the peroxide substrate or product.

Epoxide hydrolases (i.e., the enzymes which synthesize LTB_4 from LTA_4) have received considerable attention. In addition to the enzymes in leukocytes, erythrocytes have been found to be a rich source of this activity (Fitzpatrick *et al.*, 1984), and plasma has also been shown to contain such an activity (Fitzpatrick *et al.*, 1983). LTA_4 is not a substrate for the epoxide hydrolases of liver, which are capable of hydrolyzing a variety of arachidonate epoxides, however.

8.5. LTA_4: GLUTATHIONE *S*-TRANSFERASE

Depletion of glutathione, either by supplying a substrate for glutathione *S*-transferase (Parker *et al.*, 1980b) or by inhibiting glutathione biosynthesis (Rouzer *et al.*, 1981), resulted in the inhibition of LTC_4 biosynthesis. The mixed glutathione *S*-transferases of rat liver are capable of synthesizing LTC_4 from LTA_4 (Bach *et al.*, 1982a). However, it appears that the different subforms of liver glutathione *S*-transferase differ in their capacity to accept LTA_4 as a substrate, and this difference appears to be unrelated to the specificity of the same enzyme for other more traditional substrates (Bach and Brashler, 1985). The synthesis of LTC_4 in cells and tissues which are ordinarily recognized as sources of these substances occurs in a high-speed particulate fraction of the cell homogenates (Jakschik *et al.*, 1982; Bach *et al.*, 1984b). In RBL-1 cells, the LTC_4-synthesizing activity is not associated with any classical glutathione *S*-transferase activity but rather appears to be the function of a unique enzyme. This conclusion is also supported by extensive studies with inhibitors of the classical glutathione *S*-transferases which, on the whole, had little inhibitory activity on this new enzyme whereas diethylcarbamazine, which is inactive on liver glutathione *S*-transferases, is an inhibitor of the LTC synthetase of RBL cells (Bach *et al.*, 1986).

8.6. γ-GLUTAMYL TRANSPEPTIDASE, LTD_4 DIPEPTIDASE

Kinetic evidence for a precursor–product relationship between the two spasmogenic components of SRS actually preceded the determination of the structures of these components (Bach *et al.*, 1980c) and the conversion of LTC_4 to LTD_4 by commercial γ-glutamyl transpeptidase was a critical step in the proof of the structure of LTD_4 (Örning *et al.*, 1980). The conversion in RBL-1 cells was inhibited by several inhibitors of this enzyme (Örning and Hammarström, 1980). The reaction of the γ-glutamyl transpeptidase is an equilibrium reaction. Thus, in the presence of excess glutathione, a condition which exists in liver, LTD_3 was converted to LTC_3. Similarly, a γ-glutamyl residue can be added to LTE_4 to form the γ-glutamyl-cysteinyl derivative, LTF_4 (Anderson *et al.*, 1982). As will be discussed in more detail below, the kinetics of the contractile response of guinea pig ileum to LTC and LTD are different; there is a lag of approximately 1 min in duration before contraction caused by LTC can be discerned, whereas the contraction due to LTD begins as soon as the agonist is added to the tissue bath. The conversion of LTC to LTD takes place in the mucosal lining of the guinea pig ileum and thus it was found that, once this layer is removed, the kinetics of the contractile response to LTC change, requiring a distinct lag phase which was not seen when the mucosal layer was present (Krilis *et al.*, 1983a).

Two distinct routes have been recognized for the "inactivation" of the thiol ether leukotrienes. A variety of cells and body fluids are capable of converting LTD_4 to LTE_4 which, in general, has only approximately 10% of the spasmogenic activity of LTD_4 (Parker *et al.*, 1980a,c; Örning *et al.*, 1981; Sok *et al.*, 1981). The "LTDase" which is responsible for this conversion has been shown to reside in the specific granule fraction from human PMNs (Lee *et al.*, 1982b). The dipeptidase action can also be carried out by cathepsin H (Yokota *et al.*, 1983). A recent report by Hammarström indicated that LTE_4 is metabolized further, at least in the rat, to form *N*-acetyl LTE_4 which was found in the urine. This metabolite had approximately 10% of the smooth muscle contractile activity on the guinea pig ileum as did LTC_4 (Hammarström, *et al.*, 1985).

8.7. ω-OXIDATION, PEROXIDATIVE INACTIVATION

The second route of inactivation of the thiol ether leukotrienes is peroxidative. It has been reported that the peroxidase from eosinophils, in the presence of halide ions, can cause the inactivation of the thiol ether leukotrienes (Henderson *et al.*, 1982). A similar reaction also takes place in the presence of PMNs if the cells are stimulated with relatively high concentrations of the phorbol ester. In that case, three pairs of closely related metabolites have been found. One pair is biologically active and immunologically cross-reactive with LTC. These have been identified as the two stereoisomers of the sulfoxide of LTC_4. Another pair has very low cross-reactivity with LTC in the radioimmunoassay and has no biologic activity. The third pair is both biologically and immunochemically inactive and has been shown to be 5*S*,12*S*- and the 5*S*,12*R*-6-*trans*-LTB_4 (Lee *et al.*, 1982a). The enzymatic reaction can be mimicked by the Haber–Weiss reaction, even in the absence of any enzymes (Henderson and Klebanoff, 1983b). Furthermore, neutrophils from both patients with chronic granulomatous disease and those with a genetically defined myeloperoxidase deficiency were capable of this oxidative degradation of the sulfidopeptide leukotrienes, indicating that both a myeloperoxidase and an ·OH radical-dependent process are responsible for the oxidative metabolism of LTC in normal neutrophils (Henderson and Klebanoff, 1983a; Henderson *et al.*, 1984).

LTB_4 is rapidly metabolized to the ω-hydroxy compound and the corresponding dicarboxylic acid. The 5,12,20-trihydroxy-eicosatetraenoic acid was the major urinary metabolite found when LTB_4 was infused into monkeys (Serafin *et al.*, 1984). The oxidation is apparently catalyzed by a cytochrome P_{450} since it can be inhibited by carbon monoxide (Shak and Goldstein, 1984).

9. ACTIONS OF LTB_4

9.1. LEUKOCYTE ACTIVATION

The best-studied and -documented actions of LTB_4 involve the activation of leukocytes. These actions range from the chemokinetic and chemotactic effects, which were the first to be subjected to detailed analysis, to the biochemical events leading to leukocyte activation ranging from receptor occupancy, through effects on second messengers, to energy uptake and, finally, to the expression of functional changes. The studies of the receptor for LTB_4 have been discussed in an earlier section.

The mechanism of transmembrane signal transmission to the inside of the cell following the activation of a cell by LTB_4 has been examined. Activation of human leukocytes by LTB_4 in the presence of an NSAIC caused a dose-dependent rise in cAMP which peaked by 30 sec (Claesson, 1982).

At concentrations in the range of 3×10^{-8} to 3×10^{-6} M, LTB_4 caused dose-dependent and specific changes of calcium flux in rabbit PMNs which were time-dependent and were largely analogous to the changes which could also be induced with the chemotactic peptide. The evidence suggested that LTB_4 caused a rapid increase in membrane permeability to calcium and a release of calcium from previously unexchangeable intracellular pools leading to a transient increase in intracellular calcium concentration. Sodium ion fluxes were also increased by LTB_4 (Sha'afi *et al.*, 1981a; Naccache *et al.*, 1982). The effects of LTB_4 resembled those of fMLP. However, cross-deactivation studies revealed that whereas fMLP caused deactivation to itself and to LTB_4, LTB_4 only deactivated the cells to further exposures to LTB_4 (Sha'afi *et al.*, 1981b; O'Flaherty *et al.*, 1981). A corollary of cell activation is an increased requirement for metabolic energy. Thus, it is interesting that another effect of LTB_4 on PMNs is to increase the rate at which they take up 2-deoxyglucose from the medium (Bass *et al.*, 1981).

9.2. *IN VITRO* EFFECTS

Collectively, the actions of LTB_4 are consistent with a role of this mediator in the expression of the inflammatory response (Smith *et al.*, 1980). Thus, it attracts inflammatory cells to the affected site, causes their adhesion to endothelial cells at the site, and causes their aggregation. Furthermore, it stimulates these cells to release their lysosomal enzymes which, in turn, can cause an inflammatory response. In addition to acting on pro-inflammatory leukocytes, LTB_4 induces human T suppressor cells in a dose-dependent fashion and in the same range of concentrations which are effective in activating the inflammation-causing cells (Rola-Pleszczynski *et al.*, 1982).

9.2.1. Chemotaxis and Chemokinesis

Although other hydroxy acids have chemotactic activity, LTB_4 is some 100 times more potent as a chemoattractant and chemostimulator than any of the monohydroxy acids which had been tested (Bray *et al.*, 1981a,b). Consideration of the actions of the lipoxins is both premature at this time and outside the sphere of this review. As little as 10 pg of LTB_4 could be readily detected in the assays for chemotaxis, making this compound equipotent to fMLP and anaphylatoxin, the two most potent chemotactic agents known. The marked structural specificity of LTB_4 was confirmed in studies using leukocytes from rabbits, rats, and humans (R. M. J. Palmer *et al.*, 1980). In this regard, it is essential to distinguish LTB from all the other stereoisomers of 5,12 DHETE, some of which are quite difficult to resolve from LTB on reverse phase HPLC (Jubiz, 1983). The chemotactic and chemokinetic activities of LTB_4 have been shown for PMNs, mononuclear cells, and eosinophils (Czarnetzki *et al.*, 1983).

A number of animal models have been described in which the chemotactic activity of LTB_4 can be demonstrated *in vivo*. These include measurement of the accumulation of PMNs in the peritoneal or pleural cavity upon the implantation of a sponge or similar material containing LTB_4 (Smith *et al.*, 1980), measurement of leukocyte accumulation at injected skin sites in the rabbit (Higgs *et al.*, 1982), and similar measurements in the humor of the rabbit eye (Bhattacherjee *et al.*, 1981). In the last case, the contralateral eye served as a control, and 25 ng of LTB_4 were readily detected.

9.2.2. Leukocyte Aggregation

The dose-dependence of leukocyte aggregation upon exposure to leukotrienes is quite similar to that for chemotaxis and chemokinesis. It is not clear if the mechanism of aggregation is different from that of chemotaxis or if aggregation is merely an expression of the desensitization to the chemotactic stimulus added to the membrane depolarization which is known to be caused by this mediator. Despite some earlier papers which suggested the opposite position, the current weight of evidence indicates that the activation of PMNs by other agents, such as fMLP or PAF, is mediated by the induction of LTB_4 formation in the PMNs by these agents (Lin *et al.*, 1982; Chilton *et al.*, 1982). Thus, the role of LTB_4 appears to be central to the activation of the cells.

9.2.3. Leukocyte Adherence

When LTB_4 was injected intravenously into hamsters, there was a marked increase in the tendency of the leukocytes to adhere to the walls of the capillaries in the cheek pouch. This adherence-inducing activity was dose-dependent and occurred at the same concentrations of LTB_4 which induced increased vascular permeability in this model (see Section 9.3.1; Dahlén *et al.*, 1981).

9.2.4. Lysosomal Enzyme Release

LTB_4, at nanomolar to micromolar concentrations, caused a dose-dependent though modest increase in the secretion of lysosomal enzymes from PMNs. The effect was stereospecific in that LTB_4 was at least 50-fold more potent than the closest analogue (5*S*,12*S*-6-*trans*-DHETE). While fMLP caused up to a 10-fold increase in the release of lysosomal enzymes over untreated controls, LTB_4 caused only a doubling. However, the concentrations of the two secretagogues which were required for maximal stimulation were comparable. The release which was induced by LTB_4 was inhibitable by ETYA. The nature of this inhibition remains unclear; it does not appear to involve an inhibition of *de novo* synthesis of more leukotrienes in the target cells (Hafström *et al.*, 1981; Rae and Smith, 1981; Showell *et al.*, 1982a).

9.3. *IN VIVO* EFFECTS

9.3.1. Vascular Permeability

A standard model for measuring the effects of drugs on vascular permeability is to measure the extravasation of a colloidal dye or a radiolabeled serum protein into the dermis of animals which have been intravenously injected with the tracer and challenged intradermally with the suspected permeability-modifying substance. When LTB_4 was tested in such a model in guinea pigs, it was without effect. However, the simultaneous injection of LTB_4 and a dose of PGE_2, which was inactive by itself, into guinea pig, rabbit, or rat skin caused a dose-dependent extravasation of radioactive serum albumin (Bray *et al.*, 1981a,b). Similar results were obtained in human volunteers; however, PGD_2 was used in place of PGE_2 (Camp *et al.*, 1981). This is consistent with the two-signal hypothesis for the induction of inflammation in which PGE_2 acts as the vasodilator and LTB_4 acts as the vascular permeability-inducing substance (Williams *et al.*, 1983).

9.3.2. Effects on Smooth Muscle

While there was no effect of injected LTB_4 on blood flow (Bray *et al.*, 1981a), the intrajugular administration of this compound caused profound increases in blood pressure which peaked in 2 min (Sirois *et al.*, 1981a). Changes in heart rate were minor.

Comparison of the spasmogenic activity of LTB_4 on various smooth muscle preparations showed that LTB_4 had no contractile activity on the guinea pig ileum (Bray *et al.*, 1981b). Of other organs of the guinea pig, the duodenum was somewhat active and the lung parenchyma was quite active. None of the following showed any activity: rat aorta or gastrointestinal tract, rabbit aorta, trachea, or parenchyma (Sirois *et al.*, 1981a,b). Contractions of parenchymal strips from guinea pig lung have been proposed for the bioassay of LTB_4 (Sirois *et al.*, 1981d). The actions of LTB_4 on the parenchymal strip are indirect in that indomethacin can effectively antagonize them whereas the same is not true for contractions induced by the sulfidopeptide leukotrienes (Dahlén *et al.*, 1983c).

9.3.3. Contribution to the Inflammatory Response

In line with the actions of LTB_4 which have been described above, it is widely believed that LTB_4 may play a pivotal role in the elicitation of inflammatory responses. However, while it is relatively easy to demonstrate the formation of LTB_4 by PMNs upon the addition of a proinflammatory stimulus (Palmer and Salmon, 1983), there is no correlation between the peak LTB production *in vivo* and the time course of PMN accumulation (Simmons *et al.*, 1983; Ford-Hutchinson *et al.*, 1984) nor is there a correlation between the ability of inhibitors to inhibit LTB formation and their capacity to inhibit cell migration into an inflammatory site (Salmon *et al.*, 1984). On the other hand, evidence has been presented that LTB_4, at low concentrations, can produce hyperalgesia independent of its action on PMNs (Levine *et al.*, 1984).

10. ACTIONS OF THIOL ETHER LEUKOTRIENES

10.1. SMOOTH MUSCLE CONTRACTION

The relative potencies of the thiol ether leukotrienes were compared on a variety of smooth muscle preparations from a large number of animal species (Holme *et al.*, 1980; Drazen *et al.*, 1980; Hedqvist *et al.*, 1980; Sirois *et al.*, 1981e; Hand *et al.*, 1981). The guinea pig is the species for which the most information is available. It also appears to be uniquely sensitive to the action of the leukotrienes. Parenchymal strips from guinea pig lung are particularly sensitive to the action of the leukotrienes (Drazen *et al.*, 1979). While guinea pig trachea and ileum were approximately equisensitive to the leukotrienes, parenchymal tissue was at least 10 times more sensitive and segments of the pulmonary artery were between 10 and 100 times less responsive. This selective sensitivity of lung parenchyma to the leukotrienes is consistent with the unique preferential action of these compounds on the peripheral portion of the lung *in vivo* (Drazen and Austen, 1974).

Dose–response curves for the effect of synthetic leukotrienes on parenchymal strips showed that LTD_4 had a uniquely biphasic response curve (Drazen *et al.*, 1980) which was not seen when the effect of LTD_4 on tracheal spirals was tested. Interestingly, only the low-dose, high-affinity component of this biphasic curve was inhibitable by the specific end-organ antagonist for leukotrienes, FPL 55712 (Augstein *et al.*, 1973), suggesting the existence of low- and high-affinity receptors for LTD_4 in lung parenchyma.

Exposure of guinea pig lung parenchyma to leukotrienes results in the induced formation of thromboxane in this tissue. Depending on variables in the assay procedure, the thromboxane which is formed may contribute significantly to the overall contractile response (Piper and Samhoun, 1981; Weichman *et al.*, 1982b; Leitch *et al.*, 1982; Dahlén *et al.*, 1983c; Iacopino *et al.*, 1984). It has been possible to demonstrate that thromboxane formation and contraction of the tissue can be dissociated both by use of inhibitors (e.g., indomethacin) and by studying the time course of the two events (Creese and Bach, 1983; Creese *et al.*, 1984). However, in addition to the apparent contribution to thromboxane to the magnitude of the contractile response, other products of the cyclooxygenase pathway appear to ameliorate the response so that the inclusion of indomethacin in smooth muscle baths or its administration to guinea pigs *in vivo* has been reported to cause a hyperreactivity to the leukotrienes (Austen *et al.*, 1983; Leitch *et al.*, 1983a).

The relatively high sensitivity of parenchymal tissue to leukotrienes, compared to that of tracheal preparations, was also found in human lung (Jones *et al.*, 1982a). Pulmonary vascular muscle was relatively insensitive and thus is not likely to have contributed to this differential response (Hanna *et al.*, 1981). There is incomplete agreement at the moment regarding the contribution of products of the cyclooxygenase pathway to the response of human tissue to leukotrienes. Some reports state that the addition of NSAICs did not affect the response (Dahlén *et al.*, 1980) while others report a hyperreactivity (Gardiner and Stock, 1984). Both the guinea pig and the human preparations showed tachyphylaxis to repeated dosing with leukotrienes (Dahlén *et al.*, 1980; Hansson *et al.*, 1983).

Leukotrienes have also been investigated for their actions in the cow (Burka, 1977), rat (Welton *et al.*, 1981a; Goldenberg and Subers, 1982), rabbit (Piper and Samhoun, 1981; Kito *et al.*, 1981), and monkey (Michoud *et al.*, 1977; Patterson *et al.*, 1978, 1983; Bach *et al.*, 1981) in addition to guinea pig and human. Pulmonary tissue from both rat and rabbit was relatively insensitive to the leukotrienes. The small response which was reported by some investigators was insensitive to NSAICs, suggesting that thromboxane formation may not be involved. On the other hand, gastrointestinal tissues from both rat and rabbit, as well as certain vascular preparations, have been reported to have high sensitivity.

10.2. CARDIOVASCULAR EFFECTS

The thiol ether leukotrienes have both peripheral and central effects on the cardiovascular system. Injection of LTD_4 intradermally into human volunteers caused a dilatation of the deep dermal venules which persisted for at least 4 hr. This was augmented by the simultaneous injection of PGD_2 (Lewis *et al.*, 1981c). By contrast, LTD_4 and, more potently, LTC_4 acted as vasoconstrictors in guinea pig skin and the vasoconstriction was accompanied by a decrease in blood flow (Peck *et al.*, 1981). Vasoconstriction was also seen in the isolated perfused rat kidney (Rosenthal and Pace-Asciak, 1983) and in the hamster cheek pouch where the leukotrienes were judged to be approximately equipotent to angiotensin in this respect (Dahlén *et al.*, 1981).

Intravenous administration of LTC_4 and LTD_4 into guinea pigs has been reported to cause varying and sometimes complex changes in blood pressure (Schiantarelli *et al.*, 1981; Lewis *et al.*, 1982a; Folco *et al.*, 1982b). These changes in pressure were accompanied by tachycardia, rhythm disturbances, convulsions, and, at times, death. Indomethacin, in a repeat administration to the same animals, reduced the severity of the response although no control for tachyphylaxis appears to have been included in this study. Direct injection of LTD_4 (10^{-13} to 10^{-9} mole) into the left circumflex artery of the sheep caused profound vasoconstriction and reduced coronary blood flow peaking at 20 sec and resolving by 3 min. NSAICs had no effect on this response while FPL 55712 abolished it (Michelassi *et al.*, 1982). A short-lived hypertensive response was seen when LTD_4 was injected, intravenously, into normal rats (Iacopino *et al.*, 1983). However, a similar injection into spontaneously hypertensive (SHR) rats resulted in abrupt hypotension, followed by a hypertension which lasted for about 1 min and which was followed by a longer hypotensive period. Heart rate increased to 400 beats/min at the peak of the hypertensive period (G. Feuerstein *et al.*, 1981). The hypotension which was caused by leukotrienes could be reversed by thyrotropin-releasing hormone (Feuerstein *et al.*, 1983).

Effects on blood pressure in a variety of species have been compared (Berkowitz *et al.*, 1984), and species differences have been observed. Different vascular beds appear to differ in their sensitivity to the leukotrienes. The pulmonary vein appears to be particularly sensitive (Schellenberg and Foster, 1984) as is the cerebral vasculature (Tagari *et al.*, 1983). The latter is particularly interesting in view of the recent finding that nerve tissue is capable of generating significant amounts of leukotrienes (Moskowitz *et al.*, 1984; Lindgren *et al.*, 1984).

Leukotrienes appear to play a role in cardiac anaphylaxis and in coronary function. Thus, both the tachycardia and the arrhythmia which accompany cardiac anaphylaxis could be reduced by the administration of FPL 55712 and inhibitors of leukotriene formation, suggesting that endogenously produced leukotrienes may play a role in this reaction (Levi *et al.*, 1980; Aeringhaus *et al.*, 1983). The administration of synthetic leukotrienes to Langendorf perfused heart preparations or directly into the coronary circulation of dogs (Letts *et al.*, 1983; Panzenbeck and Kaley, 1983) was shown to cause a reduced contractility, reduced coronary flow, and, in the case of LTD_4, also to increase the chronotropic effect of histamine (Burke *et al.*, 1982; Letts and Piper, 1982). Similar studies have been carried out in the rat (Pfeffer *et al.*, 1983), cat (Roth and Lefer, 1983), and pig (Greenwald *et al.*, 1984). Some of these effects of LTC_4, but not those of LTD_4, were reported to be inhibitable by indomethacin (Letts and Piper, 1982; Woodman and Dusting, 1983). Levi has constructed an elaborate schema to accommodate the functions of histamine, the prostaglandins, and the leukotrienes in the regulation of cardiac function (Levi *et al.*, 1982). The effect of the leukotrienes appears to be on the calcium-dependent response (Hattori and Levi, 1984).

10.3. EFFECTS ON VASCULAR PERMEABILITY

In addition to their effects on the smooth muscle in the vasculature, the sulfidopeptide leukotrienes also affect the tightness of the junctions between the cells lining the blood vessels and thus vascular permeability. The combination of these two actions is expressed in the apparent contribution of the leukotrienes to the edematous and hypertensive response which follows the exposure of the lungs to hypoxic conditions (Stenmark *et al.*, 1983; Ahmed and Oliver, 1983).

Intradermal injection of LTD_4, but not LTC_4, caused an increased exudation of radiolabeled serum albumin from guinea pig skin which was potentiated by the coinjection of PGE_2 at doses of the prostaglandin which were without effect by themselves (Peck *et al.*, 1981). Similar increases in vascular permeability were reported in a number of other species. In the hamster cheek pouch, the increased permeability followed the arteriolar constriction phase of the response to topically applied LTC_4 or LTD_4. The increased permeability appeared to occur in the postcapillary venules (Dahlén *et al.*, 1981). Crude bovine SRS-A was reported to increase vascular permeability in calf skin (Burka and Eyre, 1977). The leukotrienes are potent enhancers of vascular permeability in rat skin but, in contrast to the observations in guinea pig and human skin, the simultaneous administration of prostaglandins did not potentiate the action of the leukotrienes. Thus, the intradermal injection of synthetic LTC_4 or LTE_4 into rat skin, without the coinjection of a prostaglandin, resulted in a dose-dependent increase in vascular permeability. Neither pretreatment of the animals with methysergide to block serotonin-dependent reactions nor with an antihistamine abolished the response (Welton *et al.*, 1981a; Gulati *et al.*, 1983).

The injection of 0.19 nmole of LTC_4 and approximately twice as much LTD_4 into the skin of human volunteers elicited an erythema and whealing which peaked at 5 min. There was no itching or leukocyte infiltration into the sites, induration development, or delayed development of erythema although the injections were accompanied by pain and discomfort. Coinjection of 0.18 nmole PGE_2 did not synergize the response (Camp *et al.*, 1981). In another study, the injection of approximately 1 nmole of LTD_4 caused whealing in human skin which was augmented by the coinjection of 3 nmole of PGD_2 (Lewis *et al.*, 1981c).

10.4. EFFECTS ON CELL ACTIVATION AND HORMONE SECRETION

It has become apparent in the past few years that, in addition to their effects on leukocytes, vessel walls (endothelial cells), nerve cells, and smooth muscle cells, the leukotrienes also appear to activate the secretion of various hormones and even gastric juice. Thus, Magous *et al.* (1983) reported that relatively high and possibly unphysiologic concentrations of all three sulfidopeptide leukotrienes caused a marked increase in the production of acid from gastric parietal cells. The more physiologic and potent stimulation of mucus production in the airways will be discussed in the next section.

The ability of isolated pancreatic islet cells to respond to stimulation by glucose with the secretion of insulin appears to be modulated via products of the arachidonate cascade. It has been reported that cyclooxygenase inhibitors enhance the secretion and nonspecific inhibitors which inhibit lipoxygenases appear to inhibit the response (Metz *et al.*, 1983; Yamamoto *et al.*, 1983). 12-HETE, 12-HPETE, and 15-HPETE but not 5-HETE were reported to potentiate the release. More recently, Peck and Walsh (1984) reported, however, that both LTC_4 and LTB_4 and, to a much smaller degree, LTE_4 and LTD_4 caused a marked stimulation of release from perfused, isolated islet cells. Thus, the exact nature of the metabolite which regulates this function is open to further investigation.

10.5. EFFECTS ON MUCUS PRODUCTION AND MUCUS TRANSPORT

Excessive mucus production and slowed mucus transport are two events which accompany attacks of asthma. Mucus transport can be measured by blowing small X-ray-opaque disks into the trachea of an animal or a human and then measuring the rate at which these disks are transported. Using such an approach, it was found that mucus transport velocity decreased sharply when *Ascaris* antigen-sensitive dogs were challenged with the antigen by inhalation (Wanner *et al.*, 1975). This slowing of mucus transport could be prevented if the dogs were pretreated with FPL 55712. Similar results have been reported for antigen-induced bronchospasm in allergic asthmatic patients (Ahmed *et al.*, 1981). These indirect observations suggest that leukotrienes, which are presumably formed as a consequence of the anaphylactic challenge, may be responsible for the slowing of the mucus transport velocity. This has now been confirmed in studies using exogenously administered synthetic leukotrienes in sheep (Russi *et al.*, 1983; Wanner *et al.*, 1983).

Effects of the leukotrienes on mucus production have been studied with human lung fragments *in vitro*. Such fragments can be maintained for prolonged periods of time in organ culture, and their ability to produce mucus can be measured by the incorporation of radioactive precursor molecules into acid-precipitable mucus. The leukotrienes are potent inducers of mucus production in this system; doses which are required for activity are comparable to the minimal doses which cause the contraction of sensitive muscle preparations such as the guinea pig ileum (Shelhamer *et al.*, 1982; Marom *et al.*, 1982). Potentiation of mucus production in this system has been shown to reflect increased production of the gel component of the mucus rather than the sol component since secretion of lysozyme, which is secreted along with the water of mucus, was not increased in parallel with the increased production of mucus (Coles *et al.*, 1983).

Effects on mucus production can also be demonstrated *in vivo* in both cat and dog trachea. Although LTC_4 caused a dose-dependent increase in the incorporation of both $^{35}SO_4$ and [^{3}H]glucose into mucus in the tracheae of living cats (6×10^{-7} to 6×10^{-5} M), no stimulation could be elicited with the same tissue *in vitro* (Peatfield *et al.*, 1982). Mucus production in dog trachea has been quantitated as the formation of "hillocks" on the surface of exposed tracheae which had been coated with tantalum powder. In this model, LTC_4 and especially LTD_4 are potent inducers of mucus production. The effect of LTC_4 was not affected by vagotomy, but both atropine and diphenhydramine inhibited the effect, suggesting that there was a postganglionic neuroendocrine response involved (Johnson and McNee, 1983; Johnson *et al.*, 1983a,b). Indeed, there was a synergism between the effect of the leukotrienes on mucus secretion and the effect of the local ganglionic stimulator, dimethylphenylpiperazinium, on the same end point (Johnson *et al.*, 1983c).

10.6. EFFECTS ON NERVE CELLS

When rat brain Purkinje cells were pressure-injected with a 0.4 mM solution of LTC_4, there was a sustained increase in the firing rate of the cells which reached a peak several minutes after the application of the leukotriene. The increased rate persisted for more than 1 hr. This response was longer in duration than any other response which had been studied. Pressure injection of a solution containing 2 μM FPL 55712 reversed the response to LTC_4. LTD_4 was approximately half as potent as LTC_4, and LTB_4 was inactive (M. R. Palmer *et al.*, 1980, 1981). A functional correlate to these observations was the recent report that the administration of 0.2 ng of leukotriene into the cavity in cat teeth resulted in a spontaneous decrease in nerve activity and in the response of the nerves to the presence of hyperosmotic salt solutions (Gazelius *et al.*, 1984). Thus, it is clear that leukotrienes can be produced in brain tissue (Moskowitz *et al.*, 1984; Lindgren *et al.*, 1984) and can affect the action of brain tissue.

10.7. EFFECTS ON THE IMMUNE RESPONSE

Leukotrienes can induce the formation of products of the cyclooxygenase pathway of arachidonate metabolism. Some of these products, and especially PGE_2, are known to modulate the immune response. Thus, there is the possibility that endogenously produced leukotrienes may have an indirect effect on the immune response. Relatively large concentrations of exogenous LTD_4 and LTE_4 were reported to inhibit both PHA- and LPS-induced blastogenesis and also the immune response to SRBC (Webb *et al.*, 1982). LTD_4 suppressed the expression of NK cells at 10^{-8} and 10^{-10} M (Rola-Pleszczynski *et al.*, 1982). This effect was considerably smaller than that of LTB_4 in the same system.

Antigen-induced leukocyte adherence inhibition (LAI) is a common immunologic parameter for assessing the immune response to tumor antigens in cancer patients. Cells from normal individuals can be armed with sera from cancer patients, and when these armed cells are then mixed with the immunizing antigen (i.e., an extract of tumor tissue), their adherence to the substratum is inhibited. These armed mononuclear cells produce thiol ether leukotrienes upon exposure to tumor extract. Furthermore, LAI could be abrogated by minute amounts of FPL 55712 and could be elicited in a dose-dependent fashion by equally minute amounts of the thiol ether leukotrienes. These effects of the leukotrienes correlated with increases in the transmembrane potential of the cells as measured by the partitioning of the radioactive tracer [^{3}H]methyl-triphenyl phosphonium bromide. Similar changes in transmembrane potential were demonstrated upon antigen challenge of armed leukocytes (Thomson *et al.*, 1982; Shenouda *et al.*, 1982).

10.8. EFFECTS ON PRODUCTION, RELEASE, OR ACTION OF LEUKOTRIENES, PROSTAGLANDINS, AND OTHER MEDIATORS

The formation of the various products of arachidonate metabolism in the body is extensively interconnected. Thus, production of leukotrienes is enhanced in the presence of inhibitors of the cyclooxygenase or of the thromboxane synthetase. By contrast, some leukotriene-induced responses could be inhibited by NSAICs (Schiantarelli *et al.*, 1981; Welton *et al.*, 1983; Dahlén, 1983; Dahlén *et al.*, 1983b,c) which is presumably due to the induced formation of thromboxane A_2 upon the stimulation of the tissues with leukotrienes and the potentiation of the observed (smooth muscle contractile) responses by the thromboxane.

Further examination of the nature of the arachidonate metabolites which are produced in lung tissue upon stimulation with leukotrienes revealed that there was a marked difference between the products produced by parenchymal tissue and those produced by tracheal tissue. Parenchymal tissue produced thromboxane, whereas tracheal tissue produced PGI_2. Thus, the addition of indomethacin to tracheal muscle preparations resulted in an augmentation of the response to leukotrienes rather than an inhibition presumably because the relaxant action of the PGI_2 was not present under those conditions (Folco *et al.*, 1982a,b). This was confirmed by the observation that the addition of nanogram amounts of PGI_2, but not PGE_2 or 6-keto-$PGF_{1\alpha}$, was able to reverse the action of perfused indomethacin (Mugridge *et al.*, 1984). Pulmonary vascular tissue and human endothelial cells in culture also produced PGI_2 in the presence of LTC_4 (Cramer *et al.*, 1983; Benjamin *et al.*, 1983).

When rat peritoneal mononuclear cells were stimulated with LTC_4 or LTD_4, there was a dose-dependent production of PGE_2, PGI_2 (measured as 6-keto-$PGF_{1\alpha}$), and thromboxane B_2 in the dose range 10^{-8} to 10^{-4} M (N. Feuerstein *et al.*, 1981). But in addition to causing the activation of the cyclooxygenase pathway, SRS-A also caused the increased production of 12-HPETE (Berti *et al.*, 1981). Bearing in mind that 12-HPETE is reportedly a stimulator of the 5-lipoxygenase, this observation may indicate a positive feedback loop of leukotrienes stimulating their own synthesis through the intermediate action of platelet 12-lipoxygenase.

Leukotriene synthesis in immunologically or ionophore-stimulated cells is self-limiting. It was found (Bach, unpublished results; Rankin *et al.*, 1984), however, that when cells which had completed leukotriene synthesis were recovered from the leukotriene-containing medium by centrifugation, and were then resuspended in fresh medium, they produced more leukotrienes. It was not even necessary to restimulate these cells with a fresh dose of antigen and antibody (Rankin *et al.*, 1984). In a similar study, using IgE-sensitized and ragweed antigen E-challenged human lung fragments, it was found (Weichman *et al.*, 1982a) that the addition of synthetic LTC_4 or LTD_4 to the lung fragments caused an apparent dose-dependent inhibition of leukotriene synthesis in the fragments. Thus, it appears that leukotrienes can have a negative feedback regulation on their own synthesis. The mechanism of this action remains to be elucidated.

In addition to the interactions which exist in the control of the generation of the mediators, there are a variety of interactions which have been reported in their actions. A potentiation of the response to histamine by previous exposure to leukotrienes has been observed in various systems (Creese and Bach, 1983; Stewart *et al.*, 1983; Hand *et al.*, 1983). Similar potentiation of the response to histamine by pretreatment with leukotrienes was observed *in vivo* (Barnes *et al.*, 1984; Inoue *et al.*, 1984). A mechanistic explanation for this potentiation may have been found by T. H. Lee *et al.* (1984a), who demonstrated that LTE_4 specifically lowered the EC_{50} for histamine on guinea pig tracheal preparations. Other potentiations which have been reported involve PAF (Patterson *et al.*, 1984; O'Flaherty *et al.*, 1984) and even cholinergic responses (Nakahata *et al.*, 1984).

11. ROLE OF LEUKOTRIENES IN DISEASE

It should be apparent that it is very likely that the leukotrienes play a role in the etiology of a number of diseases having an inflammatory component. Among these are inflammatory bowel disease, psoriasis, and acute respiratory distress syndrome, to name a few. Evidence for the role of the leukotrienes in these diseases ranges from the observation that diseased tissue appears to produce supernormal quantities of leukotrienes to pharmacologic observations on the effects of potential inhibitors of the cyclooxygenase and/or the 5-lipoxygenase pathways on the course of the disease or a model of the disease in animals (Voorhees 1983; Littner *et al.*, 1984; Morganroth *et al.*, 1984; Brain *et al.*, 1984). For the purposes of this review, however, I shall only consider the case for the role of the leukotrienes in asthma. Using Dale's criteria, the proof for this role requires three elements:

1. It must be shown that leukotrienes are formed in the course of an asthmatic attack.
2. The administration of synthetic leukotrienes must cause symptoms which are reminiscent of those of asthma.
3. Therapeutic intervention with a drug which selectively inhibits leukotriene formation must result in beneficial effects in asthmatic patients.

Unfortunately, there are no animal models of asthma which resemble the human disease in all aspects. Monkeys can be shown to respond to instilled leukotrienes (Bach *et al.*, 1984a) just as most other animals which have been examined. However, the response is acute and does not manifest itself in hyperreactivity of the airways. Airway hyperreactivity which has many features reminiscent of human asthma has been found in basenji greyhound hybrid dogs and it has been shown that challenge of these dogs with an irritant such as citric acid resulted in the production of leukotrienes in the airways while challenge with mecholyl did not (Hirshman *et al.*, 1983b). It was also shown that these dogs were hyperresponsive to exogenous LTD_4 compared to mongrel dog controls (Hirshman *et al.*, 1983a).

Challenge of human asthmatics with an aerosol of antigen or a nonspecific irritant often results in the development of a biphasic reaction; following the acute response which has been traditionally associated with responses to such challenges, there develops a second phase of bronchospasm, which peaks several hours after the symptoms of the first phase have resolved. It is believed that this so-called late phase reaction is an expression of hyperreactivity and efforts have been made to find animal models which might reproduce this response. One such model was found in the allergic sheep which develops bronchospasm upon challenge with aerosolized *Ascaris* antigen. It was found that the administration of relatively small amounts of LTD_4 by the same route also caused the development of both the immediate and the late phase reactions which are ordinarily seen upon challenge with antigen (Abraham *et al.*, 1983).

Turning to a consideration of the elements of Dale's proof in human asthma itself, it can be shown that anaphylactically challenged human lung can produce large amounts of leukotrienes. In a unique study, using lung tissue from an extrinsic asthmatic patient, it was shown (Dahlén *et al.*, 1983a) that antigen challenge *in vitro* of this actively sensitized tissue resulted in the production of as much as 86 pmoles of leukotrienes per g lung tissue (sum of LTC_4, LTD_4, and LTE_4 found). This is 70% of the amount of leukotrienes which the same tissue produced upon challenge with the calcium ionophore. Interestingly, though the lung produced LTB_4 after challenge with the ionophore, none was produced during the anaphylactic challenge. SRS has also been found in the sputum of asthmatics and not in the sputum of normal individuals (Turnbull *et al.*, 1977). One study actually suggested that there may be an underlying shift in the balance of arachidonate metabolites which are produced in tissues from asthmatics compared to those from normal individuals. In that study, platelets from asthmatics incorporated 70% more radioactivity from arachidonate into lipoxygenase products and 40% less into cyclooxygenase products than did platelets from normal controls (Yen and Morris, 1981). These observations amply satisfy the first of Dale's criteria for a role of leukotrienes in human asthma.

The second element of the proof of the role of leukotrienes in asthma, the elicitation of symptoms upon administration of leukotrienes, has also been satisfied. Leukotrienes have been given to normal volunteers, and elicited changes in pulmonary function which appeared to be selective for the small airways and were reminiscent of the changes seen in asthma (Holroyde *et al.*, 1981; Weiss *et al.*, 1982; Davidson *et al.*, 1984). These changes were unaffected by pretreatment of the volunteers with aspirin which is consistent with the general lack of benefit resulting from attempted therapy of most asthmatics with NSAICs (Weiss *et al.*, 1983). Responses to leukotrienes have also been elicited in asthmatics (Griffin *et al.*, 1983; Smith *et al.*, 1984). In the first of these studies, surprisingly, the asthmatics were no more sensitive to the leukotrienes than were the normal volunteers while in the second study the hyperreactivity of asthmatics to LTD_4 was quite comparable to their hyperreactivity to mecholyl.

No convincing evidence dealing with the third stipulation of the proof is available thus far, largely because of the fact that no suitable inhibitor has been available. The effect of FPL 55712 on mucociliary transport in extrinsic asthmatics has been noted (Ahmed *et al.*, 1981).

It must be stressed at this point that an underlying assumption in seeking to prove the role of leukotrienes in asthma by way of satisfying Dale's criteria is that the leukotrienes of and by themselves are not only necessary but also sufficient to elicit the symptoms of disease. It should be apparent from all that has been said thus far that it is actually unlikely that this will be the case, given the complex interactions which are known to occur both during the production of the mediators and in their action. Thus, a much more likely scenario is one where the leukotrienes may play an important role in diseases such as asthma but the alleviation of the symptoms of the disease (Dale's third required criterion) may require the use of possibly even relatively complex mixtures of selective agents, each antagonizing one of the causative events or agents.

12. PHARMACOLOGIC INTERVENTION

Because of the potential utility in inhibitors of various aspects of the 5-lipoxygenase pathway in the treatment of human disease, the field is progressing rapidly and space will not permit a detailed consideration here. Much of this literature has been reviewed (Bach, 1984). Only a few examples of relatively selective agents will be briefly considered.

12.1. FATTY ACID ANALOGUES

A variety of analogues of arachidonic acid have been evaluated as potential substrate analogues of the 5-lipoxygenase. The oldest and best known of these, by far, is ETYA, the tetraacetylenic analogue of arachidonic acid. Because of the fact that all the ethylenic bonds of arachidonate are modified in that molecule, its inhibitory activity is not selective but rather extends, more or less, to all the reactions in which arachidonate serves as a substrate. A level of selectivity can be introduced by replacing only one of the ethylenic bonds by an acetylenic bond but, in general, these molecules have not been very active *in vivo*. More interesting substitutions have been reported recently, however. One of these is a series of molecules in which the methylene bridge between two ethylenic carbons (e.g., carbons 7 or 10) are substituted by two methyl groups. This would be expected to block the rearrangement of the double bonds and thus inhibit the formation of the leukotrienes and, indeed, this was observed (Cohen *et al.*, 1984). Other derivatives have included the incorporation of atoms other than oxygen into the epoxide bridge at carbons 5,6 and/or the shifting of the location of this three-atom ring. Some of these derivatives have been reported to be highly active inhibitors of the 5-lipoxygenase (Arai *et al.*, 1983).

12.2. PIRIPROST (U-60,257)

6,9-Deepoxy-6,9-(phenylimino)-$\Delta^{6,8}$-PGI_1 (piriprost, U-60,257) was recently identified as a selective inhibitor of thiol ether leukotriene synthesis in ionophore-challenged peritoneal mononuclear cells from rats and in antigen-challenged human lung fragments (Bach *et al.*, 1982b). Selectivity of action is based on the observation that neither anaphylactic release of histamine, nor the concomitant formation of products of the cyclooxygenase appear to be affected at concentrations of piriprost which completely inhibit leukotriene synthesis. In addition to inhibiting the formation of the leukotrienes, piriprost is also a competitive end-organ antagonist for these spasmogens based on evidence obtained using the guinea pig ileum in a superfusion arrangement. Piriprost also inhibited both ionophore- and fMLP-induced secretion of lysosomal enzymes from human PMNs and the production of superoxide anion by these cells (Smith *et al.*, 1982).

Low concentrations of aerosolized piriprost inhibited the pulmonary changes which accompany anaphylaxis in *Ascaris* antigen-challenged rhesus monkeys. In this *in vivo* system, the action of the compound was synergistic with that of diphenhydramine, an H_1-type antihistamine (Johnson *et al.*, 1983c). Similar concentrations also inhibited the leukotriene-dependent component of the anaphylactic response of guinea pigs which were challenged with antigen intravenously (Bach *et al.*, 1983). The compound also inhibited the production of mucus in dog tracheae *in vivo*. Interestingly, in this system, piriprost inhibited the spontaneous production of mucus as well as the production which was induced by hypoxia or by the administration of arachidonate (Johnson and McNee, 1984). This observation lends further support to the belief that the leukotrienes may exert a homeostatic control on lung mucus production. Finally, the incorporation of piriprost into the perfusion fluid of rat lungs undergoing reversible hypoxic vasoconstriction blocked the development of the vasoconstriction (Morganroth *et al.*, 1984).

When piriprost was added to fragments of the lung of a human asthmatic with known sensitivity to birch pollen, and the fragments were challenged *in vitro* with birch pollen, the formation of the thiol ether leukotrienes was inhibited. Bronchiolar smooth muscle preparations from the same lung specimen, which contracted upon exposure to pollen, were inhibited from contracting when piriprost was included in the incubations (Dahlén *et al.*, 1983a). In another study, the challenge of mechanically dissociated human lung cells with the calcium ionophore A23187 in the presence of piriprost resulted in the expected inhibition of leukotriene formation and a concomitant doubling in PGD_2 formation. Interestingly, thromboxane formation was inhibited in a dose-dependent fashion with complete inhibition at 0.1 mM piriprost (Holgate and Robinson, 1964).

Mode of action studies suggest that piriprost inhibits the 5-lipoxygenase of human PMNs with a 50% inhibitory concentration of approximately 5 μM (Sun and McGuire, 1983). However, when high concentrations of arachidonate were added to the cells, both 5-HETE and 5,12-DHETEs accumulated, suggesting that the inhibition may be overcome by excess substrate. Early studies on the synthesis of LTC_4 from LTA_4, using the high-speed supernatant fraction from rat liver homogenates as a source of glutathione *S*-transferase, indicated that this reaction could be efficiently inhibited by piriprost (Bach *et al.*, 1982b). However, the particulate LTA_4:glutathione *S*-transferase of RBL cells is insensitive to this inhibitor (Bach *et al.*, 1984c). The recent observation that the inhibition of LTB_4 formation in murine bone marrow-derived mast cells by piriprost was more readily reversed upon removal of the inhibitor than was the inhibition of LTC_4 formation (Razin *et al.*, 1984) remains to be explained. It is possible that this apparent preferential inhibition, which appears at first glance as though the compound were inhibiting the LTC synthetase, may be due to the relative dependence of the LTA hydrolase and the LTC synthetase reactions on the pool size of LTA_4 but this remains to be established. Thus, piriprost may afford the opportunity of obtaining the third element which is necessary to firmly establish the role of the leukotrienes in asthma.

12.3. AA861, QUINONES, HYDROQUINONES, AND FLAVONOIDS

It has become apparent in the last few years that lipid-soluble antioxidants as a family appear to be good inhibitors of the 5-lipoxygenase. Examples from several different classes of compounds have been described:

1. Esculetin (6,7-dihydroxycoumarin) is a relatively nonselective lipoxygenase inhibitor (Neichi *et al.*, 1983).
2. Caffeic acid has been reported to be a relatively specific inhibitor of the 5-lipoxygenase and, indeed, a series of derivatives with markedly higher potency have been prepared (Koshihara *et al.*, 1983).
3. Flavonoids have been known to inhibit the 5-lipoxygenase. Among these, quercetin has probably been recognized the longest. However, more recently cirsiliol was found to be the most active (Hope *et al.*, 1983; Yoshimoto *et al.*, 1983).
4. Perhaps the most interesting series of compounds of this class is the one represented by AA861, a 3,5,6-trimethyl-1,4-benzoquinone having a 12-carbon diacetylenic alcohol side chain at position 2, which has been reported to be a highly selective inhibitor of the 5-lipoxygenase and which has oral activity in a number of models of leukotriene-dependent responses (Shiraishi and Terao, 1983; Ishihara *et al.*, 1983; Ashida *et al.*, 1983). Similar compounds, in which the quinone

was reduced to a hydroquinone, or the side chain was modified have also been reported to be highly active (Suno and Nagaoka, 1984).

It is not clear from these examples if antioxidant properties and lipophilicity are sufficient for activity or what else may be required for optimal activity.

12.4. FPL 55712

FPL 55712, a chromone dicarboxylic acid, was reported to be a selective end-organ antagonist of SRS-A more than 10 years ago (Augstein *et al.*, 1973). Schild plots which were included in the original publication suggested that the inhibition may be competitive even though the slope of the curves was somewhat less than 1.0. Much of the early work using FPL 55712 has been summarized (Chand, 1979). With the identification of specific receptors for the different sulfidopeptide leukotrienes, the lack of true competitiveness in the Schild plots can be readily explained: As already noted, only the high-affinity portion of the dose–response curve for LTD_4 on lung parenchyma was antagonized by FPL 55712. Subsequent studies have confirmed that, indeed, this compound is a better antagonist for LTD and LTE than it is for LTC (Jones *et al.*, 1983).

Despite the early indications of selectivity of action, more recent reports have suggested that FPL 55712 has a variety of activities in addition to its end-organ antagonist action on the leukotrienes. It was reported that this compound was a potent inhibitor of the cyclic nucleotide phosphodiesterase in rat brain, with K_i values of 2.8 and 11.7 μM toward the cAMP- and the cGMP-degrading enzymes, respectively. These values compare favorably to those of the best inhibitors of these enzymes [e.g., papaverine, 1.3 and 56 μM, respectively (Chasin and Scott, 1978)]. FPL 55712 was also reported to be a potent inhibitor of thromboxane synthetase of human platelets, K_i 6.5 μM compared to 10 mM for 1-nonyl-imidazole (Welton *et al.*, 1981b). It also inhibited histamine release from IgE-sensitized and antigen-challenged human lung and from rat peritoneal cells with 50% inhibitory concentrations that were fully ten times smaller than those of disodium cromoglycate (Welton *et al.*, 1981a).

12.5. OTHER END-ORGAN ANTAGONISTS

A potent end-organ antagonist of the leukotrienes may, in principle, be as desirable as an inhibitor of their formation provided it can be established which of the many different receptors for specific leukotrienes which we now recognize really need to be blocked. Accordingly, a number of groups have recently reported the identification of different classes of end-organ antagonists for the leukotrienes:

1. The removal of the second carbon from the sulfidopeptide leukotrienes has resulted in a family of partial agonists having approximately 10% of the contractile activity of the natural leukotrienes but at the same time having significant antag-

onist activity (Gleason *et al.*, 1983; Weichman *et al.*, 1983). Unfortunately, the compounds are too labile for *in vivo* studies.
2. KC 404 is an isopropylpyrazole pyridine which is orally active and suppressed the response of several laboratory animals to administered leukotrienes. As expected, it also inhibited anaphylactic responses which are believed to be caused by endogenously produced leukotrienes (Nagai *et al.*, 1983; Nishino *et al.*, 1983).
3. A series of pyrido-quinazoline carboxylic acids have been described as selective antagonists of the leukotrienes (Tilley *et al.*, 1983) although it is not clear to what extent the best of these compounds exceeds the activity of FPL 55712.

13. CONCLUDING REMARKS

This review has attempted to highlight the areas of active research involving the leukotrienes. The selection of subject matter and of references, though lengthy, are far from exhaustive and were made in most instances in an attempt to focus the discussion on the areas in which agreement exists and on those key areas of ongoing effort in which only fragmentary information has been published. The field is progressing rapidly and many new insights are being afforded by the virtual deluge of new publications which are appearing. This rapid progress can be viewed as the driving force for much of the excitement which this area has generated in the past few years.

REFERENCES

Abraham, W. M., Delehunt, J. C., Russi, E., Wanner, A., Yerger, L., and Chapman, G. A., 1983, Inhalation of leukotriene D_4 produces both early and late bronchoconstriction in allergic sheep, *Am. Rev. Respir. Dis.* **127**(4II):65.

Aeringhaus, U., Wöbling, R. H., König, W., Patrono, C., Peskar, B. M., and Peskar, B. A., 1982, Release of leukotriene C_4 from human polymorphonuclear leukocytes as determined by radioimmunoassay, *FEBS Lett.* **146**:111.

Aeringhaus, U., Peskar, B. A., Wittenberg, H. R., and Wöbling, R. H., 1983, Effect of inhibition of synthesis and receptor antagonism of SRS-A in cardiac anaphylaxis, *Br. J. Pharmacol.* **80**:73.

Aharony, D., Dobson, P., Bernstein, P. R., Kusner, E. J., and Smith, J. B., 1983, Determination of SRS-A release from guinea pig lungs by a radioimmunoassay, *Biochem. Biophys. Res. Commun.* **117**:574.

Ahmed, T., and Oliver, W., Jr., 1983, Does slow reacting substance of anaphylaxis mediate hypoxic pulmonary vasoconstriction?, *Am. Rev. Respir. Dis.* **127**:566.

Ahmed, T., Greenblatt, D. W., Birch, S., Marchette, B., and Wanner, A., 1981, Abnormal mucociliary transport in allergic patients with antigen-induced bronchospasm: Role of slow reacting substance of anaphylaxis, *Am. Rev. Respir. Dis.* **124**:110.

Anderson, M. E., Allison, R. D., and Meister, A., 1982, Interconversion of leukotrienes catalyzed by purified γ-glutamyl transpeptidase; concomitant formation of leukotriene D_4 and γ-glutamyl amino acids, *Proc. Natl. Acad. Sci. USA* **79**:1088.

Anggård, A., and Strandberg, K., 1981, Vascular effects of slow reacting substance (SRS) in the nasal mucosa and the appearance of a SRS-like principle in cat nasal secretion on nerve stimulation, *Acta Physiol. Scand.* **111**:329.

Appelgren, L.-E., and Hammarström, S., 1982, Distribution and metabolism of ^3H-labeled leukotriene C_3 in the mouse, *J. Biol. Chem.* **257**:531.

Arai, Y., Shimoji, K., Kanno, M., Konishi, Y., Okuyama, S., Iguchi, S., Hayashi, M., Miyamoto, T., and Toda, M., 1983, Synthesis of 5-lipoxygenase inhibitory activities of eicosanoic compounds, *J. Med. Chem.* **26**:72.

Ashida, Y., Saijo, T., Kuriki, H., Makino, H., Terao, S., and Maki, Y., 1983, Pharmacological profile of AA861, a 5-lipoxygenase inhibitor, *Prostaglandins* **26**:955.

Augstein, J., Farmer, J. R., Lee, T. B., Sheard, P., and Tattersall, M. L., 1973, Selective inhibitor of slow reacting substance of anaphylaxis, *Nature New Biol.* **245**:215.

Austen, K. F., Corey, E. J., Drazen, J. M., and Leitch, A. G., 1983, The effect of indomethacin on the contractile response of the guinea pig lung parenchymal strips to leukotrienes B_4, C_4, D_4 and E_4, *Br. J. Pharmacol.* **80**:47.

Bach, M. K., 1984, Inhibitors of leukotriene synthesis and action, in: *The Leukotrienes* (L. W. Chakrin and D. M. Bailey, eds.), pp. 163–194, Academic Press, New York.

Bach, M. K., and Brashler, J. R., 1974, In vivo and in vitro production of a slow reacting substance in the rat upon treatment with calcium ionophores, *J. Immunol.* **113**:2040.

Bach, M. K., and Brashler, J. R., 1978, Ionophore A 23187-induced production of slow reacting substance of anaphylaxis (SRS-A) by rat peritoneal cells in vitro: Evidence for production by mononuclear cells, *J. Immunol.* **120**:998.

Bach, M. K., and Brashler, J. R., 1979, A family of calcium ionophore-induced slow reacting substances produced by rat mononuclear cells: Enhancement of production with different mercaptans results in formation of different products, *Int. J. Immunopharmacol.* **1**:205.

Bach, M. K., and Brashler, J. R., 1985, A comparison of the leukotriene synthesizing ability of subfractions of rat liver glutathione S-transferases, *Prostaglandins Leukotrienes Med.* **17**:125.

Bach, M. K., and Brashler, J. R., 1986, Inhibition of the LTC synthetase of rat basophil leukemia cells by diethylcarbamazine and synergism between diethylcarbamazine and piriprost, a 5-lipoxygenase inhibitor, *Biochem. Pharmacol.* **35**:425.

Bach, M. K., Brashler, J. R., and Gorman, R. R., 1977, On the structure of slow reacting substance of anaphylaxis: Evidence of biosynthesis from arachidonic acid, *Prostaglandins* **14**:21.

Bach, M. K., Brashler, J. R., Brooks, C. D., and Neerken, J. A., 1979, Slow reacting substances: Comparison of some properties of human SRS-A and two distinct fractions from ionophore-induced rat mononuclear cell SRS, *J. Immunol.* **122**:160.

Bach, M. K., Brashler, J. R., Hammarström, S., and Samuelsson, B., 1980a, Identification of a component of rat mononuclear cell SRS as leukotriene D, *Biochem. Biophys. Res. Commun.* **93**:1121.

Bach, M. K., Brashler, J. R., Hammarström, S., and Samuelsson, B., 1980b, Identification of leukotriene C-1 as a major component of slow reacting substance from rat mononuclear cells, *J. Immunol.* **125**:115.

Bach, M. K., Brashler, J. R., Johnson, M. A., and Drazen, J. M., 1980c, Two rat mononuclear cell-derived slow reacting substances: Kinetic evidence that the peripheral airways-selective spasmogen is derived from a non-selectively acting precursor, *Immunopharmacology* **2**:361.

Bach, M. K., Brashler, J. R., Johnson, H. G., and McNee, M. L., 1981, Pharmacology of the slow reacting substances: Some aspects of their selective action and modulation of their synthesis, in: *SRS-A and Other Leukotrienes* (P. J. Piper, ed.), pp. 161–180, Research Studies Press, Chichester.

Bach, M. K., Brashler, J. R., Morton, D. R., Steel, L. K., Kaliner, M. A., and Hugli, T. E., 1982a, Formation of leukotrienes C and D and pharmacologic modulation of their synthesis, *Adv. Prostaglandin Thromboxane Leukotriene Res.* **9**:103.

Bach, M. K., Brashler, J. R., Smith, H. W., Fitzpatrick, F. A., Sun, F. F., and McGuire, J. C., 1982b, 6,9-Deepoxy-6,9-(phenylimino)-$\Delta^{6,8}$-prostaglandin I_1, (U-60,257), a new inhibitor of leukotriene C and D synthesis: In vitro studies, *Prostaglandins* **23**:759.

Bach, M. K., Brashler, J. R., Fitzpatrick, F. A., Griffin, R. L., Iden, S. S., Johnson, H. G., McNee, M. L., McGuire, J. C., Smith, H. W., Sun, F. F., and Wasserman, M. A., 1983, in vivo and in vitro actions of a new selective inhibitor of leukotriene C and D synthesis, *Adv. Prostaglandin Thromboxane Leukotriene Res.* **11**:39.

Bach, M. K., Brashler, J. R., McNee, M. L., Johnson, H. G., and Johnson, M. A., 1984a, A pharmacological evaluation of the responses of the subhuman primate to instillation of leukotrienes into the lungs, in: *Advances in Clinical Immunology* (M. Condorelli, G. Marone, and L. M. Lichtenstein, eds.), pp. 91–104, OIC Medical Press, Milan.

Bach, M. K., Brashler, J. R., and Morton, D. R., Jr., 1984b, Solubilization and characterization of the leukotriene C_4 synthetase of rat basophil leukemia cells, a novel, particulate glutathione S-transferase, *Arch. Biochem. Biophys.* **230**:455.

Bach, M. K., Brashler, J. R., Peck, R. E., and Morton, D. R., Jr., 1984c, Leukotriene C synthetase, a special glutathione S-transferase: Properties of the enzyme and inhibitor studies with special reference to the mode of action of U-60,257, a selective inhibitor of leukotriene synthesis, *J. Allergy Clin. Immunol.* **74**:353.

Bach, M. K., Bowman, B. J., Brashler, J. R., Fitzpatrick, F. A., Griffin, R. L., Johnson, H. G., Major, N. J., McGuire, J. C., McNee, M. L., Richards, I. M., Smith, H. W., Smith, R. J., Speziale, S. C., and Sun, F. F., 1985, Piriprost, 6,9-deepoxy-6,9-(phenylimino)-$\Delta^{6,8}$-prostaglandin I_1 (U-60,257), a selective inhibitor of leukotriene synthesis, *Adv. Prostaglandin Thromboxane Leukotriene Res.* **15**:225.

Bailey, J. M., Bryant, R. W., Low, C. E., Pupillo, M. B., and Vanderhoek, J. Y., 1982, Regulation of T lymphocyte mitogenesis by the leukocyte product 15-hydroxy-eicosatetraenoic acid (15-HETE), *Cell. Immunol.* **67**:112.

Baker, S. R., Boot, J. R., and Osborne, D. J., 1982, A comparative study of the physical properties and enzymatic reactions of slow reacting substance of anaphylaxis and some leukotriene D_4 isomers, *Prostaglandins* **23**:569.

Baker, S. R., Boot, J. R., Morgan, S. E., Osborne, D. J., Ross, W. J., and Shrubshall, P. R., 1983, Synthesis of 8,9-leukotriene C_3 and D_3, *Tetrahedron Lett.* **24**:4469.

Barnes, N. C., Watson, A., Piper, P. J., and Costello, J. F., 1984, Action of inhaled leukotriene C and D on large and small airways: Effect of preinhalation of leukotriene D on histamine dose response curve, *Am. Rev. Respir. Dis.* **129**:A1.

Bass, D. A., Thomas, M. J., Goetzl, E. J., DeChatelet, E. R., and McCall, C. E., 1981, Lipoxygenase-derived products of arachidonic acid mediate stimulation of hexose uptake in human polymorphonuclear leukocytes, *Biochem. Biophys. Res. Commun.* **100**:1.

Benjamin, C. W., Hopkins, N. K., Oglesby, T. D., and Gorman, R. R., 1983, Agonist specific desensitization of leukotriene C_4-stimulated PGI_2 biosynthesis in human endothelial cells, *Biochem. Biophys. Res. Commun.* **117**:780.

Benveniste, J., Hadji, J., Jouvin, E., Mencia-Huerta, J. M., Pirotzky, E., and Roubin, R., 1981, Release of slow reacting substances (SRS) from various cell types and organs, *Fed. Proc.* **40**:1022.

Berkowitz, B. A., Zabko-Potapovich, B., Valocik, R., and Gleason, J. G., 1984, Effects of leukotrienes on the vasculature and blood pressure of different species, *J. Pharmacol. Exp. Ther.* **229**:105.

Berti, F., Folco, G. C., and Omini, C., 1981, An outline of the pharmacological control of the formation and release of thromboxane A_2 (TxA_2) and 12-HETE in the lungs, *Atheroscler. Rev.* **8**:109.

Bhattacherjee, P., Eakins, K. E., and Hammond, B., 1981, Chemotactic activity of arachidonic acid lipoxygenase products in the rabbit eye, *Br. J. Pharmacol.* **73**:254P.

Bisgaard, H., Ford-Hutchinson, A. W., Charleson, S., and Taudog, E., 1984, Detection of leukotriene C_4-like immunoreactivity in tear fluid from subjects challenged with specific antigen, *Prostaglandins* **27**:369.

Bokoch, G. M., and Reed, P. W., 1981, Evidence for inhibition of leukotriene A_4 synthesis by 5,8,11,14-eicosatetraynoic acid in guinea pig polymorphonuclear leukocytes, *J. Biol. Chem.* **256**:4156.

Bonney, R. J., and Humes, J. L., 1984, Physiological and pharmacological regulation of prostaglandin and leukotriene production by macrophages, *J. Leukocyte Biol.* **35**:1.

Borgeat, P., 1984, Reversed phase HPLC profiling and quantitation of lipoxygenase products, *Prostaglandins* **27**:356.

Borgeat, P., and Samuelsson, B., 1979a, Transformation of arachidonic acid by rabbit polymorphonuclear leukocytes: Formation of a novel dihydroxy-eicosatetraenoic acid, *J. Biol. Chem.* **254**:26453.

Borgeat, P., and Samuelsson, B., 1979b, Arachidonic acid metabolism in polymorphonuclear leukocytes: Unstable intermediate in formation of dihydroxyacids, *Proc. Natl. Acad. Sci. USA* **76**:3213.

Brain, S. D., Camp, R. D. R., Cunningham, F. M., Dowd, P. M., Greaves, M. W., and Black, A. K., 1984, Leukotriene B_4-like material in scale of psoriatic skin lesions, *Br. J. Pharmacol.* **83**:313.

Bray, M. A., Cunningham, F. M., Ford-Hutchinson, A. W., and Smith, M. J. H., 1981a, Leukotriene B_4: A mediator of vascular permeability, *Br. J. Pharmacol.* **72**:483.

Bray, M. A., Ford-Hutchinson, A. W., and Smith, M. J. H., 1981b, Leukotriene B_4: Biosynthesis and biological activities, in: *SRS-A and Leukotrienes* (P. J. Piper, ed.), pp. 253–270, Research Studies Press, Chichester.

Bremm, K. D., König, W., Spur, B., Crea, A., and Galanos, C., 1984, Generation of slow reacting substance (leukotrienes) by endotoxin and lipid A from human polymorphonuclear granulocytes, *Immunology* **53**:299.

Bretz, U., Dewald, B., Payne, T., and Schnyder, J., 1980, Phagocytosis stimulates the release of a slow reacting substance in cultured macrophages, *Br. J. Pharmacol.* **71**:631.

Brocklehurst, W. E., 1960, The release of histamine and formation of slow reacting substance (SRS-A) during anaphylactic shock, *J. Physiol. (London)* **151**:416.

Brocklehurst, W. E., 1962, Slow reacting substance and related compounds, *Prog. Allergy* **6**:539.

Bruns, R. F., Thomsen, W. J., and Pugsley, T. A., 1983, Binding of leukotrienes C_4 and D_4 to membranes from guinea pig lung: Regulation by ions and guanine nucleotides, *Life Sci.* **33**:645.

Burka, J. F., 1977, Pharmacological studies of bovine slow reacting substance of anaphylaxis, *Diss. Abstr. B* **37**:4421B.

Burka, J. F., and Eyre, P., 1977, A pharmacological study of SRS-A on the bovine cutaneous vasculature, *Can. J. Physiol. Pharmacol.* **55**:904.

Burke, J. A., Levi, R., Guo, Z.-G., and Corey, E. J., 1982, Leukotrienes C_4, D_4, and E_4: Effect on human and guinea pig cardiac preparations in vitro, *J. Pharmacol. Exp. Ther.* **221**:235.

Camp, R. D. R., Coutts, A. A., Greaves, M. W., Kay, A. B., and Walport, M. J., 1981, Responses of human skin to intradermal injection of leukotrienes C_4, D_4 and B_4, *Br. J. Pharmacol.* **75**:168P.

Chand, N., 1979, FPL 55712—An antagonist of slow reacting substance of anaphylaxis (SRS-A): A review, *Agents Actions* **9**:133.

Chasin, M., and Scott, C., 1978, Inhibition of cyclic nucleotide phosphodiesterase by FPL 55712, an SRS-A antagonist, *Biochem. Pharmacol.* **27**:2065.

Cheng, J. B., and Townley, R. G., 1984a, Identification of leukotriene D_4 receptor binding sites in guinea pig lung homogenates using [^{3}H] leukotriene D_4, *Biochem. Biophys. Res. Commun.* **118**:20.

Cheng, J. B., and Townley, R. G., 1984b, Effect of serine borate complex on the relative ability of leukotrienes C_4, D_4 and E_4 to inhibit lung and brain [^{3}H]-leukotriene C_4 binding: Demonstration of the agonists' potency order for leukotriene D_4 and leukotriene C_4 receptors, *Biochem. Biophys. Res. Commun.* **119**:612.

Cheng, J. B., and Townley, R. G., 1984c, Evidence for a similar receptor site binding [^{3}H]-leukotriene E_4 and [^{3}H]-leukotriene D_4 to the guinea pig crude lung membrane, *Biochem. Biophys. Res. Commun.* **122**:949.

Chilton, F. H., O'Flaherty, J. T., Walsh, C. E., Thomas, M. J., Wykle, R. L., DeChatelet, L. R., and Waite, B. M., 1982, Platelet activating factor: Stimulation of lipoxygenase pathway in polymorphonuclear leukocytes by 1-O-alkyl-2-O-acetyl-*sn*-glycerol-3-phosphocholine, *J. Biol. Chem.* **257**:5402.

Cho, H., Ueda, M., Funahashi, K., and Koda, K., 1983, Smooth muscle contractile activities of leukotriene analogues, *Chem. Pharm. Bull.* **31**:3326.

Claesson, H.-E., 1982, Leukotrienes A_4 and B_4 stimulate the formation of cyclic AMP in human leukocytes, *FEBS Lett.* **139**:305.

Claesson, H.-E., Lundberg, U., and Malmsten, C., 1981, Serum-coated zymosan stimulates the synthesis of leukotriene B_4 in human polymorphonuclear leukocytes, *Biochem. Biophys. Res. Commun.* **99**:1230.

Clancy, R. M., and Hugli, T. E., 1983, The extraction of leukotrienes (LTC_4, LTD_4 and LTE_4) from tissue fluids: The metabolism of these mediators during IgE-dependent hypersensitivity reactions in lung, *Anal. Biochem.* **133**:30.

Clancy, R. M., Dahinden, C. A., and Hugli, T. E., 1983, Arachidonate metabolism by human polymorphonuclear leukocytes stimulated by N-formyl-met-leu-phe or complement component C5a is independent of phospholipase activation, *Proc. Natl. Acad. Sci. USA* **80**:7200.

Clark, M. A., Cook, M., Mong, S., Hogaboom, G. K., Shor, R., Stradel, J., and Crooke, S. T., 1984, Leukotriene C_4 ([^{3}H]-LTC_4) binding to membranes isolated from hamster smooth muscle cell line DDT_1MF2, *Life Sci.* **35**:441.

Cohen, N., Weber, G., Banner, B. L., Welton, A. F., Hope, W. C., Crowley, H., Anderson, W. A., Simko, B. A., O'Donnell, M., Coffey, J. W., Fiedler-Nagy, C., and Batula-Bernardo, C., 1984, Analogs of arachidonic acid methylated at C-7 and C-10 as inhibitors of leukotriene biosynthesis, *Prostaglandins* **27**:553.

Coles, S. J., Neill, K. H., Reid, L. M., Austen, K. F., Nii, Y., Corey, E. J., and Lewis, R. A., 1983, Effects of leukotrienes C_4 and D_4 on glycoprotein and lysozyme secretion by human bronchial mucosa, *Prostaglandins* **25**:155.

Conroy, M. C., Orange, R. P., and Lichtenstein, L. M., 1976, Release of slow reacting substance of anaphylaxis (SRS-A) from human leukocytes by the calcium ionophore A23187, *J. Immunol.* **116**:1677.

Corey, E. J., and Munroe, J. E., 1982, Irreversible inhibition of prostaglandin and leukotriene biosynthesis from arachidonic acid by 11,12-dehydro- and 5,6-dehydro-arachidonic acids, respectively, *J. Am. Chem. Soc.* **104**:1752.

Corey, E. J., Marfat, A., Goto, G., and Brion, F., 1980, Leukotriene B: Total synthesis and assignment of stereochemistry, *J. Am. Chem. Soc.* **102**:7984.

Corey, E. J., Shih, C., and Cashman, J. R., 1983, Docosahexaenoic acid is a strong inhibitor of prostaglandin but not leukotriene synthesis, *Proc. Natl. Acad. Sci. USA* **80**:3581.

Cramer, E. B., Pologe, L., Pawlowski, N. A., Cohn, Z. A., and Scott, W. A., 1983, Leukotriene C promotes prostacyclin synthesis by human endothelial cells, *Proc. Natl. Acad. Sci. USA* **80**:4109.

Creese, B. R., and Bach, M. K., 1983, Hyperreactivity of airways smooth muscle produced in vitro by leukotrienes, *Prostaglandins Leukotrienes Med.* **11**:161.

Creese, B. R., Bach, M. K., Fitzpatrick, F. A., and Bothwell, W. M., 1984, Leukotriene responses and arachidonic acid metabolism of guinea pig lung parenchyma, *Eur. J. Pharmacol.* **102**:197.

Czarnetzki, B. M., 1981, In vitro generation of eosinophil chemotactic factor from human and murine mononuclear phagocytes, *Scand. J. Immunol.* **13**:511.

Czarnetzki, B. M., and Benveniste, J., 1981, Effect of synthetic PAF-Acether on human neutrophil function, *Agents Actions* **11**:549.

Czarnetzki, B. M., Frosch, P. J., Mardin, M., and Macher, E., 1983, Leukotriene B_4 and neutrophil-derived eosinophil chemotactic factor (ECF_{PMN}): Comparison of their biologic activities, *Monographs in Allergy* **18**:193.

Dahinden, C. A., Clancy, R. M., and Hugli, T. E., 1984, Stereospecificity of leukotriene B_4 and structure–function relationships for chemotaxis of human neutrophils, *J. Immunol.* **133**:1477.

Dahlén, S.-E., 1983, The significance of liberated cyclooxygenase products for the pulmonary and cardiovascular actions of leukotriene C_4 in the guinea pig depends upon the route of administration, *Acta Physiol. Scand.* **118**:415.

Dahlén, S.-E., Björk, T., Hedqvist, P., Arfors, K.-E., Hammarström, S., Lindgren, J.-A., and Samuelsson, B., 1981, Leukotrienes promote plasma leakage and leukocyte adhesion in postcapillary venules: In vivo effects with relevance to the acute inflammatory response, *Proc. Natl. Acad. Sci. USA* **78**:3887.

Dahlén, S.-E., Hansson, G., Hedqvist, P., Björk, T., Granström, E., and Dahlén, B., 1983a, Allergen challenge of lung tissue from asthmatics elicits bronchial contraction that correlates with the release of leukotrienes C_4, D_4 and E_4, *Proc. Natl. Acad. Sci. USA* **80**:1712.

Dahlén, S.-E., Hedqvist, P., and Hammarström, S., 1983b, Contractile activities of several cysteine-containing leukotrienes in the guinea pig lung strip, *Eur. J. Pharmacol.* **86**:208.

Dahlén, S.-E., Hedqvist, P., Hammarström, S., and Samuelsson, B., 1980, Leukotrienes are potent constrictors of human bronchi, *Nature* (Lond) **288**:484.

Dahlén, S.-E., Hedqvist, P., Westlund, P., Granström, E., Hammarström, S., Lindgren, J.-A., and Råmark, O., 1983c, Mechanisms of leukotriene induced contractions of guinea pig airways: Leukotriene C_4 has a potent direct action whereas leukotriene B_4 acts indirectly, *Acta Physiol. Scand.* **118**:393.

Damon, M., Chavis, C., Godard, P., Michel, F. B., and Crastes de Paulet, A., 1983, Purification and mass spectrometry identification of leukotriene D_4 synthesized by human alveolar macrophages, *Biochem. Biophys. Res. Commun.* **111**:518.

Davidson, A. B., McFadden, E. R., Jr., Drazen, J. M., Lee, T. H., Austen, K. F., and Corey, E. J., 1984, Bronchoconstrictor effects of leukotriene E in normal adults, *Am. Rev. Respir. Dis.* **129**:A1.

Dembinska-Kieć, A., Simmet, T., and Peskar, B., 1984, Formation of leukotriene C_4-like material by rat brain tissue, *Eur. J. Pharmacol.* **99**:57.
Drazen, J. M., and Austen, K. F., 1974, Effects of intravenous administration of slow-reacting substance of anaphylaxis, histamine, bradykinin, and prostaglandin $F_{2\alpha}$ on pulmonary mechanics in the guinea pig, *J. Clin. Invest.* **53**:1679.
Drazen, J. M., Lewis, R. A., Wasserman, S. I., Orange, R. P., and Austen, K. F., 1979, Differential effects of a partially purified preparation of slow reacting substance of anaphylaxis on guinea pig tracheal spirals and parenchymal strips, *J. Clin. Invest.* **63**:1.
Drazen, J. M., Austen, K. F., Lewis, R. A., Clark, G. A., Goto, G., Marfat, A., and Corey, E. J., 1980, Comparative airway and vascular activities of leukotrienes C-1 and D in vivo and in vitro, *Proc. Natl. Acad. Sci. USA* **77**:4354.
Drazen, J. M., Lewis, R. A., Austen, K. F., Toda, M., Brion, F., Marfat, A., and Corey, E. J., 1981, Contractile activities of structural analogs of leukotriene C and D: Necessity of a hydrophobic region, *Proc. Natl. Acad. Sci. USA* **78**:3195.
Du, J. T., Foegh, M., Maddox, Y., and Ramwell, P. W., 1983, Human peritoneal macrophages synthesize leukotrienes B_4 and C_4, *Biochim. Biophys. Acta* **753**:159.
Engineer, D. M., Piper, P. J., and Sirois, P., 1976, Interaction between the release of SRS-A and of prostaglandins, *Br. J. Pharmacol.* **57**:460P.
Engineer, D. M., Niederhauser, U., Piper, P. J., and Sirois, P., 1978, Release of mediators of anaphylaxis inhibition of prostaglandin synthesis and the modification of release of slow reacting substance of anaphylaxis and histamine, *Br. J. Pharmacol.* **62**:61.
Falkenhein, S. F., MacDonald, H., Huber, M. M., Koch, D., and Parker, C. W., 1980, Effect of the 5-hydroperoxide of eicosatetraenoic-acid and inhibitors of the lipoxygenase pathway on the formation of slow reacting substance by rat basophilic leukemia cells; direct evidence that slow reacting substance is a product of the lipoxygenase pathway, *J. Immunol.* **125**:163.
Farmer, J. B., Richards, I. M., Sheard, P., and Woods, A. M., 1975, Mediators of passive lung anaphylaxis in the rat, *Br. J. Pharmacol.* **55**:57.
Feldberg, W., and Kellaway, C. H., 1938, Liberation of histamine and formation of lysocithin-like substances by cobra venom, *J. Physiol. (London)* **94**:187.
Feuerstein, G., Zukowska-Grojec, Z., and Kopin, I. J., 1981, Cardiovascular effects of leukotriene D_4 in SHR and WKY rats, *Eur. J. Pharmacol.* **76**:107.
Feuerstein, G., Zukowska-Grojec, Z., Bayorh, M., Kopin, I. J., and Faden, A. I., 1983, Leukotriene D_4-induced hypotension is reversed by thyrotropin-releasing hormone, *Prostaglandins* **26**:711.
Feuerstein, N., Foegh, M., and Ramwell, P. W., 1981, Leukotrienes C_4 and D_4 induce prostaglandin and thromboxane release from rat peritoneal macrophages, *Br. J. Pharmacol.* **72**:389.
Fitzpatrick, F., Haeggström, J., Granström, E., and Samuelsson, B., 1983, Metabolism of leukotriene A_4 by an enzyme in blood plasma: A possible leukotactic mechanism, *Proc. Natl. Acad. Sci. USA* **80**:5425.
Fitzpatrick, F., Liggett, W., McGee, J., Bunting, S., Morton, D., and Samuelsson, B., 1984, Metabolism of leukotriene A_4 by human erythrocytes: A novel cellular source of leukotriene B_4, *J. Biol. Chem.* **259**:11403.
Folco, G., Omini, C., Rossoni, G., Vigano, T., and Berti, F., 1982a, Anticholinergic agents prevent guinea-pig airway constriction induced by histamine, bradykinin, and leukotriene C_4: Relationship to circulating TXA_2, *Eur. J. Pharmacol.* **78**:159.
Folco, G. C., Omini, C., Vigano, T., Brunelli, G., Rossoni, G., and Berti, F., 1982b, Biological activity of leukotriene C_4 in guinea pigs: In vitro and in vivo studies, *Adv. Prostaglandin Thromboxane Leukotriene Res.* **9**:153.
Ford-Hutchinson, A. W., Bray, M. A., Doig, M. V., Shipley, M. E., and Smith, M. J. H., 1980, Leukotriene B, a potent chemokinetic and aggregating substance released from polymorphonuclear leukocytes, *Nature (London)* **286**:264.
Ford-Hutchinson, A. W., Bray, M. A., Cunningham, F. M., Davidson, E. M., and Smith, M. J. H., 1981, Isomers of leukotriene B_4 possess different biological potencies, *Prostaglandins* **21**:143.
Ford-Hutchinson, A. W., Piper, P. J., and Samhoun, M. N., 1982, Generation of leukotriene B_4, its all *trans* isomers, and 5-hydroxyeicosatetraenoic acid by rat basophilic leukemia cells, *Br. J. Pharmacol.* **76**:215.

Ford-Hutchinson, A. W., Brunet, G., Savard, P., and Charleson, S., 1984, Leukotriene B_4, polymorphonuclear leukocytes and inflammatory exudates in the rat, *Prostaglandins* **28**:13.
Fruteau de Laclos, B., Braquet, P., and Borgeat, P., 1984, Characteristics of leukotriene (LT) and hydroxyeicosatetraenoic acid (Hete) synthesis in human leukocytes *in vitro:* Effect of arachidonic acid, *Prostaglandins Leukotrienes Med.* **13**:47.
Furukawa, M., Yoshimoto, T., Ochi, K., and Yamamoto, S., 1984, Studies on arachidonate 5-lipoxygenase of rat basophilic leukemia cells, *Biochim. Biophys. Acta* **795**:458.
Gardiner, P. J., and Stock, D. M., 1984, Indomethacin-induced hyperreactivity of human small airways to LTD_4, *Br. J. Pharmacol.* **81**:66P.
Gazelius, B., Panopoulos, P., Orlander, B., and Claesson, H.-E., 1984, Inhibition of intradental nerve excitability by leukotriene B_4 and C_4, *Acta Physiol. Scand.* **120**:141.
Girard, Y., Larue, M., Jones, T. R., and Rokach, J., 1982, Synthesis of the sulfones of leukotriene C_4, D_4 and E_4, *Tetrahedron Lett.* **23**:1023.
Gleason, J. G., Ku, T. W., McCarthy, M. E., Weichman, B. M., Holden, D., Osborn, R. R., Zabko-Potapovich, B., Berkowitz, B., and Wasserman, M. A., 1983, 2-Nor leukotriene analogs: Antagonists of the airway and vascular smooth muscle effects of leukotriene C_4, D_4, and E_4, *Biochem. Biophys. Res. Commun.* **117**:732.
Godard, P., Damon, M., Chavis, C., Kervran, C., and Michel, F. B., 1983, Leukotriene (LT) D_4 generation by human alveolar macrophages (AM): Higher synthetic ability in allergic asthmatic patients, *Am. Rev. Respir. Dis.* **127**(4II):73.
Godard, P., Damon, M., Chavis, C., Bognie, M., Crastes de Paulet, A., and Michel, F. B., 1984, Arachidonic acid metabolism in alveolar macrophages: Comparison of cells from healthy subjects, patients with chronic bronchitis, and allergic asthmatic patients, *Am. Rev. Respir. Dis.* **129**:81.
Goetzl, E. J., 1981, Oxygenation products of arachidonic acid as mediators of hypersensitivity and inflammation, *Med. Clin. North Am.* **65**:809.
Goetzl, E. J., 1982, Selective feed-back inhibition of the 5-lipoxygenation of arachidonic acid in human T-lymphocytes, *Biochem. Biophys. Res. Commun.* **101**:344.
Goetzl, E. J., and Pickett, W. C., 1981, Novel structural determinants of the human neutrophil chemotactic activity of leukotriene B, *J. Exp. Med.* **153**:482.
Goetzl, E. J., Phillips, M. J., and Gold, W. M., 1983, Stimulus specificity of the generation of leukotrienes by dog mastocytoma cells, *J. Exp. Med.* **158**:731.
Goetzl, E. J., Payan, D. G., and Goldman, D. W., 1984, Immunopathogenic roles of leukotrienes in human diseases, *J. Clin. Immunol.* **4**:79.
Goldenberg, M. M., and Subers, E. M., 1982, The reactivity of rat isolated gastrointestinal tissues to leukotrienes, *Eur. J. Pharmacol.* **78**:463.
Goldman, D. W., and Goetzl, E. J., 1982, Specific binding of leukotriene B_4 to receptors on human polymorphonuclear leukocytes, *J. Immunol.* **129**:1600.
Goldman, D. W., and Goetzl, E. J., 1984, Heterogeneity of human polymorphonuclear leukocyte receptors for leukotriene B_4: Identification of a subset of high affinity receptors that transduce the chemotactic response, *J. Exp. Med.* **159**:1027.
Goldman, D. W., Pickett, W. C., and Goetzl, E. J., 1983, Human neutrophil chemotactic and degranulating activities of leukotriene B_5 (LTB_5) derived from eicosapentaenoic acid, *Biochem. Biophys. Res. Commun.* **117**:282.
Goldyne, M. E., Burrish, G. F., Poubelle, P., and Borgeat, P., 1984, Arachidonic acid metabolism among human mononuclear leukocytes: Lipoxygenase-related pathways, *J. Biol. Chem.* **259**:8815.
Greenwald, S. E., Letts, L. G., Newman, D. L., and Piper, P. J., 1984, Actions of cysteinyl leukotrienes in coronary, femoral and carotid vessels of the pig, *Eur. J. Pharmacol.* **103**:107.
Griffin, M., Weiss, J. W., Leitch, A. G., McFadden, E. R., Jr., Corey, E. J., Austen, K. F., and Drazen, J. M., 1983, Effects of leukotriene D on the airways in asthma, *N. Engl. J. Med.* **308**:436.
Gulati, O. P., Malmsten, C., Ponard, G., and Gulati, N., 1983, The local edemogenic effects of leukotriene C_4 and prostaglandin E_2 in rats, *Prostaglandins Leukotrienes Med.* **10**:11.
Gunter, K., and Stechschulte, D. J., 1980, Generation of slow reacting substance (SRS) termed leukotriene C1 from human peripheral blood monocytes, *Clin. Res.* **28**:707a.
Hafström, I., Palmbald, J., Malmsten, C. L., Rådmark, O., and Samuelsson, B., 1981, Leukotriene

B_4—A stereospecific stimulator for release of lysosomal enzymes from neutrophils, *FEBS Lett.* **130**:146.
Ham, E. A., Soderman, D. D., Zanetti, M. E., Dougherty, H. W., McCauley, E., and Kuehl, F. A., Jr., 1983, Inhibition by prostaglandins of leukotriene B_4 release from activated neutrophils, *Proc. Natl. Acad. Sci. USA* **80**:4349.
Hammarström, S., 1977, Selective inhibition of platelet n-8 lipoxygenase by 5,8,11-eicosatriynoic acid, *Biochim. Biophys. Acta* **487**:517.
Hammarström, S., 1980, Leukotriene C_5: A slow-reacting substance derived from eicosapentaenoic acid, *J. Biol. Chem.* **255**:7093.
Hammarström, S., 1981, Conversion of dihomo-γ-linolenic acid to an isomer of leukotriene C_3, oxygenated at C-8, *J. Biol. Chem.* **256**:7712.
Hammarström, S., 1982, Metabolism of leukotriene C_3 in the guinea pig, *Adv. Prostaglandin Thromboxane Leukotriene Res.* **9**:83.
Hammarström, S., and Samuelsson, B., 1980, Detection of leukotriene A_4 as an intermediate in the biosynthesis of leukotrienes C_4 and D_4, *FEBS Lett.* **122**:83.
Hammarström, S., Murphy, R. C., Samuelsson, B., Clark, D. A., Mioskowski, C., and Corey, E. J., 1979, Structure of leukotriene C: Identification of the amino acid part, *Biochem. Biophys. Res. Commun.* **91**:1266.
Hammarström, S., Samuelsson, B., Clark, D. A., Goto, G., Marfat, A., Mioskowski, C., and Corey, E. J., 1980, Stereochemistry of leukotriene C-1, *Biochem. Biophys. Res. Commun.* **92**:946.
Hammarström, S., Bernström, K., Örning, L., Dahlén, S.-E., Hedqvist, P., Smedegard, G., and Revenas, B., 1981, Rapid *in vivo* metabolism of leukotriene C_3 in the monkey *Macaca irus*, *Biochem. Biophys. Res. Commun.* **101**:1109.
Hammarström, S., Örning, L., Bernström, K., Gustafsson, B., Norin, E., and Kaijser, L., 1985, Metabolism of leukotriene C_4 in rats and humans, *Adv. Prostaglandin Thrombaxane Leukotriene Res.* **15**:185.
Hand, J. M., Will, J. A., and Buckner, C. K., 1981, Effects of leukotrienes on isolated guinea pig pulmonary arteries, *Eur. J. Pharmacol.* **76**:439.
Hand, J. M., Will, J. A., and Buckner, C. K., 1983, Enhancement by histamine of the contractile effects of leukotriene D_4 on extralobar pulmonary arteries isolated from the guinea pig, *Eur. J. Pharmacol.* **87**:341.
Hanna, C. J., Bach, M. K., Pare, P. D., and Schellenberg, R. R., 1981, Slow reacting substances (leukotrienes) contract human airway and pulmonary vascular smooth muscle in vitro, *Nature (London)* **290**:343.
Hansson, G., and Rådmark, O., 1980, Leukotriene C_4: Isolation from human polymorphonuclear leukocytes, *FEBS Lett.* **122**:87.
Hansson, G., Björk, T., Dahlén, S.-E., Hedqvist, P., Granström, E., and Dahlén, B., 1983, Specific allergen induces contraction of bronchi and formation of leukotrienes C_4, D_4 and E_4 in human asthmatic lung, *Adv. Prostaglandin Thromboxane Leukotriene Res.* **12**:153.
Harkavy, J., 1930, Spasm-producing substance in the sputum of patients with bronchial asthma, *Arch. Intern. Med.* **45**:641.
Hattori, Y., and Levi, R., 1984, Negative inotropic effect of leukotrienes: Leukotrienes C_4 and D_4 inhibit calcium-dependent contractile responses in potassium-depolarized guinea-pig myocardium, *J. Pharmacol. Exp. Ther.* **230**:646.
Hayes, E. C., Girard, Y., Lombardo, D. L., Macock, A. L., Rokach, J., Young, R. N., and Zweerink, H. J., 1983, Measuring leukotrienes of slow reacting substance of anaphylaxis: Development of a specific radioimmunoassay, *J. Immunol.* **131**:429.
Hedqvist, P., Dahlén, S.-E., Gustafson, L., Hammarström, S., and Samuelsson, B., 1980, Biological profile of leukotrienes C_4 and D_4, *Acta Physiol. Scand.* **110**:331.
Henderson, W. R., and Klebanoff, S. J., 1983a, Leukotriene production and inactivation by normal, chronic granulomatous diseases and myeloperoxidase-deficient neutrophils, *J. Biol. Chem.* **258**:13522.
Henderson, W. R., and Klebanoff, S. J., 1983b, Leukotriene B_4, C_4, D_4 and E_4 inactivation by hydroxyl radicals, *Biochem. Biophys. Res. Commun.* **110**:266.
Henderson, W. R., Jörg, A., and Klebanoff, S. J., 1982, Eosinophil peroxidase-mediated inactivation of leukotrienes B_4, C_4 and D_4, *J. Immunol.* **128**:2609.

Henderson, W. R., Harley, J. B., and Fauci, A. S., 1984, Arachidonic acid metabolism in normal and hypereosinophilic syndrome human eosinophils: Generation of leukotrienes B_4, C_4, D_4 and 15-lipoxygenase products, *Immunology* **51**:679.
Higgs, G. A., Bax, C. M. P., and Moncada, S., 1982, Inflammatory properties of lipoxygenase products and the effects of indomethacin and BW 755c on prostaglandin production, leukocyte migration and plasma exudation in rabbit skin, *Adv. Prostaglandin Thromboxane Leukotriene Res.* **9**:331.
Hirshman, C. A., Darnell, M., Brugman, T., and Peters, J., 1983a, Airway constrictor effects of leukotriene D_4 in dogs with hyperreactive airways, *Prostaglandins* **25**:481.
Hirshman, C. A., Peters, J., Butler, J., Hanifin, J. M., and Downes, H., 1983b, Role of mediators in allergic and nonallergic asthma in dogs with hyperreactive airways, *J. Appl. Physiol.* **54**:1108.
Hofmann, S. L., Prescott, S. M., and Majerus, P. W., 1982, The effects of mepacrine and p-bromophenacylbromide on arachidonic acid release in human platelets, *Arch. Biochem. Biophys.* **215**:237.
Hogaboom, G. K., Mong, S., Wu, H.-L., and Crooke, S. T., 1983, Peptidoleukotrienes: Distinct receptors for leukotriene C_4 and D_4 in the guinea pig lung, *Biochem. Biophys. Res. Commun.* **116**:1136.
Holgate, S. T., and Robinson, C., 1984, 6,9-deepoxy-6,9-(phenylimino)-$\Delta^{6,8}$-prostacyclin I_1, (U-60,257), stimulates prostaglandin D_2 and inhibits thromboxane B_2 release from ionophore challenged human dispersed lung cells, *Br. J. Pharmacol.* **83**:603.
Holme, G., Brunet, G., Piechuta, H., Masson, P., Girard, Y., and Rokach, J., 1980, The activity of synthetic leukotriene C-1 on guinea pig trachea and ileum, *Prostaglandins* **20**:717.
Holroyde, M. C., Altounyan, R. E. C., Cole, M., Dixon, M., and Elliott, E. V., 1981, Bronchoconstriction produced in man by leukotrienes C and D, *Lancet* **2**:17.
Homma, Y., Onozaki, K., Hashimoto, T., Nagai, Y., and Takenawa, T., 1982, Differential activation of phospholipids metabolism by formylated peptide and ionophore A 23187 in guinea pig peritoneal macrophages, *J. Immunol.* **129**:1619.
Hope, W. C., Welton, A. F., Fiedler-Nagy, C., Batula-Bernardo, C., and Coffey, J. W., 1983, Quercetin inhibits the biosynthesis of slow reacting substance of anaphylaxis (SRS-A) and lipoxygenase activity in vitro, *Biochem. Pharmacol.* **32**:367.
Houglum, J., Pai, J.-K., Atrache, V., Sok, D.-E., and Sih, C. J., 1980, Identification of the slow-reacting substances from cat paws, *Proc. Natl. Acad. Sci. USA* **77**:5688.
Humes, J. L., Sadowski, S., Galavage, M., Goldenberg, M., Subers, E., Bonney, R. J., and Kuehl, F. A., Jr., 1982, Evidence for two sources of arachidonic acid for oxidative metabolism by mouse peritoneal macrophages, *J. Biol. Chem.* **257**:1591.
Iacopino, V. J., Fitzpatrick, T. M., Ramwell, P. W., Rose, J. C., and Kot, P. A., 1983, Cardiovascular responses to leukotriene C_4 in the rat, *J. Pharmacol. Exp. Ther.* **227**:244.
Iacopino, V. J., Compton, S., Fitzpatrick, T., Ramwell, P., Rose, J., and Kot, P., 1984, Responses to leukotriene C_4 in the perfused rat lung, *J. Pharmacol. Exp. Ther.* **229**:654.
Inoue, H., Ichinose, M., Horio, S., Suzuki, S., Hida, W., and Takashima, T., 1984, Interaction between leukotriene and histamine, $PGF_{2\alpha}$-stimulant on lung mechanics in vivo, *Am. Rev. Respir. Dis.* **129**:A2.
Irvine, R. F., 1982, How is the level of free arachidonic acid controlled in mammalian cells?, *Biochem. J.* **204**:3.
Ishihara, Y., Kitamura, S., and Takaku, F., 1983, Effect of lipoxygenase inhibitors, AA861 and T22083, on chemical mediators released from sensitized guinea pig lung tissue, *Prostaglandins* **26**:623.
Jakschik, B. A., and Kuo, C. G., 1983, Characterization of leukotriene A_4 and B_4 biosynthesis, *Prostaglandins* **25**:767.
Jakschik, B. A., Falkenhein, S., and Parker, C. W., 1977a, Precursor role of arachidonic acid in release of slow reacting substance from rat basophilic leukemia cells, *Proc. Natl. Acad. Sci. USA* **74**:4577.
Jakschik, B. A., Kulczycki, A., Jr., MacDonald, H. H., and Parker, C. W., 1977b, Release of slow reacting substance (SRS) from rat basophilic leukemia (RBL-1) cells, *J. Immunol.* **119**:618.
Jakschik, B. A., Sun, F. F., Lee, L.-H., and Steinhoff, M. M., 1980, Calcium stimulation of a novel lipoxygenase, *Biochem. Biophys. Res. Commun.* **95**:103.

Jakschik, B. A., DiSantis, D. M., Sankarappa, S. K., and Sprecher, H., 1981, Delta four acetylenic acids—Selective inhibitors of the formation of slow reacting substance, *Biochem. Biophys. Res. Commun.* **102**:624.

Jakschik, B. A., Harper, T., and Murphy, R. C., 1982, Leukotriene C_4 and D_4 formation by particulate enzymes, *J. Biol. Chem.* **257**:5346.

Jakschik, B. A., Morrison, A. R., and Sprecher, H., 1983, Products derived from 5,8,11-eicosatrienoic acid by the 5-lipoxygenase leukotriene pathway, *J. Biol. Chem.* **258**:12797.

Johnson, H. G., and McNee, M. L., 1983, Secretagogue responses of leukotrienes C_4 and D_4: A comparison of potency in canine trachea in vivo, *Prostaglandins* **25**:237.

Johnson, H. G., and McNee, M. L., 1984, Regulation of canine mucus secretion by a novel leukotriene synthesis inhibitor, *Int. Arch. Allergy Appl. Immunol.* **75**:97.

Johnson, H. G., Chin, R. A., Chow, A. W., Bach, M. K., and Nadel, J. A., 1983a, Leukotriene C_4 enhances mucus production from submucosal glands in canine trachea in vivo, *Int. J. Immunopharmacol.* **5**:391.

Johnson, H. G., McNee, M. L., Bach, M. K., and Smith, H. W., 1983b, The activity of a new, novel inhibitor of leukotriene synthesis in rhesus monkey Ascaris reactors, *Int. Arch. Allergy Appl. Immunol.* **70**:169.

Johnson, H. G., McNee, M. L., Johnson, M. A., and Miller, M. D., 1983c, Leukotriene C_4 and dimethyphenylpiperazinium-induced responses in canine airway tracheal muscle contraction and fluid secretion, *Int. Arch. Allergy Appl. Immunol.* **71**:214.

Jones, T. R., Davis, C., and Daniel, E. E., 1982a, Pharmacological study of the contractile activity of leukotriene C_4 and D_4 on isolated human airway smooth muscle, *Can. J. Physiol. Pharmacol.* **60**:638.

Jones, T., Masson, P., Hamel, R., Brunet, G., Holme, G., Girard, Y., Larue, M., and Rokach, J., 1982b, Biological activity of leukotriene sulfones on respiratory tissues, *Prostaglandins* **24**:279.

Jones, T., Denis, D., Hall, R., and Ethier, D., 1983, Pharmacological study of leukotrienes C_4, D_4, E_4 and F_4 on guinea pig trachealis: Interaction with FPL 55712, *Prostaglandins* **26**:833.

Jörg, A., Henderson, W. R., Murphy, R. C., and Klebanoff, S. J., 1982, Leukotriene generation by eosinophils, *J. Exp. Med.* **155**:390.

Jubiz, W., 1983, A reevaluation of the chemotactic potency of leukotriene B_4, *Biochem. Biophys. Res. Commun.* **110**:842.

Kellaway, C. V. H., and Trethewie, E. R., 1940, Liberation of a slow reacting smooth muscle stimulating substance in anaphylaxis, *Q. J. Exp. Physiol.* **30**:121.

Kito, G., Okuda, H., Ohkawa, S., Terao, S., and Kikuchi, K., 1981, Contractile activities of leukotrienes C_4 and D_4 on vascular strips from rabbits, *Life Sci.* **29**:1325.

Koide, T., Neichi, T., Takato, M., Matsushita, H., Sugioka, K., Nakano, M., and Hata, S.-I., 1982, Possible mechanisms of 15-hydroperoxyarachidonic acid-induced contraction of the canine basilar artery in vitro, *J. Pharmacol. Exp. Ther.* **221**:481.

Koshihara, Y., Neichi, T., Murota, S.-I., Lao, A.-N., Fujimoto, Y., and Tatsuno, T., 1983, Selective inhibition of 5-lipoxygenase by natural compounds isolated from Chinese plants, *Artemisia rubripes* Nakai, *FEBS Lett.* **158**:41.

Kreisle, R. A., and Parker, C. W., 1983, Specific binding of leukotriene B_4 to a receptor on human polymorphonuclear leukocytes, *J. Exp. Med.* **157**:628.

Krell, R. D., Tsai, B. S., and Giles, R. E., 1983, Pharmacologic aspects of leukotriene receptors, in: *Leukotrienes and Other Lipoxygenase Products* (P. J. Piper, ed.), pp. 222–233, Research Studies Press, Chichester.

Krilis, S., Lewis, R. A., Corey, E. J., and Austen, K. F., 1983a, Bioconversion of C-6 sulfidopeptide leukotrienes by the responding guinea pig ileum determines the time course of its contraction, *J. Clin. Invest.* **71**:909.

Krilis, S., Lewis, R. A., Corey, E. J., and Austen, K. F., 1983b, Specific receptors for leukotriene C_4 on a smooth muscle cell line, *J. Clin. Invest.* **72**:1516.

Krilis, S., Lewis, R. A., Corey, E. J., and Austen, K. F., 1984, Specific binding of leukotriene C_4 to ileal segments and subcellular fractions of ileal smooth muscle cells, *Proc. Natl. Acad. Sci. USA* **81**:4529.

Lapetina, E. G., Billah, M. M., and Cuatrecasas, P., 1981, The phosphatidyl inositol cycle and the regulation of arachidonic acid production, *Nature (London)* **292**:367.

Lee, C. W., Lewis, R. A., Corey, E. J., Barton, A., Oh, H., Tauber, A. I., and Austen, K. F., 1982a, Oxidative inactivation of leukotriene C_4 by stimulated human polymorphonuclear leukocytes, *Proc. Natl. Acad. Sci. USA* **79**:4166.
Lee, C. W., Lewis, R. A., Austen, K. F., and Corey, E. J., 1982b, Metabolism of leukotriene D (LTD) by human polymorphonuclear leukocytes (PMN's), *Fed. Proc.* **41**:487.
Lee, J. Y., Chernov, T., and Goetzl, E. J., 1984, Characteristics of the epitope of leukotriene B_4 recognized by a highly specific mouse monoclonal antibody, *Biochem. Biophys. Res. Commun.* **123**:944.
Lee, T. H., Austen, K. F., Corey, E. J., and Drazen, J. M., 1984a, Leukotriene E_4-induced airway hyperresponsiveness of guinea pig tracheal smooth muscle to histamine and evidence for three separate sulfidopeptide leukotriene receptors, *Proc. Natl. Acad. Sci. USA* **81**:4922.
Lee, T. H., Mencia-Huerta, J.-M., Shih, C., Corey, E. J., Lewis, R. A., and Austen, K. F., 1984b, Characterization and biologic properties of 5,12-dihydroxyderivatives of eicosapentaenoic acid, including leukotriene B_5 and the double lipoxygenase product, *J. Biol. Chem.* **259**:2383.
Leitch, A. G., Austen, K. F., Corey, E. J., and Drazen, J. M., 1982, Effects of indomethacin (INDO) on the contractile response of guinea pig lung parenchymal strips (GPLS) to leukotrienes (LT) B_4, C_4, D_4 and E_4, *Fed. Proc.* **41**:1047.
Leitch, A. G., Corey, E. J., Austen, K. F., and Drazen, J. M., 1983a, Indomethacin potentiates the pulmonary response to aerosol leukotriene C_4 in the guinea pig, *Am. Rev. Respir. Dis.* **128**:639.
Leitch, A. G., Lewis, R. A., Corey, E. J., and Austen, K. F., 1983b, Quantitation and characterization of leukotriene products released following immunologic stimulation of human lung cells in vitro, *Chest* **83**:81S.
Letts, L. G., and Piper, P. J., 1982, The actions of leukotrienes C_4 and D_4 on guinea pig isolated hearts, *Br. J. Pharmacol.* **76**:169.
Letts, L. G., Newman, D. L., Greenwald, S. E., and Piper, P. J., 1983, Effects of intracoronary administration of leukotriene D_4 in the anesthetized dog, *Prostaglandins* **26**:563.
Levi, R., Burke, J. A., Holland, B. A., and Green, F. J., 1980, Immediate hypersensitivity reactions of the heart: Reduction of tachycardia and arrhythmias by the SRS-A antagonist FPL 55712, *Immunopharmacology* **2**:173.
Levi, R., Burke, J. A., and Corey, E. J., 1982, SRS-A, leukotrienes, and immediate hypersensitivity reactions of the heart, *Adv. Prostaglandin Thromboxane Leukotriene Res.* **9**:215.
Levine, J. D., Lau, W., Kwiat, G., and Goetzl, E. J., 1984, Leukotriene B_4 produces hyperalgesia that is dependent on polymorphonuclear leukocytes, *Science* **225**:743.
Levine, L., Morgan, R. A., Lewis, R. A., Austen, K. F., Clark, D. A., Marfat, A., and Corey, J., 1981, Radioimmunoassay of the leukotrienes of slow reacting substance of anaphylaxis, *Proc. Natl. Acad. Sci. USA* **78**:7692.
Lewis, R. A., and Austen, K. F., 1984, The biologically active leukotrienes: Biosynthesis, metabolism, receptors, functions, and pharmacology, *J. Clin. Invest.* **73**:889.
Lewis, R. A., Goetzl, E. J., Wasserman, S. I., Valone, F. H., Rubin, R.H., and Austen, K. F., 1975, The release of four mediators of immediate hypersensitivity from human leukemic basophils, *J. Immunol.* **114**:87.
Lewis, R. A., Drazen, J. M., Austen, K. F., Clark, D. A., and Corey, E. J., 1980, Identification of the C(6)-S-conjugate of leukotriene A with cysteine as a naturally occurring slow reacting substance of anaphylaxis (SRS-A): Importance of the 11-*cis*-geometry for biologic activity, *Biochem. Biophys. Res. Commun.* **96**:271.
Lewis, R. A., Drazen, J. M., Austen, K. F., Toda, M., Brion, F., Marfat, A., and Corey, E. J., 1981a, Contractile activities of structural analogs of leukotrienes C and D: Role of the polar substituents, *Proc. Natl. Acad. Sci. USA* **78**:4579.
Lewis, R. A., Goetzl, E. J., Drazen, J. M., Soter, N. A., Austen, K. F., and Corey, E. J., 1981b, Functional characterization of synthetic leukotriene B and its stereochemical isomers, *J. Exp. Med.* **154**:1243.
Lewis, R. A., Soter, N. A., Corey, E. J., and Austen, K. F., 1981c, Local effects of synthetic leukotrienes (LT's) on monkey (M) and human (H) skin, *Clin. Res.* **29**:492a.
Lewis, R. A., Drazen, J. M., Figueiredo, J. C., Corey, E. J., and Austen, K. F., 1982a, A review of recent contributions on biologically active products of arachidonate conversion, *Int. J. Immunopharmacol.* **4**:85.

Lewis, R. A., Mencia-Huerta, J.-M., Soberman, R. J., Hoover, D., Marfat, A., Corey, E. J., and Austen, K. F., 1982b, Radioimmunoassay for leukotriene B_4, *Proc. Natl. Acad. Sci. USA* **79**:7904.

Lin, A. H., Morton, D. R., and Gorman, R. R., 1982, Acetyl glyceryl ether phosphoryl-choline stimulates leukotriene B_4 synthesis in human polymorphonuclear leukocytes, *J. Clin. Invest.* **70**:1058.

Lindgren, J.-A., Hammarström, S., and Goetzl, E. J., 1983, A sensitive and specific radioimmunoassay for leukotriene C_4, *FEBS Lett.* **152**:83.

Lindgren, J.-A., Höfelt, T., and Dahlén, S.-E., 1984, Leukotrienes in the rat central nervous system, *Proc. Natl. Acad. Sci. USA* **81**:6212.

Littner, M. R., Kazmi, G. M., and Lott, F. D., 1984, Inhibition of cyclooxygenase production does not prevent arachidonate from increasing extravascular lung water and albumin in an isolated dog lung, *Prostaglandins Leukotrienes Med.* **15**:53.

Lundberg, U., Rådmark, O., Malmsten, C., and Samuelsson, B., 1981, Transformation of 15-hydroperoxy-5,9,11,13-eicosatetraenoic acid into novel leukotrienes, *FEBS Lett.* **126**:127.

Maas, R. L., Brash, A. R., and Oates, J. A., 1981, A second pathway of leukotriene biosynthesis in porcine leukocytes, *Proc. Natl. Acad. Sci. USA* **78**:5523.

MacDermot, J., Kelsey, C. R., Waddell, K. A., Richmond, R., Knight, R. K., Cole, P. J., Dollery, C. T., Landon, D. N., and Blair, I. A., 1984, Synthesis of leukotriene B_4 and prostanoids by human alveolar macrophages: Analysis by gas chromatography/mass spectrometry, *Prostaglandins* **27**:163.

MacGlashan, D. W., Jr., Schleimer, R. P., Peters, S. P., Schulman, E. S., Adams, G. K., III, Newball, H. H., and Lichtenstein, L. M., 1982, Generation of leukotrienes by purified human lung mast cells, *J. Clin. Invest.* **70**:747.

Maclouf, J., Fruteau de Laclos, B., and Borgeat, P., 1982, Stimulation of leukotrienes biosynthesis in human blood leukocytes by platelet-derived 12-hydroperoxy-eicosatetraenoic acid, *Proc. Natl. Acad. Sci. USA* **79**:6042.

Magous, R., Bali, J.-P., Rossi, J.-C., and Girard, J.-P., 1983, Leukotrienes stimulate acid secretion from isolated gastric parietal cells, *Biochem. Biophys. Res. Commun.* **114**:897.

Malik, K. U., and Wang, P. Y.-K., 1981, Leukotriene C_4: The major lipoxygenase metabolite of arachidonic acid in dog spleen, *Biochem. Biophys. Res. Commun.* **103**:511.

Marcus, A. J., Safier, L. B., Broekman, J., Ullman, H. L., Islam, N., Sorell, T. C., Serhan, C. N., Weissmann, G., Oblesby, T. D., and Gorman, R. R., 1984, Production of metabolic products of arachidonic acid during cell–cell interaction, *J. Allergy Clin. Immunol.* **74**:338.

Marom, Z., Shelhamer, J. H., Bach, M. K., Morton, D. R., and Kaliner, M., 1982, Slow-reacting substances, leukotrienes C_4 and D_4 increase the release of mucus from human airways in vitro, *Am. Rev. Respir. Dis.* **126**:449.

Martin, T. R., Altman, L. C., Albert, R. K., and Henderson, W. R., 1984, Leukotriene B_4 production by the human alveolar macrophage: A potential mechanism for amplifying inflammation in the lung, *Am. Rev. Respir. Dis.* **129**:106.

Mathews, W. R., and Murphy, R. C., 1982, Inhibition of leukotriene biosynthesis in mastocytoma cells by diethylcarbamazine, *Biochem. Pharmacol.* **31**:2129.

Mencia-Huerta, J.-M., Hadji, L., and Benveniste, J., 1981, Release of a slow-reacting substance from rabbit platelets, *J. Clin. Invest.* **68**:1586.

Mencia-Huerta, J.-M., Razin, E., Ringel, E. W., Corey, E. J., Hoover, D., Austen, K. F., and Lewis, R. A., 1983, Immunologic and ionophore-induced generation of leukotriene B_4 from mouse bone marrow-derived mast cells, *J. Immunol.* **130**:1885.

Metz, S. A., Hall, M. E., Harper, T. W., and Murphy, R. C., 1982, Rapid extraction of leukotrienes from biologic fluids and quantitation by high performance liquid chromatography, *J. Chromatogr. Biomed. Appl.* **233**:193.

Metz, S., Van Rollins, M., Strife, R., Fujimoto, W., and Robertson, R. P., 1983, Lipoxygenase pathway in islet endocrine cells, *J. Clin. Invest.* **71**:1191.

Michelassi, F., Landa, L., Hill, R. D., Lowenstein, E., Watkins, W. D., Petkau, A. J., and Zapol, W. M., 1982, Leukotriene D_4: A potent coronary artery vasoconstrictor associated with impaired ventricular contraction, *Science* **217**:841.

Michoud, M.-C., Pare, P. D., Orange, R. P., and Hogg, J. C., 1979, Airway sensitivity of slow reacting substance of anaphylaxis, histamine and antigen in ascaris sensitive monkeys, *Am. Rev. Respir. Dis.* **115**:66.

Mong, S., Wu, H.-L., Hogaboom, G. K., Clark, M. A., and Crooke, S. T., 1984, Characterization of the leukotriene D_4 receptor in guinea-pig lung, *Eur. J. Pharmacol.* **102**:1.

Morganroth, M. L., Reeves, J. T., Murphy, R. C., and Voelkel, N. F., 1984, Leukotriene synthesis and receptor blockers block hypoxic pulmonary vasoconstriction, *J. Appl. Physiol.* **56**:1340.

Morris, H. R., Piper, P. J., Taylor, G. W., and Tippins, J. R., 1979, Comparative studies on immunologically and non-immunologically produced slow reacting substances from man, guinea-pig and rat, *Br. J. Pharmacol.* **67**:79.

Morris, H. R., Taylor, G. W., Piper, P. J., and Tippins, J. R., 1980a, Structure of slow reacting substance of anaphylaxis from guinea pig lung, *Nature (London)* **285**:104.

Morris, H. R., Taylor, G. W., Piper, P. J., Samhoun, M. N., and Tippins, J. R., 1980b, Slow reacting substances (SRS's): The structure identification of SRS's from rat basophil leukaemia (RBL-1) cells, *Prostaglandins* **19**:185.

Morris, H. R., Taylor, G. W., Rokach, J., Girard, Y., Piper, P. J., Tippins, J. R., and Samhoun, M. N., 1980c, Slow reacting substance of anaphylaxis (SRS-A): Assignment of the stereochemistry, *Prostaglandins* **20**:601.

Moskowitz, M. A., Kiwak, K. J., Hekimian, K., and Levine, L., 1984, Synthesis of compounds with properties of leukotriene C_4 and D_4 in gerbil brains after ischemia and reperfusion, *Science* **224**:886.

Mugridge, K. G., Higgs, G. A., and Moncada, S., 1984, Prostacyclin modulates the responses to leukotrienes C_4 and D_4 of guinea-pig airway smooth muscle, *Eur. J. Pharmacol.* **104**:1.

Murphy, R. C., Hammarström, S., and Samuelsson, B., 1979, Leukotriene C: A slow-reacting substance from murine mastocytoma cells, *Proc. Natl. Acad. Sci. USA* **76**:4275.

Naccache, P. H., Molski, T. F. P., Becker, E. L., Borgeat, P., Picard, S., Vallerand, P., and Sha'afi, R. I., 1982, Specificity of the effect of lipoxygenase metabolites of arachidonic acid on calcium homeostasis in neutrophils, *J. Biol. Chem.* **257**:8608.

Nagai, H., Iwamoto, T., Nishiyori, T., Takizawa, T., Tsuchiya, H., and Koda, A., 1983, Pharmacological studies on newly synthesized anti-allergic agents, 2-methyl-3-piperidino-β-propionaphthone hydrochloride (KZ 111) and 3-isobutyl-(2-isopropylpyrazole-[1,5a] pyridine (KC 404), *Jpn. J. Pharmacol.* **33**:1215.

Nakahata, N., Ono, T., and Nakanishi, H., 1984, Possible involvement of a product of the 5-lipoxygenase pathway in mediation of indomethacin-induced inhibition of cholinergic transmission in guinea pig ileum, *Eur. J. Pharmacol.* **104**:133.

Neichi, T., Koshihara, Y., and Murota, S.-I., 1983, Inhibitory effect of esculetin on 5-lipoxygenase and leukotriene biosynthesis, *Biochim. Biophys. Acta* **753**:130.

Nicolaou, K. C., Petasis, N. A., Li, W. S., Ladduwahetty, T., Randall, J. L., Webber, S. E., and Hernandez, P. E., 1983, Ethanoarachidonic acids: A new class of arachidonic acid cascade modulators. Monoethano compounds, *J. Org. Chem.* **48**:5400.

Nishino, K., Ohkubo, H., Ohashi, M., Hara, S., Kito, J., and Irikura, T., 1983, KC 404: A potential anti-allergic agent with antagonistic action against slow reacting substance of anaphylaxis, *Jpn. J. Pharmacol.* **33**:267.

O'Driscoll, B. R. C., Cromwell, O., and Kay, A. B., 1984, Sputum leukotrienes in obstructive airways diseases, *Clin. Exp. Immunol.* **55**:397.

O'Flaherty, J. T., Wykle, R. L., McCall, C. E., Shewmake, T. B., Lees, C. J., and Thomas, M., 1981, Desensitization of the human neutrophil degranulation response: Studies with 5,12-dihydroxy-6,8,10,14-eicosatetraenoic acid, *Biochem. Biophys. Res. Commun.* **101**:1290.

O'Flaherty, J. T., Wykle, R. L., Thomas, M. J., and McCall, C. E., 1984, Neutrophil degranulation responses to combinations of arachidonate metabolites and platelet activating factor, *Res. Commun. Chem. Pathol.* **43**:3.

Orange, R. P., and Austen, K. F., 1968, Pharmacologic dissociation of immunologic release of histamine and slow-reacting substance of anaphylaxis in rats, *Proc. Soc. Exp. Biol. Med.* **129**:836.

Orange, R. P., and Austen, K. F., 1969a, Slow-reacting substance of anaphylaxis, *Adv. Immunol.* **10**:105.

Orange, R. P., and Austen, K. F., 1969b, Slow reacting substance in the rat, in: *Cellular and Humoral Mechanisms in Anaphylaxis and Allergy* (H. Z. Movat, ed.), pp. 196–206, Karger, Basel.

Orange, R. P., and Austen, K. F., 1971, The immunological release of chemical mediators of the immediate type hypersensitivity from human lung, in: *Advances in Immunology* (B. Amos, ed.), pp. 173–186, Academic Press, New York.

Orange, R. P., and Austen, K. F., 1976, The biological assay of slow-reacting substances—SRS-A, bradykinin, prostaglandins, *Methods Immunol. Immunochem.* **5**:145.

Orange, R. P., and Chang, P. L., 1975, The effect of thiols on immunologic release of slow-reacting substance of anaphylaxis. I. Human lung, *J. Immunnol.* **115**:1072.

Orange, R. P., Valentine, M. D., and Austen, K. F., 1967, Release of slow-reacting substance of anaphylaxis in the rat: Polymorphonuclear leukocyte, *Science* **157**:318.

Orange, R. P., Murphy, R. C., and Austen, K. F., 1974, Inactivation of slow-reacting substance of anaphylaxis (SRS-A) by arylsulfatases, *J. Immunol.* **113**:316.

Orange, R. P., Moore, E. G., and Gelfand, E. W., 1980, The formation and release of slow-reacting substance of anaphylaxis (SRS-A) by rat and mouse peritoneal mononuclear cells induced by ionophore A23187, *J. Immunol.* **124**:2264.

Ormstad, K., Uehara, N., Orrenius, S., Örning, L., and Hammarström, S., 1982, Uptake and metabolism of leukotriene C_3 by isolated rat organs and cells, *Biochem. Biophys. Res. Commun.* **104**:1434.

Örning, L., and Hammarström, S., 1980, Inhibitors of leukotriene C and leukotriene D synthesis, *J. Biol. Chem.* **255**:8023.

Örning, L., Hammarström, S., and Samuelsson, B., 1980, Leukotriene D: A slow-reacting substance from rat basophilic leukemia cells, *Proc. Natl. Acad. Sci. USA* **77**:2014.

Örning, L., Bernström, K., and Hammarström, S., 1981, Formation of leukotrienes E_3, E_4 and E_5 in rat basophilic leukemia cells, *Eur. J. Biochem.* **120**:41.

Palmblad, J., Udén, A.-M., Lindgren, J.-A., Rådmark, O., Hansson, G., and Malmsten, C. L., 1982, Effects of novel leukotrienes on neutrophil migration, *FEBS Lett.* **144**:81.

Palmer, M. R., Mathews, W. R., Murphy, R. C., and Hoffer, B. J., 1980, Leukotriene C elicits a prolonged excitation of cerebellar Purkinje neurons, *Neurosci. Lett.* **18**:173.

Palmer, M. R., Mathews, W. R., Hoffer, B. J., and Murphy, R. C., 1981, Electrophysiological response of cerebellar Purkinje neurons to leukotriene D_4 and B_4, *J. Pharmacol. Exp. Ther.* **219**:91.

Palmer, R. M. J., and Salmon, J. A., 1983, Release of leukotriene B_4 from human neutrophils and its relationship to degranulation induced by N-formyl-methionyl-leucyl-phenylalanine, serum-treated zymosan, and the ionophore A23187, *Immunology* **50**:65.

Palmer, R. M. J., Stepney, R. J., Higgs, G. A., and Eakins, K. E., 1980, Chemokinetic activity of arachidonic acid lipoxygenase products on leukocytes of different species, *Prostaglandins* **20**:411.

Panzenbeck, M. J., and Kaley, G., 1983, Leukotriene D_4 reduces coronary blood flow in the anesthetized dog, *Prostaglandins* **25**:661.

Parker, C. W., Jakschik, B. A., Huber, M. M., and Falkenhein, S. F., 1979, Characterization of slow-reacting substance as a family of thiolipids derived from arachidonic acid, *Biochem. Biophys. Res. Commun.* **89**:1186.

Parker, C. W., Falkenhein, S. F., and Huber, M. M., 1980a, Sequential conversion of the glutathionyl side chain of slow-reacting substance (SRS) to cysteinyl glycine and cysteine in rat basophil leukemia cells stimulated with A 23187, *Prostaglandins* **20**:863.

Parker, C. W., Fischman, C. M., and Wedner, H. J., 1980b, Relationship of biosynthesis of slow-reacting substance to intracellular glutathione concentrations, *Proc. Natl. Acad. Sci. USA* **77**:6870.

Parker, C. W., Koch, D., Huber, M. M., and Falkenhein, S. F., 1980c, Formation of the cysteinyl form of slow-reacting substance (leukotriene E_4) in human plasma, *Biochem. Biophys. Res. Commun.* **97**:1038.

Patterson, R., Orange, R., and Harris, K. E., 1978, A study of the effect of slow reacting substances of anaphylaxis on the rhesus monkey airway, *J. Allergy Clin. Immunol.* **62**:371.

Patterson, R., Harris, K. E., Smith, L. J., Greenberger, P. A., Shaughnessy, M. A., Bernstein, P. R., and Krell, R. D., 1983, Airway responses to leukotriene D_4 in rhesus monkeys, *Int. Arch. Allergy Appl. Immunol.* **71**:156.

Patterson, R., Bernstein, P. R., Harris, K. E., and Krell, R. D., 1984, Airway responses to sequential challenges with platelet activating factor and leukotriene D_4 in rhesus monkeys, *J. Lab. Clin. Med.* **104**:340.

Pawlowski, N. A., Kaplan, G., Hamill, A. L., Cohn, Z. A., and Scott, W. A., 1983, Arachidonic acid metabolism by human monocytes: Studies with platelet-depleted cultures, *J. Exp. Med.* **158**:393.
Payan, D. G., and Goetzl, E. J., 1983, Specific suppression of human T lymphocyte function by leukotriene B_4, *J. Immunol.* **131**:551.
Payan, D. G., Missirian-Bastian, A., and Goetzl, E. J., 1984, Human T-lymphocyte subset specificity of the regulatory effects of leukotriene B_4, *Proc. Natl. Acad. Sci. USA* **81**:3501.
Peatfield, A. C., Piper, P. J., and Richardson, P. S., 1982, The effect of leukotriene C_4 on mucin output in the cat trachea, *J. Physiol. (London)* **325**:56P.
Peck, M. J., Piper, P. J., and Williams, T. J., 1981, The effect of leukotrienes C_4 and D_4 on the microvasculature of guinea pig skin, *Prostaglandins* **21**:315.
Peck, S. B., and Walsh, M. F., 1984, Leukotrienes stimulate insulin release from the rat pancreas, *Proc. Natl. Acad. Sci. USA* **81**:2199.
Peters, S. P., Schulman, E. S., Lin, M. C., Hayes, E. C., and Lichtenstein, L. M., 1983, Separation of major prostaglandins, leukotrienes and monohetes by high performance liquid chromatography, *J. Immunol. Methods* **64**:335.
Pfeffer, M. A., Pfeffer, J. M., Lewis, R. A., Braunwald, E., Corey, E. J., and Austen, K. F., 1983, Systemic hemodynamic effects of leukotrienes C_4 and D_4 in the rat, *Am. J. Physiol.* **244**:H628.
Phillips, M. J., Gold, W. M., and Goetzl, E. J., 1983, IgE-dependent and ionophore-induced generation of leukotrienes by dog mastocytoma cells, *J. Immunol.* **131**:906.
Piper, P. J., and Samhoun, M. N., 1981, The mechanism of action of leukotrienes C_4 and D_4 in guinea pig isolated perfused lung and parenchymal strips of guinea-pig, rabbit and rat, *Prostaglandins* **21**:793.
Piper, P. J., and Seale, J. P., 1978, Release of slow-reacting substance from guinea pig and human lung by calcium ionophore A 23187, *Br. J. Pharmacol.* **63**:364P.
Piper, P. J., Letts, L. G., and Galton, S. A., 1983, Generation of leukotriene-like substances from porcine vascular and other tissues, *Prostaglandins* **25**:591.
Pong, S.-S., and DeHaven, R. N., 1983, Characterization of a leukotriene D_4 receptor in guinea pig lung, *Proc. Natl. Acad. Sci. USA* **80**:7415.
Pong, S.-S., DeHaven, R. N., Kuehl, F. A., Jr., and Egan, R. W., 1983, Leukotriene C_4 binding to rat lung membranes, *J. Biol. Chem.* **258**:9616.
Prescott, S. M., 1984, The effect of eicosapentaenoic acid on leukotriene B production by human neutrophils, *J. Biol. Chem.* **259**:7615.
Rådmark, O., Malmsten, C., Samuelsson, B., Goto, G., Marfat, A., and Corey, E. J., 1980a, Leukotriene A: Isolation from human polymorphonuclear leukocytes, *J. Biol. Chem.* **255**:11828.
Rådmark, O., Malmsten, C., Samuelsson, B., Clark, D. A., Goto, G., Marfat, A., and Corey, E. J., 1980b, Leukotriene A: Stereochemistry and enzymatic conversion to leukotriene B, *Biochem. Biophys. Res. Commun.* **92**:954.
Rådmark, O., Malmsten, C., and Samuelsson, B., 1980c, Leukotriene A_4: Enzymatic conversion to leukotriene C_4, *Biochem. Biophys. Res. Commun.* **96**:1679.
Rådmark, O., Lundberg, U., Jubiz, W., Malmsten, C., and Samuelsson, B., 1982, New group of leukotrienes formed by initial oxygenation at C15, *Adv. Prostaglandin Thromboxane Leukotriene Res.* **9**:61.
Rae, S. A., and Smith, M. J. H., 1981, The stimulation of lysosomal enzyme secretion from human polymorphonuclear leukocytes by leukotriene B_4, *J. Pharm. Pharmacol.* **33**:616.
Rankin, J. A., Hitchcock, M., Merrill, W., Bach, M. K., Brashler, J. R., and Askenase, P. W., 1982, IgE-dependent release of leukotriene C_4 from alveolar macrophages, *Nature (London)* **297**:329.
Rankin, J. A., Hitchcock, M., Merrill, W. W., Huang, S. S., Brashler, J. R., Bach, M. K., and Askenase, P. W., 1984, IgE immune complexes induce immediate and prolonged release of leukotriene C_4 (LTC_4) from rat alveolar macrophages, *J. Immunol.* **132**:1993.
Razin, E., Mencia-Huerta, J.-M., Lewis, R. A., Corey, E. J., and Austen, K. F., 1982, Generation of leukotriene C_4 from a subclass of mast cells differentiated in vitro from mouse bone marrow, *Proc. Natl. Acad. Sci. USA* **79**:4665.
Razin, E., Romeo, L. C., Krilis, S., Liu, F.-T., Lewis, R. A., Corey, E. J., and Austen, K. F., 1984, An analysis of the relationship between 5-lipoxygenase product generation and the secretion of preformed mediators from mouse bond marrow-derived mast cells, *J. Immunol.* **133**:938.

Regal, J. F., and Pickering, R. J., 1981, C5a-induced tracheal contraction: Effect of an SRS-A antagonist and inhibitors of arachidonate metabolism, *J. Immunol.* **126**:313.
Rola-Pleszczynski, M., Borgeat, P., and Sirois, P., 1982, Leukotriene B_4 induces human suppressor lymphocytes, *Biochem. Biophys. Res. Commun.* **108**:1531.
Rosenthal, A. and Pace-Asciak, C. R., 1983, Potent vasoconstriction of the isolated perfused rat kidney by leukotrienes C_4 and D_4, *Can. J. Physiol. Pharmacol.* **61**:323.
Roth, D. M., and Lefer, A. M., 1983, Studies on the mechanism of leukotriene-induced coronary artery constriction, *Prostaglandins* **26**:573.
Rouzer, C. A., Scott, W. A., Cohn, Z. A., Blackburn, P., and Manning, J. M., 1980, Mouse peritoneal macrophages release leukotriene C in response to a phagocytic stimulus, *Proc. Natl. Acad. Sci. USA* **77**:4928.
Rouzer, C. A., Scott, W. A., Griffith, O. W., Hamill, A. L., and Cohn, Z. A., 1981, Depletion of glutathione selectively inhibits synthesis of leukotriene C by macrophages, *Proc. Natl. Acad. Sci. USA* **78**:2532.
Rouzer, C. A., Scott, W. A., Hamill, A. L., and Cohn, A. Z., 1982a, Synthesis of leukotriene C and other arachidonic acid metabolites by mouse pulmonary macrophages, *J. Exp. Med.* **155**:720.
Rouzer, C. A., Scott, W. A., Hamill, A. L., Liu, F.-T., Katz, D. H., and Cohn, Z. A., 1982b, Secretion of leukotriene C and other arachidonic acid metabolites by macrophages challenged with immunoglobulin E immune complexes, *J. Exp. Med.* **156**:1077.
Russi, E., Chapman, G. A., Marchette, B., Abraham, W. M., and Wanner, A., 1983, Effects of leukotriene D_4 on mucociliary function and airway mechanics in allergic sheep, *Am. Rev. Respir. Dis.* **127**(4II):70.
Saad, M. H., and Burka, J. F., 1983, Isolation of leukotriene C_4 from human seminal fluid, *Prostaglandins* **26**:943.
Salmon, J. A., 1984, Analysis of cyclooxygenase and lipoxygenase products in inflammatory exudates, *Prostaglandins* **27**:364.
Salmon, J. A., Simmons, P. M., and Palmer, R. M. J., 1982, A radioimmunoassay for leukotriene B_4, *Prostaglandins* **24**:225.
Samuelsson, B., and Hammarström, S., 1980, Nomenclature for leukotrienes, *Prostaglandins* **19**:645.
Samuelsson, B., Borgeat, P., Hammarström, S., and Murphy, R. C., 1979, Introduction of a nomenclature: Leukotrienes, *Prostaglandins* **17**:785.
Samuelsson, B., Hammarström, S., Murphy, R. C., and Borgeat, P., 1980, Leukotrienes and slow reacting substance of anaphylaxis (SRS-A), *Allergy* **35**:375.
Schellenberg, R. R., and Foster, A., 1984, Differential activity of leukotrienes upon human pulmonary vein and artery, *Prostaglandins* **27**:475.
Schiantarelli, P., Bongrani, S., and Folco, G., 1981, Bronchospasm and pressor effects induced in the guinea pig by leukotriene C_4 are probably due to release of cyclooxygenase products, *Eur. J. Pharmacol.* **73**:363.
Serafin, W. E., Oates, J. A., and Hubbard, W. C., 1984, Metabolism of leukotriene B_4 in the monkey: Identification of the principal nonvolatile metabolites in urine, *Prostaglandins* **27**:899.
Sha'afi, R. I., Naccache, P. H., Molski, T. F. P., Borgeat, P., and Goetzl, E. J., 1981a, Cellular regulatory role of leukotriene B_4: Its effects on cation homeostasis in rabbit neutrophils, *J. Cell. Physiol.* **108**:401.
Sha'afi, R. I., Molski, T. F. P., Borgeat, P., and Naccache, P. H., 1981b, Deactivation of the effects of f-met-leu-phe and leukotriene B_4 on calcium mobilization in rabbit neutrophils, *Biochem. Biophys. Res. Commun.* **103**:766.
Shak, S., and Goldstein, I. M., 1984, Carbon monoxide inhibits ω-oxidation of leukotriene B_4 by human polymorphonuclear leukocytes: Evidence that catabolism of leukotriene B_4 is mediated by a cytochrome P_{450} enzyme, *Biochem. Biophys. Res. Commun.* **123**:475.
Shaw, R. J., Cromwell, O., and Kay, A. B., 1984, Preferential generation of leukotriene C_4 by human eosinophils, *Clin. Exp. Immunol.* **56**:716.
Shelhamer, J. H., Marom, Z., Sun, F., Bach, M. K., and Kaliner, M., 1982, The effects of arachinoids and leukotrienes on the release of mucus from human airways, *Chest* **81**:36S.
Shenouda, G., Thomson, D. M. P., and Bach, M. K., 1982, Tumor antigen induced changes in transmembrane potential of leukocytes of cancer patients, in: *Prostaglandins and Cancer: First International Conference*, pp. 679–684, Liss, New York.

Shimizu, T., Rådmark, O., and Samuelsson, B., 1984, Enzyme with dual lipoxygenase activities catalyzes leukotriene AS_4 synthesis from arachidonic acid, *Proc. Natl. Acad. Sci. USA* **81**:689.

Shiraishi, M., and Terao, S., 1983, Quinones. Part 3. Synthesis of quinone derivatives having ethylenic and acetylenic bonds: Specific inhibitors of the formation of leukotrienes and 5-hydroxyeicosa-6,8,11,14-tetraenoic acid (5-HETE), *J. Chem. Soc. Perkin Trans. 1* **1983**:1591.

Showell, H. J., Naccache, P. H., Borgeat, P., Picard, S., Vallerand, P., Becker, E. L., and Sha'afi, R. I., 1982a, Characterization of the secretory activity of LTB_4 toward rabbit neutrophils, *J. Immunol.* **128**:811.

Showell, H. J., Otterness, I. G., Marfat, A., and Corey, E. J., 1982b, Inhibition of LTB_4-induced neutrophil degranulation by leukotriene B_4-dimethylamide, *Biochem. Biophys. Res. Commun.* **106**:741.

Siegel, M. I., McConnell, R. T., Bonser, R. W., and Cuatrecasas, P., 1981, The production of 5-HETE and leukotriene B in rat neutrophils from carrageenan pleural exudates, *Prostaglandins* **21**:123.

Simmons, P. M., Salmon, J. A., and Moncada, S., 1983, The release of leukotriene B_4 during experimental inflammation, *Biochem. Pharmacol.* **32**:1353.

Sirois, P., 1979, Inactivation of slow-reacting substance of anaphylaxis (SRS-A) by lipoxidase, *Prostaglandins* **17**:395.

Sirois, P., Borgeat, P., and Jeanson, A., 1981a, Comparative effects of leukotriene B_4, prostaglandins I_2 and E_2, 6-keto-$PGF_{1\alpha}$, thromboxane B_2 and histamine on selected smooth muscle preparations, *J. Pharm. Pharmacol.* **33**:466.

Sirois, P., Kerouac, R., Roy, S., Borgeat, P., Picard, S., and Rioux, F., 1981b, In vivo effects of leukotriene B_4, C_4 and D_4: Evidence that changes in blood pressure are mediated by prostaglandins, *Prostaglandins Med.* **7**:363.

Sirois, P., Roy, S., Borgeat, P., Picard, S., and Corey, E. J., 1981c, Structural requirement for the action of leukotriene B_4 on the guinea pig lung: Importance of double bond geometry in the 6,8,10 triene unit, *Biochem. Biophys. Res. Commun.* **99**:385.

Sirois, P., Roy, S., and Borgeat, P., 1981d, The lung parenchymal strip as a sensitive assay for leukotriene B_4, *Prostaglandins Med.* **6**:153.

Sirois, P., Roy, S., Tetrault, J. P., Borgeat, P., Picard, S., and Corey, E. J., 1981e, Pharmacological activity of leukotrienes A_4, B_4, C_4 and D_4 on selected guinea-pig, rat, rabbit, and human smooth muscles, *Prostaglandins Med.* **7**:327.

Smith, L. J., Patterson, R., Greenberger, P., Krell, R., and Bernstein, P., 1984, Airway response to inhaled leukotriene D_4 in man, *Am. Rev. Respir. Dis.* **129**:A1.

Smith, M. J. H., Ford-Hutchinson, A. W., and Bray, M. A., 1980, Leukotriene B: A potential mediator of inflammation, *J. Pharm. Pharmacol.* **32**:517.

Smith, R. J., Sun, F. F., Bowman, B. J., Iden, S. S., Smith, H. W., and McGuire, J. C., 1982, Effect of 6,9-deepoxy-6,9-(phenylimino)-$\Delta^{6,8}$-prostaglandin I_1 (U-60,257), an inhibitor of leukotriene synthesis, on human neutrophil function, *Biochem. Biophys. Res. Commun.* **109**:943.

Sok, D.-E., Pai, J.-K., Atrache, V., and Sih, C. J., 1980, Characterization of slow-reacting substances (SRSs) of rat basophilic leukemia (RBL-1) cells: Effect of cysteine on SRS profile, *Proc. Natl. Acad. Sci. USA* **77**:6481.

Sok, D.-E., Pai, J.-K., Atrache, V., Kang, Y.-C., and Sih, C. J., 1981, Enzymatic inactivation of SRS-cys-gly (leukotriene D), *Biochem. Biophys. Res. Commun.* **101**:222.

Sok, D.-E., Han, C.-O., Shieh, W.-R., Zhou, B.-N., and Sih, C. J., 1982, Enzymatic formation of 14,15 leukotriene A and C(14)-sulfur-linked peptides, *Biochem. Biophys. Res. Commun.* **104**:1363.

Stenmark, K. R., James S. L., Voelkel, N. F., Toews, W. H., Reeves, J. T., and Murphy, R. C., 1983, Leukotrienes C_4 and D_4 in neonates with hypoxemia and pulmonary hypertension, *N. Engl. J. Med.* **309**:77.

Stewart, A. G., Thompson, D. C., and Fennessy, M. R., 1983, Interaction of leukotriene D_4 and histamine on bronchomotor tone in the guinea pig, *J. Pharm. Pharmacol.* **35**:755.

Stimler, N. P., Bach, M. K., Bloor, C. M., and Hugli, T. E., 1982, Release of leukotrienes from guinea pig lung stimulated by $C5a_{des\ arg}$ anaphylatoxin, *J. Immunol.* **128**:2247.

Sun, F. F., and McGuire, J. C., 1983, Inhibition of human neutrophil arachidonate 5-lipoxygenase by 6,9-deepoxy-6,9-(phenylimino)-$\Delta^{6,8}$-prostaglandin I_1 (U-60,257), *Prostaglandins* **26**:211.

Suno, M., and Nagaoka, A., 1984, Inhibition of lipid peroxidation by a novel compound, idebenone (CV 2619), *Jpn. J. Pharmacol.* **35**:196.

Tagari, P., Du Boulay, G. H., Aitken, V., and Boullin, D. J., 1983, Leukotriene D_4 and the cerebral vasculature in vivo and in vitro, *Prostaglandins Leukotrienes Med.* **11**:281.

Terano, T., Salmon, J. A., and Moncada, S., 1984, Biosynthesis and biological activity of leukotriene B_5, *Prostaglandins* **27**:217.

Thomson, D. M. P., Phelan, K., Morton, D. R., and Bach, M. K., 1982, Armed human monocytes challenged with a sensitizing cancer extract release substances pharmacologically similar to leukotrienes, *Int. J. Cancer* **30**:299.

Tilley, J. W., Levitan, P., Welton, A. F., and Crowley, H. J., 1983, Antagonists of slow reacting substance of anaphylaxis. 1. Pyrido-[2,1-b] quinazolinecarboxylic acid derivatives, *J. Med. Chem.* **26**:1638.

Tsai, B. S., Bernstein, P., Macia, R. A., Conaty, J., and Krell, R. D., 1982, Comparative potency and pharmacology of isomers of leukotriene D_4 on guinea pig trachea: Requirement for a 5(S), 6(R) configuration, *Prostaglandins* **23**:489.

Turk, J., Maas, M. L., Brash, A. R., Roberts, L. J., II, and Oates, J. A., 1982, Arachidonic acid 15-lipoxygenase products from human eosinophils, *J. Biol. Chem.* **257**:7068.

Turnbull, L. S., Jones, D. G., and Kay, A. B., 1976, Slow-reacting substance as a preformed mediator from human lung, *Immunology* **31**:813.

Turnbull, L. S., Turnbull, L. W., Leitch, A. G., Crofton, J. W., and Kay, A. B., 1977, Mediators of immediate-type hypersensitivity in sputum from patients with chronic bronchitis and asthma, *Lancet* **2**:526.

Vanderhoek, J. Y., Bryant, R. W., and Bailey, J. M., 1980, Inhibition of leukotriene biosynthesis by the leukocyte product 15 hydroxy-5,8,11,13-eicosatetraenoic acid, *J. Biol. Chem.* **255**:10064.

Van Praag, D., and Farber, S. J., 1981, Leukotriene synthesis in rabbit kidney, *Fed. Proc.* **40**:1713.

Van Praag, D., and Farber, S. J., 1983, Biosynthesis of cyclooxygenase and lipoxygenase metabolites of arachidonic acid by rabbit renal microsomes, *Prostaglandins Leukotrienes Med.* **12**:29.

Verhagen, J., Bruynzeel, P. L. B., Koedam, J. A., Wassink, G. A., DeBoer, M., Trepstra, G. K., Kruikniet, J., Veldink, G. A., and Vlieghart, J. F. G., 1984a, Specific leukotriene formation by purified human eosinophils and neutrophils, *FEBS Lett.* **168**:23.

Verhagen, J., Walstra, P., Veldink, G. A., Vliegenthart, J. F. G., and Bruynzeel, P. L. B., 1984b, Separation and quantitation of leukotrienes by reversed phase high performance liquid chromatography, *Prostaglandins Leukotrienes Med.* **13**:15.

Voelkel, N. F., Worthen, S., Reeves, J. T., Henson, P. M., and Murphy, R. C., 1982, Nonimmunological production of leukotrienes induced by platelet activating factor, *Science* **218**:286.

Voorhees, J. J., 1983, Leukotrienes and other lipoxygenase products in the pathogenesis and therapy of psoriasis and other dermatoses, *Arch. Dermatol.* **119**:541.

Walstra, P., Verhagen, J., Veldink, G. A., and Vliegwnthart, J. F. G., 1984, Leukotriene formation by bovine polymorphonuclear leukocytes, *Biochim. Biophys. Acta* **795**:499.

Wanner, A., Zarzecki, S., Hirsch, J., and Epstein, S., 1975, Tracheal mucous transport in experimental canine asthma, *J. Appl. Physiol.* **39**:950.

Wanner, A., Maurer, D., Abraham, W. M., Szepfalusi, Z., and Sielczak, M., 1983, Effects of chemical mediators of anaphylaxis on ciliary function, *J. Allergy Clin. Immunol.* **72**:663.

Webb, D. R., Nowowiejski, I., Healy, C., and Rogers, T. J., 1982, Immunosuppressive properties of leukotriene D_4 and E_4 in vitro, *Biochem. Biophys. Res. Commun.* **104**:1617.

Weichman, B. M., Hostelley, L. S., Bostick, S. P., Muccitelli, R. M., Krell, R. D., and Gleason, J. G., 1982a, Regulation of the synthesis and release of slow-reacting substance of anaphylaxis from sensitized monkey lung, *J. Pharmacol. Exp. Ther.* **221**:295.

Weichman, B. M., Muccitelli, R. M., Osborn, R. R., Holden, D. A., Gleason, J. G., and Wasserman, M. A., 1982b, In vitro and in vivo mechanisms of leukotriene-mediated bronchoconstriction in the guinea pig, *J. Pharmacol. Exp. Ther.* **222**:202.

Weichman, B. M., Wasserman, M. A., Holden, D. A., Osborn, R. R., Woodward, D. F., Ku, T. W., and Gleason, J. G., 1983, Antagonism of the pulmonary effects of the peptidoleukotrienes by a leukotriene D_4 analog, *J. Pharmacol. Exp. Ther.* **227**:700.

Weiss, J. W., Drazen, J. M., Coles, N., McFadden, E. R., Jr., Weller, P. F., Corey, E. J., Lewis, R. A., and Austen, K. F., 1982, Bronchoconstrictor effects of leukotriene C in humans, *Science* **216**:196.

Weiss, J. W., Drazen, J. M., McFadden, E. R., Jr., Weller, P. F., Corey, E. J., Lewis, R. A., and

Austen, K. F., 1983, Airway constriction in normal humans produced by inhalation of leukotriene D: Potency, time course, and effect of aspirin therapy, *J. Am. Med. Assoc.* **249:**2814.

Weller, P. F., Lewis, R. A., Corey, E. J., and Austen, K. F., 1981, The interaction of purified human eosinophil arylsulfatase B with synthetic leukotrienes, *Fed. Proc.* **40:**1023.

Weller, P. F., Lee, C. W., Foster, D. W., Corey, E. J., Austen, K. F., and Lewis, R. A., 1983, Generation and metabolism of 5-lipoxygenase pathway leukotrienes by human eosinophils: Predominant production of leukotriene C_4, *Proc. Natl. Acad. Sci. USA* **80:**7626.

Welton, A. F., Crowley, H. J., Miller, D. A., and Yaremko, B., 1981a, Biological activities of a chemically synthesized form of leukotriene E_4, *Prostaglandins* **21:**287.

Welton, A. F., Hope, W. C., Tobias, L. D., and Hamilton, J. G., 1981b, Inhibition of antigen-induced histamine release and thromboxane synthetase by FPL 55712, a specific SRS-A antagonist, *Biochem. Pharmacol.* **30:**1378.

Welton, A. F., O'Donnell, M., Anderson, W., Crowley, H., Medford, A., Simko, B., and Yaremko, B., 1983, The role of cyclooxygenase products in some of the biological effects of chemically synthesized leukotrienes (LTB_4, LTC_4, LTD_4 and LTE_4), *Adv. Prostaglandin Thromboxane Leukotriene Res.* **12:**145.

Welton, A. F., Hope, W. C., Fiedler-Nagy, C., Batula-Bernardo, C., and Coffey, J. W., 1984, Analogs of arachidonic acid methylated at C-7 and C-10 inhibit leukotriene biosynthesis and phospholipase A_2 activity, *Fed. Proc.* **43:**845.

Wightman, P. D., Humes, J. L., Davies, P., and Bonney, R. J., 1981a, Identification and characterization of two phospholipase A_2 activities in resident mouse peritoneal macrophages, *Biochem. J.* **195:**427.

Wightman, P. D., Dahlgren, M. E., Hall, J. C., Davies, P., and Bonney, R. J., 1981b, Identification and characterization of a phospholipase C activity in resident mouse peritoneal macrophages, *Biochem. J.* **197:**523.

Williams, J. D., Czop, J. K., and Austen, K. F., 1984, Release of leukotrienes by human monocytes on stimulation of their phagocytic receptor for particulate activators, *J. Immunol.* **132:**3034.

Williams, T. J., Jose, P. J., Wedmore, C. V., and Peck, M. J., 1983, Mechanisms underlying inflammatory oedema: The importance of synergism between prostaglandins, leukotrienes, and complement derived peptides, *Adv. Prostaglandin Thromboxane Leukotriene Res.* **11:**33.

Wöbling, R. H., Aeringhaus, U., Morgenroth, K., Peskar, B. M., and Peskar, B. A., 1983a, Release of slow reacting substance of anaphylaxis (SRS-A) from guinea pig colon, *N. S. Arch. Pharmacol.* **322**(Suppl.):R115.

Wöbling, R. H., Aeringhaus, U., Peskar, B. A., Morgenroth, K., and Peskar, B. M., 1983b, Release of slow reacting substance of anaphylaxis and leukotriene C_4-like immunoreactivity from guinea pig colonic tissue, *Prostaglandins* **25:**809.

Woodman, O. L., and Dusting, G. J., 1983, Coronary vasoconstriction induced by leukotrienes in the anesthetized dog, *Eur. J. Pharmacol.* **86:**125.

Wynalda, M. A., Brashler, J. R., Bach, M. K., Morton, D. R., and Fitzpatrick, F. A., 1984, Determination of leukotriene C_4 by radioimmunoassay with a specific antiserum generated from a synthetic hapten mimic, *Anal. Chem.* **56:**1862.

Yamamoto, S., Ishii, M., Nakadate, T., Nakaki, T., and Kato, R., 1983, Modulation of insulin secretion by lipoxygenase products of arachidonic acid: Relation to lipoxygenase activity of pancreatic islets, *J. Biol. Chem.* **258:**12149.

Yecies, L. D., Parker, C. W., and Watanabe, S., 1979a, Slow reacting substances from ionophore A 23187-stimulated basophilic leukemia cells and peritoneal mast cells in the rat. I. Purification and comparison during sequential Sephadex LH-20 and thin layer chromatography, *Life Sci.* **25:**1909.

Yecies, L. D., Wedner, H. J., Johnson, S. M., Jakschik, B. A., and Parker, C. W., 1979b, Slow reacting substance (SRS) from ionophore A 23187-stimulated peritoneal mast cells of the normal rat. I. Conditions of generation and initial characterization, *J. Immunol.* **122:**2083.

Yen, S. S., and Morris, H. G., 1981, An imbalance of arachidonic acid metabolism in asthma, *Biochem. Biophys. Res. Commun.* **103:**774.

Yokota, K., Shono, F., Yamamoto, S., Kominami, E., and Katunuma, N., 1983, Transformation of leukotriene D_4 catalyzed by lysosomal cathepsin H of rat liver, *J. Biochem. (Tokyo)* **94:**1173.

Yoshimoto, T., Furukawa, M., Yamamoto, S., Horie, T., and Watanabe-Kohno, S., 1983, Flavonoids: Potent inhibitors of arachidonate 5-lipoxygenase, *Biochem. Biophys. Res. Commun.* **116**:612.
Young, R. N., Coombs, W., Guindon, Y., Rokach, J., Ethier, D., and Hall, R., 1981, The preparation of octahydroleukotrienes C, D, and E via a stereoselective sulfenyl lactonization reaction, *Tetrahedron Lett.* **22**:4933.
Young, R. N., Zamboni, R., and Rokach, J., 1983, Studies on the conjugations of leukotriene B_4 with proteins for the development of a radioimmunoassay for leukotriene B_4, *Prostaglandins* **26**:605.
Ziltener, H. J., Chavaillaz, P.-A., and Jörg, A., 1983, Leukotriene formation by eosinophil leukocytes: Analysis with ion-pair high pressure liquid chromatography and effect of the respiratory burst, *Hoppe Seylers Z. Physiol. Chem.* **364**:1029.

12

Cutaneous Basophil Hypersensitivity

STEPHEN J. GALLI and PHILIP W. ASKENASE

1. INTRODUCTION

The term *cutaneous basophil hypersensitivity* (CBH) initially referred to a systemic but evanescent pattern of delayed-onset reactivity in guinea pigs sensitized to protein antigens administered *without* mycobacterial adjuvants (Richerson *et al.*, 1969, 1970; H. Dvorak *et al.*, 1970). When animals immunized in this fashion were skin tested 5–7 days later, sites of antigenic challenge developed intradermal infiltrates containing large numbers of basophilic granulocytes, as well as lymphocytes (H. Dvorak *et al.*, 1970). Although the participation of basophils in certain delayed-onset reactions had been recognized for some time (Rinogoen, 1923; Wolf-Jurgensen, 1965, 1967), the description of CBH focused renewed attention on the occurrence of basophils in delayed-onset immunologic reactions, particularly in the guinea pig. And with the application of improved methods of tissue fixation and processing, basophils soon were discovered in the inflammatory infiltrates elicited by a large number of biologically significant antigens (reviewed in Askenase, 1977, 1980; Galli and Dvorak, 1979; Galli *et al.*, 1983a, 1984).

The identification of CBH and other immunologic reactions containing basophils represented more than a morphologic curiosity. First, it dramatically illustrated that cell-mediated immunity may be heterogeneous in its tissue expression and that certain delayed-onset reactions include large numbers of basophilic granulocytes, in addition to lymphocytes and monocytes. Second, efforts to understand the pathogenesis of CBH and similar responses uncovered new and previously unsuspected examples of synergism between T cells and antibody in the expression of immune reactions with a delayed time course.

STEPHEN J. GALLI • Departments of Pathology, Beth Israel Hospital and Harvard Medical School; and the Charles A. Dana Research Institute, Beth Israel Hospital, Boston, Massachusetts 02215. PHILIP W. ASKENASE • Department of Medicine, Yale University School of Medicine, New Haven, Connecticut 06510.

Third, analysis of one example of a basophil-rich delayed reaction, the cutaneous immune response of guinea pigs to the feeding of ixodid ticks, provided the most convincing evidence to date for a specific biologic function of the basophilic granulocyte. Finally, the discovery and analysis of CBH highlighted the important contribution of modern morphologic techniques to investigations of the pathogenesis of immunologic reactions *in vivo*.

In this review, we will attempt to summarize current understanding of CBH and other delayed-onset reactions containing large numbers of basophils, concentrating primarily on the mechanisms regulating the participation of basophils in these processes. We will also attempt to point out significant gaps in our knowledge of these responses. Finally, we will offer some suggestions about the nomenclature of immunologic reactions with a basophil component. By necessity we have focused largely on work performed in the guinea pig, since this species has been used for most of the experimental studies in this field. The discussion will concentrate primarily on cutaneous reactions, which have been analyzed in detail, although the involvement of basophils in inflammatory reactions in other organs will be considered briefly where appropriate. Attaining these goals represented a challenging exercise for the authors, whose views on these matters do not perfectly overlap, and we have welcomed this opportunity to attempt a synthesis of our individual perspectives. A few areas where a synthesis may not yet be possible are also discussed, an approach we hope will stimulate further experimental efforts to clarify the underlying issues.

Although it is axiomatic that any attempt to elucidate the pathogenesis of an immunologic reaction *in vivo* first requires the accurate identification of the cells whose activities together comprise the response, most routine methods of tissue fixation and processing are poorly suited for this purpose. We therefore will begin with a description of the morphologic techniques appropriate for studying cell-mediated immune responses *in vivo*. We will then briefly review the ontogeny and biology of basophils and mast cells, emphasizing the essential point that in humans and the guinea pig these cells represent distinct populations that can be separately identified in tissues analyzed by appropriate morphologic techniques.

2. TECHNIQUES FOR MORPHOLOGIC ANALYSIS OF DELAYED-ONSET IMMUNOLOGIC REACTIONS CONTAINING BASOPHILS

Routine histologic approaches employ 7- to 10-μm-thick sections of paraffin-embedded tissues fixed according to a variety of protocols and stained with hematoxylin and eosin or similar dyes. These methods have several serious deficiencies for morphologic studies of immunologic processes. Because of section thickness, they do not allow exploitation of the full resolving capacity of the light microscope. They also introduce numerous artifacts, including shrinkage and distortion that accompany paraffin embedding, and loss of structural detail that results from many common fixatives; as a consequence, important morphologic features of inflammatory processes are blurred. Finally, most routine

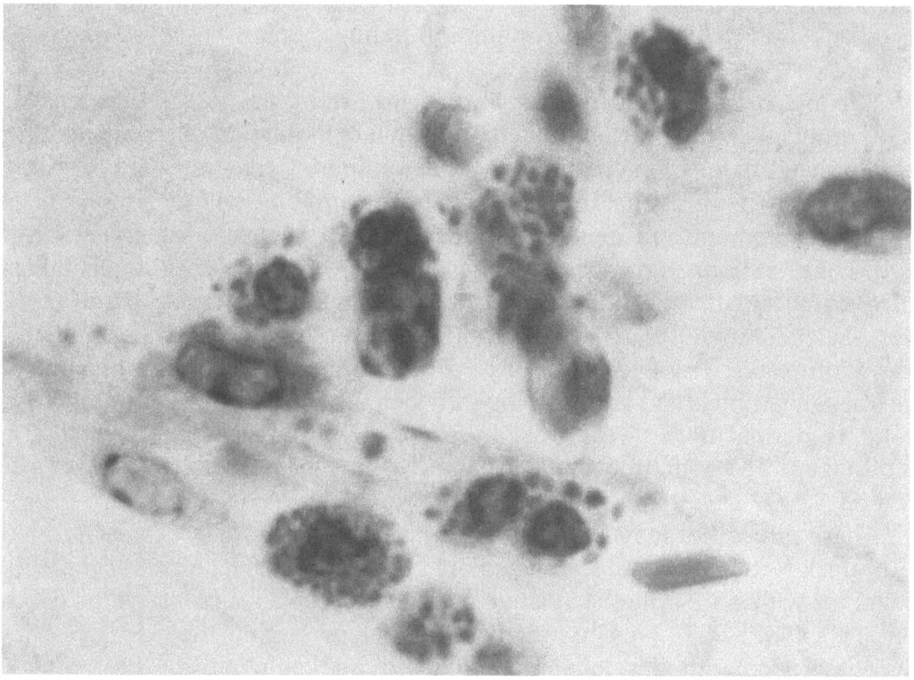

FIGURE 1. Photomicrograph of Giemsa-stained paraffin section from 24-hr cutaneous basophil reaction in flank skin of a guinea pig sensitized with picryl (trinitrophenyl)-keyhole limpet hemocyanin and challenged with picryl (trinitrophenyl)-human serum albumin. The large dark granules of the basophils are widely spaced compared to the smaller and more numerous granules of the eosinophils. A basophil is at the bottom right and an eosinophil is at the bottom on the left. ×1250. (From Askenase *et al.*, 1976a, with permission.)

histologic techniques do not permit the unequivocal identification of basophils or mast cells. In the guinea pig, for example, formalin fixation and hematoxylin and eosin staining produces slides in which both basophils and eosinophils appear red (H. Dvorak *et al.*, 1970).

With Helly's fixative and Giemsa staining (reviewed in Askenase, 1977), by contrast, slides can be produced from paraffin-embedded tissue (Fig. 1) which do permit the separate identification of guinea pig basophils (which stain blue-purple) and eosinophils (which stain red-orange). Similar results can be obtained with Bouin's fixative and Giemsa staining (Brown and Knapp, 1980), and in the hands of some investigators with formalin fixation and Giemsa staining (Yoshida *et al.*, 1979). Because relatively large pieces of tissue can be processed by these simple techniques, basophils and eosinophils infiltrating widely separated sites of inflammation may be rapidly identified and enumerated in a large number of samples (Askenase *et al.*, 1976a; Brown *et al.*, 1982a). But despite their advantages, these techniques suffer from the other problems of routine histologic methods including shrinkage, distortion, and lack of resolution of features such as changes affecting individual inflammatory cells (e.g., degranulation of basophils), endothelial cells (e.g., "activation" or damage), or the interstitium (e.g., fibrin deposition). In addition, Helly's fixative and Giemsa staining does not adequately demonstrate basophils infiltrating human tissues (Askenase, 1977).

In contrast to light microscopic approaches, transmission electron microscopy permits both the unequivocal identification of individual cells participating in immune responses, including basophils and mast cells, and the detailed analysis of their ultrastructure. But electron microscopy has limitations. It is costly in equipment and demanding in technical skills. In addition, electron microscopic sections encompass very small areas (~ 1 mm^2). Although this is not a significant handicap in studies of relatively homogeneous tissue (e.g., liver, skeletal muscle), it may be a formidable obstacle to the analysis of inflammatory processes. These reactions often consist of small foci of infiltrating leukocytes separated by broad and histologically heterogeneous zones of uninvolved tissue (epidermis, dermis, subcutis, and so on). Thorough sampling and analysis of such processes entirely by electron microscopy may require many weeks or months.

These difficulties may be overcome with Epon-embedded, Giemsa-stained 1-μm-thick sections (Fig. 2), so-called giant 1-μm sections (H. Dvorak *et al.*, 1970, 1977a), or with similar methacrylate-embedded, Giemsa-stained sections (Godfrey *et al.*, 1981, 1983). This intermediate technology permits the extensive sampling associated with light microscopy but with markedly improved cell preservation and resolution, and is suitable for studies of both experimental animal (mouse, guinea pig) and human tissues. Giant 1-μm-thick, Epon-embedded sections measuring up to 5 mm on edge, i.e., about 25 times the surface area of a typical electron microscope section face, generally provide a satisfactory overview of tissue reaction patterns. These sections also permit identification of the individual infiltrating leukocytes and connective tissue cells, as well as other features, for example, small fibrin deposits, that together comprise the inflammatory response. Because basophils, eosinophils, and other leukocytes can be identified and reliably distinguished from fixed tissue cells such as histiocytes, mast cells, and pericytes, both the absolute and relative frequency of each type of inflammatory cell can be enumerated. In addition, this method allows identification of changes reflecting basophil activation, such as degranulation sacs (Brown *et al.*, 1982a), and appreciation of other intracellular alterations such as the loss of mast cell or basophil cytoplasmic granule contents (H. Dvorak *et al.*, 1974a; A. Dvorak *et al.*, 1976a,b).

Although either 1-μm sections of Epon-embedded tissue or 1- to 2-μm sections of glycol methacrylate-embedded tissue are greatly superior to conventional sections of paraffin-embedded tissue for morphologic evaluation of inflammatory processes, the techniques are not strictly equivalent. For some observers, Giemsa-stained glycol methacrylate-embedded sections facilitate discrimination between basophils (whose granules stain purple in the guinea pig) and eosinophils (granules stained bright red), particularly for such purposes as low-power quantitation. By contrast, in Epon-embedded sections of guinea pig skin stained with alkaline Giemsa, basophil granules appear a dull blue-purple and eosinophil granules appear bright blue-green. On the other hand, the ap-

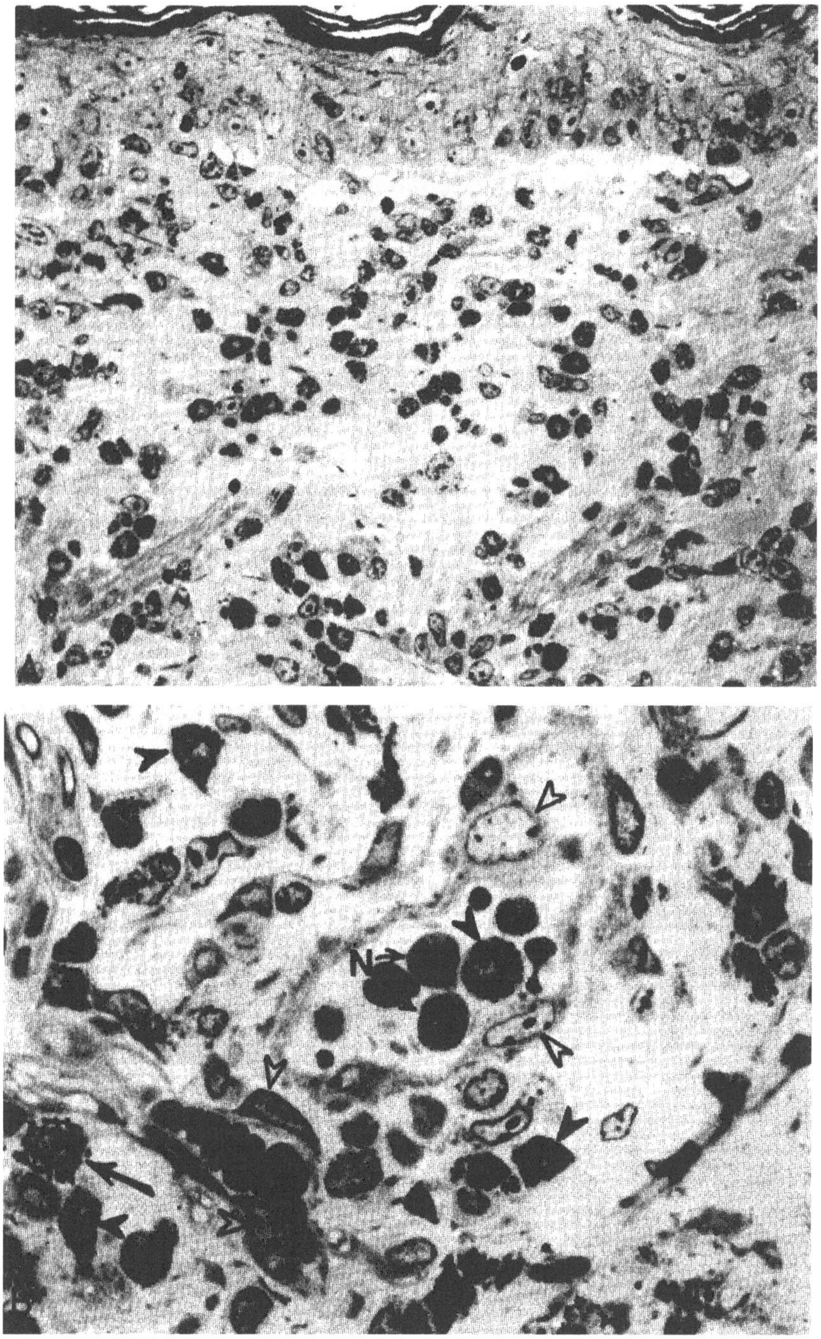

FIGURE 2. A 24-hr CBH reaction elicited in a guinea pig 1 week after sensitization with vaccinia virus. (A) The superficial dermis is infiltrated with many basophils. (B) At higher magnification, several basophils (some of them indicated by solid arrowheads) exhibit two or more nuclear lobes and prominent, darkly stained granules. By contrast, a mast cell (solid arrow, lower left) has a round nucleus and contains cytoplasmic granules that are smaller than those of basophils. Activated endothelial cells (open arrowheads) are also evident, as are a few neutrophils (e.g., N). (A) ×380; (B) ×960. Both are taken from 1-μm-thick, Giemsa-stained Epon sections. (From S. J. Galli, A. M. Dvorak, and H. F. Dvorak, 1984, with permission.)

pearance of basophils and eosinophils in guinea pig tissues embedded in Epon are so different that the cells are readily discriminated at higher magnifications. Furthermore, 1-μm sections of Epon-embedded tissues generally provide better resolutions of fine details of infiltrating cells and tissue structures than do 1- to 2-μm sections of glycol-methacrylate-embedded material.

But to examine the fine structure of inflammatory reactions in detail, the 1- or 2-μm plastic sections must be supplemented with electron microscopy, and optimal electron microscopy is best performed on tissues processed according to somewhat different protocols (A. Dvorak *et al.*, 1976b). As a result, it is often useful to bisect some important specimens during fixation and to process half for giant 1-μm sections and the other half for electron microscopy. With modifications, tissue processed for 1-μm sections also may be used for immunoperoxidase or immunofluorescent localization of many antigens, further enhancing the general attractiveness of this approach for histologic studies of immunity and inflammation. The preparation of giant 1-μm sections is much less demanding technically than the preparation of "thin sections" for electron microscopy, putting this methodology within reach of any investigator with access to the assistance of a capable, conventionally trained histotechnologist.

3. ORIGIN AND NORMAL DISTRIBUTION OF BASOPHILS AND MAST CELLS

Mast cells and basophils share several notable properties. As recognized by Ehrlich in the late 19th century (Ehrlich, 1878, 1879), both cell types contain prominent cytoplasmic granules with affinity for certain basic dyes. Both cells represent a major source of potent chemical mediators implicated in a wide spectrum of inflammatory and immunologic processes (reviewed in Sobotka *et al.*, 1978; Lewis and Austen, 1981; Metcalfe *et al.*, 1981b; Samuelsson, 1983; Schwartz and Austen, 1984). In all mammalian species analyzed, both mast cells and basophils express plasma membrane receptors ($Fc_{\epsilon}R$) that specifically bind the Fc portion of IgE antibody with high affinity (reviewed in K. Ishizaka and Ishizaka, 1979; Froese, 1984). After active or passive sensitization with IgE (or IgG1 in the case of guinea pigs), exposure to specific multivalent antigen triggers both cell types to undergo an integrated, noncytolytic series of biochemical (Kagey-Sobotka *et al.*, 1982; Siraganian *et al.*, 1982; T. Ishizaka and Ishizaka, 1984; Winslow and Austen, 1984) and ultrastructural alterations (reviewed in A. Dvorak *et al.*, 1983; Galli *et al.*, 1984), often referred to as anaphylactic degranulation, that results in the release of preformed, granule-associated mediators such as histamine, as well as newly generated mediators such as the oxidative products of arachidonic acid metabolism. Basophil and/or mast cell degranulation thus may be regarded as a critical step in the biologic events underlying disorders of immediate hypersensitivity. In addition, evidence to be described in detail below establishes that basophils and mast cells also participate in a variety of immune responses in which IgE (or IgG1) antibody is thought to have little if any role.

Although the factors regulating basophil or mast cell function in delayed-onset immune responses are incompletely understood, *in vitro* studies have identified multiple agents which may contribute to basophil and mast cell activation independently of IgE or IgG1 antibody. These secretagogues, whose effectiveness may vary according to species and cell type, include certain complement fragments and lymphokines, many naturally occurring basic peptides, and some neurotransmitters (Ho *et al.*, 1979; Thueson *et al.*, 1979a,b; Metcalfe *et al.*, 1981b; Castro, 1982; Fantozzi *et al.*, 1983). Recent work in mice indicates that limited release of mast cell mediators such as the vasoactive amine serotonin can be induced by antigen-specific, but T-cell-derived factors that appear to be distinct from IgE (Ptak *et al.*, 1982; Askenase *et al.*, 1983a,b). These factors also differ from other lymphocyte products generated in response to antigen activation but which act on vasoactive amine-containing target cells (e.g., basophils) nonspecifically (Thueson *et al.*, 1979a,b; A. Dvorak *et al.*, 1984). The antigen-specific T-cell factors activate mast cells only in the presence of antigen, possibly by an immune aggregate mechanism (Kops *et al.*, 1984a; Ratzlaff *et al.*, 1984), and may be responsible for the mast cell activation observed during the initiation of contact hypersensitivity reactions in the mouse (Kops *et al.*, 1984b).

Despite their remarkable similarities, mammalian mast cells and basophils are clearly not identical. Basophils, like other granulocytes, differentiate and mature in the bone marrow, circulate in mature form in the blood, and are not normally found in the connective tissues. In contrast to basophils, mast cells are ordinarily distributed throughout normal connective tissues, where they are often situated adjacent to blood and lymphatic vessels, near or within peripheral nerves, in the fibrous capsules of internal organs, and beneath epithelial surfaces, such as those of the respiratory and gastrointestinal systems and the skin, that are exposed to environmental antigens (Selye, 1965; Metcalfe *et al.*, 1981b; Galli *et al.*, 1984). In some species, such as mice and rats, mast cells are also present, as a minor component ($\leq$ 10%) of cells normally found in peritoneal and pleural transudates. In man, by contrast, accumulations of peritoneal fluid contain few or no recognizable mast cells but may contain small numbers of basophils [e.g., in peritoneal fluid of patients undergoing peritoneal dialysis (Fox *et al.*, 1984)]. Mast cells are also a normal, if numerically minor, component of bone marrow and lymphoid tissues.

However, unlike mature basophils, mature mast cells do not normally circulate in the blood. Instead, studies in mice and rats strongly suggest that mast cells mature locally in the connective tissue from mitotically active precursor cells that contain few cytoplasmic granules (Combs *et al.*, 1965; Combs, 1966, 1971). These precursors differ from mature mast cells in histochemical staining, suggesting that their granules contain little or no heparin (Combs *et al.*, 1965). Morphologic studies suggest that as rat mast cells mature in the connective tissue (Combs *et al.*, 1965) or in the peritoneal cavity (Pretlow and Cassady, 1970), their proliferative capacity *in situ* declines and they acquire large numbers of cytoplasmic granules. However, apparently mature mast cells can be observed in mitosis during inflammatory reactions (reviewed in Galli *et al.*, 1984). And recent work by Sonoda *et al.* (1984) indicates that at least some fully granulated mouse peritoneal mast cells can express considerable replicative ability when transferred to the skin of congenitally mast cell-deficient W/W^v mice. The granules of mature murine mast cells acquire the histochemical reactivity of heparin (Combs *et al.*, 1965; Pretlow and Cassady, 1970; Tas and Berndsen, 1977), and biochemical studies have confirmed that rat (Yurt *et al.*, 1977) and mouse (Galli *et al.*, 1982b; Razin *et al.*, 1982) peritoneal mast cells contain heparin, as do the mast cells of the skin and many other tissues (Schwartz and Austen, 1984).

The origin of the immature mast cells that populate the connective tissue or serous spaces has been widely discussed. Ginsburg and his associates were the first to provide convincing *in vitro* evidence that mast cell precursors are present in mouse thymus and other lymphoid tissues (Ginsburg, 1963; Ginsburg and Lagunoff, 1967; Ginsburg *et al.*, 1978, 1979, 1981, 1982). T. Ishizaka *et al.* (1976, 1977) reported similar findings in the rat. In a series of elegant and convincing experiments, Kitamura *et al.* (1979b) and Sonoda *et al.* (1982) confirmed the presence of mast cell precursors in mouse lymphoid tissue, but reported the important finding that much larger numbers of mast cell precursors reside in the bone marrow. This group also established that small numbers of mast cell precursors ($\sim 1/10^5$ peripheral blood leukocytes) circulate in the blood of mice (Kitamura *et al.*, 1979a), a finding later confirmed by Zucker-Franklin *et al.* (1981) in the rat.

The precise nature of the circulating mouse mast cell precursor is poorly understood. Kitamura *et al.* (1981) reported evidence that mast cells are derived from a primitive hematopoietic stem cell which can also differentiate into leukocytes of the granulocytic series. Schrader *et al.* (1981) and Nakahata *et al.* (1982) drew similar conclusions based on *in vitro* studies. This evidence, and the recent finding that the mouse granulocytic series includes a cell with the morphologic features of basophils (Urbina *et al.*, 1981; A. Dvorak *et al.*, 1982), supports the plausible notion (Zucker-Franklin, 1980) that mast cells and basophils share a common, bone marrow-derived precursor cell. But until patterns of hematopoietic cell differentiation are understood in detail, it will be difficult to ascertain whether mast cells have a closer ontogenetic relationship to basophils than to the other granulocytes.

Although circulating basophils and tissue mast cells differ in natural history and mediator content, these cells may be able to express similar or complementary functions. The broad similarities of basophil and mast cell mediators, and the responsiveness of these cells to similar secretagogues (notably IgE and antigen), have already been mentioned. In addition, mammalian species generally exhibit an inverse relationship between the frequency of tissue mast cells and circulating basophils (reviewed in Michels, 1938; Galli and Dvorak, 1979). Furthermore, in many chronic inflammatory responses, including some provoked by parasite infections and neoplasms, basophil infiltration may occur during early phases of the reactions, whereas local mast cell hyperplasia occurs later (reviewed by Askenase, 1977, 1980; Galli and Dvorak, 1979; Galli *et al.*, 1984). In processes such as these, the quickly mobilized but short-lived basophil may represent the primary initial source of vasoactive amines (and other basophil/mast cell-derived mediators), particularly in species or anatomic sites relatively poor in mast cells. In the more chronic phases of these responses, vasoactive amines may be derived predominantly from mast cells, as this population expands by replication and differentiation from local and/or recruited precursors.

Certain mast cell populations appear to be capable of particularly rapid proliferation during immune responses. These cells, often designated "atypical" or "mucosal" mast cells, are especially prominent in the intestines of mice and rats during T-cell-dependent immune response to certain intestinal parasites. Mucosal mast cells were first recognized owing to their distinctive staining qualities and light microscopic appearance (Maximow, 1905; Enerbäck, 1966a,b,c; Csaba and Kovacs, 1975), but a large body of evidence now indicates that mucosal mast cells also differ from peritoneal (or "connective tissue") mast cells in mediator content and susceptibility to secretagogues (reviewed in Enerbäck and Lundin, 1974; Askenase, 1977, 1980; Kaliner, 1980; Enerbäck, 1981; Jarrett and Miller, 1982; Befus and Bienenstock, 1982; Bienenstock *et al.*, 1982; Galli *et al.*, 1984). The precise relationship between "mucosal" and "connective tissue" type mast cells remains to be determined. Present evidence, much of it derived from recent work *in vitro* and *in vivo* (reviewed in Schrader *et al.*, 1983; Crowle and Reed, 1984; Galli *et al.*, 1984; Guy-Grand *et al.*, 1984), strongly suggests that mucosal mast cells arise from bone marrow-derived precursors whose local expansion is regulated by macromolecular products elaborated by inducer T cells and perhaps certain other leukocytes. One important regulatory factor appears to be the macromolecule originally reported as "mast cell growth factor" (Nabel *et al.*, 1981; Yung *et al.*, 1981) or "P-cell growth factor" (Schrader *et al.*, 1981) but now most frequently designated interleukin-3 (IL-3) (Ihle *et al.*, 1982; Fung *et al.*, 1984; Yokota *et al.*, 1984). The identity of additional molecules involved in regulating mucosal mast cell proliferation and differentiation remains to be determined.

Local proliferation of morphologically identifiable mast cells also has been noted in association with T-cell-dependent immune responses in humans (H. Dvorak *et al.*, 1974a; A. Dvorak *et al.*, 1976b) and guinea pigs (Rothwell, 1975; Rothwell and Love, 1975), but such effects appear to be quantitatively modest when compared to the striking increases in mucosal mast cells seen in mice and rats. In addition, it is not yet clear to what extent the mitotically active mast cells in human or guinea pig tissues are similar to the rat or mouse "mucosal" mast cell subset. Furthermore, in mice (Galli *et al.*, 1982b, 1984; Schrader *et al.*, 1983; Sredni *et al.*, 1983; Razin *et al.*, 1984) and rats (Woodbury *et al.*, 1978; Haig *et al.*, 1982), normal hematopoietic cells stimulated *in vitro* with growth factors derived from lectin- or antigen-activated lymphoid cells give rise, over a period of weeks, to populations consisting largely of mast cells. While maintained under usual conditions of culture, these cells express many of the features thought to characterize mucosal mast cells *in vivo*. By contrast, when similar tissue culture methods are used in the guinea pig (Denburg *et al.*, 1980) or humans (Ogawa *et al.*, 1983; Tadokoro *et al.*, 1983), they favor the proliferation of cells with ultrastructural features typical of basophils (T. Ishizaka *et al.*, 1985; A. Dvorak *et al.*, unpublished data). Whether the factors promoting the development of guinea pig or human basophils *in vitro*, or stimulating basophilia *in vivo* (Wingvist, 1960; H. Dvorak *et al.*, 1974b; Denburg *et al.*, 1980), are the analogues of mouse IL-3 remains to be determined.

Beside sharing responsiveness to T-cell-derived growth factors, both murine mucosal mast cells and guinea pig basophils differ from typical connective tissue mast cells in lacking heparin (Orenstein *et al.*, 1978; Galli *et al.*, 1982; Razin *et al.*, 1982; Sredni *et al.*, 1983). The apparent similarities in their growth regulation, mediator content, and patterns of involvement in the effector limb of certain immune responses raise the possibility that murine mucosal mast cells and guinea pig basophils share some similar functions (e.g., in parasite immunity, see Sections 6.3 and 6.4). But whatever current *in vitro* studies ultimately reveal about the relationships among "connective tissue-type" (heparin-containing) mast cells, mucosal mast cells and basophils, the tissue culture findings reviewed above illustrate why conclusions based on *in vitro* work with basophils and mast cells in a single species cannot be extrapolated uncritically to others.

Given the rapid pace of current investigation in these areas, the potential for semantic difficulties would seem to be great, particularly in rats and mice. In order to avoid confusion, we will now define the terms used in this chapter to discuss cells with high-affinity IgE receptors on their surface and prominent, metachromatic cytoplasmic granules. *Basophils* refers to cells with the kinetics and natural history of granulocytes. Basophils normally mature in the bone marrow from myelocytic precursors, circulate in mature form in the blood, and exhibit certain characteristic ultrastructural features, even after migrating into the tissues during inflammatory or immunologic processes. The ultrastructure of mature mammalian basophils (reviewed in detail in A. Dvorak *et al.*, 1983; Galli *et al.*, 1984) varies according to species but generally includes electron-dense cytoplasmic granules which appear distinct by ultrastructure from those of the mast cells in the same species, a nuclear chromatin pattern similar to that of other granulocytes, prominent aggregates of cytoplasmic glycogen (which can be visualized in appropriately processed cells), and short, blunt, irregularly distributed plasma membrane processes. There is no convincing evidence that mature mammalian basophils, whether in the circulation or in the tissues, retain significant mitotic capability, or that basophils metamorphose into mast cells upon entering the tissues.

Mast cells, by contrast, designates cells that also arise from bone marrow precursors, but whose major population differentiates or matures outside of the bone marrow or circulation, generally in the connective tissues or serous cavities. (Some mature mast cells may normally be found in the bone marrow as well.) Mast cell populations, wherever distributed, may retain at least limited or latent proliferative capacity. Thus, mast cells in mitoses may be observed morphologically in a variety of tissues expressing inflammatory or immunological processes (reviewed in Galli *et al.*, 1984). In addition, approximately 1 of 17 differentiated mast cells in the peritoneal cavities of normal mice is capable of extensive proliferation when injected into the skin of mast cell-deficient W/W^v mice (Sonoda *et al.*, 1984). In all mammalian species where they have been carefully studied, the granules of mast cells and basophils differ distinctively in ultrastructure (reviewed in A. Dvorak *et al.*, 1983; Galli *et al.*, 1984). Mast cells also differ from basophils in lacking electron-dense aggregates of cytoplasmic glycogen, and in having a plasma membrane surface with uniformly distributed, thin, elongate folds and processes. Mast cell nuclei may appear bilobed in individual electron micrographs, but they generally lack the pattern of peripherally condensed nuclear chromatin characteristic of basophils and other granulocytes.

To conform to current usage, we refer to some murine mast cell populations as "atypical" or "mucosal" mast cells (as reviewed above) and point out that one of their major distinctions from conventional connective tissue mast cells is the critical sensitivity of this population to T-cell-derived growth factors (such as IL-3). It should be emphasized that in the two species which represent the main focus of this chapter, humans and the guinea pig, connective tissue mast cells and basophilic granulocytes are readily distinguishable from one another at both light and electron microscopic levels, provided appropriate methods of fixation, processing, and staining have been used.

4. PROTOTYPIC REACTIONS OF CELL-MEDIATED IMMUNITY

4.1. HISTORICAL PERSPECTIVE

Before turning to a detailed analysis of basophil-rich delayed-onset reactions, we will first briefly review the evolution of the concept of delayed hypersensitivity. We will then compare the features of one of the prototypic reactions of delayed hypersensitivity, the response to mycobacterial antigens, to those of CBH as it was initially described. This will help place the discovery of CBH in a historical context, and will serve to establish a point of reference useful in discussions of other delayed reactions containing basophils.

Hans Zinsser (1921) applied the term *delayed hypersensitivity* to the erythematous, indurated reactions elicited in specifically sensitized subjects by intracutaneous challenge with bacterial antigens. The reactions were called "delayed" to describe what then appeared to be their most significant property: their development at a substantially slower pace than antibody-mediated reactions, first appearing 4–6 hr after antigen challenge and not achieving maximum intensity for 1 or more days. A more fundamental property of these reactions, their dependence on sensitized lymphocytes, was later identified by Landsteiner and Chase (1942) in one of the landmark experiments in immunology. In contrast to other forms of immunologic reactivity, which may be passively transferred to unsensitized hosts with antibodies, the passive transfer of delayed hypersensitivity required lymphoid cells obtained from sensitized donors. Generally speaking, such transfers are most readily effected by living cells. As a result, this group of reactions is commonly referred to as "cellular immunity," "cell-mediated hypersensitivity," or "cell-mediated immunity" (CMI), rather than delayed hypersensitivity. They now can be more precisely defined as T-cell mediated immunity, to emphasize the fact that these responses require thymus-dependent lymphocytes and not antibodies (products of lymphocytes in the B-cell lineage).

Although Zinsser's work applied exclusively to bacterial antigens, of which tuberculin hypersensitivity was the prototype, the concept of cellular immunity was later extended to other immunogens by the work of Dienes (1929), who elicited typical delayed-type reactions in guinea pigs with several purified proteins. Subsequently, the concept of CMI has grown to encompass host responses to a wide variety of antigens, including microbial organisms, many viruses, simple reactive chemicals (haptens), foreign proteins, tumors, and allografts of foreign tissue, as well as some of the host's own antigens in instances of autoimmunity.

By the mid-1960s, most investigators believed that the morphology of delayed hypersensitivity was fully understood and that it consisted essentially of perivascular accumulations of mononuclear cells, primarily lymphocytes but also including monocytes/macrophages, at sites of antigen challenge (Dienes and Mallory, 1932; Gell and Minde, 1954; Waksman, 1960; Boughton and Spector, 1963; Martins and Raffel, 1964). Indeed, this characteristic but relatively nonspecific histologic pattern represented one of the features that delineated delayed hypersensitivity from other types of inflammation. It was thought that the "round cells" (i.e., lymphocytes and monocytes) of these reactions migrated away from the perivascular region, invaded the surrounding tissues, and there effected the various expressions of CMI such as the destruction of allografts or the neutralization of viruses or other antigenic substances. Ultrastructural examination of delayed hypersensitivity reactions generally confirmed the results of previous light microscopic work, suggesting that morphologic studies had little further to add to our understanding of these processes (Goldberg *et al.*, 1962; Flax and Caulfield, 1963; Wyllie *et al.*, 1964; Norton and Ziff, 1965; Wiener *et al.*, 1965; Willms-Kretschmer *et al.*, 1967).

However, the prevailing opinion that the expression of CMI *in vivo* was completely understood was called into question with the discovery of CBH (Richerson *et al.*, 1969; 1970; H. Dvorak *et al.*, 1970). Although others had identified occasional basophils in tuberculin reactions, the finding that these granulocytes comprised 50–90% of total infiltrating cells in what had previously been regarded as exclusively mononuclear cell reactions revealed prominent gaps in our knowledge of the expression of CMI *in vivo*. It also illustrated the potential for further important discoveries that were possible using morphologic techniques. During the past 15 years, this potential has been realized with the description of a number of additional intrinsic features of the cellular immune response *in vivo*. These features both accompany and succeed the lymphocytic exudations common to all forms of CMI so far studied and include the prominent role played by granulocytes, particularly basophilic leukocytes; degranulation of both basophils and mast cells; mast cell replication; increased vascular permeability with consequent vascular compaction and tissue edema; microvascular alterations affecting endothelial cells and pericytes, as well as the basal lamina; and activation of the clotting system with deposition of fibrin (reviewed in H. Dvorak *et al.*, 1980).

Each of these features of cell-mediated reactions may be expressed to a differing extent, depending on many variables but particularly on the species, the mode of sensitization, the interval following initial contact with antigen, the nature of the antigen itself, and the site of antigen challenge. To establish a frame of reference in the midst of this complexity, we will first describe the distinctive properties of two relatively stereotyped examples of CMI as they occur in guinea pig skin: classic delayed (tuberculin-type) hypersensitivity (DH) and classic CBH. These two reactions may be regarded as prototypes of CMI and together illustrate most of the features of cellular immunity that have been recognized in humans and other species (H. Dvorak *et al.*, 1980).

4.2. CLASSIC DELAYED (TUBERCULIN-TYPE) HYPERSENSITIVITY

Classic DH may be induced by immunizing guinea pigs with intradermal or footpad injections containing microgram quantities of many foreign proteins (e.g., ovalbumin) in an emulsion of Freund's complete adjuvant (A. Dvorak and Dvorak, 1974). Alternatively, DH may be induced by infection with mycobacteria such as BCG. Although DH represents a systemic form of altered reactivity to antigen, its presence is most conveniently demonstrated *in vivo* by local skin test with a few micrograms of antigen. Delayed skin reactions first appear 4–6 hr after test as local areas of erythema which become progressively larger and more intense and which acquire induration. Induration, a characteristic feature of DH, refers to a local firmness that frequently may be more sensitively palpated than visualized and that often is accompanied by elevation, giving the lesion a three-dimensional (papular) configuration. In an exquisitely reactive guinea pig, or in one challenged with a large dose of antigen, unusually intense reactions may develop in which central portions of the skin test site become pale and subsequently may undergo necrosis. Reactions are maximal at about 24 hr in the guinea pig (at 48 hr or later in humans) and then regress gradually over the course of several days.

One of the earliest microscopic features of DH reactions is the emigration of lymphocytes from dermal venules at 4–6 hr after skin testing, at about the time that local erythema first becomes evident. Lymphocyte infiltration subsequently increases in parallel with erythema, reaching maximum intensity at about 24 hr. As the reaction evolves, occasional lymphocytes invade epithelial structures such as the epidermis or hair follicles. Bloodborne monocytes, some of which differentiate into macrophages at the site, also participate extensively in these reactions in the guinea pig but are less frequent in humans; by contrast, basophils are exceedingly uncommon if skin tests are performed beyond a week after sensitization. Other important features include a generalized increase in microvascular permeability with kinetics roughly paralleling those of cellular infiltration and erythema; substantial deposition of fibrin in the intervascular reticular dermis; and striking microvascular alterations consisting of endothelial cell and pericyte activation, basement membrane thickening, and variable amounts of endothelial cell injury (reviewed in H. Dvorak *et al.*, 1980). Mast cells are not overly prominent in guinea pig flank skin, and definitive ultrastructural studies of these cells in guinea pig DH have not been performed. At the light microscopic level, however, mast cells often appear hypogranulated after about 6 hr.

The picture presented above may be additionally complicated in the case of antigens that elicit antibody formation as well as CMI. This is a common occurrence at 3 weeks after immunization, when skin tests are performed to elicit the most classic examples of DH. For example, there may be features of anaphylaxis with early (i.e., 20 min after skin test) massive degranulation of mast cells and a substantial but transient (lasting less than 1 hr) increase in microvascular permeability. Alternatively (or additionally, because Freund's complete adjuvant induces a strong response of IgG2, complement-fixing antibodies), features of the Arthus reaction (Cochrane and Janoff, 1974) may be present with antigen–antibody complex formation within vessel walls, leading to local endothelial cell damage quite apart from that induced by CMI and accompanied by varying amounts of microvascular thrombosis.

4.3. CUTANEOUS BASOPHIL HYPERSENSITIVITY AS ORIGINALLY DESCRIBED (CLASSIC CBH)

The participation of basophils in guinea pig cutaneous immunologic reactions with a delayed time course was first recognized in studies of animals immunized *without* mycobacterial adjuvants (Richerson *et al.*, 1969, 1970; H. Dvorak *et al.*, 1970). [Rinogoen (1923) almost certainly elicited similar reactions in the subcutis of guinea pigs, although he referred to the infiltrating cells as "mast leukocytes."] Guinea pigs sensitized to protein antigens emulsified with Freund's incomplete adjuvant developed a systemic but often evanescent form of delayed-onset reactivity that was elicited optimally at about 1 week after immunization. Reactions similar in their transience and limited intensity had been studied earlier in human volunteers by Jones and Mote (1934), leading to the name *Jones–Mote reaction* (Raffel and Newel, 1958). However, human Jones–Mote reactions had not been studied in either immunologic or morphologic detail and their relation to the guinea pig reactions was uncertain. The new term *cutaneous basophilic hypersensitivity* therefore was introduced (Richerson *et al.*, 1969) [later changed to *cutaneous basophil hypersensitivity* or simply CBH (Richerson *et al.*, 1970; H. Dvorak *et al.*, 1970)] to call attention to one of the principal and essential characteristics of these reactions, namely, the striking infiltration of basophilic leukocytes.

In typical guinea pig CBH reactions, 300–600 basophils are present per linear millimeter of skin surface and account for 40–90% of leukocytes in the papillary or superficial dermis, the cutaneous zone where basophil infiltration is generally most abundant (Table 1; H. Dvorak *et al.*, 1970). This concentration is particularly notable in that basophils normally account for 1% of circulating guinea pig leukocytes (Winqvist, 1960; H. Dvorak *et al.*, 1974b), a figure that may transiently increase to ~ 2% of leukocytes as a result of sensitization and testing for CBH (Lett-Brown *et al.*, 1982). Like classic DH responses, CBH reactions have a delayed time course (becoming apparent at 4–6 hr and maximal at 18–24 hr after skin test) and the cellular infiltrate includes a major lymphocyte component. However, classic DH reactions that are elicited optimally at 3–4 weeks after immunization, typically contain very few basophils [< 2% of *total* infiltrat-

TABLE 1. DIFFERENTIAL CELL COUNTS OF LEUKOCYTES INFILTRATING CBH, DELAYED, AND LATE REACTIONS[a,b]

Reaction	Sensitizing antigen	24-hr skin reaction[c]	Time of biopsy (hr)	No. of animals	Mean cell counts ± S.E.[d] Mononuclear	Neutrophils	Basophils	Eosinophils	Total	% basophils
CBH	ABA–BSA	14 (0)	24	6	92	6	68 ± 17	3	169 ± 37	40 ± 3
		13 (0)	48	6	204	3	119 ± 18.8	2	328 ± 41	36 ± 4
	Peroxidase	12 (0)	48	3	217	4	105 ± 14.8	3	329 ± 27	32 ± 7
	KLH	10 (0)	48	4	83	<1	68 ± 26.9	2	154 ± 44	44 ± 9
Delayed	ABA–tyrosine	14 (++–+++)	24	9	491	36	9 ± 2.7	<1	536 ± 52	1.7 ± 0.6
		13 (+++)	48	2	547	50	6 ± 2.1	0	603 ± 11	1.0 ± 0.1
	Peroxidase	14 (+++)	24	3	479	74	5 ± 0.9	3	561 ± 83	0.8 ± 0.9
Late	ABA–BSA	11 (0)	24	8	126	89	13 ± 5	48	276 ± 55	5 ± 2.0
		11 (0)	48	7	208	20	26 ± 8.6	47	301 ± 45	9 ± 1.7

[a]From H. Dvorak *et al.* (1970), with permission.
[b]CBH reactions were elicited in animals sensitized either with 100 μg of azabenzene-arsanilic acid (ABA) conjugated to bovine serum albumin (ABA–BSA) in Freund's incomplete adjuvant (FIA), with 100 μg of peroxidase in FIA, or with 100 μg of KLH in saline, and skin was tested 7 days later with 50 μg of ABA conjugated to a polymer of L-glutamic acid, L-alanine, and L-tyrosine (ABA–L-GAT), peroxidase, or KLH. Delayed reactions were elicited in animals sensitized with 2×10^{-6} mole of ABA conjugated to *N*-acetyltyrosine (ABA–L-tyrosine) or 20 μg of peroxidase in Freund's complete adjuvant and skin tested 2–3 weeks later with 50 μg of ABA–L-GAT or 6 μg of peroxidase. Late reactions were studied in animals sensitized with 100 μg of ABA–BSA in FIA and skin tested 5–6 weeks later with 50 μg of ABA–L-GAT.
[c]Gross reactions expressed as diameter of erythema (millimeters) and, in parentheses, degree of induration (0–++++).
[d]Differential cell counts derived by enumerating all infiltrating leukocytes encountered in two vertical swaths (each 100 μm wide) extending through the reactions from the dermal–epidermal junction to the deep limit of the panniculus carnosus. Counts were performed in 1-μm, Epon-embedded, Giemsa-stained sections.

ing cells, although numbers of basophils in the papillary (high) dermis may be greater (5–6.5%); see Table 2, from H. Dvorak *et al.* (1970)]. Even higher numbers of basophils are found in reactions elicited 1 week after sensitization for DH (Table 3, from H. Dvorak *et al.*, 1970; H. Dvorak *et al.*, 1971). Also, DH reactions are characterized by extensive local activation of clotting with abundant dermal fibrin deposition, leading to induration as a cardinal feature (Colvin and Dvorak, 1975). By contrast, classic CBH reactions exhibit little fibrin deposition and are typically flat, nonindurated erythematous lesions (Colvin and Dvorak, 1975). Central pallor and necrosis, features of severe DH reactions, are not observed in CBH.

As originally described, CBH thus differed from classic DH in several respects. CBH was induced by antigens emulsified with Freund's *incomplete* adjuvant or even administered in saline buffers; erythematous cutaneous reactions elicited by antigen challenge generally were flat and *nonindurated;* in some instances, reactivity was more *transient* than that seen after immunization for classic DH; and skin reactions to relatively high doses of antigen contained large numbers of basophilic leukocytes. Data from an early series of experiments (H. Dvorak *et al.*, 1970) illustrating these differences are shown in Tables 1–3. Note that if antigen challenge of animals primed for CBH was delayed until 5–6 weeks after immunization ("late reactions" in Table 1), the number of basophils in the reactions was substantially less than in classic "CBH" reactions, but still more than was present in classic DH. Although we will discuss the "late reaction" phenomenon in more detail in subsequent sections of the chapter, it is introduced now to make a single point: the number of basophils recruited to sites of antigen challenge varies not only with the type of immunization protocol and antigen employed, but also with the interval between sensitization and cutaneous testing. These early data suggested that the control of basophil traffic into the tissues may be subject to complex regulatory mechanisms. As we shall see, this notion has been confirmed.

In addition to depending on the interval after sensitization, the degree of basophil infiltration also varied according to anatomic site (Table 2). Basophils had a particular predilection for the upper or papillary dermis, where they accounted for up to 90% of infiltrating cells in individual animals primed for CBH. Basophils also represented up to 6.5% of the leukocytes in the papillary dermis of DH reactions, a figure three to five times greater than that computed for the DH lesions as a whole.

Finally, the number of basophils infiltrating sites of antigen challenge varied according to the dose of antigen employed, as well as with the time of testing after immunization. We have already pointed out that immunization with protein antigens in Freund's incomplete adjuvant sensitizes for CBH reactions which are elicited optimally at 1 week by high doses of antigen. With ovalbumin, a 50-μg challenge dose given at 7 days elicited an infiltrate containing 41% basophils, whereas the corresponding figure for a 10-μg dose was only 8% (Table 3). A 10-μg challenge dose of antigen also produced an infiltrate with 8% basophils in animals primed for a *DH* response to ovalbumin and tested 7 days later. And remarkably, actual cell counts showed that more basophils were

TABLE 2. DIFFERENTIAL CELL COUNTS OF LEUKOCYTES INFILTRATING THE UPPER DERMIS IN CBH AND DELAYED REACTIONS[a,b]

Reaction	Sensitizing antigen	Time of biopsy (hr)	No. of animals	Mean cell counts ± S.E.[c]					% basophils
				Mononuclear	Neutrophils	Basophils	Eosinophils	Total	
CBH	ABA–BSA	24	6	73	2	116 ± 9	9	200	58 ± 4.3
		48	6	54	2	140 ± 7.5	4	200	70 ± 4.9
	Peroxidase	48	3	71	0	125 ± 3.3	4	200	63 ± 1.8
Delayed	ABA–tyrosine	24	7	167	20	13 ± 6	<1	200	6.5 ± 2.8
•		48	2	173	17	10 ± 5.0	0	200	5.0 ± 2.5

[a]From H. Dvorak *et al.* (1970), with permission.
[b]Animals were sensitized and skin tested as in Table 1.
[c]Differential cell counts of the first 200 leukocytes encountered in the papillary dermis of reaction sites within 150 μm of the dermal–epidermal junction.

TABLE 3. EFFECT OF VEHICLE AND SKIN TEST DOSE ON CELLULAR COMPOSITION OF SKIN REACTIONS AT EARLY INTERVALS AFTER SENSITIZATION WITH EGG ALBUMIN[a,b]

Sensitizing vehicle	Skin test dose (μg)	24-hr skin reaction[c]	Time of biopsy (hr)	No. of animals	Mean cell counts ± S.E.[d]					
					Mononuclear	Neutrophils	Basophils	Eosinophils	Total	% basophils
FCA	10	20 (+)	24	4	195	80	21 ± 7.2	5	301 ± 44	7 ± 2.4
		20 (+)	48	3	312	48	29 ± 11.2	5	394 ± 45.8	8 ± 1.9
FIA	10	15 (0)	24	4	110	17	12 ± 3.2	2	141 ± 17.4	8 ± 2.4
	50	16 (0)	24	4	113	5	82 ± 10.4	2	202 ± 7.3	41 ± 5.7
		16 (0)	48	4	151	<1	105 ± 42.5	<1	257 ± 54.5	41 ± 8.1

[a]From H. Dvorak *et al.* (1970), with permission.
[b]Animals were sensitized with 10 μg egg albumin in Freund's complete (FCA) or incomplete (FIA) adjuvant, and skin was tested 7 days later with 10 or 50 μg egg albumin in saline.
[c]Scored as in Table 1.
[d]Cell counts performed as in Table 1.

present in "DH" reactions to 10 μg of ovalbumin than in "CBH" reactions to the same dose. However, unlike CBH reactions, the 7-day DH reactions to ovalbumin were indurated (Table 3), although not as indurated as DH reactions elicited 2–3 weeks after sensitization with ABA–L-tyrosine or peroxidase (Table 1). DH reactions elicited over succeeding weeks typically increase markedly in induration and their basophil component disappears. These findings indicate that the features of delayed reactions elicited 1 week after immunization with proteins emulsified in Freund's complete adjuvant represent mixtures of the features of classic DH and classic CBH. Reactions elicited at later times become purer and stronger examples of DH, and the "CBH component" may become suppressed (see also Section 5.3).

In summary, the striking infiltrations of basophils regarded as characteristic of "CBH" in the guinea pig are optimally elicited at early intervals after sensitization and with relatively high doses of antigen. With certain doses of antigen, the difference in basophil counts in "CBH" and "DH" reactions can be obscured, particularly when skin testing is performed at early intervals after immunization for DH. As a result, differences in patterns of reactivity of animals primed for classic DH or classic CBH are best established by considering *multiple* features of the reactions (e.g., induration, as well as the composition of the leukocytic infiltrate) and by testing the animals at multiple intervals after sensitization and with different doses of antigen.

5. IMMUNOLOGY OF THE INDUCTION AND EXPRESSION OF DELAYED REACTIONS CONTAINING BASOPHILS

5.1. THE ROLE OF T LYMPHOCYTES

Despite the striking differences in the appearance of typical DH and CBH reactions in the tissues, these patterns of delayed-onset responsiveness exhibit several similarities in requirements for induction and expression. Both are systemic forms of hypersensitivity under the control of histocompatibility-linked immune response genes that regulate T-cell responses (Stadecker and Leskowitz, 1976b). Hapten- and carrier-specific forms of both CBH and DH have been reported. For example, strong "hapten-specific" DH and CBH can be elicited with appropriate conjugates of azobenzene-arsanilic acid (ABA) (Richerson *et al.*, 1970). But ABA is an unusual hapten in that both T-cell and antibody responses are "hapten specific," perhaps because the ABA hapten can act as a "carrier" in the sense that ABA-specific helper T cells can be induced that aid B cells to produce antibodies to macromolecules linked to ABA (Alkan *et al.*, 1971). When protein conjugates of more conventional haptens such as dinitrophenyl (DNP), trinitrophenyl (TNP), or oxazolone (OX) are conjugated to proteins and used for immunization in Freund's complete or incomplete adjuvant, DH and CBH responses demonstrate carrier specificity in that the strongest reactions are elicited by challenge with the hapten–carrier combination used for immunization. However, hapten-specific delayed-time-course responses can also be elicited in these animals. In animals primed for DH or for CBH to certain antigens (e.g., DNP conjugated to several different carriers), these hapten-specific delayed-onset reactions generally are weak (H. Dvorak *et al.*, 1975). However, strong hapten-specific delayed reactions can be elicited in guinea pigs primed with several other hapten–carrier combinations administered in Freund's complete or incomplete adjuvant (Askenase, 1973; Askenase *et al.*, 1976a). These latter reactions, which contain prominent basophil infiltrates with a tissue distribution similar to that in carrier-specific CBH responses, most probably are dependent on IgG1 antibodies (see also Section 5.2). A similar mechanism may also contribute to the basophil infiltrates of certain CBH responses elicited at sites of challenge with the hapten–protein conjugate used for immunization.

Animals primed for CBH demonstrate expansion of the thymus-dependent paracortical zones of draining lymph nodes (A. Dvorak *et al.*, 1971), and reactions containing both basophils and lymphocytes can be transferred with lymph node cell preparations (H. Dvorak *et al.*, 1971) or T-cell-enriched lymphocyte populations (Askenase, 1976; Stashenko *et al.*, 1977) obtained from animals actively sensitized for CBH. Both systemic (Askenase, 1976; H. Dvorak *et al.*, 1971) and intradermal (Stashenko *et al.*, 1977) T-cell transfers of CBH-like responses have been reported. Whether the delayed reactivity that can be transferred passively from immunized to nonsensitized animals can properly be designated CBH is a difficult issue since the full definition of this entity is based on multiple gross and microscopic findings in actively sensitized animals. By contrast, the transferred reactions generally have been evaluated only by their macroscopic characteristics and pattern of leukocyte infiltration, with less attention given to associated alterations of vessels or the presence or absence of local clotting. Also, it is likely that some components contributing to the overall response in the actively sensitized donor might not be transferred to recipients qualitatively or quantitatively and/or that the recipient could alter expression of the reactivity transferred with sensitized cells (or antibodies). For example, the reactivity transferred by T-cell-containing preparations derived from animals primed for CBH resembles classic CBH in its lack of induration, but the elicited reactions generally are weak responses containing many fewer basophils than are normally present in the corresponding infiltrates of actively sensitized animals. On the other hand, many workers have noted that reactions transferred with lymphocytes from animals immunized for DH are distinctly less indurated than those of actively sensitized donors, and are often more evident macroscopically if larger skin test antigen doses are employed. Furthermore, the systemic administration of lymphocyte preparations or purified T cells from guinea pigs actively sensitized for *classic DH* results in passive transfer of a CBH-like, delayed-onset cutaneous reactivity with a prominent basophil component (Askenase, 1976). This finding may be related to the fact that lymphocytes from guinea pigs primed for classic DH (DNP–BSA in Freund's complete adjuvant for 3 weeks) produced as much basophil chemotactic activity *in vitro* as did lymphocytes from animals primed for classic CBH [DNP–BSA in Freund's incomplete adjuvant for 1 week (Ward *et al.*, 1975)].

Taken together, passive transfer studies have established the following important points: (1) antigen-dependent delayed-onset reactivity may be passively conferred by the transfer of purified T cells derived from animals primed for *either* DH or CBH; but (2) the delayed reactions "transferred" by preparations containing sensitized T cells may differ in certain important respects from those elicited in actively sensitized animals. Considered with other evidence to be discussed in detail below (Sections 5.2, 5.3), these findings suggest that the *full* expression of CBH (or DH) in the skin, as defined by patterns of reactivity in actively sensitized animals, involves the participation of multiple cellular and/or humoral factors.

While the results of passive transfer experiments indicate that sensitized T cells derived from guinea pigs primed for CBH can mediate antigen-specific delayed-onset accumulations of basophils, the precise T-cell subset or product involved in the expression of CBH remains to be defined. For example, in certain experimental protocols, expression of CBH reactivity, like expression of DH reactivity, is inhibited by antilymphocyte serum (Richerson *et al.*, 1970) or anti-T-lymphocyte serum (Stadecker and Leskowitz, 1976a) administered 4 or 1 hr before skin testing, respectively. However, a recent preliminary report indicates that CBH reactivity to ovalbumin may remain intact in guinea pigs treated at the time of skin testing with a mouse monoclonal antibody reactive with guinea pig T cells (Sobel *et al.*, 1984). By contrast, this maneuver ablated expression of DH to PPD in the same animals, strongly suggesting that the CBH and DH reactions differed in their T-cell requirement, and raising the possibility that expression of these CBH reactions was actually independent of intact T cells.

One hypothesis which could reconcile the findings of Sobel *et al.* (1984) with the results of previous work would be to postulate that the recruitment of basophils in classic CBH depends on a T-cell subset not affected by the monoclonal antibodies used by Sobel *et al.* Another possibility is that basophil recruitment in CBH can be effected by soluble antigen-specific factors produced by T cells. In this scheme, ablation of the T cells responsible for the production of the antigen-specific soluble product might be ineffective in blocking the expression of CBH reactions elicited soon after administration of the anti-T-cell antibodies. Evidence for the existence of such a putative antigen-specific factor has recently been provided by Katayama *et al.* (1982, 1984). This factor, which appeared in the serum of guinea pigs sensitized with DNFB 1–2 weeks after topical application of the antigen, passed through an anti-guinea pig IgG1 affinity column, absorbed to an affinity column coupled with DNP–bovine gamma globulin, and was approximately the size of bovine serum albumin according to Sephadex G-150 gel filtration (Katayama *et al.*, 1982). *In vivo* studies indicated that this activity passively transferred antigen-specific cutaneous reactivity with a prominent (33%) basophil component (Katayama *et al.*, 1982). Furthermore, more recent experiments by this group have identified a similar or identical factor (CBH-factor or CBH-F) which, when incubated with guinea pig bone marrow basophils, apparently primed them to exhibit antigen-specific chemotaxis *in vitro* (Katayama *et al.*, 1984). By contrast, CBH-F did not prime induced peritoneal macrophages for antigen-specific chemotaxis. As discussed by Katayama *et al.* (1982, 1984), this factor(s) may represent a fragment of immunoglobulin, a T-cell product analogous to the mouse T-cell-derived antigen-binding factor described by Ptak *et al.* (1982) and Askenase *et al.* (1983a,b), or some other class of molecule. If the activities responsible for antigen-specific basophil recruitment *in vivo* and basophil chemotaxis *in vitro* are shown to be the same (as seems likely), and if the activity is shown to be derived from T cells, this factor obviously would represent a major candidate for explaining the mechanism of basophil recruitment in antibody-independent CBH reactions.

5.2. THE ROLE OF ANTIBODIES

Although we have mentioned that T-cell-dependent antibodies are involved in some important examples of "CBH," initial analyses suggested that antibodies were not necessary for the CBH responses originally described. First, in many CBH reactions, including those induced by soluble protein antigens, only a few participating basophils (generally < 5%) demonstrated morphologic evidence of anaphylactic degranulation, such as might be expected to occur if infiltrating basophils expressed cell-bound antibody with specificity for the eliciting antigen (H. Dvorak *et al.*, 1970, 1973). Second, typical CBH reactions could be elicited 4 days after sensitization, before systemic antibodies were detectable by passive or active cutaneous anaphylasix or by serologic methods (reviewed in H. Dvorak, 1974). Third, intravenous or local injections of specific antigen into guinea pigs bearing mature (24 hr) CBH reactions to several protein antigens failed to elicit either anaphylactic degranulation of basophils or the augmented vascular permeability characteristic of histamine release (H. Dvorak *et al.*, 1971, 1977a; H. Dvorak, 1974). Fourth, fewer than 10% of either circulating basophils, or basophils infiltrating the skin in animals primed for CBH to any of three different antigens, expressed specific cell-bound antibody by a sensitive rosetting technique (H. Dvorak *et al.*, 1975). Fifth, skin test of guinea pigs performed at late intervals (6 weeks) after immunization for CBH elicited relatively basophil-poor "late reactions": complex lesions with sequential, overlapping components of cutaneous anaphylaxis and Arthus reactivity (Colvin *et al.*, 1973). However, unlike the basophils infiltrating typical CBH reactions, substantial fractions (~ one-third) of the basophils teased from "late reactions" had specific antibody as determined by the rosetting assay (H. Dvorak *et al.*, 1975). Finally, attempts to transfer CBH reactivity using serum from animals sensitized to certain soluble protein antigens administered in Freund's incomplete adjuvant were unsuccessful, even though *lymphocyte* preparations from these same guinea pigs were able to transfer CBH-like reactivity (H. Dvorak *et al.*, 1974b). Taken together, these studies suggested that neither the attraction of basophils nor their subsequent behavior in CBH reactions is *necessarily* mediated by cell-bound antibodies.

On the other hand, guinea pig basophils clearly have the capacity to bind IgG1 immunoglobulins to their surface with high affinity, and the results of rosetting assays suggested that specific antibody was present on a minority of basophils even in classic CBH reactions (H. Dvorak, 1974). This fact, the finding of occasional basophils in classic CBH reactions that exhibited evidence of anaphylactic degranulation (H. Dvorak *et al.*, 1970, 1973), and the observation that a substantial fraction of basophils in "late reactions" elicited weeks after sensitization for CBH acquired antibodies specific for sensitizing antigen (H. Dvorak *et al.*, 1975), supported the view that *some* examples of delayed reactions containing basophils might represent a cooperation (H. Dvorak *et al.*, 1971) or synergism (H. Dvorak, 1974) between cellular immunity and immediate hypersensitivity.

The notion that antibodies may mediate the accumulation of basophils in delayed-onset reactions was placed on a firm experimental footing by Askenase (1973), Askenase *et al.* (1975a, 1976a), Haynes and Askenase (1977), and Haynes *et al.* (1978) who demonstrated convincingly that systemic administration of serum containing specific IgG1 antibody, harvested from donors immunized for CBH by antigens administered in Freund's incomplete adjuvant or buffer, or by contact sensitization, confers on previously untreated guinea pigs the ability to develop cutaneous infiltrates including basophils at sites of local antigen challenge. More recent work has shown that serum fractions rich in IgE antibodies and depleted of IgG1 antibodies can also transfer cutaneous basophil reactivity to naive recipients (Graziano *et al.*, 1983). Studies of reactions in serum recipients with 1-μm sections and immunofluorescence (S. Galli, H. Dvorak, and P. Askenase, unpublished data) have revealed fibrin deposits (normally absent from classic, T-cell-dependent CBH reactions) that may account for some of the mild induration of reactions transferred with serum.

The precise mechanisms by which passive administration of IgG1 antibody confers reactivity are not yet clear. Askenase (1977) originally hypothesized that the IgG1 antibodies may function in part by sensitizing cutaneous mast cells. This alone probably is not sufficient, since local passive transfer of these antibodies by intradermal administration does not result in a CBH reaction but instead is followed by a passive cutaneous anaphylaxis (PCA) response (Ovary *et al.*, 1963) rich in eosinophils, but with few basophils (Askenase, 1977). However, basophils also bind IgG1 and IgE antibodies via Fc receptors, and in certain CBH responses both circulating and skin test site basophils are sensitized with specific antibodies (Askenase *et al.*, 1978a; Mitchell *et al.*, 1982; Lett-Brown *et al.*, 1983). Furthermore, expression of antibody-mediated cutaneous basophil responses can be blocked by administration of irrelevant rabbit IgG or its Fc fragments, an effect attributed to the ability of rabbit IgG or its Fc fragments to block Fc receptors on mast cells and basophils (Graziano and Askenase, 1979). These findings have led to the formulation of an "anaphylactic cascade" hypothesis (Graziano and Askenase, 1979). It is envisioned that local challenge results in antigen-induced activation of mast cells, which in turn leads to localization of circulating basophils that bear specific IgG1 or IgE antibodies. These recruited sensitized basophils release mediators upon activation by local antigen, leading to accumulation of additional basophils and other leukocytes.

As suggested previously, an alternative explanation is that the basophil-rich reactions transferred with serum involve the participation of lymphocytes or monocytes (Graziano and Askenase, 1979; H. Dvorak *et al.*, 1980). Mononuclear cells, as well as basophils, are prominent in reactions transferred with serum, as they are in classic CBH. While there is as yet no evidence to suggest that basophil products are chemotactic for lymphocytes, the converse is true. Thus, administration of sensitized lymphocytes, whether locally or systemically, does result in the attraction of basophils to sites of specific antigen challenge. Moreover, leukocyte products chemotactic for guinea pig (Ward *et al.*, 1975) or human (Kay and Austin, 1972; Lett-Brown *et al.*, 1981) basophils have been identified and partially characterized *in vitro*. [See also the discussion of the work of Katayama *et al.* (1982, 1984) in Section 5.1.]

Another mechanism that could explain the antigen-dependent mobilization of basophils to sites of antigen challenge has been suggested by the work of Orida *et al.* (1983). These investigators showed that certain rat basophilic leukemia sublines (two of four tested) exhibited a chemotactic response to antigen after passive sensitization with specific IgE antibody. Whether normal basophil populations can respond in a similar fashion remains to be established. Initial attempts to demonstrate antigen-dependent chemotaxis of basophils obtained from guinea pigs primed to develop a basophilia in response to any of three different antigens (SRBC, ovalbumin, bovine IgG) were unsuccessful (Ward *et al.*, 1975). Indeed, in these experiments the addition of specific antigen *inhibited* basophil chemotaxis to lymphokine or C5 fragment preparations. Other work has shown that basophils obtained under these conditions bear anaphylactic antibodies (IgG1) on their surface (e.g., see H. Dvorak *et al.*, 1975), suggesting that at some concentrations specific antigens can retard rather than promote chemotaxis of antibody-coated basophils. One possible mechanism for this effect, suggested by several lines of evidence (reviewed in A. Dvorak *et al.*, 1984), is that basophil motility is impaired during anaphylactic-type degranulation induced by antigen or other agents.

Although the precise mechanism responsible for the *tissue localization* of basophils in reactions elicited after passive transfer of IgG1 (or IgE) antibodies remains unsettled, several lines of evidence strongly suggest that IgG1 antibodies can influence the *function* of basophils infiltrating some CBH responses. First, the systemic transfer of several types of immune sera both permits the elicitation of antibody-dependent basophil accumulations in the skin and passively sensitizes the circulating basophils of these guinea pigs for antigen-dependent histamine release *in vitro* (Lett-Brown *et al.*, 1983). Second, in CBH elicited to certain antigens such as KLH, the injection of additional small amounts of specific antigen into mature (24 hr) CBH reaction sites results in the degranulation of many of the basophils in these reactions (Askenase *et al.*, 1978a). This phenomenon, which results in a transient agumentation of local vascular permeability, has been designated *cutaneous basophil anaphylaxis*. Finally, Mitchell *et al.* (1982) demonstrated that basophils elicited by a CBH response to one antigen (KLH) may be passively sensitized *in vivo* by intravenous administration of IgG1 antibodies derived from antipicryl immune serum. Basophils sensitized in this manner respond to intradermal challenge with picryl–human serum albumin by undergoing anaphylactic-type degranulation, a reaction associated with a transient local augmentation of vascular permeability.

Passive transfer studies thus indicate that IgG1 (and IgE) antibodies can influence basophil recruitment and function *in vivo*. However, it appears likely that the pathogenesis of many delayed reactions containing basophils requires the orchestration of both T-cell- and IgG1-dependent mechanisms. This concept is illustrated graphically by the experiments of Lett-Brown *et al.* (1982). These investigators primed guinea pigs to develop CBH responses to KLH or ovalbumin. Six or eight days later, they skin tested the animals with KLH or ovalbumin and also determined whether circulating blood basophils released histamine in response to these antigens *in vitro*. Although animals sensitized with either antigen exhibited CBH reactions both on day 6 and on day 8, circulating basophils had detectable surface antibody (as judged by their ability to release histamine in response to challenge with specific antigen *in vitro*) only at the later time.

This study, taken together with earlier work, including the demonstration of basophils coated with specific antibody in "late reactions" elicited several weeks after priming for CBH (H. Dvorak *et al.*, 1975), illustrates three important points. First, at early intervals (4–7 days) after sensitization for CBH, antibodies with specificity for the immunizing antigen are detectable neither in the circulation nor on the basophil surface, yet characteristic CBH reactions develop in the skin. These observations suggest that the CBH reactions elicited at early intervals after sensitization may be dependent entirely on T cells. Second, at later times following sensitization (beyond 6 days), antibodies capable of specifically sensitizing basophils may appear in the circulation and prime basophils for the local release of mediators at sites of repeated antigen challenge. Third, the results of Lett-Brown *et al.* (1982) suggest that the appearance of specific antibodies capable of binding to the surface of guinea pig basophils may be associated with either an increase or a decrease in the expression of cutaneous reactivity. Thus, CBH reactions to KLH were larger on day 8 than on day 6 (15 ± 2 versus 11 ± 1 mm of erythema), and apparently also contained more basophils than those elicited 2 days earlier (186 ± 18 basophils/five high-power fields on day 8 versus 80 ± 8 on day 6). By contrast, reactions to ovalbumin were *smaller* on day 8 than on day 6 (7 ± 1 versus 11 ± 3 mm of erythema), although the numbers of basophils were similar in both groups. While the mechanism underlying the reduction in CBH reactivity between days 6 and 8 was not directly investigated, Katayama and Nishioka (1981) have reported that under certain circumstances, IgG1 antihapten antibody can suppress contact sensitivity in the guinea pig.

In summary, the experiments reviewed in this section indicate that while *some* grossly and histologically typical CBH reactions may be elicited in the apparent absence of specific antibody, *other* basophil-rich delayed responses, including some reactions elicited at late intervals after sensitization for CBH, may incorporate both T-cell- and antibody (IgG1)-dependent mechanisms. With certain antigens, for example KLH, the appearance of specific IgG1 antibodies in the circulation and on the basophil surface appears to be associated with increased cutaneous reactivity. For other antigens, for example ovalbumin, the appearance of specific antibodies capable of sensitizing basophils may be associated with decreased cutaneous reactivity.

5.3. THE REGULATION OF BASOPHIL RECRUITMENT TO DELAYED REACTIONS IN THE GUINEA PIG

The regulation of basophil traffic into individual T-cell- and/or antibody-dependent delayed reactions can be influenced by a variety of systemic or local factors. While a detailed consideration of all the factors contributing to the regulation of delayed reactions in the guinea pig is beyond the scope of this chapter, we will present a brief review of those experiments which have focused on the basophil component of the reactions.

Although certain antibodies (of IgG1 and IgE classes) have been shown to *promote* the delayed-onset accumulation of basophils at sites of cutaneous immune responses (see Section 5.2), antibodies also have been implicated in the suppression of these responses. Turk and his collaborators studied CBH-like reactions ("Jones–Mote reactions") in guinea pigs treated before immunization with doses of cyclophosphamide known to deplete B cells (300 mg/kg) (Turk and Parker, 1973, 1974; Katz *et al.*, 1974; Parker *et al.*, 1974). The reactions in animals treated with cyclophosphamide became macroscopically intensified and indurated, and the basophil component was reduced. In view of these results and other lines of evidence, these investigators have proposed that B lymphocytes suppress the expression of DH, and that CBH in this scheme may represent a suppressed form of DH (reviewed in Turk and Parker, 1983). However, cyclophosphamide can also diminish suppressor T-cell function (Askenase *et al.*, 1975b; Turk and Parker, 1983; Drössler *et al.*, 1983) and may impair basophil production (Katz *et al.*, 1974), suggesting additional or alternative mechanism by which this drug could affect the expression of CBH.

Askenase (1976) and Mitchell and Askenase (1982) studied the regulation of cutaneous basophil responses by employing passive transfers. Askenase (1976) initially showed that T cells from animals primed exclusively for classic DH (using antigens emulsified with Freund's complete adjuvant) passively transferred CBH-like cutaneous reactivity that exhibited a delayed onset, weak induration, hapten–conjugate specificity, and leukocytic infiltrates containing many basophils. Mitchell and Askenase (1982) later demonstrated that when immune serum *and* cells are transferred from animals sensitized to express DH, the basophil component of the reactions elicited in recipients is suppressed, mononuclear cell infiltration is increased, and the macroscopic reactions are augmented in size and induration. These responses thus much more closely resemble the classic DH responses elicited in actively sensitized animals than do the responses elicited in recipients of T cells alone. One serum factor is an antigen-specific IgG2 antibody that augments the macroscopic reaction; another serum factor is antigen-nonspecific, approximately the size of albumin by chromatography, and suppresses T-cell-dependent basophil recruitment (Mitchell and Askenase, 1982). These experiments confirm the T-cell dependence of DH but suggest that the characteristic features of classic DH reflect the influence of at least two different humoral factors whose appearance is favored by immunization protocols employing Freund's complete adjuvant.

These findings also raise the possibility that "CBH-like" reactivity is a more fundamental and pathogenetically simpler form of T-cell-mediated immunity than classic DH (Mitchell and Askenase, 1982). Studies with neonatal guinea pigs also support this notion. In neonatal guinea pigs, cutaneous reactions to antigen elicited 1 week after immunization for classic DH resembled CBH in their lack of induration and high content of basophils (Haynes and Askenase, 1977). While the mechanisms responsible for this finding have not been defined, it would be of interest to evaluate whether it reflects a deficiency in neonates of the two humoral factors implicated in regulating the composition of the leukocytic infiltrate and the macroscopic features of DH reactions in adult animals.

Finally, it is important to emphasize that the factors responsible for the regulation of the gross and microscopic features of classic DH and CBH reactions can operate simultaneously in the same animal. Perhaps the most graphic illustration of this point is that after appropriate sensitization protocols, individual guinea pigs can express classic DH reactions and classic CBH reactions at the same time at different cutaneous sites (Galli *et al.*, 1978; Leonard *et al.*, 1979).

In addition to its sensitivity to systemic factors, the regulation of basophil traffic may also depend on structural features of the tissue microenvironment. Basophils in cutaneous reactions are particularly abundant in the upper or papillary dermis, even when the major source of antigen is in the deep dermis [e.g., certain tumor cells (H. Dvorak *et al.*, 1973; Galli *et al.*, 1982a)]. Cutaneous infiltrates of basophils also develop when antigen is administered intravenously to CBH-primed guinea pigs (H. Dvorak *et al.*, 1977a). By contrast, examination of the lungs of these animals reveals accumulations of leukocytes containing many eosinophils but few basophils. Similarly, when guinea pigs are infected with *Schistosoma mansoni,* skin tests with schistosome eggs elicit CBH responses, and basophil-rich responses occur around eggs that lodge in the intestines. However, eggs that embolize to the liver, or are injected intravenously and lodge in the lung, elicit a strong granulomatous response that is rich in eosinophils but virtually lacks basophils (Askenase, 1977). Finally, when guinea pigs primed to express classic CBH to soluble protein antigens are tested by antigen challenge to the cornea, the elicited reactions resemble cutaneous CBH in their time course and character (minimal edema and fibrin deposition), but differ from cutaneous CBH in having relatively fewer basophils and more eosinophils (Friedlaender and Dvorak, 1977). These differences in patterns of cellular infiltration according to anatomic site may reflect local variation in the production of chemotactic or other regulatory factors, regional differences in the ability of endothelial cells to support basophil emigration, and/or other factors. But whatever the explanation for its occurrence, the predilection of the skin, the cornea, and the intestines (Askenase *et al.*, 1978b) for the development of basophilic infiltrates supports the notion, developed in detail below (Section 6), that one special role of basophils in host immunity may be the expression of resistance to parasite infection at cutaneous and mucosal surfaces.

6. BIOLOGIC FUNCTION OF BASOPHILS IN DELAYED REACTIONS

6.1. INTRODUCTION

In the guinea pig, a wide variety of biologically significant antigens induce delayed-onset cutaneous reactivity with a prominent basophil component. These responses may be assigned to two broad categories. The first comprises those reactions in which infiltrating basophils exhibit little if any evidence of anaphylactic degranulation. Such reactions include classic CBH elicited at early intervals (4–7 days) after immunization with many soluble proteins, as well as the immune responses to certain viruses (H. Dvorak and Hirsch, 1971) and tumors (H. Dvorak *et al.*, 1973, 1979; Galli *et al.*, 1982a), skin allografts (H. Dvorak, 1971), antigen–antibody complexes (H. Dvorak *et al.*, 1971), and some contact allergens (H. Dvorak *et al.*, 1971; Askenase, 1973). Although these responses have not all been studied in the same degree of detail, some of them clearly differ from classic CBH [the reactions induced by administration of protein antigens in Freund's incomplete adjuvant and elicited optimally at early intervals (1 week) after immunization]. For example, the response to certain syngeneic tumor cells contains a moderate number of basophils (10–20% of infiltrating cells), develops after immunization with tumor cells emulsified with Freund's *complete* adjuvant, and may be elicited at long intervals (months) after initial sensitization (H. Dvorak *et al.*, 1973). Other long-lasting examples of cutaneous reactivity with a prominent basophil component include certain contact hypersensitivity reactions (H. Dvorak *et al.*, 1971; Godfrey *et al.*, 1983), and some responses to viruses (H. Dvorak and Hirsch, 1971; Griffith *et al.*, 1982) or parasites (Askenase *et al.*, 1976b, 1982; Brown and Askenase, 1981, 1983).

In many CBH responses lacking evidence of significant anaphylactic-type basophil degranulation, the infiltrating basophils may exhibit light and electron microscopic evidence of a progressive, partial loss of cytoplasmic granule content (H. Dvorak *et al.*, 1973). As originally proposed by A. Dvorak *et al.* (1972) and reviewed in detail elsewhere (Galli *et al.*, 1984), this finding may reflect a slow, "piecemeal" loss of granule-associated mediators by a mechanism distinct from conventional anaphylactic degranulation. This model of mediator release, which hypothesizes the transport of small quanta of granule-associated mediators to the plasma membrane within cytoplasmic vesicles, is supported by *in vitro* evidence in guinea pig basophils (A. Dvorak *et al.*, 1980; Galli *et al.*, 1984). A similar mechanism may also occur in mast cells (A. Dvorak *et al.*, 1976b; Tamir *et al.*, 1982; Kops *et al.*, 1984b). While the actual transport of basophil mediators by traffic of cytoplasmic vesicles awaits confirmation, this mechanism theoretically could result in the slow, sustained release of histamine and other mediators in

CBH reactions. But guinea pig basophils contain several mediators of inflammation, each of which may have multiple, and sometimes contradictory, biologic effects (reviewed in Galli and Dvorak, 1979; see also H. Dvorak *et al.*, 1977b, 1978; Brendel and Graziano, 1984); it is therefore difficult to predict what the net effect of a slow release and/or protracted expression of basophil mediators might be. Similarly, the factors which regulate basophil secretory activity during CBH remain to be fully defined. For example, it is tempting to speculate that T-lymphocyte products ("lymphokines") could be responsible for nonanaphylactic release of basophil mediators in T-cell-dependent CBH responses. Although T-cell products capable of provoking guinea pig basophil mediator release have not been described, Thueson *et al.* (1979a,b) and A. Dvorak *et al.* (1984) have identified a mononuclear cell product that provokes *anaphylactic-type* degranulation of human basophils in an antigen-nonspecific fashion.

6.2. USE OF ANTIBASOPHIL SERUM

One approach to understanding the role of basophils in delayed reactions is to attempt to elicit the responses after ablation of basophils *in vivo*. Galli *et al.* (1978) performed this experiment by using a highly specific rabbit antiserum to destroy bone marrow and circulating basophils in guinea pigs primed to express simultaneously both classic CBH to ovalbumin and classic DH to PPD. Antibasophil serum (ABS) or normal rabbit serum (NRS) was administered to guinea pigs 1 hr before challenging separate cutaneous sites with ovalbumin (to elicit CBH), PPD (to elicit DH), or Con A (to elicit a nonspecific inflammatory reaction that contained only small numbers of basophils). ABS treatment reduced bone marrow and blood basophil levels to 7 and 30% of control levels, respectively. In addition, CBH reactions in ABS-treated animals had 88% fewer basophils than those in NRS-treated controls, while the numbers of other leukocytes were not significantly affected. However, CBH reactions in ABS-treated animals depleted of basophils were significantly *more* erythematous by rank order testing than those in NRS-treated controls. By contrast, ABS treatment had no significant effects on either the gross or the microscopic features of DH reactions to PPD or nonspecific reactions to Con A.

The experiment with ABS offers little support for the idea that basophils *promote* erythema in typical experimental CBH reactions but, on the contrary, is consistent with the possibility, also suggested by Katz *et al.* (1974), that under certain circumstances basophils may have a net anti-inflammatory role. The mediator(s) responsible for such an effect has not been defined, but histamine has a number of activities that can reasonably be considered anti-inflammatory, and several lines of evidence support the hypothesis that histamine could act *in vivo* to inhibit certain delayed-type hypersensitivity reactions (Avella *et al.*, 1978; Askenase *et al.*, 1981; Plaut and Lichtenstein, 1982; Siegel *et al.*, 1982; Rocklin and Beer, 1983).

6.3. COMPLEX CBH RESPONSES TO AN INTESTINAL NEMATODE

The second major category of basophil-rich delayed-onset reactions comprises those in which many of the infiltrating basophils exhibit evidence of conventional *anaphylactic* degranulation. This morphologic pattern of basophil activation can be provoked by the interaction of multivalent antigen with antibody bound to the basophil surface (A. Dvorak *et al.*, 1981a), or by other factors such as complement components (A. Dvorak *et al.*, 1981b). Reactions featuring prominent evidence of anaphylactic-type basophil degranulation can be elicited in the guinea pig small intestine during infections with the nematode *Trichostrongylus colubriformis* (Rothwell and Dineen, 1972; Huxtable and Rothwell, 1975; Rothwell and Love, 1975), and in the skin by the feeding of certain ticks (Allen, 1973; Brown *et al.*, 1982a; McLaren *et al.*, 1983). The participation of basophils in the response to *T. colubriformis* will be considered first.

After primary infection with *T. colubriformis* administered by mouth, guinea pigs developed a thymus-dependent bone marrow and peripheral blood basophilia, and basophils infiltrated the lamina propria of the infected small intestine (Rothwell and Love, 1975). Although the focus of the immune response to *T. colubriformis* was the gut, infected animals exhibited delayed reactions with a basophil component at skin sites challenged with small amounts of material solubilized from *T. colubriformis* fourth-stage larvae (Rothwell and Huxtable, 1976). Primary infection also provoked: (1) a systemic eosinophilia, with infiltration of eosinophils into the lamina propria of the involved intestine; and (2) increased numbers of intestinal basophils and mast cells. [Although gut basophils and mast cells were combined for initial quantitative studies, the increased numbers of cells with metachromatically stained granules observed in the gut of *T. colubriformis*-infected guinea pigs appeared to reflect predominantly basophil accumulation (Rothwell, 1975).] The worms were expelled 2 to 4 weeks after initial infection. This event was associated with an elevation of histamine levels in the intestinal lumen, probably due at least in part to anaphylactic degranulation of intestinal basophils (Huxtable and Rothwell, 1975). Infection was also associated with an elevation of gut levels of serotonin and, afterwards, at approximately the time of worm expulsion, by significant serotonin release. These events may reflect local mast cell proliferation and activation during the response (and/or release of serotonin by enterochromaffin cells), as we are unaware of reports that guinea pig basophils contain serotonin. After worm expulsion, basophil levels in the gut declined, whereas the number of intestinal mast cells remain elevated for at least 6 months (Rothwell, 1975; Rothwell and Love, 1975; Handlinger and Rothwell, 1984).

Several experiments were performed to evaluate the role of histamine and serotonin in worm expulsion. These involved both the intra-intestinal administration of these amines, alone or in combination, as well as the use of pharmacologic agents to block amine activity. While the interpretation of experiments such as these is not simple (see Keller and Ogilvie, 1972; Kelley and Dineen, 1972; Jones *et al.*, 1974; Askenase, 1977, 1980), taken in aggregate this work suggests that histamine and serotinin may contribute to worm expulsion, particularly when the two amines are released together at certain susceptible stages of the organism's life cycle (Rothwell *et al.*, 1974; Jones *et al.*, 1974). These experiments thus are consistent with the notion that worm expulsion may require the coordinated activation of and mediator release from basophils and/or mast cells, but do not define separate roles for these two cell types.

Identification of a specific role for basophils, as opposed to mast cells, in the immune response to *T. colubriformis* is further complicated by the change in the pattern of response that occurs during *secondary* infection. In reinfected guinea pigs, blood and bone marrow basophilia and eosinophilia occurred earlier than in primary infections, as did both infiltration of the intestinal mucosa by these cells and parasite expulsion (Rothwell and Dineen, 1972; Rothwell, 1975). Furthermore, light microscopic observations suggested that some infiltrating basophils degranulated at sites of secondary worm infection. But challenge infection provoked little (Rothwell and Dineen, 1972) or no (Rothwell, 1975) change in the total number of basophils and mast cells in the intestinal mucosa of previously infected guinea pigs, largely because primary infection resulted in persistently elevated gut mast cell levels.

In summary, the work with *T. colubriformis* suggests that basophil anaphylactic degranulation may be an important element in a protective immune response to this intestinal parasite. But the prominent parallel changes involving mast cells, and the potential involvement of other amine-containing cells, complicate determination of whether basophils have a unique or critical role in resistance to this parasite. We will therefore now consider another, and arguably less complex, example of a delayed response to parasites involving the attraction and degranulation of basophils: the immune response to ticks.

6.4. COMPLEX CBH RESPONSES TO INFESTATION WITH IXODID TICKS

A number of investigators have reported that the feeding of ixodid ticks on guinea pigs subjected to a previous cycle of infestation elicits among the most basophil-rich cutaneous reactions yet observed in any species; and that the reactions may include abscesses containing up to 90% basophils (Allen, 1973; Wikel and Allen, 1976; Allen *et al.*, 1977; Brown and Askenase, 1981; Brown and Knapp, 1981). These responses represent an expression of immune resistance, in that ticks placed on previously sensitized hosts either die or feed inadequately (Allen, 1973; Brown and Askenase, 1981; Brown and Knapp, 1981; Askenase *et al.*, 1982; Brown *et al.*, 1982a). Furthermore, these reactions appear to represent an example of synergism between T cells and antibodies: passive transfer of resistance to tick challenge can be achieved either when cell preparations including T lymphocytes, or when serum IgG1 antibodies, are transferred from actively sensitized hosts to naive recipients (reviewed in Askenase, 1977, 1980; Brown and Askenase, 1981, 1983; Askenase *et al.*, 1982; Brown *et al.*, 1982b). One role of antibodies in these reactions is participation in the mechanism of basophil recruitment to the sites of tick feeding. Another role may be to sensitize the basophils which infiltrate tick feeding sites, so that basophil degranulation is induced by release of tick antigens. Such a mechanism may explain why in certain immune responses to ticks the great majority of basophils infiltrating feeding sites undergo anaphylactic degranulation (Allen, 1973; Brown *et al.*, 1982a; McLaren *et al.*, 1983). Basophil degranulation in these reactions appears similar ultrastructurally (McLaren *et al.*, 1983) to that induced by systemic (Chan, 1972; Mitchell *et al.*, 1982) or local (Askenase *et al.*, 1978a) challenge with protein antigens *in vivo* or by antigen or Con A *in vitro* (A. Dvorak, 1978; A. Dvorak *et al.*, 1979, 1981a).

Several lines of evidence suggest that the effector limb of this form of tick immunity is dependent on the infiltration and function of basophils. Tick rejection coincides with the arrival and degranulation of basophils (Allen, 1973; Wikel and Allen, 1976; Brown and Askenase, 1981; Brown and Knapp, 1981; Brown *et al.*, 1982a), and, in several tick species, the level of resistance correlates with the magnitude of the cutaneous basophil response (Brown and Askenase, 1981). Furthermore, ablation of circulating and skin reaction site basophils by administration of a specific antibasophil serum (ABS) (Galli *et al.*, 1978) totally abrogates immunity to the feeding of larval *Amblyomma americanum* ticks (Brown *et al.*, 1982a). For these experiments, guinea pigs were sensitized by a primary infestation of *A. americanum,* then challenged 26 days later with a second infestation of 100 larvae. Larvae from the same batch were also used to infect a matched control group of unimmunized (naive) guinea pigs (group A). The previously sensitized guinea pigs were divided into four groups: group B received no serum, group C received normal rabbit serum (NRS), group D received a rabbit ABS, and group E received a rabbit antieosinophil serum (AES). The effects of the different sera were assessed by monitoring blood leukocyte levels. On day 4, tick feeding success was measured by counting the number of fully engorged or actively feeding ticks and by determining the mean weight of engorged ticks. In contrast to feeding ticks, which appeared robust and responded actively to probing with forceps, dead ticks were small, desiccated, unresponsive to touch, and either loosely attached to the skin or free within the hair coat.

As shown in Fig. 3 and 4, ABS eliminated basophils at tick feeding sites and completely abrogated immunity to *A. americanum:* hosts treated with ABS had tick yields and mean fed-tick weights that were not statistically different from those of naive animals. ABS also virtually eliminated blood and bone marrow basophils but neither reacted with eosinophils *in vitro* (Galli *et al.*, 1978) nor significantly reduced eosinophils in the blood or bone marrow *in vivo* (Galli *et al.*, 1978; Brown *et al.*, 1982a). However, ABS-treated animals had diminished eosinophils at tick feeding sites, perhaps because these sites were deficient in basophil-derived chemotactic factors. AES markedly reduced blood and feeding site eosinophils, did not affect basophil levels, and partially impaired tick resistance. In addition to abrogating immunity to *A. americanum,* depletion of basophils by ABS was also associated with *increased* erythema of tick feeding sites and increased mononuclear cell infiltration. This result is similar to that obtained in a study of the effect of ABS on classic CBH reactions to ovalbumin

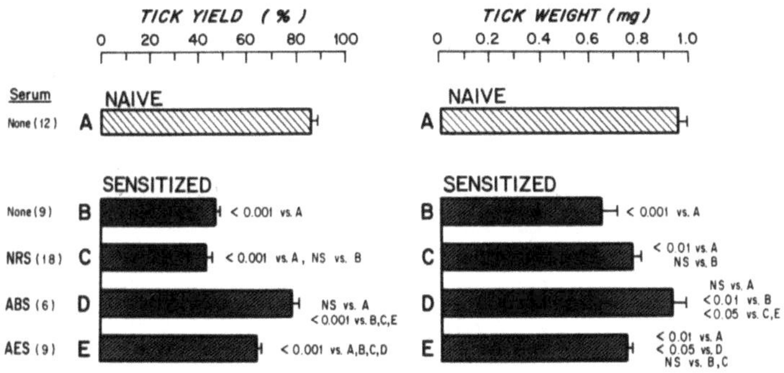

FIGURE 3. One hundred larval *Amblyomma americanum* ticks were placed on the flanks of nonsensitized (naive) guinea pigs or hosts sensitized 26 days earlier by a primary infestation of *A. americanum*. The naive guinea pigs and one group of sensitized hosts received no serum. Other sensitized animals were treated with NRS, ABS, or AES, as described in the text. The number {tick yield, left: % tick yield = [number of ticks successfully fed/number applied (100)] × 100} and weight (right) of engorged ticks were determined at 90 hr of infestation. ABS (D) completely ablated immunity; AES (E) partially impaired resistance; NRS (C) had no effect. Data (mean ± S.E.) were pooled from three separate experiments, with the total number of animals in each group shown in parentheses. NS, not significant ($p > 0.05$), Newman–Keuls multiple sample comparison test. (From Brown *et al.*, 1982a, with permission.)

(see Section 6.2), and suggests that basophils may *inhibit* the development of erythema and the infiltration of mononuclear cells in tick feeding sites.

The experiments with ABS, taken together with the work cited above linking the arrival and degranulation of basophils with resistance to tick feeding, strongly suggest that basophils have a significant role in the expression of protective immunity to ticks. The ability to transfer immunity to *A. americanum* with

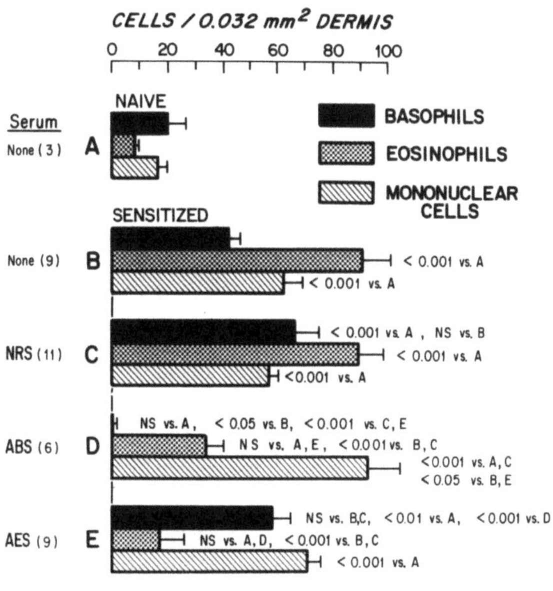

FIGURE 4. Leukocytes (mean ± S.E.) infiltrating tick feeding sites were enumerated as described in Brown *et al.* (1982a). ABS-treated hosts (D) had virtually no feeding site basophils, reduced eosinophils, and increased mononuclear cells. AES (E) reduced eosinophils but did not affect basophils or mononuclear cells. NRS (C) had no effect. There were no significant differences (Newman–Keuls multiple sample comparison test) in the number of feeding site neutrophils (0.7–9.4 cells/0.032 mm^2). NS, not significant ($p > 0.05$). (From Brown, *et al.*, 1982a, with permission.)

lymphocytes and/or serum IgG1 antibodies (Brown and Askenase, 1981; Brown *et al.*, 1982b) suggests that basophil recruitment *and function* in these reactions may be mediated by *both* T cells and IgG1 antibody. However, the precise mechanism by which basophil degranulation results in tick rejection/resistance is still unclear. For example, recent work by Brown and Askenase (1985) failed to document a role for histamine in basophil-mediated rejection of *A. americanum*, but indicates that other basophil mediators are involved.

In addition to questions about the identity of the basophil mediators influencing tick immunity, it is not yet certain whether basophils or their products directly injure ticks, interfere with their feeding, or act through an effect on other infiltrating cells. Kazura *et al.* (1981) reported that human leukemic basophils, in the presence of complement and specific antibody, were as effective as human eosinophils in mediating cytotoxicity to schistosomula *in vitro*. This result suggests one mechanism by which basophils and eosinophils may act independently as effector cells in tick immunity *in vivo*. In addition, work with mast cells suggests at least two potential mechanisms of basophil–eosinophil cooperation in immunity to ticks. (1) Eosinophil peroxidase expresses greater iodinating and cytocidal activity when bound to mast cell granules than when free in solution (reviewed in Henderson *et al.*, 1981; Jong *et al.*, 1981). Like mast cell granules, guinea pig basophil granules contain poorly soluble negatively charged sulfated glycosaminoglycans (Orenstein *et al.*, 1978) that may bind and potentiate the activity of eosinophil peroxidase. (2) Mast cells, or mast cell-derived ECF-A tetrapeptides, appear to enhance eosinophil-mediated damage of schistosomula by antibody-dependent cell-mediated cytotoxicity (ADCC) (Capron *et al.*, 1981a,b) or complement-dependent (Anwar *et al.*, 1980) mechanisms *in vitro*. Since specific antibody appears to participate in resistance to ticks as well (Brown and Askenase, 1981; Brown *et al.*, 1982b), it is possible that basophil-derived mediators serve a function analogous to that established for mast cell-derived factors in augmenting eosinophil-mediated cytotoxicity.

6.5. CBH RESPONSES TO TUMORS

Basophils have been identified as a prominent component of the inflammatory infiltrate elicited in association with the immune responses to the line 1 or line 10 guinea pig hepatomas (H. Dvorak *et al.*, 1973, 1979; Leonard *et al.*, 1979; Galli *et al.*, 1982a). The immune response to intradermal or subcutaneous line 1 tumor is associated with the elimination of this neoplasm from the host, and morphologic studies have shown basophils in the vicinity of dead or damaged tumor cells (H. Dvorak *et al.*, 1973). In addition, *in vitro* studies have shown that guinea pig basophils stimulated to undergo anaphylactic degranulation attach to line 1 or line 10 tumor cells and that a small number of tumor cells in these preparations exhibit morphologic evidence of injury and/or death (A. Dvorak *et al.*, 1979). It is unclear, however, to what extent this *in vitro* model is relevant to the process of tumor rejection *in vivo*. First, basophils at sites of tumor rejection *in vivo* generally do not exhibit evidence of anaphylactic-type degranulation. Second, although tumor cells in the *in vitro* preparations exhibited evidence of damage, the effect seemed only modest at the effector/target ratio tested (10 : 1), even though this ratio probably exceeds those occurring *in vivo*. Third, although basophils represented approximately half the leukocytes in the coculture experiments, the other leukocytes present, including varying proportions of eosinophils and lymphocytes, may have contributed to the modest tumor cytotoxicity observed when specific antigen was added to the reaction mixture.

6.6. SUMMARY

Basophils have been identified as prominent participants in guinea pig immune responses to many biologically significant antigens. These reactions may be divided into two categories, based on whether or not the infiltrating basophils regularly exhibit morphologic evidence of anaphylactic-type degranulation. In reactions characterized by extensive basophil degranulation, such as those elicited in the lamina propria of the intestine by the nematode *T. colubriformis* and in the skin by ixodid ticks, basophils may be important for the expression of immune resistance. The clearest evidence for such a role is provided by the secondary immune response to the feeding of *A. americanum* ticks. Studies with ABS indicate that basophils are required for expression of immune rejection of the ticks, although the mechanism by which basophils exert this effect is unknown. In the other category of basophil-rich delayed reactions, such as classic CBH responses or reactions to certain solid tumors, in which few basophils exhibit evidence of anaphylactic degranulation, the function of basophils is perhaps best characterized as "obscure." Studies with ABS in one such reaction, the classic CBH response to ovalbumin (Galli *et al.*, 1978), like similar work in the more complex response to ticks (Brown *et al.*, 1982a), suggest that basophils may serve, in part, to limit the extent of inflammation.

7. DOES CBH OCCUR IN SPECIES OTHER THAN THE GUINEA PIG?

The application of appropriate methods of tissue fixation and processing has revealed basophils in a number of delayed-onset immunologic reactions in several mammalian species including humans (reviewed in Askenase, 1977, 1980; Galli and Dvorak, 1979; Mitchell and Askenase, 1983; Galli *et al.*, 1984). Mice had long been thought to virtually lack basophils, but recent light (Urbina *et al.*, 1981) and electron microscopic (A. Dvorak *et al.*, 1982) evidence indicates that small numbers of basophils may be found in mouse blood and bone marrow. Occasional cells with the morphologic features of basophils also may be identified in delayed-onset responses in mice (Hurtado and Urbina, 1981; Galli *et al.*, unpublished data).

But in studies of delayed reactions in most mammalian species other than the guinea pig [certain responses in the rabbit may constitute interesting exceptions (Clark *et al.*, 1977)], basophils have represented a minor fraction of the infiltrating cells. In one study of human allergic contact dermatitis, for example, basophils rarely accounted for > 15% (often < 5%) of infiltrating cells (H. Dvorak *et al.*, 1974a). Basophils were also present in some human DH reactions provoked by intradermal Old Tuberculin, *Candida albicans*, or streptokinase/streptodornase. However, when these reactions were compared to those of allergic contact dermatitis, basophils appeared more frequently and in greater numbers in the latter (basophils were identified in 61% of delayed reactions to intradermal microbial antigens and were abundant in 7%; the values for the contact reactions were 88 and 35%, respectively). In another study (Askenase and Atwood, 1976), the Rebuck skin window technique (Rebuck, 1958) was used to examine basophil responses to KLH in human volunteers and to tuberculin in patients with tuberculosis. Small numbers of basophils (9–12% of infiltrating cells) were present in virtually all (18 of 20) immune reactions to KLH. By contrast, basophils were prominent (> 2%) in the tuberculin reactions of only 4 of 9 patients with tuberculosis. But in the 4 patients with tuberculosis whose reactions contained basophils, these cells were present in percentages similar to those in KLH reaction sites.

Taken together, these results illustrate that the sharp distinction between the basophil content of classic CBH responses and classic DH responses in the guinea pig may generally be blurred in humans. This may be due to the fact that the KLH-induced responses were elicited 1 week after sensitization of normal individuals and thus resemble classic CBH responses of guinea pigs, while the delayed reaction elicited with bacterial antigens occurred at various times after sensitization and thus may reflect the influence of complex regulatory factors that vary between individuals. In the case of species other than humans, there is too little information available to permit any conclusions about the existence of a pattern of delayed immunologic reactivity strictly analogous to guinea pig CBH.

8. NOMENCLATURE OF CBH: SUMMARY

Delayed-onset immunologic reactions, whose leukocytic infiltrates for many years were thought to be comprised largely of mononuclear cells, may also contain variable but sometimes striking infiltrates of basophilic granulocytes. Such reactions have been studied most extensively in the guinea pig, where they may be elicited by a wide variety of biologically significant antigens. The identification of these processes overturned conventional notions about the histopathology of delayed-onset immunologic reactions and also showed that a morphologic approach employing appropriate methods of tissue fixation and processing could contribute significantly to our knowledge of the expression of CMI *in vivo*.

Present evidence indicates that guinea pig delayed-onset responses including prominent infiltrates of basophils are themselves heterogeneous in pathogenesis. Perhaps the most graphic evidence of this is derived from attempts to transfer such reactions passively. For example, delayed accumulations of leukocytes including ≥ 5% basophils may occur in experimental reactions transferred by T-cell, B-cell, or immunoglobulin (IgG1 or IgE) preparations derived from actively sensitized guinea pigs. But none of these transferred reactions contain the striking infiltrates of basophils seen in the lesions of actively immunized animals. This may simply reflect quantitative deficiencies in the adoptively transferred materials. However, we feel that a more likely explanation is that the recruitment of basophils to sites of delayed-onset reactivity *in vivo* often reflects a complex interplay of local and systemic factors. The separate contributing parts may be identified using the reductionist approach of passive transfers, but the orchestrated whole, as occurs in the actively sensitized host, is difficult to mimic precisely.

This notion is also supported by the analysis of reactions adoptively transferred with sensitized lymphocytes derived from guinea pigs primed to express DH reactions, the classical experiment of CMI (Landsteiner and Chase, 1942). Although DH reactions in actively sensitized animals contain very few basophils, and are strongly indurated, sensitized T cells from these animals passively transfer a form of cutaneous reactivity, which, like CBH, has a prominent basophil component (Askenase, 1976). By contrast, if recipient animals receive both serum *and* lymphocytes of guinea pigs sensitized for DH, the basophil component of the transferred reaction is suppressed, and the macroscopic size and degree of induration are increased (Mitchell and Askenase, 1982).

In view of these considerations, it may be appropriate to reassess the nomenclature of delayed-onset immunologic reactions in the guinea pig, particularly those with a prominent basophil component. Because many mechanisms may result in the delayed-onset accumulation of basophils at sites of antigen challenge, lumping all of these processes under the rubric "CBH" would be justified only if "CBH" were construed to indicate simply the presence of a certain minimum number of basophils in a cutaneous immune reaction. An alternative approach, which we favor, would be to use "classic DH" and "classic CBH" as descriptive terms to refer to rather extreme points on what may well represent a continuous spectrum of delayed-onset reactivity. In this scheme, classic DH refers to the pattern of T-cell-dependent systemic reactivity, modified by several humoral factors, which is best revealed by skin tests applied a few weeks after immunizing adult guinea pigs with antigen emulsified in Freund's complete adjuvant. Classic CBH (Richerson *et al.*, 1969, 1970; H. Dvorak *et al.*, 1970) refers to another pattern of T-cell-mediated systemic reactivity, which appears to differ from classic DH in its dependence on humoral factors, and which is best revealed by skin tests with large doses of antigen placed on adult guinea pigs at early intervals (4–7 days) after sensitization with certain proteins (e.g., ovalbumin) in Fruend's incomplete adjuvant. These two prototypic T-cell-dependent reactions differ in a number of characteristics, as described in detail in Section 4. Note particularly that when complement-fixing (IgG2, IgM) antibody responses occur at late intervals after sensitization for "classic CBH," they are associated with a *decline* in the gross evidence of cutaneous reactivity. In classic CBH, cutaneous reactions elicited *after* the evolution of a complement-fixing antibody response ("late reactions") contain fewer basophils than reactions elicited on day 7. In addition, a relatively high proportion of basophils in "late reactions" may bear surface IgG1 antibody with specificity for immunizing antigen (Colvin *et al.*, 1973; H. Dvorak *et al.*, 1975).

In guinea pigs sensitized to certain protein antigens, infected with certain worms, ticks, or viruses, or sensitized to some contactants, both sensitized T cells and strong anaphylactic (principally IgG1) antibody responses are generated. In this setting, intense basophil-rich cutaneous reactions often can be elicited by antigen challenge performed *before* or *after* the development of a significant antibody response. In these reactions, many basophils infiltrating the tissues after the development of the IgG1 response bear surface immunoglobulin with specificity for immunizing antigen. Furthermore, these basophils may undergo anaphylactic-type degranulation in response to additional local antigen challenge, a reaction designated "cutaneous basophil anaphylaxis" (Askenase *et al.*, 1978a). The cutaneous reactions to certain ticks (Brown *et al.*, 1982a; Brown and Askenase, 1983) probably represent examples of this pattern of reactivity, in which basophil degranulation may be due to shed tick antigens triggering antibody-coated basophils. For responses such as these, which appear to represent fascinating mixtures of delayed and immediate hypersensitivity, we suggest the term "complex CBH."

Passive transfer of lymphocytes or antibodies from guinea pigs actively sensitized for DH or CBH has revealed the complexity of these reactions. As described in detail above, this is perhaps best illustrated by the observation that systemic transfer of T cells derived from guinea pigs actively sensitized for classic DH renders *recipients* able to express CBH-like reactions at sites of cutaneous antigen challenge. While these reactions have a histopathology different from that of classic DH, to refer to them as "CBH" may serve to confuse rather than to clarify the important issues raised by the initial observation. On the other hand, the use of "CBH" to designate basophil-containing lesions elicited in animals which have received adoptive transfer of T-cell or antibody preparations from donor animals primed to develop *basophil-rich* delayed responses is now commonplace. A reasonable compromise may be to call passively transferred reactions with a prominent (? > 5–10%) basophil component "CBH-like."

The nomenclature suggested here is provisional and operational in nature, and more than one of the terms may be applicable during the analysis of a single experimental model. For example, animals primed by the feeding of *A. americanum* ticks appear to express "complex CBH" at sites of secondary tick challenge (Brown *et al.*, 1982a), and also may serve as donors for either T-cell- or antibody-mediated "CBH-like" reactivity (Brown and Askenase, 1981; Brown *et al.*, 1982b). A more important consideration is that some reaction patterns appear to occupy a position intermediate between classic CBH and classic DH. The ability to mount a delayed response to challenge with line 10 tumor cells in guinea pigs primed with line 10 tumor in Freund's *complete* adjuvant persists indefinitely after sensitization (like classic DH) and elicits an inflammatory infiltrate containing 10–20% basophils, a value more than that observed in classic DH but less than that seen in many examples of classic CBH (H. Dvorak *et al.*, 1973; Leonard *et al.*, 1979). Put differently, classic CBH and complex CBH may represent only two of many possible reaction patterns drawn from a broad spectrum of delayed reactivity regulated by T cells and antibodies and featuring a prominent basophil component. Such a spectrum may be defined at one extreme by classic CBH to a simple protein like ovalbumin and at the other by the basophil-rich responses to the feeding of ixodid ticks, with the response to KLH occupying an intermediate position.

But these semantic considerations should not be permitted to obscure more important issues. For example, whether they are viewed as pathogenetically distinct entities or extreme points on a spectrum of T cell dependent delayed cutaneous responses, classic DH reactions and classic CBH reactions differ in a number of features in addition to the composition of their leukocytic infiltrates. These include the extent of changes in vascular permeability (Colvin and Dvorak, 1975), the pattern of activation of the clotting or fibrinolytic systems (Colvin and Dvorak, 1975), and the nature of alterations involving endothelial cells (Galli *et al.*, 1982b, 1983b) and connective tissue elements (reviewed in H. Dvorak *et al.*, 1980). The factors determining these characteristics remain largely undefined. Even when the regulation of basophil infiltration is the only feature considered, many specific and nonspecific modulating factors have been identified but important questions remain unanswered. Thus, chemotaxis studies *in vitro* and attempts to transfer DH or CBH passively *in vivo* have suggested the existence of multiple factors that could influence basophil recruitment. Which of these mechanisms actually operate during the evolution of reactions elicited in actively sensitized animals remains a difficult issue that may not be resolved easily. Finally, even though a large body of work has confirmed the participation of basophils in a variety of delayed-onset reactions in the guinea pig, the function of basophils in most of these responses is unknown.

Studies with a specific anti-guinea pig basophil serum suggest two roles for basophils in delayed reactions *in vivo:* the local anaphylactic degranulation of basophils appears to be essential for the expression of immune resistance to the feeding of *A. americanum* ticks; both in complex, tick-induced CBH reactions and in classic CBH to ovalbumin, basophils may serve, in part, to limit the extent of inflammation (Galli *et al.*, 1978; Brown *et al.*, 1982a). While these represent tantalizing observations, the basophil mediators directly or indirectly responsible for the expression of tick resistance or suppression of inflammation have not been identified. In species other than the guinea pig, the role of basophils in delayed reactions is even more obscure. One possibility is that a major biologic function of the basophil is to represent an essential participant (? effector cell) in complex immune responses to parasitic infection of cutaneous and mucosal surfaces. If this is the case, then the occurrence of variable numbers of basophils in other immunologic reactions, including classic CBH or responses to certain tumors in the guinea pig, and a number of clinically significant allergic responses in humans (Mitchell and Askenase, 1983; Galli *et al.*, 1984), may reflect the "inappropriate" activation of immunologic circuits or mediator pathways ordinarily mobilized during parasitic infections. This notion is akin to the suggestion that pathologic allergic responses dependent on IgE and mast cells mimic host reactions which confer benefit in the context of parasitic infections (e.g., see Jarrett and Miller, 1982).

Interpreted narrowly, these ideas have little room for a significant beneficial role of basophils and mast cells in the biology of healthy populations free of parasitic infections. On the other hand, the possibility that basophils and mast cells have important responsibilities beyond dealing with parasites and providing work for allergists (and some pathologists) cannot be dismissed out of hand. Consideration of this possibility has fueled much imaginative and provocative speculation, but definitive proof of such roles has proven elusive.

ACKNOWLEDGMENT. The authors thank Dr. Harold F. Dvorak for review of the manuscript and for permission to use original data from H. F. Dvorak *et al.* (1970).

REFERENCES

Alkan, S. S., Nitecki, D. E., and Goodman, J. W., 1971, Antigen recognition and the immune response: The capacity of 1-tyrosine-azobenzenearsonate to serve as a carrier for a macromolecular hapten, *J. Immunol.* **107**:353–358.

Allen, J. R., 1973, Tick resistance: Basophils in skin reactions of resistant guinea pigs, *Int. J. Parasitol.* **3**:195–200.

Allen, J. R., Doube, B. M., and Kemp, D. H., 1977, Histology of bovine skin reactions of *Ixodes holocyclus*, Neuman, *Can. J. Comp. Med.* **41**:26–35.

Anwar, A. R. E., McKean, J. R., Smithers, S. R., and Kay, A. B., 1980, Human eosinophil- and neutrophil-mediated killing of schistosomula of *Schistosoma mansoni* in vitro. I. Enhancement of complement-dependent damage by mast cell-derived mediators and formyl methionyl peptides, *J. Immunol.* **124**:1122–1129.

Askenase, P. W., 1973, Cutaneous basophil hypersensitivity in contact-sensitized guinea pigs. I. Transfer with immune serum, *J. Exp. Med.* **128**:1144–1155.

Askenase, P. W., 1976, Cutaneous basophil hypersensitivity uncovered in the cell transfer of classical tuberculin hypersensitivity, *J. Immunol.* **117**:741–747.

Askenase, P. W., 1977, The role of basophils, mast cells and vasoamines in hypersensitivity reactions with a delayed time course, *Prog. Allergy* **23**:199–320.

Askenase, P. W., 1980, Immunopathology of parasitic diseases: Involvement of basophils and mast cells, *Semin. Immunopathol.* **2**:417–442.

Askenase, P. W., and Atwood, J. E., 1976, Basophils in tuberculin and "Jones–Mote" delayed reactions of humans, *J. Clin. Invest.* **58**:1145–1154.

Askenase, P. W., Haynes, J. D., Tauben, D., and DeBernardo, R., 1975a, Specific basophil hypersensitivity induced by skin testing and transferred using immune serum, *Nature (London)* **256**:52–54.

Askenase, P. W., Hayden, B. J., and Gershon, R. K., 1975b, Augmentation of delayed-type hypersensitivity by doses of cyclophosphamide which do not effect antibody responses, *J. Exp. Med.* **141**:697–702.

Askenase, P. W., Haynes, J. D., and Hayden, B. J., 1976a, Antibody-mediated basophil accumulations in cutaneous hypersensitivity reactions in guinea pigs, *J. Immunol.* **117**:1722–1730.

Askenase, P. W., Hayden, B., and Higashi, G. I., 1976b, Cutaneous basophil hypersensitivity and inhibited macrophage migration in guinea pigs with schistosomiasis, *Clin. Exp. Immunol.* **23**:318–327.

Askenase, P. W., DeBernardo, R., Tauben, D., and Kashgarian, M., 1978a, Cutaneous basophil anaphylaxis: Immediate vasopermeability increases and anaphylactic degranulation of basophils at delayed hypersensitivity reactions challenged with additional antigen, *Immunology* **35**:741–755.

Askenase, P. W., Boone, W. T., and Binder, H. J., 1978b, Colonic basophil hypersensitivity, *J. Immunol.* **120**:198–201.

Askenase, P. W., Schwartz, A., Siegel, J. N., and Gershon, R. K., 1981, Role of histamine in the regulation of cell-mediated immunity, *Int. Arch. Allergy Appl. Immunol.* **66**(Suppl. 1):225–233.

Askenase, P. W., Bagnall, B. G., and Worms, M. J., 1982, Cutaneous basophil associated resistance to ectoparasites (ticks). I. Transfer with immune serum or immune cells, *Immunology* **45**:501–511.

Askenase, P. W., Rosenstein, R. W., and Ptak, W., 1983a, T cells produce an antigen binding factor with in vivo activity analogous to IgE antibody, *J. Exp. Med.* **157**:862–873.

Askenase, P. W., Van Loveren, H., Rosenstein, R. W., and Ptak, W., 1983b, Immunologic specificity of antigen-binding T cell-derived factors that transfer mast cell-dependent, immediate hypersensitivity-like reactions, *Monogr. Allergy* **18**:249–255.

Avella, J., Binder, H., Madsen, J. E., and Askenase, P. W., 1978, Effect of histamine H2-receptor antagonists on delayed hypersensitivity, *Lancet* **1**:624–626.

Befus, D., and Bienenstock, J., 1982, Factors involved in symbiosis and host resistance at the mucosa–parasite interface, *Prog. Allergy* **31**:76–177.

Bienenstock, J., Befus, A. D., Pearce, F., Denburg, J., and Goodacre, R., 1982, Mast cell heterogeneity: Derivation and function, with emphasis on the intestine, *J. Allergy Clin. Immunol.* **70**:407–412.

Boughton, B., and Spector, W. G., 1963, Histology of the tuberculin reaction in guinea pigs. *J. Pathol. Bacteriol.* **85**:371–381.

Brendel, J., and Graziano, F. M., 1984, Basophil granules provoke an inflammatory response in guinea pig skin, *Fed. Proc.* **43**:1759a.

Brown, S. J., and Askenase, P. W., 1981, Cutaneous basophil responses and immune resistance of guinea pigs to ticks: Passive transfer with peritoneal exudate cells or serum, *J. Immunol.* **127**:2163–2167.

Brown, S. J., and Askenase, P. W., 1983, Immune rejection of ectoparasites (ticks) by T cell and IgG_1 antibody recruitment of basophils and eosinophils, *Fed. Proc.* **42**:1744–1749.

Brown, S. J., and Askenase, P. W., 1985, Rejection of ticks from guinea pigs by anti-hapten-antibody mediated degranulation of basophils at cutaneous basophil hypersensitivity sites: Role of mediators other than histamine, *J. Immunol.* **134**:1160–1165.

Brown, S. J., and Knapp, F. W., 1980, *Amblyomma americanum:* Sequential histological analysis of larval and nymphal feeding sites on guinea pigs, *Exp. Parasitol.* **49**:188–205.

Brown, S. J., and Knapp, F. W., 1981, Response of hypersensitized guinea pigs to the feeding of *Amblyomma americanum* ticks, *Parasitology* **83**:213–223.

Brown, S. J., Galli, S. J., Gleich, G. J., and Askenase, P. W., 1982a, Ablation of immunity of *Amblyomma americanum* bv anti-basophil serum: Cooperation between basophils and eosinophils in expression of immunity to ectoparasites (ticks) in guinea pigs, *J. Immunol.* **129**:790–796.

Brown, S. J., Graziano, F. M., and Askenase, P. W., 1982b, Immune serum transfer of cutaneous basophil associated resistance to ticks: Mediation by 7S IgG_1 antibodies, *J. Immunol.* **129**:2407–2412.

Capron, M., Bazin, H., Joseph, M., and Capron, A., 1981a, Evidence for IgE-dependent cytotoxicity by rat eosinophils, *J. Immunol.* **126**:1764–1768.

Capron, M., Capron, A., Goetzl, E. J., and Austen, K. F., 1981b, Tetrapeptides of the eosinophil chemotactic factor of anaphylaxis (ECF-A) enhance eosinophil Fc receptor, *Nature (London)* **289**:71–73.

Castro, G. A., 1982, Immunological regulation of epithelial function, *Am. J. Physiol.* **G243**:321–329.

Chan, B. S. T., 1972, Ultrastructural changes in guinea pig bone marrow basophils during anaphylaxis, *Immunology* **23**:215–224.

Clark, J. M., Altman, G., and Fromowitz, F. B., 1977, Basophil hypersensitivity response in rabbits, *Infect. Immun.* **15**:305–312.

Cochrane, C. G., and Janoff, A., 1974, The arthus reaction: A model of neutrophil and complement-mediated injury, in: *The Inflammatory Process,* 2nd ed. (B. W. Zweifach, L. Grant, and R. T. McCluskey, eds.), Volume 3, pp. 85–162, Academic Press, New York.

Colvin, R. B., and Dvorak, H. F., 1975, Role of the clotting system in cell-mediated hypersensitivity. II. Kinetics of fibrinogen/fibrin accumulation and vascular permeability. Changes in tuberculin and cutaneous basophil hypersensitivity reactions, *J. Immunol.* **114**:377–387.

Colvin, R. B., Pinn, V. W., Simpson, B. A., and Dvorak, H. F., 1973, Cutaneous basophil hypersensitivity. IV. The late reaction. Sequel to Jones–Mote type hypersensitivity. Comparison with rabbit Arthus reaction. Effect of passive antibody on induration and expressions of Jones–Mote hypersensitivity, *J. Immunol.* **110**:1279–1289.

Combs, J. W., 1966, Maturation of rat mast cells: An electron microscopic study, *J. Cell Biol.* **31:**563–575.

Combs, J. W., 1971, An electron microscopic study of mouse mast cells arising *in vivo* and *in vitro, J. Cell Biol.* **48:**676–684.

Combs, J. W., Lagunoff, D., and Benditt, E. P., 1965, Differentiation and proliferation of embryonic mast cells of the rat, *J. Cell Biol.* **25:**577–592.

Crowle, P. K., and Reed, N. D., 1984, Bone marrow origin of mucosal mast cells, *Int. Arch. Allergy Appl. Immunol.* **73:**242–247.

Csaba, G., and Kovacs, P., 1975, Demonstration of the heterogeneity of the mast cell population on the basis of the mucopolysaccharide content, *Acta Morphol. Acad. Sci. Hung.* **23:**227–233.

Denburg, J. A., Davison, M., and Bienenstock, J., 1980, Basophil production: Stimulation by factors derived from guinea pig splenic T-lymphocytes, *J. Clin. Invest.* **65:**390–399.

Dienes, L., 1929, The technic of producing the tuberculin type of sensitization with eggwhite in tuberculous guinea pigs, *J. Immunol.* **17:**531–538.

Dienes, L., and Mallory, T. B., 1932, Histological studies of the hypersensitive reactions, *Am. J. Pathol.* **8:**689–710.

Drössler, K., Parker, D., and Turk, J. L., 1983, The effect of cyclophosphamide on immunological memory, *Immunopharmacology* **6:**123–132.

Dvorak, A. M., 1978, Biology and morphology of basophilic leukocytes, in: *Immediate Hypersensitivity: Modern Concepts and Developments* (M. K. Bach, ed.), pp. 369–405, Dekker, New York.

Dvorak, A. M., and Dvorak, H. F., 1974, Structure of Freund's complete and incomplete adjuvants: Relation of adjuvanticity to structure, *Immunology* **27:**99–114.

Dvorak, A. M., Bast, R. C., Jr., and Dvorak, H. F., 1971, Morphologic changes in draining lymph nodes and in lymphocyte cultures after sensitization with complete or incomplete Freund's adjuvant: Correlation with immunologic events *in vivo* and in culture, *J. Immunol.* **107:**422–435.

Dvorak, A. M., Dvorak, H. F., and Karnovsky, M. J., 1972, Uptake of horseradish peroxidase by guinea pig basophilic leukocytes, *Lab. Invest.* **26:**27–39.

Dvorak, A. M., Mihm, M. C., Jr., and Dvorak, H. F., 1976a, Degranulation of basophilic leukocytes in allergic contact dermatitis reactions in man, *J. Immunol.* **116:**687–695.

Dvorak, A. M., Mihm, M. C., Jr., and Dvorak, H. F., 1976b, Morphology of delayed-type hypersensitivity reactions in man. II. Ultrastructural alterations affecting the microvasculature and the tissue mast cells, *Lab. Invest.* **34:**179–191.

Dvorak, A. M., Galli, S. J., Galli, A. S., Hammond, M. E., Churchill, W. H., Jr., and Dvorak, H. F., 1979, Tumor–basophil interactions *in vitro:* Scanning and transmission electron microscopic study, *J. Immunol.* **122:**2447–2457.

Dvorak, A. M., Hammond, M. E., Morgan, E., Orenstein, N. S., Galli, S. J., and Dvorak, H. F., 1980, Evidence for a vesicular transport mechanism in guinea pig basophilic leukocytes, *Lab. Invest.* **42:**263–276.

Dvorak, A. M., Galli, S. J., Morgan, E., Galli, A. S., and Dvorak, H. F., 1981a, Anaphylactic degranulation of guinea pig basophilic leukocytes. I. Fusion of granule membranes and cytoplasmic vesicles. Formation and resolution of degranulation sacs, *Lab. Invest.* **44:**174–191.

Dvorak, A. M., Lett-Brown, M., Thueson, D., and Grant, J. A., 1981b, Complement-induced degranulation of human basophils, *J. Immunol.* **126:**523–528.

Dvorak, A. M., Nabel, G., Pyne, K., Cantor, H., Dvorak, H. F., and Galli, S. J., 1982, Ultrastructural identification of the mouse basophil, *Blood* **59:**1279–1285.

Dvorak, A. M., Galli, S. J., Schulman, E. S., Lichtenstein, L. M., and Dvorak, H. F., 1983, Basophil and mast cell degranulation: Ultrastructural analysis of mechanisms of mediator release, *Fed. Proc.* **42:**2510–2515.

Dvorak, A. M., Lett-Brown, M. A., Thueson, D. O., Pyne, K., Paghuprasad, P. K., Galli, S. J., and Grant, J. A., 1984, Histamine-releasing activity (HRA). III. HRA induces human basophil histamine release by provoking noncytotoxic granule exocytosis, *Clin. Immunopathol.* **32:**142–150.

Dvorak, H. F., 1971, Role of the basophilic leukocyte in allograft rejection, *J. Immunol.* **106:**279–281.

Dvorak, H. F., 1974, Delayed hypersensitivity, in: *The Inflammatory Process*, 2nd ed. (B. W. Zweifach, L. Grant, and R. T. McCluskey, eds.), Volume 3, pp. 291–345, Academic Press, New York.

Dvorak, H. F., and Hirsch, M. S., 1971, Role of basophilic leukocytes in cellular immunity to vaccinia virus infection. *J. Immunol.* **107:**1576–1582.

Dvorak, H. F., Dvorak, A. M., Simpson, B. A., Richerson, H. M., Leskowitz, S., and Karnovsky, M. J., 1970, Cutaneous basophil hypersensitivity. II. A light and electron microscopic description. *J. Exp. Med.* **132**:558–582.

Dvorak, H. F., Simpson, B. A., Bast, R. C., Jr., and Leskowitz, S., 1971, Cutaneous basophil hypersensitivity. III. Participation of the basophil in hypersensitivity to antigen–antibody complexes, delayed hypersensitivity and contact allergy. Passive transfer, *J. Immunol.* **107**:138–148.

Dvorak, H. F., Dvorak, A. M., and Churchill, W. H., 1973, Immunologic rejection of diethylnitrosamine induced hepatomas in strain 2 guinea pigs: Participation of basophilic leukocytes and macrophage aggregates, *J. Exp. Med.* **137**:751–775.

Dvorak, H. F., Mihm, M. C., Jr., Dvorak, A. M., Johnson, R. A., Manseau, E. J., Morgan, E., and Colvin, R. B., 1974a, Morphology of delayed-type hypersensitivity reactions in man. I. Quantitative description of the inflammatory response, *Lab. Invest.* **31**:111–130.

Dvorak, H. F., Selvaggio, S. S., Dvorak, A. M., Colvin, R. B., Lean, D. B., and Rypysc, J., 1974b, Purification of basophilic leukocytes from guinea pig blood and bone marrow, *J. Immunol.* **113**:1694–1702.

Dvorak, H. F., Colvin, R. B., and Churchill, W. H., 1975, Specificity of basophils and lymphocytes in cutaneous basophil hypersensitivity, *J. Immunol.* **114**:507–511.

Dvorak, H. F., Hammond, M. E., Colvin, R. B., Manseau, E. J., and Goodwin, J., 1977a, Systemic expression of cutaneous basophil hypersensitivity, *J. Immunol.* **118**:1549–1557.

Dvorak, H. F., Orenstein, N. S., Dvorak, A. M., Hammond, M. E., Roblin, R. O., Feder, J., Schott, C. F., Goodwin, J., and Morgan, E., 1977b, Isolation of the cytoplasmic granules of guinea pig basophilic leukocytes: Identification of esterase and protease activities, *J. Immunol.* **119**:38–46.

Dvorak, H. F., Orenstein, N. S., Rypysc, J., Colvin, R. B., and Dvorak, A. M., 1978, Plasminogen activator of guinea pig basophilic leukocytes: Probable localization to the plasma membrane, *J. Immunol.* **120**:766–773.

Dvorak, H. F., Dvorak, A. M., Manseau, E. J., Wiberg, L., and Churchill, W. H., 1979, Fibrin gel investment associated with line 1 and line 10 solid tumor growth, angiogenesis, and fibroplasia in guinea pigs: Role of cellular immunity, myofibroblasts, microvascular damage and infarction in line 1 tumor regression, *J. Natl. Cancer Inst.* **62**:1459–1472.

Dvorak, H. F., Galli, S. J., and Dvorak, A. M., 1980, Expression of cell-mediated hypersensitivity in vivo: Recent advances, *Int. Rev. Exp. Pathol.* **21**:119–194.

Ehrlich, P., 1878, Beitrage zur Theorie und Praxis der histologischen Farbung, thesis, Leipzig.

Ehrlich, P., 1879, Uber die specifischen Granulationen des Blutes, *Arch. Anat. Physiol. Physiol. Abt.* 571–577.

Enerbäck, L., 1966a, Mast cells in rat gastrointestinal mucosa. 1. Effects of fixation, *Acta Pathol. Microbiol. Scand.* **66**:289–302.

Enerbäck, L., 1966b, Mast cells in rat gastrointestinal mucosa. 2. Dye-binding and metachromatic properties, *Acta Pathol. Microbiol. Scand.* **66**:303–312.

Enerbäck, L., 1966c, Mast cells in rat gastrointestinal mucosa. 3. Reactivity towards compound 48/80, *Acta Pathol. Microbiol. Scand.* **66**:313–322.

Enerbäck, L., 1981, The gut mucosal mast cell, *Monogr. Allergy* **17**:222–232.

Enerbäck, L., and Lundin, P. M., 1974, Ultrastructure of mucosal mast cells in normal and compound 48/80-treated rats, *Cell Tissue Res.* **150**:95–105.

Fantozzi, R., Masini, E., Blandina, P., Brunelleschi, S., and Mannaioni, P. F., 1983, Receptor modulation of histamine release in isolated purified rat mast cells, *Monogr. Allergy* **18**:190–192.

Flax, M. H., and Caulfield, J. B., 1963, Cellular and vascular components of allergic contact dermatitis, *Am. J. Pathol.* **43**:1031–1053.

Fox, C. C., Dvorak, A. M., MacGlashan, D. W., Jr., and Lichtenstein, L. M., 1984, Histamine-containing cells in human peritoneal fluid, *J. Immunol.* **132**:2177–2179.

Friedlaender, M. H., and Dvorak, H. F., 1977, Morphology of delayed-type hypersensitivity reactions in the guinea pig cornea, *J. Immunol.* **118**:1558–1563.

Froese, A., 1984, The receptors for IgE on mast cells and basophils, *Prog. Allergy* **34**:142–187.

Fung, M. C., Hapel, A. J., Ymer, S., Cohen, D. R., Johnson, R. M., Campbell, H. D., and Young, I. G., 1984, Molecular cloning of cDNA for murine interleukin-3, *Nature (London)* **307**:233–237.

Galli, S. J., and Dvorak, H. F., 1979, Basophils and mast cells: Structure, function, and role in hypersensitivity, in: *Cellular, Molecular, and Clinical Aspects of Allergic Diseases* (S. Gupta and R. A. Good, eds.), pp. 1–53, Plenum Press, New York.

Galli, S. J., Colvin, R. B., Verderber, E., Galli, A. S., Monahan, R., Dvorak, A. M., and Dvorak, H. F., 1978, Preparation of a rabbit anti-guinea pig basophil serum: *In vitro* and *in vivo* characterization, *J. Immunol.* **121**:1157–1166.

Galli, S. J., Bast, R. C., Jr., Bast, B. S., Isomura, T., Zbar, B., Rapp, H. J., and Dvorak, H. F., 1982a, Bystander suppression of tumor growth: Evidence that specific targets and bystanders are damaged by injury to a common microvasculature, *J. Immunol.* **129**:1790–1799.

Galli, S. J., Dvorak, A. M., Marcum, J. A., Ishizaka, T., Nabel, G., Der Simonian, H., Pyne, K., Goldin, J. M., Rosenberg, R. D., Cantor, H., and Dvorak, H. F., 1982b, Mast cell clones: A model for the analysis of cellular maturation. *J. Cell Biol.* **95**:435–444.

Galli, S. J., Dvorak, A. M., and Dvorak, H. F., 1983a, Morphology, biochemistry, and function of basophils, in: *Hematology*, 3rd ed. (W. J. Williams, E. Beutler, A. J. Erslev, and M. A. Litchman eds.), pp. 820–825, McGraw–Hill, New York.

Galli, S. J., Isomura, T., Dvorak, A. M., and Dvorak, H. F., 1983b, Lymph node cells (LNC) activated by antigen (Ag) nonspecifically damage cloned strain 2 guinea pig (S2GP) aortic endothelial cells *in vitro*, *Fed. Proc.* **42**:662a.

Galli, S. J., Dvorak, A. M., and Dvorak, H. F., 1984, Basophils and mast cells: Morphologic insights into their biology, secretory patterns, and function, *Prog. Allergy* **34**:1–141.

Gell, P. G. H., and Minde, I. T., 1954, Observations on the histology of the Arthus reaction and its relation to other known types of skin hypersensitivity, *Int. Arch. Allergy Appl. Immunol.* **5**:23–46.

Ginsburg, H., 1963, The *in vitro* differentiation and culture of normal mast cells from mouse thymus, *Ann. N.Y. Acad. Sci.* **103**:20–39.

Ginsburg, H., and Lagunoff, D., 1967, The *in vitro* differentiation of mast cells: Cultures of cells from immunized mouse lymph nodes and thoracic duct lymph on fibroblast monolayers, *J. Cell Biol.* **35**:685–697.

Ginsburg, H., Nir, I., Hammel, I., Eren, R., Weissman, B.-A., and Naot, Y., 1978, Differentiation and activity of mast cells following immunization in cultures of lymph node cells, *Immunology* **35**:485–502.

Ginsburg, H., Ben-Shahar, D., Hammel, I., and Ben-David, E., 1979, Degranulation capacity of mast cells grown in cell culture in the presence of histamine releaser, *Nature (London)* **280**:151–153.

Ginsburg, H., Olson, E. C., Huff, T., Okudaira, H., and Ishizaka, T., 1981, Enhancement of mast cell differentiation *in vitro* by T-cell factor(s), *Int. Arch. Allergy Appl. Immunol.* **66**:447–458.

Ginsburg, H., Ben-Shahar, D., and Ben-David, E., 1982, Mast cell growth on fibroblast monolayers: Two-cell entities, *Immunology* **45**:371–380.

Godfrey, H. P., Pastore, J., and Askenase, P. W., 1981, A modified Giemsa stain for demonstrating basophils in glycol methacrylate-embedded tissue sections, *J. Histotechnol.* **4**:80–82.

Godfrey, H. P., Phillips, M. E., and Askenase, P. W., 1983, Histopathology of delayed-onset hypersensitivities in contact-sensitive guinea pigs, *Int. Arch. Allergy Appl. Immunol.* **70**:50–58.

Goldberg, B., Kantor, F. S., and Benacerraf, B., 1962, An electron microscopic study of delayed sensitivity to ferritin in guinea pigs, *Br. J. Exp. Pathol.* **43**:621–626.

Graziano, F. M., and Askenase, P. W., 1979, Involvement of host F_c receptors in antibody-mediated cutaneous basophil hypersensitivity reactions, *J. Immunol.* **123**:1645–1652.

Graziano, F. M., Gunderson, L., Larson, L., and Askenase, P. W., 1983, IgE antibody-mediated cutaneous basophil hypersensitivity reactions in guinea pigs, *J. Immunol.* **131**:2675–2681.

Griffith, B. P., Askenase, P. W., and Hsiung, G. D., 1982, Serum and cell-mediated viral-specific delayed cutaneous basophil reactions during cytomegalovirus infection of guinea pigs, *Cell. Immunol.* **69**:138–149.

Guy-Grand, D., Dy, M., Luffau, G., and Vassalli, P., 1984, Gut mucosal mast cells: Origin, traffic, and differentiation, *J. Exp. Med.* **160**:12–28.

Haig, D. M., McKee, T. A., Jarrett, E. E. E., Woodbury, R., and Miller, H. R. P., 1982, Generation of mucosal mast cells is stimulated *in vitro* by factors derived from T cells of helminth-infected rats, *Nature (London)* **300**:188–190.

Handlinger, J. H., and Rothwell, T. L. W., 1984, Intestinal mast cell changes in guinea pigs infected with the nematode *Trichostrongylus colubriformis*, *Int. Arch. Allergy Appl. Immunol.* **74**:165–171.

Haynes, J. D., and Askenase, P. W., 1977, Cutaneous basophil responses in neonatal guinea pigs: Active immunization, hapten specific transfer with small amounts of serum, and preferential elicitation with phytohemagglutinin skin testing, *J. Immunol.* **118:**1063–1069.

Haynes, J. D., Rosenstein, R. W., and Askenase, P. W., 1978, A newly described activity of guinea pig IgG_1 antibodies: Transfer of cutaneous basophil reactions, *J. Immunol.* **120:**886–894.

Henderson, W. R., Chi, E. Y., Jong, E. C., and Kelbanoff, S. J., 1981, Mast cell mediated tumor cell cytotoxicity. Role of the peroxidase system. *J. Exp. Med.* **153:**520–533.

Ho, P. C., Lewis, R. A., Austen, K. F., and Orange, R. P., 1979, Mediators of immediate hypersensitivity, in: *Cellular, Molecular, and Clinical Aspects of Allergic Disorders* (S. Gupta and R. A. Good, eds.), pp. 179–228, Plenum Press, New York.

Hurtado, I., and Urbina, C., 1981, Cutaneous basophil hypersensitivity (CBH) in C3H/HeJ mice contact sensitized with dinitrochlorobenzene, *Fed. Proc.* **40:**989a.

Huxtable, C. R., and Rothwell, T. L. W., 1975, Studies of the responses of basophil and eosinophil leucocytes and mast cells to the nematode *Trichostrongylus colubriformis:* Ultrastructural changes in basophils and eosinophils at the site of infection. *Aust. J. Exp. Biol. Med. Sci.* **53:**437–445.

Ihle, J. N., Rebar, L., Keller, J., Lee, J. C., and Hapel, A. J., 1982, Interleukin 3: Possible roles in the regulation of lymphocyte differentiation and growth, *Immunol. Rev.* **63:**5–32.

Ishizaka, K., and Ishizaka, T., 1979, Immunoglobulin E: Biosynthesis and immunological mechanisms of IgE-mediated hypersensitivity, in: *Cellular, Molecular, and Clinical Aspects of Allergic Disorders* (S. Gupta and R. A. Good, eds.), pp. 153–178, Plenum Press, New York.

Ishizaka, T., and Ishizaka, K., 1984, Activation of mast cells for mediator release through IgE receptors, *Prog. Allergy* **34:**188–235.

Ishizaka, T., Okudaira, H., Mauser, L. E., and Ishizaka, K., 1976, Development of rat mast cells *in vitro*. I. Differentiation of mast cells from thymus cells, *J. Immunol.* **116:**747–754.

Ishizaka, T., Adachi, T., Chang, T.-H., and Ishizaka, K., 1977, Development of mast cells *in vitro*. II. Biologic function of culture mast cells. *J. Immunol.* **118:**211–217.

Ishizaka, T., Dvorak, A. M., Conrad, D. H., Niebyl, J. R., and Ishizaka, K., 1985, Morphological and immunological characterization of human basophils developed in cultures of cord blood mononuclear cells, *J. Immunol.* **134:**532–540.

Jarrett, E. E. E., and Miller, H. R. P., 1982, Production and activities of IgE in helminth infections. *Prog. Allergy* **31:**178–233.

Jones, T. D., and Mote, J. R., 1934, The phases of foreign protein sensitization in human beings, *N. Engl. J. Med.* **210:**120–123.

Jones, V. O., Rothwell, T. L. W., Dineen, J. K., and Griffiths, D. A., 1974, Studies on the role of histamine and 5-hydroxytryptamine in immunity against the nematode *Trichostrongylus colubriformis.* III. Amine levels in the intestine of infected guinea pigs. *Int. Arch. Allergy Appl. Immunol.* **64:**14–17.

Jong, E. C., Mahmoud, A. A., and Klebanoff, S. J., 1981, Peroxidase-mediated toxicity to schistosomula of *Schistosoma mansoni*, *J. Immunol.* **126:**468–471.

Kagey-Sobotka, A., MacGlashan, D. W., and Lichtenstein, L. M., 1982, Role of receptor aggregation in triggering IgE-mediated reactions, *Fed. Proc.* **41:**12–16.

Kaliner, M. A., 1980, Is a mast cell a mast cell a mast cell?, *J. Allergy Clin. Immunol.* **66:**1–4.

Katayama, I., and Nishioka, K., 1981, Suppression of contact sensitivity by IgG_1 antihapten antibody in contact-sensitized guinea pigs, *J. Invest. Dermatol.* **76:**197–201.

Katayama, I., Nishioka, K., Nishida, K., and Doi, T., 1982, Non-IgG_1 nature of cutaneous basophil hypersensitivity factor in contact sensitivity, *Int. Arch. Allergy Appl. Immunol.* **68:**164–168.

Katayama, I., Nishioka, K., Nishida, K., and Doi, T., 1984, Non-IgG_1 nature of cutaneous basophil hypersensitivity factor in contact sensitivity, *Int. Arch. Allergy Appl. Immunol.* **73:**289–295.

Katz, S. I., Heather, C. J., Parker, D., and Turk, J. L., 1974, Basophilic leukocytes in delayed hypersensitivity reactions, *J. Immunol.* **113:**1073–1078.

Kay, A. B., and Austen, K. F., 1972, Chemotaxis of human basophil leucocytes, *Clin. Exp. Immunol.* **11:**557–563.

Kazura, J. W., Fanning, M. M., Beumer, J. L., and Mahmoud, A. A., 1981, Role of cell-generated hydrogen peroxide in granulocyte-mediated killing of schistosomula of *Schistosoma mansoni* in vitro, *J. Clin. Invest.* **67:**93–102.

Keller, R., and Ogilvie, B. M., 1972, The effects of drugs on worm expulsion in the *Nippostrongylus brasiliensis* infected rat: A discussion of the interpretation of drug action, *Parasitology* **64**:217–227.

Kelly, J. D., and Dineen, J. K., 1972, The suppression of rejection of *Nippostrongylus brasiliensis* in the rat by promethazine hydrochloride: The site of immunological impairment. *Immunology* **22**:361–370.

Kitamura, Y., Hatanaka, K., Murakami, M., and Shibata, H., 1979a, Presence of mast cell precursors in peripheral blood of mice demonstrated by parabiosis, *Blood* **53**:1085–1088.

Kitamura, Y., Shimada, M., Go, S., Matsuda, H., Hatanaka, K., and Seki, M., 1979b, Distribution of mast-cell precursors in hematopoietic and lymphopoietic tissues of mice, *J. Exp. Med.* **150**:482–490.

Kitamura, Y., Yokoyama, M., Matsuda, H., Ohno, T., and Mori, K. F., 1981, Spleen colony-forming cell as common precursor for tissue mast cells and granulocytes, *Nature (London)* **291**:159–160.

Kops, S. K., Ratzlaff, R. E., and Askenase, P. W., 1984a, Mast cell rosettes mediated by an antigen-specific T-cell factor, *Fed. Proc.* **43**:1973a.

Kops, S. K., Van Loveren, H., Rosenstein, R. W., Ptak, W., and Askenase, P. W., 1984b, Mast cell activation and vascular alterations in immediate hypersensitivity-like reactions induced by a T cell-derived antigen-binding factor, *Lab. Invest.* **50**:421–434.

Landsteiner, K., and Chase, M. N., 1942, Experiments on transfer of cutaneous sensitivity to simple chemical compounds, *Proc. Soc. Exp. Biol. Med.* **49**:688–690.

Leonard, E. J., Lett-Brown, M. A., and Askenase, P. W., 1979, Simultaneous generation of tuberculin-type and cutaneous basophilic hypersensitivity at separate sites in the guinea pig, *Int. Arch. Allergy Appl. Immunol.* **58**:460–469.

Lett-Brown, M. A., Aelvoet, M., Hooks, J. J., Georgiades, J. A., Thueson, D. O., and Grant, J. A., 1981, Enhancement of basophil chemotaxis *in vitro* by virus induced interferon, *J. Clin. Invest.* **67**:547–552.

Lett-Brown, M. A., Thueson, D. O., and Grant, J. A., 1982, Dissociation between cutaneous basophil hypersensitivity and basophil sensitization. *J. Immunol.* **129**:1262–1266.

Lett-Brown, M. A., Mitchell, E. B., and Askenase, P. W., 1983, Sensitization of circulating basophils in guinea pig recipients of passive transfer of cutaneous basophil hypersensitivity (CBH) with immune serum: Antigen-specific histamine release *in vitro*, *J. Immunol.* **130**:2104–2108.

Lewis, R. A., and Austen, K. F., 1981, Mediators of local homeostasis and inflammation by leukotrienes and other mast cell-dependent compounds, *Nature (London)* **293**:103–108.

McLaren, D. J., Worms, M. J., and Askenase, P. W., 1983, Cutaneous basophil associated resistance to ectoparasites (ticks). III. Electron microscopy of *Rhipicephalus appendiculatus* larval feeding sites in actively sensitized guinea pigs and recipients of immune serum, *J. Pathol.* **139**:291–308.

Martins, A. B., and Raffel, S., 1964, Cellular activities in hypersensitive reactions. I. Comparative cytology of delayed, "Jones–Mote" and Arthus reactions, *J. Immunol.* **93**:937–947.

Maximow, A., 1905, Über die Zellformen des lockeren Bindegewebes, *Arch. Mikrosk. Anat. Entwicklungsmech.* **67**:680–757.

Metcalfe, D. D., Kaliner, M., and Donlon, M. A., 1981b, The mast cell, *CRC Crit. Rev. Immunol.* **3**:23–74.

Michels, N. A., 1938, The mast cells, in: *Handbook of Haematology*, Volume 1 (H. Downey, ed.), pp. 232–372, Hoeber, New York. Reprinted in *Ann. N.Y. Acad. Sci.* **103**:Appendix, 1–372 (1963).

Mitchell, E. B., and Askenase, P. W., 1982, Suppression of T cell mediated cutaneous basophil hypersensitivity by serum from guinea pigs immunized with mycobacterial adjuvant, *J. Exp. Med.* **156**:159–172.

Mitchell, E. B., and Askenase, P. W., 1983, Basophils in human disease, *Clin. Rev. Allergy* **1**:427–448.

Mitchell, E. B., Brown, S. J., and Askenase, P. W., 1982, IgG_1 antibody dependent mediator release following passive systemic sensitization of basophils arriving at cutaneous basophil hypersensitivity reactions, *J. Immunol.* **129**:1663–1669.

Nabel, G., Galli, S. J., Dvorak, A. M., Dvorak, H. F., and Cantor, H., 1981, Inducer T lymphocytes synthesize a factor that stimulates proliferation of cloned mast cells, *Nature (London)* **291**:332–334.

Nakahata, T., Spicer, S. S., Cantey, J. R., and Ogawa, M., 1982, Clonal assay of mouse mast cell colonies in methylcellulose, *Blood* **60**:352–361.

Norton, W., and Ziff, M., 1965, Electron microscopic observations on the localization of antigen in the tuberculin reaction, *Immunology* **8**:81–87.
Ogawa, M., Nakahata, T., Leary, A. G., Sterk, A. R., Ishizaka, K., and Ishizaka, T., 1983, Suspension culture of human mast cells/basophils from umbilical cord blood mononuclear cells, *Proc. Natl. Acad. Sci. USA* **80**:4494–4498.
Orenstein, N. S., Galli, S. J., Dvorak, A. M., Silbert, J. E., and Dvorak, H. F., 1978, Sulfated glycosaminoglycans of guinea pig basophilic leukocytes, *J. Immunol.* **121**:586–592.
Orida, N., Feldman, J. D., Katz, D. H., and Liu, F.-T., 1983, IgE-mediated chemotaxis of rat basophilic leukemia cells towards specific antigen, *J. Exp. Med.* **157**:2166–2171.
Ovary, Z., Benacerraf, B., and Block, K. J., 1963, Properties of guinea pig 7S antibodies. II. Identification of antibodies involved in passive cutaneous and systemic anaphylaxis, *J. Exp. Med.* **117**:951–964.
Parker, D., Katz, S. I., and Turk, J. L., 1974, Demonstration of suppressor cells in delayed hypersensitivity by B lymphocytes, *Int. Arch. Allergy Appl. Immunol.* **49**:276–280.
Plaut, M., and Lichtenstein, L. M., 1982, Histamine and immune responses, in: *Pharmacology of Histamine Receptors* (C. R. Ganellin and M. E. Parsons, eds), pp. 392–435, Wright, London.
Pretlow, T. G., and Cassady, I. M., 1970, Separation of mast cells in successive stages of differentiation using programmed gradient sedimentation, *Am. J. Pathol.* **61**:323–338.
Ptak, W., Askenase, P. W., Rosenstein, R. W., and Gershon, R. K., 1982, Transfer of an antigen specific immediate hypersensitivity-like reaction with an antigen binding factor produced by T-cells, *Proc. Natl. Acad. Sci. USA* **79**:1969–1973.
Raffel, S., and Newel, J. M., 1958, The delayed hypersensitivity induced by antigen–antibody complexes, *J. Exp. Med.* **108**:823–841.
Ratzlaff, R. E., Meade, R., and Askenase, P. W., 1984, An immune aggregate mechanism in mast cell activation by an antigen-binding T cell factor, *Fed. Proc.* **43**:1626a.
Razin, E., Stevens, R. L., Akiyama, F., Schmid, K., and Austen, K. F., 1982, Culture from mouse bone marrow of a subclass of mast cells possessing a distinct chondroitin sulfate proteoglycan with glycosaminoglycans rich in N-acetylgalactosamine-4,6-disulfate, *J. Biol. Chem.* **257**:7229–7236.
Razin, E., Ihle, J. N., Seldin, D., Mencia-Heurta, J.-M., Katz, H. R., Leblanc, P. A., Hein, A., Caulfield, J. P., Austen, K. F., and Stevens, R. L., 1984, Interleukin 3: A differentiation and growth factor for the mouse mast cell that contains chondroitin sulfate E proteoglycan, *J. Immunol.* **132**:1479–1486.
Rebuck, J. W., 1958, Technic for the study of leukocytic functions in man, *Methods Med. Res.* **7**:161–164.
Richerson, H. B., Dvorak, H. F., and Leskowitz, S., 1969, Cutaneous basophilic hypersensitivity: A new interpretation of the Jones–Mote reaction, *J. Immunol.* **103**:1431–1434.
Richerson, H. B., Dvorak, H. F., and Leskowitz, S., 1970, Cutaneous basophilic hypersensitivity. I. A new look at the Jones–Mote reaction, general characteristics, *J. Exp. Med.* **132**:546–551.
Rinogoen, A. R., 1923, The mast leukocytes in the adult guinea-pig under experimental conditions, *Am. J. Anat.* **31**:319–337.
Rocklin, R. E., and Beer, D. J., 1983, Histamine and immune modulation, *Adv. Intern. Med.* **28**:225–251.
Rothwell, T. L. W., 1975, Studies of the responses of basophil and eosinophil leukocytes and mast cells to the nematode *Trichostrongylus colubriformis*. I. Observations during the expulsion of first and second infections by guinea pigs, *J. Pathol.* **116**:51–60.
Rothwell, T. L. W., and Dineen, J. K., 1972, Cellular reactions in guinea pigs following primary and challenge infection with *Trichostrongylus colubriformis* with special reference to the roles played by eosinophils and basophils in rejection of the parasite, *Immunology* **22**:733–745.
Rothwell, T. L. W., and Huxtable, C. R., 1976, Basophil leukocytes in cutaneous hypersensitivity reactions in nematode-infected guinea pigs, *Aust. J. Exp. Biol. Med. Sci.* **54**:329–335.
Rothwell, T. L. W., and Love, R. J., 1975, Studies on the response of basophil and eosinophil leukocytes and mast cells to the nematode *Trichostrongylus colubriformis*. II. Changes in cell numbers following infection of thymectomized and adaptively or passively immunized guinea pigs, *J. Pathol.* **116**:183–194.

Rothwell, T. L. W., Prichard, R. K., and Love, R. J., 1974, Studies on the role of histamine and 5-hydroxytryptamine in immunity against the nematode *Trichostrongylus colubriformis*. I. *In vivo* and *in vitro* effects of the amines, *Int. Arch. Allergy Appl. Immunol.* **46**:1–13.
Samuelsson, B., 1983, Leukotrienes: Mediators of immediate hypersensitivity reactions and inflammation; *Science* **220**:568–575.
Schrader, J. W., Lewis, S. J., Clark-Lewis, I., and Culvenor, J. G., 1981, The persisting (P) cell: Histamine content, regulation by a T cell-derived factor, origin from a bone marrow precursor, and relationship to mast cells. *Proc. Natl. Acad. Sci. USA* **78**:323–327.
Schrader, J. W., Clark-Lewis, I., Crapper, R. M., and Wong, G. W., 1983, P-cell stimulating factor: Characterization, action on multiple lineages of bone-marrow-derived cells and role in oncogenesis, *Immunol. Rev.* **76**:79–104.
Schwartz, L. B., and Austen, K. F., 1984, Structure and function of the chemical mediators of mast cells, *Prog. Allergy* **34**:271–321.
Selye, H., 1965, *The Mast Cells*, Butterworths, Washington, D.C.
Siegel, J. N., Schwartz, A., Askenase, P. W., and Gershon, R. K., 1982, T-cell suppression and contrasuppression induced by histamine H_2 and H_1 agonists, respectively, *Proc. Natl. Acad. Sci. USA* **79**:5052–5056.
Siraganian, R. P., McGivney, A., Barsumian, E. L., Crews, F. T., Hirata, F., and Axelrod, J., 1982, Variants of the rat basophilic leukemia cell line for the study of histamine release, *Fed. Proc.* **41**:30–34.
Sobel, R. A., Hauzakos, J. L., Blanchette, B. W., and Colvin, R. B., 1984, Anti-T cell monoclonal antibodies (MA) in vivo: Inhibition of delayed hypersensitivity (DH) but not cutaneous basophil hypersensitivity (CBH) in guinea pigs (GP), *Fed. Proc.* **43**:1804a.
Sobotka, A. K., Malveaux, F. J., Marone, G., Thomas, L. L., and Lichtenstein, L. M., 1978, IgE-mediated basophil phenomena: quantitation, control, inflammatory interactions, *Immunol. Rev.* **41**:171–185.
Sonoda, T., Ohno, T., and Kitamura, Y., 1982, Concentration of mast-cell progenitors in bone marrow, spleen, and blood of mice determined by limiting dilution analysis, *J. Cell. Physiol.* **112**:136–140.
Sonoda, T., Kanayama, Y., Hara, H., Hayashi, C., Tadokoro, M., Yonezawa, T., and Kitamura, Y., 1984, Proliferation of peritoneal mast cells in the skin of W/W^v mice that genetically lack mast cells, *J. Exp. Med.* **160**:138–151.
Sredni, B., Friedman, M. M., Bland, C. E., and Metcalfe, D. D., 1983, Ultrastructural, biochemical and functional characteristics of histamine-containing cells cloned from mouse bone marrow: Tentative identification as mucosal mast cells. *J. Immunol.* **131**:915–922.
Stadecker, M. J., and Leskowitz, S., 1976a, The inhibition of basophil rich delayed skin reactions by a heterologous anti guinea pig T cell serum, *J. Immunol.* **116**:1646–1651.
Stadecker, M. J., and Leskowitz, S., 1976b, Genetic control of cutaneous basophil hypersensitivity, *J. Exp. Med.* **143**:206–210.
Stashenko, R. P., Bhan, A. K., Schlossman, S. F., and McCluskey, R. T., 1977, Local transfer of delayed hypersensitivity and cutaneous basophil hypersensitivity, *J. Immunol.* **119**:1987–1993.
Tadokoro, K., Stadler, B. M., and de Weck, A. L., 1983, Factor-dependent growth of human bone-marrow-derived basophil-like cells, *J. Exp. Med.* **158**:857–871.
Tamir, H., Theoharides, T. C., Gershon, M. D., and Askenase, P. W., 1982, Serotonin storage pools in basophil leukemia and mast cells: Characterization of two types of serotonin binding protein and radioautographic analysis of the intracellular distribution of (^{3}H) serotonin, *J. Cell Biol.* **93**:638–647.
Tas, J., and Berndsen, R. G., 1977, Does heparin occur in mucosal mast cells of the rat small intestine?, *J. Histochem. Cytochem.* **25**:1058–1062.
Thueson, D. O., Speck, L. S., Lett-Brown, M. A., and Grant, J. A., 1979a, Histamine-releasing activity (HRA). I. Production by mitogen- or antigen-stimulated human mononuclear cells, *J. Immunol.* **123**:626–632.
Thueson, D. O., Speck, L. S., Lett-Brown, M. A., and Grant, J. A., 1979b, Histamine-releasing activity (HRA). II. Interaction with basophils and physiocochemical characterization, *J. Immunol.* **123**:633–639.

Turk, J. L., and Parker, D., 1973, Further studies on B-lymphocyte suppression in delayed hypersensitivity indicating a possible mechanism for Jones–Mote hypersensitivity, *Immunology* **24:**751–758.

Turk, J. L., and Parker, D., 1974, Modulation of T-lymphocyte function by B lymphocytes in delayed hypersensitivity, *Int. Arch. Allergy Appl. Immunol.* **49:**241–246.

Turk, J. L., and Parker, D., 1983, The effect of cyclophosphamide on immunological memory, *Immunopharmacology* **6:**123–132.

Urbina, C., Ortiz, C., and Hurtado, I., 1981, A new look at basophils in mice, *Int. Arch. Allergy Appl. Immunol.* **66:**158–160.

Waksman, B. M., 1960, in: *Cellular Aspects of Immunity* (G. E. W. Wolstenholme and M. O'Connor, eds.), pp. 280–322, Little, Brown, Boston.

Ward, P. A., Dvorak, H. F., Cohen, S., Yoshida, T., Data, R., and Selvaggio, S., 1975, Chemotaxis of basophils by lymphocyte-dependent and lymphocyte-independent mechanisms, *J. Immunol.* **114:**1523–1531.

Wiener, J., Spiro, D., and Zunker, H. O., 1965, A cellular study of tuberculin sensitivity, *Am. J. Pathol.* **47:**723–763.

Wikel, S. K., and Allen, J. R., 1976, Acquired resistance to ticks. I. Passive transfer of resistance, *Immunology* **30:**311–326.

Willms-Kretschmer, K., Flax, M. H., and Cotran, R. S., 1967, The fine structure of the vascular response in hapten-specific delayed hypersensitivity and contact dermatitis, *Lab. Invest.* **17:**334–349.

Winqvist, G., 1960, Experimental production of basophil granulocytes in the guinea pig, *Exp. Cell Res.* **19:**7–12.

Winslow, C. M., and Austen, K. F., 1984, Role of cyclic nucleotides in the activation–secretion response, *Prog. Allergy* **34:**236–270.

Wolf-Jurgensen, P., 1965, Basophilic leukocytes in experimental tuberculin reactions studied by the skin-window technique, *Acta Allergol.* **20:**38–46.

Wolf-Jurgensen, P., 1967, Basophilic leukocytes in delayed hypersensitivity: Experimental studies in man using the skin window technique, *Dan. Med. Bull.* **14:**249–253.

Woodbury, R. G., Gruzenski, G. M., and Lagunoff, D., 1978, Immunofluorescent localization of a serine protease in rat small intestine, *Proc. Natl. Acad. Sci. USA* **75:**2785–2789.

Wyllie, J. C., More, T. H., and Haust, M. D., 1964, Electron microscopy of epidermal lesions elicited during delayed hypersensitivity, *Lab. Invest.* **13:**137–151.

Yokota, T., Frank, L., Rennick, D., Hall, C., Arai, N., Mosmann, T., Nabel, G., Cantor, H., and Arai, K.-I., 1984, Isolation and characterization of a mouse cDNA clone that expresses mast-cell growth-factor activity in monkey cells, *Proc. Natl. Acad. Sci. USA* **81:**1070–1074.

Yoshida, S., Nomoto, K., and Himeno, K., 1979, Dissociation of erythema and basophil accumulation in Jones–Mote type hypersensitivity in the guinea pig, *Br. J. Exp. Pathol.* **60:**596–603.

Yung, Y.-P., Eger, R., Tertian, G., and Moore, M. A. S., 1981, Long-term *in vitro* culture of murine mast cells. II. Purification of a mast cell growth factor and its dissociation from TCGF, *J. Immunol.* **127:**794–799.

Yurt, R. W., Leid, R. W., Austen, K. F., and Silbert, J. E., 1977, Native heparin from rat peritoneal mast cells, *J. Biol. Chem.* **252:**518–521.

Zinsser, H., 1921, Studies on the tuberculin reaction and on specific hypersensitiveness in bacterial infection, *J. Exp. Med.* **34:**495–524.

Zucker-Franklin, D., 1980, Ultrastructural evidence for the common origin of human mast cells and basophils, *Blood* **56:**534–540.

Zucker-Franklin, D., Grusky, G., Hirayama, N., and Schnipper, E., 1981, The presence of mast cell precursors in rat peripheral blood, *Blood* **58:**544–551.

13

The Nephritogenic Immune Response and Renal Immunopathogenesis
A Progress Report

MICHAEL D. CLAYMAN, CARL S. GOLDSTEIN, and ERIC G. NEILSON

1. INTRODUCTION

It is generally agreed that the immune system plays a central role in many forms of renal injury. Hallmark studies of Dixon *et al.* (1961) emphasized the importance of antibodies and immune complex deposition in glomerular disease processes. A wealth of data have subsequently accumulated which lend support to the concept that humoral immunity is an intrinsic component of many nephritogenic immune responses. The conceptual framework for understanding renal immunology, as a result, has been largely defined in terms of antibody production, renal deposition, and subsequent immune amplification processes.

In this review, we choose to provide a slightly broader perspective on basic immunologic mechanisms as they might relate to renal disease; one in which antibody and its subsequent interactions are seen as only one expression of a pleiomorphic system. We furthermore propose that such immune responses can be viewed as a profile divided into three arbitrary phases as outlined in Table 1. The first phase of this profile is the afferent, induction, or antigen-recognition limb which initiates the nephritogenic immune response. This induction process is dependent on several variables, including the molecular form and route of antigen administration (Waksman, 1979), as well as the genetically determined capacity to respond to the antigen. In many cases a response to self-alloantigens must also be considered, and this is an area which has undergone a conceptual

MICHAEL D. CLAYMAN, CARL S. GOLDSTEIN, and ERIC G. NEILSON • Renal Electrolyte Section, Department of Medicine, University of Pennsylvania School of Medicine, Philadelphia, Pennsylvania 19104.

TABLE 1. THE NEPHRITOGENIC IMMUNE RESPONSE PROFILE

Antigen-recognition	Immunoregulation	Effector limb
A. Antigen 1. Size 2. Form (including vehicle) 3. Route of administration 4. Self as antigen a. Self-tolerance and autoimmunity b. MHC gene products c. Idiotypic network B. Immune response genes	A. Suppressor T cells and their factors B. Idiotypic network C. Pharmacologic mediators D. Other—nutritional status, systemic illness	A. Humoral response 1. Tissue-specific antibody 2. Immune complexes B. Cell-mediated response 1. Cytotoxic T cells 2. Delayed-type hypersensitivity T cells 3. Monocytes/macrophages 4. Polymorphonuclear leukocytes C. Complement

revolution in the past decade. While accepting the importance of tolerance to self, and appreciating that its abrogation is permissive for many autoimmune diseases, data from many sources suggest that the immune system necessarily and routinely recognizes some self-antigens in the form of major histocompatibility complex (MHC) gene products and idiotypic determinants. Further, this recognition of self involves accessory cells which function according to immunogenetically defined rules. Thus, the induction of the nephritogenic immune response has an important dependence on and an interface with the RES.

Immunoregulation, the second phase, is concerned with the modulation of already initiated nephritogenic immune responses. Its function depends on cell-mediated interactions, soluble lymphokines, and anti-idiotypic antibodies in addition to the well-documented but poorly understood effects of nutritional and hormonal changes (Neilson and Phillips, 1980). The idiotypic network has also emerged as an important feature of immune regulation (Jerne, 1974), and studies in this area have documented that its function can affect the nephritogenic immune response. Suppression by T cells and the regulatory effects of lymphokines, and other soluble mediators, seem to have an important role as well.

The third and final limb of the nephritogenic immune response profile is that of the effector pathways producing local renal injury. In addition to the well-described and now-classic roles for antibody, recent data also suggest that certain disease states are primarily dependent on T cells and/or monocytes for their pathologic evolution. The potential effects of soluble histokines have also been under investigative evaluation as new mediators of nephritis.

In summary, then, this review will be structured to include the above components of the nephritogenic immune response profile. Such a profile can serve both as a conceptual framework and as an integrator of the varied facets of renal immunopathogenesis. Because the fine immunologic details have not been completely elucidated for every form of renal injury, this review can only serve as a progress report.

2. ANTIGEN-RECOGNITION, IMMUNOGENETICS, AND THE AFFERENT PHASE OF THE NEPHRITOGENIC IMMUNE RESPONSE

2.1. BASIC CONSIDERATIONS

One hallmark of the immune response is its capability for exquisite specificity. This, in general, is dictated by T- and B-lymphocyte surface receptors which process antigenic determinants having maximal complementarity or "fit" with the binding portion of receptor molecules. These lymphocytes each express a receptor for only one antigenic specificity, and cell activation by antigen results in clonal proliferation and the differentiation of other cells which will eventually effect a humoral or cell-mediated immune response. While the immune system is capable of distinguishing molecules that differ in very small ways, and while proteins, lipids, and carbohydrates can all elicit an immune response, only molecules above a certain minimum size can serve as an immunogen.

Complete antigens are distinguishable from haptens. Whereas the former are defined by their ability to initiate an immune response, the latter (from the Greek *haptein*, to grasp) while capable of combining specifically with the antigen-binding region of immune receptors, do not meet the minimum size to qualify as immunogens. This is because most complex immune responses require T-cell activation, and hapten determinants are simply not large enough to stimulate T cells. In order to generate an antihapten immune response, the hapten must be physically linked to a carrier which serves to initiate cooperative T-cell function. The logical corollary of the hapten–carrier description is that compounds large enough to act as complete antigens must be perceived by the immune system as possessing one or more carrier determinants (in addition to any smaller hapten portions).

It should be noted that not all immunogens require T-cell cooperation to effect an immune response. T-independent antigens, most commonly of repeating polysaccharide subunits (e.g., pneumococcal polysaccharide), can only stimulate B-cell proliferation and their differentiation to IgM-secreting plasma cells. While not completely understood, it seems that T-independent antigens can activate B cells by virtue of multiple identical subunits which may allow a single molecule to simultaneously occupy numerous receptor sites on one B cell. Of interest, T-independent antigens are associated with IgM antibody that attends the primary immune response. T-dependent antigens, on the other hand, stimulate an IgM response followed by an IgG response which represents "switching" from IgM to IgG production by cells which have undergone clonal proliferation. It appears that T-cell cooperation is required for the IgM-to-IgG switch; without this, T-independent antigens are destined to elicit only IgM responses.

The macrophage plays an important role in the afferent phase of the immune response, as this cell is required to process antigen and "present" it in an appropriate form to T cells. Without these accessory cells it is very difficult to elicit a T-dependent response. It should be noted that recent data on humans (van Voorhis *et al.*, 1983) call into question the notion of the antigen-presenting cell being of monocyte/macrophage origin. An Fc-receptor-negative, esterase-negative cell with distinctive parenchymal distribution which can be distinguished from monocytes/macrophages by monoclonal antibodies and glass adherence properties has been named the dendritic cell and would also appear to be the primary antigen-presenting cell.

Other significant variables associated with the antigen-recognition limb of the immune response include the form in which antigen is introduced into the system. Whereas particulate or aggregated compounds are highly immunogenic, in purely soluble form these same compounds can induce immune paralysis, or tolerance, specific for that antigen. This is seen, for example, in the case of bovine serum albumin, where removal of aggregated molecules from a suspension transforms an antigen into a tolerogen. The basis for this is not completely understood, but it appears that macrophages more readily ingest and process aggregated compounds; either suboptimal or deficient processing of antigen by macrophages may contribute to the induction of tolerance by soluble antigen.

Various vehicles have also been used to enhance the immunogenicity of experimentally administered antigens. The most common of these is Freund's adjuvant (primarily paraffin oil), either with PPD (referred to as complete) or without (incomplete). The route of antigen administration plays an important role in the subsequent immune response. Whereas intravenous administration tends to elicit tolerance, subcutaneous or intraperitoneal injections are usually immunogenic (Waksman, 1979). In addition to form, vehicle, and site of administration, quantity of antigen is an integral variable in the immune response equation. Exceedingly low or high doses of antigen can result in specific tolerance. Hence, one might expect a maximal immune response using an appropriately adjusted dose of particulate antigen in Freund's complete adjuvant administered subcutaneously or intraperitoneally.

Having provided a preliminary background regarding the immune response to exogenous antigen, we can now focus on self-antigens. Here, tolerance plays an obvious critical role as the organism's general well-being is dependent on unresponsiveness to the vast majority of self-antigens. Mechanisms involved in self-tolerance are not completely delineated, but may temporally depend on exposure of the immature immune system to self-antigen, and functionally involve clonal deletion (of B and/or T lymphocytes), active T-cell suppression, and/or receptor blockade (Nossal and Pike, 1975). It has become more clear in the past 10 years, however, that immune responses to self not only occur regularly, but are of physiologic import and are required for the expression of a vast array of immunologic functions. At the present time, the concept of self-recognition relates to two broad categories of antigen: MHC determinants and idiotypes. Interestingly, the pivotal proposals regarding these were made within a year of each other in 1973 (Rosenthal and Shevach) and 1974 (Jerne).

Regarding MHC determinants, it would appear that the organism's ability to recognize self is intimately linked to the simultaneous recognition of foreign or exogenous antigens. Thus, the antigen responsiveness of helper T lymphocytes is dependent on antigen being seen in association with MHC determinants on the surface of antigen-presenting cells (e.g., macrophages or dendritic cells) (McDevitt *et al.*, 1972; Rosenthal and Shevach, 1973). Models to explain T-cell recognition of self and exogenous antigen invoke hypotheses of either dual recognition of antigen and MHC determinants, or single receptor recognition of a neoantigen comprised of foreign antigen and MHC product, also referred to as altered-self. A recently proposed single receptor hypothesis alternatively suggests that appropriate T-cell clones possess a receptor with affinity for nominal antigen and also for MHC antigen, and that the T-lymphocyte response is dependent on its recognition of a critical number of antigen and/or MHC determinants (Siliciano *et al.*, 1983).

The requirement for recognition of appropriate MHC products has been deemed restrictive and requires antigen interaction with class II MHC determinants (*vide infra*) for accessory cell presentation to T-helper lymphocytes, and for one pathway of cooperation between helper T cells and B cells. Optimal target cell killing by cytotoxic T lymphocytes, on the other hand, depends on antigen-recognition in the context of class I (or serologically defined) determinants (Doherty and Zinkernagel, 1974). Thus, it would appear that the immune system not only routinely recognizes itself, but must do so in order to function optimally.

In 1974, Jerne proposed that self-recognition of idiotype (those unique protein determinants intimately related to the antigen-binding site of immunoglobulins and T-cell receptors) also occurs routinely and that the anti-idiotypic response involves both B and T cells. The stimulus to the system occurs when idiotypes peculiar to a particular antigenic response increase in concentration in the course of a response to that antigen. Once a critical-threshold value is reached, an immune response is elicited to idiotype, an "antigen" which had previously existed at such low levels as to be functionally hidden. The anti-idiotypic response, particularly at the level of idiotype-specific T lymphocytes, is generally a down-regulating influence which appears to perform an important immunoregulatory function in experimental animal systems (Neilson and Zakheim, 1983) and in humans (Geha, 1983). In summary, then, it appears that MHC molecules and idiotypes are two groups of determinants which modulate the course of an immune response, and that this self-recognition has important implications for induction, effector, and immunoregulatory limbs of the immune system.

Because the ability to respond to specific antigen is genetically determined, it is appropriate to consider, under the heading of induction, the area of immunogenetics. Not unlike the conceptual revolutions attendant to the description of MHC-restricted responses and the idiotypic network, the understanding of immunogenetics has recently advanced dramatically, particularly with the utilization of gene cloning techniques and congenic and/or mutant inbred mouse strains. While a detailed description of the structure and function of genes related to immune function is beyond the scope of this chapter, several excellent reviews have been published recently (Hood *et al.*, 1983; Klein *et al.*, 1983; Dorf, 1981). What follows is an overview of an area which seems to be playing an increasingly important role in developing immunopathogenesis.

The genetically determined ability to respond to certain antigens was first described over 20 years ago by Levine *et al.* (1963) when they showed that the response to a particular polymer (poly-L-lysine) occurred in one guinea pig strain but not in another. Subsequently, it was shown that genetically determined antigen responsiveness was linked to the MHC (McDevitt and Chinitz, 1969). The MHC derives its name from the fact that it encodes so-called transplantation antigens, which provide the strongest barriers to tissue transplantation across strain and species lines. It is referred to by different names, according to mammalian species, as follows: human—HLA (human leukocyte antigen) on chromosome 6; rat—AgB, H-1, and, currently, RT1; and mouse—H-2 on chromosome 17.

The chromosomal segment encompassing H-2 has been elucidated in great detail and is presented in schematic form in Fig. 1A, along with HLA for reasons of comparison. Class I, or serologically defined antigens, correspond to K and D regions of H-2 and A and B regions of HLA. Gene products of these regions are expressed on the surface of most cells and are associatively recognized with nominal antigen in target cell killing by T lymphocytes. Class II molecules are encoded for by genes in the I region of H-2 and the D region of HLA (note that class I/class II organization of HLA is actually typical of most mammalian species; H-2 appears to have undergone an extra-MHC rearrangement such that the class II genes are sandwiched between the two class I loci). Class II gene products are found on macrophages, dendritic cells, B cells, and activated T cells (in humans), and on some tissues of nonlymphoid origin. These antigens serve as restriction elements in T-helper cell–antigen-presenting cell and T-helper cell–B cell interactions as discussed above. Proteins encoded by the genes in the I region appear also to dictate the ability of T lymphocytes to recognize certain foreign antigens, and consequently responder or nonresponder phenotype. These genes are referred to as immune response (Ir) genes and their products as I-region-associated (Ia) antigens.

Using the above discussion of the afferent phase of the immune response as background, an integration of this area with different facets of experimental and human renal immunopathogenesis can now be presented.

2.2. ANTIGEN

2.2.1. Introduction

As one might anticipate, the afferent phase of the nephritogenic immune response, particularly as it relates to antigen characterization, exists at a relatively primitive level of understanding in human disease. This is due to the fact that, for most human immune renal diseases, the responsible antigen has not been isolated, and further, ethical constraints preclude deliberate challenge with suspected antigen. Poststreptococcal glomerulonephritis is one human renal disease which might lend itself to a detailed afferent phase evaluation (Fillit *et al.*, 1983), but even here observations on potential nephritogenic antigens are preliminary (see Section 22.2.). In contrast, animal studies allow a more detailed analysis of afferent phase immunopathogenesis, and provide for the possibility

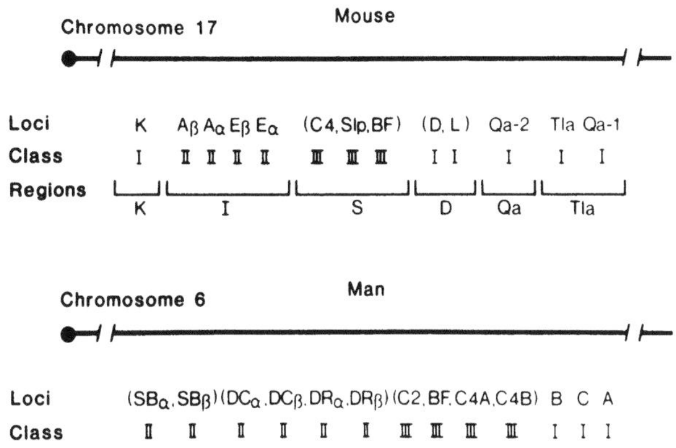

FIGURE 1. Schematic representation of the major histocompatibility complex genes in mouse and man. Parentheses enclose loci of unknown order. See text for further explanation. (From Steinmetz and Hood, 1983.)

that responsible antigens may eventually be isolated in pure form (Wilson, 1983). What follows is a selected discussion of some nephritogenic antigens, with an emphasis on experimental models of disease induction, commensurate with the large proportion of data in nonhuman systems (see Table 2). Observations on antigens relevant to human disease will be integrated where appropriate.

2.2.2. Glomerular Disease Associated with an Immune Response to Antigens of Renal Origin

2.2.2a. Glomerular Basement Membrane (GBM) Antigens. For over 30 years it has been known that the GMB contains antigens intimately involved in experimental models of renal disease (Krakower and Greenspon, 1951). Nephrotoxic serum nephritis (earlier called Masugi nephritis) refers to an animal model of disease which entails passive transfer of heteroantisera containing antibodies with anti-GBM specificity. Disease occurs in two phases, depending on the quantity of anti-GBM antibodies administered. The immediate (heterologous) phase involves binding of transferred anti-GBM antibodies to the glomerulus. If enough anti-GBM antibody is bound [75 μg/g kidney in rat (Unanue and Dixon, 1965), 15 μg/g kidney in rabbit (Unanue *et al.*, 1967), and 5 μg/g kidney in sheep (Lerner and Dixon, 1966)], then a proliferative lesion of the glomerulus develops with destruction of the GBM. Should less than requisite quantities of anti-GBM antibody be bound initially, a nephrotoxic insult can still be imparted in the autologous phase of the disease. Here, heterogenic anti-GBM immunoglobulin, which is now attached to GBM structures, is recognized by the host as foreign. Consequently, an anti-immunoglobulin antibody response is initiated with targeting of these antibodies to the GBM via the previously deposited anti-GBM immunoglobulin. Linear immunofluorescent staining of the GBM ensues and renal injury occurs as described for the heterologous phase. Hence, nephrotoxic serum nephritis can be thought of as a two-phase disease-state following the administration of heteroantisera containing anti-GBM antibodies.

TABLE 2. ANTIGENS OF THE NEPHRITOGENIC IMMUNE RESPONSE PROFILE

I. Associated with glomerular disease

Glomerular disease (species)	Antigen/origin	Molecular weight/ composition	Immunopathogenesis	Reference
Anti-GBM disease (human)	GBM	53,000; 27,000/noncollagen protein	Autoimmune	Holdsworth *et al.* (1979b)
Heymann's nephritis (rat)	Fx1A; RTE 5; GP330/tubule brush border	330,000/glycoprotein	Autoimmune	Kerjaschki and Farquhar (1982)
Proliferative GN with mesangial immunoglobulin deposition (rat)	Nephritogenoside/glomerulus	Branched oligosaccharide	Autoimmune	Shibata (1978)
Membranous and proliferative GN (rabbit)	Glomerular capillary wall antigen	Not available	Autoimmune	Neale and Wilson (1978), Woodroffe *et al.* (1978)
Proliferative GN, acute phase; membranous GN, chronic phase (rabbit)	Bovine or human serum albumin or gamma globulin	Albumin—68,000; gamma globulin (immunoglobulin)—150,000	Circulating immune complex deposition	Dixon *et al.* (1961)
Immune complex GN (mouse)	DNA	26,000 per monomer (dimer or tetramer)	Antigen-directed *in situ* immune complex deposition	Izui *et al.* (1976, 1977a)

	Con A		Antigen-directed *in situ* immune complex deposition	Golbus and Wilson (1979), Sitrin *et al.* (1982)
Poststreptococcal GN (human)	Endostreptosin/*Streptococcus*	48,000/protein	Antigen-directed *in situ* immune complex deposition	Seligson *et al.* (1982), Lange *et al.* (1983)
	Nephritis strain-associated protein (NSAP)/*Streptococcus*	46,000/protein	? Cross-reactive immune response	Fillit *et al.* (1983)

II. Associated with tubulointerstitial disease

Tubulointerstitial disease (species)	Antigen/origin	Molecular weight/ composition	Immunopathogenesis	Reference
Anti-TBM disease (rat)	TBM	42,000/noncollagen protein	Autoimmune	Zanetti and Wilson (1983)
Anti-TBM disease (mouse)	TBM	30,000/protein	Autoimmune	Wakashin *et al.* (1981)
Immune complex interstitial nephritis (rat)	Tamm–Horsfall protein/thick ascending limb of the loop of Henle	80,000/glycoprotein	*In situ* immune complex deposition	Hoyer (1980), Fletcher *et al.* (1970)

A more recently described model of anti-GBM disease entails active immunization of sheep with human GBM in Freund's complete adjuvant (Steblay, 1962). In this circumstance, an autoimmune anti-GBM response is elicited, and severe anti-GBM glomerulonephritis ensues, with animals dying of end-stage renal disease. Finally, spontaneously occurring anti-GBM disease has been documented in horses (Banks and Henson, 1972). The only other spontaneous autoimmune form of the disease is seen in humans.

In summary, the GBM can clearly act as the focus for an immunologically mediated glomerulonephritis under a wide spectrum of inciting circumstances and in a variety of mammalian species. Concerning the molecular nature of this glomerular structure, biochemical, functional, and immunologic studies have been performed, and provide an emerging picture of the GBM. By electron microscopy (reviewed in Farquhar and Kanwar, 1983), the GBM can be divided into three zones, two relatively electron-lucent layers, the lamina rara interna (immediately adjacent to the glomerular endothelial cells) and the lamina rara externa (applied to the epithelial foot processes), and an intervening middle layer, the lamina densa. Of functional note, the laminae rarae have been demonstrated to be the primary repository of negatively charged molecular determinants (probably sulfated glycosaminoglycans) and are the dominant structural correlate for the permselectivity properties of the glomerulus.

The GBM, on the molecular level (Kefalides, 1983), appears to consist predominantly of two distinguishable species. A procollagen core consisting of the triple-helical hydroxyproline and hydroxylysine-rich collagen molecule with N- and C-terminal nonhelical peptide appendages is surrounded by glycoprotein moieties. (Interestingly, these two species appear to be synthesized and secreted separately with association occurring as a postsecretion event.) These noncollagenous glycoproteins are linked by disulfide bonds. Approximately 75% of the GBM is procollagen, while the remainder is noncollagenous glycoprotein. It is generally contended, as it relates to the specificity of antibodies in both experimental and human forms of anti-GBM disease, that the relevant antigenic properties of the GBM are contained within the glycoprotein moieties. In this context, it is interesting that immunization with whole GBM preparations results in the production of antibodies which usually do not bind to the collagen components of this structure. This is felt to be related to the fact that collagen is hidden within the internal structure of the GBM and hence not available to stimulate an immune response [Fig. 2 from Kefalides (1983) provides a structural representation of procollagen and glycoprotein moieties in the GBM].

A large, noncollagenous glycoprotein component of the GBM has been named laminin (Timpl *et al.*, 1979). When antilaminin antibodies prepared in sheep were used in passive transfer experiments in rats, there was excellent linear immunofluorescent anti-GBM staining, but the pathologic sequelae were less, and clearly distinguishable from classic nephrotoxic serum nephritis models. The possibility that this might reflect deficient quantitative antilaminin antibody deposition has not been excluded. Further distinguishing, however, was the observation that no autologous phase followed the passive transfer of antilaminin antibodies (Abrahamson and Caulfield, 1982).

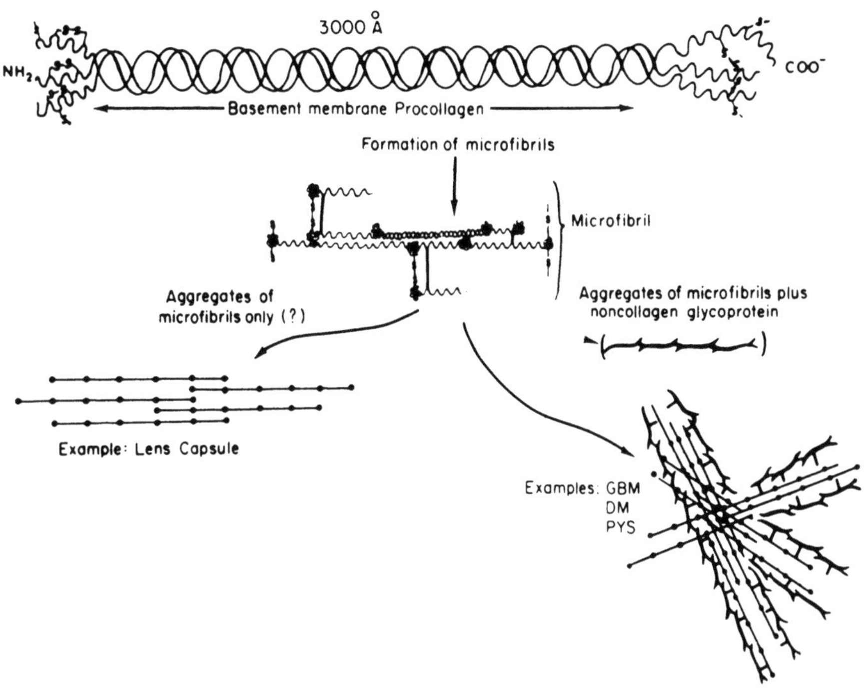

FIGURE 2. Schematic diagram of proposed basement membrane organization. Procollagen (triple-helical collagen core with glycoprotein ends) molecules associate to produce microfibrils. These may exist in disulfide linkage with distinct noncollagenous glycoproteins in the structure of glomerular basement membrane (GBM), Descemet's membrane (DM), and rat embryo parietal yolk sac (PYS). Alternatively, aggregates of these microfibrils without noncollagenous glycoproteins may form the substance of lens capsule. (From Kefalides *et al.*, 1983.)

A high-resolution immunofluorescence study has attempted to address the issue of which portions of the GBM possess relevant epitopes as far as human anti-GBM disease is concerned (Fish *et al.*, 1983). Specifically, this study compared localization of antibodies eluted from kidneys of patients with anti-GBM disease to that of antisera from rabbits immunized with human GBM. In the former, staining was limited to the inner aspect of the GBM (lamina rara interna), whereas in the latter, double parallel stripes of GBM immunofluorescence were seen, corresponding to both laminae rarae. It is interesting to correlate these findings with the above-discussed suggestion that the laminae rarae are the sites of the anionic glycoprotein moieties which impart permselectivity to the glomerulus, and that GBM preparations elicit predominantly antiglycoprotein antibodies. These findings are consistent with the fact that proteinuria is an early manifestation of nephrotoxic serum nephritis, as the negative charges of the laminae rarae are neutralized by anti-GBM antibodies. Further, in the spontaneous human form of the disease, it is interesting that while proteinuria is an almost invariable concomitant, nephrotic range proteinuria is unusual, suggesting some preservation of the charge-selective properties. This correlates with the unilaminar immunofluorescent staining of the above study. It should be noted, however, that a critical evaluation of the permselective properties of the glomerulus in human anti-GBM disease has not been performed, and it is conceivable that the rapid deterioration in glomerular filtration rate seen in this disease precludes the development of massive proteinuria.

It should further be noted that Wakashin *et al.* (1981) have isolated a trypsinized fraction of human GBM which reacts with rabbit anti-human GBM antisera. Using a series of analytical techniques, they estimated the size of the relevant GBM antigen at 40,000 daltons. Unfortunately, no molecular characterization of the compound was provided, and its ability to induce disease was not addressed. This study stands in contrast with another (Holdsworth *et al.*, 1979a) which isolated collagenase-solubilized fractions of GBM (53,000 and 27,000 daltons) which were reactive with human anti-GBM antisera. In concert with the immunofluorescence data from above, one preliminary interpretation would be that the important GBM epitope in human anti-GBM disease is a noncollagenous glycoprotein of 27,000 and/or 53,000 daltons which is located in or near the lamina rara interna.

The distribution of GBM from one person to another is not uniform, and in the condition referred to as hereditary nephritis, or Alport's syndrome, it appears that one concomitant of the disease state in some patients is a deficiency of certain relevant GBM antigens. Whether this is a developmental abnormality or a congenital one has not been addressed directly, but there is indirect evidence to suggest the latter. Specifically, early in the disease, when pathologic abnormalities are rare, there are no demonstrable nephritogenic GBM antigens (as defined by the cross-reactive binding of eluted antibodies from nephritic kidneys of patients with anti-GBM disease). Second, it appears that some patients with Alport's syndrome are not tolerant to the nephritogenic GBM. This has been illustrated in the case of a patient with end-stage renal disease secondary to Alport's syndrome who was transplanted with a cadaveric kidney. Shortly thereafter, he rejected the kidney and evaluation revealed classic anti-GBM disease in the transplanted kidney with no GBM binding in the native kidney (McCoy *et al.*, 1982). This suggests immunization with cadaveric GBM as a new antigen. One possible explanation for this observation is that congenital absence of the relevant GBM epitopes resulted in a failure to induce tolerance.

It should be noted that while the kidneys of patients with Alport's syndrome do not appear to react with anti-GBM antibody eluted from kidneys of patients with anti-GBM disease, they will react with heteroantisera raised to human GBM. This is consistent with the immunofluorescence data from Fish *et al.* (1983), indicating that spontaneous human anti-GBM antibodies bind to determinants distinct from those recognized by rabbit antisera to human GBM.

The Mayo Clinic has recently reviewed its kidney transplantation experience in 11 patients with Alport's syndrome who had reached end-stage renal disease (Milliner *et al.*, 1982). In 2 of 3 patients who lost allografts to rejection, the underlying pathologic process was anti-GBM disease. Interestingly, both of these patients were successfully retransplanted, and with a mean follow-up time of 6 years, 10 of the patients had functioning grafts and none had serum creatinine levels greater than 2 mg/dl. Further, only 1 patient had developed detectable circulating anti-GBM antibodies. The seeming lack of an anti-GBM response in the majority of Alport's patients might be related to impaired afferent mechanisms associated with either immunosuppressive regimens attendant to renal transplantation and/or persistent GBM tolerance in the host. Finally, since renal biopsies in those transplanted patients with good renal function are not available, one cannot exclude an undetectably low anti-GBM response and some renal anti-GBM antibody deposition, with abrogation of disease expression associated with immunosuppressive therapy.

Finally, whether an immune response to GBM or GBM-like material may play a role in renal lesions other than classic anti-GBM disease is unknown, but there are data to indicate that proliferative glomerular lesions are associated with lymphocyte reactivity to GBM. This has recently been studied by several groups (Fillit *et al.*, 1983; Schmitt *et al.*, 1982; Matsumoto *et al.*, 1978) and will be discussed in more detail later.

2.2.2b. Brush Border Antigens. Almost 25 years ago, Heymann *et al.* (1959) first described an experimental model of renal disease which structurally resembles membranous glomerulonephritis. Using a proximal brush border antigen emulsified in Freund's complete adjuvant, they were able to immunize rats and induce nephrotic syndrome associated with a glomerular lesion which revealed subepithelial deposits and, by immunofluorescence, showed subepithelial immune complexes. The crude antigen is referred to as fraction 1A (Fx1A), the more isolated nephritogenic component is called renal tubular epithelial antigen or fraction α5 (RTEα5) (Edgington *et al.*, 1968), and the highly purified glycoprotein is now referred to as GP330 (Kerjaschki and Farquhar, 1982). In addition to the autologous form of the disease (also referred to as autologous immune complex glomerulonephritis), an identical lesion can be induced by the administration of heterologous antisera containing anti-Fx1A antibodies (Barabas and Lannigan, 1974; Salant *et al.*, 1980).

While originally considered to be an example of glomerulonephritis secondary to the passive deposition of circulating immune complexes, the localization of the nephritogenic antigen in the glomerulus (Van Damme *et al.*, 1978) and the demonstration that perfusion of an isolated rat kidney with anti-Fx1A antibodies alone can result in the pathologic picture of immune complexes, have led to the newer concept of *in situ* immune complex deposition (Couser and Salant, 1980; Couser, 1981). Further, the fact that infused immune complexes preferentially localize in the subendothelial GBM and not in the subepithelial GBM, more directly speaks against a role for circulating immune complexes in the pathogenesis of this experimental disease (Couser *et al.*, 1978). The relevant glomerular antigen is a glycoprotein of 330,000 daltons (GP330) which can be localized on the epithelial side of the GBM at the base of the epithelial podocytes (Kerjaschki and Farquhar, 1983). Absorption of Fx1A heterologous antisera with mannan abolishes glomerular binding of the antisera, suggesting that mannan may represent part of the nephritogenic epitope. The absence of linear immunofluorescent staining in the disease is indicative of the discontinuous arrangement of this antigen in the GBM.

While antibodies to Fx1A have been sought in patients with membranous glomerulonephritis, their documentation has been extremely unusual (Douglas *et al.*, 1981; Collins *et al.*, 1981). This suggests the possibility of other destructive immunopathologic processes and/or nephritogenic antigens in most cases of human membranous glomerulonephritis. Before this issue is decided, however, it may be valuable to screen these antisera with the recently purified nephritogenic antigen, GP330.

2.2.2c. Nephritogenic Glycoprotein. Over approximately the past 15 years, Shibata and colleagues (Shibata, 1978; Shibata *et al.*, 1966, 1971, 1972, 1976) have isolated fractions of GBM with the selective capacity to induce several patterns of renal injury. First, using trypsin digestion of GBM, they were able to induce anti-GBM antibodies, and further simulate the nephrotoxic serum nephritis model of glomerulonephritis by passive administration of these heterologous antibodies. When this trypsinized segment of GBM was then subjected to prolonged (3 days) pronase digestion, a glycoprotein (GP-P) was then derived which, while not inducing anti-GBM antibodies, could cause (following a single hindfoot injection in rats) a severe proliferative glomerulonephritis and associated granular mesangial immunofluorescence which, over the course of several months, evolved into typical membranous glomerulonephritis. Compared with the trypsinized fraction, this glycoprotein had a higher carbohydrate content, particularly in glucose moieties. Finally, when this GP-P was treated with trichloroacetic acid (TCA), to precipitate free amino acids and small peptides, a glycoprotein, referred to as nephritogenoside, was isolated. In rats this causes a proliferative glomerulonephritis associated with granular mesangial immunofluorescence, and evolves over 7 to 8 months into a chronic glomerulonephritis with contraction of renal parenchyma. At no time is a membranous-like lesion observed following immunization with the TCA-treated glycoprotein. Biochemical analysis suggests that nephritogenoside is a branched oligosaccharide with an α-D-glucopyranosyl unit at the nonreducing terminus. It appears to be free of disulfide linkages and contains hydroxylysine and hydroxyproline in small quantities (Shibata, 1978).

It should be noted that the pathogenesis of this experimental disease is poorly understood. Shibata (1978) initially suggested a nonimmunologic explanation for the glomerular injury which involves the binding of injected nephritogenoside to circulating Con A-like substances in blood. The unoccupied carbohydrate-binding moiety of Con A would then bind glomerular capillary wall, allowing deposition of complexes between circulating Con A-like compounds and nephritogenoside. Unfortunately, this leaves the mesangial immunofluorescence seen in the model unexplained. No detailed immunologic study of afferent, efferent, or immunoregulatory limbs of this renal disease is presently available.

2.2.2d. Glomerular Antigen Distinguishable from Classic Nephritogenic GBM Determinants. A spontaneously occurring glomerulopathy in NZW rabbits associated with antibody which recognizes a glomerular capillary wall antigen distinguishable from typical GBM has been reported (Neale and Wilson, 1978; Woodroffe *et al.*, 1978). The lesion bears both membranous and proliferative features, is associated with proteinuria and epithelial foot process fusion, and is seen in 4% of younger animals. While immunofluorescence studies demonstrate glomerular specificity of eluted antibody and antisera from affected animals, glomerular capillary wall staining is granular, distinguishing reactivity of these antibodies from those of classical anti-GBM antibody. Further, electron microscopic investigation reveals atypical, ill-defined dense deposits within and along the subepithelial side of the GBM, different from that which has been associated with granular glomerular immunofluorescence. *In toto,* these observations suggest the possibility that such an antibody may give rise to granular immunofluorescence reflecting a mixed membranous and proliferative glomerulonephritis.

2.2.3. Glomerular Disease Associated with an Immune Response to Nonrenal Antigens

At least three pathophysiologic processes may result in the glomerular localization of antibody whose induction appears unrelated to the presence of a fixed glomerular antigen. These include (1) deposition of circulating immune complexes, (2) *in situ* immune complex deposition secondary to sequential nonimmunologic adsorption of antigen to the glomerular capillary wall followed by specific recognition of that antigen by antibodies (antigen-directed *in situ* immune complex formation), and (3) a humoral immune response to nonglomerular antigens which are cross-reactive with glomerular moieties (cross-reactive immune response). Although initiated by nonrenal exogenous antigen, this third process implies the development of an autoimmune response. Specifically, those mechanisms responsible for precluding a response to self must be superseded. Such pathways of potential operational significance have recently been reviewed (Cook *et al.*, 1983) and include an enhancing cross-reactive idiotype which leads to activation and expansion of normally silent self-reactive T- and B-cell clones, and exogenous antigen which is viewed by the immune system as self-hapten linked to foreign carrier, thereby allowing T-cell help to autoimmune B cells.

Since there are data to support the presence of each of these immunopathologic processes in different experimental animal systems or human disease, this section will approach the topic of nonglomerular antigens by briefly reviewing each of them.

2.2.3a. Circulating Immune Complex Deposition. A rabbit model of serum-sickness-induced glomerular disease has been known for over 30 years (Germuth, 1953). In essence, the injection of foreign serum proteins leads to either a proliferative lesion in the acute phase, or a membranous-like glomerulopathy (with GBM thickening and subepithelial deposits) with or without proliferative changes after chronic exogenous antigen administration. While bovine serum albumin appeared most nephritogenic in terms of reliably inducing renal injury, Dixon *et al.* (1961) demonstrated that bovine gamma globulin, human serum albumin, and human gamma globulin all possessed nephritogenic potential. These investigations drew the parallel between the pathologic alterations seen in the rabbit model and the spectrum of human glomerular lesions, and further suggested a central pathogenic role for immune complex deposition. This was based on the observation that both antigen and antibody could be demonstrated in affected glomeruli using immunohistochemical techniques, and further that the strongest predictor of chronic disease was the presence of antigen–antibody equivalence in the circulation, indicating the likelihood of soluble immune complex formation. Animals that had a marked antibody response to antigen probably developed insoluble immune complexes which were removed by the RES. Renal histopathology in these animals revealed a transient proliferative glomerulonephritis. Animals that failed to respond to antigen (and *a priori* could not have immune complexes) experienced no renal injury.

While it is usually assumed that an immune response to exogenous antigen or an autoimmune response to self-determinants forms the basis for renal immune complex formation, Lambert and colleagues (Rose *et al.*, 1982; Goldmann *et al.*, 1982) have also demonstrated that a syngeneic, broadly cross-reactive idiotype may provide an antigenic stimulus for circulating immune complex production and, further, that these idiotype–anti-idiotype immune complexes could localize in glomeruli. In a series of experiments, they showed that polyclonal B-cell activation in BALB/c mice was associated with the formation of antiphosphorylcholine antibodies which share idiotypic determinants with the mouse myeloma protein, TEPC-15. Antibody to TEPC-15 idiotype along with idiotype–anti-idiotype circulating immune complexes could then be demonstrated as deposits.

The above information generally supports the concept that circulating immune complexes can play a central role in the immunopathogenesis of renal disease. This has provided insight into a variety of human conditions associated with circulating immune complexes and renal injury. These include lupus nephropathy, postinfectious proliferative glomerulonephritis, cryoglobulinemia, and tumor-associated glomerulonephritis. It should be noted, however, that as our understanding of antigen characteristics and antibody specificity advances, alternative explanations for the presence of glomerular immune complexes emerge. This was seen in Heymann's nephritis model of *in situ* immune complex deposition (see Section 2.2.2b) and will be described in several other systems. In addition, the fact that as many as 80% of solid tumor patients have demonstrable circulating immune complexes (Rossen *et al.*, 1976), but far less have evidence of glomerular pathology (Eagen and Lewis, 1977), suggests that circulating immune complexes may be a necessary but not sufficient condition for induction of immune renal injury. In addition, alternative pathogenic schemata, such as tumor antigen-directed *in situ* immune complex formation and/or tumor-associated antigenic cross-reactivity, may be important.

2.2.3b. Antigen-Directed *in Situ* Immune Complex Deposition. There are several well-described experimental situations in which nonimmunologic adsorption of antigen is associated with glomerular focusing of the humoral immune response, *in situ* immune complex deposition, and subsequent glomerular pathology.

Izui *et al.* (1976, 1977a) showed that DNA has a particular affinity for GBM, and can be demonstrated to localize there, nonimmunologically, after it is seen in the circulation. Further, in a murine model of renal disease in which DNA is released into the intravascular compartment and anti-DNA antibodies are subsequently formed, typical immune complex-type glomerular lesions evolved in the absence of demonstrable circulating DNA–anti-DNA complexes and eluted glomerular antibody was shown to have anti-DNA activity. These authors suggested that intravascular DNA was filtered at the level of the glomerulus, adhered to GBM structures, and directed anti-DNA antibodies to the glomerular capillary wall, producing glomerular lesions. Their failure to find circulating DNA–anti-DNA immune complexes in patients with SLE (Izui *et al.*, 1976) provides provocative, albeit preliminary and controversial (Bruneau and Benveniste, 1979) evidence that such a mechanism may be operational in humans.

Other moieties which appear to nonimmunologically adsorb to glomerular structures and then direct *in situ* immune complex formation include Con A which binds to GBM nephritogenoside (Shibata *et al.*, 1977) through carbohydrate moieties with subsequent localization of anti-Con A antibodies producing glomerulonephritis (Golbus and Wilson, 1979) and endostreptosin, a cationic group A streptococcal cytoplasmic antigen seen in the glomeruli (subendothelial and mesangial regions) in some patients with poststreptococcal glomerulonephritis (Seligson *et al.*, 1982).

2.2.3c. Glomerular Disease Related to a Cross-Reactive Immune Response. While there is no unequivocal evidence, some data in the poststreptococcal glomerulonephritis literature would suggest that at least one mechanism for poststreptococcal glomerular disease involves renal antigens cross-reactive with glomerular moieties. Rapaport *et al.* (1969) reported that antisera raised to streptococcal membranes can induce glomerular pathology. Fillit *et al.* (1983) isolated a 46,000-dalton protein which is specifically associated with streptococcal strains recovered from patients with acute poststreptococcal glomerulonephritis. This nephritic-strain-associated protein could not be identified in other streptococcal strains. The intriguing question of whether this protein exhibits determinants cross-reactive with glomerular constituents remains unanswered. Along these lines, however, it is interesting to note that patients with chronic glomerulonephritis (particularly with proliferative histology) possess immune reactivity to both streptococcal antigens and altered GBM antigens (Fillit *et al.*, 1983; Schmitt *et al.*, 1982; Matsumoto *et al.*, 1978). Taken together, there are several indirect lines of evidence which would indicate that the immunopathogenesis of acute poststreptococcal glomerulonephritis could be related to serologic cross-reactivity between glomerular and streptococcal antigens. This may also apply to those patients with acute poststreptococcal glomerulonephritis who progress to chronic disease, in that an endogenous glomerular antigen may provide an ongoing stimulus to immune injury.

At present, it is unknown whether cross-reactive immune responses play a role in other renal diseases. Given the high incidence of membranous glomerulonephritis in those patients with solid tumors and nephrotic syndrome (Eagen and Lewis, 1977) plus the examples of the immunopathogenesis of Heymann's nephritis discussed above, it would be potentially revealing to ascertain whether renal eluates from such patients are reactive with normal glomerular tissue.

2.2.4. Tubulointerstitial Nephropathy Associated with an Immune Response to Renal Antigens

2.2.4a. Tubular Basement Membrane. Over a decade ago, Steblay and Rudofsky (1971) first described the induction of anti-tubular basement membrane (anti-TBM) disease in guinea pigs following immunization with xenogeneic renal tubular antigen in Freund's complete adjuvant. The lesions produced are characterized by extensive cortical interstitial infiltrates. Subsequently, this model has been extended to rats (Lehman *et al.*, 1974) and mice (Neilson and Phillips, 1982a; Rudofsky *et al.*, 1980).

In all of the above rodent systems, susceptibility is genetically determined among strains and, in susceptible animals, anti-TBM antibodies are observed usually within 2 weeks of immunization. Interstitial infiltrates are seen 10 to 40 days later. Immunofluorescent evaluation of affected kidneys early in the course of disease (before significant tubular destruction has occurred) demonstrates linear staining of TBM (especially, but not exclusively of proximal tubules). While the TBM alone is stained in rats and guinea pigs, both the GBM and the TBM are stained in mice (Rudofsky *et al.*, 1980). Thus, it would appear that the nephritogenic TBM antigen is peculiar to the TBM of guinea pigs and rats, but is shared by or cross-reactive with the GBM in mice.

The molecular nature of the relevant TBM antigen has not been fully characterized. Zanetti and Wilson (1983) have provided a detailed characterization of the anti-TBM antibody response in rats, following immunization with a bovine preparation of TBM. From this, they have been able to derive meaningful information regarding the nature of the nephritogenic TBM antigen. First, it would appear that the relevant disease-inducing antigenic determinants in the TBM preparation are not solubilized by collagenase. This observation parallels that for the GBM, as discussed above. Second, using immunoprecipitation and analytic separative techniques, they were able to correlate serologic response to a 42,000-dalton component of collagenase-solubilized Brown Norway (BN) rat (a susceptible strain) TBM with disease susceptibility. While all seven rat strains evaluated made a similar antibody response to the xenogeneic TBM preparation, BN antibody levels to the BN collagenase-solubilized TBM were clearly of greatest magnitude. The remaining six strains made less anti-BN collagenase-solubilized TBM antibody and did not get disease.

Unlike guinea pig and mouse, where all strains thus far examined appear to possess TBM, which will bind anti-TBM antibodies in linear fashion by immunofluorescence, there are rat strains which appear to be deficient in these antigens. Thus, the Lewis rat, which makes a demonstrable anti-TBM response to xenogeneic preparations, does not appear to possess sufficient autologous TBM to bind these antibodies. Consequently, while the sera of these immunized animals will cause linear immunofluorescent staining of BN TBM, the kidneys of these same animals are immunofluorescence-negative. Zanetti and Wilson (1983) have provided data suggesting that this is a relative, not an absolute deficiency of relevant antigen. It is worthy of note, however, that Paul and Carpenter (1982), utilizing immunofluorescent staining patterns of different antisera, have suggested the presence of at least two distinct epitopes intrinsic to the TBM.

Two considerations regarding the above work are worthy of note. First, it will be interesting to see whether the isolated 42,000-dalton fraction of BN collagenase-solubilized TBM is capable of inducing anti-TBM disease itself, as this must remain the ultimate criterion for a nephritogenic antigen. Second, in our experience there is variability in the quantitative antibody response to autologous collagenase-solubilized TBM among susceptible rat strains, suggesting that the magnitude of the serologic response can be dissociated from disease predisposition. Hence, it should be revealing to evaluate the humoral response to BN collagenase-solubilized TBM in "low"-responder susceptible rat strains, to determine whether this response remains an invariant correlate of disease induction.

Further supporting the concept of dissociation between a serologic response to a given nephritogenic antigen and susceptibility to disease, are recent data from our laboratory in the murine system (Zakheim *et al.*, 1984) which document transfer of typical tubulointerstitial nephropathy (without anti-TBM antibody deposition) to naive recipients with lymphoid cells from animals immunized with xenogeneic renal tubular antigen. It is noteworthy that tubulointerstitial nephropathy associated with anti-TBM antibodies has been documented in humans in a variety of clinical settings. This has occurred spontaneously (Bergstein and Litman, 1975) and has also been observed in association with methicillin-induced interstitial nephritis (Border *et al.*, 1974) and renal allograft rejection (Wilson *et al.*, 1974). In addition, as many as 70% of patients with anti-GBM disease will have concomitant anti-TBM antibodies (Wilson and Dixon, 1973), the presence of which can be correlated with tubulointerstitial pathology (Andres *et al.*, 1978).

Interestingly, there is preliminary evidence to suggest that the pathologic equivalent of Alport's syndrome may exist in the case of selected TBM antigens as well. Wilson *et al.* (1974) described a renal transplant patient who rejected two successive allografts related to the development of anti-TBM antibodies which displayed reactivity with the grafts but not with the patient's own kidneys. These anti-TBM antibodies were then eluted from the rejected grafts and were shown to have consistent reactivity with a broad spectrum of allogeneic kidney tissue from both normal and diseased patients. In addition, the antibody displayed no GBM specificity and by immunofluorescence criteria bound to BN but not Lewis rat renal tubules. Thus, this would appear to be an example of deficiency of immunogenic TBM epitopes, with subsequent immune recognition of these moieties when presented on allogeneic renal tissues. As discussed above, this parallels the situation for GBM in Alport's syndrome, and for TBM in Lewis versus BN rats. It should be noted that in the patient discussed above, congenital deficiency of TBM antigens versus developmental loss of such constituents cannot be formally distinguished.

Wakashin *et al.* (1981) have isolated a 30,000-dalton component of trypsin-solubilized TBM which does not cross-react with human GBM but does cross-react with TBM from goat, guinea pig, and mouse. In addition, immunization of BALB/c mice (an anti-TBM disease-susceptible strain) with this preparation resulted in anti-TBM antibodies in all of five animals. Thus, it would appear that this component could fulfill criteria for a nephritogenic antigen. It would be especially intriguing to determine whether this is the relevant antigen in those renal allograft patients who develop anti-TBM antibody-related rejection.

2.2.4b. Tamm–Horsfall Protein. Produced by cells of the thick ascending limb of the loop of Henle, this high-molecular-weight glycoprotein represents the major protein derived from renal parenchyma which is excreted in the urine on a daily basis. Hoyer and colleagues (Friedman *et al.*, 1982; Seiler and Hoyer, 1981; Hoyer, 1980) have developed a rat model of what, morphologically, is immune complex tubulointerstitial nephropathy. Granular immunofluorescence on the basal aspect of the cells of the thick ascending limb is accompanied by focal leukocytic infiltrates. Electron microscopy reveals dense deposits in the region between the TBM and the basal limit of the renal tubular cells in this segment. An autoimmune form of the disease is seen in rats chronically challenged with autologous Tamm–Horsfall protein in Freund's complete adjuvant (Seiler and Hoyer, 1981; Hoyer, 1980). Conversely, disease can be induced by passive transfer of anti-Tamm–Horsfall antibody (Friedman *et al.*, 1982). An *in situ* mechanism for the formation of peritubular immune complexes has been postulated in this experimental model (Hoyer *et al.*, 1983). While autoantibodies to Tamm–Horsfall protein have been observed in patients with pyelonephritis (Mayrer *et al.*, 1983), there are presently no firm data which demonstrate a primary immunopathogenic role for this protein in human tubulointerstitial nephropathy.

2.3. IMMUNOGENETICS OF RENAL DISEASE

2.3.1. Introduction

The genetics of any immunologically mediated disease relates ultimately to the capability of specifying given gene sequences or segments as requirements for disease induction and/or expression. As with inherited differences in immune responses to nominal antigens seen in various animal systems immunologic disease associations most commonly relate to Ir genes of the MHC. It seems that many Ir genes function at the level of the interaction between antigen-presenting cells and helper T cells, in that nonresponders fail to make an immune response to nominal antigen because of an inability to associate with available class II molecules (determinant selection model) (Rosenthal, 1982) or because of a "hole" in the T-cell repertoire (clonal deletion model) (Schwartz,

1978). Suppressor cell mechanisms have also been invoked as an additional basis for unresponsiveness. By extension, in those instances where susceptibility to disease states is genetically determined and is linked to the MHC, induction of pathologic (frequently autoimmune) responses may relate to the ability of the immune system to recognize normally inert, but cross-reactive with self, moieties because of (1) a unique association with class II molecules on antigen-presenting cells, (2) the presence of self-reactive T-cell clones, or possibly (3) the deficiency of an I–J-region-restricted, antigen-specific suppressor cell. This third explanation associates the phenomenon of induction of pathologic responses with alterations in the immunoregulatory limb of the nephritogenic immune response, and, similarly, there are data (see Section 2.1) to suggest that in some instances disease susceptibility may be linked to MHC genes which code for class I molecules and restrict cytotoxic T-cell responses (Neilson and Phillips, 1982a). Thus, it is apparent that the immunogenetics of disease is theoretically not limited to the afferent phase, but also to immunoregulatory and efferent limbs of the immune response. Further, as discussed below, disease susceptibility may also be linked to immunoglobulin heavy-chain allotypes, which suggests that one determinant of immune responsiveness is the paratypic and idiotypic repertoire of variable-region genes which share the same chromosomal segment as heavy-chain genes.

Because any discussion of immunogenetics frequently relates to the MHC and antigen-presentation, it is appropriate to include this in the afferent phase of the immune response and relevant to begin this section with a discussion of the data which examine renal MHC antigens, renal cells bearing macrophage markers, and the ability of the kidney to act as an antigen-presenting organ to the lymphoid system. Subsequently, immunogenetic data will be explored as they relate to selected experimental animal models, and to human renal disease.

2.3.2. Renal MHC Antigens and the Kidney as an Antigen-Presenting Organ

A body of literature has appeared which documents the distribution of class I and class II MHC antigens in the kidney. In the rat, resident renal cells appear to express both classes of antigen (Davies and Butcher, 1978). Detailed localization studies suggest diversely distributed class I antigens while class II antigens are found primarily on tubular and dendritic cells of the cortex (Hart and Fabre, 1981). Mayrhofer and Schon-Hegrad (1983), using monoclonal antibodies to framework determinants on the rat class II molecules, have extended these studies to show that Ia products are found on proximal convoluted tubular epithelium and dendritic cells of the interstitium, glomeruli, and mucosa of the renal pelvis in the rat. Of note, while strain variation was observed in the epithelial expression of RT1.B-region gene products, B lymphocytes and dendritic cells appeared to have a uniform density of these molecules, suggesting non-MHC gene regulation of epithelial Ia antigen expression.

The demonstration of *in vivo* Ia^+ cells in the glomerulus supports the study of Schreiner *et al.* (1981), which documents that approximately 2% of rat glomerular cells are Ia$^+$. These observations were complemented by functional investigations which showed that Ia$^+$ glomerular cells were capable of presenting nominal antigen in genetically restricted fashion and could provide allogeneic stimulation in a one-way mixed lymphocyte reaction.

Taken together, the above rat data suggests: (1) widely distributed class I antigens in the kidney, (2) localization of class II antigens predominantly to renal dendritic and tubuloepithelial cells, and (3) functional competence of renal antigen-presenting cells. Hence, these findings provide a feasible basis for understanding the induction of immune responses within the kidney, and also indicate that resident renal cells may be sufficient to induce allogeneic stimulation after renal transplantation, thus obviating the need to invoke the passenger-leukocyte theory of allograft rejection. Additional relevant data in the mouse demonstrate that interstitial cells of the kidney bear a macrophage-specific marker (Hume and Gordon, 1983) and that in the human, in contrast to the rat, class II MHC antigens are found on peritubular capillaries, glomerular endothelium, and mesangium (Hart *et al.*, 1981; Platt *et al.*, 1983).

2.3.3. Immunogenetics of Experimental Renal Disease

2.3.3a. Heymann's Nephritis. In this model of membranous glomerulopathy, Lewis rats are the prototypic susceptible strain, whereas BN rats are resistant. Further elucidation by Stenglein and colleagues (Stenglein *et al.*, 1975, 1978) has shown linkage of susceptibility to RT1 but that susceptibility is a relative term, and that background (i.e., non-MHC) genes can exert an important influence. Thus, the congenic strain Lew.BN, which carries a "nonsusceptible" BN haplotype (RT1^n), does in fact develop glomerular disease (as judged by proteinuria and positive immunofluorescence) but requires a prolonged period to do so, and further has a lower incidence of disease with less proteinuria in affected animals. Similarly, (BN × Lewis)F_1 hybrids develop glomerulonephritis but in less severe fashion and in lower incidence than their Lewis parent. The studies in Lew.BN animals point to the enhancing effect of Lewis background genes and the possibility of complementation of genes in disease induction as seen in other experimental disease systems in rats (Gasser *et al.*, 1978).

2.3.3b. Mercuric Chloride. Chronic injection of mercuric chloride ($HgC1_2$) in susceptible strains of rats results in a two-phase immunologic renal lesion with anti-GBM antibody production followed by an immune complex-type glomerulonephritis. The BN rat is typically susceptible to both phases of this response. Druet *et al.* (1977) have shown that the anti-GBM response is linked to RT1 and appears to be inherited in dominant fashion. While F_1 hybrids, Lewis (nonsusceptible) and BN parents all produce anti-GBM antibodies, Lew.BN congenics are entirely resistant, suggesting that the RT1^n haplotype is a necessary but not sufficient condition for antibody production and that non-MHC background gene products also contribute to susceptibility (or resistance). Subsequently, Sapin *et al.* (1980), using (Lew × BN)$F_1 \rightarrow$ P(Lew) chimeras, have demonstrated that susceptibility to the induction of both anti-GBM antibodies and immune complex glomerulonephritis is conferred by immunocompetent cells of bone marrow or splenic origin. This suggests that the evolution of $HgCl_2$-induced renal lesions can occur in a resistant (Lew) environment if susceptible lymphoid cells are employed. Hence, it would appear that the non-MHC gene(s) which exert an influence on susceptibility in this model, do so at the level of the lymphoid population.

2.3.3c. Circulating Immune Complex Glomerulonephritis. From the time of one of the earliest descriptions of this disease model in outbred rabbits (Dixon *et al.*, 1961), it has been evident that the nature of the antibody response to heterologous proteins plays an important role in the ultimate development of glomerular pathology. Thus, rabbits that made no antibody had no disease, and those that made large amounts of antibody and, hence, readily cleared insoluble immune complexes, developed a transient proliferative glomerulonephritis. Those animals, on the other hand, whose quantitative antibody response dictated antigen–antibody equivalence, experienced an immune complex glomerulonephritis by histopathologic and immunofluorescent criteria. While the quantitative antibody response is generally accorded MHC-linked gene control (Benacerraf and McDevitt, 1972), the qualitative paratypic profile is largely dictated by variable (V)-region immunoglobulin genes. The V-region library for a given antigen receptor could conceivably influence disease induction, and Steward and colleagues (Steward *et al.*, 1979; Steward, 1979; 1981) have explored this issue by examining the role of antibody affinity in a murine model of immune complex glomerulonephritis, induced with repeated injections of human serum albumin or transferrin. Random-bred mice were mated on the basis of antibody affinity to these two proteins, resulting in two lines of mice, one which produced high-affinity antibody, and the other which produced low-affinity antibody. As would be predicted, breeding experiments indicated independent segregation of genes controlling antibody affinity and those regulating antibody level. Further, both high- and low-affinity lines of mice possessed similar distributions of MHC antigens (Steward *et al.*, 1979). When these two lines were then subjected to chronic human serum albumin injections, there was increased immune complex deposition in low-affinity mice, primarily localized to the GBM. High-affinity mice, in contrast, had less immune complex deposition with enhanced mesangial localization. Antibody levels were similar between the two groups, as were renal function and renal deaths. The differing immunopathologic pictures, however, suggest that the qualitative antibody response, as dictated by non-MHC genes, may influence the expression of renal disease.

2.3.4. Immunogenetics of Experimental Tubulointerstitial Nephropathy

The immunogenetics of this disease has been investigated in guinea pigs (Hyman *et al.*, 1976a,b), rats (Lehman *et al.*, 1974; Krieger *et al.*, 1981; Neilson *et al.*, 1983), and mice (Rudofsky *et al.*, 1980; Neilson and Phillips, 1982a).

Hyman *et al.* (1976a) showed that strain XIII guinea pigs are exquisitely sensitive to disease induction and will develop anti-TBM antibodies and severe interstitial infiltrates after as little as 10 μg of xenogeneic renal tubular antigen (RTA) in Freund's complete adjuvant, or following the passive transfer of immune serum. Strain II guinea pigs, on the other hand, are resistant to anti-TBM disease, requiring 1000 times more RTA than their strain XIII counterparts in order to make an anti-TBM autoimmune response. Even then, linear immunofluorescence of renal cortical TBM occurs without accompanying histopathologic derangement. Furthermore, administration of immune serum fails to induce a renal lesion in these animals. Absorption studies suggest a similar content of TBM epitopes in strain II compared with strain XIII renal tissue. Thus, it appears that both susceptible and resistant strains can be induced to make anti-TBM antibodies and seem to have similar endogenous TBM antigen profiles, suggesting that a factor unrelated to either antibody production or antigen presence may influence disease expression. Additional genetic evaluation in the species (Hyman *et al.*, 1976b) revealed MHC linkage to susceptibility with codominance of gene inheritance.

Immunogenetic evaluation in rats (Lehman *et al.*, 1974; Krieger *et al.*, 1981; Neilson *et al.*, 1983) suggests that ultimate disease expression is dependent on a complex set of requirements. Specifically, susceptible rat strains [e.g., BN ($RT1^{n}$)] must have not only the ability to make an antibody and cell-mediated response to TBM determinants, but also the nephritogenic RTA in sufficient quantity to allow immune injury to occur. Consequently, resistant strains may be so on the basis of any one of several variables. For example, F344 ($RT1^{av1}$) rats have the appropriate TBM antigen, but do not make an antibody response and do not get disease. Maxx ($RT1^{n}$) and Lewis ($RT1^{l}$), conversely, make an antibody response, but are deficient in the cross-reacting renal antigens, and also do not develop interstitial lesions. When F344 and Maxx rats are mated, the immunized F_1 hybrids ($RT1^{n/av1}$) make an anti-TBM antibody response, possess the necessary TBM epitopes (for linear immunofluorescence of cortical tubules by anti-TBM-containing antisera), and develop disease, whereas (Lewis × F344)F_1 ($RT1^{l/av1}$) rats also develop anti-TBM antibodies and have readily recognizable TBM determinants, but do not develop disease. These findings indicate the possible failure of a cell-mediated effector response in the (Lewis × F344)F_1 hybrids, which would appear to be under the control of RT1. They further suggest the potential for a multifactorial basis for resistance, with roles for genes which control relevant antigen synthesis, antibody response, and cell-mediated immune injury. Neilson *et al.* (1983) and Krieger *et al.* (1981) have documented the independent segregation of RT1 haplotype and TBM antigen, with the former group showing that inheritance of the genes specifying TBM antigen are linked to the gene for albinism in the first linkage group. Because adoptive transfer of serum or cells fails to induce disease in rats (Neilson, unpublished observation), it is impossible to state with certainty the relative importance of humoral versus cell-mediated immune responses in anti-TBM disease in the rat. The example of (Lewis × F344)F_1 animals suggests that the presence of TBM antigen and an anti-TBM antibody response are necessary but not sufficient conditions for disease induction. Data from work by Zanetti and Wilson (1983) would allow an alternative explanation for resistance in such animals, i.e., the failure of an antibody response to the appropriate nephritogenic TBM. The importance of this latter finding, however, has been challenged by the observation that F.BN rats make very little antibody but produce substantial interstitial lesions (Neilson *et al.*, 1983). Further clarification of these issues regarding the immunogenetics of anti-TBM disease awaits further studies.

The murine model of anti-TBM disease is particularly intriguing on two counts: (1) the potential for refined immunogenetic evaluation because of the availability of mutant strains, congenic strains, and a detailed molecular understanding of the H-2 complex, and (2) the observation that, thus far, all mouse strains, following immunization with xenogeneic RTA, make anti-TBM antibodies which bind autologous renal cortical TBM, while only a fraction of these develop interstitial lesions (Neilson and Phillips, 1982; Rudofsky *et al.*, 1980). Because susceptibility to disease maps to the H-2K region of the MHC [which is positively correlated with T-cell cytotoxicity against tubular antigen-coated targets (Neilson and Phillips, 1982a)], and because immune T lymphocytes can transfer disease to naive recipients (Zakheim *et al.*, 1984), it would appear that genetic control of anti-TBM disease may be exerted at the level of the effector T-cell repertoire.

While the immunogenetic evaluation of anti-TBM disease has provided a wealth of valuable data, there still remain many gaps in our understanding of this model. It is hoped that efforts currently underway to characterize the clonotypic heterogeneity of the anti-TBM antibody response in mice, and to purify the relevant nephritogenic TBM antigen will allow even more refined immunogenetic analysis.

2.3.5. Immunogenetics of Human Renal Disease

As discussed earlier, the approach to the genetic aspects of immune-mediated renal injury in humans has *a priori* limitations related to difficulties in isolating purified nephritogenic antigens, and ethical constraints that generally preclude experimental evaluation until disease induction has already occurred. Thus, human data in this field are, for the most part, population studies which demonstrate positive or negative associations between HLA alleles and renal disease.

Because our understanding of the molecular nature, function, and polymorphism of human class II gene products is evolving rapidly (Gonwa *et al.*, 1983), it is conceivable that in the not too distant future an HLA association may be translatable into a distinctive amino acid sequence on an Ia molecule. In conjuction with the observation that determinant selection may be an important control element in immune responsiveness, the possibility of blocking pathologic immune response by an anti-Ia monoclonal antibody that prevents immunogenetic association between nominal antigen and Ia becomes a distinct consideration. The obvious concern would be that of precluding normal inductive immune responses to other antigens (i.e., bacterial and viral antigen) which might share the same epitope-binding region of the Ia molecule. Anti-Ia monoclonal antibodies should provide an ideal probe. Especially provocative in this regard is the preliminary report of Adelman *et al.* (1983), in which treatment of NZB mice with monoclonal anti-Ia antibodies appeared to abrogate the typically fatal immune nephritis of these animals. This approach has also been utilized successfully to obviate individual antigen responses in mice (Rosenbaum *et al.*, 1981) and has recently been reviewed by Gorzynski and David (1983). It represents an exciting potential avenue for therapeutic intervention and portends the possibility of moving immunogenetics from the realm of correlative documentation to that of functional manipulation.

2.3.5a. HLA Associations. The first description of a renal disease in which there was an increased incidence of a given human MHC antigen was provided 15 years ago by Patel *et al.* (1969) for chronic glomerulonephritis. That such a lesion is a final common pathway for a number of glomerulopathies, makes the interpretation of this finding somewhat obscure. Nevertheless, in the ensuing years, many primary and secondary glomerulopathies have been shown to be associated with other HLA alleles (reviewed in Schur and Carpenter, 1979; Garovoy, 1982). For the purpose of succinctness, this information is presented in tabular form (Table 3), and the interested reader is referred to the original references for further information.

TABLE 3. HLA ASSOCIATIONS WITH HUMAN GLOMERULOPATHIES[a]

	HLA association	Reference
Primary glomerulopathies		
Minimal change glomerulonephritis	B12 (England)	Trompeter *et al.* (1980)
	B8; B18 (Ireland)	O'Regan *et al.* (1980)
Membranous glomerulonephritis	DR3/MT2[b]	Garovoy (1980)
Berger's disease (IgA nephropathy)	BW35 (Australia)	MacDonald *et al.* (1976)
	BW 35/DR4 (France)	Noel *et al.* (1978), Berthoux *et al.* (1978)
	DR4 (Japan)	Kashwabara *et al.* (1980)
	B12 (England, USA)	Brettle *et al.* (1978)
		Richman *et al.* (1979)
Anti-GBM glomerulonephritis	DR2	Rees *et al.* (1978)
Secondary glomerulopathies		
SLE glomerulonephritis	DR3	Garovoy (1982)
Gold-induced membranous glomerulonephritis	DR3	Wooley *et al.* (1980)
Penicillamine-induced membranous glomerulonephritis	DR3	Wooley *et al.* (1980)
Poststreptococcal glomerulonephritis	D"EN"[c] (Japan)	Sasazuki *et al.* (1979)
	None	Read *et al.* (1977), Layrisse *et al.* (1983)

[a]Modified from Garovoy (1982).
[b]MT2 = a probably public D-region domain, shared by several DR alleles, including DR3.
[c]D"EN" = a new D-locus allele closely related to DRw6.

2.3.5b. Heavy-Chain Allotype Associations. Immunoglobulin isotypes bear several regions which display allogeneic polymorphism that are referred to as immunoglobulin allotypes. Since heavy-chain genes are linked to V-region genes, it is reasonable to believe that immunoglobulin heavy-chain allotypes might specify limited families of V-region genes, to the extent that linkage disequilibrium can be demonstrated to exist between the two. Since it is generally agreed that germ-line V-region genes provide a significant degree of eventual paratypic and idiotypic diversity, it is possible that the study of immunoglobulin heavy-chain allotypes may provide insight into individual antigen-recognition capabilities and the presence or absence of given idiotypes. By extension, and analogy to MHC associations discussed above, certain allotypes may bear indirectly on the likelihood of making an autoimmune response either because of inclusion of the pathologic reactivity in the B-cell repertoire of paratopes or because of an idiotype-related immunoregulatory defect (e.g., one idiotype may stimulate less of an anti-idiotypic suppressive immune response than another). This investigation of allotypes may allow further dissection of the immunogenetics of various disease states. In this context, Demain *et al.* (1982) have documented a statistically significant association between renal failure and the Gm2 heavy-chain allotype. Nakao *et al.* (1980) in Japan have examined the Gm heavy-chain allotype for possible associations with a variety of autoimmune disorders. They found a significant increase in the Gm2 allele among patients with myasthenia gravis, Grave's disease, and Hashimoto's thyroiditis. Further characterization of paratypes and idiotypic repertoires in such patients, as well as similar studies in patients with immunologically mediated renal disease, should provide valuable immunogenetic insights.

3. IMMUNE REGULATION AND THE NEPHRITOGENIC IMMUNE RESPONSE

Immune regulation can be viewed as a complex network of antibody and cell-mediated circuits which modulate and balance communication among immune processes responding to antigenic perturbation. These regulatory responses are coordinated by recognition of complementary cell surface receptors so that under normal circumstances activated immune processes can be efficiently controlled to minimize the chronic expression of detrimental mediators of immunopathology. As a consequence, the effectiveness of immune regulation measurably influences, among other things, the overall level of activity of the mediators that may produce chronic inflammation and hypersensitivity. The effects of immune regulation can also be viewed in the context of developing renal injury. That is, regulatory processes can theoretically influence the development of immune-mediated renal disease in two general ways. First, an expected regulatory response to a nephritogenic antigen could be delayed or fail to appear (Hoffsten *et al.*, 1977; Neilson *et al.*, 1983b). In this instance, the immune response cannot be terminated and, under such circumstances, a progressively destructive lesion would be expected. Second, a component of the normal regulatory process could cease to operate or fail to develop such that any number of autoreactive immune responses would spontaneously arise to produce mediators of renal injury (Cantor and Gershon, 1979; Cohen and Eisenberg, 1982; Steinberg *et al.*, 1982). In the context of renal disease, and for the purposes of this progress report, immune regulation can be viewed as a complex network of antibody and cell-mediated circuits usually operating to produce feedback suppression and down-regulation leading to the eventual control of ongoing renal injury. Because of its pivotal influence, it would also not be surprising to find that regulatory networks may provide new and additional pathogenic pathways for immune-mediated renal disease. That is, defects or alterations in regulatory systems may create a permissive environment for the uncontrolled expression of effector responses producing renal injury. Correcting these deficiencies in immune regulation may furthermore provide another goal for therapeutic interventions tending to reduce host morbidity. The subject of immune regulation in renal disease has recently been reviewed (Neilson and Zakheim, 1983b). A great deal of progress has been made in this area of investigation since our previous survey (Neilson and Phillips, 1980), but unfortunately this field of research is so new that much of the information available is still incomplete.

3.1. IMMUNE REGULATION AND EXPERIMENTAL RENAL DISEASE

With the availability and widespread use of antibodies to T-cell subpopulations, it has now been possible to critically evaluate the role of selected T lymphocytes as modulators of a nephritogenic immune response. Experiments in T-cell-reconstituted mice directly address this point (Cantor and Gershon, 1979). In these experiments, it was observed that B-cell mice devoid of T lymphocytes did not develop autoantibodies or immune-mediated renal disease. If these mice, however, were further reconstituted with Lyt-1^+ T helper cells, a variety of autoantibodies appeared and their kidneys accumulated immune deposits, producing glomerulonephritis. Further repopulation with Lyt-2^+ T suppressor cells abrogated this autoimmune condition and reduced host morbidity. In another set of experiments, antigen-specific suppressor cells generated to Fx1A and transferred into naive recipients were able to prevent the development of experimental membranous nephropathy following immunization (Harmon *et al.*, 1980). Finally, young mice carrying lymphocytic choriomeningitis virus develop a glomerulonephritis in association with focal hepatitis and generalized lymphoproliferation (Morrison and Wright, 1977). When T lymphocytes from adult mice sensitized to the virus are adoptively transferred into the young carrier animals, the viremia is weakened and the renal disease improves (Hoffsten *et al.*, 1977). These three experiments directly illustrate that suppressor T cells can prevent or interrupt an ongoing nephritogenic immune response.

The best-studied regulatory network of all the models of experimental immune-mediated renal disease is that of spontaneous immune complex glomerulonephritis in NZB/W, MRL, and BXSB mice (Steinberg *et al.*, 1982). Our understanding of the autoimmune disturbances in these mice has undergone several phases of evolution. It now seems that a number of immunologic aberrations in T- and B-cell function have been associated with the development of immunopathology. Hyperactive B-cell function and suppressor T-cell deficiencies have been observed under a variety of experimental conditions (Steinberg *et al.*, 1981; Schwartz, 1982). While all of this information collectively implies that several different mechanisms may underlie the regulatory disturbances found in these mice, none of the findings clearly defines a principal basis for the autoimmune response. Interestingly enough, it now seems that autoantibody production can be genetically dissociated from B-cell hyperactivity (Datta *et al.*, 1982). Because of this finding, an alternative explanation for autoantibody production is required. In several recent studies, it has been observed that NZB mice do not make anti-idiotypic immune responses to selected antigens, including some autoantibodies (Cohen and Eisenberg, 1982). This finding implicates a regulatory disturbance which under normal circumstances might control autoreactive immune responses by suppressing the expression of certain V-region gene products. Because the anti-DNA antibody repertoire in NZB/W and MRL mice contains cross-reactive idiotypes, it has been hypothesized that defective regulation of autoantibody expression in these mice may be a result of a failure to provide an appropriate anti-idiotypic repertoire (Carnegie and Mackay, 1975). In additional experiments, NZB/W mice injected with monoclonal anti-DNA antibodies expressing high-frequency idiotypes had significantly delayed nephritis and demonstrated lower titers of antibodies to double-stranded DNA (Hahn and Ebling, 1983). These injections of monoclonal anti-DNA antibody furthermore induced an anti-idiotypic immune response which may have been the critical modulatory event. Studies looking at T-cell interaction in association with anti-idiotypic immune responses are yet to be performed in these autoimmune mice, and it could be speculated that the results of such studies will provide interesting insights into the molecular disturbance in this disease process.

The regulatory events in a new model of experimental autoimmune interstitial nephritis have also been evaluated (Neilson *et al.*, 1983). *kdkd* mice have normal kidneys at birth but after 8 weeks of life develop a progressively fatal interstitial disease. Humoral immune products are not present in the kidney nor thought to contribute to the nephritogenic immune response. Disease can be transferred through bone marrow chimeras and prevented by chimeras in the reverse direction using donor bone marrow from nonnephritic congenic control strains (Neilson *et al.*, 1983). Of interest, the interstitial nephritis in *kdkd* mice can also be substantially prevented by treating these mice with T lymphocytes from congenic substrain controls. These findings suggest that the defect in *kdkd* mice may be related to a defective suppressor T-cell mechanism.

In experimental anti-TBM disease, anti-idiotypic antisera directed against idiotypes expressed by kidney-bound anti-TBM antibodies can prevent the development of disease in naive recipients (Brown *et al.*, 1979; Zanetti *et al.*, 1983). Immunization with tubular antigen-reactive T lymphoblasts in adjuvant or the intravenous administration of tubular antigen-reactive T cells can also induce an anti-idiotypic immune response which prevents the development of interstitial nephritis in BN rats (Neilson and Phillips, 1982b; Neilson *et al.*, 1983). In more recent studies, tubular antigen-derivatized lymphocytes have been used to induce a suppressor T-cell network which can interfere with the development of experimental interstitial nephritis in SJL mice (Neilson *et al.*, 1984b). Induction of these suppressor cells is I–J-restricted and the cells themselves are Thy-1.2$^+$, Lyt-1.2$^+$, I–J$^+$ lymphocytes. These suppressor cells appear to be phenotypically similar to the Ts_1 suppressor cells described in other complex suppressor cell systems. The generation of these suppressor cells and the dissection of the genetic rules governing their function should prove useful to our further understanding of immune regulation in experimental kidney disease.

3.2. IMMUNE REGULATION AND HUMAN RENAL DISEASE

Studies of immunoregulation in human renal disease are at their earliest stages of development, and many of the observations that have been made are only initial, isolated evaluations. Furthermore, the problems of conducting these kinds of investigations in humans are compounded by the genetic heterogeneity within the population, the lack of time controls for the disease expression, and a poorly understood analysis of nephritogenic effector functions. In spite of these qualifications, several interesting observations can be reviewed. Abnormalities in T-cell regulation have been observed in patients who have what is commonly referred to as nil lesion (Shalhoub, 1974). A lymphokine has been demonstrated which can alter glomerular permeability to serum proteins (Lagrue *et al.*, 1975), and it has been proposed that the lymphokine is secreted as part of an imbalance in thymic function. Nonspecific suppressor T cells have also been observed in patients with nil lesions (Osakabe and Matsumoto, 1981), as well as in other patients with nephrotic range proteinuria (Beale *et al.*, 1980). A metabolic defect related to the nephrotic syndrome has been proposed as a common denominator in these patients. A nonspecific suppressor monocyte has also been observed in patients with membranous glomerulopathy (Ooi *et al.*, 1980). This suppressive monocyte effect is thought to have a modulatory influence on the character of autoantibody formation and, thus, is considered to influence either immune complex size or antibody affinity. In other human studies, it has been observed that serum from patients with IgA–IgG nephropathy often contained elevated amounts of IgA (Egido *et al.*, 1983). An increase in IgA-secreting B cells has also been observed (Egido *et al.*, 1982, 1983; Sakai *et al.*, 1979) in conjunction with a deficiency in IgA-specific suppressor T cells (Nomoto and Arimori, 1979). The absence of this suppressor cell subpopulation could prevent patients with IgA–IgG nephropathy from effectively modulating a pathogenic IgA immune response, and in effect, then, the disease process would remain uncontrolled.

Patients with systemic lupus erythematosus have been an important source of information on altered immune regulation in immune complex glomerulonephritis. Hyperactivity of human lupus B cells (Steinberg *et al.*, 1982), altered suppressor T-cell function (Steinberg *et al.*, 1981) and circulating anti-suppressor

T-cell antibodies (Sakane *et al.*, 1979) have all been described. These findings, however, are not present in all patients and this regulatory defect does not clearly point to a fundamental or pathogenically specific abnormality in regulatory circuits which would predict the wide array of disturbances that have been observed in the setting of lupus nephritis. Several years ago, Carnegie and Mackay (1975) hypothesized that autoimmune disorders could represent a failure in the regulation of antireceptor immunity. In a more recent study in lupus patients, an inverse relationship between the presence of anti-idiotypic antibodies to anti-DNA antibodies and the activity of clinical disease was observed (Abdou *et al.*, 1981). Remission serum containing anti-idiotypic antibodies preferentially reacted with autologous anti-DNA antibodies, suggesting that anti-idiotypic antibodies from one lupus patient may not recognize the same idiotypic repertoire in other patients. Of further interest, however, serum from normal individuals who had been exposed to lupus patients or lupus blood products could cross-reactively inhibit the binding of anti-DNA antibodies from unrelated patients. These latter findings suggest that normal individuals may have a broader anti-idiotypic repertoire than lupus patients in remission.

Thus, in summary, the study of altered immune regulation in human renal disease is now in its earliest stages of development. The future of this area of research, however, depends heavily on additional studies in basic animal models. An examination of regulatory functions at this level and in these terms can only promise new and potentially exciting information.

4. EFFECTOR LIMB FUNCTION AND THE NEPHRITOGENIC IMMUNE RESPONSE

The phenotypic expression of many renal diseases is highly variable. Many circumstances of common pathogenesis may furthermore result in different histopathologic lesions in different patients. Several histologic parameters are subject to this variability: The character and distribution of cellular infiltrates, the identity and pattern of immunoglobulin deposition within the kidney, and the degree of end-stage fibrosis can all vary in severity among glomeruli, interstitium, and vascular structures. Even within a single structure, the glomerulus, we recognize forms of disease which affect some but not all glomeruli (focal), and forms which involve all glomeruli (diffuse). Some lesions will affect only parts of each glomerulus (segmental) or the entire glomerulus (global). In a very large sense, then, we only understand selected factors which can lead to glomerular and tubulointerstitial injury. The fact is that most of the variables which modulate the phenotypic expression of disease have not been cohesively integrated or fully accepted on a cause-and-effect basis. There are a number of issues relating to genetic background, concurrent disease, vehicle of induction, and unique immune reactivity which are not presently known. The effector limb of the nephritogenic immune response has historically been subdivided into humoral and cell-mediated processes. In a large majority of renal lesions, this subdivision is somewhat arbitrary, as the final complexion of the pathologic injury pattern often involves both of these arms of the effector limb. We have nevertheless chosen to review the humoral immune response, T-cell-mediated effector function, and the macrophage response to renal injury as separate entities in an effort to preserve clarity and organization.

4.1. AN OVERVIEW OF THE HUMORAL IMMUNE RESPONSE IN RENAL DISEASE

The humoral immune response in renal disease, from a historical perspective, has held a preeminent position as the principal effector mechanism of immune-mediated disease. This position is largely unchallenged, even today. Because antibody formation to complex antigens is dependent on the T-cell repertoire, because antibodies lodged in parenchymal tissues often attract infiltrating cells, and because the removal of immune deposits depends a great deal on reticuloendothelial function, it has become increasingly necessary to view the humoral immune response in renal disease as being closely related to and interdependent on other inflammatory processes (Neilson and Phillips, 1980).

The actual location of immunoglobulin within the kidney architecture has always been assumed to be of some pathogenic importance. With the routine use of immunofluorescence, various renal diseases have also been subdivided into potential mechanisms based on immune deposit morphology. It has not been easy, however, to fully understand a particular pathogenic mechanism simply on the basis of antigen/antibody location (McCluskey, 1974). Antibodies directed to kidney-specific determinants will be distributed according to the location of these determinants. If antigenic epitopes are repeated with high frequency along basement membranes, a fluorescent pattern of linear staining will appear. If, on the other hand, the epitopic density is interrupted, then a lumpy–bumpy pattern of immunofluorescence will be observed. In recent experiments, it has also been observed that exogenous antigens which have lectinlike properties can selectively bind to glycoproteins in the kidney. Only later will antibody find these antigens in the kidney. The distribution of antibody will depend largely on previous glycoprotein interactions with the antigen, and the morphologic distribution will not reflect additional pathogenic events. Many of these recent views have contributed to the idea that immune deposits, in some cases, may be formed by an "*in situ* mechanism" as compared to the fallout and deposition of circulating preformed immune complexes (Couser and Salant, 1980). It now seems that preformed or *in situ*-formed complexes are probably not an either/or proposition. These immune reactants exist in a molecular equilibrium and may remain in a particular tissue location as a function of other biochemical constraints. As a general rule, however, it does not seem that preformed immune complexes will deposit in the subepithelial regions of renal basement membranes (Couser and Salant, 1980). Preformed immune complexes may and probably do contribute to the inflammation observed following subendothelial and mesangial deposition. A variety of factors have been described which have some influence on the glomerular and interstitial localization of

TABLE 4. DETERMINANTS OF RENAL IMMUNE DEPOSIT FORMATION[a]

A. Quantitative conditions of immune complex formation
   1. Antigen/antibody ratios
   2. Equivalence point
B. Intrinsic qualities of immune reactants
   1. Affinity and avidity characteristics
   2. Antigen valence properties
   3. Temperature dependence
C. Physiopharmacologic factors
   1. Blood pressure and flow
   2. Complement fixation
   3. Vasoactive amine production
D. Properties of the glomerulus
   1. Basement membrane charge
   2. IgG and C3b receptors
E. Mesangial and phagocytic systems
F. Complement clearance mechanisms

[a]Modified from Neilson and Phillips (1980).

preformed immune complexes (see Table 4). The ratio of antigen to antibody in the circulation partially determines optimal immune complex formation. Large circulating aggregates formed during antibody excess are usually removed by phagocytic mechanisms before they reach parenchymal tissues. Very small immune complexes formed in antigen excess remain in the circulation for extensive periods of time and do not seem to contribute to tissue damage. Thus, the medium-size complexes formed in slight antigen excess but near equivalence, appear to have an important pathogenic role (Wilson and Dixon, 1981). The quality of the antibody and the valence properties of the antigen are also important considerations in the formation of pathogenic circulating immune complexes (Neilson and Phillips, 1980). Antibody affinity is genetically controlled, and experimental animals that produce low-affinity antibodies are at particular risk for significant glomerular injury (Steward, 1981). Temperature dependence and complement fixation are additional determinants of pathogenicity (Neilson and Phillips, 1980). Blood pressure and flow as well as the indirect effects of vasoactive amines released by inflammatory cells can all influence immune complex localization. The presence of C3 and IgG receptors in the glomerular and interstitial structures of some species may also influence immune complex localization (Gelfand *et al.*, 1976, 1979). Finally, the integrity and function of the mesangium may regulate many aspects of immune complex localization (Sterzel *et al.*, 1982). Alterations in mesangial morphology may be the direct result of immune complex formation, and mechanical or physiologic dysfunction of this specialized pericapillary tissue may alter the rate of removal of deposited material. Further studies of the humoral immune response leading to immune deposit formation, however, must also take into account a variety of biochemical and cellular functions which collectively modulate the formation and determinants of glomerular injury.

4.2. THE ROLE OF T LYMPHOCYTES IN THE EFFECTOR PATHWAY

T lymphocytes traditionally have been given little or no role in actual effector events producing immune-mediated glomerulonephritis. Even today the evidence that T lymphocytes are important mediators of glomerular injury is still somewhat tenuous. With the advent of monoclonal antibody technology, it has been possible to determine the phenotype of the cells present in renal inflammatory infiltrates (Platt *et al.*, 1982). Initial studies, yet to be confirmed, suggest that T lymphocytes are rarely, if ever, found in a wide variety of glomerular infiltrates (Atkins *et al.*, 1982). These findings are in distinction to the well-documented presence of T cells in the lesions of human transplant rejection (Atkins *et al.*, 1982; Platt *et al.*, 1982), and in the lesions of both human (Platt *et al.*, 1982) and experimental tubulointerstitial nephritis (Mampaso and Wilson, 1983; Zakheim *et al.*, 1984). In addition to the presence of T lymphocytes in these infiltrates, a number of investigators have also reported a variety of T helper to suppressor cell ratios in these infiltrates. The functional significance of these findings has yet to be determined. Unfortunately, as has recently been demonstrated in mice, phenotyping of T-cell subpopulations does not clearly correlate or predict immune function (Swain, 1981).

Because T lymphocytes are not directly present in a variety of glomerular lesions, it has been proposed that such cells might indirectly influence the composition of renal histopathology by the effects of soluble lymphokines. In nil lesion, for example, a vascular permeability factor released by T lymphocytes has been observed to induce proteinuria in rats (Lagrue *et al.*, 1975). This lymphokine effect, of course, is not specific to the glomerular capillary. A number of studies have reported that renal antigen-activated T lymphocytes secrete migration inhibition factor (MIF) which is thought to act by trapping monocytes in areas of antigen deposition (Neilson and Phillips, 1980; Fillit and Zabriskie, 1982). It is hard to know, however, what role this lymphokine plays in the development of glomerular lesions if the T cell is not physically present to expand the monocyte infiltration. In spite of the lack of evidence supporting a presence of T lymphocytes in the lesions of immune-mediated glomerulonephritis, there are provocative experimental studies which implicate T cells in the development of glomerular injury. In a model of accelerated nephrotoxic serum nephritis, rats pretreated with lymphocytes responsive to rabbit immunoglobulin are subsequently given subnephritogenic doses of rabbit anti-GBM antiserum (Schreiner *et al.*, 1982). Recipients of the sensitized lymphocytes demonstrated active hypercellular lesions in approximately 40% of glomeruli compared to only about 10% of glomeruli in recipients receiving antisera alone. Bhan *et al.* (1978) in a similar model also observed segmental glomerular hypercellularity using purified T lymphocytes. In both of these studies, the transferred antibody could not account for the development of the glomerular lesion, although the presence of T lymphocytes was not demonstrated in the areas of hypercellularity. Using preformed immune complexes, Bhan *et al.* (1979) also reported that the transfer of sensitized T lymphocytes could induce glomerular proliferation. This proliferation, however, could only be demonstrated by radioactive thymidine labeling techniques.

In the case of experimental interstitial nephritis, the role of T lymphocytes in the effector limb of the nephritogenic immune response is more convincingly established (Neilson and Phillips, 1979). Using the model of anti-TBM disease in rodents, a variety of studies support this conclusion. From experimental work in rats, it has been observed that antibody titer, subclass, and affinity cannot predict disease susceptibility (Neilson *et al.*, 1983). In the SJL mouse, the activation of helper T cells (Lyt 1^+) depends on I–A-region gene products. These helper cells subsequently induce a cell-mediated (Lyt 2^+) effector response defined by gene products in H-2K (Neilson *et al.*, 1982a, 1983b). These Lyt-1^+ helper cells are sufficient to adoptively transfer the interstitial lesions to a naive recipient (Zakheim *et al.*, 1984). The lesions themselves contain many different kinds of mononuclear cells including macrophages, NK cells, and T cells of either helper or suppressor/cytotoxic phenotype (Zakheim *et al.*, 1984). Cytotoxic T cells and cells participating in delayed-type hypersensitivity responses to tubular antigen are also restricted by genes in H-2K. T cells producing delayed-type hypersensitivity can also be harvested from nephritic kidneys. When Lyt-1^+ T cells are transferred, disease does not appear for 4–6 weeks. This finding suggests that the transferred cells are not effector cells. Thus, it seems that Lyt-1^+ cells are responsible for disease activation and, in the presence of appropriate Ir genes, they subsequently induce Lyt-2^+ T cells for disease expression. Interestingly enough, T cells from guinea pigs with anti-TBM disease also secrete lymphokines which stimulate fibroblast proliferation and collagen synthesis (Neilson *et al.*, 1980). In this experimental model, the role of anti-TBM antibodies cannot be completely excluded as a contributor to the development of the renal lesion. It can be stated, however, that in some species, the T-cell repertoire is sufficient to induce many of the characteristic findings of this interstitial injury pattern. A role for T cells in chronic pyelonephritis in animals and in humans has also been proposed in a recent investigation (Mayrer *et al.*, 1983).

In summary, the role of the T lymphocyte as an effector cell has been expanded in recent years. While its influence in the development of glomerular lesions is tentative, its influence on the development of interstitial nephritis seems fully established.

4.3. ANALYSIS OF MACROPHAGE EFFECTOR FUNCTION

Mononuclear phagocytes are reported to have an important role in the effector limb of the immune response during immunologically mediated renal disease. Information from morphologic, immunohistochemical, and ultrastructural studies also advances the contention that monocytes are normally present in the kidney. Experiments to be reviewed below show that the number and character of glomerular and interstitial monocytes can change in response to disease or antigen challenge.

Complementing the identification of monocytes in the intact kidney are those *in vitro* studies of monocyte effector function which may be active during renal injury. These include the interaction of mononuclear phagocytes with the coagulation cascade (as a producer of procoagulant activity, as a source of plasminogen activator, and as a mediator in the synthesis of polymerized fibrin), fibroblast activity (positive and negative modulation of fibroblast proliferation and collagen synthesis), and as responders to chemotactic mediators. The monocyte may also be involved in vascular endothelial proliferation and the clearance of immune complexes from glomerular structures.

4.3.1. Renal Monocyte-Macrophage Populations during Disease

Several reports have documented the presence of monocytes and macrophages in the glomeruli from biopsy specimens of human renal disease. Magil and associates (Magil and Wadsworth, 1981, 1982; Magil *et al.*, 1981) have evaluated more than 300 renal biopsies for monocyte involvement. Using nonspecific esterase staining, morphologic and ultrastructural analysis, they found monocytes in several conditions. Monocytes were present in crescents from patients with anti-GBM disease, but much less so in immune complex disease. In a separate study, they showed monocytes insinuated between electron-dense deposits and the endothelium, and argued that monocytes were involved in the clearance of immune deposits. Finally, these investigators showed that intraglomerular monocytes were present in relatively high numbers in immune complex disease expressing subendothelial electron-dense deposits, whereas fewer numbers of monocytes were seen in diseases with subepithelial, mesangial, intramembraneous, or no deposits. These are helpful data in that they demonstrate the presence of intraglomerular monocytes during disease, perhaps with some relationship to the histopathologic lesion. More detailed interpretation is not possible without additional information on case selection, extrarenal disease, patient drug use, and blood immunochemistry.

Harry *et al.* (1982) found nonspecific esterase-positive glomerular cells (presumably monocytes) in 8 of 68 unselected renal biopsies accumulated prospectively over a 15-month period. No distinguishing associated findings were discovered. Jothy and Sawka (1981) studied 13 biopsy specimens from 10 patients with systemic lupus erythematosus. They identified monocytes by immunoperoxidase positivity in 8 specimens from 6 patients. A positive correlation existed between immunoperoxidase staining and mesangial hypercellularity as well as proteinuria. Shigematsu *et al.* (1976) showed monocytes in association with glomerular hypercellularity in proliferative glomerulonephritis patients who had completely recovered. A similar finding was shown in a rat Masugi nephritis model. These were exclusively morphologic studies. A divergent view was given by Monga *et al.* (1981), who failed to show a significant association between glomerular hypercellularity and monocyte infiltration. These human studies share the problem of specific case-disease definition and do not carefully control for systemic factors which may affect monocyte traffic into the kidney. They do, however, demonstrate the frequent presence of mononuclear phagocytes in the glomerulus during the active phase of a wide spectrum of immune-mediated diseases.

Animal models of experimental renal disease have also been employed to explore the effector role of the mononuclear phagocyte in the kidney (Schreiner *et al.*, 1982). Hunsicker *et al.* (1979) mapped the time course of glomerular hypercellularity by [^{3}H]thymidine incorporation, morphology, and monocyte infiltration using nonspecific esterase staining in a bovine serum albumin-induced serum sickness model. In this system, they found a parallel increase in hypercellularity and monocyte infiltration. Several groups have also examined the relationship of macrophage accumulation during bovine serum albumin-induced serum sickness in rabbits to the development of proteinuria and renal insufficiency. Holdsworth *et al.* (1980b) and Becker *et al.* (1982) report similar observations. The histologic lesions induced by repeated albumin injection were variable, ranging from mesangial proliferation to crescentic proliferative glomerulonephritis. Macrophage accumulation was detected by light and electron microscopy between the second and third week of immunization, and persisted through the ninth week. In both studies, the onset of proteinuria was coincident with monocytic infiltration. Glomerular culture techniques were also used to quantify macrophage accumulation in the nephron. Holdsworth *et al.* (1980a) found that there was variability in the amount of mononuclear phagocytes per glomerulus among animals, but that the mononuclear cell number/glomerulus ratio correlated positively with the timing and degree of proteinuria and glomerular hypercellularity. These reports confirm the coincidence among several markers of disease but fail to specifically implicate the macrophage as the effector cell in the system. The serum sickness model has also been used by two groups (Holdsworth *et al.*, 1981; Lavelle *et al.*, 1981) to show that treatment of experimental animals with antimacrophage serum following induction of renal disease will attenuate or abrogate its pathologic expression. This information is more exacting in its support of the descriptive serum sickness data above.

The presence of macrophages in autoimmune renal disease has been shown by Hunt *et al.* (1980) and McGiven *et al.* (1981). Outgrowths from murine glomerular cultures were characterized by morphology, ultrastructure, and histochemistry. A motile and phagocytic cell type was recovered from NZB/NZW mice but not from controls. These cells stained positively for nonspecific esterase, had surface Fc receptors, stained positively by indirect immunofluorescence with specific rabbit anti-mouse macrophage serum, and had macrophage ultrastructural characteristics.

The most frequently studied model of macrophage accumulation is nephrotoxic serum nephritis. Sterzel and Pabst (1982) used an accelerated rat anti-GBM nephritis to study the time course of glomerular hypercellularity and monocyte infiltration. Histologic measures of both parameters proved insensitive. However, cellular incorporation of [^{3}H]thymidine as a measure of cell proliferation, and nonspecific esterase staining showed a rise in endogenously proliferating cells on day 1, but a rise in monocytic infiltration on days 2 through 4 following induction of disease. These results, based on more sensitive probes than the human studies reviewed above, suggest that the initial hypercellularity in this disease is not due to renotropic migration of circulating monocytes. The exact role of the invading monocyte in terms of disease resolution or interaction with intrinsic renal structures was not elucidated by this study. Differences in the mechanism of injury between this model and that of Hunsicker described earlier may contribute to their divergent results. A role for the interaction between renotropic bloodborne monocytes and intrinsic renal structures has been proposed by Sraer *et al.* (1984). In admixture experiments involving rodent peripheral blood monocytes and isolated glomeruli from syngeneic animals, a stimulation of C20:4 fatty acid metabolism in macrophages was seen with a subsequent inhibition of C20:4 fatty acid metabolism in glomeruli. These changes were not seen when monocytes or glomeruli were studied separately. The pathways by which this metabolic interaction/regulation affects disease development and expression remain to be elucidated.

Holdsworth *et al.* (1980b) induced nephrotoxic nephritis in NZW rabbits using sheep anti-rabbit GBM antibody. This produced a crescentic, proliferative lesion. Using histochemical stains for phagolysosomal enzymes, they showed that over half of all crescentic cells were monocytic in origin. Only trace amounts of enzyme were found in control animal kidneys. Monocytic infiltration was demonstrated microscopically at 48 to 96 hr after induction of rabbit anti-rat serum nephrotoxic nephritis by Schreiner *et al.* (1978). An extrarenal origin for these mononuclear cells was suggested by [^{3}H]thymidine incorporation and systemic irradiation (with kidney shields) experiments. These observations were verified by Cattell and Arlidge (1981) in the rabbit. Parallel results were also obtained by Dubois *et al.* (1981), who additionally showed an increased outgrowth of endothelial cells and macrophages from glomerular cultures established 2 to 4 days after induction of nephrotoxic nephritis in the rat. These culture data were interpreted to represent a growth-stimulating effect of monocytes on endothelial cells (*vide infra*).

Unusual experimental models have recently been examined to provide support for the recognition that mononuclear phagocytes are probably involved in a wide spectrum of renal diseases. Nagle *et al.* (1982), in a detailed pathologic study, infected rabbits with a human isolate of *Trypanosoma rhodesiense*. Animals developed trypanosomiasis, and the kidney lesion was characterized by monocytic accumulation as detected by nonspecific esterase staining.

4.3.2. Mechanisms of Macrophage Effector Function

4.3.2a. Mononuclear Phagocytes and the Coagulation Cascade. There is ample pathologic evidence to include blood coagulation as part of the mechanism of injury in many glomerular diseases. Capillary thrombosis, aggregated and degranulated platelets, and fibrin tactoids are prominent histologic features of the glomerulus in the hemolytic–uremic syndrome, thrombotic thrombocytopenic purpura, the defibrination–intravascular coagulation syndromes of infection or malignancy as well as many forms of proliferative glomerulonephritis (Spargo *et al.*, 1980). Prompted by these observations, many clinicians have attempted to treat these glomerular diseases with anticoagulant drugs. The mononuclear phagocytic system appears to be involved in both procoagulant and anticoagulant aspects of these disease processes.

The role of a monocyte procoagulant tissue factor in the immune response has recently been reviewed (Edwards and Rickles, 1980). This material [tissue factor (TF)] is a membrane-associated, lipid-dependent glycoprotein of 56,000 daltons. TF apparently exists in an inactive form on cell surfaces and is activated by cell membrane perturbation. Mononuclear cells harvested from peripheral blood of patients anticoagulated with warfarin sodium (prothrombin time twice normal) demonstrated significantly decreased TF production following antigenic or mitogenic stimulation (Edwards and Rickles, 1978), suggesting that TF or an intermediate protein in the activation sequence may be vitamin K-dependent. Monocyte subpopulations separated on albumin gradients have been shown to vary in their ability to generate procoagulant activity in response to stimulation with bacterial endotoxin (Edwards *et al.*,). These experiments also demonstrate that granulocytes are not required for TF activity and that only a subset of monocytes may be responsible for the production of physiologically important quantities of TF. Edwards and Rickles (1980) have summarized experiments which show cultured mononuclear cell TF generation in response to a wide variety of immunologically relevant stimuli, including bacterial endotoxin, mitogens, antigens, antigen–antibody complexes, aggregated immunoglobulins, and complement components. Many of these activators have also been identified as stimuli or components of immune-mediated injury in experimental and clinical renal disease.

The possible role of a monocyte TF-like substance in the mediation of renal disease was evaluated by Hattler *et al.* (1973), who studied the procoagulant activity of mononuclear cells recovered from a human renal allograft which had been rejected. They showed that these mononuclear cells, karyotypically of recipient origin, generated increased procoagulant activity as compared to peripheral blood mononuclear cells or kidney parenchymal cells from the same patient, using freshly prepared cell sonicates. The procoagulant activity of the mononuclear cells was eradicated by incubation of the cell sonicate with antibody to human TF. There was, interestingly, intense fibrin deposition present in the rejected organ. Evidence also exists demonstrating the production of TF by human neonatal foreskin fibroblasts in culture (Korn *et al.*, 1982). The generation of TF by these cells was inhibited by the addition of supernatants of phytohemagglutinin (PHA)-stimulated peripheral human blood monocytes. This inhibition was independent of effects on cell viability and proliferation (determined by cell protein measurements). The inhibition was associated with a 20-fold increase in prostaglandin E_2 production which was indomethacin inhibitable. Several lines of evidence have been reviewed which argue that procoagulant activity of mononuclear cells is amplified and promoted by the soluble mediators released from activated lymphocytes (Hogg, 1983). This induction is extremely rapid and may be genetically restricted by HLA-DR gene products (Helin and Edgington, 1983).

Further work which explores the way in which coagulation and immune responses may be associated has recently been reported. A subpopulation of human peripheral blood monocytes have been shown to surround themselves with a radial array of needlelike projections after 1.5 sec when allowed to adhere to glass in the presence of human plasma (Sherman, 1983). These needles were shown to be fibrin by gel electrophoresis and reaction with a well-characterized monoclonal antibody. This fibrin, however, may differ from that found in conventional clots (Hogg, 1983). The glycoprotein fibronectin may mediate fibrin attachment, as suggested by cocapping experiments. Macrophages from various tissue sites differ in their ability to form fibrin needles. Within the peritoneal cavity, only 25–50% of exudate macrophages display surface fibrin or fibrinogen (Colvin, 1983). The question remains whether different stable subpopulations of mononuclear cells are responsible for fibrin elaboration, procoagulant activity, and plasminogen activator secretion or whether this diversity reflects altered states of macrophage activation. Additional evidence linking fibrin formation, monocytes, and expression of renal disease was presented by Holdsworth *et al.* (1979a). Nephrotoxic serum nephritis was induced in rabbits and shown to produce a proliferative glomerulonephritis with extensive fibrin deposition in Bowman's space, renal failure, and macrophage infiltration. Defibrination of animals with ancrod prevented renal failure, abrogated the fibrin deposition in Bowman's space, and limited the degree of crescent formation. *In vitro* studies suggested that the number of intraglomerular macrophages were the same in the defibrinated and non-defibrinated group, but a much lower rate of macrophage egress toward Bowman's space with evolution into epithelial cells and crescents was seen among the defibrinated group.

A second regulatory pathway through which the mononuclear phagocyte system may mediate the pathologic influence of blood coagulation is through the fibrinolytic system. Mononuclear phagocytes have been shown to secrete a plasminogen activator. Unkeless *et al.* (1974) have shown that mononuclear cells isolated from inflammatory exudates secrete substantial quantities of this enzyme whereas mononuclear cells isolated from sterile peritoneal cavities create little or none. Colchicine, vinblastine, and glucocorticoids block plasminogen activator secretion. Similar data describing enzyme secretion and inhibition have been presented for neutrophils. Granelli-Piperno *et al.* (1977) have advanced a hypothesis that plasminogen activator secretion alters the cellular microenvironment, facilitating mononuclear cell and polymorphonuclear leukocyte migration. There are, however, other considerations on this point. Hau *et al.* (1979) have shown that fibrinolytic activity of peritoneal biopsies from operated, infected (peritonitis) dogs was dramatically suppressed compared to sham-operated animals. They postulated a suppression of mesothelial cell procoagulant activity in this model. The contribution of local factors may alter the role of mononuclear cell plasminogen activation during inflammation occurring at different tissue sites. In the kidney, there are little or no data addressing the fibrinolytic activity of glomerular, tubular, or vascular cell types.

Clearly, factors promoting coagulation which produce a network of fibrin strands trapping immune competent cells and providing a substrate for attachment and collagen synthesis for tissue repair must be intricately balanced against anticoagulant factors which may facilitate cellular migration into sites of inflammation. The regulatory network which operates to guarantee this balance in the kidney is not fully known.

4.3.2b. Immune Modulation of Fibrogenesis. The end-stage kidney which has lost functional capacities for ultrafiltration, tubular secretion and reclamation, blood pressure regulation, and mineral homeostasis is pathologically characterized by extensive fibrosis. The process by which this state is achieved involves the accumulation, activation, proliferation, and synthetic performance of fibroblasts. Many aspects of this sequence of "fibrogenesis" have been demonstrated to be responsive to immune regulation.

Monocytes Elaborate Factors Which Stimulate Fibroblast Proliferation: An early report suggesting that macrophages produce a fibroblast growth factor was that of Leibovich and Ross (1976). They observed that fibroblasts from guinea pig skin wounds which were grown in platelet-poor plasma serum (PPPS) were truly quiescent, with less than 3% of the cells synthesizing DNA and dividing. Peritoneal exudate cells from guinea pigs were prepared from saline lavage and mineral oil elicitation. Pooled macrophages were cultured in PPPS and the supernatant was found to contain a factor(s) which stimulated fibroblast proliferation as determined by hemocytometer cell counts following trypsinization of target monolayers. This factor(s) was dubbed "macrophage-dependent fibroblast-stimulating activity" (MFSA), and found to be nondialyzable and heat stable (56°C for 30 min). Unstimulated lymphocytes from mesenteric lymph nodes failed to show fibroblast-stimulating activity.

Glenn and Ross (1981) explored fibroblast growth factor elaboration from monocyte-derived macrophages compared to tissue macrophages. They found various differences among monocytes, *in vitro* monocyte-derived macrophages, and peritoneal exudate macrophages. They did, however, reliably generate a monocyte-derived growth factor(s) (MDGF) from 5-day-old monocyte cultures (maintained in serum-free medium) exposed to endotoxin or Con A. This MDGF stimulated both fibroblast DNA synthesis and an increase in fibroblast cell number, was trypsin sensitive, heat stable (56°C for 30 min), and nondialyzable. Although there are several similarities between MDGF and MFSA, no conclusion can yet be drawn about their possible structural homologies.

Rutherford *et al.* (1982) confirmed earlier work by demonstrating that human peripheral monocytes, upon stimulation with *N*-acetylmuramyl-L-alanyl-D-isoglutamine (MDP), produced a factor(s) which stimulated fibroblast proliferation determined by counting nuclei *in situ*. This ability to induce proliferation appeared to decrease with monocyte culture age, but was restored with the addition of plasma-derived serum. The authors drew analogy to platelet-derived growth factor and suggested that the monocyte-elaborated factor may act coordinately with substances in plasma-derived serum to effect growth stimulation.

Diegelmann *et al.* (1982) analyzed both a murine and a human *in vitro* system to detect monocyte–fibroblast interactions. They showed that coincubation of mouse peritoneal exudate lymphocytes (PEL) and 3T3 fibroblast low-density monolayers; or mouse PEL conditioned medium and 3T3 fibroblast low-density monolayers; or human peripheral blood monocyte conditioned medium and human dermal fibroblast monolayers all succeeded in stimulating fibroblast proliferation as determined by $[^{125}I]$-UdR incorporation. Cell conditioned media were made without stimulating the conditioning cells and all media contained 10% fetal calf serum. Optimal dilution of conditioned media differed for the murine and human systems. In fact, at concentrations greater than 1:5 in the mouse system, inhibition of proliferation was observed. The murine cells were elicited with thioglycollate which may account for this difference, as compared to the "resting" peripheral blood monocytes. Comparison to other growth factors is restricted by insufficient characterization and purification. De Lustro *et al.* (1980) identified a nondialyzable soluble factor from human peripheral blood monocytes which promoted human foreskin fibroblast growth. The fibroblast growth-stimulating activity was demonstrated in the presence of 10% fetal calf serum and was not inhibited by indomethacin. In a later report, De Lustro and LeRoy (1982) further characterized these monocyte-derived mediators. They reported that adherence of peripheral blood monocytes promotes rapid release of these factors and that treatment of adherent peripheral blood monocyte cultures with lipopolysaccharide enhances the level of fibroblast-stimulating activity in the cell-free supernatant. Phagocytosis or pinocytosis had no effect on mediator release, whereas blockage by pretreatment of monocytes with puromycin resulted in decreased production of these mediators. Monocyte serum proteases were felt to be essential in mediator synthesis or release since tosyl-lysine-chloromethyl ketone (TLCK) and phenylmethyl-sulfonyl fluoride (PMSF), irreversible inhibitors of serum esterase activity, diminished the release of fibroblast-stimulating factors.

The direct application of this principle to renal disease is hinted by the preliminary observation of Melcion *et al.* (1982), who showed increased [^{3}H]thymidine incorporation by dermal fibroblasts and mesangial cell cultures when exposed to a medium conditioned by human peritoneal macrophages isolated from peritoneal dialysis fluid. There are many uncontrolled variables in this report, but the interaction among peripheral blood monocyte, fibroblast, and mesangial cell deserves further study.

Lymphocytes Elaborate Factors Which Stimulate Fibroblast Proliferation: It has been recognized for many years that in certain disease states, fibrosis develops in areas which become heavily invested with mononuclear cells. For example, scleroderma or graft-versus-host disease frequently follows this pattern. Several laboratories have attempted to demonstrate immune mediation in this sequence.

Wahl *et al.* (1978) described a fibroblast growth-promoting agent elaborated by T lymphocytes. Using guinea pig skin explants in culture as a target, these investigators harvested peritoneal exudate cells (PEL) from guinea pigs sensitized *in vivo* to either tetanus toxoid or DNP–ovalbumin. Nonadherent PEL, eluted through a glass bead column, were cultured for 24 to 48 hr in antigen-free medium and then stimulated *in vitro* by the same antigen used for *in vivo* sensitization. Stimulated lymphocyte supernatants were shown to increase fibroblast proliferation, determined by [^{3}H]thymidine incorporation and actual cell counts. Fibroblast cell density was important in the expression of this response which was maximal at a fibroblast density of 1×10^5 per vial.

Schmidt *et al.* (1982) pursued the possibility of a lymphocyte stimulant by assessing and characterizing the effect of supernatant fluid from a secondary allogeneic mixed-lymphocyte reaction on fibroblast behavior. Augmented fibroblast proliferation was demonstrated by [^{3}H]thymidine incorporation and by counting nuclei. Biochemical characterization of their supernatant fluid by gel filtration chromatography and isoelectric focusing demonstrated that the fibroblast proliferation factors had molecular weights of 75,000 and 13,000 and isoelectric points of 6.8 to 7.2, 5.8 to 6.2, and 5.2 to 5.4. The activity profiles of thymocyte and fibroblast proliferation by this supernatant were superimposable and fractions with peak proliferation-stimulating activity were demonstrated to stimulate interleukin-2 (IL-2) production by a murine T-cell lymphoma line. These data were interpreted to suggest that the active component of the secondary allogeneic mixed-lymphocyte reaction supernatant was IL-1, and that local production of IL-1 by mononuclear cells may serve to regulate fibroblast proliferation and fibrosis in chronic inflammatory states.

Postlethwaite and Kang (1983) have identified similar IL-1-like substances in the supernatants of unfractionated mononuclear leukocytes stimulated with T-cell antigen and mitogens. This supernatant also increased [^{3}H]thymidine uptake in nonconfluent fibroblast cultures, was heat stable (56°C for 45 min), trypsin sensitive, and yielded two major activity peaks on Sephadex G-100 gel filtration of 60,000 and 16,000 daltons.

Neilson *et al.* (1979, 1980) in their series of experiments on cell-mediated immunity in interstitial nephritis, examined the question of lymphocyte modulation of fibrogenesis. Interstitial nephritis was induced in guinea pigs with an injection of renal tubular antigen (RTA). T lymphocytes recovered from nephritic animals were stimulated *in vitro* with soluble tubular antigen. Dialyzed supernatants prepared early in disease were found to contain a factor which enhanced fibroblast proliferation as determined by [^{3}H]thymidine incorporation. The potency of this stimulation was found to vary with the duration of disease. In contrast, dialyzed supernatants from T lymphocytes harvested from control animals were found to contain a factor which inhibited fibroblast proliferation.

Monocytes Elaborate Factors Which Suppress Fibroblast Growth: Korn *et al.* (1980, 1983) studied the effects of supernatants from mixed (monocyte–lymphocyte) cell cultures, either unstimulated or activated by PHA. Supernatants from both cultures suppressed fibroblast proliferation (up to 90%) as determined by [^{3}H]thymidine incorporation. Concomitantly, fibroblast prostaglandin E (PGE) synthesis increased 20- to 70-fold. Fibroblast growth suppression paralleled the increase in fibroblast PGE, was inhibited by indomethacin and meclofenamate, and was reproduced by the addition of exogenous PGE. This activity was recovered from the non-T-lymphocyte, adherent cell fraction and appeared within 6 hr of culturing. The activity was heat (56°C) and trypsin sensitive, nondialyzable, and appeared in the 12,000- to 20,000-dalton fractions on Sephadex G-100 chromatography. At low dilutions there also appeared an indomethacin-resistant suppressive activity. In Rutherford and colleagues' (1982) experiments using MDP-stimulated peripheral blood monocytes (*vide supra*), the fibroblast-stimulating effect was lost in the absence of plasma-derived serum. Finally, in Diegelmann and colleagues' (1982) report of macrophage effects on fibroblast DNA synthesis and proliferation, distinct inhibition of DNA synthesis was observed in a murine 3T3 fibroblast/peritoneal macrophage system at low peritoneal macrophage supernatant dilutions.

Lymphocytes Elaborate Factors Which Suppress Fibroblast Proliferation: Suppression of fibroblast proliferation is not a unique feature of any specific experimental model. Many mononuclear cell population/fibroblast target systems can show stimulation, no effect, or suppression as a function of prevailing experimental conditions. In the same model, this variability may have several explanations. Necessary cofactors may be absent; critical effector cell subpopulations may be absent or insufficiently matured or activated; different methods of stimulation may yield opposing regulatory actions; transformation of cell surface properties by disease may alter the response pattern to a constant activator; some mononuclear cell–fibroblast interactions may be genetically restricted, or relative cell densities or mediator concentrations may promote positive or negative responses.

Neilson *et al.* (1980) showed that the supernatants from cultured T lymphocytes from control guinea pigs produced a suppression of fibroblast proliferation as measured by [^{3}H]thymidine incorporation. In the experimental group with interstitial nephritis, as described earlier, the opposite effect was seen early in the disease, but a suppressive lymphokine was also secreted later.

Mononuclear Cell Fractions Influence Protein Synthesis, Including Collagen and Collagenase: The immune modulation of fibrosis may involve effects on fibroblast synthetic and secretory activity, in addition to the proliferative or anti-proliferative effects described above. The reports on this subject indicate that positive or negative regulation of these fibroblast functions is complex, probably determined by several variables which may include specific immunocompetent cell types, tissue site and cell densities of immunocyte-mesenchymal cell interaction, availability of serum proteins, and the net effects on the target cell of multiple (perhaps competing) mediators.

Wahl *et al.* (1978), using guinea pig skin explants as target tissue, harvested peritoneal exudate lymphocytes elicited with mineral oil from guinea pigs previously immunized with DNP–ovalbumin. They showed that *in vitro* stimulation of these lymphocytes with DNP–ovalbumin produced a supernatant which induced increased PGE_2 production and increased collagen synthesis (measured by protein [^{3}H]proline incorporation and subsequent collagenase sensitivity). These synthetic changes preceded the fibroblast proliferative response.

Melcion *et al.* (1982), using peritoneal macrophages harvested from the first exchange of patients undergoing peritoneal dialysis and stimulated *in vitro* with zymosan-activated serum, also showed an increase in [^{3}H]proline incorporation by fibroblast target cells when exposed to macrophage supernatant. Their cells also showed a positive proliferative response, but detailed time controls were not included. Supporting the view that normal mononuclear cells can secrete substances which depress human dermal fibroblast activity are the data of Jimenez *et al.* (1979), who studied human peripheral blood monocytes and lymphocytes isolated by Ficoll–Hypaque gradient centrifugation. Supernatants from mononuclear cell suspensions, either unstimulated or activated by PHA, were incubated with confluent human foreskin fibroblasts, using [^{14}C]proline incorporation into [^{14}C]hydroxyproline as a measure of collagen synthesis. Experiments were conducted in 1% serum-containing medium, based on the authors' observation that serum-free medium gave variable results and that noncollagen [^{14}C]proline incorporation at higher serum concentrations could mask an inhibiting effect of their supernatants on collagen synthesis. Adequate ascorbic acid availability in the reaction medium was carefully controlled. These experiments showed a consistent, significant, and selective inhibition of collagen synthesis by the target fibroblasts when incubated with the supernatant from stimulated mononuclear cells. The role of serum deprivation or excess in the reaction medium was reviewed and invoked as a possible explanation for the discrepant effects of mononuclear cell supernatants on collagen synthesis reported by others. The contribution of monocytes to the remodeling or immunoregulatory dimension of fibrogenesis was recently considered by Postlethwaite *et al.* (1983). They showed that highly purified human peripheral blood monocyte IL-1 was able to stimulate collagenase production by the human infant foreskin fibroblast line HFL-1. These experiments were conducted in serum-free medium with a 3-day period of target cell serum starvation.

Neilson *et al.* (1980), working with a model of interstitial nephritis in guinea pigs, showed that collagen synthesis (measured by [^{14}C]proline incorporation into [^{14}C]hydroxyproline) was stimulated by dialyzed T-lymphocyte supernatants from nephritic animals but was suppressed by dialyzed T-lymphocyte supernatants from control animals. These experiments were conducted at fibroblast confluency to eliminate the possible complicating effects of cell growth.

Recent work by Korn (1983) significantly alters the interpretative framework for cytokine–fibroblast studies. He has shown that fibroblast PGE_2 synthesis was increased by exposure to supernatant from PHA-stimulated human peripheral blood mononuclear cells. This effect persisted for 20 weeks and 19 cell generations after the original exposure to mononuclear cell products. Experimentally, it seems critical to know the long-term exposure history of target cells in these other study systems. A growing body of recent work has also explored possible mechanisms by which monocytes and fibroblasts are recruited to the kidney during certain phases of selected diseases. A widely held belief is that a chemoattractant is released at the initial site of injury which promotes migration of immune competent cells to a renal site of injury (Norman *et al.*, 1978). As these effector cells accumulate, they amplify production of the attractant(s) or initiate the manufacture of the substance(s) autonomously. One such substance which has been studied in detail with regard to monocyte–fibroblast interaction in the kidney is fibronectin (cold-insoluble globulin).

Fibronectin is a high-molecular-weight glycoprotein which was discovered as a constituent of "conditioned media" in early tissue culture experiments. Its biologic role is diverse. Fibronectin is an effective cell agglutinin and cell-surface adhesive. In reconstitution experiments with transformed fibroblasts, fibronectin restores normal "social" cell–cell interaction (overlap and alignment), restores normal fibroblast morphology including the cell-surface features of microvilli and membrane ruffling, and restores normal cell-surface adhesion. Fibronectin is chemotactic for fibroblasts *in vitro*. There are cellular and soluble (plasma) forms of fibronectin, each of which has been shown to possess a cell-surface binding site and a collagen binding site (Yamada, 1980).

Fibronectin has been demonstrated in the healthy kidney, primarily in the mesangium. Measurements of immunofluorescence show an increase in fibronectin in those diseases associated with mesangial proliferation (Stenman and Vaheri, 1978; Linder *et al.*, 1980; Weiss *et al.*, 1979; Pettersson and Colvin, 1978; Oberley *et al.*, 1979; Dixon *et al.*, 1980; Courtoy *et al.*, 1980; Scheinman *et al.*, 1978). Murphy-Ullrich *et al.* (1982) have also shown that the immunization of mice with autologous or heterologous fibronectin produces a mesangial proliferative glomerulonephritis.

A role for fibronectin in immune renal disease has indirectly been suggested by recent reports which show that fibronectin and fibronectin fragments are chemotactic for monocytes (Norris *et al.*, 1982) and fibroblasts (Tsukamoto *et al.*, 1981). Fibroblasts, monocytes, and endothelial cells are also able to elaborate fibronectin, in addition to the occurrence of this glycoprotein in the mesangium. A possible scenario involving monocytes, fibronectin, and fibrin in the inflammatory process (which may be operative in kidney disease) has been proposed by Bianco (1983). Plasma protein-derived chemotactic factors are theorized to attract monocytes into the area of tissue injury, where surfaces containing denatured collagen and fibrin-containing clots will be coated with fibronectin. Monocytes would bind via their fibronectin receptors. This binding event would facilitate fibrinogen and fibrin net formation retaining the mononuclear cells at the site of injury and could stimulate fibronectin elaboration by arriving monocytes. Binding also may trigger monocyte differentiation into macrophages with increased expression of Fc and C3b receptors, facilitated phagocytic ability, and subsequent neutral protease secretion. This sequence may also explain the variable cell composition (neutrophils or monocytes) of different inflammatory lesions since monocytes are the only leukocytes recognized to express fibronectin receptors.

Fibronectin and other macrophage-related substances are not the only fibroblast chemotactic factors. Postlethwaite *et al.* (1976; Postlethwaite and Kang, 1980) have identified a lymphocyte product [lymphocyte-derived chemotactic factor for fibroblasts (LDCF-F)], produced by T cells in response to antigen or mitogen stimulation, which can stimulate directed migration of fibroblasts *in vitro*.

4.3.2c. Endothelial Proliferation and Angiogenesis. Many glomerulonephritidies involve a vascular proliferative component. Diabetic nephropathy may develop in this fashion. The intimal hyperplasia of malignant hypertension and progressive systemic sclerosis may also be in this category. Clearly, the dramatic endothelial proliferation in the kidney during eclampsia is part of this consideration.

The possibility that activated macrophages are responsible for this endothelial response was first described by Polverini *et al.* (1977), who showed that media conditioned by activated guinea pig peritoneal macrophages (elicited by thioglycollate) were able to induce neovascularization when injected into guinea pig corneas. These experiments were performed in a syngeneic system and the vascular reaction was not inflammatory (no edema or neutrophils). Media conditioned by lymphocytes or neutrophils did not produce this result. In the following year, Greenberg and Hunt (1978) demonstrated that wound-derived macrophages and cell-free wound fluid were able to stimulate endothelial and smooth muscle cell proliferation determined by cell counts of trypsinized monolayers. They hypothesized that the macrophages were the source of the mitogenic activity found in the wound fluid. Recently, Banda *et al.* (1982) isolated and characterized a nonmitogenic angiogenesis factor from rabbit wound fluid. This substance was shown to produce angiogenesis in the corneal implant assay and was shown *in vitro* to be chemotactic for endothelial cells. This angiogenesis factor has also been reported to be produced in response to hypoxia (Knighton *et al.*, 1982). Dubois *et al.* (1981) reported findings in support of this effector function. They induced nephrotoxic serum nephritis in rabbits, some of which were subjected to systemic but not renal irradiation. Outgrowths from *in vitro* glomerular culture, isolated from nonirradiated animals, included endothelial cells and macrophages, whereas cultures of glomeruli from irradiated animals contained only epithelial and mesangial cells. If one accepts the postulate that infiltrating macrophages in glomerulonephritis are marrow-derived (Striker *et al.*, 1979), then the presence of endothelial outgrowths from nonirradiated animals may reasonably be linked to the growth-promoting presence of marrow-derived macrophages. Although speculative at present, the concept of a macrophage-derived endothelial chemoattractant/proliferative angiogenesis factor is appealing. This monokine could mediate certain endothelial reactions which have been well characterized pathologically (Spargo *et al.*, 1980).

4.3.2d. Monocyte-Macrophage Growth-Promoting Effects on Mesangial and Epithelial Cells. The mesangium cooperates in several ways with the RES. In the preceding sections, findings have been reviewed which illustrate the way in which the mesangium serves as a repository for infiltrating monocytes from the circulating blood. In this section the interactions between mesangial cells and monocytes will be discussed as well as the effect of the extrarenal RES activity on mesangial immune clearance.

Monocytes exert both a metabolic and an immunologic influence over the mesangium. Circumstantial evidence suggests that the presence of infiltrating monocytes can lead to mesangial proliferation (Dubois *et al.*, 1981). Dialyzed macrophage supernatants (mouse peritoneal or human peripheral monocytes) can stimulate mesangial cell DNA synthesis *in vitro*, supporting the contention that monocytes may induce mesangial proliferation *in vivo* (Wagner *et al.*, 1983). In a system, perhaps analogous to that which mediates macrophage–fibroblast interactions, these macrophage–mesangial cell interactions may be stimulatory or suppressive. Ooi *et al.* (1983) reported the identification of two factors from resident and endotoxin-stimulated mouse peritoneal macrophages which exerted a suppressive effect on [^{3}H]thymidine incorporation of cultured mesangial cells from syngeneic animals. One G-100 column chromatography fraction of 14,600 to 29,000 daltons was shown to stimulate endogenous PGE synthesis in the mesangial cells, whereas a second factor of 29,000 to 68,000 daltons did not.

Exogenous PGE exerted a suppressive effect on mesangial cell growth. The supernatant suppression, however, was not totally abrogated by the addition of indomethacin, suggesting that one fraction acted through a prostaglandin-independent pathway and one through a prostaglandin-dependent pathway. Although no specific mechanism for ligand–receptor–effector linkage can be derived from these preliminary observations, the findings definitely support a role for the macrophage in the pathogenesis of glomerular mesangial hypercellularity.

4.3.2e. Immune Clearance by Mononuclear Phagocytes. There are numerous studies which characterize the role of phagocytosis as a mechanism by which immune complexes are cleared from the circulation. Receptor–ligand binding, pinocytosis, and membrane turnover dynamics have a demonstrated importance in this process. Phagocytic defects account for a spectrum of human disease. Although a complete review of this topic is beyond the scope of the present chapter, there are selected aspects of the "immune clearance" function of macrophages which are germane to glomerular disease.

It has long been recognized that glomerular, in particular mesangial, structures were responsible for the removal of material from afferent arteriolar blood. Particulate matter (thorium dioxide, colloidal gold, carbon, and iron dextran), polysaccharides (dextran, polyvinyl alcohol polymers), proteins (horseradish peroxidase, myeloperoxidase, catalase, and ferritin), and aggregated proteins or complexes (aggregated globulin, aggregated IgG, aggregated albumin, and antigen–antibody complexes) have been identified in the mesangium following intravenous injection (Michael *et al.*, 1980).

Mesangial cells, while morphologically distinct from monocytes, have several features in common with phagocytic macrophages, including endocytosis of immune complexes, high lysosomal enzyme content, secretion of extracellularly active neutral proteases, and PGE_2 elaboration. Furthermore, mesangial cells have been shown to produce a cytokine ("mesangial cell thymocyte-activating factor") which appears very similar to IL-1 (Lovett *et al.*, 1983).

The mechanism by which macrophages are recruited to the mesangium has been explored from the perspective of "immune clearance," specifically the removal of deposited immunoproteins from the glomerulus. A role for complement mediation has been considered, based on several lines of evidence. Schreiner *et al.* (1982) showed that cellular infiltrates, proteinuria, and C3 immunofluorescence were abolished in rats with accelerated nephrotoxic nephritis, if the animals were decomplemented by treatment with cobra venom factor. Glomerular C3b receptors were found to be decreased in human biopsy specimens from patients with a variety of renal diseases (Colasanti *et al.*, 1980). No correlation was found between C3b receptor activity and disease processes involving C3b-bearing immune complexes. Rather, C3b receptors became undetectable in many immune and nonimmune glomerular diseases with lesions of capillary walls, due to loss of integrity of the C3b receptor-bearing visceral epithelial cells. Kazatchkine *et al.* (1982) verified the epithelial location of C3b receptors in normal human kidney sections using a monospecific rabbit antibody to the 205,000-dalton glycoprotein that is the C3b receptor of human peripheral blood erythrocytes. In contrast to Colasanti *et al.* (1980), these investigators found a relationship between complete loss of C3b antigen and diffuse proliferative lupus nephritis. Normal podocyte staining was found in mesangial proliferative and membranous lupus nephritis. There are, however, many studies which underscore the need for C3b and Fc receptor cooperation to allow and promote phagocyte attachment and ingestion of opsonized material. Furthermore, there are data to suggest a role for a lymphokine in the conversion of nonphagocytic C3b receptors on macrophages to phagocytic ones. This information, reviewed by Griffin (1982), predicts that a defect in either complement or immunoglobulin receptors can impair effective phagocytosis by macrophages.

The essential role for Fc-mediated immune adherence of monocytes has recently been argued by Holdsworth (1983). Several observations were made using a model of passive nephrotoxic nephritis in rabbits. Leukocyte depletion with nitrogen mustard prevented macrophage accumulation, proteinuria, and histologic injury. Decomplementation with cobra venom factor had no effect. Effects of administration of intact anti-GBM IgG and $F(ab')_2$ fraction of the same antibody were compared. Intact IgG was associated with glomerular deposition, macrophage infiltration, heavy proteinuria, and diffuse endocapillary proliferative glomerulonephritis. Deposition of $F(ab')_2$ fraction was associated with only minimal proteinuria. These experiments support the critical role for Fc-mediated adherence in the macrophage infiltration and phenotypic expression of disease in this model.

Impaired Fc-mediated clearance has been suggested as a possible mechanism for disease severity and renal involvement in systemic lupus. Using opsonized, ^{51}Cr-labeled patient erythrocytes, Frank *et al.* (1979) and Parris *et al.* (1982) demonstrated impaired clearance following intravenous injection in lupus patients. The degree of clearance impairment appeared to be independently related to immune complex levels in the blood, clinical activity of disease, and presence of renal injury. Similar clearance defects have been identified in disease-free adults whose leukocytes bear HLA haplotypes DR2, MT1, or B8/DR3. Each of these HLA antigens is also preferentially expressed by patients with lupus.

The observation that impaired Fc receptor-mediated clearance occurs among patients with lupus who have renal disease could be interpreted to mean that inadequate clearance of immune complexes leads to greater amounts of circulating immunoproteins which then deposit in glomeruli. In fact, there are many studies which suggest that glomerular deposition of immune complexes is functionally modulated by an extrarenal component of the mononuclear phagocyte system. Activation of the RES by systemic administration of *Corynebacterium parvum* (Barcelli *et al.*, 1981), zymosan (Raij *et al.*, 1981), and endotoxin (Ford, 1975) can reduce the amount of BSA–anti-BSA complexes which deposit in the kidney of experimental animals following intravenous BSA administration. Conversely, RES blockade in rabbits by injection of thorotrast or horse serum produces augmented immune complex deposition in glomeruli as measured by increased renal radioactivity following an intravenous injection of heat-aggregated–radioiodinated human gamma globulin. RE clearance function was shown to improve with resolution of the horse serum blockade, but was promptly rendered defective with repeat horse serum injections.

These clearance studies emphasize the importance of several factors relevant to the role of macrophages in renal disease. From the pathologic studies reviewed earlier, it appears that the macrophage enters the glomerulus in response, directly or indirectly, to immunoprotein deposition. Factors which affect this deposition (immune complex formation, antibody production, extrarenal clearance, or sequestration) will influence macrophage accumulation in the kidney. These studies do not, however, carefully consider nonimmune factors which may alter disease expression, such as hemodynamic, metabolic, or nutritional parameters.

4.4. NK CELLS IN RENAL DISEASE

NK cells are lymphoid cells which are genetically programmed to spontaneously recognize selected viruses, tumors, and some altered self-antigens in an immunosurveillance capacity (Herberman *et al.*, 1979). The evaluation of NK cells has also implicated their role as effector cells in antibody-dependent cellular cytotoxicity reactions. Very little is known about the role of NK cells in immune-mediated renal disease. In experimental interstitial nephritis, however, these cells have been observed among the infiltrates of nephritic kidneys (Zakheim *et al.*, 1984), and with anti-TBM antibodies may form an informational bridge in antibody-dependent cellular cytotoxicity reactions (Neilson and Phillips, 1981).

4.5. INTERFERON AND RENAL HISTOPATHOGENESIS

There are at least two types of interferon: type 1 is the traditional interferon secreted by most somatic cells in response to viruses, toxins, and polynucleotides (Baron and Diazani, 1977); type 2 interferon, sometimes called immune interferon, is a soluble lymphokine produced in response to sensitizing antigens and immune complexes (Valle *et al.*, 1975; Fujibayashi *et al.*, 1975). Both of these interferons are potent antiviral substances with a variety of immunoregulatory properties (DeMaeyer and DeMayer-Guignard, 1980). Gresser *et al.* (1976) observed that newborn mice treated with interferon for the first week of life developed a progressively fatal glomerulonephritis. The lesions were characterized by mesangial proliferation, thickening of the capillary loops, presence of epithelial deposits, and eventual hyalinization and sclerosis with diffuse tubular atrophy. Interestingly enough, this same lesion was observed in nude mice. The mechanism of this interferon effect is not entirely understood, but the results of its action are quite impressive, and its role in the pathogenesis of virus-induced glomerulonephritis may be significant. For example, it has been observed that lymphocytic choriomeningitis (LCM) virus infections in mice induce the production of interferon (Riviere *et al.*, 1977; Gresser *et al.*, 1978). When these mice were treated with antibodies to mouse interferon, the LCM virus titers dramatically rose, but the lesions of LCM-induced glomerulonephritis did not develop. These latter findings suggest that interferon may have the unique capability of inducing renal lesions following viral infections in susceptible hosts. Interferon has also been observed to enhance the glomerular lesions seen in NZB/W mice (Adam *et al.*, 1980). Such observations further suggest that interferon and/or viral infections may augment an ongoing nephritogenic immune response to the subsequent detriment of the host.

5. SUMMARY

The nephritogenic immune response can be analyzed from several different perspectives. A larger collective view would suggest that the effector pathways of renal injury depend on the complex interplay of a variety of other immunologic phenomena. The role of immunogenetics, the RES, and the cellular influence on the development of antigen-recognition and immunoregulation have accordingly assumed important positions within the study of renal immunopathogenesis. These observations have made it clear that many factors outside the local environment of the kidney can affect the differentiation of pathways leading to renal injury. The further exploration of these issues forms the substance of more recent research initiatives within this field of investigation. The continued elucidation of these components provides some hope that our ability to better understand the mechanisms of renal injury will also lead to the development of novel strategies for specific therapeutic interventions.

REFERENCES

Abdou, N. I., Wall, H., Lindsley, H. B., Halsey, J. F., and Suzuki, T., 1981, Network theory in autoimmunity: In vitro suppression of serum anti-DNA antibody binding to DNA by anti-idiotype antibody in systemic lupus erythematosus, *J. Clin. Invest.* **67**:1297.

Abrahamson, D. R., and Caulfield, J. P., 1982, Proteinuria and structural alterations in rat glomerular basement membranes induced by intravenously injected anti-laminin immunoglobulin G, *J. Exp. Med.* **156**:128.

Adams, C., Thova, Y., Ronco, P., Verroust, P., Tovey, M., and Motel-Maroger, L., 1980, The effect of exogenous interferon: acceleration of autoimmune and renal diseases in (N2B/W)F_1 mice, *Clin. Exp. Immunol.* **40**:373.

Adelman, N. B., Watleng, D. L., and McDevitt, H. O., 1983, Treatment of (NZB × NZW)F_1 disease with anti-I-A monoclonal antibodies, *J. Exp. Med.* **158**:1350.

Andres, G., Brentjens, J., Kohli, R., Anthone, R., Anthone, S., Baliah, T., Montes, M., Mookerjee, B. K., Prezyna, A., Sepulveda, M., Venuto, R., and Elwood, C., 1978, Histology of human tubulo-interstitial nephritis associated with antibodies to renal basement membranes, *Kidney Int.* **13**:480.

Atkins, R. C., Holdsworth, S. R., Hancock, W. W., Thompson, N. M., and Glasgow, E. F., 1982, Cellular immune mechanisms in human glomerulonephritis: The role of mononuclear leukocytes, *Springer Semin. Immunopathol.* **5**:269.

Banda, M. J., Knighton, D. R., Hunt, T. K., and Werb, Z., 1982, Isolation of a non-mitogenic angiogenesis factor from wound fluid, *Proc. Natl. Acad. Sci. USA* **79**:7773.

Banks, K. L., and Henson, J. B., 1972, Immunologically mediated glomerulitis of horses. II. Anti-glomerular basement membrane antibody and other mechanisms in spontaneous disease, *Lab. Invest.* **26:**708.

Barabas, A. Z., and Lannigan, R., 1974, Induction of an autologous immune-complex glomerulonephritis in the rat by intravenous injection of heterologous anti-rat kidney tubular antibody. I. Production of chronic progressive immune complex glomerulonephritis, *Br. J. Exp. Pathol.* **55:**47.

Barcelli, U., Rademacher, R., Ooi, Y. M., and Ooi, B. S., 1981, Modification of glomerular immune complex deposition in mice by activation of the reticuloendothelial system, *J. Clin. Invest.* **67:**20.

Baron, S., and Diazani, F., 1977, General considerations of the interferon system, *Tex. Rep. Biol. Med.* **35:**1.

Beale, M. G., Hoffstein, P. E., Robson, A. M., and MacDermott, R. P., 1980, Inhibitory factors of lymphocyte transformation in sera from patients with minimal change nephrotic syndrome, *Clin. Naphrol.* **13:**271.

Becker, G. J., Hancock, W. W., Stow, J. L., Glasgow, E. F., Atkins, R. C., and Thomson, N. M., 1982, Involvement of the macrophage in experimental chronic immune complex glomerulonephritis, *Nephron* **32:**227.

Benacerraf, B., and McDevitt, H. O., 1972, Histocompatibility defined immune response genes, *Science* **175:**273.

Bergstein, J. M., and Litman, N., 1975, Interstitial nephritis with anti-tubular basement membrane antibody, *N. Engl. J. Med.* **292:**875.

Berthoux, F. C., Gagne, A., Sabatier, J. C., Ducret, F., LePetit, J. C., Marcellin, M., Mercier, B., and Prezard, C. P., 1978, HLA-BW 35 and mesangial IgA glomerulonephritis, *N. Engl. J. Med.* **298:**1034.

Bhan, A. K., Schneeberger, E., Collins, A., and McCluskey, R. T., 1978, Evidence for a pathogenic role of a cell-mediated immune mechanism in experimental glomerulonephritis, *J. Exp. Med.* **148:**246.

Bhan, A. K., Collins, A., Schneeberger, E., and McCluskey, R. T., 1979, A cell-mediated reaction against glomerular-bound immune complexes, *J. Exp. Med.* **150:**1410.

Bianco, C., 1983, Fibrin, fibronectin and macrophages, *Ann. N.Y. Acad. Sci.* **408:**602.

Border, W. A., Lehman, D. H., Egan, J. D., Sass, H. J., Glade, J. E., and Wilson, C. B., 1974, Antitubular basement membrane antibodies in methicillin-associated interstitial nephritis, *N. Engl. J. Med.* **291:**381.

Brettle, R., Peters, D. K., and Batchelor, J. R., 1978, Mesangial IgA glomerulonephritis and HLA antigens, *N. Engl. J. Med.* **299:**200.

Brown, C. A., Carey, K., and Colvin, R. B., 1979, Inhibition of autoimmune tubulointerstitial nephritis in guinea pigs by heterologous antisera containing anti-idiotype antibodies, *J. Immunol.* **123:**2102.

Bruneau, C., and Benveniste, J., 1979, Circulating DNA:anti-DNA complexes in systemic lupus erythematosus: Detection and characterization by ultracentrifugation, *J. Clin. Invest.* **64:**191.

Cantor, H., and Gershon, R. K., 1979, Immunological circuits: Cellular composition, *Fed. Proc.* **38:**2058.

Carnegie, P. R., and Mackay, J. R., 1975, Vulnerability of cell surface receptors to autoimmune reactions, *Lancet* **2:**684.

Cattell, V., and Arlidge, S., 1981, The origin of proliferating cells in the glomerulus and Bowman's capsule in nephrotoxic serum nephritis: Effects of unilateral renal irradiation, *Br. J. Exp. Pathol.* **62:**669.

Cohen, P. L., and Eisenberg, R. A., 1982, Anti-idiotypic antibodies to the Coombs antibody in NZB F_1 mice, *J. Exp. Med.* **156:**173.

Colasanti, G., Moran, J., Bellini, A., and D'Amico, G., 1980, Significance of glomerular C3b receptors in human renal diseases, *Renal Physiol.* **3:**387.

Collins, A. B., Andres, G. A., and McCluskey, R. T., 1981, Lack of evidence for a role of renal tubular antigen in human membranous glomerulonephritis, *Nephron* **27:**297.

Colvin, R. B., 1983, Fibrinogen–fibrin interactions with fibroblasts and macrophages, *Ann. N.Y. Acad. Sci.* **408:**621.

Cook, A., Lydyard, P. M., and Roitt, I. M., 1983, Mechanisms of autoimmunity: a role for cross-reactive idiotypes, *Immunol. Today* **4:**170.

Courtoy, P. J., Kanwar, Y. S., Hynes, R. O., and Farquhar, M. G., 1980, Fibronectin localization in the rat glomerulus, *J. Cell Biol.* **87**:691.
Couser, W. G., 1981, What are circulating immune complexes doing in glomerulonephritis?, *N. Engl. J. Med.* **304**:1230.
Couser, W. G., and Salant, D. J., 1980, In situ immune complex formation and glomerular injury, *Kidney Int.* **17**:1.
Couser, W. G., Steinmuller, D. R., Stilmant, M. M., Salant, D. J., and Lowenstein, L. M., 1978, Experimental glomerulonephritis in the isolated perfused rat kidney, *J. Clin. Invest.* **62**:1275.
Datta, S. K., Owen, F. L., Womack, J. E., and Riblet, R. J., 1982, Analysis of recombinant inbred lines derived from autoimmune (NZB) and "high leukemia" (C58) strains: Independent multigenic systems control B cell hyperactivity, retrovirus expression, and autoimmunity, *J. Immunol.* **129**:1539.
Davies, H. S., and Butcher, G. W., 1978, Kidney alloantigens determined by two regions of the rat major histocompatibility complex, *Immunogenetics* **6**:171.
DeLustro, F., and LeRoy, E. C., 1982, Characterization of the release of human monocyte regulators of fibroblast proliferation, *J. Reticuloendothel. Soc.* **31**:295.
DeLustro, F., Sherer, G. K., and LeRoy, E. C., 1980, Human monocyte stimulation of fibroblast growth by a soluble mediator(s), *J. Reticuloendothel. Soc.* **28**:519.
DeMaeyer, E., and DeMayer-Guignard, J., 1980, Interferons and the immune system, in: *The Immune System: Function and Therapy of Dysfunction* (G. Doria and A. Eshkol, eds.), pp. 213–225, Academic Press, New York.
Demaine, A. G., Vaughan, R. W., and Welsn, K. I., 1982, Immunoglobulin (GM) allotype frequencies in patients with renal failure, *Transplantation* **34**:153.
Diegelmann, R. F., Cohen, I. K., and Kaplan, A. M., 1982, Effect of macrophages on fibroblast DNA synthesis and proliferation, *Proc. Soc. Exp. Biol. Med.* **169**:445.
Dixon, A. J., Burns, J., Dunnill, M. S., and McGee, J., 1980, Distribution of fibronectin in normal and diseased human kidneys, *J. Clin. Pathol.* **33**:1021.
Dixon, F. J., Feldman, J. D., and Vasquez, J. J., 1961, Experimental glomerulonephritis: The pathogenesis of a laboratory model resembling the spectrum of human glomerulonephritis, *J. Exp. Med.* **113**:899.
Doherty, P. C., and Zinkernagel, R. M., 1974, Restriction of in vitro T cell-mediated cytotoxicity in lymphocytic choriomeningitis within a syngeneic or semiallogeneic system, *Nature (London)* **248**:701.
Dorf, M. E., 1981, Genetic control of immune responsiveness, in: *The Role of the Major Histocompatibility Complex in Immunobiology* (M. E. Dorf, ed.), pp. 221–254, Garland STPM Press, New York.
Douglas, M. F. S., Rabideau, D. P., Schwartz, M. M., and Lewis, E. J., 1981, Evidence of autologous immune-complex nephritis, *N. Engl. J. Med.* **305**:1326.
Druet, E., Sapin, C., Gunther, E., Feingold, N., and Druet, P., 1977, Mercuric chloride-induced antiglomerular basement membrane antibodies in the rat, *Eur. J. Immunol.* **7**:348.
Dubois, C. H., Foidart, J. B., Hautier, M. B., Dechenne, C. A., Lemaire, M. J., and Mahieu, P. R., 1981, Proliferative glomerulonephritis in rats: Evidence that mononuclear phagocytes infiltrating the glomeruli stimulate the proliferation of endothelial and mesangial cells, *Eur. J. Clin. Invest.* **11**:91.
Eagen, J. W., and Lewis, E. J., 1977, Glomerulopathies of neoplasia, *Kidney Int.* **11**:297.
Edgington, T. S., Glassock, R. J., and Dixon, F. J., 1968, Autologous immune complex nephritis induced with renal tubular antigen. I. Identification and isolation of the pathogenetic antigen, *J. Exp. Med.* **127**:555.
Edwards, R. L., and Rickles, F. R., 1978, Delayed hypersensitivity in man: Effects of systemic anticoagulation, *Science* **200**:541.
Edwards, R. L., and Rickles, F. R., 1980, The role of monocyte tissue factor in the immune response, *Lymphokine Rep.* **1**:181.
Edwards, R. L., Rickles, F. R., and Bogrove, A. M., 1979, Mononuclear cell tissue factor: Cell of origin and requirements for activation, *Blood* **54**:359.
Egido, J., Blasco, R., Sancho, J., Lozaro, L., Sanchez-Crespo, M., and Hernando, L., 1982, Increased

rates of polymeric IgA synthesis by circulating lymphoid cells in IgA mesangial glomerulonephritis, *Clin. Exp. Immunol.* **47**:309.

Egido, J., Sancho, J., Blasco, R., Rivera, F., and Hernando, L., 1983, The immunopathogenetic aspects of IgA nephropathy, *Adv. Naphrol.* **12**:103.

Farquhar, M. G., and Kanwar, Y. S., 1983, Functional organization of the glomerulus: Presence of glycosaminoglycans (proteoglycans) in the glomerular basement membrane, in: *Immune Mechanisms in Renal Disease* (N. B. Cummings, A. F. Michael, and C. B. Wilson, eds.), pp. 1–35, Plenum Medical Book Co., New York.

Fillit, H. M., and Zabriskie, J. B., 1982, Cellular immunity in glomerulonephritis, *Am. J. Pathol.* **109**: 227.

Fillit, H. M., Villareal, H. J., and Zabriskie, J. B., 1983, The role of streptococcal and glomerular basement membrane antigens in glomerulonephritis, in: *Immune Mechanisms in Renal Disease* (N. B. Cummings, A. F. Michael, and C. B. Wilson, eds.), pp. 339–418, Plenum Medical Book Co., New York.

Finbloom, D. S., and Plotz, P. H., 1979, Studies of reticuloendothelial function in the mouse with model immune complexes. I. Serum clearance and tissue uptake in normal C3H mice, *J. Immunol.* **123**:1594.

Fish, A. J., Carmody, K. M., Schiffer, M. S., and Michael, A. F., 1983, Glomerular basement membrane antigens, in: *Immune Mechanisms in Renal Disease* (N. B. Cummings, A. F. Michael, and C. B. Wilson, eds.), pp. 71–81, Plenum Medical Book Co., New York.

Fletcher, A. P., Neuberger, A., and Ratcliff, W. A., 1970, Tamm–Horsfall urinary glycoprotein: The subunit structure, *Biochem. J.* **120**:425.

Ford, P. M., 1975, The effect of manipulation of reticuloendothelial system activity on glomerular deposition of aggregated protein and immune complexes in two different strains of mice, *Br. J. Exp. Pathol.* **56**:523.

Frank, M. M., Hamburger, M. I., Lawley, T. J., Kimberly, R. P., and Plotz, P. H., 1979, Defective reticuloendothelial system Fc-receptor function in systemic lupus erythematosus, *N. Engl. J. Med.* **300**:518.

Friedman, J., Hoyer, J. R., and Seiler, M. W., 1982, Formation and clearance of tubulointerstitial immune complexes in kidneys of rats immunized with heterologous antisera to Tamm–Horsfall protein, *Kidney Int.* **21**:575.

Fujibayashi, T., Hooks, J. J., and Notkins, A. L., 1975, Production of interferon by immune lymphocytes exposed to herpes simplex virus–antibody complexes, *J. Immunol.* **115**:1191.

Garovoy, M. R., 1980, Idiopathic membranous glomerulopathy: An HLA associated disease, in: *Histocompatibility Testing* (P. I. Terasaki, ed.), pp. 673–680, UCLA Press, Los Angeles.

Garovoy, M. R., 1982, Immunogenetic associations in nephrotic states, in: *Contemporary Issues in Nephrology*, Volume 9 (B. M. Brenner and J. H. Stein, eds.), pp. 259–282, Churchill Livingstone, Edinburgh.

Gasser, D. L., Baloyannus, S. J., and Gonatas, N. K., 1978, Influence of MHC-linked and non-MHC linked genes on susceptibility of rats to experimental allergic encephalomyelitis, in: *Genetic Control of Autoimmune Disease* (N. R. Rose, P. E. Bigazzi, and N. L. Warner, eds.), pp. 355–364, Elsevier/North-Holland, Amsterdam.

Geha, R., 1983, Presence of circulating anti-idiotype bearing cells after booster immunization with tetanus toxoid (TT) and inhibition of anti-TT antibody synthesis by auto-anti-idiotypic antibody, *J. Immunol.* **130**:1634.

Gelfand, M. C., Shin, M. L., Nagle, R. B., Green, I., and Frank, M. M., 1976, The glomerular complement receptor in immunologically mediated renal glomerular injury, *N. Engl. J. Med.* **295**:10.

Gelfand, M. C., Frank, M. M., Green, I., and Shin, M. L., 1979, Binding sites for immune complexes containing IgG in the renal interstitium, *Clin. Immunol. Immunopathol.* **13**:19.

Germuth, F. G., 1953, A comparative and immunologic study in rabbits of induced hypersensitivity of the serum sickness type, *J. Exp. Med.* **97**:257.

Glenn, K. C., and Ross, R. 1981, Human monocyte-derived growth factor(s) for mesenchymal cells: Activation of secretion by endotoxin and concanavalin A, *Cell* **25**:603.

Golbus, S., and Wilson, C. B., 1979, Experimental glomerulonephritis induced by in situ formation of immune complexes in glomerular capillary wall, *Kidney Int.* **16**:148.

Goldmann, M., Rose, L. M., Hochmann, A., and Lambert, P. H., 1982, Deposition of idiotype–anti-idiotype immune complexes in renal glomeruli after polyclonal B cell activation, *J. Exp. Med.* **155**:1385.

Gonwa, T. A., Picker, L. J., Raff, H. V., Goyert, S. M., Silver, S., and Stobo, J. D., 1983, Antigen-presenting capabilities of human monocytes correlate with their expression of HLA-DS, an Ia determinant distinct from *HLA-DR, J. Immunol.* **130**:706.

Gorzynski, T. J., and David, C. S., 1983, Immune-response gene-associated antigens (Ia/DR): Structure and function in immunologically related diseases, *Mayo Clin. Proc.* **58**:547.

Granelli-Piperno, A., Vasoalli, J. D., Perch, E., 1977, Secretion of plasminogen activator by human polymorphonuclear leukocytes. Modulation by glucocorticoids and other effectors. *J. Exp. Med.* **146**:1693.

Greenberg, G. B., and Hunt, T. K., 1978, The proliferative response in vitro of vascular endothelial and smooth muscle cells exposed to wound fluids and macrophages, *J. Cell. Physiol.* **97**:353.

Gresser, I., Morel-Maroger, L., Maury, C., Tovey, M. G., and Pontillon, F., 1976, Progressive glomerulonephritis in mice treated with interferon preparations at birth, *Nature (London)* **263**:420.

Gresser, I., Morel-Maroger, L., Verroust, P., Riviere, Y., and Guillon, J. C., 1978, Anti-interferon globulin inhibits the development of glomerulonephritis in mice infected at birth with lymphocytic choriomeningitis virus, *Proc. Natl. Acad. Sci. USA* **75**:3413.

Griffin, F. M., 1982, Mononuclear cell phagocytic mechanisms and host defense, in: *Advances in Host Defense Mechanisms,* Volume 1 (J. I. Gallin and A. S. Fauci, eds.), pp. 31–55, Raven Press, New York.

Hahn, B. H., and Ebling, F. M., 1983, Suppression of NZB/NZW murine nephritis by administration of a syngeneic monoclonal antibody to DNA: Possible role of anti-idiotypic antibodies, *J. Clin. Invest.* **71**:1728.

Harmon, W. E., Grupe, W. E., and Parkman, R., 1980, Control of autologous immune complex nephritis. I. Suppression of the disease in the presence of T cell sensitization, *J. Immunol.* **124**:1034.

Harry, T., Bryant, D., Coles, G. A., Davies, M., and Fortt, W., 1982, The detection of monocytes in human renal biopsies: A prospective study, *Clin. Nephrol.* **18**:29.

Hart, D. N. J., and Fabre, J. W., 1981, Major histocompatibility complex antigens in rat kidney, ureter, and bladder, *Transplantation* **31**:318.

Hart, D. N. J., Fuggle, S. V., Williams, K. A., Fabre, J. W., Trug, A., and Mours, P. J., 1981, Localization of HLA-ABC and DR antigens in human kidney, *Transplantation* **31**:428.

Hattler, B. G., Rocklin, R. E., Ward, P. A., and Rickles, F. R., 1973, Functional features of lymphocytes recovered from a human renal allograft, *Cell. Immunol.* **9**:289.

Hau, T., Adrenholz, D. H., Simmons, R. L., 1979, Secondary bacterial peutourtis: The biological basis on treatment, *Curr. Prob. Surg.* **16**:5.

Helin, H., and Edgington, T. S., 1983, Allogeneic induction of the human T cell-restricted monocyte procoagulant response is rapid and is elicited by HLA-DR, *J. Exp. Med.* **158**:962.

Herberman, R. B., Djeu, J. Y., Kay, H. D., Ortaldo, J. R., Riccardi, C., Bonnard, G. D., Holden, H. T., Fognani, R., Santoni, A., and Puccetti, P., 1979, Natural killer cells: Characteristics and regulation of activity, *Immunol. Rev.* **44**:43.

Heymann, W., Hackel, D. B., Harwood, S., Wilson, S. G. F., and Hunter, J. L. P., 1959, Production of nephrotic syndrome in rats by Freund's adjuvant and kidney suspensions, *Proc. Soc. Exp. Biol. Med.* **110**:660.

Hoffsten, P. E., Villalobos, R., Hill, C., and Klahr, S., 1977, T-cell deficiency in immune complex glomerulonephritis, *Kidney Int.* **11**:318.

Hoffsten, P. E., Swerdlin, A., Bartell, M., Hill, C. L., Venverloh, J., Brotherson, K., and Klahr, S., 1979, Reticuloendothelial and mesangial function in murine immune complex glomerulonephritis, *Kidney Int.* **15**:144.

Hogg, N., 1983, Human monocytes are associated with the formation of fibrin, *J. Exp. Med.* **157**:473.

Holdsworth, S. R., 1983, Fc-dependence of macrophage accumulation and subsequent injury in experimental glomerulonephritis, *J. Immunol.* **130**:735.

Holdsworth, S. R., Thompson, N. M., Glasgow, E. F., and Atkins, R. C., 1979a, The effect of defibrination on macrophage participation in rabbit nephrotoxic nephritis: Studies using glomerular culture and electron microscopy, *Clin. Exp. Immunol.* **37**:38.

Holdsworth, S. R., Golbus, S. M., and Wilson, C. B., 1979b, Characterization of collagenase solubilized human glomerular basement membrane antigens reacting with human antibodies, *Kidney Int.* **16**:797a.

Holdsworth, S. R., Neale, T. J., and Wilson, C. B., 1980a, The participation of macrophages and monocytes in experimental immune complex glomerulonephritis, *Clin. Immunol. Immunopathol.* **15**:510.

Holdsworth, S. R., Allen, D. E., Thompson, N. M., Glasgow, E. F., and Atkins, R. C., 1980b, Histochemistry of glomerular cells in animal models of crescentic glomerulonephritis, *Pathology* **12**:339.

Holdsworth, S. R., Neale, T. J., and Wilson, C. B., 1981, Abrogation of macrophage-dependent injury in experimental glomerulonephritis in the rabbit, *J. Clin. Invest.* **68**:686.

Hood, L., Steinmetz, M., and Malissen, B., 1983, Genes of the major histocompatibility complex of the mouse, *Annu. Rev. Immunol.* **1**:529.

Hoyer, J. R., 1980, Tubulointerstitial immune nephritis in rats immunized with Tamm–Horsfall protein, *Kidney Int.* **17**:284.

Hoyer, J. R., Friedman, J., and Seiler, M. W., 1983, Autoimmunity to Tamm–Horsfall protein, in: *Immune Mechanisms in Renal Disease* (N. B. Cummings, A. F. Michael, and C. B. Wilson, eds.), pp. 295–304, Plenum Medical Book Co., New York.

Hume, D. A., and Gordon, S., 1983, Mononuclear phagocyte system of the mouse defined by immunohistochemical localization of antigen F4/80: Identification of resident macrophages in renal medullary and cortical interstitium and the juxtaglomerular complex, *J. Exp. Med.* **157**:1704.

Hunsicker, L. G., Shearer, T. P., Plattner, S. B., and Weisenburger, D., 1979, The role of monocytes in serum sickness nephritis, *J. Exp. Med.* **150**:413.

Hunt, J. S., McGiven, A. R., Jackson, A. E., and Day, W. A., 1980, Macrophages in NZB/NZW mouse glomerular disease, *Renal Physiol.* **3**:368.

Hyman, L. R., Colvin, R. B., and Steinberg, A. D., 1976a, Immunopathogenesis of autoimmune tubulo-interstitial nephritis. I. Demonstration of differential susceptibility in strain II and strain XIII guinea pigs, *J. Immunol.* **116**:327.

Hyman, L. R., Steinberg, A. D., Colvin, R. B., and Bernard, E. F., 1976b, Immunopathogenesis of autoimmune tubulointerstitial nephritis. II. Role of an immune response gene linked to the major histocompatibility complex, *J. Immunol.* **117**:1894.

Izui, S., Lambert, P. H., and Miescher, P. A., 1976, In vitro demonstration of a particular affinity of glomerular basement membrane and collagen for DNA: A possible basis for a local formation of DNA–anti-DNA complexes in systemic lupus erythematosus, *J. Exp. Med.* **144**:428.

Izui, S., Lambert, P. H., Fournie, G. J., Turler, H., and Miescher, P. A., 1977, Features of systemic lupus erythematosus in mice injected with bacterial lipopolysaccharides: Identification of circulating DNA and renal localization of DNA–anti-DNA complexes, *J. Exp. Med.* **145**:1115.

Jerne, N. K., 1974, Towards a network theory of the immune system, *Ann. Immunol. (Inst. Pasteur)* **125C**:373.

Jimenez, S. A., McArthur, W., and Rosenbloom, J., 1979, Inhibition of collagen synthesis by mononuclear cell supernatants, *J. Exp. Med.* **150**:1421.

Jothy, S., and Sawka, R. J., 1981, Presence of monocytes in systemic lupus erythematosus-associated glomerulonephritis: Marker study and significance, *Arch. Pathol. Lab. Med.* **105**:590.

Kashwabara, H., Shishido, H., Yokoyama, T., and Mryayuna, T., 1980, HLA in IgA nephropathy, *Tissue Antigen* **16**:411.

Kazatchkine, M. D., Fearon, D. T., Appay, M. D., Mandet, C., and Bariety, J., 1982, Immunohistochemical study of the human glomerular C3b receptor in normal kidney and in seventy-five cases of renal diseases, *J. Clin. Invest.* **69**:900.

Kefalides, N. A., 1983, The molecular structure of basement membranes as it relates to function, in:

Immune Mechanisms in Renal Disease (N. B. Cummings, A. F. Michael, and C. B. Wilson, eds.), pp. 47–69, Plenum Medical Book Co., New York.
Kefalides, N. A., Alper, R., and Clark, C. C., 1979, Biochemistry and metabolism of basement membranes, *Int. Rev. Cytol.* **60**:167.
Kerjaschki, D., and Farquhar, M. G., 1982, The pathogenic antigen of Heymann nephritis is a membrane glycoprotein of the renal proximal tubule brush border, *Proc. Natl. Acad. Sci. USA* **79**:5557.
Kerjaschki, D., and Farquhar, M. G., 1983, Immunocytochemical localization of the Heymann nephritis antigen (GP 330) in glomerular epithelial cells of normal Lewis rats, *J. Exp. Med.* **157**:667.
Klein, J., Figueroa, F., and Nagy, Z., 1983, Genetics of the major histocompatibility complex: The final act, *Annu. Rev. Immunol.* **1**:119.
Knighton, D. R., Oredsson, S., Banda, M. J., Werb, Z., and Hunt, T. K., 1982, Hypoxia stimulates production of angiogenesis factor, plasminogen activator and growth factor by rabbit bone marrow macrophages, *Fed. Proc.* **41**:270.
Korn, J. H., 1983, Fibroblast prostaglandin E2 synthesis, *J. Clin. Invest.* **71**:1240.
Korn, J. H., Halushka, P. V., and LeRoy, E. C., 1980, Mononuclear cell modulation of connective tissue function: Suppression of fibroblast growth by stimulation of endogenous prostaglandin production, *J. Clin. Invest.* **65**:543.
Korn, J. H., Rickles, F. R., Ewan, V. A., Edwards, R. L., Myers, T. J., and Onder, O., 1982, Mononuclear cell modulation of fibroblast procoagulant activity, *J. Lab. Clin. Med.* **99**:657.
Krakower, C. A., and Greenspon, S. A., 1951, Localization of the nephrotoxic antigen within the isolated renal glomerulus, *Arch. Pathol.* **51**:629.
Krieger, A., Thoenes, G. H., and Gunther, E., 1981, Genetic control of autoimmune tubulointerstitial nephritis in rats, *Clin. Immunol. Immunopathol.* **21**:301.
Lagrue, G., Xheneumont, S., Branellec, A., and Weil, B., 1975, Lymphokines and nephrotic syndrome, *Lancet* **1**:271.
Lange, K., Seligson, G., and Cronin, W., 1983, Evidence for the in situ origin of poststreptococcal glomerulonephritis: Glomerular localization of endostreptosin and the clinical significance of the subsequent antibody response, *Clin. Nephrol.* **19**:3.
Lavelle, K. J., Durland, B. D., and Yum, M. N., 1981, The effect of anti-macrophage antiserum on immune complex glomerulonephritis, *J. Lab. Clin. Med.* **98**:195.
Layrisse, Z., Rodriquez-Iturbe, B., Garcia-Ramirez, R., Rodriquez, A., and Tuvori, J., 1983, Family studies of the HLA system in acute post-streptococcal glomerulonephritis, *Hum. Immunol.* **7**:177.
Lehman, D. H., Wilson, C. B., and Dixon, F. J., 1974, Interstitial nephritis in rats immunized with heterologous tubular basement membrane, *Kidney Int.* **5**:187.
Leibovich, S. J., and Ross, R., 1976, A macrophage-dependent factor that stimulates the proliferation of fibroblasts in vitro, *Am. J. Pathol.* **84**:501.
Lerner, R., and Dixon, F. J., 1966, Transfer of ovine experimental allergic glomerulonephritis (EAG) with serum, *J. Exp. Med.* **124**:431.
Levine, B. B., Ojeda, A., and Benacerraf, B., 1963, Studies on artificial antigens. III. The genetic control of the immune response to hapten poly-L-lysine conjugates in guinea pigs, *J. Exp. Med.* **118**:953.
Linder, E., Miettinen, A., and Tornroth, T., 1980, Fibronectin as a marker for the glomerular mesangium in immunohistology of kidney biopsies, *Lab. Invest.* **42**:70.
Lovett, D. H., Ryan, J. L., and Sterzel, R. B., 1983, A thymocyte-activating factor derived from glomerular mesangial cells, *J. Immunol.* **130**:1796.
McCluskey, R. T., 1974, Immunologic mechanisms in renal disease, in: *Pathology of the Kidney* (R. H. Heptinstall, ed.), pp. 273–317, Little, Brown, Boston.
McCoy, R. C., Johnson, H. K., Stone, W. J., and Wilson, C. B., 1982, Absence of nephritogenic GBM antigens in some patients with hereditary nephritis, *Kidney Int.* **21**:642.
McDevitt, H. O., and Chinitz, A., 1969, Genetic control of antibody response: Relationship between immune response and histocompatibility (H-2) type, *Science* **163**:1207.
McDevitt, H. O., Deak, B. D., Shreffler, D. C., Klein, J., Stimpfling, J. H., and Snell, G. D., 1972, Genetic control of the immune response mapping to the Ir.1 locus, *J. Exp. Med.* **135**:1259.

McGiven, A. R., Hunt, J. S., Day, W. A., and Jackson, A. E., 1981, Phagocytic cells in glomerular cultures from NZB/W mice, *Br. J. Exp. Pathol.* **62**:59.

Magil, A. B., and Wadsworth, L. D., 1981, Monocytes in human glomerulonephritis—An electron microscopic study, *Lab. Invest.* **45**:77.

Magil, A. B., and Wadsworth, L. D., 1982, Monocyte involvement in glomerular crescents, *Lab. Invest.* **47**:160.

Magil, A. B., Wadsworth, L. D., and Loewen, M., 1981, Monocytes and human renal glomerular disease: A quantitative evaluation, *Lab. Invest.* **44**:27.

Mampaso, F. M., and Wilson, C. B., 1983, Characterization of inflammatory cells in autoimmune tubulointerstitial nephritis in rats, *Kidney Int.* **23**:448.

Matsumoto, K., Yoslugana, N., and Hatano, M., 1978, Studies of cell-mediated immunity in human glomerulonephritis by macrophage migration inhibition test, *Nephron* **21**:192.

Mayrer, A. R., Miniter, P., and Andriole, V. T., 1983, Immunopathogenesis of chronic pyelonephritis, *Am. J. Med.* (Suppl.) **75**:59.

Mayrhofer, G., and Schon-Hegrad, M. A., 1983, Ia antigens in rat kidney with special reference to their expression in tubular epithelium, *J. Exp. Med.* **157**:2097.

Melcion, C., Lachman, L., Killen, P. D., Marel-Maroger, L., and Striker, G. E., 1982, Mesangial cells and the effect of monocyte products on proliferation and matrix synthesis, *Transplant. Proc.* **14**:559.

Michael, A., Keane, W. F., Raij, L., Vernier, R. L., and Mauer, S. M., 1980, The glomerular mesangium, *Kidney Int.* **17**:141.

Milliner, D. S., Pierides, A. M., and Holley, K. E., 1982, Renal transplantation in Alport's syndrome: Anti-glomerular basement membrane glomerulonephritis in the allograft, *Mayo Clin. Proc.* **57**:35.

Monga, G., Mazzucco, G., DiBelgiojoso, G. B., and Busnach, G., 1981, Monocyte infiltration and glomerular hypercellularity in human acute and persistent glomerulonephritis, *Lab. Invest.* **44**:381.

Morrison, W. I., and Wright, N. G., 1977, Viruses associated with renal disease of man and animals, *Prog. Med. Virol.* **23**:22.

Murphy-Ullrich, J. E., Oberley, T. D., and Mosher, D. F., 1982, Glomerular and vascular injury in mice following immunization with heterologous and autologous fibronectin, *Virchows Arch. B* **39**:305.

Nagle, R. B., Dong, S., Janacek, L. L., Guillot, J. M., and Lindsley, H. B., 1982, Glomerular accumulation of monocytes and macrophages in experimental glomerulonephritis associated with *Trypanosoma rhodesiense* infection, *Lab. Invest.* **46**:365.

Nakao, Y., Matsumoto, H., Miyazaki, T., Nishitani, H., Takatsuki, K., Kasukawa, R., Nakayama, S., Izumi, S., Fujita, T., and Tsuji, K., 1980, IgG heavy chain allotypes (GM) in autoimmune diseases, *Clin. Exp. Immunol.* **42**:20.

Neale, T. J., and Wilson, C. B., 1978, Non-GBM glomerular antigen in spontaneous nephritis in rabbits, *Kidney Int.* **14**:715.

Neilson, E. G., and Phillips, S. M., 1979, Cell-mediated immunity in interstitial nephritis. II. T lymphocyte effector mechanisms in nephritic guinea pigs: Analysis of the renotropic migration and cytotoxic response. *J. Immunol.* **123**:2381.

Neilson, E. G., and Phillips, S. M., 1980, The immunobiology of nephritis, *Prog. Allergy* **27**:167.

Neilson, E. G., and Phillips, S. M., 1981, Cell-mediated immunity in interstitial nephritis. IV. Antitubular basement membrane antibodies can function in antibody-dependent cellular cytotoxicity reactions: Observations on a nephritogenic effector mechanism acting as an informational bridge between the humoral and cellular immune response, *J. Immunol.* **126**:1990.

Neilson, E. G., and Phillips, S. M., 1982a, Murine interstitial nephritis. I. Analysis of disease susceptibility and its relationship to pleiomorphic gene products defining both immune response genes and a restrictive requirement for cytolytic T cells at H-2K, *J. Exp. Med.* **155**:1075.

Neilson, E. G., and Phillips, S. M., 1982b, Suppression of interstitial nephritis by auto-antiidiotypic immunity, *J. Exp. Med.* **155**:179.

Neilson, E. G., and Zakheim, B., 1983, T cell regulation, anti-idiotypic immunity, and the nephritogenic immune response, *Kidney Int.* **24**:289.

Neilson, E. G., Jimenez, S. A., and Phillips, S. M., 1980, Cell-mediated immunity in interstitial nephritis. III. T lymphocyte-mediated fibroblast proliferation and collagen synthesis: An immune mechanism for renal fibrogenesis, *J. Immunol.* **125**:1708.

Neilson, E. G., Gasser, D. L., McCafferty, E., Zakheim, B., and Phillips, S. M., 1983, Polymorphism of genes involved in anti-tubular basement membrane disease in rats, *Immunogenetics* **17**:55.

Neilson, E. G., McCafferty, E., Feldman, A., Clayman, M., Zakheim, B., and Korngold, R., 1984, Spontaneous interstitial nephritis in kdkd mice. I. An experimental model of autoimmune renal disease, *J. Immunol.* **133**:2560.

Noel, L. H., Descamps, B., Junges, P., Bach, J. R., Basson, M., Suet, C., Hors, J., and Dausset, J., 1978, HLA antigens in three types of glomerulonephritis, *Clin. Immunol. Immunopathol.* **10**:19.

Nomoto, S. Y., and Arimori, S., 1979, Decrease in IgA-specific suppressor T cell activity in patients with IgA nephropathy, *Clin. Exp. Immunol.* **38**:243.

Norman, M. E., Mandle, R., and Nilsson, U. R., 1978, Further studies of a humoral chemotactic abnormality in glomerulonephritis, *Int. Archs. Allergy Appl. Immun.* **56**:144.

Norris, D. A., Clark, R. A. F., Swigart, L. M., Huff, C., Weston, W. L., and Howell, S. E., 1982, Fibronectin fragment(s) are chemotactic for human peripheral blood monocytes, *J. Immunol.* **129**:1612.

Nossal, G. J. V., and Pike, B. L., 1975, Evidence for the clonal abortion theory of B lymphocyte tolerance, *J. Exp. Med.* **141**:904.

Oberley, T. D., Mosher, D. F., and Mills, M. D., 1979, Localization of fibronectin within the renal glomerulus and its production by cultured glomerular cells, *Am. J. Pathol.* **96**:651.

Ooi, B. S., Ooi, Y. M., Hsu, A., and Hurtubise, P. E., 1980, Diminished synthesis of immunoglobulin by peripheral lymphocytes in patients with idiopathic membranous glomerulonephropathy, *J. Clin. Invest.* **65**:789.

Ooi, Y. M., Weiss, M. A., Hsu, A., and Ooi, B. S., 1983, Mechanisms of suppression of mouse mesangial cell proliferation by macrophage supernatants, *J. Immunol.* **130**:1790.

O'Regan, D., O'Callaghan, U., Dundon, S., and Rean, D. J., 1980, HLA antigens and steroid responsive nephrotic syndrome of childhood, *Tissue Antigens* **16**:147.

Osakabe, K., and Matsumoto, K., 1981, Concanavalin A-induced suppressor cell activity in lipoid nephrosis, *Scand. J. Immunol.* **14**:161.

Parris, T. M., Kimberly, R. P., Inman, R. D., McDougal, S., Gibofsky, A., and Christian, C. L., 1982, Defective Fc receptor-mediated function of the mononuclear phagocyte system in lupus nephritis, *Ann. Intern. Med.* **97**:526.

Patel, P., Mickey, M. R., and Terasaki, P. I., 1969, Leukocyte antigens and disease; association of HLA-A2 and chronic glomerulonephritis, *Br. Med. J.* **2**:424.

Paul, L. C., and Carpenter, C. B., 1982, Antigenic determinants of tubular basement membranes and Bowman's capsule in rats, *Kidney Int.* **21**:800.

Pettersson, E. E., and Colvin, R. B., 1978, Cold-insoluble globulin (fibronectin, LETS protein) in normal and diseased human glomeruli: Papain-sensitive attachment to normal glomeruli and deposition in crescents, *Clin. Immunol. Immunopathol.* **11**:425.

Platt, J. L., LeBien, T. W., and Michael, A. F., 1982, Interstitial mononuclear cell population in renal graft rejection, *J. Exp. Med.* **155**:17.

Platt, J. L., LeBien, T. W., and Michael, A. F., 1983, Stages of renal ontogenesis identified by monoclonal antibodies reactive with lympho-hemopoietic differentiation antigens, *J. Exp. Med.* **157**:155.

Polverini, P. J., Cotran, R. S., Gimbrone, M. A., and Unanue, E. R., 1977, Activated macrophages induce vascular proliferation. *Nature (London)* **269**:804.

Postlethwaite, A. E., and Kang, A. H., 1980, Characterization of guinea pig lymphocyte-derived chemotactic factor for fibroblasts, *J. Immunol.* **124**:1462.

Postlethwaite, A. E., and Kang, A. H., 1983, Induction of fibroblast proliferation by human mononuclear leukocyte-derived proteins, *Arthritis Rheum.* **26**:22.

Postlethwaite, A. E., Snyderman, R., and Kang, A. H., 1976, The chemotactic attraction of human fibroblasts to a lymphocyte-derived factor, *J. Exp. Med.* **144**:1188.

Postlethwaite, A. E., Lachman, L. B., Mainardi, C. L., and Kang, A. H., 1983, Interleukin 1 stimulation of collagenase production by cultured fibroblasts, *J. Exp. Med.* **157**:801.

Raij, L., Sibley, R. K., and Keane, W. F., 1981, Mononuclear phagocytic system stimulation: Protective role from glomerular immune complex deposition, *J. Lab. Clin. Med.* **98:**558.

Rapaport, F. T., Markowitz, A. S., McCluskey, R. T., Hanaoka, T., and Shumada, T., 1969, Induction of renal disease with antisera to group A streptococcal membranes, *Transplant. Proc.* **1:**981.

Read, S. E., Reid, H., Poon-King, T., Fischetti, V. A., Zabrieskie, J. B., and Rapaport, F. T., 1977, HLA and predisposition to the non-suppurative sequelae of group A streptococcal infections, *Transplant. Proc.* **9:**543.

Rees, A. J., Peters, D. K., Compston, D. A. S., and Batchelor, J. R., 1978, Strong association between HLA-DR2 and antibody mediated Goodpasture's syndrome, *Lancet* **1:**966.

Richman, A. V., Mahoney, J. J., and Filler, T. J., 1979, Higher prevalence of HLA-B12 in patients with IgA nephropathy, *Ann. Intern. Med.* **90:**201.

Riviere, Y., Gresser, I., Guillon, J. C., and Tovey, M. G., 1977, Inhibition by anti-interferon serum of lymphocytic choriomeningitis virus disease in suckling mice, *Proc. Natl. Acad. Sci. USA* **74:**2135.

Rose, L. M., Goldmann, M., and Lambert, P. H., 1982, The production of anti-idiotypic antibodies and of idiotype–anti-idiotype immune complexes after polyclonal activation induced by bacterial LPS, *J. Immunol.* **128:**2126.

Rosenbaum, J. T., Adelman, N. E., and McDevitt, H. O., 1981, In vivo effects of antibodies to immune response gene products. I. Haplotype-specific suppression of humoral immune response with a monoclonal anti-I-A, *J. Exp. Med.* **154:**1694.

Rosenthal, A. S., 1982, Determinant selection and macrophage function, *Immunol. Today* **3:**33.

Rosenthal, A. S., and Shevach, E. M., 1973, Function of macrophages in antigen recognition by guinea pig T lymphocytes. I. Requirement for histocompatible macrophages and lymphocytes, *J. Exp. Med.* **138:**1194.

Rossen, R. D., Reisberg, M. A., Hersh, E. M., and Gutterman, J. U., 1976, Measurement of soluble immune complexes: A guide to prognosis in cancer patients, *Clin. Res.* **24:**462a.

Rudofsky, U. H., Dilwith, R. L., and Tung, K. S. K., 1980, Susceptibility differences of inbred mice to induction of autoimmune renal tubulointerstitial lesions, *Lab. Invest.* **43:**463.

Rutherford, B., Stefflin, K., and Sexton, J., 1982, Activated human mononuclear phagocytes release a substance(s) that reduces replication of quiescent human fibroblasts *J. Reticuloendothel. Soc.* **31:**281.

Sakane, T., Steinberg, A. D., Reeves, J. P., and Green, I., 1979, Studies of immune functions of patients with systemic lupus euythematosus: complement-dependent immunoglobulin M anti-thymus-derived cell antibodies preferentially inactive suppressor cells, *J. Clin. Invest.* **63:**954.

Salant, D. J., Darby, C., and Couser, W. G., 1980, Experimental membranous glomerulonephritis in rats. Quantitative Studies of glomerular immune deposit formation in isolated glomeruli and whole animals, *J. Clin. Invest.* **66:**71.

Sapin, C., Druet, P., and Mandet, C., 1980, Induction of susceptibility to $HgCl_2$ immune glomerulonephritis in the Lewis rat by immunocompetent cells from susceptible hybrids, *Eur. J. Immunol.* **10:**371.

Sasazuki, T., Hayase, R., Iwamoto, I., and Tsuchida, H., 1979, HLA and poststreptococcal glomerulonephritis, *N. Engl. J. Med.* **301:**1184.

Scheinman, J. I., Fish, A. J., Matas, A. J., and Michael, A. F., 1978, The immunohistopathology of glomerular antigens. II. The glomerular basement membrane, actin, myosin, and fibroblast surface antigens in normal, diseased, and transplanted human kidneys, *Am. J. Pathol.* **90:**71.

Schmidt, J. A., Mizel, S. B., Cohen, D., and Green, I., 1982, Interleukin 1, a potential regulator of fibroblast proliferation, *J. Immunol.* **128:**2177.

Schmitt, E., Seyfarth, M., Werner, H., and Klinkmann, H., 1982, Cellular immunity in glomerulonephritis, *Clin. Nephrol.* **17:**271.

Schreiner, G. F., Cotran, R. S., Pardo, V., and Unanue, E. R., 1978, A mononuclear cell component in experimental immunological glomerulonephritis, *J. Exp. Med.* **147:**369.

Schreiner, G., Kiely, J. M., Cotran, R. S., and Unanue, E. R., 1981, Characterization of resident glomerular cells in the rat expressing Ia determinants and manifesting genetically restricted interactions with lymphocytes, *J. Clin. Invest.* **68:**920.

Schreiner, G. F., Cotran, R. S., and Unanue, E. R., 1982, Macrophages and cellular immunity in experimental glomerulonephritis, *Springer Semin. Immunopathol.* **5:**251.

Schur, P. S., and Carpenter, C. B., 1979, Host and genetic factors contributing to immunologic renal disease, in: *Contemporary Issues in Nephrology*, Volume 3 (C. B. Wilson, B. M. Brenner, and J. H. Stein, eds.), pp. 144–166, Churchill Livingstone, Edinburgh.
Schwartz, R. H., 1978, A clonal deletion model for Ir gene control of the immune response, *Scand. J. Immunol.* **7**:3.
Schwartz, R. S., 1982, Recent trends in systemic lupus erythematosus, *Pharmacol. Rev.* **34**:105.
Seiler, M. W., and Hoyer, J. R., 1981, Ultrastructural studies of tubulointerstitial immune complex nephritis in rats immunized with Tamm-Horsfall protein, *Lab. Invest.* **45**:321.
Seligson, G., Large, K., and Cronui, W., 1982, Evidence for the in situ origin of post-streptococcal glomerulonephritis, *Kidney Int.* **21**:213.
Shalhoub, R. J., 1974, Pathogenesis of lipoid nephrosis: A disorder of T cell function, *Lancet* **2**:556.
Sherman, L. A., 1983, Binding of soluble fibrin in macrophages, *Ann. N.Y. Acad. Sci.* **408**:610.
Shibata, S., 1978, Immunologic and non-immunologic aspects of glomerulonephritis, in: *Biology and Chemistry of Basement Membranes* (N. A. Kefalides, ed.), pp. 535–560, Academic Press, New York.
Shibata, S., Nagasawa, T., Takuma, T., Naruse, T., and Muyakawa, Y., 1966, Isolation and properties of the soluble antigen specific for the production of nephrotoxic glomerulonephritis. I. Immunopathological demonstration of the complete antigenicity of the soluble antigen, *Jpn. J. Exp. Med.* **36**:127.
Shibata, S., Nagasawa, T., Muyakawa, Y., and Naruse, T., 1971, Nephritogenic glycoprotein. I. Proliferative glomerulonephritis induced in rats by a single injection of the soluble glycoprotein isolated from homologous glomerular basement membrane, *J. Immunol.* **106**:1284.
Shibata, S., Sakaguchi, H., Nagasawa, T., and Naruse, T., 1972, Nephritogenic glycoprotein. I. Experimental production of membranous glomerulonephritis in rats by a single injection of homologous renal glycopeptide, *Lab. Invest.* **27**:457.
Shibata, S., Sakaguchi, H., and Nagasawa, T., 1976, Induction of chronic progressive glomerulonephritis with immunofluorescent "mesangial pattern" in rats, *Nephron* **16**:241.
Shibata, S., Nagasawa, T., and Muira, K., 1977, Nephritogenoside, the receptor glycoprotein for concanavalin A in rat glomerular basement membrane: Demonstration of α-D-glucopyranosyl unit at the non-reducing terminus, *Biochim. Biophys. Acta* **499**:392.
Shigematsu, H., Shishido, H., Sano, M., Kogayashi, Y., Kondo, Y., and Okabayashi, A., 1976, Participation of monocytes in transient glomerular hypercellularity in acute glomerulonephritis of experimental animals and man, *Contrib. Nephrol.* **2**:41.
Siliciano, R. F., Brookmeyer, R., and Shin, H. S., 1983, The diversity of T cell receptors specific for self MHC gene products, *J. Immunol.* **130**:1512.
Sitrin, R. D., Antell, L., Griswold, D. E., Bender, P. E., Greig, R. G., and Poste, G., 1982, A high performance liquid chromatography assay for the rapid analysis of the subunit content of concanavalin A, *Biochim. Biophys. Acta* **717**:175.
Spargo, B. H., Seymore, A. E., and Ordonez, N. G., 1980, *Renal Biopsy Pathology with Diagnostic and Therapeutic Implication*, Wiley, New York.
Sraer, J., Baud, L., Bens, M., Podjarny, E., Schlondorff, D., Ardaillou, R., and Sraer, J. D., 1984, Glomeruli cooperate with macrophages in converting arachidonic acid to prostaglandins and hydroxyeicosatetraenoic acids, *Kidney Int.* **25**:337a.
Steblay, R. W., 1962, Glomerulonephritis induced in sheep by injections of heterologous glomerular basement membrane and Freund's complete adjuvant, *J. Exp. Med.* **116**:253.
Steblay, R. W., and Rudofsky, U., 1971, Renal tubular disease and autoantibodies of tubular basement membrane induced in guinea pigs, *J. Immunol.* **107**:589.
Steinberg, A. D., Huston, D. P., Taurog, J. D., Cowdery, J. S., and Raveche, E. S., 1981, The cellular and genetic basis of murine lupus, *Immunol. Rev.* **55**:121.
Steinberg, A. D., Smith, H. R., Laskin, C. A., Steinberg, B. J., and Smolen, J. S., 1982, Studies of immune abnormalities in systemic lupus erythematosus, *Am. J. Kidney Dis.* **2**:S101.
Steinmetz, M., and Hood, L., 1983, Genes of the major histocompatibility complex in mouse and man, *Science* **222**:727.
Stenglein, B., Thoenes, G. W., and Gunther, E., 1975, Genetically controlled autologous immune complex glomerulonephritis in rats, *J. Immunol.* **115**:895.
Stenglein, B., Thoenes, G. W., and Gunther, E., 1978, Genetic control of susceptibility to autologous immune complex glomerulonephritis in inbred rat strains, *Clin. Exp. Immunol.* **33**:88.

Stenman, S., and Vaheri, A., 1978, Distribution of a major connective tissue protein, fibronectin, in normal human tissues, *J. Exp. Med.* **147**:1054.

Sterzel, R. B., and Pabst, R., 1982, The temporal relationship between glomerular cell proliferation and monocyte infiltration in experimental glomerulonephritis, *Virchows Arch. B* **38**:337.

Sterzel, R. B., Lovett, D. H., Stein, H. D., and Kashgarian, M., 1982, The mesangium and glomerulonephritis, *Klin. Wochenschr.* **60**:1077.

Steward, M. W., 1979, Chronic immune complex disease in mice: The role of antibody affinity, *Clin. Exp. Immunol.* **38**:414.

Steward, M. W., 1981, The biological significance of antibody affinity, *Immunol. Today* **2**:134.

Steward, M. W., Reinhardt, M. C., and Staines, N. A., 1979, The genetic control of antibody affinity: Evidence from breeding studies with mice selectively bred for either high or low affinity production, *Immunology* **37**:697.

Striker, G. E., Mannik, M., and Tung, M. Y., 1979, Role of marrow-derived monocytes and mesangial cells in removal of immune complexes from renal glomeruli, *J. Exp. Med.* **149**:127.

Swain, S. L., 1981, Significance of Lyt phenotypes: Lyt 2 antibodies block activities of T cells that recognize class I major histocompatibility complex antigens regardless of function, *Proc. Natl. Acad. Sci. USA* **78**:7101.

Timpl, R., Rolide, H., Gehron, R. P., Renard, S.I., Foidart, J. M., and Martin, G. R., 1979, Laminin glycoprotein from basement membranes, *J. Biol. Chem.* **254**:9933.

Trompeter, R. S., Barratt, T. M., Kay, R., Turner, M. W., and Soothall, J. F., 1980, HLA, atopy and cyclophosphamide in steroid responsive childhood nephrotic syndrome, *Kidney Int.* **17**:113.

Tsukamoto, Y., Helsel, W. E., and Wahl, S. M., 1981, Macrophage production of fibronectin, a chemoattractant for fibroblasts, *J. Immunol.* **127**:673.

Unanue, E. R., and Dixon, F. J., 1965, Experimental glomerulonephritis. V. Studies on the interaction of nephrotoxic antibodies with the tissues of the rat, *J. Exp. Med.* **121**:697.

Unanue, E. R., Dixon, F. J., and Feldman, J. D., 1967, Experimental allergic glomerulonephritis induced in the rabbit with homologous renal antigens, *J. Exp. Med.* **125**:163.

Unkeless, J. C., Gordon, S., and Reich, E., 1974, Secretion of plasminogen activator by stimulated macrophages, *J. Exp. Med.* **139**:834.

Valle, M. J., Jordan, G. W., Haahr, S., and Merigan, T. C., 1975, Characteristics of immune interferon produced by human lymphocyte cultures compared to other human interferons, *J. Immunol.* **115**:230.

Van Damme, B. J. C., Fleuren, G. J., Bakker, W. W., Vernude, R. L., and Hoedemaker, P. J., 1978, Experimental glomerulonephritis in the rat induced by antibodies directed against tubular antigens. V. Fixed glomerular antigens in the pathogenesis of heterologous immune complex glomerulonephritis, *Lab. Invest.* **38**:502.

van Voorhis, W. C., Volinsky, J., Hoffman, E., Luban, J., Hair, L. S., and Steenman, R. M., 1983, Relative efficacy of human monocytes and dendritic cells as accessory cells for T cell replication, *J. Exp. Med.* **158**:714.

Wagner, C. M., Lucas, D. O., and Nagle, R. B., 1983, The effect of macrophages on the metabolism of glomerular cells: Preliminary studies, *J. Reticuloendothel. Soc.* **33**:93.

Wahl, S. M., Wahl, L. M., and McCarthy, J. B., 1978, Lymphocyte-mediated activation of fibroblast proliferation and collagen production, *J. Immunol.* **121**:942.

Wakashin, Y., Takei, I., Ueda, S., Mori, Y., Jesato, K., Wakashin, M., and Ekuda, K., 1981, Autoimmune interstitial disease of the kidney and associated antigens: Purification and characterization of a soluble tubular basement membrane antigen, *Clin. Immunol. Immunopathol.* **19**:360.

Waksman, B. H., 1979, Adjuvants and immune regulation by lymphoid cells, *Springer Semin. Immunopathol.* **2**:5.

Weiss, M. A., Ooi, B. S., Ooi, Y. M., Engvall, E., and Ruoslahti, E., 1979, Immunofluorescent localization of fibronectin in the human kidney, *Lab. Invest.* **41**:340.

Wilson, C. B., 1983, Nephritogenic immune response involving basement membrane and other antigens in or of the glomerulus, in: *Immune Mechanisms in Renal Disease* (N. B. Cummings, A. F. Michael, and C. B. Wilson, eds.), pp. 233–248, Plenum Medical Book Co., New York.

Wilson, C. B., and Dixon, F. J., 1973, Anti-glomerular basement membrane antibody induced glomerulonephritis, *Kidney Int.* **3**:74.

Wilson, C. B., and Dixon, F. J., 1981, The renal response to immunologic injury, in: *The Kidney* (B. M. Brenner and F. C. Rector, Jr., eds.), pp. 1237–1350, Saunders, Philadelphia.

Wilson, C. B., Lehman, D. H., McCoy, R. C., Gunnels, J. C., and Stickel, D. L., 1974, Antitubular basement membrane antibodies after renal transplantation, *Transplantation* **18**:447.

Woodroffe, A. J., Neale, T. J., and Wilson, C. B., 1978, Spontaneous glomerulonephritis (GN) in New Zealand White (NZW) rabbits, in: *VIIIth International Congress of Nephrology*, Montreal, (Abstract).

Wooley, P. H., Griffin, J., Panayi, G. S., Batchelor, J. R., Welsh, K. I., and Gibson, T. J., 1980, HLA-DR antigens and toxin reactions to sodium aurothiomalate and penicillamine in patients with rheumatoid arthritis, *N. Engl. J. Med.* **303**:300.

Yamada, K. M., 1980, Fibronectin: Transformation-sensitive cell surface protein, *Lymphokine Rep.* **1**:231.

Zakheim, B., McCafferty, E., Phillips, S. M., Clayman, M., and Neilson, E. G., 1984, Murine interstitial nephritis. II. The adoptive transfer of disease with immune T lymphocytes produces a phenotypically complex interstitial lesion, *J. Immunol.* **133**:234.

Zanetti, M., and Wilson, C. B., 1983, Characterization of anti-tubular basement membrane antibodies in rats, *J. Immunol.* **130**:2173.

Zanetti, M., Mampaso, F., and Wilson, C. B., 1983, Anti-idiotype as a probe in the analysis of autoimmune tubulointerstitial nephritis in the Brown Norway rat, *J. Immunol.* **131**:1268.

14

Macrophages in Immunoparasitology

GRAHAM F. MITCHELL and PAUL R. WOOD

1. INTRODUCTION

The discussion herein on macrophage activities in the immunology of host–parasite relationships is selective. It embraces a range of topics that can be termed, in the broadest sense, "extracellular macrophage-related events" in the immunology of parasitism and, in particular, rodent parasitic infections. Aspects of the cell biology of recognition, entry, persistence, and intramacrophage destruction or inhibition of parasitic protozoa will not be covered, reviews on these topics being plentiful (e.g., Trager, 1974; Jones, 1974, 1981; Smith, 1979; Chang, 1981; Mauel, 1982; Sethi, 1982; Bray, 1982; Edelson, 1982; Pearson *et al.*, 1983). Phagocytosis will only be covered in outline. Rather, attention will focus on the macrophage in, or from, parasitized hosts as a *regulator* of antiparasite immune responses and as an *effector* cell of host resistance involving mediator *release*. Large sections in the review cover two aspects that are research interests of the authors, viz., the immunoregulatory role of parasitized macrophages in murine cutaneous leishmaniasis and the role of macrophage products in the control of infection with plasmodia and related intraerythrocytic protozoa. Experimental results are detailed in these two sections more as an indication of approaches currently being pursued and certainly not to indicate that a clear picture currently exists.

Emphasis is placed in some parts of this discussion (e.g., Section 2) on difficulties in interpretation of observed phenomena and the sort of conclusion that is not warranted on the basis of available data. This does not reflect a negative attitude but rather the fact that very little is "cut-and-dried." With immunologists still grappling with the intricacies of immunoregulation and the role played by macrophages in response to the simplest "nonliving" laboratory

GRAHAM F. MITCHELL • Immunoparasitology Unit, The Walter and Eliza Hall Institute of Medical Research, Melbourne, Victoria 3050, Australia. PAUL R. WOOD • Department of Microbiology, University of Melbourne, Parkville, Victoria 3052, Australia.

antigen, the immunoparasitologist interested in the same events is really only able to make quite general statements.

As might be expected, most attention on the macrophage in parasitology has focused on this cell population as a host for parasitic protozoa, i.e., the cell biology of intramacrophage parasitism. Protozoa infecting macrophages in at least some stage of the cycle in mammals include *Toxoplasma gondii, Trypanosoma cruzi,* and *Leishmania* spp. Intriguing observations have emerged on the location of resident protozoa in macrophages—*Toxoplasma gondii* residing in phagosomes and inhibiting fusion with lysosomes (Jones and Hirsch, 1972), *Trypanosoma cruzi* residing in the cytoplasm upon having escaped from the phagosome (Peters, 1974; Kress *et al.*, 1975; Nogueira and Cohn, 1976; Milder and Kloetzel, 1980), and *Leishmania* spp. residing within a phagolysosome apparently resistant to the action of digestive enzymes and other inhibitors (Alexander and Vickerman, 1975; Chang and Dwyer, 1976; Lewis and Peters, 1977; Berman *et al.*, 1981; Shepherd *et al.*, 1983). In the past few years, an enormous effort has been directed to lymphokine-mediated activation of parasitized macrophages for inhibition or destruction of intramacrophage protozoa *in vitro* and the role of oxygen metabolites in the killing event (Hoff and Frenkel, 1974; Borges and Johnson, 1975; Jones *et al.*, 1975; Sethi *et al.*, 1975; Anderson *et al.*, 1976; Shirahata *et al.*, 1977; Nogueira and Cohn, 1978; Chinchilla and Frenkel, 1978; Sethi and Brandes, 1978; Nathan *et al.*, 1979; Behin *et al.*, 1979; Handman and Burgess, 1979; Murray and Cohn, 1979; Wilson *et al.*, 1980; Murray, 181a,b, 1982; Nacy *et al.*, 1981, 1983a; Büchmüller and Mauel, 1981; Pearson *et al.*, 1981; Chang and Chiao, 1981; Tanaka *et al.*, 1982; Haidaris and Bonventre, 1982; Murray *et al.*, 1982; Nogueira *et al.*, 1982; also Beaman *et al.*, 1983).

2. MACROPHAGES AND THE REGULATION OF IMMUNE RESPONSES TO PARASITE ANTIGENS

Historically, the macrophage in parasitic infections was viewed as a key phagocytic cell in the uptake and destruction of opsonized and nonopsonized extracellular parasitic protozoa and infected erythrocytes. Quantitative studies on phagocytosis of parasites and parasitized erythrocytes are not common in parasitology although a huge literature exists on macrophage-mediated clearance or ingestion. A detailed review is available on phagocytosis of *Trypanosoma* spp., *Toxoplasma gondii,* and erythrocytes in malaria (Sethi, 1982). Studies on phagocytosis and blood clearance of parasites have been greatly facilitated by the use of ^{75}Se-labeled parasites (e.g., Mascaskill *et al.*, 1980; Scott and Moyes, 1982; Dempsey and Mansfield, 1983) and the use of defined amounts, isotypes, and specificities of opsonizing (and/or cytophilic) monoclonal antiparasite antibodies (e.g., Hauser and Remington, 1981; Sethi *et al.*, 1981). Observations of particular interest that are highlighted in the excellent review by Sethi (1982) are: (1) the demonstration by Ferrante and Jenkin (1978, 1979) that even relatively large extracellular trypanosomes such as *Trypanosoma lewisi* are killed *following*, rather than prior to, ingestion. Numerous early studies in rodents implicated an involvement of phagocytosis in blood clearance of trypanosomes such as *T. lewisi* and *T. musculi* (e.g., Takayanagi *et al.*, 1974; Dusanic, 1975; Targett and Viens, 1975). Even in hypothymic nude mice, macrophage activities can exert a degree of control over *T. musculi* parasitemias. Thus, injection of trypan blue to nude (and intact) mice markedly increases parasitemia early in infection with *T. musculi* (Brooks and Reed, 1979). Moreover, T-independent IgM antibodies, possibly with opsonizing activity, have been implicated in control of *T. rhodesiense* infections in mice (Campbell *et al.*, 1978). (2) The difference in the fate of opsonized *T. cruzi* compared with opsonized *Toxoplasma gondii* after uptake by normal macrophages—the latter are killed but not the former; and (3) the enhanced phagocytosis of both parasitized and uninfected erythrocytes in malaria (see Section 3.). What is not known is whether *any* parasite-encoded antigens are displayed or released by the *phagocytic* macrophage for induction of antiparasite immune responses.

Systemic helminthic and protozoan parasites confront the immune system of the vertebrate host with a diverse and changing array of antigens. Evidence that the macrophage population is centrally involved in induction and regulation of host immune responses is comprehensive (e.g., Möller, 1978; Unanue, 1981). Functions for the macrophage population include regulating the availability of antigen as well as the processing and presentation of antigen for antigen-reactive lymphocytes, particularly T cells. Not that these are by any means exclusive functions of the macrophage. When it comes to analysis of the role played by macrophages in induction and regulation of antiparasite immune responses, great difficulties arise. Apart from aspects such as macrophage heterogeneity (e.g., McIntyre *et al.*, 1967; Nelson, 1976; Wing *et al.*, 1977; Cowing *et al.*, 1978; Forster and Landy, 1981), the question of the role played by other specialized antigen-presenting cells (e.g., M cells, Langerhans cells, dendritic reticulum cells) remains undefined. The fact that antigen presentation is shown to be defective in macrophages *from* parasitized hosts *in vitro* may or may not bear on antigen presentation by specialized cells *in* parasitized hosts. The fact that natural parasites are living and genetically diverse poses major problems for the immunoparasitologist interested in dissecting events of immune induction and regulation in parasitized hosts. The following discussion may serve to illustrate this.

If it is accepted that recognition of antigen by T cells involved in help for antibody production, cytotoxicity, and delayed-type hypersensitivity (DTH) is dependent upon associative (restrictive) recognition of MHC gene products, then the only antigens of extracellular parasites recognized efficiently by these T-cell types will be those associated with antigen-presenting cells, including macrophages. In addition, parasitized macrophages will only stimulate such cells if they display antigens on the surface in association with host cell MHC molecules. Parasitized erythrocytes may also stimulate these T cells if surface MHC antigens are present (e.g., in mouse compared with humans). Thus, in the case of extracellular parasites, the only antigens stimulating T_D, T_C, or T_H cells will be antigens *released* from living parasites (i.e., exoantigens) plus dead and dying parasites (Mitchell and Anders, 1982; Sendashonga and Black, 1982). A prediction is that somatic antigens of parasites that establish efficiently without substantial parasite death, or that appear as differentiation antigens in established parasites, will not be efficient inducers of T-cell-dependent antibody production and DTH. Far greater sensitization will occur to liberated antigens of parasites. If this notion has validity, then it will be appreciated that interpretation problems could arise from situations such as the following:

(1) *Nonspecific in vivo* activation of macrophages prior to parasite challenge. Increased destruction of challenge parasites and liberation of certain antigens could very dramatically alter specificities and "titers" of immune responses generated subsequently. Greater emphasis could thus be placed on the role of nonspecific activation of macrophages in mediating the degree of resistance measured than might be warranted. Activated macrophages may increase (e.g., Wiener and Bandieri, 1975) or reduce (see Playfair and deSouza, 1979) the availability of certain antigens for immune induction.

(2) Challenge of a host with a cloned genetically homogeneous parasite line compared with a heterogeneous parasite population. If, for example, the latter contains a proportion of avirulent parasites that are killed, liberated antigens could markedly alter "titers" and types of immune response induced to shared somatic antigens.

Although largely theoretical, these points are raised to emphasize the potential degree of complexity of *in vivo* immunological events in parasitic infections, the need for the reductionist approach through *in vitro* systems, and yet the great difficulty in extrapolating events *in vitro* to the *in vivo* situation.

Mice with various genetically based defects are useful in the analysis of host–parasite interactions, the prime example being the hypothymic *nu/nu* (nude) mouse (Mitchell, 1982a; Owen, 1982). Such mice provide a convenient means to assess the broad *in vivo* T-cell dependence of host resistance as well as various antiparasite responses such as antibody production, eosinophilia, macrophage activation, hypergammaglobulinemia, granuloma formation, fibrosis, and so on (Mitchell, 1982a). A comparable B-cell-deficient mouse has been sought for some time. One means to reduce numbers of B cells and thus to reduce antibody production *in vivo* is to inject the host from birth with heterologous anti-immunoglobulin sera, in particular anti-μ antibodies, prior to parasite challenge (e.g., Jacobson *et al.*, 1977; Dessein *et al.*, 1981).

Observations made on the alteration of infection characteristics in anti-immunoglobulin-injected hosts could be used to implicate an involvement of antibodies in expression of these characteristics in intact mice. However, a potential difficulty arises in interpretation of results using this strategy of B-cell deprivation and the difficulty relates to macrophages. Multiple injections of anti-immunoglobulin, even when commenced very early in life when the recipient is relatively hypogammaglobulinemic, must result in the formation of immune complexes with consequences such as chronic macrophage activation and possibly hyperplasia. Thus, an observation such as unaltered resistance or even increased resistance to parasitic infection or disease in such mice (e.g., Rosenberg and Evans, 1979) may reflect not T-cell activities in the absence of circulating antibodies, but an additional contribution from a preexisting hyperactive macrophage axis.

Naturally, the above problem will arise in those parasitic infections in which macrophages are parasitized and where activated macrophages are involved in inhibitory effects on parasites through mediator release. On the same note, macrophage hyperactivity may also be a feature of nude mice (Cheers and Waller, 1975), particularly those bred or maintained for prolonged periods in a conventional mouse colony where there must be selection for survivors of environmentally acquired infections.

Lines of mice, selected from outbred Swiss albino mice according to high (Ab/H) and low (Ab/L) antibody production, have been developed by Biozzi and colleagues (Biozzi *et al.*, 1979). At least under "selection method I," macrophages in Ab/L mice are activated relative to those in Ab/H mice (Wiener and Bandieri, 1974). These strains of mice have been used with various protozoan and helminthic parasites to examine the consequences of low antibody production plus macrophage activation compared with high antibody production (Perrudet-Badeux *et al.*, 1975, 1978; Kierszenbaum and Howard, 1976; Blum and Cioli, 1978; Heumann *et al.*, 1979; Biozzi *et al.*, 1979; Hale and Howard, 1981; Mitchell *et al.*, 1981a, 1982; Jenkins and Carrington, 1981; Mitchell, 1982b). More precise definition of the activation status of macrophages in Ab/L relative to Ab/H (preferably raised in an SPF facility) is urgently required. It has been proposed that in mice with a genetically-based or induced hyperactive RES with increased antigen degradation, there is little opportunity for, and little point in, induction of high titers of antibody (Mitchell, 1977). In the future, quantitative studies in Ab/L and Ab/H mice should provide useful information on the contributions of opsonizing and cytophilic antibodies in clearance of hemoprotozoa, and hyperactive macrophages as antiparasite effector cells and regulators of antiparasite immune responses.

Immunopathologic responses to egg antigens underlie numerous disease manifestations of *chronic* schistosomiasis mansoni and japonica (reviewed in Phillips and Colley, 1978; Warren, 1982; Smithers and Doenhoff, 1982; Damian, 1984). Antigens released from *Schistosoma mansoni* and *S. japonicum* eggs trapped in organs, principally the liver but also the lungs and the intestinal wall, induce T-cell-dependent granulomatous responses with fibrosis. With increasing duration of infection, new granulomas formed around eggs are reduced in size, a phenomenon referred to as endogenous desensitization or modulation of granuloma formation. A major effort has been directed toward the cellular mechanisms of this apparent anti-immunopathology and to ascribe it to immunoregulatory T-cell effects (Chensue and Boros, 1979; Chensue *et al.*, 1981; Colley, 1981; Green and Colley, 1981; Doughty and Phillips, 1982). Alternative mechanisms, such as immunologically based accelerated destruction of eggs (James and Colley, 1976, 1978; Olds and Mahmoud, 1980) and inhibition of embryonation of eggs (Garcia and Mitchell, 1982; Mitchell and Cruise, 1986) in chronically infected hosts, have also been raised. Impetus in pursuing these alternatives to anti-immunopathology (immunoregulation) in granuloma modulation derives from evidence that it is the mature egg which is the richest source of soluble, exported immunopathologic antigens and that egg maturation occurs over several days, if not weeks, after oviposition by the female schistosome. Eosinophil-mediated, antibody-dependent destruction of eggs has been demonstrated

(James and Colley, 1976, 1978) and whether activated macrophages can do the same thing remains unknown.

Granulomas contain macrophages (Adams, 1976; Boros, 1978), these cells being more plentiful in intestinal granulomas than in hepatic granulomas in *S. mansoni*-infected mice (Weinstock and Boros, 1983). Because of the evidence for T-cell dependency of both immunopathologic and anti-immunopathologic responses in granuloma formation and its modulation, a search has been made for changes in the macrophage population in granulomas (Wellhausen and Boros, 1981; Stadecker *et al.*, 1982; Sunday *et al.*, 1983). Nothing remarkable has yet emerged from these studies regarding differences in macrophages from early and late (modulated) granulomas in mice. Reduced phagocytosis of opsonized red cells (Wellhausen and Boros, 1981) and tumoricidal activity (Loveless *et al.*, 1982) have been demonstrated in macrophages from modulated granulomas, and a reduced display of Fc receptors for IgG1 has been described in lung (Amsden and Boros, 1979) relative to liver granulomas. Activated macrophages (Wellhausen and Boros, 1981), presumably under the influence of T-cell lymphokines, have been implicated in stimulation of fibroblasts for collagen synthesis (Wyler *et al.*, 1981; Wyler and Tracy, 1982; see also Pollacco *et al.*, 1978). Besides these indirect effects of egg antigens in collagen biosynthesis, direct effects of egg products on fibroblasts have also been examined (Wyler and Tracy, 1982; Boros and Lande, 1983; Boros *et al.*, 1983).

Studies on the immunology of intestinal parasitism are severely handicapped by a general lack of knowledge on the nature of antigens gaining access from the intestinal lumen to sites of immune induction and regulation (e.g., lymphoid aggregates in the intestinal wall, mesenteric lymph nodes, liver, spleen, and so on). Histological studies by Owen *et al.* (1981) suggest that macrophages of Peyer's patches in mice infected with the luminal protozoan parasite, *Giardia muris,* may be involved in induction of anti-*Giardia* immune responses. *G. muris* trophozoites and their fragments have been observed in macrophages associated with the epithelium of Peyer's patches. Lymphoblasts have been reported to be closely apposed to such macrophages in infected intact but not infected nude mice. The latter develop chronic giardiasis, rejection of this parasite being highly T cell dependent (Roberts-Thomson and Mitchell, 1978; Stevens *et al.*, 1978). It would be of interest to explore this macrophage–lymphocyte interaction in Peyer's patches of young male C3H/He mice, these mice differing from many other inbred mouse strains in developing persistent infection with *G. muris* (Roberts-Thomson and Mitchell, 1978; Underdown *et al.*, 1981). Moreover, differences in antitrophozoite antibody specificities have been detected in chronically infected C3H/He relative to recovered genetically resistant mice (Erlich *et al.*, 1983).

Altered immunoresponsiveness in parasitized compared with uninfected hosts can be profound, secondary immunodeficiency being particularly pronounced in certain *Plasmodium* and *Trypanosoma* infections (World Health Organization, 1978; reviewed in Clayton, 1978; Weidanz, 1982). Reduced responses to homologous parasite antigens, or injected antigens structurally unrelated to those of the resident parasite population, can be described as immunodepression or immunosuppression (specific or nonspecific). By convention, the term *immunosuppression* usually implies an active suppressor event (e.g., *in vitro* inhibition of one host cell population by another or its released products). One of the likely major contributors to reduced immunoresponsiveness in parasitized hosts is the macrophage population, possible macrophage-related activities being:

1. Accelerated clearance and destruction of antigen by activated and/or increased numbers of macrophages
2. Impaired accessory cell functions, i.e., defective antigen presentation and/or processing
3. Failure to clear circulating immune complexes (see Dessaint and Capron, this volume) with several possible immunological consequences as well as consequences for efficient expression of antiparasite effector functions (e.g., Karavodin and Ash, 1982)
4. Elaboration of suppressive molecules

At the reactive lymphocyte level, reduced immunoresponsiveness in parasitized hosts has not been analyzed using clonal approaches. Moreover, at the macrophage level, use is generally made of adherent cell populations of a particular organ. It will be appreciated therefore that few unequivocal statements can be made and that this area of endeavor has not progressed far. Though of great potential in analysis of mechanisms, additional complications are added by the quite different effects of closely related parasites in the one host species (discussed extensively by Weidanz, 1982).

There is evidence for macrophage dysfunction in malaria, and particularly in the spleen macrophage population (Parashar *et al.*, 1983) from malarious mice. Murphy (1981b) reported that plasmodial infections caused defects in macrophage microbicidal capacity and that this defect could be mimicked by intravenous injection of SRBC. Recently, Lee Hand and King Thompson (1983) suggested that erythrophagocytosis inhibits macrophage antibacterial function because of interactions between erythrocyte components and reactive products of phagocyte oxygen metabolism. These papers suggest mechanisms whereby the increased erythrophagocytosis seen in malaria infections may result in suppression of both antibacterial and antimalarial (see below) functions of macrophages.

Adherent cells derived from spleens of mice infected with certain mouse malarias are poor accessory cells in *in vitro* antibody formation to heterologous antigens (Warren and Weidanz, 1976; Brown *et al.*, 1977; Correa *et al.*, 1980). A change in the adherent cell population from one with net enhancing activities with respect to lymphocyte stimulation to one with net suppressive activities has been demonstrated (Wyler *et al.*, 1979). Active suppression by adherent cells (Allison, 1978) has been reported using cells from hosts infected with *Plasmodium* spp. (Weinbaum *et al.*, 1978; Lelchuk and Playfair, 1980; Correa *et al.*, 1980) and others such as *Schistosoma mansoni* (Coulis *et al.*, 1978; Todd *et al.*, 1979; Ottesen, 1979; Powell and Damian, 1982), *Trypanosoma cruzi* (Cunningham *et al.*, 1980; Cunningham and Kuhn, 1980; Mansfield *et al.*, 1981; Kierszenbaum, 1982), *Leishmania tropica major* (Scott and Farrell, 1981), African trypanosomes (Eardley and Jayawardena, 1977; Wellhausen and Mansfield, 1979, 1980), and *Toxoplasma gondii* (Suzuki *et al.*, 1981; Suzuki and Kobayashi, 1983). Macrophage-derived suppressive prostaglandins (Passwell *et al.*,1979, Metzger *et al.*, 1980; Stenson and Parker, 1980) have been implicated in several systems (Lelchuk and Playfair, 1980; Scott and Farrell, 1981; Kierszenbaum, 1982).

In the case of African trypanosomes, parasite-derived inhibitory substances have been described (Cunningham *et al.*, 1980; Albright and Albright, 1981) and as far as phagocytic cells are concerned it is not yet clear what immunosuppressive effects are mediated by products coded for by host macrophages and what are mediated by macrophage-derived or processed parasite products (Grosskinsky and Askonas, 1981; Sacks *et al.*, 1982; Grosskinsky *et al.*, 1983; see also Liew *et al.*, 1979). Defects in responding lymphocyte *and* macrophage populations have been demonstrated using proliferation assays with cells from mice infected with trypanosomes (e.g., Gasbarre *et al.*, 1981; Harel-Bellon *et al.*, 1983).

In malarious mice, the loss of *in vitro* accessory cell function of splenic macrophages is T cell dependent (Morges and Weidanz, 1980). However, various macrophage changes appear to be T cell independent (Lelchuk *et al.*, 1983; see also Mansfield *et al.*, 1981). Coupled with evidence of T-cell-dependent splenic recruitment and activation of macrophages in malaria (Jayawardena *et al.*, 1977; Wyler and Gallin, 1977; Roberts and Weidanz, 1978), the possibility exists that cell movement into and proliferation within the spleen may dilute various cell populations, the activities of which are measured *in vitro* using spleen cell suspensions. Thus, Ia^- macrophages may dilute out those Ia^+ macrophages that purportedly have accessory functions *in vitro* (Weidanz, 1982; see also Freeman and Parish, 1978; Mayor-Withey *et al.*, 1978; Djoko-Tamnou *et al.*, 1981). Dilution of mitogen-responsive cells may contribute to reduced *in vitro* responsiveness of spleen cells to mitogens. Although blood clearance of heterologous erythrocytes is increased (Loose *et al.*, 1972), spleen localization may be reduced (Loose and DiLuzio, 1976); increased antigen elimination by macrophage hyperplasia/activation (Yoshikai *et al.*, 1979) has been implicated in defective induction of DTH responses in malarious mice (Michel *et al.*, 1979). Liver macrophages in plasmodium-infected mice (Dockrell *et al.*, 1980) may be involved in both antigen clearance and effects on intracellular parasites (see Section 3.) and are worthy of more detailed examination.

MACROPHAGES AND THE IMMUNOLOGY OF MURINE CUTANEOUS LEISHMANIASIS

Four models of human leishmaniasis that have been exploited to analyze the genetics and mechanisms of expression of resistance are the *Leishmania tropica major*/mouse, *L. mexicana*/mouse, *L. donovani*/mouse, and *L. enrietti*/guinea pig systems. Concerning *immunological* aspects, most information is available on cutaneous (± systemic) leishmaniasis caused by *L. t. major* in mice. After cutaneous injection of promastigotes, lesions develop and either resolve over a period of some months (in healer strains of mice) or persist with visceralization of infection usually being obvious (in nonhealer strains of mice such as BALB/c and its close relatives). This genetically based variability (literature cited in Rosenstreich *et al.*, 1982; Mitchell *et al.*, 1982) has greatly assisted the analysis of the immunobiology of host resistance and susceptibility to this intramacrophage protozoan parasite.

In cutaneous leishmaniasis in humans and animal models, the bulk of the evidence points to a decisive role for cell-mediated DTH responses, rather than antibodies, in the resolution of natural infection (reviewed by Garnham and Humphrey, 1969; Bryceson, 1970; Zuckerman, 1975; Mauel and Behin, 1982). This is not to suggest that antileishmania antibodies [perhaps concentrated in lesions (Moriearty *et al.*, 1982)] cannot mediate resistance in vaccinated or manipulated hosts—just that there is as yet no unequivocal demonstration that circulating antibodies appearing during the course of "natural" infection and disease resolution, have host-protective effects (e.g., Mitchell, 1983).

Parasite antigens are present on the surface of *L. t. major*-infected macrophages *in vitro* as detected by indirect immunofluorescence using polyspecific (Farah *et al.*, 1975; Handman *et al.*, 1979; Berman and Dwyer, 1981) or monoclonal (Handman and Hocking, 1982; Lebrija de Ibarra *et al.*, 1982) antileishmania antibodies. It is reasonable to assume that some such antigens are the targets of reactive T cells of the DTH type (Ly 1^+2^- in the mouse) and that focusing of aggressive immune attack at the parasitized macrophage surface would involve associative (restrictive) recognition of host cell MHC antigens. It is further assumed that released products (lymphokines) from activated T cells acting at close quarters, in turn activate macrophages in the complex process leading to intracellular parasite destruction. How "close quartered" this T-cell activation need be is conjectural; it is of some interest that irrelevant DTH responses elicited and sustained in the vicinity of infected macrophages will inhibit cutaneous lesion development (Behin *et al.*, 1977). What is most unlikely is that T cells destroy infected macrophages via cytotoxic mechanisms at least *in vivo* (Bray and Bryceson, 1968; Bryceson *et al.*, 1970). T cells which efficiently reconstitute resistance to cutaneous leishmaniasis in nude mice are clearly Ly1^+2^- (Mitchell *et al.*, 1980).

Peculiarities in *L. t. major*-infected macrophages seem to underlie the high susceptibility of intact BALB/c mice to leishmaniasis (reviewed in Mitchell *et al.*, 1982) expressed as absence of healing of cutaneous lesions and visceralization of infection (Handman *et al.*, 1979; Behin *et al.*, 1979; Howard *et al.*, 1980a; Gorczynski and MacRae, 1982). Expression of genetically based high and low resistance to chronic cutaneous disease has been shown to be associated with donor hematopoietic cells in radiation chimeras (Howard *et al.*, 1980a). BALB/c macrophages are more supportive of *L. t. major* amastigote proliferation *in vitro* at 37°C than are macrophages from resistant (healer) strains of mice such as C57BL/6. Addition of lymphokines (Büchmüller and Mauel, 1979; Nacy *et al.*, 1981) to infected macrophages from susceptible mice will inhibit intracellular parasite growth or persistence *in vitro* (Handman and Burgess, 1979; Behin *et al.*, 1979; see also Murray *et al.*, 1982). However, other evidence suggests that lymphokine-initiated killing by such macrophages is defective (Nacy *et al.*, 1983a). It would seem that less attention than is warranted has been paid to control of temperature in some of the *in vitro* killing systems with *L. t. major* (Berman and Neva, 1981).

Handman *et al.* (1979) have presented evidence that one defect in infected BALB/c macrophages is a decrease in the availability of *H-2*-encoded antigens at the cell surface. The experiments which led to this conclusion involved the use of infected and uninfected macrophages of nonhealer strain BALB/c and healer strain CBA/H as cold "blockers" in anti-H-2 cytotoxic T-cell assays. It was found that infected BALB/c macrophages were inefficient inhibitors of target cell lysis. Serological evidence for defective H-2 expression has not been sought as yet. Moreover, immunochemical studies involving surface labeling techniques with infected and uninfected macrophages have been inconclusive (E. Handman, unpublished observations). Thus, the notion that *L. t. major* infection of BALB/c macrophages leads to loss of *H-2*- including *I*-region-encoded molecules, and thus *inefficient* recognition of infected macrophages by Ly-1^+2^- T_D cells, remains an attractive yet unproven mechanism to account for a component of high genetically based susceptibility to cutaneous leishmaniasis in BALB/c mice (see below).

Gorczynski and MacRae (1982) examined *in vitro* proliferative responses of Ly-1^+ T cells from peripheral lymphoid organs of *L. t. major*-challenged and recovered CBA/J $\rightarrow$ (BALB/c $\times$ CBA/J)F_1 and BALB/c $\rightarrow$ F_1 radiation chimeras. Proliferative responses in an allied system are known to be generally *I–A* restricted (Louis *et al.*, 1981, 1982). Stimulator cells used by Gorczynski and MacRae (1982) were adherent cells (macrophages?) harvested from the dorsal skin of CBA/J, BALB/c, and F_1 mice (which are as resistant as CBA/J) and either infected with *L. t. major* or *additionally* pulsed with a crude *L. t. major* antigen preparation. Proliferative responses were obtained using, as appropriate stimulators, infected CBA/J skin cells or antigen-pulsed infected CBA/J and BALB/c skin cells. In contrast, proliferative responses were not induced by *infected* BALB/c skin cells. The presumed defect in antigen presentation by infected BALB/c cells was not shared by another infected *H-2^d^* cell population, B10.D2, mice of this strain being more resistant than BALB/c to chronic cutaneous leishmaniasis. A discrepancy exists between these data and *in vivo* DTH studies of Handman *et al.* (1979) in which antigen-pulsed infected BALB/c macrophages (from the peritoneal cavity) were inefficient at sensitization for DTH to crude *L. t. major* antigen. These two studies support the notion of a pivotal role for infected macrophages in dictating the immunological consequences and outcome of infection with *L. t. major* but also raise the issue of macrophage heterogeneity (Gorczynski *et al.*, 1981) in the interplay between T cells and macrophages that are infected or that present released antigen to reactive T cells.

Gorczynski and MacRae (1982) also demonstrated that infected BALB/c skin cells can cause induction of cells that have inhibitory effects on *in vitro* lymphocyte proliferation although the surface marker status of the "suppressor" cells remains uncertain. Moreover, indirect evidence was obtained that defective stimulating activity in this system was associated with low Ia expression (see also Bagasra *et al.*, 1981). The presence of suppressor T cells (Howard *et al.*, 1980b, 1981; Liew *et al.*, 1982), alternatively known as disease-promoting T cells (Mitchell *et al.*, 1981b, 1982; Mitchell, 1983), has been demonstrated in peripheral lymphoid organs of diseased BALB/c mice, the surface marker phenotype of T cells with such activities being Ly $1^{+}2^{-}$ (Mitchell *et al.*, 1981b; Liew *et al.*, 1982). In the studies by Howard *et al.* (1980b) on impairment of DTH responses to crude promastigote antigen in diseased BALB/c mice, activity of the T_S cell population was demonstrated in the induction phase of DTH. This is in keeping with observations by Louis *et al.* (1982) that *in vitro* generated BALB/c T-cell lines with antileishmania specificity could transfer specific DTH responses to both infected and uninfected BALB/c recipients. Suppression of the lymphokine-producing capacity of spleen cells has been demonstrated in *L. donovani*-infected BALB/c mice (Murray *et al.*, 1982). A potential complication in interpretation of activities of cell populations harvested from organs of chronically diseased BALB/c mice is that large numbers of parasites are generally also present in these cell suspensions and antigens must also be present (Nasseri and Modabber, 1979; Mitchell *et al.*, 1981b; Leclerc *et al.*, 1981; Mitchell, 1983; Hill *et al.*, 1983). Cloned T-cell lines with defined antileishmania specificities will prove particularly useful in analysis of T-cell-mediated suppression in this system.

With defective antigen presentation, yet the capacity to kill parasites in the presence of added lymphokines (but see Nacy *et al.*, 1983a), infected BALB/c macrophages resemble those of immature mice (Lu and Unanue, 1982). As emphasized, the evidence remains circumstantial for defective Ia antigen expression on at least some *L. t. major*-infected BALB/c macrophages being responsible for poor "accessory cell" function and for inefficiency of recognition by aggressive host-protective Ly-$1^{+}2^{-}$ T cells. Lymphokines, and in particular γ-interferon (King and Jones, 1983; Basham and Merigan, 1983) from activated T cells, will increase Ia expression on macrophages (Scher *et al.*, 1980; Steeg *et al.*, 1980, 1982; Steinman *et al.*, 1980) whereas prostaglandins may decrease Ia expression (Snyder *et al.*, 1982; Bouta and Parnham, 1982). As mentioned previously, an *in vitro* inhibitory effect of activated macrophages from *L. t. major*-infected BALB/c mice on lymphocyte proliferative responses has been ascribed to prostaglandin effects (Scott and Farrell, 1981). If Ia antigen expression on at least some infected BALB/c macrophages is defective, then this could result from a direct effect of parasitization of the cell (see below), refractoriness to the effects of γ-interferon, or a direct inhibitory effect of prostaglandins.

A most peculiar phenomenon has been described in reconstituted BALB/c nude mice infected with *L. t. major.* As expected, nude mice of CBA/H and C57BL/6 genotypes differ from intact CBA/H and C57BL/6 mice in developing chronic rather than resolving cutaneous leishmaniasis. Resistance (i.e., healing) is reconstituted by injection of a small number of syngeneic or *H-2*-compatible/non-*H-2*-incompatible T cells (Mitchell *et al.*, 1980; Mitchell, 1983). In C57BL/6.*nu*/*nu*, cells from bm12 mutant mice which differ from C57BL/6 in the *I* region of *H-2*, do not reconstitute resistance (Mitchell, 1983). BALB/c.*nu*/*nu* mice are also highly susceptible to chronic leishmaniasis as expected and die prematurely with large cutaneous lesions and visceralization of infection after cutaneous injection of *L. t. major*. However, injection of a *small* number of syngeneic T cells results in resistance. Thus, cutaneous lesions in BALB/c.*nu*/*nu* mice injected with 10^6 or 10^7 syngeneic cells resolve rapidly and completely. T cells from chronically infected BALB/c.*nu*/*nu* mice do not have resistance-promoting effects at any cell dose in infected BALB/c.*nu*/*nu* mice and small numbers of Ly-1$^+$2$^-$ T cells from such donors readily abrogate resistance in nudes given a small number of normal T cells. Extensive studies by Howard and colleagues in resistant sublethally irradiated BALB/c mice injected with T cells from diseased mice have led to similar conclusions (Howard *et al.*, 1980b, 1981; Liew *et al.*, 1982). That is, a dominant effect of suppressor or disease-promoting T cells from diseased BALB/c mice can be demonstrated readily in syngeneic mice that are resistant as a result of reduced total T-cell involvement.

Recent experiments in the mouse model of cutaneous leishmaniasis suggest that peripheral lymphoid organs of diseased BALB/c and related mice contain a mixture of disease-promoting and resistance-promoting T cells. Thus, cells from diseased (genetically susceptible) BALB/c.*H-2*^{b} and BALB/c.*H-2*^{k} mice can have resistance-promoting effects when injected at certain cell doses into C57BL/6.*nu*/*nu* and CBA/H.*nu*/*nu* mice, respectively. It appears that nude mice of *resistant* genotype are required to demonstrate the presence of the resistance-promoting subset of (Ly 1$^+$2$^-$) T cells in chronically diseased *H*-2-compatible mice (Mitchell, 1983).

A sequence of events that could account for the immunological and infection consequences of a primary defect in *L. t. major*-infected BALB/c macrophages has been proposed (Mitchell, 1983). This intricate scheme would not be necessary if Ly-1$^+$2$^-$ T-cell-dependent "blocking antibodies" (Garnham and Humphrey, 1969) were shown to be responsible for inhibition of host-protective Ly-1$^+$2$^-$ T cells of the DTH, macrophage-activating type. It is proposed that loss of MHC-encoded molecules on infected macrophages results in increased release of parasite-encoded antigens normally associated with such molecules. Uptake of this antigen by noninfected macrophages or specialized antigen-presenting cells leads to induction of Ly-1$^+$2$^-$ T cells with, in general, specificity for epitopes of the antigen that are normally cryptic when the antigen is presented on the *infected* macrophage. The effect of these expanded and activated T cells is to recruit and activate uninfected macrophages for accelerated destruction of released antigen and therefore a further reduction in the *efficiency of induction* (see Dye and North, 1981) of host-protective Ly-1$^+$2$^-$ T cells with specificity for displayed epitopes on infected macrophages. The essential feature of the hypothesis is that host-protective and disease-promoting Ly-1$^+$2$^-$ T cells have specificity for macrophage-associated antigen but react with different epitopes of this antigen. Defective expression of Ia antigens on infected macrophages not only reduces the efficiency of infected target cell recognition, but also increases the availability of (released) antigen for induction and expression of inappropriate T cells. In order to account for the host-protective effects of low but not high numbers of normal T cells in BALB/c.*nu*/*nu* mice, it is necessary to postulate that the frequency of T cells with specificity for "organized" epitopes of parasite antigen on infected macrophages is higher than the frequency of T cells with specificity for epitopes of the antigen when taken up and presented by uninfected cells. This ratio of frequencies of T-cell specificities is reversed in chronically diseased mice following clonal expansion of the inappropriate specificities. Clearly the issue of T-cell specificities will only be resolved through the use of biologically active T-cell clones and isolated antigens in this system.

BALB/c and related genetically susceptible mice can be vaccinated quite effectively against cutaneous leishmaniasis by infection of irradiated promastigotes (Alexander, 1982; Howard *et al.*, 1982), or crude infected macrophage antigens plus the adjuvant *Corynebacterium parvum* (Mitchell *et al.*, 1981a; Mitchell and Handman, 1983). Nonpathogenic (avirulent) cloned promastigotes derived from a pathogenic isolate (Kellina *et al.*, 1981; Handman *et al.*, 1984) will also induce a degree of resistance in genetically susceptible mice (Mitchell *et al.*, 1984). Dramatic restraints are imposed by route of injection in all these systems, subcutaneous injections being totally without protective effects. The nature of the resistance demonstrated in vaccinated mice remains unclear. In the system in which crude antigen plus *C. parvum* is injected intraperitoneally, infected macrophage antigens appear to be more effective than promastigote or even amastigote preparations. Whether or not "host-protective antigens" of infected macrophages are present in promastigotes can only be answered after the identity of the antigens is known. Antigenic sharing as determined by high-resolution immunochemical techniques is extensive between the life cycle stages of *L. t. major* (Handman *et al.*, 1984; see also Chang and Fong, 1982).

Concerning the putative reduction in MHC expression on *L. t. major*-infected BALB/c macrophages, interesting possibilities regarding mechanisms are raised by recent studies on MHC expression on SV40-transformed cells, in this case fibroblasts and mammary tumor cells (Swaminathan and Gooding, 1983). Tunicamycin treatment of such cells results in loss of susceptibility to anti-H-2 cytotoxicity mediated by alloreactive cells or antibodies plus complement. Conceivably, prevention of glycosylation leads to accelerated destruction of the unglycosylated H-2 antigens. In the discussion of this paper, mention is also made of the increased production of proteases such as plasminogen activator by transformed cell lines and the generation of plasmin with further opportunities for protein cleavage at the cell surface. It could be rewarding to examine whether infection of BALB/c macrophages with *L. t. major* disturbs glycosylation of host cell H-2^d antigens, or results in increased protease production with resultant decrease in surface H-2 display. An active, parasite-mediated effect of this type could also protect other organisms within the macrophage from *immunologically* initiated destruction. Products released from *Leishmania* spp. and with "proparasitic" effects have to date been studied more from the viewpoint of inhibiting intramacrophage digestive enzymes (e.g., Handman and Greenblatt, 1977; El-On *et al.*, 1980).

3. MACROPHAGES AND DESTRUCTION OF PARASITES NOT WITHIN MACROPHAGES

Antibody-dependent, cell-mediated (cellular) cytotoxicity (ADCC) is an *in vitro* effector mechanism of helminth destruction (Dean *et al.*, 1974; Butterworth *et al.*, 1975; Capron *et al.*, 1975) that has been studied using targets such as young schistosomules of *Schistosoma mansoni,* microfilariae and infective larvae of filarial nematodes, and larvae of nematodes such as *Trichinella spiralis, Nematospiroides dubius,* and *Nippostrongylus brasiliensis* (see Kazura, this volume). In tumor cell killing involving both antibody-dependent and non-antibody-dependent cytotoxicity mediated by macrophages, there has been a rapid accumulation of information of effector roles of cytolytic proteases, hydrogen peroxide, and others of the 60 or so different substances secreted by macrophages (reviewed in Adams and Nathan, 1983).

Any helminth with accessible surface antigens, when incubated with appropriate antibody specificities, should adsorb any or all of the various accessory cells such as macrophages, neutrophils, and eosinophils (e.g., Higashi and Chowdury, 1970; Leventhal and Soulsby, 1976; Mackenzie *et al.*, 1981) and mast cells and platelets, provided:

1. Surface antigens are not being turned over rapidly, or the parasite antigen–antibody complexes are not sloughed off
2. The "indicator cells" have appropriate numbers, types, and distribution of surface Fc receptors

Concerning efficiency and/or stability of binding and as in the case of opsonization and phagocytosis of smaller entities, complement components can be expected to increase cell binding to antibodies on the helminth surface. Moreover, efficiency of any subsequent parasite destruction can be expected to increase when the mononuclear phagocytic effector cell is "activated." *Comparative* studies have only recently commenced on activated macrophages as effector cells in tumor cell killing compared with extracellular helminth destruction and intramacrophage killing of microbes (Haidaris and Bonventre, 1981; James *et al.*, 1982a). Reports are available on fractionation of lymphokines involved in these activated macrophage functions (Nacy *et al.*, 1983b; Tabor and Saluk, 1983).

Controversy still surrounds the *relative* contributions of various accessory cells involved in the helminthotoxic component of ADCC. What is not controversial any longer is that virtually all formed elements of the blood can have effector and/or amplification functions in these *in vitro* events. One persistent concern is that at least some *in vitro* systems used for analysis of ADCC may be more a form of accelerated *in vitro* demise of a compromised parasite than a reflection of a major *in vivo* effector mechanism. Culture conditions usually employed have been developed for optimal mammalian cell survival or activity. Parasite repair functions *in vitro* may be severely impaired compared with these same functions *in vivo.*

Systems in which macrophages have been implicated, or demonstrated to be involved in ADCC are the *Schistosoma mansoni* schistosomule/rat (see below), the *Litomosoides carinii* microfilariae/rat (Subrahmanyam *et al.*, 1976; Mehta *et al.*, 1980, 1982), the *Dipetalonema viteae* microfilariae/hamster (Weiss and Tanner, 1979; Haque *et al.*, 1981; see also Worms and McLaren, 1982), and the *Nematospiroides dubius* infective larvae/mouse systems (Chaicumpa and Jenkin, 1978). Most data have come from the *S. mansoni* rat system analyzed by Capron and colleagues (Capron *et al.*, 1975, 1977, 1982a,b). Using the schistosomules of *S. mansoni,* this group has established that macrophages are involved in IgE-dependent ADCC, the participation of IgE being something of an initial surprise (see also Mehta *et al.*, 1980). The bulk of the evidence has been obtained in a rat system although a similar phenomenon has been described using human immune sera and human monocytes or baboon macrophages (Joseph *et al.*, 1978). Apparently, IgE-containing immune complexes from serum, and in antibody excess, bind to macrophages through Fc receptors for aggregated IgE. This leads to activation as well as focusing of the activated macrophage to the surface of the newly transformed schistosomule with subsequent parasite destruction. Activation involves release of lysosomal enzymes (β-glucuronidase) and plasminogen activator, increased glucosamine incorporation and production of superoxide anion (Capron *et al.*, 1980). Interestingly, certain indices of activation mediated by aggregated IgE were inhibited by peptide fragments of IgG cleaved by proteolytic enzymes derived from schistosomules; schistosomule killing was reduced in the presence of IgG peptides (Auriault *et al.*, 1980, 1981). The contribution of IgE-dependent, macrophage-mediated killing of schistosomules to expression of *in vivo* resistance to *S. mansoni* infection in rat and more "permissive hosts" remains unknown.

In vitro schistosomulocidal effects of *mouse* macrophages in the presence (Kassis *et al.*, 1979) or absence of added specific antibody (Mahmoud *et al.*, 1979; Olds *et al.*, 1980; Bout *et al.*, 1981; James *et al.*, 1982b,c; see also Olds *et al.*, 1981) have been demonstrated. Soluble mediators derived from freshly harvested activated peritoneal macrophages of BCG- or *C. parvum*-injected mice have been implicated in *S. mansoni* schistosomule killing (Mahmoud *et al.*, 1979; Ellner and Mahmoud, 1981) with arginase as a possible effector molecule (Olds *et al.*, 1980). Inflammatory macrophages, which are generally inactive (Mahmoud *et al.*, 1979; James *et al.*, 1982b), have been shown to kill *S. mansoni* schistosomules when lymphokines ("Con A supernatants") are added (Bout *et al.*, 1981). Factors involved in activation of macrophages for extracellular killing of schistosomules are likely to differ from those that activate macrophages for killing of intramacrophage target organisms (Nacy *et al.*, 1983b; see also Tabor and Saluk, 1983; Ralph *et al.*, 1983). Plastic-adherent peritoneal cells from *S. mansoni*-infected mice can kill skin stage (James *et al.*, 1982b) but not lung schistosomules (Sher *et al.*, 1982), the latter being particularly resistant to a variety of antiparasite effector mechanisms (literature cited in Mitchell and Cruise, 1986). A requirement for attachment of activated macrophages from infected mice in increasing the efficiency of killing of schistosomules has been emphasized (James *et al.*, 1982b). No evidence is available that activated macrophages in any location actually mediate schistosomulocidal effects *in vivo* but it is of some interest that Biozzi Ab/L mice with activated macrophages (see Section 2) are marginally more resistant to *S. mansoni* infection than Biozzi Ab/H mice (Blum and Cioli, 1978).

3.1. MACROPHAGES IN MURINE MALARIA AND BABESIA INFECTIONS

A role for antibody in resistance to the erythrocytic stage of malaria has been well established by the successful passive immunization of humans and experimental animals (reviewed by Zuckerman, 1977; Cohen, 1979; Playfair, 1982) including the *in vivo* use of monoclonal antibodies (e.g., Freeman *et al.*, 1980). However, it has become clear that T-cell-dependent, non-antibody-mediated mechanisms are also important in expression of resistance to malaria infections.

B-cell-deficient chickens, cured of lethal *Plasmodium gallinaceum* infection by drug therapy, are resistant to homologous challenge (Rank and Weidanz, 1976). B-cell-deficient mice (i.e., anti-μ treated and thus not only B cell deficient; see Section 2) will resist a challenge infection with *P. yoelii* (Roberts and Weidanz, 1979) after drug cure of a primary infection. Moreover, B-cell-deficient mice can resolve an acute primary infection with the nonlethal parasite *P. chabaudi adami* and subsequently resist challenge with both homologous and some heterologous infections [*P. vinckei*, but not *P. yoelii* or *P. berghei* (Grun and Weidanz, 1981)].

These non-antibody-mediated mechanisms of resistance which are presumably T cell dependent (Brown *et al.*, 1968; Clark and Allison, 1974; Weinbaum *et al.*, 1976b) are not likely to result from direct cytolysis of infected red blood cells. There is only one report of a cytotoxic lymphocyte lysing *P. berghei*-infected red cells *in vitro* (Coleman *et al.*, 1975), but this finding has not been generally confirmed. Playfair *et al.* (1977) found that mice made resistant to infection by recovery from a homologous or heterologous hemoprotozoan infection had normal or reduced helper T-cell responses, and concluded that resistance to infection with those parasites may not involve the helper T-cell subpopulation. However, in such studies, meaningful data can only come from clonal analyses of various antigen-reactive T-cell types.

T cells also mediate DTH responses to malarial antigens. Mice protected or sensitized against *P. yoelii* and *P. berghei* exhibit both pronounced T-cell blastogenesis *in vitro* (Jayawardena *et al.*, 1975; Weinbaum *et al.*, 1976a) and DTH response (Finerty and Krehl, 1977; Whitelaw *et al.*, 1977) to malarial antigens. Mice vaccinated with a formalin-fixed preparation of either *P. berghei* or *P. yoelii* also exhibit DTH responses to homologous parasite antigens (Cottrell *et al.*, 1978), and the DTH responses of these vaccinated mice to antigen challenge correlated well with the degree of host resistance. These authors also found evidence of a "systemic DTH response" with radiolabeled bone marrow cells homing into the spleen and liver of vaccinated mice challenged with live *P. yeolii*.

McDonald and Sherman (1980) reported a lack of correlation between DTH response and host resistance to *P. chabaudi* infection. They found that DTH responses to parasite antigens were depressed in immunized mice after challenge with live *P. chabaudi*. A possible explanation for this depression is the heavy antigen burden that the challenge infection imposes on the host. In a different antigen system, such a burden has been reported to suppress peripheral DTH expression (Collins and Mackaness, 1970). The expression rather than induction of DTH responses to heterologous antigens has also been found to be depressed in malarious mice (Freeman, 1978).

3.1.1. Phagocytosis of Parasite-Infected Red Cells

It was originally thought that the death of malaria parasites during both natural and acquired immunity was associated with phagocytosis of infected red cells (Taliaferro, 1941). Hyperplasia of the RES, particularly in the spleen and

TABLE 1. INCREASED PHAGOCYTIC ACTIVITY DURING INTRAERYTHROCYTE PROTOZOAN INFECTIONS

Parasite	Animal species	Agent phagocytosed	Reference
Plasmodium vinckei	Mice	Carbon	Cox *et al.* (1964)
P. vinckei, P. chabaudi	Mice	Carbon	Lucia and Nussenzweig (1969)
Plasmodium spp.	Humans	^{125}I-labeled aggregated human serum albumin	Sheagren *et al.* (1970)
P. berghei	Rats	Carbon	Cantrell *et al.* (1970)
	Rats	Carbon, gelatinized ^{131}I-labeled triolein lipid	Kitchen and DiLuzio (1971)
P. berghei yoelii	Mice	Carbon	Rivera-Ortiz and Nussenzweig (1972)
	Mice	^{51}Cr-labeled SRBC	Loose *et al.* (1972), Loose and DiLuzio (1976)
	Mice	Carbon, ^{51}Cr-labeled SRBC	Greenwood *et al.* (1971)
Babesia microti, B. hylomysci	Mice	Carbon	Hussein (1979)

liver, was characteristic of malaria and early workers found that blockade of the RES with foreign red cells (Gingrich, 1941) or India ink (Goble and Singer, 1960) interfered with the ability of animals to control malaria infections. Since that time, many groups have examined the effect of malaria and babesia infections on the phagocytic activity of the RES of animals (Table 1). In all these studies, phagocytic activity in animals was increased during infection, although in one study, while phagocytosis of colloidal carbon was enhanced, a gelatinized lipid was taken up normally (Kitchen and DiLuzio, 1971).

Two groups (Cox *et al.*, 1964; Cantrell *et al.*, 1970) reported that animals lost their ability to clear carbon rapidly from their circulation at high parasitemias. This loss of hyperactivity did not, however, adversely affect the recovery of rats from *P. berghei* infection, indicating that recovery was not primarily due to enhanced phagocytosis. Lucia and Nussenzweig (1969) reported normal blood clearance rates during latent chronic *P. chabaudi* infection when animals were actively suppressing this infection and were resistant to an otherwise lethal *P. vinckei* superinfection. They also concluded that nonspecific phagocytosis does not have a major role in resistance to rodent malaria.

In 1971 Criswell *et al.* found that humoral factors and macrophages acted synergistically to eliminate parasitized erythrocytes from chambers placed in mice with chronic malaria. This finding was confirmed *in vitro* by Chow and Kreier (1972), who reported that the combination of immune macrophages and immune serum was more effective in phagocytosis of *P. berghei* parasites than either component alone. The cytophilic antibody responsible was later characterized as IgG which, on passive transfer, acted synergistically with an opsonizing antibody to protect against *P. berghei* infection (Green and Kreier, 1978). The presence of opsonizing antibody has also been demonstrated by Shear *et al.* (1979), who found that splenic macrophages from malarious mice could ingest *P. berghei*-parasitized erythrocytes *in vitro* and that this uptake was mediated by disease-associated immunoglobulin bound to the surface of parasitized cells. They also reported that splenic, but not peritoneal, macrophages appeared to be activated, in terms of increased spreading and greater phagocytosis of opsonized and nonopsonized SRBC, during infection with *P. berghei.*

Erythrophagocytosis has also been implicated in the anemia produced during *P. berghei* infections (Roth and Herman, 1979; Howard and Mitchell, 1979) as the degree of anemia is apparently greater than that accounted for by direct destruction of infected erythrocytes by the eruption of mature schizonts. Although opsonizing antibodies have been found to be involved in this erythrophagocytosis, antibody alone in the absence of macrophage activation could not explain the phagocytosis of noninfected erythrocytes involved in the development of anemia (Tosta and Wederburn, 1980).

Immune serum also enhances phagocytosis of *B. microti*-infected hamster red cells by macrophages (Bautista and Kreier, 1979, 1980). However, macrophages from immune animals had an inhibitory effect on parasite growth even in the absence of immune serum, although this inhibition was greater in cultures containing both immune serum and immune macrophages. Cell-free supernatants from these cultures also inhibited parasite growth (Bautista and Kreier, 1980). This finding supports the hypothesis of Clark *et al.* (1975) that soluble factors released by macrophages from immune animals inhibit the growth of *B. microti* (see below).

There are also several reports of increased ADCC activity and/or K cell activity during malaria infections (Greenwood *et al.*, 1977; McDonald and Phillips, 1978). Brown and Smalley (1980) found that lymphoid cells from *P. falciparum*-infected humans could kill parasites *in vitro.* They also reported that polymorphonuclear neutrophils from children with acute *P. falciparum* inhibit parasite growth in homologous and nonimmune sera (Brown and Smalley, 1981). The mechanism responsible for these effects is, however, not clear as lymphoid cells from uninfected animals have been shown to be cytotoxic for *P. yoelii in vitro* (Taverne *et al.*, 1982b). G. A. Butcher (personal communication) also found that normal human peripheral blood mononuclear cells could inhibit *P. falciparum* growth *in vitro* and that the macrophage cell line U937 significantly inhibited *P. falciparum* growth at cell/parasite ratios as low as 1 : 5.

3.1.2. NK Cells and Malaria

Eugui and Allison (1980, 1982) proposed a role for NK cells in resistance to *P. chabaudi* on the basis of increased splenic NK activity and a correlation between susceptibility to plasmodium infection and low NK activity in mice. Several other groups have also reported increased NK activity during other malaria infections, *P. yoelii* (Hunter *et al.*, 1981), and *P. falciparum* (Ojo-Amaize *et al.*, 1981). This increase in NK activity early after infection correlated well with increased interferon levels in serum. However, three groups have now concluded from several lines of evidence, including the use of Nk-defective beige mutant mice, that the conventional NK cell (i.e., that capable of lysing YAC target cells in 4 hr in culture) is probably not involved in resistance to *P. vinckei petteri* and *B. microti* infections (Wood and Clark, 1982a; Ruebush and Burgess, 1982). Skamene *et al.* (1983) also found that NK cell activity and resistance to *P. chabaudi* segregated independently in genetic studies and that beige mice were as resistant as their heterozygous littermates to infection.

Although the "conventional" NK cell now appears not to be involved in resistance to malaria infections, NK cells are known to be an extremely heterogeneous group (Bartlett and Burton, 1982) and their lineages are still in dispute. Cells of the monocyte-macrophage lineage are also known to express natural cytotoxic activity (Roder *et al.*, 1981). Therefore, Allison and Eugui (1982) have modified their earlier hypothesis to suggest that promacrophages are the effectors in the postulated NK-mediated resistance to malaria parasites.

3.1.3. Evidence for Macrophage Activation during Malaria and Babesia Infections

Mice are known to become hypersensitive to the lethal effects of endotoxin (LPS) during malaria and babesia infections (Loose *et al.*, 1971; Clark, 1978, 1982). This increased LPS sensitivity of malaria-infected animals has also been demonstrated by the release of several monokines *in vivo* after injection of LPS (Clark *et al.*, 1981; Taverne *et al.*, 1981, 1982a). The susceptibility of animals to the lethal effects of LPS *in vivo* has been shown to be paralleled by an increase in the sensitivity of their macrophages to LPS *in vitro* (Shands *et al.*, 1974; Glode *et al.*, 1977; Peavy *et al.*, 1978, 1979).

Wood and Clark (1984) showed that the increased LPS sensitivity of *P. vinckei petteri*- and *B. microti*-infected mice was accompanied by an increase in the ability of their macrophages to release one of the soluble mediators (LAF) *in vitro*. Thus, increased LPS sensitivity of animals and release of these mediators reflects macrophage activation during these infections. If these mediators are of biological relevance *in vivo*, then some endogenous trigger for their release is required. It is tempting to speculate that this trigger shares functional properties with LPS, since LPS protects mice against *P. berghei* (Martin *et al.*, 1967; McGregor *et al.*, 1969), and inhibits metabolism in *P. vinckei petteri* parasites (Rzepczyk and Clark, 1981). The LPS-induced cytostatic effect on *P. vinckei* has also been shown to be mediated via macrophages (Rzepczyk, 1982). In addition, the pathology of severe malaria can be induced in healthy mice with *P. vinckei petteri* infections of low parasitemia by injection of a small dose of LPS (Clark and Clouston, 1980).

There is some evidence that an endotoxinlike material may be present in malaria parasites and malarious sera. Tubbs (1980) detected small amounts of *Limulus*-positive material, presumed to be endotoxin, in plasma from all of 16 patients acutely infected with *P. falciparum* but from only 2 of 18 individuals who were either healthy or had other diseases. Tubbs (1980) also reported this activity in plasma of *P. berghei*-infected mice but not controls. This result has been confirmed by Felton *et al.* (1980), although these authors considered that the amounts they could detect were insufficient to harm the mice. However, as referred to above, parasitized mice show increased sensitivity to the effects of LPS. It should be noted that a range of substances other than LPS have been reported to give positive readings in *Limulus* assays (Bartl *et al.*, 1975).

More recently, Gauci *et al.* (1982), using an indirect assay for endotoxinlike activity [the induction of tyrosine aminotransferase (TAT) activity], found that a cell-free preparation of *P. vinckei petteri* parasites increased TAT activity when injected into endotoxin-sensitive mice. These authors calculated that when a mouse became ill with this infection, its parasite load had a TAT-inducing activity equivalent to that of 17 ng of LPS.

There is good circumstantial evidence that the trigger to the illness in human malaria is something released from parasitized red cells when they burst at schizogony. Untreated human infections, particularly with *P. vivax* and *P. malariae*, quickly become synchronous, so that all the schizonts of a particular "brood" rupture within a short period. Many studies clearly establish that this synchronous schizogony is followed, at a fixed interval of some 4 hr, by the onset of the fever-associated illness characteristic of this disease (Neva *et al.*, 1974). Not only is this disease recognized as very similar to the sepsis of gram-negative bacterial infections (Kitchen, 1949), which is generally acknowledged to be associated with release of LPS, but this 4-hr lag also characterizes the pyrogenic response which follows the injection of LPS (Snell, 1971). This suggests that material released at schizogony, presumably of parasite origin, can function *in vivo* like LPS.

From a functional point of view, it matters little whether the trigger released at schizogony is chemically similar to bacterial endotoxin or is providing antigen for the formation of circulating immune complexes, since both can have very similar triggering effects on macrophages (Passwell *et al.*, 1979). The presence of these immune complexes has been associated with the onset of cerebral symptoms in falciparum malaria (Adam *et al.*, 1981), and has been discussed in the literature for some years (reviewed by D'Amelio, 1980).

Other indications of macrophage activation during the course of malaria infections are enhanced macrophage spreading and increased phagocytosis of opsonized and nonopsonized SRBC (Shear *et al.*, 1979; Roubin *et al.*, 1981), increased macrophage numbers and acid phosphatase activity in the liver (Dockrell *et al.*, 1980), and more recently, secretion of plasminogen activator (Lelchuk *et al.*, 1983). There is also some evidence of activation of adherent cells with increased tumoricidal activity (Wood and Clark, 1982a) and prostaglandin synthesis (Clark and Hunt, personal communication) from peritoneal cells of *P. vinckei petteri-* and *B. microti*-infected mice. The T-cell dependence of resistance to these infections is also compatible with a role for activated macrophages as many macrophage functions are amplified in the presence of T-cell products (Lazdins *et al.*, 1978; Allison, 1978; but see Lelchuk *et al.*, 1983).

3.1.4. Nonantibody Soluble Mediators and Resistance

Nonspecifically induced resistance to malaria and babesia infections is a well-described phenomenon. A range of *in vivo* immunostimulants is known to

TABLE 2. IMMUNOSTIMULANTS KNOWN TO PROTECT MICE AGAINST HEMOPROTOZOAN PARASITES[a]

Immunostimulant	Parasite	Reference
C. parvum	*P. berghei*	Nussenzweig (1967), Murphy (1981a)
	B. microti	Wood and Clark (1982b)
	B. rodhaini	
	P. vinckei	Clark *et al.* (1977a)
	P. chabaudi	
	P. yoelii	Cottrell *et al.* (1977)
	P. vinckei	
Brucella abortus	*B. microti*	Herod *et al.* (1978)
Newcastle disease virus	*P. berghei*	Jahiel *et al.* (1969)
Coxiella burnetti	*B. microti*	Clark (1979b)
	B. rodhaini	
	P. vinckei petteri	
BCG	*B. microti*	Clark (1979b)
	B. rodhaini	
	P. yoelii	
	P. vinckei	
	B. rodhaini	Mitchell *et al.* (1978)
	P. berghei	Mitchell *et al.* (1978), Smrkovski and Strickland (1978), Murphy (1981a)
Cord factor, Coam, Zymosan, Glucan, Salmonella, Listeria	*B. microti*	Clark (1979a)

[a]Reviews: Allison and Clark (1977), Cox (1978).

protect against these intraerythrocytic protozoan parasites (Table 2), and concurrent infections with viruses and bacteria will also induce resistance (reviewed by Cox, 1978). Clark *et al.* (1977a,c; Clark, 1979b) have argued that neither enhanced phagocytosis nor augmented antibody responses could account for this protection, and that a nonantibody soluble mediator was most likely involved. By examining the genetic control of *C. parvum*-induced protection against *B. microti*, it has been shown that this nonspecifically induced resistance to infection is not merely the result of an amplification of the normal immune response to *B. microti*. Thus, innate and nonspecifically induced resistance to infection are controlled by different genes (Wood and Clark, 1982b). *C. parvum* most likely functions by the recruitment of macrophage precursors from the bone marrow into the RES where, with additional "priming," they are able to kill parasitized red cells coming into intimate contact with them. Jahiel *et al.* (1970) found that exogenous mouse serum interferon protected mice against sporozoite-induced *P. berghei* malaria, but Clark *et al.* (1981) reported that repeated doses of higher-titer serum interferon did not affect *P. vinckei petteri* infections. Other reported differences relate to the apparent T-cell independence of nonspecific immunity induced by immunostimulants in nude mice, particularly *C. parvum* versus BCG (Clark *et al.*, 1977a; Cottrell *et al.*, 1977; cf. Mitchell *et al.*, 1978).

Nonspecifically induced protection results in intraerythrocytic death of the parasites, an event which is characteristic of normal recovery from infection with *B. microti* and some *Plasmodium* spp. (Taliaferro and Taliaferro, 1944; Clark *et al.*, 1977b; Quinn and Wyler, 1980). The period of "parasite crisis" in peripheral blood, in which abnormal forms appear, is a reversible process that requires the presence of the spleen for its maintenance (Quinn and Wyler, 1980). Clark *et al.* (1975) suggested that soluble mediators liberated by host cells (probably macrophages) were responsible for the morphological degeneration of the parasites, and serum containing a range of these mediators has now been found to inhibit the growth of murine *Plasmodium* spp. *in vitro* and *in vivo* (Taverne *et al.*, 1981, 1982b; Clark *et al.*, 1981).

3.1.5. Free Oxygen Radicals in Malaria

The most recent work on the mechanisms by which macrophages might kill malaria parasites suggests a role for reactive oxygen intermediates (Clark and Hunt, 1983; Clark *et al.*, 1983). Injection of alloxan, a compound that on reduction to dialuric acid generates free oxygen radicals, results in a rapid and dramatic reduction in *P. vinckei* parasitemia. The finding that these effects could be reversed by prior treatment with certain iron chelators suggests that hydroxyl radicals rather than superoxide are mediating this damage:

$$O_2^{\cdot -} + H_2O_2 \xrightarrow{\text{Fe salt}} O_2 + OH^{\cdot} + OH^-$$

Dockrell and Playfair (1983) found that both *P. yoelii* and *P. berghei* were killed *in vitro* by H_2O_2. Injection of H_2O_2 *in vivo* also resulted in a similar effect, although *P. berghei* parasites appear to be less sensitive to its effects *in vivo*. Polyamine oxidase, another enzyme system released by activated macrophages, has also been shown to abolish infectivity of *P. chabaudi* and *P. yoelii* when added to cultures containing the substrate spermine. The products of oxidized polyamines are proposed to be responsible for this effect (Morgan and Christensen, 1983).

REFERENCES

Adam, C. M., Geniteau, M., Gougerat-Pacidalo, M., Verroust, P., Lebras, J., Gilbert, C., and Morel-Maroger, L., 1981, Cryoglobulins, circulating immune complexes, and complement activation in cerebral malaria, *Infect. Immun.* **31**:530–535.

Adams, D. O., 1976, The granulomatous inflammatory response: A review, *Am. J. Pathol.* **84**:164–191.

Adams, D. O., and Nathan, C. F., 1983, Molecular mechanisms in tumor-cell killing by activated macrophages, *Immunol. Today* **4**:166–170.

Albright, J. W., and Albright, J. F., 1981, Inhibition of murine humoral immune responses by substances derived from trypanosomes, *J. Immunol.* **126**:300–303.

Alexander, J., 1982, A radioattenuated *Leishmania major* vaccine markedly increases the resistance of CBA mice to subsequent infection with *Leishmania mexicana mexicana, Trans. R. Soc. Trop. Med. Hyg.* **58**:274–281.

Alexander, J., and Vickerman, K., 1975, Fusion of host cell secondary lysosomes with the parasitophorous vacuoles of *Leishmania mexicana* infected macrophages, *J. Protozool.* **22**:502–508.

Allison, A. C., 1978, Mechanisms by which activated macrophages inhibit lymphocyte responses, *Immunol. Rev.* **40**:3–27.

Allison, A. C., and Clark, I. A., 1977, Specific and non-specific immunity to haemoprotozoa, *Am. J. Trop. Med. Hyg.* **26**:216–222.

Allison, A. C., and Eugui, E. M., 1982, A radical interpretation of immunity to malaria parasites, *Lancet* **2**:1431–1433.

Amsden, A. F., and Boros, D. L., 1979, Fc-receptor-bearing macrophages isolated from hypersensitivity and foreign-body granulomas: Delineation of macrophage dynamics, Fc-receptor density/avidity and specificity, *Am. J. Pathol.* **96**:457–473.

Anderson, S. E., Bautista, S., and Remington, J. S., 1976, Induction of resistance to *Toxoplasma gondii* in human macrophages by soluble lymphocyte products, *J. Immunol.* **117**:381–387.

Auriault, C., Joseph, M., Dessaint, J.-P., and Capron, A., 1980, Inactivation of rat macrophages by peptides resulting from cleavage of IgG by schistosoma larvae proteases, *Immunol. Lett.* **2**:135–139.

Auriault, C., Pestel, J., Joseph, M., Dessaint, J.-P., and Capron, A., 1981, Interaction between macrophages and *Schistosoma mansoni* schistosomula: Role of IgG peptides and aggregates on the modulation of β-glucuronidase release and the cytotoxicity against schistosomula, *Cell. Immunol.* **62**:15–27.

Bagasra, O., Schell, R. F., and LeFrock, J. L., 1981, Evidence for depletion of Ia^+ macrophages and associated immunosuppression in African trypanosomiasis, *Infect. Immun.* **32**:188–193.

Bartl, R., Schauer, A., Hubner, G., and Burkhardt, R., 1975, Changes of the bone marrow in endotoxin shock, in: *Gram-Negative Bacterial Infections* (B. Urbaschek, R. Urbaschek, and E. Neter, eds.), pp. 206–288, Springer-Verlag, Berlin.

Bartlett, S. P., and Burton, R. C., 1982, Studies on natural killer (NK) cells: Heterogeneity of NK cells in beige mutant mice, *Aust. J. Exp. Biol. Med. Sci.* **60**:571–579.

Basham, T. Y., and Merigan, T. C., 1983, Recombinant interferon γ increases HLA-DR synthesis and expression, *J. Immunol.* **130**:1492–1494.

Bautista, C. R., and Kreier, J. P., 1979, Effect of immune serum on the growth of *Babesia microti* in hamster erythrocytes in short-term culture, *Infect. Immun.* **25**:470–472.

Bautista, C. R., and Kreier, J. P., 1980, The action of macrophages and immune serum on growth of *Babesia microti* in short-term cultures, *Tropenmed. Parasitol.* **31**:313–324.

Beaman, L., Benjamini, E., and Pappagianis, D., 1983, Activation of macrophages by lymphokines: Enhancement of phagosome–lysosome fusion and killing of *Coccidioides immitis, Infect. Immun.* **39**:1201–1207.

Behin, R., Mauel, J., and Rowe, D. S., 1977, Mechanisms of protective immunity in experimental cutaneous leishmaniasis of the guinea pig. III. Inhibition of leishmanial lesions in the guinea pig by delayed hypersensitivity reaction to unrelated antigen, *Clin. Exp. Immunol.* **29**:320–325.

Behin, R., Mauel, J., and Sordat, B., 1979, *Leishmania tropica:* Pathogenicity and *in vitro* macrophage function in strains of inbred mice, *Exp. Parasitol.* **48**:81–91.

Berman, J. D., and Dwyer, D. M., 1981, Expression of leishmania antigen on the surface membrane of infected human macrophages *in vitro, Clin. Exp. Immunol.* **44**:342–348.

Berman, J. D,, and Neva, F. A., 1981, Effect of temperature on multiplication of *Leishmania* amastigotes within human monocyte-derived macrophages *in vitro, Am. J. Trop. Med. Hyg.* **30**:318–321.

Berman, J. D., Fioretti, T. B., and Dwyer, D. M., 1981, *In vivo* and *in vitro* localization of *Leishmania* within macrophage phagolysosomes: Use of colloidal gold as a lysosomal label, *J. Protozool.* **28**:239–242.

Biozzi, G., Mouton, D., Sant'Anna, O. A., Passos, H. C., Gennari, M., Reis, M. H., Ferreira, V. C., Heumann, A. M., Bouthillier, Y., Ibanez, O. M., Stiffel, C., and Siqueira, M., 1979, Genetics of immunoresponsiveness to natural antigens in the mouse, *Curr. Top. Microbiol. Immunol.* **85**:31–98.

Blum, K., and Cioli, D., 1978, Behaviour of Biozzi high and low responder mice upon infection with *Schistosoma mansoni, Eur. J. Immunol.* **8**:52–56.

Borges, J. S., and Johnson, W. D., 1975, Inhibition of multiplication of *Toxoplasma gondii* by human monocytes exposed to T-lymphocyte products, *J. Exp. Med.* **141**:483–496.

Boros, D. L., 1978, Granulomatous inflammations, *Prog. Allergy* **24**:183–267.

Boros, D. L., and Lande, M. A., 1983, Induction of collagen synthesis in cultured human fibroblasts by live *Schistosoma mansoni* eggs and soluble egg antigens (SEA), *Am. J. Trop. Med. Hyg.* **32**:78–82.

Boros, D. L., Lande, M. A., and Carrick, L., Jr., 1983, The influence of host and parasite factors on *Schistosoma mansoni* egg-induced tissue fibrosis, in: *From Parasitic Infection to Parasitic Disease* (P. L. Gigase and E. A. E. Van Marck, eds.), pp. 230–236, Karger, Basel.

Bout, D. T., Joseph, M., David, J. R., and Capron, A., 1981, *In vitro* killing of *S. mansoni* schistosomula by lymphokine-activated mouse macrophages, *J. Immunol.* **127**:1–5.

Bouta, I. L., and Parnham, M. J., 1982, Immunomodulatory–anti inflammatory functions of E-type prostaglandins: Mini review with emphasis on macrophage-mediated effects, *Int. J. Immunopharmacol.* **4**:103–109.

Bray, R. S., 1982, Parasitic protozoa in macrophages *in vitro,* in: *Animal Models in Parasitology* (D. G. Owen, ed.), pp. 187–203, Macmillan & Co., London.

Bray, R. S., and Bryceson, A. D., 1968, Cutaneous leishmaniasis of the guinea pig: Action of sensitised lymphocytes on infected macrophages, *Lancet* **2**:898–899.

Brooks, B. O., and Reed, N. D., 1979, The effect of trypan blue on the early control of *Trypanosoma musculi* parasitemia in mice, *J. Reticuloendothel. Soc.* **25**:325–328.

Brown, I. N., Allison, A. C., and Taylor, R. B., 1968, *Plasmodium berghei* infection in thymectomized rats, *Nature (London)* **219**:292–293.

Brown, I. N. Watson, S. R., and Sljivic, V. S., 1977, Antibody response *in vitro* of spleen cells from *Plasmodium yoelii*-infected mice, *Infect. Immun.* **16**:456–460.

Brown, J., and Smalley, M. E., 1980, Specific antibody-dependent cellular cytotoxicity in human malaria, *Clin. Exp. Immunol.* **41**:423–429.

Brown, J., and Smalley, M. E., 1981, Inhibition of the *in vitro* growth of *Plasmodium falciparum* by human polymorphonuclear neutrophil leucocytes, *Clin. Exp. Immunol.* **46**:106–109.

Bryceson, A. D. M., 1970, Immunological aspects of clinical leishmaniasis, *Proc. R. Soc. Med,* **63**:1056–1060.

Bryceson, A. D. M., Bray, R. S., Wolstencroft, R. A., and Dumonde, D. C., 1970, Immunity in cutaneous leishmaniasis of the guinea-pig, *Clin. Exp. Immunol.* **7**:301–431.

Büchmüller, Y., and Mauel, J., 1979, Studies on the mechanisms of macrophage activation. II. Parasite destruction in macrophages activated by supernates from concanavalin A-stimulated lymphocytes, *J. Exp. Med.* **150**:359–370.

Büchmüller, Y., and Mauel, J., 1981, Studies on the mechanisms of macrophage activation: Possible involvement of oxygen metabolites in killing of *Leishmania enrietti* by activated mouse macrophages, *J. Reticuloendothel. Soc.* **29**:181–192.

Butterworth, A. E., Sturrock, R. F., Houba, V., Mahmoud, A. A. F., Sher, A., and Rees, P. H., 1975, Eosinophils as mediators of antibody-dependent damage to schistosomula, *Nature (London)* **256**:727–729.

Campbell, G. H., Esser, K. M., and Phillips, S. M., 1978, *Trypanosoma rhodesiense* infection in cogenitally athymic (nude) mice, *Infect. Immun.* **20**:714–720.

Cantrell, W., Elko, E. E., and Hopff, B. M., 1970, *Plasmodium berghei:* Phagocytic hyperactivity of infected rats, *Exp. Parasitol.* **28**:291–297.

Capron, A., Dessaint, J.-P., Capron, M., and Bazin, H., 1975, Specific IgE antibodies in immune adherence of normal macrophages to *Schistosoma mansoni* schistosomules, *Nature (London)* **253**:474–475.

Capron, A., Dessaint, J.-P., Joseph, M., Rousseaux, R., Capron, M., and Bazin, H., 1977, Interaction between IgE complexes and macrophages in the rat: A new mechanism of macrophage activation, *Eur. J. Immunol.* **7**:315–322.

Capron, A., Dessaint, J.-P., Capron, M., Joseph, M., and Pestel, J., 1980, Role of anaphylactic antibodies in immunity to schistosomes, *Am. J. Trop. Med. Hyg.* **29:**849–857.

Capron, A., Dessaint, J.-P., Capron, M., Joseph, M., and Torpier, G., 1982a, Effector mechanisms of immunity to schistosomes and their regulation, *Behring Inst. Mitt.* **71:**51–70.

Capron, A., Dessaint, J.-P., Capron, M., Joseph, M., and Torpier, G., 1982b, Effector mechanisms of immunity to schistosomes and their regulation, *Immunol. Rev.* **61:**41–66.

Chaicumpa, V., and Jenkin, C. R., 1978, Studies *in vitro* on the reaction of peritoneal exudate cells from mice immune to infection with *Nematospiroides dubius* with the infective third stage larvae of this parasite, *Aust. J. Exp. Biol. Med. Sci.* **56:**61–68.

Chang, K.-P., 1981, Cellular and molecular mechanisms of intracellular symbiosis in leishmaniasis, *Int. Rev. Cytol. Suppl.* **14:**267–305.

Chang, K.-P., and Chiao, J. W., 1981, Cellular immunity of mice to *Leishmania donovani in vitro:* Lymphokine-mediated killing of intracellular parasites in macrophages, *Proc. Natl. Acad. Sci. USA* **78:**7083–7087.

Chang, K.-P., and Dwyer, D. M., 1976, Multiplication of a human parasite (*Leishmania donovani*) in phagolysosomes of hamster macrophages *in vitro, Science* **193:**678–680.

Chang, K.-P., and Fong, D., 1982, Antigenic changes during intracellular differentiation of *Leishmania mexicana* in cultured macrophages, *Infect. Immun.* **36:**430–431.

Cheers, C., and Waller, R., 1975, Activated macrophages in congenitally athymic "nude" mice and in lethally irradiated mice, *J. Immunol.* **115:**844–847.

Chensue, S. W., and Boros, D. L., 1979, Modulation of granulomatous hypersensitivity. I. Characterization of T lymphocytes involved in the adoptive suppression of granuloma formation in *Schistosoma mansoni*-infected mice, *J. Immunol.* **123:**1409–1414.

Chensue, S. W., Wellhausen, S. R., and Boros, D. L., 1981, Modulation of granulomatous hypersensitivity. II. Participation of $Ly1^+$ and $Ly2^+$ T lymphocytes in the suppression of granuloma formation and lymphokine production in *Schistosoma mansoni*-infected mice, *J. Immunol.* **127:**363–367.

Chinchilla, M., and Frenkel, J. K., 1978, Mediation of immunity to intracellular infection (Toxoplasma and Besnoitia) within somatic cells, *Infect. Immun.* **19:**999–1012.

Chow, J. S., and Kreier, J. P., 1972, *Plasmodium berghei:* Adherence and phagocytosis by a rat macrophage *in vitro, Exp. Parasitol.* **31:**13–18.

Clark, I. A., 1978, Does endotoxin cause both the disease and parasite death in acute malaria and babesiosis?, *Lancet* **2:**75–77.

Clark, I. A., 1979a, Protection of mice against *Babesia microti* with cord factor, coam, zymosan, glucan, salmonella and listeria, *Parasite Immunol.* **1:**179–196.

Clark, I. A., 1979b, Resistance to *Babesia* sp. and *Plasmodium* sp. in mice pretreated with an extract of *Coxiella burnetti, Infect. Immun.* **24:**319–325.

Clark, I. A., 1982, Correlation between susceptibility to malaria and babesia parasites and to endotoxicity, *Trans, R. Soc. Trop. Med. Hyg.* **76:**4–7.

Clark, I. A., and Allison, A. C., 1974, *Babesia microti* and *Plasmodium berghei* infections in nude mice, *Nature (London)* **252:**328–329.

Clark, I. A., and Clouston, W. M., 1980, Effects of endotoxin on the histology of intact and athymic mice infected with *Plasmodium vinckei petteri, J. Pathol.* **131:**221–233.

Clark, I. A., and Hunt, N. H., 1983, Evidence for reactive oxygen intermediates causing hemolysis and parasite death in malaria, *Infect. Immun.* **39:**1–6.

Clark, I. A., Richmond, J. E., Wills, E. J., and Allison, A. C., 1975, Immunity to intra-erythrocytic protozoa, *Lancet* **2:**1128–1129.

Clark, I. A., Cox, F. E., and Allison, A. C., 1977a, Protection of mice against *Babesia* sp. and *Plasmodium* sp. with killed *C. parvum, Parasitology* **74:**9–18.

Clark, I. A., Richmond, J. E., Wills, E. J., and Allison, A. C., 1977b, Intra-erythrocytic death of the parasite in mice recovering from infection with *Babesia microti, Parasitology* **75:**189–196.

Clark, I. A., Wills, E. J., Richmond, J. E., and Allison, A. C., 1977c, Suppression of babesiosis in BCG-infected mice and its correlation with tumour inhibition, *Infect. Immun.* **17:**430–438.

Clark, I. A., Virelizier, J.-L., Carswell, E. A., and Wood, P. R., 1981, Possible importance of macrophage-derived mediators in acute malaria, *Infect. Immun.* **32:**1058–1066.

Clark, I. A., Cowden, W. B., and Butcher, G. A., 1983, Free oxygen radical generators as antimalarial drugs, *Lancet* **1**:234.

Clayton, C. E., 1978, Immunosuppression in trypanosomiasis and malaria, in: *The Role of the Spleen in the Immunology of Parasitic Diseases, Tropical Diseases Research Series* 1, pp. 97–119, Schwabe, Basel.

Cohen, S., 1979, Review lecture: Immunity to malaria, *Proc. R. Soc. London Ser. B* **203**:323–345.

Coleman, R. M., Rencricca, N. J., Stout, J. P., Brissette, W. H., and Smith, D. M., 1975, Splenic mediated erythrocyte cytotoxicity in malaria, *Immunology* **29**:49–54.

Colley, D. G., 1981, T lymphocytes that contribute to the immunoregulation of granuloma formation in chronic murine schistosomiasis, *J. Immunol.* **126**:1465–1468.

Collins, F. M., and Mackaness, G. B., 1970, The relationship of delayed hypersensitivity to acquired antituberculous immunity. I. Tubercular sensitivity and resistance to reinfection in BCG-vaccinated mice, *Cell. Immunol.* **1**:253–265.

Correa, M., Narayanan, P. R., and Miller, H. C., 1980, Suppressive activity of splenic adherent cells from *Plasmodium chabaudi*-infected mice, *J. Immunol.* **125**:749–754.

Cottrell, B. J., Playfair, J. H. L., and DeSouza, J. B., 1977, *Plasmodium yoellii* and *Plasmodium vinckei:* The effect of nonspecific immunostimulation on murine malaria, *Exp. Parasitol.* **43**:45–53.

Cottrell, B. J., Playfair, J. H. L., and DeSouza, J. B., 1978, Cell-mediated immunity in mice vaccinated against malaria, *Clin. Exp. Immunol.* **34**:147–158.

Coulis, P. A., Lewert, R. M., and Fitch, F. W., 1978, Splenic suppressor cells and cell-mediated cytotoxicity in murine schistosomiasis, *J. Immunol.* **120**:1074–1076.

Cowing, C., Schwartz, B. D., and Dickler, H. B., 1978, Macrophage Ia antigens. I. Macrophage populations differ in their expression of Ia antigens, *J. Immunol.* **120**:378–384.

Cox, F. E. G., 1978, Concomitant infections, in: *Rodent Malaria* (R. Killick-Kendrick and W. Peters, eds.), pp. 309–343, Academic Press, New York.

Cox, F. E. G., Bilbey, D. L. J., and Nicol, T., 1964, Reticuloendothelial activity in mice infected with *Plasmodium vinckei, J. Protozool.* **11**:229–230.

Criswell, B. S., Butler, W. T., Rossen, R. D., and Knight, V., 1971, Murine malaria: The role of humoral factors and macrophages in destruction of parasitized erythrocytes, *J. Immunol.* **107**:212–221.

Cunningham, D. S., and Kuhn, R. E., 1980, *Trypanosoma cruzi*-induced suppression of the primary immune response in murine cell cultures to T-cell-dependent and -independent antigens, *J. Parasitol.* **66**:16–27.

Cunningham, D. S., Benavidez, C. R., and Kuhn, R. E., 1980, Suppression of mitogen-induced blastogenesis by the *Trypansoma cruzi*-induced suppressor substance, *J. Parasitol.* **66**:722–729.

D'Amelio, R., 1980, Circulating immune complexes in infectious tropical diseases, *Immunol. Today* **1**:112–113.

Damian, R. T., 1984, Immunity in schistosomiasis: A holistic view, *Contemp. Top. Immunobiol.* **12**:359–420.

Dean, D. A., Wistar, R., and Murrell, K. D., 1974, Combined *in vitro* effects of rat antibody and neutrophilic leukocytes on schistosomula of *Schistosoma mansoni, Am. J. Trop. Med. Hyg.* **23**:420–428.

Dempsey, W. L., and Mansfield, J. M., 1983, Lymphocyte function in experimental African trypanosomiasis. V. Role of antibody and the mononuclear phagocyte system in variant specific immunity, *J. Immunol.* **130**:405–411.

Dessein, A. J., Parker, W. L., James, S. L., and David, J. R., 1981, IgE antibody and resistance to infection. I. Selective suppression of the IgE antibody response in rats diminishes the resistance and the eosinophil response to *Trichinella spiralis* infection, *J. Exp. Med.* **153**:423–436.

Djoko-Tamnou, J., Leclerc, C., Modabber, F., and Chedid, L., 1981, Studies on visceral *Leishmania tropica* infection in BALB/c mice. I. Clinical features and cellular changes, *Clin. Exp. Immunol.* **46**:493–498.

Dockrell, H. M., and Playfair, J. H. L., 1983, Killing of blood-stage murine malaria parasites by hydrogen peroxide, *Infect. Immun.* **39**:456–459.

Dockrell, H. M., DeSouza, J. B., and Playfair, J. H. L., 1980, The role of the liver in immunity to blood-stage murine malaria, *Immunology* **41**:421–430.

Doughty, B. L., and Phillips, S. M., 1982, Delayed hypersensitivity granuloma formation and modulation around *Schistosoma mansoni* eggs *in vitro*. II. Regulatory T cell subsets, *J. Immunol.* **128:**37–42.

Dusanic, D. G., 1975, *Trypanosoma musculi* infections in complement-deficient mice, *Exp. Parasitol.* **37:**205–210.

Dye, E. S., and North, R. J., 1981, T cell-mediated immunosuppression as an obstacle to adoptive immunotherapy of the P815 mastocytoma and its metastases, *J. Exp. Med.* **154:**1033–1042.

Eardley, D. D., and Jayawardena, A. N., 1977, Suppressor cells in mice infected with *Trypanosoma brucei, J. Immunol.* **119:**1029–1033.

Edelson, P. J., 1982, Intracellular parasites and phagocytic cells: Cell biology and pathophysiology, *Rev. Infect. Dis.* **4:**124–135.

Ellner, T. J., and Mahmoud, A. A. F., 1981, Cytotoxicity of activated macrophages for the multicellular parasite, *Schistosoma mansoni, Lymphokine* **3:**231–256.

El-On, J., Bradley, D. J., and Freeman, J. C., 1980, *Leishmania donovani:* Action of excreted factor on hydrolytic enzyme activity of macrophages from mice with genetically different resistance to infection, *Exp. Parasitol.* **49:**167–174.

Erlich, J. H., Anders, R. F., Roberts-Thomson, I. C., Schrader, J. W., and Mitchell, G. F., 1983, An examination of differences in serum antibody specificities and hypersensitivity reactions as contributing factors to chronic infection with the intestinal protozoan parasite *Giardia muris* in mice, *Aust. J. Exp. Biol. Med. Sci.* **61:**599–615.

Eugui, E. M., and Allison, A. C., 1980, Differences in susceptibility of various mouse strains to haemoprotozoan infections: Possible correlation with natural killer activity, *Parasite Immunol.* **2:**277–292.

Eugui, E. M., and Allison, A. C., 1982, Natural cell-mediated immunity and interferon in malaria and babesia infections, in: *NK Cells and Other Natural Effector Cells* (R. Herberman, ed.), pp. 1491–1502, Academic Press, New York.

Farah, F. S., Samra, S. A., and Nuwayri-Salti, N., 1975, The role of the macrophage in cutaneous leishmaniasis, *Immunology* **29:**755–764.

Felton, S. C., Prior, R. B., Spagna, V. A., and Kreier, J. P., 1980, Evaluation of *Plasmodium berghei* for endotoxin by the Limulus lysate assay, *J. Parasitol.* **66:**846–847.

Ferrante, A., and Jenkin, C. R., 1978, Evidence implicating the mononuclear phagocytic system of the rat in immunity to infections with *Trypanosoma lewisi, Aust. J. Exp. Biol. Med. Sci.* **56:**201–209.

Ferrante, A., and Jenkin, C. R., 1979, The role of the macrophage in immunity to *Trypanosoma lewisi* infections in the rat, *Cell. Immunol.* **42:**327–335.

Finerty, J. F., and Krehl, E. P., 1977, Delayed immune reactions in mice immunized with malarial antigens, *Am. J. Trop. Med. Hyg.* **26:**377–381.

Forster, O., and Landy, M. (eds.), 1981, *Heterogeneity of Mononuclear Phagocytes,* Academic Press, New York.

Freeman, R. R., 1978, T cell function during fatal and self-limiting malarial infections in mice, *Cell. Immunol.* **41:**373–379.

Freeman, R. R., and Parish, C. R., 1978, Spleen cell changes during fatal and self-limiting malarial infections in mice, *Immunology* **35:**479–484.

Freeman, R. R., Trejdosiewicz, A. J., and Cross, G. A. M., 1980, Protective monoclonal antibodies recognizing stage-specific merozoite antigens of a rodent malaria parasite, *Nature (London)* **284:**366–368.

Garcia, E. G., and Mitchell, G. F., 1982, Vaccination against severe hepato-splenic disease in schistosomiasis japonica: An hypothesis, *Acta Med. Philipp.* **18:**107–112.

Garnham, P. C., and Humphrey, J. H., 1969, Problems in leishmaniasis related to immunology, *Contemp. Top. Microbiol. Immunol.* **48:**29–42.

Gasbarre, L. C., Hug, K., and Louis, J., 1981, Murine T lymphocyte specificity for African trypanosomes. II. Suppression of the T lymphocyte proliferative response to *Trypanosoma brucei* by systemic trypanosome infection, *Clin. Exp. Immunol.* **45:**165–172.

Gauci, R., Bennett, D., Clark, I. A., and Bryant, C., 1982, The induction of tyrosine aminotransferase activity and its use as an indirect assay for endotoxin in mice infected with *Plasmodium vinckei petteri, Int. J. Parasitol.* **12:**279–284.

Gingrich, W. D., 1941, The role of phagocytosis in natural and acquired immunity in avian malaria, *J. Infect. Dis.* **68**:37–45.

Glode, L. M., Jacques, A., Mergenhagen, S. E., and Rosenstreich, D. L., 1977, Resistance of macrophages from C3H/HeJ mice to the *in vitro* cytotoxic effects of endotoxin, *J. Immunol.* **119**:162–166.

Goble, F. C., and Singer, I., 1960, The reticuloendothelial system in experimental malaria and trypanosomiasis, *Ann. N.Y. Acad. Sci.* **88**:149–171.

Gorczynski, R. M., and MacRae, S., 1982, Analysis of subpopulations of glass-adherent mouse skin cells controlling resistance or susceptibility to infection with *Leishmania tropica*, and correlation with the development of independent proliferative signals to Lyt-1$^+$/Lyt-2$^+$ T lymphocytes, *Cell. Immunol.* **67**:74–89.

Gorczynski, R. M., MacRae, S., Kuba, R., and Price, G. B., 1981, Macrophage heterogeneity and Ir-gene control as factors involved in the immune response of guinea pigs to infection with *Leishmania enriettii*, *Cell. Immunol.* **60**:367–375.

Green, T. J., and Kreier, J. P., 1978, Demonstration of the role of cytophilic antibody in resistance to malaria parasites (*Plasmodium berghei*) in rats, *Infect. Immun.* **19**:138–145.

Green, W. F., and Colley, D. G., 1981, Modulation of *Schistosoma mansoni* egg-induced granuloma formation: I-J restriction of T cell-mediated suppression in a chronic parasitic infection, *Proc. Natl. Acad. Sci. USA* **78**:1152–1156.

Greenwood, B. M., Brown, J. C., DeJesus, D. G., and Holborow, E. J., 1971, Immunosuppression in murine malaria. II. The effect of reticuloendothelial and germinal center function, *Clin. Exp. Immunol.* **9**:345–354.

Greenwood, B. M., Oduloju, A. J., and Stratton, D., 1977, Lymphocyte changes in acute malaria, *Trans. R. Soc. Trop. Med. Hyg.* **71**:408–410.

Grosskinsky, C. M., and Askonas, B. A., 1981, Macrophages as primary target cells and mediators of immune dysfunction in African trypanosomiasis, *Infect. Immun.* **33**:149–155.

Grosskinsky, C. M., Ezekowitz, R. A. B., Berton, G., Gordon, S., and Askonas, B. A., 1983, Macrophage activation in murine African trypanosomiasis, *Infect. Immun.* **39**:1080–1086.

Grun, J. L., and Weidanz, W. P., 1981, Immunity to *Plasmodium chabaudi adami* in the B-cell deficient mouse, *Nature (London)* **290**:143–145.

Haidaris, C. G., and Bonventre, P. F., 1981, Elimination of *Leishmania donovani* amastigotes by activated macrophages, *Infect. Immun.* **33**:918–926.

Haidaris, C. G., and Bonventre, P. F., 1982, A role for oxygen-dependent mechanisms in killing of *Leishmania donovani* tissue forms by activated macrophages, *J. Immunol.* **129**:850–855.

Hale, C., and Howard, J. G., 1981, Immunological regulation of experimental cutaneous leishmaniasis. II. Studies with Biozzi high and low responder lines of mice, *Parasite Immunol.* **3**:45–55.

Handman, E., and Burgess, A. W., 1979, Stimulation by granulocyte-macrophage colony-stimulating factor of *Leishmania tropica* killing by macrophages, *J. Immunol.* **122**:1134–1137.

Handman, E., and Greenblatt, C. L., 1977, Promotion of leishmanial infections in nonpermissive host macrophages by conditioned medium, *Z. Parasitenkd.* **53**:143–147.

Handman, E., and Hocking, R. E., 1982, Stage-specific, strain-specific and cross-reactive antigens of *Leishmania* identified by monoclonal antibodies, *Infect. Immun.* **37**:28–33.

Handman, E., Ceredig, R., and Mitchell, G. F., 1979, Murine cutaneous leishmaniasis: Disease patterns in intact and nude mice of various genotypes and examination of some differences between normal and infected macrophages, *Aust. J. Exp. Biol. Med. Sci.* **57**:9–29.

Handman, E., Hocking, R. E., Mitchell, G. F., and Spithill, T. W., 1983, Isolation and characterization of infective and non-infective clones of *Leishmania tropica*, *Mol. Biochem. Parasitol.* **7**:111–126.

Handman, E., Jarvis, H. M., and Mitchell, G. F., 1984, *Leishmania major:* Identification of stage-specific antigens and antigens shared by promastigotes and amastigotes, *Parasite Immunol.* **6**:223–233.

Haque, A., Ouaissi, A., Joseph, M., Capron, M., and Capron, A., 1981, IgE antibody and eosinophil- and macrophage-mediated *in vitro* killing of *Dipetalonema viteae* microfilariae, *J. Immunol.* **127**:716–725.

Harel-Bellon, A., Joskowicz, M., Fradelezi, D., and Eisen, H., 1983, Modification of T-cell proliferation and interleukin 2 production in mice infected with *Trypanosoma cruzi, Proc. Natl. Acad. Sci. USA* **80:**3466–3469.

Hauser, W. E., Jr., and Remington, J. S., 1981, Effect of monoclonal antibodies on phagocytosis and killing of *Toxoplasma gondii* by normal macrophages, *Infect. Immun.* **32:**637–640.

Herod, E., Clark, I. A., and Allison, A. C., 1978, Protection of mice against the haemoprotozoan *Babesia microti* with *Brucella abortus* strain 19, *Clin. Exp. Immunol.* **31:**518–523.

Heumann, A. M., Stiffel, C., Monjour, L., Bucci, A., and Biozzi, G., 1979, Correlation between genetic regulation of antibody responsiveness and protective immunity induced by *Plasmodium berghei* vaccination, *Infect. Immun.* **24:**829–836.

Higashi, G. E., and Chowdury, A. B., 1970, *In vitro* adhesion of eosinophils to infective larvae of *Wuchereria bancrofti, Immunology* **19:**65–83.

Hill, J. O., North, R. J., and Collins, F. M., 1983, Advantages of measuring changes in the number of viable parasites in murine models of experimental cutaneous leishmaniasis, *Infect. Immun.* **39:**1087–1094.

Hoff, R. L., and Frenkel, J. K., 1974, Cell-mediated immunity against *Besnoitia* and *Toxoplasma* in specifically and cross-immunized hamsters and in culture, *J. Exp. Med.* **139:**560–580.

Howard, J. G., Hale, C., and Liew, F. Y., 1980a, Genetically determined susceptibility to *Leishmania tropica* infection is expressed by haemopoietic donor cells in mouse radiation chimaeras, *Nature (London)* **288:**161–162.

Howard, J. G., Hale, C., and Liew, F. Y., 1980b, Immunological regulation of experimental cutaneous leishmaniasis. III. The nature and significance of specific suppression of cell-mediated immunity in mice highly susceptible to *Leishmania tropica, J. Exp. Med.* **152:**594–607.

Howard, J. G., Hale, C., and Liew, F. Y., 1981, Immunological regulation of experimental cutaneous leishmaniasis. IV. Prophylactic effect of sublethal irradiation as a result of abrogation of suppressor T cell generation in mice genetically susceptible to *Leishmania tropica, J. Exp. Med.* **153:**557–568.

Howard, J. G., Nicklin, S., Hale, C., and Liew, F. Y., 1982, Prophylactic immunization against experimental leishmaniasis. I. Protection induced in mice genetically vulnerable to fatal *Leishmania tropica* infection, *J. Immunol.* **129:**2206–2212.

Howard, R. J., and Mitchell, G. F., 1979, Accelerated clearance of uninfected red cells from *Plasmodium berghei*-infected mouse blood in normal mice, *Aust. J. Exp. Biol. Med. Sci.* **57:**455–457.

Hunter, K. W., Folks, T. M., Sayles, P. C., and Strickland, G. T., 1981, Early enhancement followed by suppression of natural killer cell activity during murine malarial infections, *Immunol. Lett.* **2:**209–212.

Hussein, H. S., 1979, *Babesia microti* and *Babesia hylomysci:* Spleen and phagocytosis in infected mice, *Exp. Parasitol.* **47:**1–12.

Jacobson, R. H., Reed, N. D., and Manning, D. D., 1977, Expulsion of *Nippostrongylus brasiliensis* from mice lacking antibody production potential, *Immunology* **32:**867–874.

Jahiel, R. I., Nussenzweig, R. S., Vilcek, J., and Vanderberg, J., 1969, Protective effect of interferon inducers on *Plasmodium berghei* malaria, *Am. J. Trop. Med. Hyg.* **18:**823–835.

Jahiel, R. I., Vilcek, J., and Nussenzweig, R. S., 1970, Exogenous interferon protects mice against *Plasmodium berghei* malaria, *Nature (London)* **227:**1350–1351.

James, S. L., and Colley, D. G., 1976, Eosinophil-mediated destruction of *Schistosoma mansoni* eggs, *J. Reticuloendothel. Soc.* **20:**359–374.

James, S. L., and Colley, D. G., 1978, Eosiniphil-mediated destruction of *Schistosoma mansoni* eggs *in vitro.* II. The role of cytophilic antibody, *Cell. Immunol.* **38:**35–47.

James, S. L., Leonard, E. J., and Meltzer, M. S., 1982a, Macrophages as effector cells of protective immunity in murine schistosomiasis. IV. Coincident induction of macrophage activation for extracellular killing of schistosomula and tumor cells, *Cell. Immunol.* **74:**86–96.

James, S. L., Sher, A., Lazdins, J. K., and Meltzer, M. S., 1982b, Macrophages as effector cells of protective immunity in murine schistosomiasis. II. Killing of newly transformed schistosomula *in vitro* by macrophages activated as a consequence of *Schistosoma mansoni* infection, *J. Immunol.* **128:**1535–1540.

James, S. L., Lazdins, J. K., Meltzer, M. S., and Sher, A., 1982c, Macrophages as effector cells of protective immunity in murine schistosomiasis. I. Activation of peritoneal macrophages during natural infection, *Cell. Immunol.* **67**:255–266.

Jayawardena, A. N., Targett, G. A. T., Leuchars, E., Carter, R. L., Doenhoff, M. J., and Davies, A. J. S., 1975, T cell activation in murine malaria, *Nature (London)* **258**:149–151.

Jayawardena, A. N., Targett, G. A. T., Carter, R. L., Leuchars, E., and Davies, A. J. S., 1977, The immunological response of CBA mice to *P. yoelii*. I. General characteristics, the effects of T cell deprivation and reconstitution with thymus grafts, *Immunology* **32**:849–859.

Jenkins, D. C., and Carrington, T. S., 1981, *Nematospiroides dubius:* The course of primary, secondary and tertiary infections in high and low responder Biozzi mice, *Parasitology* **82**:311–318.

Jones, T. C., 1974, Macrophages and intracellular parasitism, *J. Reticuloendothel. Soc.* **15**:439–450.

Jones, T. C., 1981, Interactions between murine macrophages and obligate intracellular protozoa, *Am. J. Pathol.* **102**:127–132.

Jones, T. C., and Hirsch, J. G., 1972, The interaction between *Toxoplasma gondii* and mammalian cells. II. The absence of lysosomal fusion with phagocytic vacuoles containing living parasites, *J. Exp. Med.* **136**:1173–1194.

Jones, T. C., Len, L., and Hirsch, J. G., 1975, Assessment *in vitro* of immunity against *Toxoplasma gondii, J. Exp. Med.* **141**:466–482.

Joseph, M., Capron, A., Butterworth, A. E., Sturrock, R. F., and Houba, V., 1978, Cytotoxicity of human and baboon mononuclear phagocytes against schistosomula *in vitro:* Induction by immune complexes containing IgE and *Schistosoma mansoni* antigens, *Clin. Exp. Immunol.* **23**: 48–56.

Karavodin, L. M., and Ash, L. R., 1982, Inhibition of adherence and cytotoxicity by circulating immune complexes found in experimental filariasis, *Parasite Immunol.* **4**:1–12.

Kassis, A. I., Aikawa, M., and Mahmoud, A. A. F., 1979, Mouse antibody-dependent eosinophil and macrophage adherence and damage to schistosomula of *Schistosoma mansoni, J. Immunol.* **122**:398–405.

Kellina, O. I., Passova, O. M., and Alekseer, A. N., 1981, Experimental proof of heterogeneous composition of natural populations of *Leishmania major* with regard to virulence, *Med. Parazitol. Farazit. Bolezni* **50**:4–11 (see *Trop. Dis. Bull.* 1982, **79**:469).

Kierszenbaum, F., 1982, Immunologic deficiency during experimental Chagas' disease (*Trypanosoma cruzi* infection): Role of adherent nonspecific esterase positive splenic cells, *J. Immunol.* **129**:2202–2205.

Kierszenbaum, F., and Howard, J. G., 1976, Mechanisms of resistance against experimental *Trypanosoma cruzi* infection: The importance of antibodies and antibody-forming capacity in the Biozzi high and low responder mice, *J. Immunol.* **116**:1208–1211.

King, D. P., and Jones, P. P., 1983, Induction of Ia and H-2 antigens on a macrophage cell line by immune interferon, *J. Immunol.* **131**:315–318.

Kitchen, A. G., and DiLuzio, N. R., 1971, Influence of *Plasmodium berghei* infections on phagocytic and humoral recognition factor activity, *J. Reticuloendothel. Soc.* **9**:237–247.

Kitchen, S. F., 1949, Falciparum malaria, in: *Malariology* (M. F. Boyd, ed.), pp. 995–1016, Saunders, Philadelphia.

Kress, Y., Bloom, B. R., Wittner, M., Rowen, A., and Tanowitz, H., 1975, Resistance of *Trypanosoma cruzi* to killing by macrophages, *Nature (London)* **257**:394–396.

Lazdins, J. K., Kuhner, A. L., David, J. R., and Karnovsky, M. J., 1978, Alteration of some functional and metabolic characteristics of resident mouse peritoneal macrophages by lymphocyte mediators, *J. Exp. Med.* **148**:746–758.

Lebrija de Ibarra, A. A., Howard, J. G., and Snary, D., 1982, Monoclonal antibodies to *Leishmania tropica major:* Specificities and antigen localisation, *Parasitology* **85**:523–531.

Leclerc, C., Modabber, F., Deriaud, E., and Chedid, L., 1981, Systemic infection of *Leishmania tropica (major)* in various strains of mice, *Trans. R. Soc. Trop. Med. Hyg.* **75**:851.

Lee Hand, W. L., and King-Thompson, M. L., 1983, Effect of erythrocyte ingestion on macrophage antibacterial functions, *Infect. Immun.* **40**:917–923.

Lelchuk, R., and Playfair, J. H. L., 1980, Two distinct types of non-specific immunosuppression in murine malaria, *Clin. Exp. Immunol.* **42**:428–435.

Lelchuk, R., Dockrell, H. M., and Playfair, J. H. L., 1983, T-independent macrophage changes in murine malaria, *Clin. Exp. Immunol.* **51**:487–493.
Leventhal, R., and Soulsby, E. J., 1976, Cuticular reactivity of the early larval stages of *Ascaris suum:* adhesion and degranulation of polymorphonuclear leukocytes of the surface of opsonized larvae of *A. suum, Int. J. Parasitol.* **6**:279–283.
Lewis, D. H., and Peters, W., 1977, The resistance of intracellular *Leishmania* parasites to digestion by lysosomal enzymes, *Ann. Trop. Med. Parasitol.* **71**:295–312.
Liew, F. Y., Dhaliwal, S. S., and Teh, K. L., 1979, Dissociative effects of malarial infection on humoral and cell-mediated immunity in mice, *Immunology* **37**:35–44.
Liew, F. Y., Hale, C., and Howard, J. G., 1982, Immunological regulation of experimental cutaneous leishmaniasis. V. Characterization of effector and specific suppressor T cells, *J. Immunol.* **128**:1917–1922.
Loose, L. D., and DiLuzio, N. R., 1976, A temporal relationship between reticuloendothelial system phagocytic alterations and antibody responses in mice infected with *Plasmodium berghei* (NYU-2 strain), *Am. J. Trop. Med. Hyg.* **25**:221–228.
Loose, L. D., Trejo, R., and DiLuzio, N. R., 1971, Impaired endotoxin detoxification as a factor in enhanced endotoxin sensitivity of malaria infected mice, *Proc. Soc. Exp. Biol. Med.* **137**:794–797.
Loose, L. D., Cook, J. A., and DiLuzio, N. R., 1972, Malarial immunosuppression—A macrophage mediated defect, *Proc. Helminthol. Soc. Wash.* **39**:484–491.
Louis, J. A., Moedder, E., MacDonald, H. R., and Engers, H. D., 1981, Recognition of protozoan parasites by murine T lymphocytes. II. Role of the H-2 gene complex in interactions between antigen-presenting macrophages and leishmania-immune T lymphocytes, *J. Immunol.* **126**:1661–1666.
Louis, J. A., Zubler, R. H., Coutinho, S. G., Lima, G., Behin, R., Mauel, J., and Engers, H. D., 1982, The *in vitro* generation and functional analysis of murine T cell populations and clones specific for a protozoan parasite, *Leishmania tropica, Immunol. Rev.* **61**:215–243.
Loveless, S. E., Wellhausen, S. R., Boros, D. L., and Heppner, G. H., 1982, Tumoricidal macrophages isolated from liver granulomas of *Schistosoma mansoni*-infected mice, *J. Immunol.* **128**:284–288.
Lu, C. Y., and Unanue, E. R., 1982, Ontogeny of murine macrophages: Functions related to antigen presentation, *Infect. Immun.* **36**:169–175.
Lucia, H. L., and Nussenzweig, R. S., 1969, *Plasmodium chabaudi* and *Plasmodium vinckei* phagocytic activity of mouse reticuloendothelial system, *Exp. Parasitol.* **25**:319–323.
McDonald, V., and Phillips, R. S., 1978, Increase in non-specific antibody mediated cytotoxicity in malarious mice, *Clin. Exp. Immunol.* **34**:159–163.
McDonald, V., and Sherman, I. W., 1980, Lack of correlation between delayed-type hypersensitivity and host resistance to *Plasmodium chabaudi* infection, *Clin. Exp. Immunol.* **42**:421–427.
McGregor, R. R., Sheagren, J. N., and Wolff, S. M., 1969, Endotoxin induced modification of *Plasmodium berghei* infection in mice, *J. Immunol.* **103**:131–139.
McIntyre, J., Rowley, D., and Jenkin, C. R., 1967, The functional heterogeneity of macrophages at the single cell level, *Aust. J. Exp. Biol. Med. Sci.* **45**:675–680.
Mackenzie, C. D., Jungery, M., Taylor, P. M., and Ogilvie, H. M., 1981, The *in vitro* interaction of eosinophils, neutrophils, macrophages and mast cells with nematode surfaces in the presence of complement or antibody, *J. Pathol.* **133**:161–175.
Mahmoud, A. A., Peters, P. A. S., Civil, R. H., and Remington, J. S., 1979, *In vitro* killing of schistosomula of *Schistosoma mansoni* by BCG and *C. parvum*-activated macrophages, *J. Immunol.* **122**:1655–1657.
Mansfield, J. M., Levine, R. F., Dempsey, W. L., Wellhausen, S. R., and Hansen, C. T., 1981, Lymphocyte function in experimental African trypanosomiasis. IV. Immunosuppression and suppressor cells in the athymic nu/nu mouse, *Cell. Immunol.* **63**:210–215.
Martin, L. K., Eineber, A., Sadun, E. H., and Wren, R. E., 1967, Effect of bacterial endotoxin on the course of *Plasmodium berghei* infection, *Exp. Parasitol.* **20**:186–199.
Mascaskill, J. A., Holmes, P. H., Whitelaw, D. D., McConnell, I., Jennings, F. W., and Urquhart, G. M., 1980, Immunological clearance of ^{75}Se-labelled *Trypanosoma brucei* in mice. II. Mechanisms in immune animals, *Immunology* **40**:629–635.

Mauel, J., 1982, *In vitro* induction of intracellular killing of parasitic protozoa by macrophages, *Immunobiology* **161**:392–400.

Mauel, J., and Behin, R., 1982, Leishmaniasis: Immunity, immunopathology and immunodiagnosis, in: *Immunology of Parasitic Infections,* Volume 2 (S. Cohen and K. S. Warren, eds.), pp. 299–355, Blackwell, Oxford.

Mayor-Withey, K. S., Clayton, C. E., Roelants, G. E., and Askonas, B. A., 1978, Trypanosomiasis leads to extensive proliferation of B, T and null cells in spleen and bone marrow, *Clin. Exp. Immunol.* **34**:359–363.

Mehta, K., Sindhu, R. K., Subrahmanyam, D., and Nelson, D. S., 1980, IgE-dependent adherence and cytotoxicity of rat spleen and peritoneal cells to *Litmosoides carinii* microfilariae, *Clin. Exp. Immunol.* **41**:107–114.

Mehta, K., Sindhu, R. K., Subrahmanyam, D., Hooper, K., and Nelson, D. S., 1982, IgE-dependent cellular adhesion and cytotoxicity to *Litomosoides carinii* microfilariae—Nature of effector cell, *Clin. Exp. Immunol.* **48**:477–484.

Metzger, Z., Hoffeld, J. T., and Oppenheim, J. J., 1980, Macrophage-mediated suppression. I. Evidence for participation of both hydrogen peroxide and prostaglandins in suppression of murine lymphocyte proliferation, *J. Immunol.* **124**:983–988.

Michel, J. C., Lagrange, P. H., and Hurtel, B., 1979, Modulation by malaria infection of the induction of T lymphocyte-dependent delayed-type hypersensitivity and antibody formation to sheep erythrocytes in mice, *Parasite Immunol.* **1**:267–275.

Milder, R., and Kloetzel, J., 1980, The development of *Trypanosoma cruzi* in macrophages *in vitro:* Interaction with lysosomes and host cell fate, *Parasitology* **80**:139–145.

Mitchell, G. F., 1977, Observations and speculations on the influence of T cells in the cellular events of induction of antibody formation and tolerance *in vivo,* in: *The Lymphocyte: Structure and Function* (J. J. Marchalonis, ed.), pp. 227–256, Dekker, New York.

Mitchell, G. F., 1982a, The nude mouse in immunoparasitology, in: *The Nude Mouse in Experimental and Clinical Research,* Volume 2 (J. Fogh and B. C. Giovanelli, eds.), pp. 267–289, Academic Press, New York.

Mitchell, G. F., 1982b, Genetic variation in resistance of mice to *Taenia taeniaeformis:* Analysis of host-protective immunity and immune evasion, in: *Cysticercosis: Present State of Knowledge and Perspective* (A. Flisser, K. Willms, J. P. Laclette, C. Larralde, C. Ridaura, and F. Beltran, eds.), pp. 575–584, Academic Press, New York.

Mitchell, G. F., 1983, Murine cutaneous leishmaniasis: Resistance in reconstituted nude mice and several F_1 hybrids infected with *Leishmania tropica major, J. Immunogenet.* **5**:395–412.

Mitchell, G. F., and Anders, R. F., 1982, Parasite antigens and their immunogenicity in infected hosts, in: *The Antigens,* Volume 6 (M. Sela, ed.), pp. 69–149, Academic Press, New York.

Mitchell, G. F., and Cruise, K. M., 1986, Schistosoma antigens: Immunochemical identification of antigens of *Schistosoma mansoni* and *Schistosoma japonicum,* in: *Parasite Antigens: Towards New Strategies for Vaccines* (T. W. Pearson, ed.), pp. 275–316, Dekker, New York.

Mitchell, G. F., and Handman, E., 1983, *Leishmania tropica major* in mice: Vaccination against cutaneous leishmaniasis in mice of high genetic susceptibility, *Aust. J. Exp. Biol. Med. Sci.* **61**:411–424.

Mitchell, G. F., Handman, E., and Howard, R. J., 1978, Protection of mice against plasmodium and babesia infections: Attempts to raise host-protective sera, *Aust. J. Exp. Biol. Med. Sci.* **56**:553–559.

Mitchell, G. F., Curtis, J. M., Handman, E., and McKenzie, I. F. C., 1980, Cutaneous leishmaniasis in mice: Disease patterns in reconstituted nude mice of several genotypes infected with *Leishmania tropica, Aust. J. Exp. Biol. Med. Sci.* **58**:521–532.

Mitchell, G. F., Curtis, J. M., and Handman, E., 1981a, Resistance to cutaneous leishmaniasis in genetically-susceptible BALB/c mice, *Aust. J. Exp. Biol. Med. Sci.* **59**:555–566.

Mitchell, G. F., Curtis, J. M., Scollay, R. J., and Handman, E., 1981b, Resistance and abragation of resistance to cutaneous leishmaniasis in reconstituted BALB/c nude mice, *Aust. J. Exp. Biol. Med. Sci.* **59**:539–554.

Mitchell, G. F., Anders, R. F., Brown, G. V., Handman, E., Roberts-Thomson, I. C., Chapman, C. B., Forsyth, K. P., Kahl, K. P., and Cruise, K. M., 1982, Analysis of infection characteristics and antiparasite immune responses in resistant compared with susceptible hosts, *Immunol. Rev.* **61**:137–188.

Mitchell, G. F., Handman, E., and Spithill, T. W., 1984, Vaccination against cutaneous leishmaniasis in mice using nonpathogenic cloned promastigotes of *Leishmania tropica major* and importance of route of injection, *Aust. J. Exp. Biol. Med. Sci.* **62:**145–153.

Möller, G. (ed.), 1978, *Role of Macrophages in the Immune Response, Immunol. Rev.* **40:**1–255.

Morgan, D. M. L., and Christensen, J., 1983, Polyamine oxidation and the killing of intracellular parasites, in: *Advances in Polyamine Research,* Volume 4 (U. Bachrock, A. Kay, and R. Chayen, eds.), pp. 169–174, Raven Press, New York.

Morges, W., and Weidanz, W. P., 1980, *Plasmodium yoelii:* The thymus-dependent lymphocyte in mice immunosuppressed by malaria, *Exp. Parasitol.* **50:**188–194.

Moriearty, P. L., Grimaldi, G., Jr., Galvao-Castro, R., de Oliveira Neto, M. P., and Marzochi, M. C. A., 1982, Intralesional plasma cells and serological responses in human cutaneous leishmaniasis, *Clin. Exp. Immunol.* **47:**59–64.

Murphy, J. R., 1981a, Host defences in murine malaria: Nonspecific resistance to *Plasmodium berghei* generated in response to *Mycobacterium bovis* infection or *Corynebacterium parvum* stimulation, *Infect. Immun.* **33:**199–211.

Murphy, J. R., 1981b, Host defences in murine malaria: Analysis of plasmodial infection caused defects in macrophage microbicidal capacities, *Infect. Immu.* **31:**396–407.

Murray, H. W., 1981a, Interaction of leishmania with a macrophage cell line: Correlation between intracellular killing and generation of oxygen intermediates, *J. Exp. Med.* **153:**1690–1695.

Murray, H. W., 1981b, Susceptibility of leishmania to oxygen intermediates and killing by normal macrophages, *J. Exp. Med.* **153:**1302–1315.

Murray, H. W., 1982, Cell-mediated immune response in experimental visceral leishmaniasis. II. Oxygen-dependent killing of intracellular *Leishmania donovani, J. Immunol.* **129:**351–357.

Murray, H. W., and Cohn, Z. A., 1979, Macrophage oxygen dependent antimicrobial activity. I. Susceptibility of *Toxoplasma gondii, J. Exp. Med.* **150:**938–949.

Murray, H. W., Masur, H., and Keithly, J. S., 1982, Cell-mediated immune response in experimental visceral leishmaniasis. I. Correlation between resistance to *Leishmania donovani* and lymphokine-generating capacity, *J. Immunol.* **129:**344–350.

Nacy, C. A., Meltzer, M. S., Leonard, E. J., and Wyler, D. J., 1981, Intracellular replication and lymphokine-induced destruction of *Leishmania tropica* in C3H/HeN mouse macrophages, *J. Immunol.* **127:**2381–2386.

Nacy, C. A., Fortier, A. H., Pappas, M. G., and Henry, R. R., 1983a, Susceptibility of inbred mice to *Leishmania tropica* infection: Correlation of susceptibility with *in vitro* defective macrophage microbicidal activities, *Cell. Immunol.* **77:**298–307.

Nacy, C. A., James, S. L., Benjamin, W. R., Farrer, J. J., Hockmeyer, W. T., and Meltzer, M. S., 1983b, Activation of macrophages for microbicidal and tumoricidal effector functions by soluble factors from EL-4, a continuous T cell line, *Infect. Immun.* **40:**820–824.

Nasseri, M., and Modabber, F. Z., 1979, Generalized infection and lack of delayed hypersensitivity in BALB/c mice infected with *Leishmania tropica major, Infect. Immun.* **26:**611–620.

Nathan, C., Nogueira, N., Juangbhanich, C., Ellis, J., and Cohn, Z. A., 1979, Activation of macrophages *in vivo* and *in vitro:* Correlation between hydrogen peroxide release and killing of *Trypanosoma cruzi, J. Exp. Med.* **149:**1056–1068.

Nelson, D. S. (ed.), 1976, *Immunobiology of the Macrophage,* Academic Press, New York.

Neva, F. A., Howard, W. A., Glew, R. H., Krotoski, W. A., Gam, A. A., Collins, W. E., Atkinson, J. P., and Frank, M. M., 1974, Relationship of serum complement levels to events of the malarial paroxysm, *J. Clin. Invest.* **54:**451–460.

Nogueira, N., and Cohn, Z. A., 1976, *Trypanosoma cruzi:* Mechanism of entry and intracellular fate in mammalian cells, *J. Exp. Med.* **143:**1402–1420.

Nogueira, N., and Cohn, Z., 1978, *Trypanosoma cruzi: In vitro* induction of macrophage microbicidal activity, *J. Exp. Med.* **148:**288–300.

Nogueira, N., Klebanoff, S. J., and Cohn, Z. A., 1982, *Trypanosoma cruzi:* Sensitization to macrophage killing by eosinophil peroxidase, *J. Immunol.* **128:**1705–1708.

Nussenzweig, R. S., 1967, Increased nonspecific resistance to malaria produced by administration of killed *Corynebacterium parvum, Exp. Parasitol.* **21:**224–231.

Ojo-Amaize, E. A., Salimonu, L. S., Williams, A. I. O., Akinwolere, O. A. O., Shabo, R., Alm, G. V., and Wigzell, H., 1981, Positive correlation between degree of parasitaemia, interferon titers, and natural killer cell activity in *Plasmodium falciparum*-infected children, *J. Immunol.* **127:**2296–2300.

Olds, G. R., and Mahmoud, A. A. F., 1980, Role of granulomatous response in murine schistosomiasis mansoni: Eosinophil-mediated destruction of eggs, *J. Clin. Invest.* **66:**1191–1199.

Olds, G. R., Ellner, J. J., Kearse, L. A., Jr., Kazura, J. W., and Mahmoud, A. A. F., 1980, Role of arginase in killing of schistosomula of *Schistosoma mansoni, J. Exp. Med.* **151:**1557–1562.

Olds, G. R., Stewart, S. J., and Ellner, J. J., 1981, Amphotericin B-induced resistance to *Schistosoma mansoni, J. Immunol.* **126:**1667–1670.

Ottesen, E. A., 1979, Modulation of the host response in human schistosomiasis. I. Adherent suppressor cells which inhibit lymphocyte proliferative responses to parasite antigens, *J. Immunol.* **123:**1639–1644.

Owen, D. G. (ed.), 1982, *Animal Models in Parasitology,* Macmillan & Co., London.

Owen, R. L., Allen, C. L., and Stevens, D. P., 1981, Phagocytosis of *Giardia muris* by macrophages in Peyer's patch epithelium in mice, *Infect. Immun.* **33:**591–601.

Parashar, A., Sehgal, S., Naik, S., and Aikat, B. K., 1983, Role of macrophage-processed antigen in a *Plasmodium berghei* model, *Parasite Immunol.* **5:**173–181.

Passwell, J. H., Dayer, J.-M., and Merler, E., 1979, Increased prostaglandin production by human monocytes after membrane receptor activation, *J. Immunol.* **123:**115–120.

Pearson, R. D., Romito, R., Symes, P. H., and Harcus, J. L., 1981, Interaction of *Leishmania donovani* promastigotes with human monocyte-derived macrophages, parasite entry, intracellular survival, and multiplication, *Infect. Immun.* **32:**1249–1253.

Pearson, R. D., Wheeler, D. A., Harrison, L. H., and Kay, H. D., 1983, The immunobiology of leishmaniasis, *Rev. Infect. Dis.* **5:**907–927.

Peavy, D. L., Baughan, R. E., and Musher, D. M., 1978, Strain dependent cytotoxic effects of endotoxin for mouse peritoneal macrophage, *Infect. Immun.* **21:**310–319.

Peavy, D. L., Baughan, R. E., and Musher, D. M., 1979, Effects of BCG infection on the susceptibility of mouse macrophages to endotoxin, *Infect. Immun.* **24:**59–64.

Perrudet-Badeux, A., Binaghi, R. A., and Biozzi, G., 1975, Trichinella infestation in mice genetically selected for high and low antibody production, *Immunology* **29:**387–390.

Perrudet-Badeux, A., Binaghi, R. A., and Boussac-Aron, Y., 1978, *Trichinella spiralis* infection in mice: Mechanism of the resistance in animals genetically selected for high and low antibody production, *Immunology* **35:**519–522.

Peters, W., 1974, Drug resistance in trypanosomiasis and leishmaniasis, in: *Trypanosomiasis and Leishmaniasis with Special Reference to Chagas' Disease* (K. Elliot, M. O'Connor, and G. E. W. Wolstenholme, eds.), pp. 309–334, Elsevier, Amsterdam.

Phillips, S. M., and Colley, D. G., 1978, Immunologic aspects of host response to schistosomiasis: Resistance, immunopathology and eosinophil involvement, *Prog. Allergy* **24:**49–182.

Playfair, J. H. L., 1982, Immunity to malaria, *Br. Med. Bull.* **38:**153–159.

Playfair, J. H. L., and DeSouza, J. B., 1979, Antibody responses in mice protected against malaria by vaccination, *Parasite Immunol.* **1:**197–208.

Playfair, J. H. L., DeSouza, J. B., and Cottrell, B. J., 1977, Reactivity and cross-reactivity of mouse helper T cells to malaria parasites, *Immunology* **32:**681–688.

Pollacco, S., Nicholas, W. L., Mitchell, G. F., and Stewart, A. C., 1978, T-cell dependent collagenous encapsulating response in the mouse liver to *Mesocestoides corti* (Cestoda), *Int. J. Parasitol.* **8:**457–462.

Powell, M. R., and Damian, R. T., 1982, Schistosomiasis mansoni in baboons. VI. Plastic-adherent cell suppression of the primary *in vitro* antibody response of peripheral blood mononuclear cells to a heterologous antigen (SRBC) during chronic infection, *Am. J. Trop. Med. Hyg.* **31:**790–795.

Quinn, T. C., and Wyler, D. J., 1980, Resolution of acute malaria (*Plasmodium berghei* in the rat): Reversibility and spleen dependence, *Am. J. Trop. Med. Hyg.* **29:**1–4.

Ralph, P., Nacy, C. A., Meltzer, M. S., Williams, N., Nakoinz, I., and Leonard, E. J., 1983, Colony-stimulating factors and regulation of macrophage tumoricidal and microbicidal activities, *Cell. Immunol.* **76:**10–21.

Rank, R. G., and Weidanz, W. P., 1976, Nonsterilizing immunity in avian malaria; an antibody-independent phenomenon, *Proc. Soc. Exp. Biol. Med.* **151:**257–259.

Rivera-Ortiz, C., and Nussenzweig, R. S., 1972, Increased phagocytic activity of the reticuloendothelial system during immunization with X-irradiated sporozoites of *Plasmodium berghei* in mice, *Proc. Helminthol. Soc. Wash.* **39**:497–506.

Roberts, D. W., and Weidanz, W. P., 1978, Splenomegaly, enhanced phagocytosis and anemia are thymus-dependent responses to malaria, *Infect. Immun.* **20**:728–731.

Roberts, D. W., and Weidanz, W. P., 1979, T cell immunity to malaria in the B cell deficient mouse, *Am. J. Trop. Med. Hyg.* **28**:1–3.

Roberts-Thomson, I. C., and Mitchell, G. F., 1978, Giardiasis in mice. I. Prolonged infections in certain mouse strains and in hypothymic (nude) mice, *Gastroenterology* **75**:42–46.

Roder, J. C., Karre, K., and Kiessling, R., 1981, Natural killer cells, *Prog. Allergy* **28**:66–159.

Rosenberg, Y. J., and Evans, C. B., 1979, Resistance of mice suppressed for IgM production to *Babesia microti* infection, *Nature (London)* **281**:302–304.

Rosenstreich, D. L., Weinblatt, A. C., and O'Brien, A. D., 1982, Genetic control of resistance to infection in mice, *CRC Crit. Rev. Immunol.* **3**:263–330.

Roth, R. L., and Herman, R., 1979, *Plasmodium berghei:* Correlation of *in vitro* erythrophagocytosis with the dynamics of early onset anemia and reticulocytosis in mice, *Exp. Parasitol.* **47**:169–179.

Roubin, R., Kennard, J., Foley, D., and Zolla-Pazner, S., 1981, Markers of macrophage heterogeneity: Altered frequency of macrophage subpopulations after various pathologic stimuli, *J. Reticuloendothel. Soc.* **29**:423–432.

Ruebush, M. J., and Burgess, D. E., 1982, Induction of natural killer cells and interferon production during infection of mice with *Babesia microti* of human origin, in: *NK Cells and Other Effector Cells* (R. Herberman, ed.), pp. 1483–1489, Academic Press, New York.

Rzepczyk, C. M., 1982, Probable macrophage origin of the LPS-induced cytostatic effect on intraerythrocytic malarial parasites (*Plasmodium vinckei*), *Immunology* **46**:261–270.

Rzepczyk, C. M., and Clark, I. A., 1981, Demonstration of a lipopolysaccharide-induced cytostatic effect on malarial parasite, *Infect. Immun.* **33**:343–347.

Sacks, D. L., Bancroft, G., Evans, W. H., and Askonas, B. A., 1982, Incubation of trypanosome-derived mitogenic and immunosuppressive products with peritoneal macrophages allows recovery of biological activities from soluble parasite fractions, *Infect. Immun.* **36**:160–168.

Scher, M. G., Beller, D. I., and Unanue, E. R., 1980, Demonstration of a soluble mediator that induces exudates rich in Ia-positive macrophages, *J. Exp. Med.* **152**:1684–1698.

Scott, M. T., and Moyes, L., 1982, ^{75}Se-methionine labelled *Trypanosoma cruzi* blood trypromastigotes: Opsonization by chronic infection serum facilitates killing in spleen and liver, *Clin. Exp. Immunol.* **48**:754–757.

Scott, P. A., and Farrell, J. P., 1981, Experimental cutaneous leishmaniasis. I. Nonspecific immunodepression in BALB/c mice infected with *Leishmania tropica*, *J. Immunol.* **127**:2395–2400.

Sendashonga, C. N., and Black, S. J., 1982, Humoral responses against *Trypanosoma brucei* variable surface antigens are induced by degenerating parasites, *Parasite Immunol.* **4**:245–257.

Sethi, K. K., 1982, Intracellular killing of parasites by macrophages, in: *Immunoparasitology, Clinics in Immunology and Allergy* (A. Capron, ed.), Vol. 2, No. 3, pp. 541–565, Saunders, Philadelphia.

Sethi, K. K., and Brandes, H., 1978, Characteristics of soluble T cell-derived factors which can induce non-immune murine macrophages to exert anti-toxoplasma activity, *Z. Immunitätsforsch.* **154**:226–242.

Sethi, K. K., Pelster, B., Suzuki, N., Piekarski, G., and Brandes, H., 1975, Immunity to *Toxoplasma gondii* induced *in vitro* in non-immune mouse macrophages with specifically immune lymphocytes, *J. Immunol.* **115**:1151–1158.

Sethi, K. K., Endo, T., and Brandes, H., 1981, *Toxoplasma gondii* trophozoites precoated with specific monoclonal antibodies cannot survive within normal murine macrophages, *Immunol. Lett.* **2**:343–346.

Shands, J. W., Peavy, D. L., Gormus, B. J., and McGraw, J., 1974, *In vitro* and *in vivo* effects of endotoxin on mouse peritoneal cells, *Infect. Immun.* **9**:106–112.

Sheagren, J. N., Tobie, J. E., Fox, L. M., and Wolff, S. M., 1970, Reticuloendothelial system phagocyte function in naturally acquired human malaria, *J. Lab. Clin. Med.* **75**:481–487.

Shear, H. L., Nussenzweig, R. S., and Bianco, C., 1979, Immune phagocytosis in murine malaria, *J. Exp. Med.* **149**:1288–1298.

Shepherd, V. L., Stahl, P. D., Bernd, P., and Rabinovitch, M., 1983, Receptor-mediated entry of β-glucuronidase into parasitophorous vacuoles of macrophages infected with *Leishmania mexicana amazonensis*, *J. Exp. Med.* **157**:1471–1482.

Sher, A., James, S. L., Simpson, A. J. G., Lazdins, J. K., and Meltzer, M. S., 1982, Macrophages as effector cells of protective immunity in murine schistosomiasis. III. Loss of susceptibility to macrophage-mediated killing during maturation of *S. mansoni* schistosomula from the skin to the lung stage, *J. Immunol.* **128**:1876–1879.

Shirahata, T., Shimiza, K., Nada, S., and Suzuki, N., 1977, Studies on production of biologically active substance which inhibits the intracellular multiplication of *Toxoplasma* within mouse macrophages, *Z. Parasitenkd.* **53**:31–40.

Skamene, E., Stevenson, M. M., and Lemieux, S., 1983, Murine malaria: Dissociation of natural killer (NK) cell activity and resistance to *Plasmodium chabaudi*, *Parasite Immunol.* **6**:557–565.

Smith, D. C., 1979, From extracellular to intracellular: The establishment of a symbosis, *Proc. R. Soc. London Ser. B* **204**:115–130.

Smithers, S. R., and Doenhoff, M. J., 1982, Schistosomiasis, in: *Immunology of Parasitic Infections* (S. Cohen and K. S. Warren, eds.), pp. 527–607, Blackwell, Oxford.

Smrkovski, L. L., and Strickland, G. T., 1978, Rodent malaria: BCG induced protection and immunosuppression, *J. Immunol.* **121**:1257–1261.

Snell, E. S., 1971, Endotoxin and the pathogenesis of fever, in: *Microbial Toxins V* (S. Kadis, G. Weinbaum, and S. J. Ajl, eds.), pp. 277–340, Academic Press, New York.

Snyder, D. S., Beller, D. I., and Unanue, E. R., 1982, Prostaglandins modulate macrophage Ia expression, *Nature (London)* **299**:163–165.

Stadecker, M., Wyler, D., and Wright, J., 1982, Ia antigen expression and antigen presenting function by macrophages isolated from hypersensitivity granulomas, *J. Immunol.* **128**:2739–2744.

Steeg, P. S., Moore, R. N., and Oppenheim, J. J., 1980, Regulation of murine macrophage Ia-antigen expression by products of activated spleen cells, *J. Exp. Med.* **152**:1734–1744.

Steeg, P. S., Johnson, H. M., and Oppenheim, J. J., 1982, Regulation of murine macrophage Ia antigen expression by an immune interferon-like lymphokine: Inhibitory effect of endotoxin, *J. Immunol.* **129**:2402–2406.

Steinman, R. M., Nogueira, N., Witmer, M. D., Tydings, J., and Mellman, I. S., 1980, Lymphokine enhances the expression and synthesis of Ia antigens on cultured mouse peritoneal macrophages, *J. Exp. Med.* **152**:1248–1261.

Stenson, W. F., and Parker, C. W., 1980, Prostaglandins, macrophages and immunity, *J. Immunol.* **125**:1–5.

Stevens, D. P., Frank, D. M., and Mahmoud, A. A. F., 1978, Thymus dependency of host resistance to *Giardia muris* infection: Studies in nude mice, *J. Immunol.* **120**:680–682.

Subrahmanyam, D., Rao, V. B. G., Mehta, K., and Nelson, D. S., 1976, Serum-dependent adhesion and cytotoxicity of cells to *Litomosoides carinii* microfilariae, *Nature (London)* **260**:529–530.

Sunday, M. E., Stadecker, M. J., Wright, J. A., Aoki, I., and Dorf, M. E., 1983, Induction of immune responses by schistosome granuloma macrophages, *J. Immunol.* **130**:2413–2417.

Suzuki, Y., and Kobayashi, A., 1983, Suppression of unprimed T and B cells in antibody responses by irradiation-resistant and plastic adherent suppressor cells in *Toxoplasma gondii*-infected mice, *Infect. Immun.* **40**:1–7.

Suzuki, Y., Watanabe, N., and Kobayashi, A., 1981, Non specific suppression of primary antibody responses and presence of plastic-adherent suppressor cells in *Toxoplasma gondii*-infected mice, *Infect. Immun.* **34**:30–35.

Swaminathan, S., and Gooding, L. R., 1983, Inhibition of glycosylation prevents H-2K and D antigen expression on SV40 virus-transformed cells, *Eur. J. Immunol.* **13**:335–340.

Tabor, D. R., and Saluk, P. H., 1983, Differential activation of resident macrophage subsets with two sources of lymphokine preparations, *Infect. Immun.* **40**:177–183.

Takayanagi, T., Nakatake, Y., and Enriquez, G. L., 1974, *Trypanosoma gambiense:* Phagocytosis *in vitro*, *Exp. Parasitol.* **36**:106–113.

Taliaferro, W. H., 1941, The cellular basis for immunity in malaria, in: *Human Malaria* (F. R. Moulton, ed.), pp. 239–249, American Association for the Advancement of Science, Washington, D.C.

Taliaferro, W. H., and Taliaferro, L. G., 1944, The effect of immunity on the asexual reproduction of *Plasmodium brasilianum*, *J. Infect. Dis.* **75**:1–32.

Tanaka, Y., Kiyotaki, C., Tanowitz, H., and Bloom, B. R., 1982, Reconstitution of a variant macrophage cell line defective in oxygen metabolism with a H_2O_2-generating system, *Proc. Natl. Acad. Sci. USA* **79**:2584–2588.
Targett, G. A. T., and Viens, P., 1975, The immunological response of CBA mice to *Trypanosoma musculi:* Elimination of the parasite from the blood, *Int. J. Parasitol.* **5**:231–234.
Taverne, J., Dockrell, H. M., and Playfair, J. H. L., 1981, Endotoxin-induced serum factor kills malarial parasites *in vitro, Infect. Immun.* **33**:83–89.
Taverne, J., Depledge, P., and Playfair, J. H. L., 1982a, Differential sensitivity *in vitro* of lethal and non-lethal malarial parasites to endotoxin-induced serum factor, *Infect. Immun.* **37**:927–934.
Taverne, J., Dockrell, H. M., and Playfair, J. H. L., 1982b, Killing of the malarial parasite *Plasmodium yoelii in vitro* by cells of myeloid origin, *Parasite Immunol.* **4**:77–91.
Todd, C. W., Goodgame, R. W., and Colley, D. G., 1979, Immune responses during human schistosomiasis mansoni. V. Suppression of schistosome antigen-specific lymphocyte blastogenesis by adherent/phagocytic cells, *J. Immunol.* **122**:1440–1446.
Tosta, C. E., and Wederburn, N., 1980, Immune phagocytosis of *Plasmodium yoelii*-infected erythrocytes by macrophages and eosinophils, *Clin. Exp. Immunol.* **42**:114–120.
Trager, W., 1974, Some aspects of intracellular parasitism, *Science* **183**:269–273.
Tubbs, H., 1980, Endotoxin in human and murine malaria, *Trans. R. Soc. Trop. Med. Hyg.* **74**:121–123.
Unanue, E. R., 1981, The regulatory role of macrophages in antigenic stimulation. Part two: symbiotic relationship between lymphocytes and macrophages, *Adv. Immunol.* **31**:1–136.
Underdown, B. J., Roberts-Thomson, I. C., Anders, R. F., and Mitchell, G. F., 1981, Giardiasis in mice: Studies on the characteristics of chronic infection in C3H/He mice, *J. Immunol.* **126**:669–672.
Warren, H. S., and Weidanz, W. P., 1976, Malarial immunosuppression *in vitro:* Adherent spleen cells are functionally defective as accessory cells in the response to horse erythrocytes, *Eur. J. Immunol.* **6**:816–819.
Warren, K. S., 1982, The secret of the immunopathogenesis of schistosomiasis: *In vivo* models, *Immunol. Rev.* **61**:189–214.
Weidanz, W. P., 1982, Malaria and alterations in immune reactivity, *Br. Med. Bull.* **38**:167–172.
Weinbaum, F. I., Evans, C. B., and Tigelaar, R. E., 1976a, An *in vitro* assay for T cell immunity to malaria in mice, *J. Immunol.* **116**:1280–1283.
Weinbaum, F. I., Evans, C. B., and Tigelaar, R. E., 1976b, Immunity to *Plasmodium berghei yoelii* in mice. I. The course of infection in T cell and B cell deficient mice, *J. Immunol.* **117**:1999–2005.
Weinbaum, F. I., Weintraub, J., Nkrumah, F. F., Evans, C. B., Tigelaar, R. E., and Rosenberg, J., 1978, Immunity to *Plasmodium berghei yoelii* in mice. II. Specific and nonspecific cellular and humoral responses during the course of infection, *J. Immunol.* **121**:629–636.
Weinstock, J. V., and Boros, D. L., 1983, Organ-dependent differences in composition and function observed in hepatic and intestinal granulomas isolated from mice with schistosomiasis mansoni, *J. Immunol.* **130**:418–422.
Weiss, N., and Tanner, M., 1979, Studies on *Dipetalonema viteae* (Filarioidea). 3. Antibody-dependent cell mediated destruction of microfilariae *in vivo, Tropenmed. Parasitol.* **30**:73–80.
Wellhausen, S. R., and Boros, D. L., 1981, Comparison of Fc, C3 receptors and Ia antigens on the inflammatory macrophage isolated from vigorous or immunomodulated liver granulomas of schistosome-infected mice, *J. Reticuloendothel. Soc.* **30**:191–203.
Wellhausen, S. R., and Mansfield, J. M., 1979, Lymphocyte function in experimental African trypanosomiasis. II. Splenic suppressor cell activity, *J. Immunol.* **122**:818–824.
Wellhausen, S. R., and Mansfield, J. M., 1980, Characteristics of the splenic suppressor cell–target cell interaction in African trypanosomiasis, *Cell. Immunol.* **54**:414–424.
Whitelaw, A., Miller, J. F. A. P., and Mitchell, G. F., 1977, Studies on immune responses to parasite antigens in mice. VI. Delayed-type hypersensitivity to blood cells in *Plasmodium berghei* infected mice, *Cell. Immunol.* **32**:216–222.
Wiener, E., and Bandieri, A., 1974, Differences in antigen handling by peritoneal macrophages from the Biozzi high and low responder lines of mice, *Eur. J. Immunol.* **4**:457–463.
Wiener, E., and Bandieri, A., 1975, Modification in the handling *in vitro* of ^{125}I-labelled keyhole limpet haemocyanin by peritoneal macrophages from mice pretreated with the adjuvant, *Corynebacterium parvum, Immunology* **29**:265–274.

Wilson, C. B., Tsai, V., and Remington, J. S., 1980, Failure to trigger the oxidative metabolic burst by normal macrophages: Possible mechanisms for survival of intracellular pathogens, *J. Exp. Med.* **151**:328–346.

Wing, E. J., Gardner, I. D. Ryning, F. W., and Remington, J. S., 1977, Dissociation of effector functions in populations of activated macrophages, *Nature (London)* **268**:642–644.

Wood, P. R., and Clark, I. A., 1982a, Apparent irrelevance of NK cells to resolution of infections with *Babesia microti* and *Plasmodium vinckei petteri* in mice, *Parasite Immunol.* **4**:319–327.

Wood, P. R., and Clark, I. A., 1982b, Genetic control of *Propionibacterium acnes*-induced protection of mice against *Babesia microti, Infect. Immun.* **35**:52–57.

Wood, P. R., and Clark, I. A., 1983, Macrophages from babesia and malaria infected mice are primed for monokine release, *Parasite Immunol.* **6**:309–317.

World Health Organization, 1978, Immunodeficiency, Technical Report Series No. 630, Geneva.

Worms, M. J., and McLaren, D. J., 1982, Macrophage-mediated damage to filarial worm *(Dipetalonema setariosum) in vivo, J. Helminthol.* **56**:235–241.

Wyler, D. J., and Gallin, J. I., 1977, Spleen-derived mononuclear cell chemotactic factor in malaria infections: A possible mechanism for splenic macrophage accumulation, *J. Immunol.* **118**:478–484.

Wyler, D. J., and Tracy, J. W., 1982, Direct and indirect effects of soluble extracts of *Schistosoma mansoni* eggs on fibroblast proliferation *in vitro, Infect. Immun.* **38**:103–108.

Wyler, D. J., Oppenheim, J. J., and Koontz, L. C., 1979, Influence of malaria infection on the elaboration of soluble mediators by adherent mononuclear cells, *Infect. Immun.* **24**:151–159.

Wyler, D. J., Wahl, S. M., Cheever, A. W., and Wahl, L. M., 1981, Fibroblast stimulation in schistosomiasis. I. Stimulation *in vitro* of fibroblasts by soluble products of egg granulomas, *J. Infect. Dis.* **144**:254–262.

Yoshikai, Y., Miakes, S., Matsumoto, T., Nomoto, K., and Takeya, K., 1979, Effect of stimulation and blockade of mononuclear phagocyte system on the delayed footpad reaction to SRBC in mice, *Immunology* **38**:577–583.

Zuckerman, A., 1975, Current status of the immunology of blood and tissue protozoa. 1. *Leishmania, Exp. Parasitol.* **38**:370–400.

Zuckerman, A., 1977, Current status of the immunology of blood and tissue protozoa. 2. *Plasmodium, Exp. Parasitol.* **42**:374–446.

Index

MIX
Papier aus verantwortungsvollen Quellen
Paper from responsible sources
FSC® C105338

If you have any concerns about our products,
you can contact us on
ProductSafety@springernature.com

In case Publisher is established outside the EU,
the EU authorized representative is:
Springer Nature Customer Service Center GmbH
Europaplatz 3, 69115 Heidelberg, Germany

Printed by Libri Plureos GmbH
in Hamburg, Germany